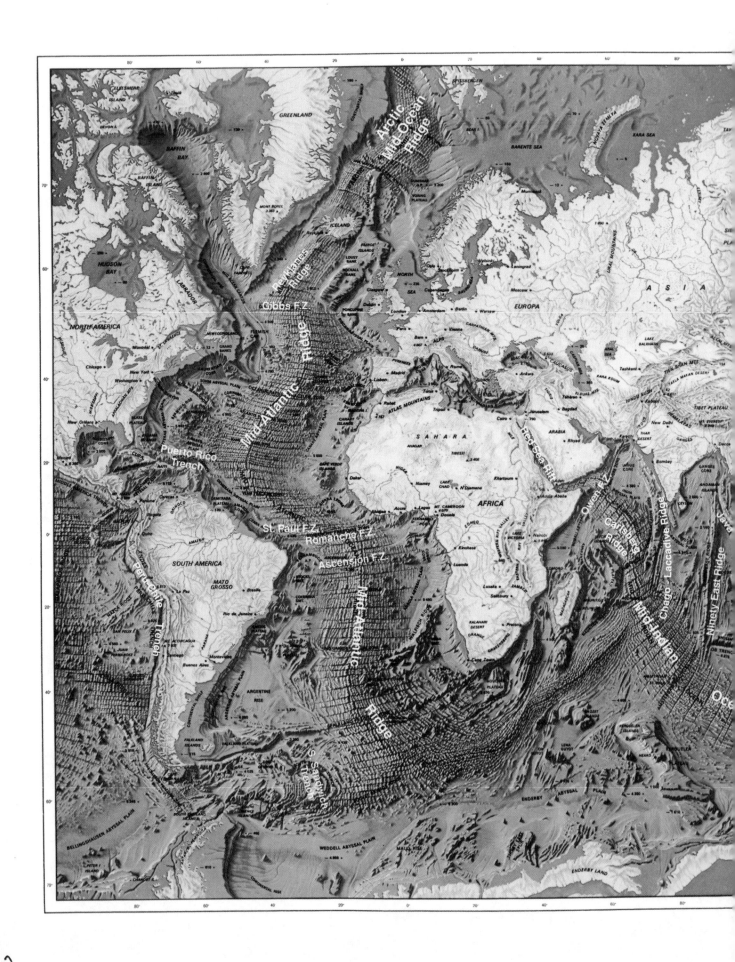

Essentials
of Oceanography

SIXTH EDITION

Essentials of Oceanography

Harold V. Thurman

Professor Emeritus
Mt. San Antonio College

Alan P. Trujillo

Associate Professor
Palomar College

PRENTICE HALL
Upper Saddle River, New Jersey 07458

Library of Congress Cataloging-in-Publication Data

Essentials of oceanography — 6th ed. / Harold V. Thurman, Alan P. Trujillo.
   p.   cm.
Includes bibliographical references index.
ISBN 0-13-727348-7
   1. Oceanography.   I. Trujillo, Alan P.
GC11.2.T49   1999
551.46—dc20   95-49157
                                      CIP

Executive Editor: Robert A. McConnin
Editor: Patrick Lynch
Editor-in-Chief: Paul Corey
Editorial Director: Tim Bozik
Assistant Vice President of Production and Manufacturing: David W. Riccardi
Executive Managing Editor: Kathleen Schiaparelli
Assistant Managing Editor: Lisa Kinne
Executive Marketing Manager: Leslie Cavaliere
Director of Creative Services: Paula Maylahn
Associate Creative Director: Amy Rosen
Art Manager: Gus Vibal
Art Editors: Karen Branson/Xiaohong Zhu
Art Director: Joseph Sengotta
Assistant to Art Director: John Christiana
Interior Design: Douglas & Gayle Limited
Manufacturing Manager: Trudy Pisciotti
Photo Researcher: Tobi Zausner
Photo Research Administrator: Melinda Reo
Copy Editor: Lynne Lackenbach
Art Studios: Academy Artworks/Geosystems Inc.
Editorial Assistant: Nancy Bauer
Production Supervision/Composition: WestWords, Inc.
Cover Designer: Joseph Sengotta
Cover Photo: View to Molokai, Lanai, Hawaii/David Muench, David Muench Photography, Inc.

©1999, 1996 by Prentice-Hall, Inc.
Simon & Schuster/A Viacom Company
Upper Saddle River, New Jersey 07458

Previous editions copyright ©1993, 1990, 1987, 1983 by Macmillan Publishing Company, a division of Macmillan, Inc.

Printed in the United States of America

10 9 8 7 6 5 4 3 2 1

ISBN 0-13-727348-7

Prentice-Hall International (UK) Limited, *London*
Prentice-Hall of Australia Pty. Limited, *Sydney*
Prentice-Hall Canada Inc., *Toronto*
Prentice-Hall Hispanoamericana, S.A., *Mexico*
Prentice-Hall of India Private Limited, *New Delhi*
Prentice-Hall of Japan, Inc., *Tokyo*
Simon & Schuster Asia Pte. Ltd., *Singapore*
Editora Prentice-Hall do Brasil Ltda., *Rio de Janeiro*

*To Karl Jay Schmid
and for Karl Edward Trujillo*

HAL THURMAN retired in May 1994, after 24 years of teaching in the Earth Sciences Department of Mt. San Antonio College in Walnut, California. Interest in geology led to a B.S. degree from Oklahoma A&M University, followed by seven years working as a petroleum geologist, mainly in the Gulf of Mexico. Here his interest in oceans developed, and he earned a M.A. degree from California State University at Los Angeles, then joined the Earth Sciences faculty at Mt. San Antonio College. Hal Thurman has also co-authored a marine biology textbook, written articles on the Pacific, Atlantic, Indian, and Arctic oceans for the 1994 edition of *World Book Encyclopedia*, and served as a consultant on the National Geographic publication, *Realms of the Sea*. He still enjoys going to sea on vacations with his wife Iantha.

AL TRUJILLO teaches at Palomar Community College in San Marcos, CA, where he is co-Director of the Oceanography Program. He received his bachelor's degree in Geology from the University of California at Davis and his master's degree in Geology from Northern Arizona University, afterwards working for several years in the industry as a developmental geologist, hydrogeologist, and computer specialist. Al began teaching at Palomar in 1990 and in 1997 was awarded Palomar's Distinguished Faculty Award in Teaching. In addition to writing and teaching, Al works as a naturalist and lecturer aboard natural history expedition vessels in Alaska and the Sea of Cortez/Baja California. His research interests include beach processes, sea cliff erosion, and computer applications in oceanography. Al and his wife, Sandy, have two children, Karl and baby Eva. (Being the child of a proud and computer-savvy parent, Eva's baby pictures can be found on Al's website at http://oceanography.palomar.edu/trujillo.htm.)

CONTENTS

PREFACE

To the Student

Welcome! You're about to embark on a journey that is far from ordinary. Over the course of this term, you will discover the central role the oceans play in the vast global system of which you are a part.

Your study of oceanography will incorporate concepts, principles and processes from virtually all the scientific disciplines—geology, chemistry, physics, and biology. However, no formal background in any of these disciplines is required to successfully master the subject matter contained within this book. Always keep in mind that we, the authors, and your instructor want you to take away from your oceanography course much more than just a collection of facts. Instead, we want the time you spend here to result in a fundamental understanding of *how the ocean works*.

To that end—and to help you make the most of your study time—we focused the presentation in this book by organizing the material into three logical and intuitive areas:

1. **Concepts**: General ideas derived or inferred from specific instances or occurrences (for instance, the concept of density can be used to explain why the ocean is layered).
2. **Processes**: Actions or occurrences that bring about a result (for instance, the process of waves breaking at an angle to the shore results in the movement of sediment along the shoreline).
3. **Principles**: Rules or laws concerning the functioning of natural phenomena or mechanical processes (for instance, the principle of sea floor spreading suggests that the geographic positions of the continents have changed through time).

Interwoven with these concepts, processes, and principles are hundreds of photographs, illustrations, real-world examples, and applications that make the material relevant and accessible (sometimes even entertaining) by bringing the science to life.

Ultimately, it is our hope that by understanding how the ocean works, you will develop a new awareness and appreciation of all aspects of the marine environment and its role in Earth systems. To this end, the book has been written for you, the student of the ocean. So enjoy and immerse yourself! You're in for an exciting ride.

Alan Trujillo
Harold Thurman

To the Instructor

The sixth edition of *Essentials of Oceanography* is designed to accompany an introductory college-level course in general or physical oceanography taught to students with no formal background in mathematics or science. Like previous editions, the goal of this edition of the textbook is to present the relationship of scientific principles to ocean phenomena in a way that can be clearly understood. In addition, this book is written in a style that is intended to engage learners and to inspire a sense of wonder and fascination about the oceans.

What's New in this Edition?

The addition of Al Trujillo—an award-winning educator—as a co-author for the sixth edition, has contributed to many improvements in this textbook. Changes in this edition are designed to increase the readability, relevance, and appeal of this book. The major changes include:

- Reorganizing the first two chapters into a new Chapter 1 ("Introduction to Planet 'Earth'"). This introduction orients the study of oceanography and the material to come within the context of Earth systems. The historical perspective on oceanography is now included as chapter-opening essays which highlight important events in the history of oceanography that are relevant to the material in specific chapters.
- Switching the order of Chapters 2 and 3 such that the chapter on plate tectonics comes before the chapter on sea floor features so that the processes of plate tectonics can be fully used to explain the origin of various sea floor features.
- Upgrading and expanding the all-important illustration program, including over 220 new line drawings and photographs (over 40 percent of all photos and line drawings are new); most other line drawings have also been revised.
- Illustrating up-to-date applications of the chapter's concepts in new feature boxes.
- Adding a new tables which better explain and summarize content.
- Adding new feature at the end of each chapter called "Students Sometimes Ask...," which are actual student questions, along with the authors' answers.
- Utilizing the international metric system (Système International or SI) of units throughout.

- Explaining word etymons (*etymon* = the origin of a word) as new terms are introduced.

- Noting of key terms with **bold print**, which are defined when they are introduced and are included in the glossary (entries have been increased by more than 30 percent).

- Increasing and updating the reference material of four new appendices.

- Creating a dedicated Web site to accompany the text at http://www.prenhall.com/thurman.

- Continuing to refine the style and clarity of the writing.

Some Notes About the Organization

The 15-chapter format of this textbook is designed for easy coverage of the material in a 15- or 16-week semester. For courses taught on a 10-week quarter system, the instructor may need to select those chapters that cover the topic and concepts of primary relevance to their course. Following the introductory chapter (Chapter 1), the four major academic disciplines of oceanography are represented in these chapters:

- Geological oceanography (Chapters 2, 3, 4, and parts of Chapters 10 and 11)

- Chemical oceanography (Chapter 5 and part of Chapter 11)

- Physical oceanography (Chapters 6–9 and parts of Chapters 10 and 11)

- Biological oceanography (Chapters 12–15)

However, we believe that oceanography is at its best when it links together several scientific disciplines and shows how they are interrelated in the oceans. Therefore, this interdisciplinary approach is a key element of every chapter.

Chapter 1, "Introduction to Planet 'Earth,'" introduces our watery world, gives some historical perspective, explains the reasoning behind the scientific method, and includes a discussion of the origin of Earth, the atmosphere, the oceans, and life itself.

The geological oceanography section begins with Chapter 2, "Plate Tectonics and the Ocean Floor," which discusses the revolutionary idea of plate tectonics, how it has changed the way we view our dynamic planet, and how it has been responsible for the continuing creation of the ocean floor. Chapter 3, "Marine Provinces," analyzes the varied features on the sea floor and includes a description of the processes of formation for these features. Chapter 4, "Marine Sediments," explains the wealth of knowledge that can be learned from the deposits on the sea floor.

The chemical and physical oceanography sections begin with Chapter 5, "Water and Seawater," where the unusual chemical and physical properties of water and seawater are explored. Chapter 6, "Air–Sea Interaction," examines the link between the ocean and the atmosphere and explores the effects those systems have on weather and world climate. Additional aspects of physical oceanography are covered in Chapters 7, 8, and 9, "Ocean Circulation," "Waves and Water Dynamics," and "Tides."

Chapter 10, "The Coast: Beaches and Shoreline Processes," examines the effect of high-energy shoreline processes which influence the coastal area. The discussion of ocean pollution is concentrated in Chapter 11, "The Coastal Ocean," which focuses on those parts of the marine environment most affected by this problem.

Chapter 12, "The Marine Habitat," introduces biological oceanography with a discussion of the changing physical conditions throughout the marine environment and some of the general adaptations required for living in the oceans. Chapter 13, "Biological Productivity and Energy Transfer," completes the introduction to biological oceanography by examining oceanic cycles and food webs. Chapters 14 and 15 focus on "Animals of the Pelagic Environment" and "Animals of the Benthic Environment," respectively.

As world population continues to expand, there is increasing evidence that human actions are having a significant and negative impact on essentially all components of Earth's environment. The ocean is at the center of many environmental issues, and thus this textbook has an environmental focus. For instance, some environmental topics that are discussed include the role of the oceans in the possible increased greenhouse effect of the atmosphere (Chapter 6) and the possible relationship between the El Niño-Southern Oscillation and changes in worldwide climate (Chapter 7). Also considered are the problems of coastal development (Chapter 10) and exploiting marine resources such as hydrocarbons and other aspects of coastal pollution (Chapter 11). Every effort has been made to include up-to-date factual information within these and other controversial topics.

Actual student questions and the authors' answers are included in a new feature at the end of each chapter called "Students Sometimes Ask...." The end-of-chapter "Summary" reviews the major concepts discussed in each chapter, and "Key Terms" are also listed. The "Questions and Exercises" section provides the student with an opportunity to test their knowledge by answering questions and performing exercises related to the major concepts. The "Suggested Readings" at the end of each chapter provide a guide to articles on relevant topics that are presented in the popular magazines *Earth* (beginning with its first issue in 1992), *Sea Frontiers* (which, unfortunately, ceased publication after the Spring 1996 issue), and *Scientific American* (where there is a somewhat more technical presentation).

To assist the student further, there are four appendixes entitled "Metric and English Units Compared," "Latitude and Longitude on Earth," "Geographic Locations," and (of particular interest to those considering a career in oceanography) "Careers in Oceanography." The Glossary includes the definitions of all the boldface key terms used throughout the book (scientific terms, excluding names of individuals), as well as additional terms in which students may be interested.

The Instructional Package

Companion Website:
http://www.prenhall.com/thurman

Designed to be integrated into a class with virtually no start-up time on the part of the instructor, the Companion Website functions as both an on-line study guide and launching pad for further exploration. The site is designed using the latest technology, written by oceanography instructors, and tied chapter-by-chapter to the text.

- **To aid in reviewing the text material,** the site contains multiple choice and true/false questions, answers to which can be submitted via the Internet to Prentice-Hall's server for a grade, which can then be submitted to your instructor via e-mail.

- **To foster critical thinking,** the site contains *Web Essay* questions—short-answer questions that require the student to use the Internet to research an issue, evaluate the information, and formulate a response. Responses to the Web Essays can also be e-mailed to your instructor for grading.

- **To encourage and enable further exploration** using the incredible array of resources now available on the Internet, every chapter contains *Web Destinations*, both general and chapter-specific links to some of the best oceanography sites on the World Wide Web. These sites are researched and annotated by oceanography instructors to insure quality and relevancy and continually checked by Prentice-Hall to insure validity.

- Finally, **to make the instructor's life easier,** the *Syllabus Manager* uses the latest technology to help professors create, store, and modify a customized on-line syllabus. Through this "living" syllabus, instructors can link assignments directly to activities on the text's Web site.

Science on the Internet: A Student's Guide (0-13-021308-x).

This brief "guidebook" is a unique resource which helps science students locate and explore myriad science resources on the World Wide Web. It also provides an overview of the Web itself, general navigational strategies, and brief student activities. *Science on the Internet* **can be packaged free with *Essentials of Oceanography*** for your students; contact your local representative for the proper ISBN.

New York Times *Themes of the Times—* Oceanography.

Show your students how dynamic oceanography really is! This newspaper-format supplement captures late-breaking news articles, drawn from the *New York Times*, that illustrate how the oceanographic concepts the students are studying in your class play out in the "real world." ***Themes of the Times* can be packaged free with *Essentials of Oceanography*** for your students; contact you local representative for the proper ISBN.

Instructor's Manual (0-13-080303-0)

Completely new and written by Al Trujillo, this manual is available to instructors and provides:

- Answers to end-of-chapter questions;

- A selection of various types of exam questions;

- A wealth of instructional resources for each chapter; and,

- Special feature articles, which cover a wide range of interesting topics. These special feature articles may be duplicated and distributed to students to add breadth and interest to coverage of material in the text or they may serve as a resource for student projects.

Slide Set (0-13-080305-7)

A set of 150 slides from the text including photographs not available with the transparencies. (All the transparencies also appear on slides.)

Transparency Set (0-13-080304-9)

Includes 100 acetates of key illustrations from the text, selected by the authors and enlarged for excellent classroom visibility.

Why a Bathysphere?

In 1934, William Beebe and Otis Barton were sealed into a hollow 2268-kilogram (5000-pound) cast iron sphere—a "bathysphere"—with walls 0.5 meter (1 1/2 feet) thick and windows of fused quartz that was lowered over the side of a ship into the deep waters off Bermuda Island. Connected to the surface by only a steel cable and a telephone line, Beebe and Barton descended to a record-setting depth of 923 meters (3028 feet). Previous to this dive, the farthest down a living human had gone was 160 meters (525 feet)! In the same unprecedented and dramatic way that Beebe and Barton's bathysphere opened the ocean depths to human exploration, the Internet has revolutionized access to the oceans and oceanography. With the Companion Website to *Essentials of Oceanography* we encourage you to explore the world's oceans through this new window!

Acknowledgments

The authors are indebted to many individuals for their helpful comments and suggestions during the revision of this book. Al Trujillo is particularly indebted to his colleagues at Palomar Community College, Patty Deen and Lisa DuBois, for their keen interest in the project and for allowing him to use some of their creative ideas in the book. They are the finest colleagues imaginable. Mary Anne Holmes of the University of Nebraska, Lincoln, deserves special recognition for thoroughly reviewing Chapters 1–9 and providing detailed suggestions for improvement. Others who reviewed certain portions of the text were Jim Pesavento at Palomar College (Chapter 9 "Tides") and Anthony Trujillo, Professor Emeritus at San Joaquin Delta College (Chapter 5 "Water and Seawater").

Many individuals at Scripps Institution of Oceanography have been particularly helpful, especially the staff at *Explorations* magazine and most notably Assistant Editor Paige Jennings. Thanks also go to the many people around the world who helped locate images or willingly donated photographs for use in the text. Without the kind help of these people, the job of tracking down experts and finding images would have been a much larger task.

Many people were instrumental in helping the text evolve from its manuscript stage. Patrick Lynch, who recently assumed the position of Geoscience Editor at Prentice-Hall, became involved with the book when it was in production and had a major role in overseeing the development of the text's accompanying Web site. Joe Sengotta, Art Editor at Prentice-Hall, artfully refined and updated the overall look of the text and even suggested using the image of a beach in Hawaii for the cover. Jennifer Maughan, Project Manager at WestWords, skillfully coordinated the production of the text and dealt with many last-minute changes with rare good humor.

Al Trujillo would also like to thank his students, whose questions provided the material for the "Students Sometimes Ask…" sections and whose input has proved invaluable for improving the manuscript. Since scientists (and teachers) are always experimenting, thanks also for allowing yourselves to be a captive audience with which to conduct my experiments.

Al Trujillo also thanks his patient and understanding wife, Sandy, for encouraging him to become involved with the project, for reading the entire manuscript, and for supporting him so thoroughly in all his endeavors. Lastly, appreciation is extended to the chocolate manufacturers Hershey, See's, and Ghiradelli for providing inspiration. A heartfelt thanks to all of you!

Many other individuals have provided advice and assistance for this work, including:

William Balsam, *University of Texas at Arlington*
Steven Benham, *Pacific Lutheran University*
Laurie Brown, *University of Massachusetts*
G. Kent Colbath, *Cerritos Community College*
Thomas Cramer, *Brookdale Community College*
Hans G. Dam, *University of Connecticut*
Wallace W. Drexler, *Shippensburg University*
Walter C. Dudley, *University of Hawaii*
Charles H. V. Ebert, *SUNY Buffalo*
Kenneth L. Finger, *Irvine Valley College*
Dave Gosse, *University of Virginia*
Joseph Holliday, *El Camino Community College*
Timothy C. Horner, *California State University, Sacramento*
Lawrence Krissek, *Ohio State University*
M. John Kocurko, *Midwestern State University*
Richard A. Laws, *University of North Carolina*
Richard D. Little, *Greenfield Community College*
Stephen A. Macko, *University of Virginia, Charlottesville*
Matthew McMackin, *San Jose University*
James M. McWhorter, *Miami-Dade Community College*
Johnnie N. Moore, *University of Montana*
B. L. Oostdam, *Millersville University*
William W. Orr, *University of Oregon*
Curt Peterson, *Portland State University*
Edward Ponto, *Onodaga Community College*
Donald L. Reed, *San Jose State University*
Cathryn L. Rhodes, *University of California, Davis*
Jill K. Singer, *SUNY College, Buffalo*
Jackie L. Watkins, *Midwestern State University*
Arthur Wegweiser, *Edinboro University of Pennsylvania*

Although this book has benefited from careful review by many individuals, the accuracy of the information rests with the authors. If you find errors in the text, please contact us at:

Al Trujillo
Department of Earth Sciences
Palomar College
1140 W. Mission Rd.
San Marcos, CA 92069
(760) 744–1150 ext. 2734
atrujillo@palomar.edu

Hal Thurman
571 Windsor Park Cove
Collierville, TN 38017
(901) 854–7129
ann_harold_thurman@classic.msn.com

San Marcos, California

INTRODUCTION

Welcome to a book about the oceans. As you read this book, we hope that it elicits a sense of wonder and a spirit of curiosity about the oceans. The oceans represent many different things to different people. To some, it is a wilderness of beauty and tranquility, a refuge from hectic civilized lives. Others see it as a vast recreational area that inspires either rest or physical challenge. To others, it is a mysterious place that is full of unknown wonders. And to others, it is a place of employment unmatched by any on land. To be sure, its splendor has inspired artists, writers, and poets for centuries (Figure I-1). Whatever your view, we hope understanding the way the oceans work will increase your appreciation of the marine environment. Above all, take time to admire the oceans.

Essentials of Oceanography was first written in the late 1970s to help students develop an *awareness about the marine environment*—that is, develop an appreciation for the oceans by learning about oceanic processes (how the ocean behaves) and their interrelationships (how

physical entities are related to one another in the ocean). In this sixth edition, our goal is the same: to give the reader the scientific background to understand the basic principles underlying oceanic phenomena. In this way, one can then make informed decisions about the oceans in the years to come. We hope that some of you will be inspired so much by the oceans that you will continue to study them formally or informally in the future (for those who may be considering a life-long career in oceanography, see Appendix IV, "Careers in Oceanography").

What Is Oceanography?

Oceanography (*ocean* = the marine environment, *graphy* = the name of a descriptive science) is quite literally the description of the marine environment. Unfortunately, this definition does not fully portray the extent of what oceanography encompasses: oceanography is much more than just *describing* marine phenomena. Oceanography

Figure I-1 The ocean environment.

could be more accurately called the scientific study of all aspects of the marine environment. Hence, the field of study called oceanography could (and maybe *should*) be called oceanology (*ocean* = the marine environment, *ology* = the study of). However, the science of studying the oceans has traditionally been called oceanography. It is often also called *marine science* and includes the study of the water of the ocean, the life within it, and the (not so) solid earth beneath it.

Since prehistoric time, people have used the oceans as a means of transportation and as a source of food. However, the importance of ocean processes has been studied technically only since the 1930s. The impetus for this study began with the search for petroleum, continued with the emphasis on ocean warfare during World War II, and more recently has been expressed in the concern for the well-being of the ocean environment. Historically, those who make their living fishing in the ocean go where the physical processes of the ocean offer good fishing. But how marine life interrelates with ocean geology, chemistry, and physics to create good fishing grounds has been more or less a mystery until only recently when scientists in these disciplines began to investigate the oceans with new technology.

Oceanography is typically divided into different academic disciplines (or subfields) of study. The four main disciplines of oceanography are:

- *Geological oceanography,* which is the study of the structure of the sea floor and how the sea floor has changed through time; the creation of sea floor features; and the history of sediments deposited on it.
- *Chemical oceanography,* which is the study of the chemical composition and properties of seawater; how to extract certain chemicals from seawater; and the effects of pollutants.
- *Physical oceanography,* which is the study of waves, tides, and currents; the ocean-atmosphere relationship that influences weather and climate; and the transmission of light and sound in the oceans.
- *Biological oceanography,* which is the study of the various oceanic life forms and their relationships to one another; adaptations to the marine environment; and developing ecologically sound methods of harvesting seafood.

Other disciplines include ocean engineering, marine archaeology, and marine policy. Since the study of oceanography often examines in detail all the different disciplines of oceanography, it is often described as being an *interdisciplinary* science, or one covering all the disciplines of science as they apply to the oceans. The content of this book includes the broad range of interdisciplinary science topics that comprises the field of oceanography. This is a book about *all* aspects of the oceans.

Perhaps one of the best desciptions of oceanography was given by geological oceanographer and explorer Willard Bascom:

> *Oceanography is not so much a science as a collection of scientists who find common cause in trying to understand the complex nature of the ocean. In the vast salty seas that encompass the earth, there is plenty of room for persons trained in physics, chemistry, biology, and engineering to practice their specialties. Thus, an oceanographer is any scientifically trained person who spends much of his [or her] career on ocean problems.*

Earth's Oceans

The oceans are the largest and most prominent feature on Earth. In fact, they are the single most defining feature of our planet. As viewed from space, our planet is a beautiful blue, white, and brown globe (Figure I-2). It is our oceans of liquid water that set us apart in the solar system. No other planet has oceans (however, a recent discovery of ice covering one of Jupiter's moons, Europa, has lead to speculation that there may be an ocean of liquid water beneath the ice!). The fact that our planet has so much water, *and in the liquid form*, is unique in the solar system.

The oceans determine where our continents end, and thus have shaped political boundaries and human history many times. The oceans conceal many features; in fact, the majority of Earth's geographic features are on the ocean floor. Remarkably, there was once more known about the surface of the moon than about the floor of the ocean! Fortunately, over the past 30 years, our knowledge of both has increased dramatically.

The oceans influence weather all over the globe, even in continental areas far from any ocean. The oceans are also the lungs of the planet, taking carbon dioxide gas (CO_2) out of the atmosphere and replacing it with oxygen gas (O_2). Some scientists have estimated that the oceans supply as much as 70 percent of the oxygen humans breathe.

The oceans are in large part responsible for the development of life on Earth, providing a stable environment for life to evolve over long periods of time. Today, the oceans contain the greatest number of living things on the planet, from microscopic bacteria and algae to the largest life form alive today (the blue whale). Interestingly, water is the major component of nearly every life form on Earth, and our own body fluid chemistry is remarkably similar to the chemistry of seawater.

The oceans hold many secrets waiting to be discovered, and new discoveries about the oceans are made nearly every day. The oceans are a source of food, minerals, and energy that remain largely untapped. Over half of the world population lives in coastal areas near the ocean, taking advantage of the mild climate, an inexpensive form

Figure I-2 Earth from space.

of transportation, and vast recreational opportunities. And unfortunately, the oceans are also the dumping ground for many of society's wastes.

Rational Use of Technology?

Many stresses have been put on the oceans by an ever-increasing human population. For instance, populations studies in the United States show that 75 percent of all Americans live within an hour's drive from an ocean or the Great Lakes. This migration to the coasts will further mar the delicate natural balance that exists in the coastal ocean (Figure I-3). Specifically, the migration is resulting in more harbor and channel dredging, industrial waste and sewage disposal at sea, chemical spills, cooling of power plants with seawater, and the destruction of wetlands that are vital to the cleansing of runoff waters and to the maintenance of coastal fisheries.

Although it may seem as if humans have severely and irreversibly damaged the oceans, our impact has been felt mostly in coastal areas. The world's oceans are a vast resource that has not yet been lethally damaged. Humans have been able to inflict only minor damage here and there along the margins of the oceans. However, as our technology makes us more powerful, the threat of irreversible harm becomes greater. For instance, in the open ocean (those areas far from shore), deep-ocean mining and nuclear waste disposal have been proposed. How do we as a society deal with these increased demands on the marine environment? How do we regulate the ocean's use?

If used wisely, our technology can actually reduce the threat of irreversible harm. Which path will we take? We all need to evaluate carefully our own actions and the effects those actions have on the environment. In addition, we need to make conscientious decisions about those we elect to public office. Some of you may even have direct responsibility for initiating legislature that affects our environment. It is our hope that you, as a student of the marine environment, will gain enough knowledge while studying oceanography to help your community (and perhaps even your nation) make rational use of technology in the oceans.

Figure I-3 Coastal population.

Heavily developed area of Miami Beach, Florida.

CHAPTER · 1

INTRODUCTION TO PLANET "EARTH"

Diving into the Marine Environment

Historically, humans have desired to submerge themselves into the marine environment to observe it directly for exploration, profit, or adventure (Figure 1A). The identity of the first divers is shrouded in the mists of antiquity. However, it has been established that as early as 4500 B.C., brave and skillful divers were reaching depths as deep as 30 meters[1] (98 feet) on one breath

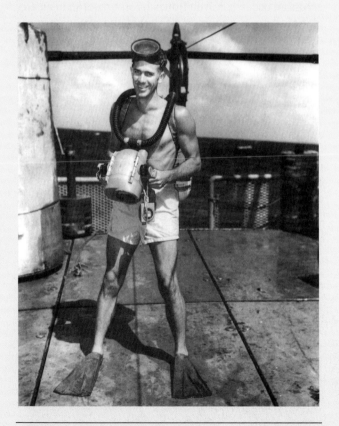

Figure 1A Oceanographer and explorer Willard Bascom.

---

[1] Throughout this book, metric measurements are used (and the corresponding English measurements are in parentheses). See Appendix I, "Metric and English Units Compared," for conversion factors between the two systems of units.

of air to retrieve such treasures as red coral and mother-of-pearl shells. Later, diving bells (bell-shaped structures full of trapped air) were lowered into the sea to provide passengers or underwater divers with an air supply. In 360 B.C., Aristotle, in his *Problematum*, recorded the use of kettles full of air lowered into the sea for use by Greek sponge divers. However, the technology to move around freely while breathing underwater was not developed until 1943, when Jacques-Yves Cousteau and Émile Gagnan invented the fully automatic, compressed-air Aqualung. The equipment was later dubbed "**scuba**," an acronym for *s*elf-*c*ontained *u*nderwater *b*reathing *a*pparatus, which is used by millions of recreational divers today. By using scuba, divers can experience the ocean first hand and in a way that leads to a full appreciation of the wonder and beauty of the marine environment.

Those who venture underwater must contend with many obstacles inherent in ocean diving. These include low temperatures, darkness, and the effects of greatly increased pressure. To combat low temperatures, specially designed clothing is worn. To combat darkness, waterproof, high-intensity diving lights are used. The only way to combat the deleterious effects of pressure is to limit the depth and duration of dives. That is why most scuba divers rarely venture below a depth of 30 meters (98 feet), where the pressure is three times that at the surface, and stay at depth less than 30 minutes.

It has always been dangerous to put humans into the marine environment because our bodies are adapted to living in a different fluid—air, a mixture of nitrogen and oxygen—at a pressure of 1 kilogram per square centimeter (15 pounds per square inch). Anyone who has been to the bottom of the deep end of a swimming pool is familiar with how quickly pressure increases with depth. Similarly, the effects of pressure changes cause problems for divers' bodies. While on a dive, the higher pressure at depth may cause them to experience a disorienting condition known as *nitrogen narcosis*, or "rapture of the deep." If divers surface too rapidly, expanding gases within the body can catastrophically rupture membranes, a condition called *barotrauma*.

In addition, when divers return to the lower pressure at the surface, they may experience a problem called *decompression illness*, which is more commonly known as the "bends." During a rapid ascent, some of the excess nitrogen dissolved in the diver's body can form tiny bubbles in the blood and other tissues (interestingly, oxygen is metabolized by tissue cells and is not usually a problem). This is analogous to the bubbles that form when a carbonated beverage is opened. Decompression illness usually results in severe joint pain (which causes divers to stoop over, hence the term the "bends"), and advanced stages lead to permanent neurologic injury and even death. To avoid it, divers must ascend slowly, allowing time for excess dissolved nitrogen to be eliminated from the blood via the lungs.

Despite these risks, divers venture to deeper and deeper depths in the ocean. In 1962, Hannes Keller and Peter Small made an open-ocean dive from a diving bell to a record-breaking depth of 304 meters (1000 feet) using a special gas mixture, but Small died once they returned to the surface. Presently, the record ocean dive is 534 meters (1752 feet), but researchers who study the physiology of deep divers have simulated a dive to 701 meters (2300 feet) in a pressure chamber using a special mix of oxygen, hydrogen, and helium gases. Researchers believe that humans will eventually be able to stay under water for extended periods of time at depths below 600 meters (1970 feet).

W hy is our planet called "Earth?" It doesn't seem to make sense that a planet with almost 71 percent of its surface covered by oceans is named after the land portion of the planet. As viewed from space, there is not much "earth" on Earth! Many early human cultures lived in the Mediterranean (*medi*=middle, *terra*=land) Sea area, which is, as its name implies, a sea surrounded by land (Figure 1–1). Historical records indicate that early Mediterranean cultures envisioned that most of the continents were similarly surrounded by marginal bodies of water. How surprised they must have been when they ventured into the larger oceans of the world. Our planet is misnamed "Earth" because we live on the land portion of the planet. If we were marine animals, our planet would probably be called "Ocean," "Water," "Hydro," "Aqua," or even "Oceanus," to indicate the prominence of Earth's oceans.

Geography of the Oceans

When looking at a map of the world (Figure 1–2), a few observations about the oceans are readily apparent. First of all, the oceans dominate the surface area of the globe. To anyone who has taken the time to travel by boat across an ocean (or even flown across an ocean in an airplane), it soon becomes clear that the oceans are remarkably large. Second, the oceans are interconnected. The oceans are actually a single continuous body of seawater. This fact has inspired some people to call it a "world ocean" (not plural) to emphasize how continuous the body of seawater really is. Notice how easily a vessel at sea can travel from one ocean to another, whereas it is impossible to travel on land from one continent to many others without crossing an ocean. Lastly, the volume of the oceans is immense. As a reservoir, the oceans contain an impressive 97.2 percent of all the water (including ice) in the world.

Figure 1–1 The world of Herodotus.

The world according to Herodotus in 450 B.C.

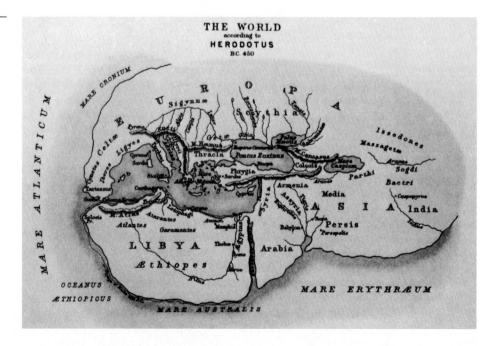

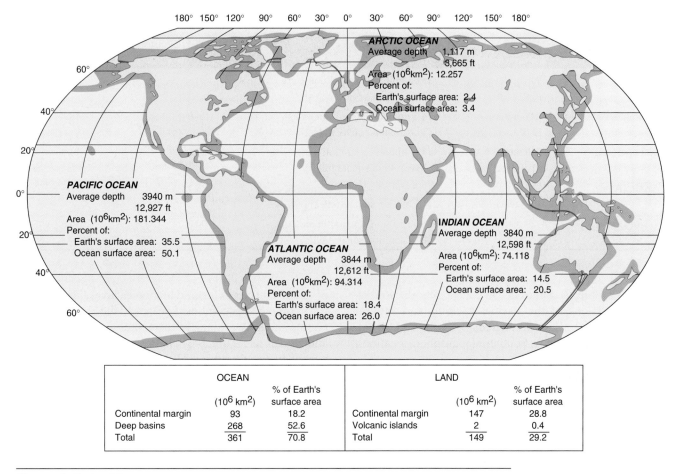

OCEAN			LAND		
	(10⁶ km²)	% of Earth's surface area		(10⁶ km²)	% of Earth's surface area
Continental margin	93	18.2	Continental margin	147	28.8
Deep basins	268	52.6	Volcanic islands	2	0.4
Total	361	70.8	Total	149	29.2

Figure 1–2 Earth's oceans.

Map showing the boundaries of the oceans, the average depths of the main ocean basins, and the percentage of Earth's surface and ocean surface they represent.

The Four Principal Oceans, Plus One

Our world ocean can easily be divided into four principal oceans (plus an additional ocean), based on the shape of the ocean basins and the positions of the continents (Table 1–1).

Pacific Ocean   The **Pacific Ocean** is the world's largest ocean and covers over half of the ocean surface area on Earth. The Pacific Ocean is, in fact, the single largest geographic feature on the planet, covering over one-third of Earth's entire surface. The Pacific Ocean is so large that, when looking at a globe, it is difficult to see any continents when this ocean faces you directly. Put another way, *all* of the continents could easily fit into the space occupied by the Pacific Ocean—with room left over! Although the Pacific Ocean is the deepest ocean in the world, it contains many small tropical islands. It was named in 1520 by explorer Ferdinand Magellan's party, for the fine weather they encountered while crossing into the Pacific (*paci*=peace) Ocean.

Atlantic Ocean   The **Atlantic Ocean** is about half the size of the Pacific Ocean and is not quite as deep. It sep-arates the Old World (Europe, Asia, and Africa) from the New World (North and South America). The Atlantic Ocean was named after the Atlas Mountains in northwest Africa.

Indian Ocean   The **Indian Ocean** is slightly smaller than the Atlantic Ocean and has about the same average depth. Note that it is mostly in the Southern Hemisphere (south of the Equator, or below 0 degrees latitude in Figure 1–2). The Indian Ocean was named for its proximity to the subcontinent of India.

Arctic Ocean   The small **Arctic Ocean** is about 7 percent the size of the Pacific Ocean and is only a little more than one-quarter as deep as the other oceans combined. Although it has a permanent layer of sea ice at the surface, the ice is only a few meters thick. The Arctic Ocean was named after the northern constellation Ursa Major, otherwise known as the Big Dipper, or the Bear (*arktos*=bear).

Southern Ocean or Antarctic Ocean   Some would say that there is an additional ocean near the continent of Antarctica in the Southern Hemisphere. Defined by the large West Wind Drift ocean current, the so-called

Table 1–1 A comparison of oceans of the world (including marginal seas).

	Average depth	Area	Percent of ocean surface area	Percent of Earth's surface area
Pacific Ocean	3940 meters (12,927 feet)	181.3 million square kilometers (70.0 million square miles)	50.1	35.5
Atlantic Ocean	3844 meters (12,612 feet)	94.3 million square kilometers (36.4 million square miles)	26.0	18.4
Indian Ocean	3840 meters (12,598 feet)	74.1 million square kilometers (28.6 million square miles)	20.5	14.5
Arctic Ocean	1117 meters (3665 feet)	12.3 million square kilometers (4.7 million square miles)	3.4	2.4
World Ocean	3729 meters (12,234 feet)	362.0 million square kilometers (139.7 million square miles)	100.0	70.8

Southern Ocean or **Antarctic Ocean** is really the southernmost portions of the Pacific, Atlantic, and Indian Oceans south of about 50 degrees south latitude. This ocean was named after its Southern Hemisphere location.

The Seven Seas?

Many have heard the evocative phrase "sailing the seven seas," but are there really seven seas? In the preceding discussion, we learned that there are four principal oceans plus an additional ocean. What is the difference between a sea and an ocean? In common use, the terms "sea" and "ocean" are often used interchangeably. For instance, a *sea* star lives in the *ocean*; the *ocean* is full of *sea*water; *sea* ice forms in the *ocean*; and one might go to the *sea*shore to live in *ocean*-front property. Technically, however, seas are defined as:

- Smaller and shallower than an ocean (interestingly, many consider the Arctic Ocean to be a sea)
- Composed of salt water (many "seas," such as the Caspian Sea in Asia, are actually large freshwater lakes)
- Somewhat enclosed by land (but some seas, such as the Sargasso Sea in the Atlantic Ocean, are defined by strong ocean currents rather than by land)

Although the definition of a sea leaves some room for interpretation, it does allow us to answer the question about the seven seas. If we count the oceans as seas and split the Pacific Ocean and the Atlantic Ocean arbitrarily at the Equator, then we have seven seas: the North Pacific, the South Pacific, the North Atlantic, the South Atlantic, the Indian, the Arctic, and the Southern or Antarctic. Somehow, "sailing the seven oceans" just doesn't sound nearly as romantic!

Comparing the Oceans to the Continents

The **hypsographic** (*hypsos*=height, *graphic*=drawn) **curve** shows the distribution of Earth's surface at elevations above and below sea level. It indicates that 70.8 percent of Earth's surface is located below sea level (Figure 1–3). Let's compare the combined oceans to the combined continents and look at some of the values given in Table 1–2 and shown in Figure 1–3. The average depth of all the oceans in the world is 3729 meters (12,234 feet). This means that there must be some extremely deep areas in the ocean to offset the shallow areas of the ocean with which most of us are familiar. Compared to the average height of all the continents in the world at 840 meters (2756 feet), it is readily apparent that the average land portion of our planet is not that far above sea level!

Let's also look at some extreme values in comparing oceans and continents (Table 1–2). The deepest depth in the oceans is a staggering 11,022 meters (36,163 feet) below sea level. Not surprisingly, this area is in the Pacific Ocean (the deepest ocean in the world), located in the Challenger Deep, part of a linear crease on the sea floor called the Mariana Trench. Could anything live at these depths, where there is crushing high pressure and absolutely no light, and the water temperature is just above freezing? Could humans ever visit this region? Surprisingly, the answer to both of these questions is yes: In 1960, Don Walsh and Jacques Piccard descended to the bottom of the Challenger Deep in the deep-diving **bathyscaphe** (*bathos*=depth, *scaphe*=a small ship) *Trieste* (Figure 1–4). The vessel was launched and began its journey to the ocean floor, getting deeper and deeper during the several hours it took for the descent. At 9906 meters (32,500 feet), the men heard a dull cracking sound that shook the cabin. They were unable to see that the Plexiglas viewing port on the entry tower had cracked. Miraculously, it held. Over five hours after leaving the surface, they reached the bottom at 10,912 meters (35,800 feet)—a record depth of human descent that still stands. Sure enough, they did see some life forms: a few small fish from the sole family and some shrimp (but hardly the giant sea monsters that some people believed would be down there).

The highest mountain in the world (that is, the mountain with the greatest height above sea level) is Mount Everest

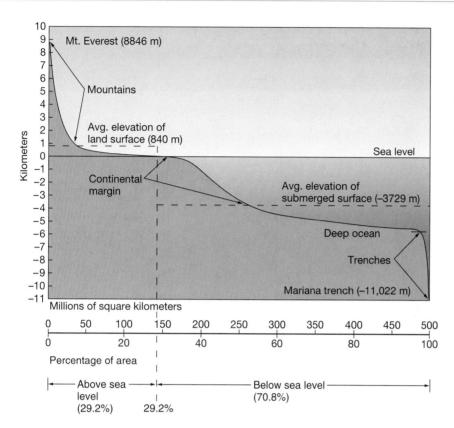

Figure 1–3 The hypso-graphic curve.

The hypsographic curve shows the elevations of Earth's crust relative to sea level. A total of 70.8 percent of Earth's surface is below sea level, while 29.2 percent is above sea level.

Table 1–2 A comparison of oceans and continents.

	Average depth/height	Extremes
Combined oceans	Average depth 3729 meters (12,234 feet)	Deepest depth 11,022 meters (36,163 feet) Challenger Deep, Mariana Trench
Combined continents	Average height 840 meters (2756 feet)	Tallest height 8846 meters (29,022 feet) Mount Everest, Himalaya Mountains

in the Himalayan Mountains of Asia at 8846 meters (29,022 feet, 3.9 inches) above sea level. Even with this impressive height, it is a full 2176 meters (7141 feet) shorter than the Mariana Trench is deep. However, there is some debate about which mountain is actually the tallest mountain in the world (that is, the mountain with the *greatest total height* from base to top). The tallest mountain in the world is Mauna Kea on the island of Hawaii in the United States, measuring 4206 meters (13,800 feet) above sea level and 5426 meters (17,800 feet) from sea level down to its base, for a total height of 9632 meters (31,600 feet). The total height of Mauna Kea is 786 meters (2578 feet) higher than Mount Everest. Even with the debate over the tallest mountain, it doesn't make much of a difference:

Mauna Kea is still 1390 meters (4563 feet) shorter than the Mariana Trench is deep. No mountain on Earth is taller than the Mariana Trench is deep.

Explorations of the Oceans: Some Historical Notes About Oceanography

Early History

Humankind probably first viewed the oceans as a source of food. At some later stage in the development of civilization, vessels were built to move upon the ocean's surface and transport ocean-going people to new fishing grounds. It was soon realized that the oceans provided an inexpensive and efficient way to move large and heavy objects, and thus trade between cultures was established and distant societies could begin to interact.

Pacific Navigators It is not known who first developed navigation, but it may have been the ancestors of Pacific Islanders. The peopling of the Pacific Islands (Oceania) is somewhat perplexing because there is no evidence that people actually evolved on these islands. Therefore, their presence required travel over hundreds or even thousands of miles of open ocean from the continents, probably in small vessels of that time (double canoes, outrigger canoes, or balsa rafts). The islands in the Pacific Ocean are widely scattered, so it is likely that only a fortunate few of the voyagers made landfall and that many others perished during the voyage. Figure 1–5 shows the three major island regions in the Pacific Ocean: Micronesia (*micro*=small,

Figure 1–4 *Trieste.*
The U.S. Navy's bathyscaphe *Trieste.*

nesia = islands), Melanesia (*melan* = black, *nesia* = islands), and Polynesia (*poly* = many, *nesia* = islands), which covers the largest area.

There are no written records of Pacific human history prior to the arrival of Europeans in the sixteenth century. However, the movement of Asian peoples into Micronesia and Melanesia is easy to visualize because the distances between islands are relatively short. This is not the case with Polynesia, where large open ocean distances separate island groups and must have presented the greatest challenge to ocean voyagers. For example, Easter Island, at the southeastern corner of the triangular-shaped Polynesian islands region, is over 1600 kilometers (1000 miles) from the nearest island, Pitcairn Island. The most challenging destination for these early voyagers must have been the Hawaiian Islands, which are a staggering 3000 kilometers (1860 miles) from the nearest inhabited islands, the Marquesas Islands (Figure 1–5).

Archeological evidence suggests that humans from New Guinea may have occupied New Ireland as early as 4000 or 5000 B.C. However, there is little evidence of human travel farther into the Pacific Ocean prior to 1500 B.C. By then, pottery makers called the "Lapita people" had traveled on to Fiji, Tonga, and Samoa. Over the next thousand years or so, the Polynesian culture developed in these islands, and Polynesian voyaging may have begun about 500 B.C.

The first archeological evidence from the eastern Pacific indicates human occupation of the Marquesas Islands by 300 A.D. Although the Maori of New Zealand, the Hawaiians, and the Easter Islanders of the sixteenth century were Polynesian, there is no clear evidence to explain *how* these peoples arrived at their present homes.

An adventuring biologist-turned-anthropologist, **Thor Heyerdahl,** believes that voyagers from South America may have reached islands of the South Pacific prior to

Figure 1–5 The peopling of the Pacific Islands.

The major island groups of the Pacific Ocean are Micronesia (*brown shading*), Melanesia (*red shading*), and Polynesia (*green shading*). The "Lapita people" present in New Ireland 5000–4000 B.C. can be traced to Fiji, Tonga, and Samoa by 1500 B.C. The route of Thor Heyerdahl's balsa raft *Kon Tiki* is also shown.

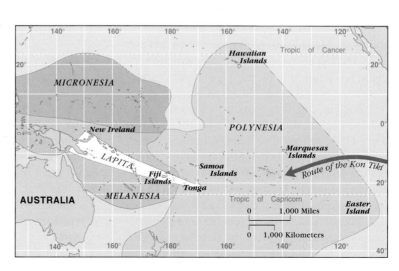

the coming of the Polynesians. He demonstrated the possibility of such voyages with his 1947 voyage of the *Kon Tiki,* a balsa raft designed like those known to have been used by South American navigators at the time of European discovery. Heyerdahl sailed the raft 11,300 kilometers (7000 miles) in the South Equatorial Current to the Tuamotu Islands. He undertook this voyage to prove that early South Americans could have traveled to Polynesia just as easily as early Asian cultures might have. Heyerdahl's views are not yet accepted by many anthropologists, but the final story of the peopling of the Pacific Islands is yet to be written.

European Navigators the first humans from the Western Hemisphere known to have developed the art of navigation were the **Phoenicians,** who lived at the eastern end of the Mediterranean Sea, in the present-day area of Syria, Lebanon, and Israel. As early as 2000 B.C., they were investigating the Mediterranean Sea, the Red Sea, and the Indian Ocean. The first recorded circumnavigation of Africa, in 590 B.C., was made by the Phoenicians, who had also sailed as far north as the British Isles.

The Greek view of the world in 450 B.C. is seen in the map made by the "father of history," **Herodotus** (see Figure 1–1). It shows the Mediterranean Sea surrounded by three continents, Europe, Asia, and Libya (Africa), and bordered by three major seas (called *mares*). The map indicates that Herodotus envisioned the oceans as a band of water that encircled all three continents.

The Greek astronomer-geographer **Pytheas** sailed northward to Iceland in 325 B.C. He had worked out a simple method for determining latitude (one's position north or south), which involved measuring the angle between an observer's line of sight to the North Star and line of sight to the northern horizon (see Appendix II, "Latitude and Longitude on Earth"). Mariners could determine the latitude of any point on the surface of Earth using the method introduced by Pytheas, but it was impossible for them to determine longitude (one's position east or west) accurately (see Appendix II, "Latitude and Longitude on Earth"). Using astronomical measurements, Pytheas also proposed that the tides were a product of lunar influence.

Eratosthenes (276-192 B.C.), a Greek librarian in the Egyptian city of Alexandria, was the first known person to determine Earth's circumference. He calculated that its circumference through the North and South Poles was 40,000 kilometers (24,840 miles), which is an error of only 0.08 percent as compared with the 40,032 kilometers (24,875 miles) determined by more precise methods in use today.

In about 150 A.D., **Ptolemy** produced a map of the world that represented Roman knowledge at that time. He introduced vertical lines of longitude and horizontal lines of latitude. Ptolemy's map indicated, as did the earlier Greek maps, the continents of Europe, Asia, and Africa. It showed the Indian Ocean to be surrounded by a partly unknown landmass. Unlike Herodotus, Ptolemy considered the major oceans to be seas similar to the Mediterranean, having boundaries defined by unknown landmasses.

The Middle Ages

After the fall of the Roman Empire in the fifth century A.D., the Mediterranean area was dominated by Arab influence; the writings of the Greeks and Romans passed into the hands of the Arabs, only to be forgotten in the following centuries by the early Christians in Europe. Subsequently, the Western concept of world geography degenerated considerably; one notion envisioned the world as a disk with Jerusalem at the center.

The Arabs, meanwhile, were trading extensively with East Africa, Southeast Asia, and India. They learned the secret of sea travel in the Indian Ocean by taking advantage of the seasonal patterns of the monsoon winds. During the summer, when the monsoon winds blow from the southwest, ships laden with goods would leave the Arabian ports and sail eastward across the Indian Ocean. The return voyage would be timed to take advantage of the reverse northeasterly trade winds that occur during winter, making the transit relatively simple for their sailing vessels.

In Europe, the nautical inactivity of the southern Europeans was offset by the vigorous exploration of the **Vikings** of Scandinavia. Late in the ninth century, aided by a period of worldwide climatic warming, the Vikings colonized Iceland (Figure 1–6). In 981, **Erik the Red** sailed westward from Greenland and discovered Baffin Island. He returned to Iceland in 984 and led the first wave of Viking colonists to Greenland. **Bjarni Herjolfsson** sailed from Iceland to join the colonists, but he sailed too far

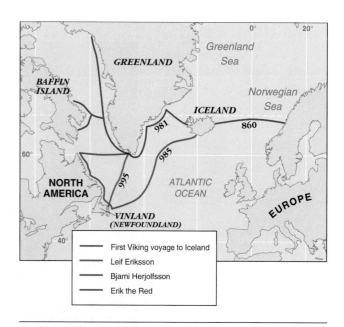

Figure 1–6 The Vikings reach North America.

Map showing the routes and dates of Viking explorations to reach North America (the New World) in 985.

Box 1–1
How Do Sailors Know Where They Are at Sea?: From Stick Charts to Satellites

How does one find where one is at sea without roads, signposts, or even any land in view—nothing but water? It is important to know where one is in the ocean to determine the distance to a destination, to find one's way back to a good fishing spot, or to relocate an area where sunken treasure has been discovered on the sea floor. Sailors have relied on a variety of navigation tools to help them locate where they are at sea.

Some of the first navigators were the Polynesians. It is surprising that the huge size of the Pacific Ocean did not prevent the Polynesians from peopling distant islands. These early navigators certainly must have been very aware of the marine environment and were able to read subtle differences in the ocean and sky. The tools they used to help them navigate between islands included the sun and moon, nighttime stars, the behavior of marine organisms, various ocean properties, and an ingenious device called a **stick chart**. These bamboo stick charts (Figure 1B) depicted islands (represented by shells at the junctions of the sticks), regular ocean wave direction (represented by the straight strips), and even waves that bent around islands (represented by the curved strips). The success of the stick charts depended on the occurrence of a regular pattern of ocean waves. By orienting their vessels relative to this regular ocean wave direction, sailors could navigate even far from land. The bent wave directions let them know when they were getting close to an island, sometimes located beyond the horizon.

The importance of knowing where you are at sea is clearly illustrated by an incident in 1707, when a British battle fleet ran aground in the Isles of Scilly near England, with the loss of four ships and 2000 men. **Latitude** (location north or south) was easy enough to determine by celestial navigation, but the

ship's crew, after several weeks at sea, had lost track of their **longitude** (location east or west; see Appendix II, "Latitude and Longitude on Earth"). To determine longitude, it was necessary to know the *time* at a reference meridian when the sun was directly overhead of a ship at sea (12:00 noon local time). By using the time difference betwen the two locations the ship's longitude could be determined because the overhead passage of the sun changes one hour for every 15 degrees of longitude traveled. However, clocks in the early 1700s were driven by pendulums and would not work for long on a rocking ship at sea. In 1714 the British government offered a £20,000 prize (about $12 million today) to whoever could develop a clock that would work well enough at sea to determine longitude within 0.5 degree after a voyage to the West Indies.

A cabinetmaker in Lincolnshire named **John Harrison** began working on such a timepiece in 1728. His **chronometer** (*chrono*=time, *meter*=measure; Figure 1C) was driven by a helical balance spring that remains horizontal independent of the motion of a ship. However, the device was complex, costly, and fragile. Harrison's first chronometer was successfully tested in 1736, but he received only £500 of the prize. Eventually, his fourth version was tested during a transAtlantic voyage in 1761. Upon reaching Jamaica, it was so accurate that the longitude error was only *0.02 degree* and it had lost only *five seconds* of time! This performance greatly exceeded the requirements of the government, and Harrison laid claim to the prize. The committee in charge of the prize was slow in paying because the majority were astronomers who wanted the solution to come from the stars. The fact that Harrison was a commoner didn't help. In addition, the committee members wanted to be convinced that reliable versions

Figure 1B Polynesian stick chart of the Marshall Islands.

(continued)

of the chronometer could be produced in large numbers. Finally, after intervention by King George III, Harrison received the balance of his prize in 1773, when he was 80 years old!

Today, navigating at sea is a far cry from what it used to be, thanks to new technology called the **Global Positioning System (GPS).** In the early 1970s, the U.S. government began launching a $13 billion system of satellites (Figure 1D) that remain in stationary orbits above Earth. Like a fixed constellation of stars, the satellites send continuous signals to the surface. When the transmissions from at least four satellites are received, a hand-held or boat-mounted receiver at the surface can be used to determine a vessel's exact latitude and longitude to within a few meters. Navigators from days gone by would be amazed how quickly a vessel's location can be determined, but they might say that it has taken all the adventure out of navigating at sea.

Figure 1C John Harrison's chronometer.

Figure 1D A Global Positioning System (GPS) satellite.

south and is thought to be the first Viking to have viewed what is now called Newfoundland. Bjarni did not land and returned to Greenland. In 995 **Leif Eriksson**, son of Erik the Red, bought Bjarni's ship and set out for the land that Bjarni had seen to the southwest. Leif spent the winter in that portion of North America and named the land Vinland (now Newfoundland) after the grapes that were found there.

The Age of Discovery in Europe

During the 30-year period from 1492 to 1522 known as the **Age of Discovery,** the Western world came to a full realization of the vastness of Earth's water-covered surface. The continents of North and South America were explored by Europeans. The globe was circumnavigated for the first time. Europeans now learned that human populations existed elsewhere in the world on newly discovered continents and islands, with cultures vastly different from those familiar to European voyagers.

Why was there such an increase in ocean exploration during the Age of Discovery? One reason was that Sultan Mohammed II captured Constantinople (the capital of eastern Christendom) in 1453. Constantinople is at the eastern crossroads of the Mediterranean Sea, providing access to the East—India, Asia, and the East Indies (modern-day Indonesia). With the capture of this city, Mediterranean port cities were isolated from the riches of the East. This caused the Western world to search for new Eastern trade routes, which opened an era of ocean exploration.

Prince **Henry the Navigator** of Portugal had established a marine observatory to improve Portuguese sailing skills, but all attempts by Portuguese to reestablish trade by ocean routes around Africa met with failure for years. The treacherous journey around the tip of Africa was a great obstacle to establishing an alternative trade route. Cape Agulhas (the southern tip of Africa) was finally rounded by **Bartholomeu Diaz** in 1486. He was followed in 1498 by **Vasco da Gama,** who continued his trip around the tip of Africa to India.

The Atlantic Ocean became familiar to European explorers such as **Christopher Columbus** in 1492 because they were looking for a new route to the East Indies by voyaging *west* (Figure 1–7; for more information about Columbus's voyage, see Box 6–1 in Chapter 6, "Air–Sea Interaction"). However, the Pacific Ocean was not seen by Europeans until 1513, when **Vasco Núñez de Balboa** attempted a land crossing of the Isthmus of Panama and sighted a large ocean to the west from atop a mountain.

The culmination of this period of discovery was the circumnavigation of the globe initiated by **Ferdinand Magellan** (Figure 1–7). In September of 1519 Magellan left Sanlúcar de Barrameda, Spain, with five ships and 280 sailors. He crossed the Atlantic Ocean and traveled through a passage to the Pacific Ocean at 52 degrees south latitude, now named the Strait of Magellan in his honor. After discovering the Philippines on March 15, 1521, Magellan was killed in a fight with the inhabitants of these islands. Juan Sebastian del Caño finally com-

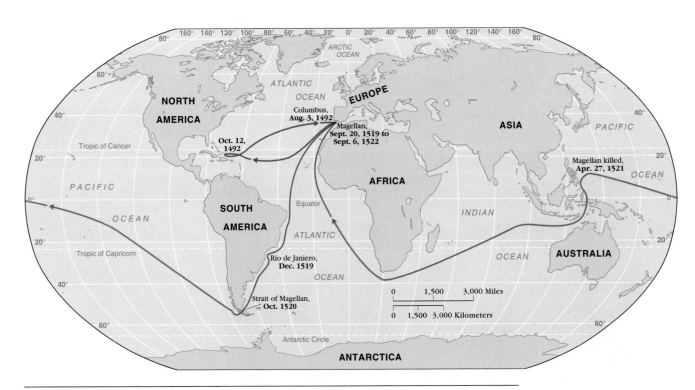

Figure 1–7 Voyages of Columbus and Magellan.
Map showing the dates and routes of Columbus's first voyage and the first circumnavigation of the globe by Magellan's party.

pleted the circumnavigation by taking the last of the ships, the *Victoria,* across the Indian Ocean, around Africa, and back to Spain in 1522. After three years, only one ship and 18 men completed the voyage.

Following these voyages, the Spanish initiated many others to take the gold that had been found in the possession of the Aztec and Inca cultures in Mexico and South America. While the Spanish were occupied with plundering Aztec and Inca wealth, the English and Dutch took advantage of the situation by robbing the Spanish. The political dominance of Spain ended with the defeat of the Spanish Armada by the British in 1588. By stopping Spain from invading the British Isles, the English became the dominant maritime power and remained so until early in the twentieth century.

Voyaging for Science

Early on, the English determined that increasing their knowledge of the oceans would help them maintain their maritime superiority. Thus, the next major focus of interest in the oceans was more scientific. One of the more successful early voyages that sought to learn about the physical nature of the oceans was conducted by the English navigator **James Cook** (1728-1779) (Figure 1–8). Cook was the son of a farm laborer who became one of the most prolific ocean explorers.

From 1768 until his death in 1779, Captain Cook undertook three voyages of discovery. He searched for the continent Terra Australis ("Southern Land," or Antarctica) and concluded that, if it existed, it lay beneath or beyond the extensive ice fields of the southern oceans. Captain Cook discovered the South Georgia and South Sandwich Islands and the Hawaiian Islands, where he was killed after searching for the fabled "northwest passage" from the Pacific Ocean to the Atlantic Ocean. Captain Cook also discovered a way to prevent his crew from contracting the

Figure 1–8 Captain James Cook, Royal Navy.

dreaded disease scurvy, which is caused by a vitamin C deficiency. Cook initiated a shipboard diet that included the German staple sauerkraut, which is a cabbage dish that is loaded with vitamin C. With this important discovery, Cook claimed that a ship's crew could exist at sea almost indefinitely.

Cook's expeditions added greatly to the scientific knowledge of the oceans. He determined the outline of the world's largest ocean, the Pacific, and was the first person known to cross the Antarctic Circle in his search for Antarctica. Cook also led the way in sampling subsurface water temperatures, measuring winds and currents, taking depth measurements (called *soundings*), and collecting data on coral reefs. By proving the value of John Harrison's chronometer as a means of determining longitude (see Box 1–1), Cook made possible the first accurate maps of Earth's surface.

History of Oceanography ... To Be Continued

Of course, there is much more to the history of oceanography. The story of the history of oceanography is continued as a chapter-opening feature at the beginning of each chapter that elaborates upon an important event in the history of oceanography related to that chapter's topic.

The Scientific Method

How can we tell if a scientific theory is right or wrong? Which of two opposing scientific theories is correct? Usually, one can distinguish between conflicting scientific theories by examining the facts. But just what are facts, and how are facts different from theories?

Let's examine an interesting example. Geologists have concluded that, based on scientific evidence, Earth is about 4.6 billion years old. Alternatively, could it be substantially younger? According to one estimate made by counting human genealogy back through time, Irish Archbishop James Ussher in 1654 concluded that Earth was created in 4004 B.C., on October 26, at 9:00 A.M., which is amazingly specific. How can we tell which of these two vastly different ages is correct? The answer may surprise you: *We can't.* Because Earth originated so long ago, no one was around to observe its creation directly. Thus, we cannot say with absolute certainty which of the two dates is correct (and there is also the possibility that neither of the two dates is right!).

This is where the **scientific method** is useful. It is based on an assumption that the physical universe "plays fair;" that is, all natural phenomena are controlled by understandable physical processes and the same physical processes operating today have been operating throughout time (although the rate of operation of these processes may have changed). The scientific method states that *science supports the theory or model that best explains all available observations.* Note that for a model

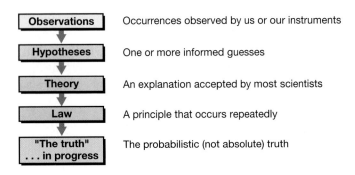

Observations	Occurrences observed by us or our instruments
Hypotheses	One or more informed guesses
Theory	An explanation accepted by most scientists
Law	A principle that occurs repeatedly
"The truth" ... in progress	The probabilistic (not absolute) truth

Figure 1–9   The scientific method.

to be supported, it must be able to take into account *all* available observations, not just some of them that tend to fit well. Also, note that as new observations become available (often with the development of new technology that allows us to observe smaller, deeper, or farther), the model may change.

Figure 1–9 illustrates the steps of the scientific method that allow one to determine which theories or models have scientific validity. **Observations** are occurrences that one can measure with one's senses. They are things one can manipulate, see, feel, hear, touch, or smell, often by experimenting with them directly or by using sophisticated tools (such a microscope or a telescope) to sense them. Other commonly used terms for "observations" include *facts*, *data*, or even *evidence*.

Through many observations (such as noticing how a whale throws its entire body out of the water, called breaching), we begin to organize those observations and develop a **hypothesis**. A hypothesis is a statement about the general nature of the phenomena observed, often labeled as an informed guess (such as the hypothesis that a breaching whale is trying to dislodge parasites from its body). Often, scientists will have multiple working hypotheses (such as a different hypothesis that the breaching whale is trying to communicate with other whales). Hypotheses are used to predict certain occurrences that lead to further research and the refinement of those hypotheses. For example, the hypothesis that a breaching whale is trying to dislodge its parasites would predict that breaching whales have more parasites than nonbreaching whales. The difficult task of collecting data and analyzing the number of parasites on breaching versus nonbreaching whales would either support or weaken that hypothesis.

An additional test that can be applied is called **Occam's razor**, named after William Occam, a thirteenth-century English philosopher. The test states that *the accepted hypothesis should be the simplest hypothesis that explains any given phenomenon*. There is an elegant simplicity that is inherent in most hypotheses. Complexity in an explanation cannot be supported without additional observations to back up that complexity. In the breaching

whale hypothesis, one explanation for the whale's behavior could be that solar flare activity causes many electric eels to discharge electricity into the water at the exact same time, startling the whale and causing it to jump out of the water. Although completely possible, this explanation seems rather unlikely and would need to be supported by observations. Any number of other complicated explanations could be devised to explain why whales breach, but the simplest explanation that is consistent with all observations is often the best one.

If a hypothesis explains something fundamental about our universe and has been strengthened by additional observations and the ability to be used as a predictive tool, then the model can be advanced to what is called a **theory**. A theory is an explanation of physical phenomena accepted by most scientists (for instance, biology's well-known theory of evolution). But just how is a theory accepted by scientists? For instance, are scientists polled on a regular basis as to which theories they accept? Actually, scientists are not polled regularly. Scientists do, however, present their theories at scientific meetings and in scientific journals that are scrutinized by the scientific community. From this, debate ensues, and usually a consensus is reached by most researchers in that particular field of study that a new theory either has validity or it does not.

Sometimes certain theories are so strongly validated that they are advanced to what is called a **law**. A law is a verbal or mathematical statement of a well-established relationship between phenomena. For instance, the law of universal gravitation states that gravity is a constant property in the universe. Drop your pencil off the table, and you know what will happen to the pencil (note that Murphy's law may prevail, but it does not have the same level of acceptance as the law of universal gravitation!). Can a law be broken? The answer to that is yes, and some people would go so far as to say that laws are *made* to be broken. For example, it appears that the law of universal gravitation does not fully describe occurrences that have been observed in some unusual parts of the universe (for example, near so-called "black holes").

The final step in the process is achieving the "*truth*." Does science ever arrive at this point? Actually, no, because we are never certain that we have all the observations, especially considering that new technology will be available in the future. As long as there can be new observations (which is always true), the nature of scientific truth is subject to change. Therefore, it might be more accurate to say that science arrives at the "*truth-in-progre*ss," or the "*probabilistic truth*" (what is *probably* true, based on the available observations).

It is not bad science to change scientific models as more observations are collected. In fact, this is the beauty of science: As new observations are gathered, there is room for the reanalysis of older models. There can be new hypotheses, new theories, and even new laws. In fact, science is littered with interesting theories that have been abandoned in favor of later theories. Hence, science is always developing. It is interesting to compare how scientists and detectives use a common approach, called deductive reasoning. Both try to determine the sequence of events that produced a certain result, often with only fragmentary evidence. The detective's analysis of a crime may change as new clues (new evidence) become available. Just as detectives try to unravel the mysteries of humankind, scientists try to unravel the mysteries of nature.

Is there really such a formal method to science as we have outlined? Actually, the work of scientists is much less formal and is not always done in a clearly logical and systematic way. It involves intellectual inventiveness, the ability to visualize models, serendipity, and even following hunches. With any physical phenomena in nature, there may always be surprises, causing scientists to re-examine old beliefs.

So how old is Earth? Based on the best available scientific evidence from radiometric age dates[2] of the oldest rocks (moon rocks and meteorites, which are thought to be remnants of the early solar system), most scientists would agree that Earth is 4.6 billion years old. However, stay tuned for future developments!

Origins

Humankind has always asked questions about the origins of physical entities in the world. For instance, where did Earth come from? Where did the atmosphere and the oceans come from? When were all of these created? This section deals with such questions and provides answers based on the fragmentary scientific evidence that remains from several billion years ago.

Origin of the Solar System and Earth

Earth is the third of nine planets in our **solar system** which revolve around the sun (Figure 1–10). The orderly nature of our solar system and the consistent age of meteorites (pieces of the early solar system) suggest to astronomers that the solar system formed from the same basic material (various gases and space dust) and at the same time as the sun. This material was a huge cloud of debris, called a **nebula** (*nebula* = a cloud).

The Nebular Hypothesis **The nebular hypothesis** (Figure 1–11) suggests that all bodies of the solar system formed from an enormous nebular cloud composed mostly of hydrogen and helium, with only a small percentage of heavier elements. As this large accumulation of gas and dust revolved around its center of rotation, the sun began to form due to the concentration of par-

Figure 1–10 The solar system.
The planets of the solar system drawn to scale.

ticles under the force of gravity. In its early stages, the diameter of the sun may have at least equaled the diameter of our entire solar system today.

As the nebular matter that formed the sun contracted, a small percentage of it was left behind in small eddies,

[2] For a discussion on radiometric age dating techniques, see "The Geologic Time Scale" section later in this chapter.

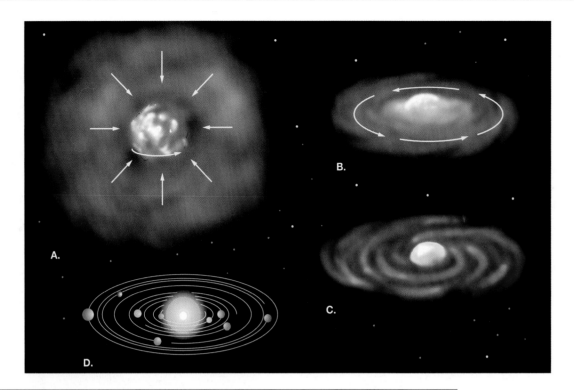

Figure 1–11 The nebular hypothesis.

A. A huge rotating cloud of dust and gases (a nebula) begins to contract. **B.** Most of the material is gravitationally swept toward the center, producing the sun. However, due to rotational motion, some dust and gases remain orbiting the central body as a flattened disk. **C.** The planets are created from the material that is orbiting the flattened disk. **D.** In time most of the remaining debris is either collected into the nine planets and their moons or swept out into space by the solar wind.

like small whirlpools in a stream. This material flattened itself into a disk that became increasingly compact. The stage was set for the formation of a suite of planets, which revolve around the sun. Because of the increasing compactness of this disk, it became gravitationally unstable and broke into separate small clouds. These were the **protoplanets** (*proto* = original) and their orbiting satellites, which later consolidated into the present planets and their moons.

The Protoearth The **Protoearth** looked very different as compared to what Earth looks like today. It was a huge mass, perhaps 1000 times greater in diameter than Earth today and 500 times more massive. There were neither oceans nor any life on the planet. In addition, the Protoearth is thought to have been **homogenous** (*homo*=alike, *genous*=producing), which means that it had a uniform composition throughout. However, this changed when the Protoearth (like all the other protoplanets) began a major period of rearrangement as the heavier constituents migrated toward the center to form a heavy core.

During this early formation of the protoplanets and their satellites, the sun condensed into such a concentrated mass that forces within its interior began releasing energy through a process known as "hydrogen burning." This is the first step in a nuclear reaction called

a **fusion reaction**. In this fusion reaction, when temperatures reach tens of millions of degrees, hydrogen **atoms** (*a*=not, *tomos*=cut) are converted to helium atoms, and large amounts of energy are released. In addition to light energy, the sun also emits ionized (electrically charged) particles. In the early stages of our solar system, this emission of ionized particles served to blow away the nebular gas that remained from the formation of the planets and their satellites.

Meanwhile, the protoplanets closest to the sun were being intensely heated by solar radiation, and their atmospheres boiled away. The combination of ionized solar particles and the internal warming of the planets drastically shrank the planets closest to the sun. Because the gaseous envelope that surrounded the planets was composed mostly of hydrogen and helium (and, most likely, this was Earth's early atmosphere), the heating easily energized atoms of these small masses to escape the gravitational attraction of their planets. As the protoplanets continued to contract, the heat that was produced deep within their cores from the spontaneous disintegration of atoms (radioactivity) became more intense.

Density Stratification The internal heat being released caused Earth ultimately to become molten, and the entire surface of the planet was covered with hot

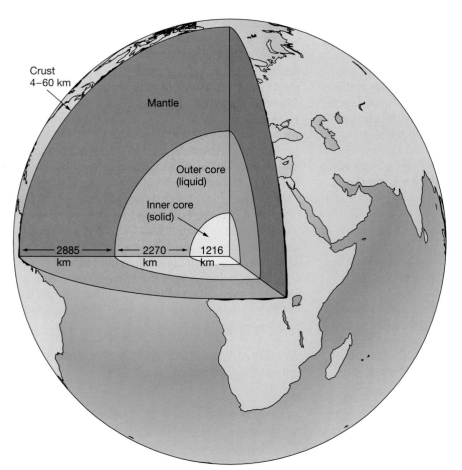

Crust
4–60 km

Mantle

Outer core
(liquid)

Inner core
(solid)

2885
km

2270
km

1216
km

Figure 1–12 View of
Earth's layered structure.
A cutaway view of Earth's interior, showing the major subdivisions of Earth's structure.
The inner core, outer core,
and mantle are drawn to scale,
but the thickness of the crust
is exaggerated about five
times.

liquid rock. This also caused the heavier components, primarily iron and nickel, to sink below the surface and concentrate in the core. The lighter-weight materials segregated according to their **densities**[3] in concentric spheres around this core. Thus, Earth is a layered sphere based on density. The process by which Earth formed these layers is called **density stratification** (*strati*=a layer), where high-density material sinks and low-density material rises (oil-and-vinegar salad dressing that has settled out is a good example of how density stratification causes separate layers to form).

The core has the highest density (Figure 1–12) and is divided into the **inner core**, a solid, and the **outer core**, a liquid. The **mantle** is a zone composed of less-dense rock. The thin **crust** overlying the mantle has an even

lower density and occurs as two types, which vary in thickness and density. One type is oceanic crust, which underlies the ocean basins and is 4 to 10 kilometers (2.5 to 6.2 miles) thick. The other type is continental crust, which underlies the continents and is 35 to 60 kilometers (22 to 37 miles) thick. Oceanic crust is more dense than continental crust. The internal structure of Earth will be discussed further in Chapter 2, "Plate Tectonics and the Ocean Floor."

Origin of the Atmosphere and the Oceans

Where did the material that makes up the present atmosphere and oceans come from? Most likely, this material came from *inside* Earth. As a result of density stratification, low-density material rises. The lowest-density material initially contained within Earth—various gases—rose to the surface, where it was expelled. What exactly was the composition of these gases? They are believed to have been similar to the gases emitted from volcanoes, geysers, and hot springs today. Studies of the composition of these gases reveal that the vast majority of gaseous emissions from volcanic vents is nothing more than water vapor (steam), with small amounts of carbon dioxide, hydrogen, and other gases. These volcanic gases produced Earth's early atmosphere. The process of expelling gases from inside Earth is appropriately called **outgassing**.

---

[3] Density is defined as mass per unit volume. An easy way to remember this is that density is a measure of *how heavy something is for its size.* For instance, an object that has a low density is light for its size (such as a dry sponge, foam packing, or a surfboard). An object that has a high density is heavy for its size (such as cement, most metals, or a large container full of water). Note that density has nothing to do with the *thickness* of an object: Some objects (such as a stack of foam packing) can be thick but have low density.

A vast amount of water vapor was released to the atmosphere during outgassing in the early stages of Earth's development. However, the water vapor didn't stay there long; it eventually condensed and fell to Earth as precipitation (rain, snow, or sleet). Evidence suggests that by at least 4 billion years ago, most of the water vapor from outgassing had condensed to form the first permanent accumulations of liquid water on Earth's surface, and thus the oceans were born (Figure 1–13). Note how the oceans, which have a lower density than the crust, occupy the area above the crust (as would be expected from density stratification), and the lowest-density material, the atmosphere, is the very topmost layer of Earth's layered sphere, farthest from the core. It is indeed surprising that the atmosphere and the oceans are thought to have come from *inside* Earth and that the oceans and the atmosphere have existed for at least 4 billion years.

The Development of Ocean Salinity   The relentless rainfall on Earth's rocky surface weathered it, eroding particles and dissolving elements and compounds, carrying them into the newly forming oceans. This has given our oceans their present chemical composition. But did the oceans start out salty? They certainly must have because many of the compounds eroded from surface rocks contained the elements that comprise salt: chlorine, sodium, magnesium, and potassium. Considering the history of the oceans, you might ask whether they have had the same salinity throughout time or whether they are growing more or less salty. By far the most important component of salinity is the chloride ion, Cl^-, which is produced by outgassing, as was the water vapor, which created the oceans. Thus, the question becomes: Has the proportion of water vapor to chloride ion that has been outgassed remained constant throughout geologic time, or has it varied? Based on studies of the chemistry of ancient marine rocks, there is no indication of any fluctuation in this ratio throughout geologic time. Therefore, based on the present evidence, it can be reasonably concluded that the oceans' salinity has been relatively constant through time. Further aspects of the ocean's salinity will be explored in Chapter 5, "Water and Seawater."

The Development of the Ocean Basins   Due to gravitational attraction, the oceans occupy lower areas

Figure 1–13 Formation of Earth's oceans.

Widespread volcanic activity released water vapor (H_2O vapor) and smaller quantities of other gases. As Earth cooled, the water vapor (**A**) condensed into clouds and (**B**) fell to Earth's surface. There (**C**) it accumulated to form the oceans.

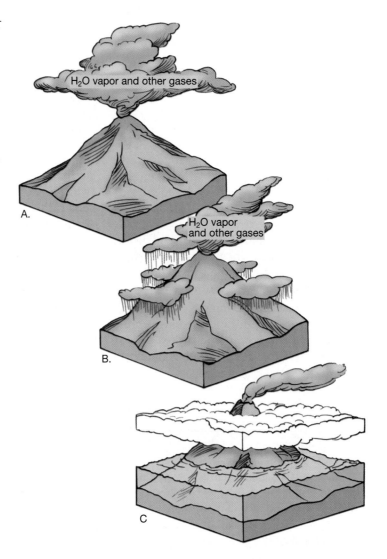

on Earth's crust. But what formed these lower areas, and why are they lower? When Earth began to form its solid crust about 4.6 billion years ago, the mass of the oceans was zero. At present, ocean mass is 14 trillion trillion grams. It is not known what the rate of volcanic release of gases from the mantle was throughout the period of 4.6 billion years. However, if the rate is assumed to have been equal to the present rate, the formation of Earth's oceans can be visualized as a gradual process.

Does Earth's continuing volcanism mean that the oceans are gradually covering more and more of the surface? Not necessarily, because the *surface area* that the oceans cover is determined directly by the *volume that the basin in which they form is able to accept.* During the initial solidification of Earth's crust, there may not have been distinct continents and ocean basins. These may have formed gradually as the oceans themselves were formed.

If it is assumed that the continents formed gradually, then the capacity of the ocean's basins could have gradually increased along with the volume of water that was being produced during the same period. Thus, it is quite possible that the ocean's surface area has been similar for a long period of geologic time and that the only major change in the ocean's character has been an increase in depth. The oceans have become deeper through the long period of time since Earth formed.

Life Begins in the Oceans

Recent scientific research has offered several new ideas on the origin of life on Earth. One idea is that the organic building blocks of life may have arrived embedded in meteors, comets, or cosmic dust (there is new enthusiasm for this idea since the discovery in 1996 of possible fossilized bacteria in a Martian meteorite found in Antarctica in 1984). Another idea is that life may have originated in association with hydrothermal vents deep in the ocean. Yet another idea is that life may have originated in rock material deep below Earth's surface. Other ideas involve the iron-sulfur mineral pyrite or various clay minerals as being an environment for early life forms.

Each idea about the origin of life on Earth has merit and is supported by diverse scientific studies. However, events that are most strongly supported by the fossil record on Earth (which goes back to the earliest life forms—fossilized primitive bacteria—that lived almost 3.5 billion years ago are emphasized in accordance with the scientific method. This record indicates that the basic building blocks for the origin of life were provided by materials already present on Earth and that the most likely place in which they could interact to produce life was in the oceans.

The Importance of Oxygen to Life

Oxygen makes up almost 21 percent of our present atmosphere. It is essential to human life for two reasons. First, oxygen is what we must inhale so that our bodies can "burn" (oxidize) food, releasing energy to our cells. Second, we depend on oxygen for protection from harmful ultraviolet radiation from the sun that constantly bombards our planet. Much of the ultraviolet radiation is absorbed in our upper atmosphere by the ozone layer (ozone molecules consist of three oxygen atoms: O_3). This allows only a small proportion of ultraviolet energy to reach Earth's surface.

Although there is some dispute about the composition of the early atmosphere, evidence suggests that it was different from the initial hydrogen-helium atmosphere on Earth and different from the one it has at present (mostly nitrogen-oxygen). This early atmosphere probably contained little free oxygen (which is oxygen that is not chemically bound to other atoms) and had large percentages of carbon dioxide and water vapor.[4] Why was there so little oxygen in the early atmosphere? Oxygen may well have been outgassed, but oxygen and iron have a strong affinity for each other (for instance, consider how common rust—iron and oxygen—is on Earth). Iron in Earth's early crust would have reacted with and chemically bonded to outgassed oxygen immediately, thus removing oxygen from the atmosphere.

Of course, if oxygen was missing from the early atmosphere, ultraviolet radiation would have readily penetrated the atmosphere and reached the surface of the young oceans. These oceans and the atmosphere contained many gases, including hydrogen (H_2), carbon dioxide (CO_2), methane (CH_4), and ammonia (NH_3). Laboratory experiments by Dr. Stanley L. Miller in 1952 showed that exposing a mixture of hydrogen, carbon dioxide, methane, ammonia, and water to ultraviolet light, plus an electrical spark (readily available in the form of lightning, which often strikes the ocean), will produce a large assortment of organic molecules (Figure 1–14).

These laboratory experiments showed that the production of organic molecules must have resulted in a vast amount of organic material present in Earth's early oceans. It is not known precisely how this organic material took on the characteristics that allowed it to become living substance. Through some process, however, the organic material became alive by being chemically self-reproductive, developing the ability to actively seek light, metabolizing food, and growing toward a characteristic size and shape dictated by internal molecular codes.

Plants and Animals Evolve

The very earliest forms of life must have been **heterotrophs** (*hetero* = different, *tropho* = nourishment). These organisms depend on an external food supply. That food supply was certainly abundant in the form of

---

[4] There is some evidence that oxides of carbon and nitrogen were released and constituted the primary component of Earth's early atmosphere.

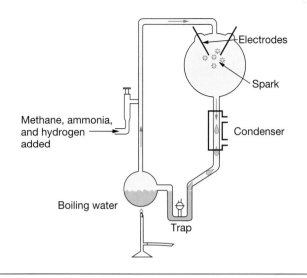

Figure 1–14 Synthesis of organic molecules.
The apparatus used by Dr. Stanley L. Miller in the 1952 experiment that resulted in the synthesis of the basic components of life (organic molecules) from a mixture resembling the composition of the early atmosphere and oceans.

molecules of the nonliving organic material from which the simplest organisms originated.

The **autotrophs** (*auto* = self, *tropho* = nourishment), which are organisms that do not depend on an external food supply but manufacture their own, eventually evolved. The first autotrophs may have been similar to our present-day bacteria of the **anaerobic** (*an* = without, *aero* = air) type, which live without atmospheric oxygen.

They may have been able to use inorganic compounds to release energy for producing their own food internally. This process is called **chemosynthesis** (*chemo* = chemistry, *syn* = with, *thesis* = an arranging).

At some later date, the more complex single-celled autotrophs probably evolved. They developed a green pigment called **chlorophyll** (*chloro* = green, *phyll* = leaf), which captures the sun's energy. Through **photosynthesis** (*photo* = light, *syn* = with, *thesis* = an arranging), these organisms produced their own food from the carbon dioxide and water that surrounded them using chlorophyll. Photosynthesis is a chemical reaction in which energy from the sun becomes stored in organic molecules (sugars) (Figure 1–15).

Fossils that may be the remains of photosynthetic bacteria have been recovered from rocks formed on the sea floor almost 3.5 billion years ago. However, an oxygen-rich atmosphere is not indicated until about 2 billion years ago, as evidenced by the oldest rocks containing iron oxide (rust). This also closely follows the appearance of complex cells with nuclei (eucaryotes) that first appear about 2.1 billion years ago. By 1.8 billion years ago, the oxygen content of the atmosphere had reached one-tenth of its present concentration. This addition of free oxygen to the atmosphere surely caused the extinction of many anaerobic single-celled species that had evolved in an oxygen-free environment. The survivors of this environmental crisis are the ancestors of Earth's present dwellers.

Photosynthesis and Respiration   Photosynthetic cells had developed security in the form of their built-in

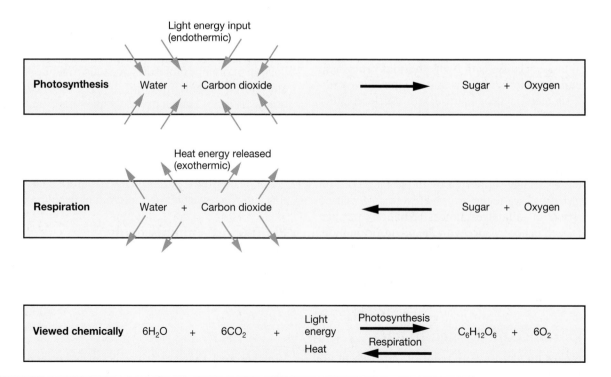

Figure 1–15 Photosynthesis and respiration.

mechanism to manufacture a food supply. Heterotrophs, not so endowed, were at a considerable disadvantage because they had to search for food. To meet this need, animals eventually evolved, with their inherent mobility and an awareness or consciousness that enabled them to obtain food.

Photosynthetic organisms and animals developed in a beautifully balanced environment where the waste products of one filled the vital needs of the other. This can be seen expressed chemically by looking at the complementary processes of photosynthesis and respiration. In the photosynthetic process, energy is captured and stored in the form of organic compounds (food sugars). This is an **endothermic** (*endo* = inside, *thermo* = heat) chemical reaction, one that *stores* energy. **Respiration** (*respir* = to breathe) is the reaction by which energy stored in food produced by photosynthesis is extracted by the plants and animals to carry on their life processes. It is the opposite of photosynthesis, making it an **exothermic** (*exo* = outside, *thermo* = heat) reaction, one that *releases* energy (these processes are shown in Figure 1–15).

Every living organism that inhabits Earth today is the result of an evolutionary process of **natural selection** that has been going on since these early times and by which various life forms (species) have been able to inhabit increasingly numerous niches within the environment. As these diverse life forms adapted to various environments, they also modified the environments in which they lived.

As plants emerged from the oceans to inhabit the terrestrial environment, they changed the landscape from a harsh, bleak panorama (which may have resembled the lunar surface) to the soft green hues that cover much of Earth's land surface today. Other changes were manifested in the ocean itself as vast quantities of the hard parts of dead organisms accumulated on the bottom as marine sediments. Some of these accumulations have been uplifted and exposed on the continents, sometimes at high elevations in mountain areas. This evidence suggests that these rocks formed as sea floor sediments and were elevated to their present position by great forces within Earth. Because over half of the rocks exposed at the surface of continents originally formed on the ocean floor, there is much to be learned about the oceans of the past by studying these rocks.

Changes to Earth's Environment Probably the most important modification of the environment by its living inhabitants involved the atmosphere. This change—the release of large quantities of oxygen from photosynthesis—made it possible for most present-day animals to develop. A by-product of photosynthesis in the early oceans was the free oxygen that makes up almost 21 percent of our present atmosphere. Carbon dioxide, which at one time must have made up a large portion of our atmosphere, was removed by photosynthesis to produce a tiny (but important) concentration of 0.035 percent in Earth's present atmosphere. These changes in the atmosphere must have had a great effect on climates and organisms (Figure 1–16).

In petroleum and coal deposits, there are remains of plant and animal life that became buried in an oxygen-free environment. This allowed their energy to remain in storage for millions of years. These deposits, which we call *fossil fuels* (coal, oil, and natural gas), provide us with over 90 percent of the energy humans consume today to do such things as heat our homes, power our lights, and manufacture goods. Not only do we, as animals, depend on the present productivity of plants to supply the energy required by our life processes, we also depend very heavily on the energy stored by plants during the geologic past.

Because of increased burning of fossil fuels for home heating, industry, power generation, and transportation over the past century, the atmospheric concentration of CO_2 and other gases that help warm the atmosphere has increased. Many people are concerned that Earth is warming to a degree that could cause serious problems now and for generations to follow. This phenomenon, referred to as the increased *greenhouse effect,* is discussed in Chapter 6, "Air-Sea Interaction."

The Geologic Time Scale

The geologic time scale (Figure 1–17) shows the names of the geologic time periods, as well as important advances in the development of plants and animals on Earth. How was this time scale developed? Initially, it was based on major extinction episodes as recorded in the fossil record. The discovery of radioactivity early in the twentieth century allowed earth scientists to measure how old certain rocks are by using small amounts of radioactive materials within the rocks[5]. The geologic time scale was then modified to reflect actual age dates (in millions or billions of years before present).

Most of the rocks that are found on the continents (as well as those from outer space) contain small amounts of radioactive elements such as uranium, thorium, and potassium. Radioactive elements break down into atoms of other elements. Each radioactive **isotope** (*iso* = equal, *topos* = place) is an atom of an element that has an atomic weight different from that of other atoms of the element. Each isotope has a specific **half-life,** the time it takes for one-half of the atoms in a sample to decay to atoms of some other element. Such dating is referred to as **radiometric** (*radio* = radioactivity, *metri* = measure) **age dating**.

By comparing the quantities of the radioactive isotope with the quantities of their decay products in rocks (Figure 1–18), the age of the rocks may be determined to within 2–3 percent, which is fairly accurate. However,

---

[5] In essence, this new technology allowed scientists to read a rock's internal "rock clock."

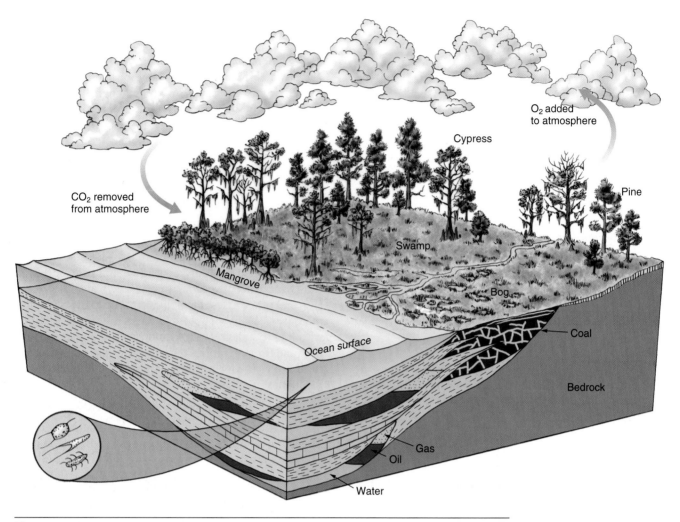

Figure 1–16 The effect of plants on Earth's environment.

As microscopic photosynthetic cells (*inset*) became established in the ocean, Earth's atmosphere was enriched in oxygen and depleted in carbon dioxide as a result of photosynthesis. As these cells and the animals that ate them died and sank to the ocean floor, some of their remains were incorporated into deposits and were eventually converted to petroleum (oil and gas). The same process occurred on land, sometimes producing coal.

in very old rocks, this seemingly small error can be quite large—for example, 2 percent of 2 billion years is 40 million years. Still, this method of age determination is an extremely powerful tool for interpreting the age of rocks.

Students Sometimes Ask...

You mentioned that the oceans came from inside Earth. However, I recently heard that the oceans came from outer space as icy comets. Which one is true?

Recent research suggests another possible mechanism for the origin of the oceans that you have heard about. Scientists have discovered what appear to be comets (each about the size of a small house) composed of ice that are entering Earth's atmosphere at a staggering rate of as many as 40,000 a day. These small comets disintegrate in the atmosphere upon entry. Still, calculations show that the comets add about 2.5 centimeters (1 inch) of water every 10,000 years to the planet, which is enough to fill the oceans over the lifetime of Earth. It is still a matter of debate whether the objects are really icy comets, but this theory suggests that the oceans originated as ice from outer space. Which theory is correct? Because the origin of the oceans occurred so long ago, there is only circumstantial evidence to help support either theory. Perhaps new developments in technology will be able to fuel discoveries that will support one or the other of these theories (or even an entirely new theory). To us, this is what makes science so interesting!

You've talked about the limits of compressed-air diving. What is the record depth for a dive from the surface with one breath?
Again, the problem with any ocean dive is the vastly increased pressure with depth. However, some people are pushing the

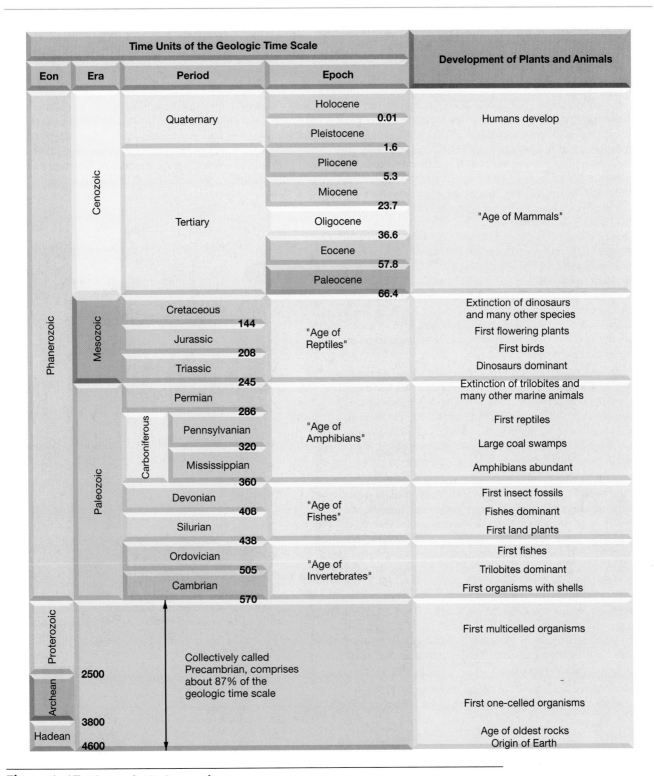

Figure 1–17 The geologic time scale.
Numbers on the time scale represent time in millions of years before the present.

limits of human endurance by holding onto a heavy weight at the surface and descending rapidly into the ocean. At the deepest depth they can tolerate, they let go of the weight and return to the surface. The world's record for these so-called "free dives" was set by Francisco Ferreras in 1996 near Cabo San Lucas, Mexico, when he dove to 130 meters (428 feet), which took 2 minutes 11 seconds.

Is technology available to allow underwater divers to breathe a substance called "liquid air"?
Yes, the technology exists, but no, it won't be available for some time. The concept of breathing an inert liquid filled with a high percentage of dissolved oxygen gas is called "liquid breathing" or "liquid ventilation." The technique uses a chemical called perfluorocarbon, which is an extremely stable and

Box 1–2
"Deep" Time

As shown in Figure 1E, Earth is 4.6 billion years old. The oceans and the atmosphere are at least 4 billion years old. The oldest fossilized life forms are 3.5 billion years old. The atmosphere became oxygen-rich about 2 billion years ago. These are impressive numbers, but it is difficult to comprehend a number as large as 4.6 billion (that's 4600 million, or 4,600,000,000). To give some idea of the immensity of a number in the billions, imagine a fresh one-dollar bill. How thick would a stack of 100 fresh one-dollar bills be? Its thickness would be about 1 centimeter (0.4 inch) high. How thick would a stack of 1000 fresh one-dollar bills be? About 10 centimeters (4 inches) high. How about a stack of 1,000,000 (one million) fresh one-dollar bills? About 100 meters (330 feet) high. How about 1,000,000,000 (one billion) in fresh one-dollar bills? As nice as this would be to imagine, it would be about 100 kilometers (62 miles) high! The problem with an example such as this is that the mind becomes immune to imagining such large amounts of money.

Let's look at another example. How long would it take to count to 4.6 billion if you were to count one number every second for 24 hours a day, seven days a week, 365 days a year? The answer to this can be worked out fairly easily, remembering that there are 60 seconds in a minute and 60 minutes in an hour. You'll be surprised by the answer (which, if you need help, is hidden below).

Another way to help visualize geologic time is by utilizing a roll of toilet paper.[6] Unrolling an entire roll of toilet paper is an interesting (and entertaining) way to envision the immensity of geologic time. Considering a standard 500-sheet roll of toilet paper (and rounding off the age of Earth to 5 billion years), each sheet on the roll represents an amazing 10 million years. When unrolling the roll (which represents the entirety of geologic time), remember that the last sheet on the roll (the last 10 million years) is about four times longer than the time humans have existed on Earth. All of recorded human history is represented by the last $^1/_{2000}$ of a sheet, and a long human lifetime of 100 years is only $^1/_{100,000}$ of a sheet! Indeed, it is humbling to realize how insignificant the length of time human existence has been.

Can one really imagine the space taken up by $1 billion, or the lifetimes it would take to count to 4.6 billion, or how many sheets of toilet paper have gone by in 4.6 billion years of Earth history? The same scale problem exists with visualizing millions or billions of years, often called "deep" time or geologic time. But keep trying, and maybe someday you will get a glimpse into the huge expanse of time represented by the geologic time scale. It will amaze you.

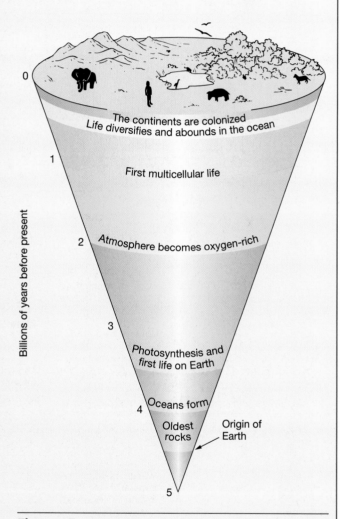

Figure 1E Major events in Earth's development.

---

[6] Toilet paper has many advantages: It is something with which most people are familiar, it is inexpensive and readily available, it is long and linear (like geologic time), and it is even perforated into individual sheets.

Answer to question posed above: 4.6 billion seconds corresponds to 145.9 years (which amounts to about two human lifetimes!).

---

nontoxic liquid used to cool electronic equipment. The first experiments were conducted successfully in 1966 to save ill animals. In 1989, it was used on a prematurely born infant (the first human) in an unsuccessful effort to save her life. The fluid is infused with oxygen and breathed into the lungs, where it exchanges oxygen directly into the lung tissue. It is currently an emergency medical technique that has never been attempted on adults. However, some believe that the use of liquid breathing may enable people to dive to record depths in the future. This technology was featured in the movie, *The Abyss*.

I've heard stories about unusual incidences at sea in the Bermuda Triangle. Where is the Bermuda Triangle and are the stories for real?

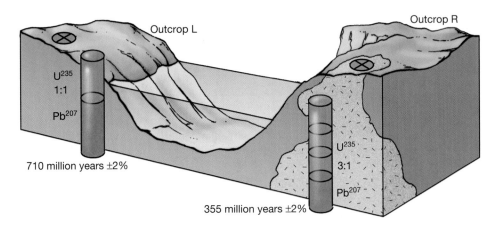

Figure 1–18 Radiometric dating.

Outcrops L and R contain the radioactive isotope uranium 235 (U^{235}). It has a half-life of 710 million years and decays to lead 207 (Pb^{207}). At outcrop L the ratio of U^{235} to Pb^{207} of 1:1 means that half of the U^{235} atoms have decayed to Pb^{207}, so the rocks are one half-life of U^{235} old, or 710 million years. In outcrop R, the ratio of U^{235} to Pb^{207} is 3:1. This means that one-fourth of the original U^{235} atoms have decayed to Pb^{207}, so the rock is one-half of a half-life old, or 355 million years.

The "Bermuda Triangle" (also known as the "Devil's Triangle" or the "Hoodoo Sea") is an imaginary triangular area located off the southeastern Atlantic coast of the United States that is noted for a high incidence of unexplained losses of ships, small boats, and aircraft. Although the U. S. Board of Geographic Names does not recognize the Bermuda Triangle as an official name, the apexes of the triangle are generally accepted to be Bermuda Island; Miami, Florida; and San Juan, Puerto Rico. Despite the popular misconception that something unusual is happening in the Bermuda Triangle, many experts (including the U.S. Coast Guard) have exposed the mystery of the Bermuda Triangle as a myth.

In the past, extensive but futile Coast Guard searches prompted by search and rescue cases such as the disappearances of an entire squadron of TBM Avengers shortly after take off from Fort Lauderdale, Florida, in 1945 or the traceless sinking of the vessel SS *Marine Sulphur Queen* in the Florida Straits in 1963 have lent credence to the popular belief in the mystery and the supernatural qualities of the Bermuda Triangle. Countless theories (including alien abductions) have been proposed to explain the many disappearances throughout the history of the area. The most practical explanations are related to environmental factors such as:

- The overlapping of the magnetic and true north directions, easily leading to confusion for navigators if they do not properly compensate for it.

- The strength and turbulent character of the Gulf Stream current.

- The unpredictable Caribbean-Atlantic weather pattern, often causing sudden local thunderstorms and water spouts.

- The abundance of navigational hazards due to extensive shoals and reefs around many islands.

- The presence of large deposits of gas hydrates (discussed in Chapter 4, "Marine Sediments") on the sea floor, which, if mobilized, may also be a hazard to navigation.

In addition, the human error factor should not be discounted as a contributing factor. For instance, a large number of pleasure boats travel the waters between Florida's Gold Coast and the Bahamas. All too often, crossings are attempted with too small a boat, insufficient knowledge of the area's hazards, unsuitable maps, and a lack of good seamanship.

The Coast Guard has publicly stated that it is not impressed with any of the supernatural explanations of disasters in the Bermuda Triangle. It has been their experience that the forces of nature combined with the unpredictability of mankind outdo even the most far-fetched science fiction that people can imagine.

I've heard of the U.S. Geological Survey and the National Biological Service. Is there a U.S. Oceanographical Survey that serves as the branch of the U.S. government dedicated to researching oceanographic phenomena?
Not quite. The branch of the U.S. government that oversees oceanographic research is the Commerce Department's National Oceanic and Atmospheric Administration (NOAA, pronounced "noah"). Scientists at NOAA work to ensure wise use of ocean resources through the National Ocean Service, the National Oceanographic Data Center, the National Marine Fisheries Service, and the National Sea Grant Office branches. Other U.S. government agencies that work with oceanographic data include the U.S. Naval Oceanographic Office, the Office of Naval Research, and the U.S. Coast Guard.

I'm considering majoring in oceanography. Which universities have graduate-level programs in oceanography?
Numerous schools both in the United States and abroad offer undergraduate and graduate degrees in oceanography or marine science. In the United States, the three most prominent oceanographic institutions are Scripps Institution of Oceanography of the University of California San Diego in La Jolla, California; Woods Hole Oceanographic Institution at Woods Hole, Massachusetts; and the Lamont-Doherty Earth Observatory of Columbia University in Palisades, New York. If you are considering working in the field of oceanography, see Appendix IV, "Careers in Oceanography."

I've seen a movie recently where Earth was flooded by the melting of the ice caps and all the continents were under water. Could this really happen?
You can't always believe what is portrayed in Hollywood movies. Since the oceans already contain over 97 percent of all the world's water, if all the ice in the world (which is only slightly more than 2 percent of the total water) melted, sea level

would rise an estimated 60 meters (200 feet), which would certainly flood coastal areas, but there would still be plenty of land above sea level on which to live. Was the movie *Waterworld*?

How has oceanography changed since the breakup of the former Soviet Union?

The major change that has occurred is the redirection of the mission of the Russian and U.S. navies from warfare to cooperative research. This has brought about the release of once-classified data. For a half-century, the Russian and U.S. navies competed for tactical advantage in the oceans of the world. This was especially true in the Arctic Ocean, where the two countries spent billions of dollars mapping the sea floor and ice

sheets and collecting data on the physical and chemical properties of Arctic seawater. In January 1997, the U.S. and Russian governments announced the release of a new atlas of the Arctic Ocean, which includes new satellite photos and an analysis of some of the 1.3 million measurements of temperature and salinity collected from 1948 to 1993. This new atlas effectively doubles the amount of data available to scientists who work on modeling the Arctic Ocean and its influence on climate change. Scientists are particularly interested in polar data because the earliest signs of phenomena such as the greenhouse effect will likely be seen in these high latitudes.

Summary

Despite our planet having 70.8 percent of its surface covered by water, it has been named "Earth." The world ocean is an interconnected body of water, which is large in size and volume. It can be divided into four principal oceans (the Pacific, Atlantic, Indian, and Arctic Oceans) plus an additional ocean (the Southern or Antarctic Ocean). Even though there is a technical distinction between a sea and an ocean, the two terms are used interchangeably. In comparing the combined oceans to the combined continents, it is apparent that the average land surface does not rise very far above sea level and that there is not a mountain on Earth that is as tall as the deepest parts of the ocean.

In the Pacific, people who populated the Pacific Islands were probably the first great navigators. In the Western world, the Phoenicians were making remarkable voyages as well. Later the Greeks, Romans, and Arabs made significant contributions and advanced oceanographic knowledge. During the Middle Ages, the Vikings colonized Greenland and made voyages to North America. The Age of Discovery in Europe renewed the Western world's interest in exploring the unknown. It began with the voyage of Columbus in 1492 and ended in 1522 with the first circumnavigation of Earth by a voyage initiated by Ferdinand Magellan. Captain James Cook was one of the first to explore the ocean for scientific purposes.

The scientific method is used to understand the occurrence of physical processes such that science supports the theory or model that best explains all available observations. This is done through making observations, forming one or more hypotheses (educated guesses), and developing a theory (an explanation accepted by most scientists). An additional test is one of simplicity, called Occam's razor. A theory can be advanced to the status of a law if the relationship stated by the theory repeatedly occurs. Science never arrives at the absolute "truth;" rather, science arrives at the "probabilistic truth," which is considered good science.

Our solar system, consisting of the sun and nine planets, may have formed from a huge cloud of gas and space dust called a nebula. According to the nebular hypothesis, the nebular matter contracted to form the sun, and the planets were formed from eddies of material that remained. The massive sun, composed of hydrogen and helium, was large enough to

become a fusion reactor and began to give off energy at high rates. The protoplanets were smaller, so they did not reach this state.

The Protoearth, more massive and larger in size than Earth today, was molten and homogenous, with an early atmosphere composed mostly of hydrogen and helium, which were later driven off into space. The Protoearth began a period of rearrangement, forming a layered structure of core, mantle, and crust through the process of density stratification. During this period, Earth also developed an atmosphere rich in water vapor and carbon dioxide produced by a process called outgassing. As Earth's surface cooled sufficiently, the water vapor condensed and accumulated in depressions on the surface (basins) to give Earth its first oceans. Rainfall on the surface dissolved elements, which were carried to the ocean and caused the ocean to be salty.

It was in these oceans that life is thought to have begun. As ultraviolet radiation fell on the oceans with their dissolved hydrogen, carbon dioxide, methane, and ammonia—along with inorganic molecules—all of which may have combined to produce carbon-containing molecules that are now formed naturally on Earth only by organisms. Chance combinations of these molecules eventually produced heterotrophic organisms (which cannot make their own food) that were probably similar to present-day anaerobic bacteria. Eventually, autotrophs evolved that had the ability to make their own food (through the process of chemosynthesis). Later, some cells developed chlorophyll, which allows photosynthesis and resulted in the development of plants.

Photosynthesis by plants caused carbon dioxide to be extracted from the atmosphere and also caused free oxygen to be released at Earth's surface. This produced an oxygen-rich atmosphere in which animals as we know them could survive. Eventually, both plants and animals evolved into forms that could survive in the stark environment of the continents, producing the lush continental environments that exist on Earth today.

Through the process of radiometric age dating, the age of certain rocks can be determined. The geologic time scale indicates that Earth has experienced a long history of changes since its origin 4.6 billion years ago.

Key Terms

Age of Discovery (p. 14)

Anaerobic (p. 22)

Antarctic Ocean (p. 8)

Arctic Ocean (p. 7)

Atlantic Ocean (p. 7)

Atom (p. 18)

Autotroph (p. 22)

Bathyscaphe (p. 8)

Chemosynthesis (p. 22)

Chlorophyll (p. 22)

Chronometer (p. 12)

Columbus, Christopher (p. 14)

Cook, James (p. 15)

Crust (p. 19)

da Gama, Vasco (p. 14)

de Balboa, Vasco Núñez (p. 14)

Density (p. 19)

Density stratification (p. 19)

Diaz, Bartholomeu (p. 14)

Endothermic (p. 23)

Eratosthenes (p. 11)

Erik the Red (p. 11)

Eriksson, Leif (p. 14)

Exothermic (p. 23)

Fusion reaction (p. 18)

Global Positioning System (GPS) (p. 13)

Half-life (p. 23)

Harrison, John (p. 12)

Henry the Navigator (p. 14)

Herodotus (p. 11)

Herjolfsson, Bjarni (p. 11)

Heterotroph (p. 21)

Heyerdahl, Thor (p. 10)

Homogenous (p. 18)

Hypothesis (p. 16)

Hypsographic curve (p. 8)

Indian Ocean (p. 7)

Inner core (p. 19)

Isotope (p. 23)

Kon Tiki (p. 11)

Latitude (p. 12)

Law (p. 16)

Longitude (p. 12)

Magellan, Ferdinand (p. 14)

Mantle (p. 19)

Natural selection (p. 23)

Nebula (p. 17)

Nebular hypothesis, the (p. 17)

Observations (p. 16)

Occam's razor (p. 16)

Outgassing (p. 19)

Outer core (p. 19)

Pacific Ocean (p. 7)

Phoenicians (p. 11)

Photosynthesis (p. 22)

Protoearth (p. 18)

Protoplanet (p. 18)

Ptolemy (p. 11)

Pytheas (p. 11)

Radiometric age dating (p. 23)

Respiration (p. 23)

Scientific method (p. 15)

Scuba (p. 5)

Solar system (p. 17)

Southern Ocean (p. 8)

Stick chart (p. 12)

Theory (p. 16)

Trieste (p. 8)

Vikings (p. 11)

Questions and Exercises

1. Discuss what problems the human body can experience as a result of diving below sea level.

2. How did the view of the ocean by early Mediterranean cultures influence the naming of planet "Earth?"

3. What is the difference between an ocean and a sea? Which ones are the seven seas?

4. Describe several differences between the combined oceans and the combined continents.

5. When humans first visited the deepest ocean floor, what was the voyage like and what did they see?

6. Describe the development of navigation techniques that have enabled sailors to navigate in the open ocean far from land.

7. Using a diagram, illustrate the method used by Pytheas to determine latitude in the Northern Hemisphere.

8. While the Arabs dominated the Mediterranean region during the Middle Ages, what were the most significant ocean-related events taking place in northern Europe?

9. Describe the important events in oceanography that occurred during the Age of Discovery in Europe.

10. List some of the major achievements of Captain James Cook.

11. In your own words, describe how the scientific method operates.

12. Does science ever reach the "truth?" Why or why not?

13. Discuss the origin of the solar system using the nebular hypothesis.

14. How was the Protoearth different from today's Earth?

15. What is density stratification, and how did it change the Protoearth?

16. What is the origin of Earth's atmosphere?

17. What is the origin of Earth's oceans?

18. Have the oceans always been salty? Why or why not?

19. New water is continually being released to the atmosphere by volcanic activity. Why does this not necessarily mean that the oceans are progressively covering an increasing percentage of Earth's surface?

20. How does the presence of oxygen in our atmosphere help reduce the amount of ultraviolet radiation that reaches Earth's surface?

21. What was Dr. Stanley Miller's experiment, and what did it help demonstrate?

22. How did the photosynthesis that produced the first organic molecules in the early oceans differ from plant photosynthesis? How does chemosynthesis differ from photosynthesis?

23. Discuss photosynthesis and respiration and explain how they are related.

24. Define in your own words the process of evolution by natural selection.

25. Describe some of the major changes to Earth's environment that were produced by plants.

26. Construct a representation of the geologic time scale, using an appropriate quantity of any substance (other than dollar bills or toilet paper). Be sure to indicate some of the major changes that have occurred on Earth since its origin.

References

Borgese, E. M., and Ginsberg, N. 1993. *Ocean yearbook 10.* Chicago: University of Chicago Press.

Boror, D. J. 1960. *Dictionary of word roots and combining forms.* Mountain View, CA: Mayfield.

Brewer, P., ed. 1983. *Oceanography: The present and future.* New York: Springer-Verlag.

Deacon, M. 1971. *Scientists and the sea 1650–1900: A study of marine science.* London: Academic Press.

Duxbury, A. C. 1971. *The earth and its oceans.* Reading, MA: Addison-Wesley.

Frank, L. A., and Sigwarth, J. B. 1993. Atmospheric holes and small comets. *Review of Geophysics* 31, 1–28.

———. 1997. Detection of atomic oxygen trails of small comets in the vicinity of Earth. *Geophysical Research Letters* 24:19, 2431–2434.

Glaessner, M. F. 1984. *The dawn of animal life: A biohistorical study.* Cambridge: Cambridge University Press.

Gregor, B. C., Garrels, R. M., Mackenzie, F. T., and Maynard, J. B., eds. 1988. *Chemical cycles in the evolution of the earth.* New York: Wiley.

Hendrickson, R. 1984. *The ocean almanac.* New York: Doubleday.

Heyerdahl, T. 1979. *Early man and the ocean.* New York: Doubleday.

Holland, J. D. 1984. *The chemical evolution of the atmosphere and oceans.* Princeton, NJ: Princeton University Press.

Idyll, C. P., ed. 1969. *The science of the sea: A history of oceanography.* London: Thomas Y. Crowell.

Kennish, M. J., ed. 1994. *Practical handbook of marine science,* 2nd ed. Boca Raton, FL: CRC Press.

Kirschvink, J. L., Ripperdan, R. L., and Evans, D. A. 1997. Evidence for a large-scale reorganization of early Cambrian continental masses by inertial interchange true polar wander. *Science* 277:5325, 541–545.

Marx, R. 1996. The early history of diving. In Pirie, R. G., ed., *Oceanography: Contemporary readings in ocean sciences,* 3rd ed. New York: Oxford University Press.

McKay, D. S., et al. 1996. Search for past life on Mars: Possible relic biogenic activity in Martian meteorite ALH84001. *Science* 273:5277, 924–930.

Melamed, Y., Shupak, A., and Bitterman, H. 1996. Medical problems associated with underwater diving. In Pirie, R. G., ed., *Oceanography: Contemporary readings in ocean sciences,* 3rd ed. New York: Oxford University Press.

Mowat, F. 1965. *Westviking.* Boston: Atlantic–Little, Brown.

Phillips, J. L. 1998. The bends: Compressed air in the history of science, diving, and engineering. Connecticut: Yale University Press.

Rubey, W. W. 1951. Geologic history of seawater: An attempt to state the problem. *Geological Society of America Bulletin* 62:1110–1119.

Schopf, J. W. 1993. Microfossils of the early archean Apex Chert: New evidence of the antiquity of life. *Science* 260:5108, 640–646.

Sears, M., and Merriman, D., eds. 1980. *Oceanography: The past.* New York: Springer-Verlag.

Sobel, D. 1995. *Longitude: The true story of a one genius who solved the greatest scientific problem of his time.* New York: Walker.

Strom, K. M., and Strom, S. E. 1982. Galactic evolution: A survey of recent progress. *Science* 216:4546, 571–580.

Sverdrup, H. U., Johnson, M. W., and Fleming, R. H. 1942, reprinted 1970. *The oceans: Their physics, chemistry, and general biology.* Englewood Cliffs, NJ: Prentice-Hall.

Various authors. 1997. Special section on small comets (5 papers). *Geophysical Research Letters* 24:24, 3105–3124.

Waldrop, M. M. 1989. The (liquid) breath of life. *Science* 245:4922, 1043–1045.

Suggested Reading

Earth

Alper, J. 1994. Earth's violent birth. 3:7, 56-63. A brief history of the changes that have occurred on Earth since its early stages of development 4.6 billion years ago.

Cone, J. 1994. Life's undersea beginnings. 3:4, 34–41. Examines the evidence supporting the idea that life's early origin may have begun at underwater volcanic hot springs, which today support unusual life forms.

O'Hanlon, L. 1996. The measure of a mountain. 5:1, 50–57. A look at the techniques used to determine the heights of mountains and a discussion of the controversy over which mountain is the tallest.

Orange, D. L. 1996. Mysteries of the deep. 5:6, 42–45. This historical view examines the advancements in technology that have brought about the discovery of many secrets of the oceans as the science of oceanography has developed.

Radetsky, P. 1998. Life's crucible. 7:1, 34–41. A debate on the origin of life on Earth.

Reid, M. 1992. Ghosts of the Burgess Shale. 1:5, 38–45. Fossils that are 530 million years old reveal some interesting early life forms.

Vogel, S. 1996. Living planet. 5:2, 26–35. A comparison of the different conditions that lead to the development of a hospitable climate on Earth, but not on Mars and Venus.

Zaburunov, S. A. 1992. Monitoring our global environment. 1:4, 46–53. Shows the kinds of oceanographic data that can be obtained by orbiting satellites and includes interesting maps.

Sea Frontiers

Baker, S. G. 1981. The continent that wasn't there. 27:2, 108–114. A history of the search for Ptolemy's Terra Australis Incognita, the unfound southern continent.

Engle, M. 1986. Oceanography's new eye in the sky. 32:1, 37–43. Photographs taken from the Space Shuttle by Paul Scully-Power, a U.S. Navy oceanographer, add to our knowledge of ocean currents and waves.

Maranto, G. 1991. Way above sea level. 37:4, 16–23. A discussion of Navstar Global Positioning System satellites and their potential for helping scientists map changes in sea level.

McClintock, J. 1987. Remote sensing: Adding to our knowledge of oceans—and Earth. 33:2, 105–113. The role of remote sensing, primarily in aiding us in understanding weather and climate, is discussed.

Schuessler, R. 1984. Ferdinand Magellan: The greatest voyager of them all. 30:5, 299–307. A brief history of the voyage initiated by Magellan to circumnavigate the globe.

Scientific American

Badash, L. 1989. The age-of-the-Earth debate. 261:2, 90–97. A history of the development of knowledge concerning Earth's age.

Bothun, G. D. 1997. The ghostliest galaxies. 276:2, 56–61. Huge galaxies too diffuse to be seen until the 1980s were found to contain mass equal to that of previously known visible galaxies. They provide clues to a better understanding of how mass is distributed throughout the universe.

De Duve, C. 1996. The birth of complex cells. 274:4, 50–59. A discussion on the evolutionary process that may have led to the development of eukaryotic cells that are from 10 to 30 times larger than the prokaryotic cells from which they evolved.

Gibson, E. K., Jr., McKay, D. S., Thomas-Keprta, K., and Romanek, C. S. 1997. The case for relic life on Mars. 227:6, 58–65. A review of the compelling evidence from a Martian meteorite found in Antarctica suggesting that Mars has had (and may still have?) microbial life.

Giles, D. L. 1997. Faster ships for the future. 277:4, 126–31. New designs for ocean freighters could double their speeds.

Herring, T. A. 1996. The global positioning system. 274:2, 44–50. A look at the technology involved in making the Global Positioning System (GPS) work.

Horgan, J. 1991. In the beginning. 264:2, 116–125. An overview of data relating to the validity of the Big Bang theory of the origin of the universe.

Kasting, J. F., Toon, O. B., and Pollack, J. B. 1988. How climate evolved on the terrestrial planets. 258:2, 90–97. A discussion of the possible sequence of events that culminated with the atmospheres that now exist on Mercury, Venus, Earth, and Mars.

Luu, J. X. and Jewitt, D. C. 1996. The Kuiper belt. 274:5, 46–53. Beyond Pluto lies a belt of objects a few tens to hundreds of kilometers across. This is where short-period comets may originate.

McMenamin, M. A. S. 1987. The emergence of animals. 256:4, 94–103. A discussion on how the explosive diversification of animal forms 570 million years ago may have been related to the breakup of a single large continental landmass.

Moon, R. E., Vann, R. D., and Bennett, P. B. 1995. The physiology of decompression illness. 273:2, 70–77. A summary of current knowledge about decompression illness and a look at research that is being conducted with pressure chambers.

Richelson, J. T. 1998. Scientists in black. 278:2, 48–55. Scientists and intelligence officials are working together to use technology that enables them to analyze features below sea level.

Stebbins, G. L., and Ayala, F. J. 1985. The evolution of Darwinism. 253:1, 72–85. New advances in molecular biology and new interpretations of the fossil record add to the knowledge of evolution.

Whitehead, H. 1985. Why whales leap. 252:3, 84–93. An attempt to answer the enigmatic question, "Why do whales breech?"

Wilson, A. C. 1985. The molecular basis of evolution. 253:4, 164–175. Mutations within the genes of organisms play an important role in evolution at the organismal level.

Oceanography on the Web

Visit the *Essentials of Oceanography* home page for on-line resources for this chapter. There you will find an on-line study guide with review exercises, and links to oceanography sites to further your exploration of the topics in this chapter. *Essentials of Oceanography* is at: **http://www.prenhall.com/thurman** (click on the Table of Contents menu and select this chapter).

CHAPTER 2

PLATE TECTONICS AND THE OCEAN FLOOR

Voyages to Inner Space: Visiting the Deep Ocean Floor via Submersibles

For as long as people have been harvesting the bounty of the oceans and traveling on the oceans in vessels, they have dreamed of plumbing the mysterious depths of the deep ocean, an area known as "inner space." The inaccessibility of the deep ocean and its environmentally different conditions as compared to the surface have limited human exploration. One way to explore the deep ocean is to build specially designed vessels that can transport surface conditions to the deep ocean. These vessels are collectively called **submersibles** and must be built to withstand the intense pressure experienced at depth. Some submersibles are suspended from ships on long cables, while others are untethered and motorized with thrusters for exploring the sea floor independently. Tradition credits Alexander the Great with the first descent in a sealed waterproof container, which reportedly took place in 332 B.C. Unfortunately, there is no record of what Alexander's submersible looked like.

Much later, warfare motivated people to develop an important submersible vessel: a **submarine**. A submarine is a submersible that can submerge, propel itself underwater, and surface under its own power. Submarines are difficult for a ship to detect and can be used to sink enemy ships. The earliest report of a submarine used in warfare is from the early sixteenth century. Greenlanders used sealskins to waterproof a three-person, oar-powered submarine, which was used to drill holes in the sides of Norwegian ships. As technology developed, so did the submarine. Today, nuclear-powered submarines are capable of cruising underwater through entire ocean basins and launching nuclear missiles.

Reaching even deeper depths was the goal for other submersibles. In 1934 naturalist William Beebe and his engineer-associate Otis Barton made a record-setting dive off Bermuda in a submersible called a **bathysphere** (*bathos* = depth, *sphere* = ball) to observe marine life.

A heavy steel ball with thick walls, the bathysphere was suspended from a ship at the end of a cable to a depth of 923 meters (3028 feet). Because the quarters were so cramped, dives were limited to no more than three-and-a-half hours. In 1964, the research submersible *Alvin* (Figure 2A) from Woods Hole Oceanographic Institution began exploring the deep ocean. The 7.6-meter- (25-foot) long *Alvin* can carry its crew of one pilot and two scientists to a depth of 4000 meters (13,120 feet) and maneuver independently along the sea floor. Since it was

Figure 2A The deep-diving submersible *Alvin*.

commissioned, it has made numerous dives to allow oceanographers to explore the deep ocean floor and retrieve samples. Some notable accomplishments of *Alvin* include locating a sunken hydrogen bomb, discovering unique life forms along hot water springs at the mid-ocean ridge, and finding the sunken wreck of the *Titanic*. The deepest-diving manned submersible currently in use is *Shinkai 6500*. This Japanese research vessel can dive to 6500 meters (21,320 feet).

Recently, there has been renewed interest in exploring inner space in a manner that would give humans the same mobility as marine mammals. Graham Hawkes has developed and tested a lightweight, single-passenger, maneuverable 3.5-meter (11.5-foot) fiberglass **microsub** called *Deep Flight I* (Figure 2B). The design of the vessel allows for optimum mobility and speed underwater for a single pilot in a position that instead of being seated is lying down, face forward, and prone. This is the nat-ural position of people and marine mammals when they swim, and, through the use of *Deep Flight's* thrusters, gives the sensation of flying underwater. Currently, the vessel has a depth capacity of 1000 meters (3281 feet). A new vessel that uses specially designed ceramics to withstand high pressure is currently in the design stage. Called *Deep Flight II*, it will be able to carry its occupant to below 11,000 meters (36,089 feet), which would make even the deepest trenches accessible.

Today, only the unmanned robotic vessel *Kaiko* has the capability to travel all the way to the deepest ocean depths. Still, there is a need for direct human observation of the deep ocean floor that robotic craft cannot duplicate. There is no substitute for having an adaptable, resourceful, inventive human for certain tasks on the deep ocean floor. In addition, manned submersibles keep alive the chance for human travel to the great ocean depths in the spirit of exploration.

Figure 2B *Deep Flight I,* with inventor Graham S. Hawkes inside the micro-sub.

How many of us have ever experienced an earth-quake or witnessed a volcanic eruption? Surprisingly, these events are fairly common (there are several thousand earthquakes and dozens of volcanic eruptions on Earth each year). Many people have experienced these events as examples of the dynamic nature of our planet. Not only do these phenomena occur on land, they also occur regularly on the ocean floor, constantly changing the surface of our planet. Even though these events have occurred throughout history, only three decades ago most scientists clung to the notion that the continents were stationary over geologic time, remaining fixed in their geographic positions. However, since that time, a bold new theory was intro-duced that helps explain why our planet looks the way that it does. Widely hailed as a modern scientific break-through, this theory has revolutionized the way in which our planet is viewed. It has received overwhelming sup-port from a variety of different scientific disciplines be-cause, for the first time, it helps explain various surface features and phenomena on Earth, including:

- The worldwide locations of volcanoes, faults, earth-quakes, and mountain building
- Why mountains on Earth haven't been eroded away
- The origin of most landforms and ocean floor fea-tures

Figure 2–1 Alfred Wegener, circa 1912–1913.

- How the continents and ocean floor formed, and why they are different
- The continuing development of Earth's surface
- The distribution of past and present life on Earth

This new theory is called **plate tectonics** (*plate* = plates of the **lithosphere**; *tekton* = to build), or "the new global geology." The basic concept of plate tectonics is this: *The outermost portion of Earth is composed of a patchwork of thin, rigid plates[1] that move horizontally with respect to one another.* Earth's outer layer is divided into about a dozen large slabs, or plates, like pieces of a cracked eggshell on a fractured egg. However, unlike a fractured egg, the pieces can move across the surface horizontally, like icebergs floating on water. The interaction of these plates as they move builds the features of Earth's crust (such as volcanoes, mountains, and ocean basins). A result of this new theory is that earth scientists now understand that the continents are *not* fixed in geographic position. The continents are mobile and move about on the Earth's surface, controlled by forces deep within Earth.

Geologists sometimes refer to the movement of the continents across Earth's surface as **continental drift**, which is the original name of the hypothesis when it was first proposed in 1912 by **Alfred Wegener** (Figure 2–1). Wegener is considered by most scientists to be the pioneer of modern plate tectonic theory. A German meteorologist, he originally proposed the concept of continents drifting across the globe to help explain the ancient climates that were evidenced in the rocks deposited in ocean basins and on the continents. However, like most new theories, Wegener's idea of continental drift encountered a great deal of skepticism from the scientific community. The main weakness of the theory was the lack of a plausible mechanism for such movement (how could a landmass as large as a continent actually move?). Unfortunately, Wegener never lived to see the widespread acceptance of his idea: He died in 1930 while gathering additional data in Greenland to help support his hypothesis.

Today, there is much more known about the internal structure of Earth and its role in controlling processes that occur on Earth's surface. Along with the features of the ocean floor, knowledge about the internal structure of Earth helped modern-day scientists advance the initial ideas about the movement of the continents.

Earth Structure

Earth is a layered sphere based on density, with the highest-density material located closest to the center of Earth. From the center of Earth toward the surface, there is lower- and lower-density material arranged in spheres or shells much like the layers of an onion. Let's examine these layers and discover their importance to plate tectonic processes.

Chemical Composition Versus Physical Properties

Figure 2–2 shows a cutaway (cross-sectional) view of Earth that illustrates Earth's internal structure. As mentioned in Chapter 1, Earth is layered based on density, with the highest-density material found near the center

---

[1] These thin, rigid plates are pieces of the lithosphere (*lithos* = rock, *sphere* = ball) which comprise Earth's outermost portion.

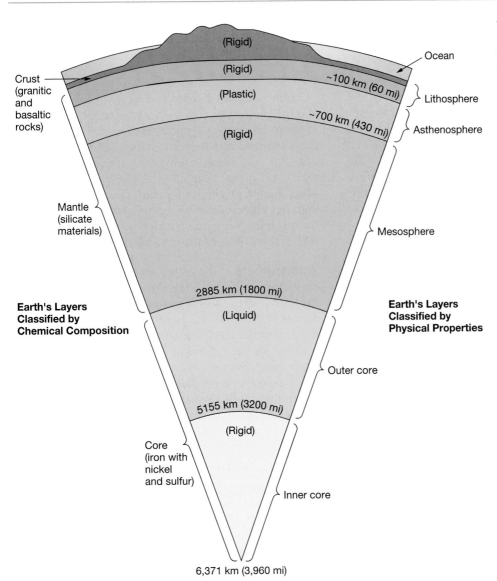

Figure 2–2 Comparison of Earth's chemical composition and physical properties.

Crust (granitic and basaltic rocks)

Mantle (silicate materials)

Earth's Layers Classified by Chemical Composition

Core (iron with nickel and sulfur)

Ocean

~100 km (60 mi)

Lithosphere

~700 km (430 mi)

Asthenosphere

Mesosphere

(Rigid)

(Rigid)

(Plastic)

(Rigid)

2885 km (1800 mi)

(Liquid)

Earth's Layers Classified by Physical Properties

Outer core

5155 km (3200 mi)

(Rigid)

Inner core

6,371 km (3,960 mi)

---

of Earth. Other layers are composed of shells or spheres around the center, each with a lower density. This creates the internal structure of Earth. There are two ways of subdividing Earth's inner structure: an older scheme based on chemical composition (what the composition of the rocks are), and a more modern scheme based on physical properties (how the rocks respond to increased temperature and pressure at depth).

Chemical Composition Based on chemical composition, Earth is composed of three layers: the **crust**, the **mantle**, and the **core** (Figure 2–2). Like the thin skin on an apple, the crust is the thinnest and outermost layer of the three, which extends from the surface to an average depth of 30 kilometers (20 miles). All of the rock material that exists at Earth's surface is considered crustal material. The crust is composed of relatively low-density rock material, consisting of aluminum, magnesium, and iron **silicates** (common rock-forming minerals with silicon and oxygen). There are two types of crust, which will be discussed in the next section.

Immediately below the crust is the mantle. It occupies the largest volume of the three layers, and extends to a depth of 2900 kilometers (1800 miles). The mantle is composed of relatively high-density rock material, consisting of iron and magnesium silicates.

Beneath the mantle is the core. It forms a large mass from 2900 kilometers (1800 miles) to the center of Earth at 6370 kilometers (3960 miles). The core is composed of high-density metal composed of iron and nickel.

Physical Properties Based on physical properties, Earth is composed of five layers: the lithosphere, the **asthenosphere** (*asthenos* = weak, *sphere* = ball), the mesosphere (*mesos* = middle, *sphere* = ball), the **outer core**, and the **inner core**. The lithosphere is Earth's cool, rigid, outermost layer, which extends from the surface to an average depth of about 100 kilometers (62 miles) and includes the crust plus the topmost portion of the mantle. The lithosphere is a **brittle** (*brytten* = to shatter) solid, meaning that it will fracture when force is applied to it. The plates involved in plate tectonic motion are the plates of the lithosphere.

Beneath the lithosphere is the asthenosphere. As its name implies, the weakness of the asthenosphere is manifested in that it is a **plastic** (*plasticus* = to mold) solid, meaning that it will flow when a gradual force is applied to it. It extends from about 100 kilometers (62 miles) to 700 kilometers (430 miles) below the surface, which is the base of the upper mantle. At these depths, it is hot enough to partially melt portions of most rocks.

The **mesosphere** is the next layer beneath the asthenosphere and extends to a depth of 2900 kilometers (1800 miles). The mesosphere is the rigid portion of the middle and lower mantle. Although it is below the asthenosphere which deforms plastically, the additional pressure probably prevents the mesosphere from flowing.

The core is broken into two subdivisions based on physical properties: the outer core, which is liquid and capable of flowing; and the inner core, which is rigid and does not flow. Again, the increased pressure keeps the inner core from flowing.

Knowledge of Earth's Inner Structure

How have scientists determined the information presented above concerning the inner structure of Earth? For instance, have scientists ever retrieved rock samples from below the crust to determine their chemical composition or physical properties? The deepest well in the world was drilled in the Kola Peninsula of Russia, where a world-record depth of 12,266 meters (40,478 feet, or nearly 8 miles) was reached in 1992. Although the temperature at the bottom of the well was a scorching 245 degrees Centigrade (475 degrees Fahrenheit), the drillers never penetrated beneath the crust. To obtain information about the deep Earth, earth scientists have had to rely on indirect observations. Like listening through a doctor's stethoscope, seismologists (*seismo* = earthquake, *ologist* = one who studies) can analyze the pattern of seismic waves bouncing around within Earth. These patterns give a wealth of information about the chemical composition and physical properties of Earth's internal layers.

Near the Surface

Let's concentrate on the layers closest to the surface. Figure 2–3 is an enlargement of the outermost 700 kilometers (430 miles) of Earth's internal structure.

Lithosphere

The lithosphere is a relatively cool, rigid shell that includes *all* of the crust and the topmost part of the mantle. In essence, the topmost part of the mantle is attached to the crust and the two act as a single unit, approximately 100 kilometers (62 miles) thick. Notice that the crust portion of the lithosphere is further subdivided into two types: **oceanic crust** and **continental crust**.

Oceanic crust is composed of the rock type **basalt**, which is dark-colored and has a relatively high density. It averages only about 8 kilometers (5 miles) in thickness. The basaltic oceanic crust has an average density of about 3.0 grams per cubic centimeter (three times the density

of water). Basalt originates as molten material beneath Earth's crust. The magma comes to the surface through underwater volcanoes on the sea floor and hardens to form new oceanic crust.

Continental crust is composed mostly of a lower-density and lighter-colored rock called **granite**.[2] It has a density of about 2.7 grams per cubic centimeter. The continental crust averages about 35 kilometers (22 miles) in thickness but may reach a maximum thickness of 60 kilometers (37 miles) beneath the highest mountain ranges. Most granite originates beneath the surface as molten material that cools and hardens within Earth's crust to form granite. No matter which type of crust is at the surface, it is still part of the lithosphere.

Astheosphere

The asthenosphere is a relatively hot, plastic region beneath the lithosphere. It extends from the base of the lithosphere to a depth of about 700 kilometers (430 miles) and is entirely contained within the upper mantle. The asthenosphere can deform without fracturing if the force is applied slowly. This means that it has the ability to flow but has high **viscosity**. Viscosity is a measure of a fluid's resistance to flow.[3] Studies indicate that the high-viscosity asthenosphere is flowing slowly through time, which has important implications in plate tectonic processes.

Isostatic Adjustment

The phenomenon known as **isostatic** (*iso* = equal, *stasis* = standing) **adjustment** refers to the vertical movement of crust. Isostatic adjustment is caused by the **buoyancy** (the tendency to float or rise in a fluid) of Earth's lithosphere as it floats on the denser, plasticlike asthenosphere below it. The concept of isostatic adjustment is comparable to a ship floating in water (Figure 2–4). An empty ship floating in water will establish its buoyancy in the water based on the weight and design of the ship. If the ship is loaded full of cargo, the ship will undergo isostatic adjustment by floating *lower* in the water (but hopefully won't sink!). Once the cargo is unloaded, the ship will again isostatically adjust itself and float *higher* in the water.

The same is true for the crust, which floats on the mantle. Continental crust is less dense than oceanic crust. Both float on the denser mantle beneath, but the less-dense continental crust floats higher. The oceanic crust floats lower in the mantle and is thin, which creates low

---

[2] At the surface, continental crust is often covered by a relatively thin layer of surface sediments. Below these, granite can be found.

[3] Substances that have high viscosity (a high resistance to flow) include toothpaste, honey, tar, and Silly Putty™; a common substance that has low viscosity is water. A substance's viscosity often changes with temperature. For instance, as honey is heated, it flows more easily.

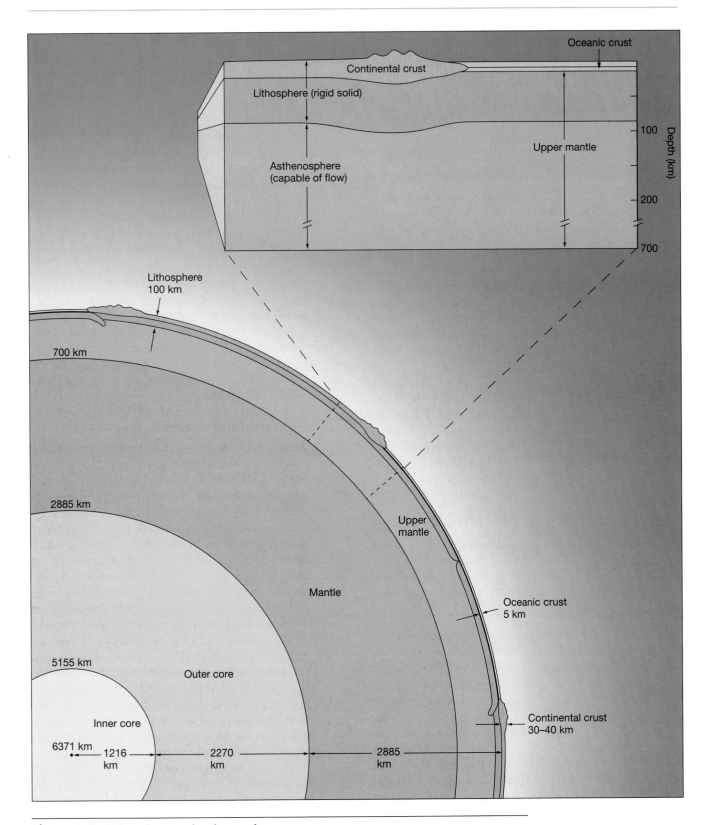

Figure 2–3 Lithosphere and asthenosphere.

areas for oceans to occupy. Areas where the continental crust is thickest (in large mountain ranges on the continents) also float higher than continental crust of normal thickness. Tall mountain ranges on earth have a thick region of crustal material beneath them that in essence keeps them afloat.

There is good evidence that during the last Ice Age, which occurred during the Pleistocene Epoch between 2 million and 10,000 years ago, massive ice sheets covered far northern continental locations such as Scandinavia and northern Canada. The additional weight of ice several kilometers (a few miles) thick caused these areas

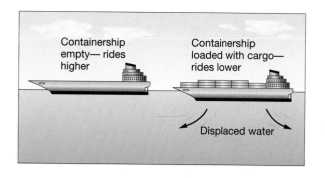

Figure 2–4 Isostatic adjustment.
A boat that rides higher in water when empty will ride lower in water when it is loaded with cargo.

to adjust themselves isostatically to lower in the mantle. Since the end of the last ice age, these areas have been relieved of the additional weight of the ice and have been isostatically adjusting themselves upward. This phenomenon is called **isostatic rebound**, which continues today in areas that have been recently deglaciated. The rate of rise that the land experiences gives scientists important information about the fluid characteristics of the mantle. In addition, it helps support one important aspect of plate tectonics: Since continents can move *vertically*, they are not firmly fixed in one position, suggesting that they might also be able to move *horizontally* across Earth's surface.

Some Principles of Plate Tectonics

Plate tectonics is unique in that so many aspects of the development of Earth's surface can be explained by plate tectonic theory. The following are examples of the phenomena and features that can be explained using some of the principles of plate tectonics

- *The plates of the lithosphere carry the continents like passive passengers.* The continents move about on Earth's surface by being carried along by the lithospheric plates. At some times in the geologic past, there were many small continents. At other times, the continents collided to form large landmasses. About 200 million years ago, all the continents were combined into one giant landmass called **Pangaea** (*pan* = all, *gaea* = Earth). At that time, Earth had only one continent (Figure 2–5; imagine what political problems this would cause if humans lived on Earth then!). Logically, if all the continents were together, then there must also have been one huge ocean surrounding them. Sure enough, there was, and this ocean is called **Panthalassa** (*pan* = all, *thalassa* = sea). Panthalassa included several smaller seas, one of which was the centrally located and shallow **Tethys** (*Tethys* = a Greek sea goddess) **Sea**. About 180 million years ago, the supercontinent began to split apart, and the various conti-

nental masses that exist today started to drift toward their present geographic positions.

- *Continents can drift at surprisingly fast rates.* How fast do plates move? Currently, the average rate of plate movement is between 2 and 12 centimeters (1 and 5 inches) per year. Surprisingly, this is about the same rate as a person's fingernails growth.[4] This may not sound very fast, but remember that plates have been moving for millions of years. Even an object moving at a low rate of speed will eventually travel a great distance over a very long time. For instance, fingernails growing at a rate of 8 centimeters (3 inches) per year for one million years would be 80 kilometers (50 miles) long.

  The movement of plates caused our planet to look very different in the geologic past. Some continents have drifted from the equator to a pole and back to the Equator again. Other continents have split apart. Still others have been created by the collision of fragments of landmasses from locations far and wide across the globe. Considering the immense span of geologic time, the continents have indeed drifted at surprisingly fast rates (but their movement is still about 10,000 times slower than the movement of an hour hand on a clock!).

- *The global mid-ocean ridge is where new crust is created.* The **mid-ocean ridge** (Figure 2–6) is one of our planet's most impressive features. It is a continuous underwater mountain range that winds through every ocean basin in the world and resembles the seam on a baseball. The mid-ocean ridge is named for the fact that in most oceans it snakes down the *middle* of the ocean basin. It is entirely volcanic in origin, wraps one-and-a-half times around the globe, and rises over 2.5 kilometers (1.5 miles) above the ocean floor (to the point where it sometimes is exposed above sea level, such as at Iceland). Studies of the mid-ocean ridge indicate that new ocean floor (new basaltic crust) is forming at the crest, or axis, of the mid-ocean ridge. It then is carried away from the axis as plates move, and the upwelling of volcanic material fills the void with new strips of ocean floor. This process is known as **sea floor spreading**, and the axis of the mid-ocean ridge, where spreading initiates, is referred to as a **spreading center**. An interesting analog for the mid-ocean ridge is a zipper that is being pulled apart. It is true, then, that the Earth's zipper (the mid-ocean ridge) is becoming unzipped!

- *The deep ocean trenches are where crust is destroyed.* **Ocean trenches** (Figure 2–6) are the deepest parts of the ocean floor and have the geometry

---

[4] Fingernail growth is dependent on many factors, including a person's heredity, gender, diet, the amount of exercise, but averages about 8 centimeters (3 inches) per year.

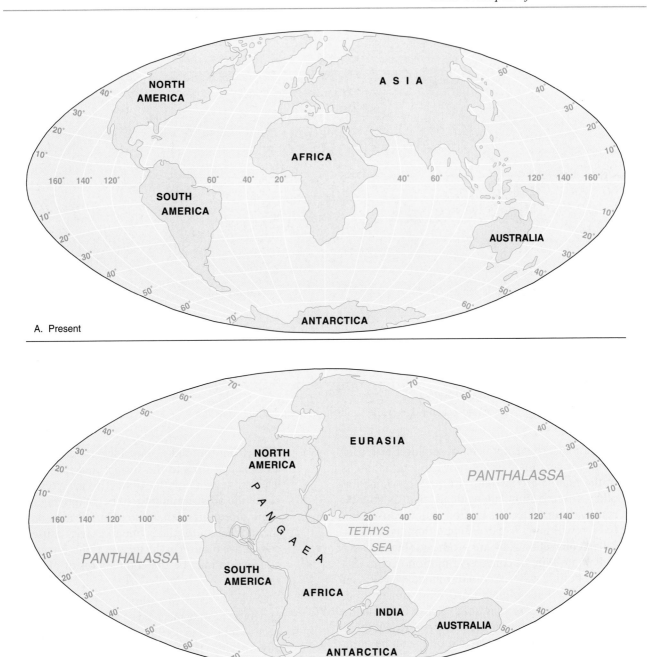

Figure 2–5 Reconstruction of Pangaea.

A. The positions of the continents today. **B.** The positions of the continents about 200 million years ago, showing the supercontinent of Pangaea.

of a narrow crease or trough. Some of the largest earthquakes in the world occur near these trenches because the lithospheric plate is bent downward and slowly plunges back into Earth's interior. This process is called **subduction** (*sub* = under, *duct* = lead), and the sloping area from the trench along the downgoing plate is called a **subduction zone**. Melting in the subduction zone causes a row of highly active and explosively erupting volcanoes to occur on the surface. This row is called a **volcanic arc** because these volcanoes occur in arc-shaped rows.

- *The driving mechanism may be gravity.* What causes the plates to move? This is not an easy question to answer, and in fact in Wegener's time opponents used this argument to cast doubts on his hypothesis of continental drift. Today, seismologists have sensitive seismic instruments that allow them to record energy released from earthquakes bouncing around inside Earth. In essence, this enables seismologists to "see" the inner workings of Earth's internal structure like a medical doctor using an X-ray or CAT scan on a patient. Although

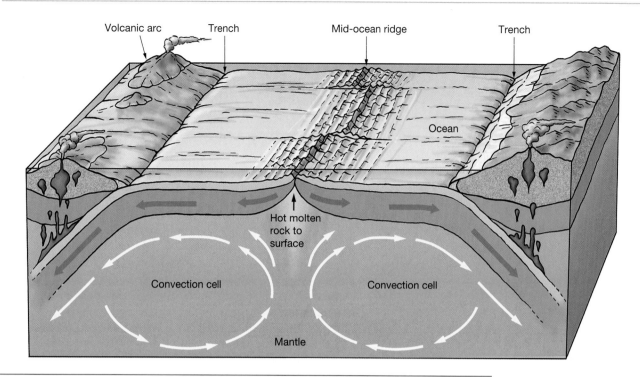

Figure 2–6 Processes of plate tectonics.

Sea floor spreading occurs at the mid-ocean ridge, while subduction occurs at the trenches.

seismologists have not been observing Earth's interior for very many years, they can tell that areas of the mantle are moving with respect to one another. There is abundant evidence that the mantle is moving in a circular motion called **convection** (*con* = with, *vect* = carried) and that giant loops, called **convection cells**, are produced (Figure 2–6). This same type of convective motion occurs in a pot of boiling water. Computer models have been developed that simulate Earth's interior motion, and these indicate that heat release from inside Earth can cause mantle convection. Further, these models show that convective motion can be quite complex, with multiple convection cells at different depths occurring inside the mantle. But is mantle convection causing the plates to move?

A new interpretation indicates that gravity *may* cause the plates to move and that the mid-ocean ridge results from the lithosphere being pulled apart by the subducting slabs at the edges of the plates. Since the mid-ocean ridge is a topographically high feature, the plates of the lithosphere slide away from this area toward the deep ocean trenches (topographically low areas). In essence, gravity may be the controlling factor in causing this movement to occur.[5] Once the plates move

away from the mid-ocean ridge, molten material from below rises to the surface and fills the gap. The movement of the plates by gravity would cause the adjacent mantle to begin moving in the same direction as the plate. Thus, it may be that the movement of plates due to gravity causes the convective motion in the mantle. If this is true, then convection may be the complex result of gravity sliding.

However, scientists are not yet convinced that gravity is the only mechanism at work. Perhaps convective motion in conjunction with gravity is actually driving the plates. There are still many unanswered questions. Part of the uncertainty is that the theory of plate tectonics is so new that scientists have been more concerned with describing plate tectonic phenomena than with defining what the controlling mechanisms are. In addition, the mantle is difficult to study because it is so far from the surface and the motions involved are quite slow. However, this has opened new areas of research that keep advancing our knowledge of Earth processes.

Evidence for Plate Tectonics

So far in this chapter, many of the features and processes of plate tectonics have been introduced. But what is the evidence that the plates are actually moving? Let's examine the wealth of data that supports the theory of plate tectonics.

[5] Similar occurrences around a house can also be attributed to the effects of gravity. For example, a comforter on a bed that is not tucked in will, when jostled, eventually slide off the bed (a high area) onto the floor (a lower area).

Fit of the Continents

Since the first accurate maps of the world were produced, many people have noticed the striking fit of the continents across certain ocean basins—for instance, the fit of South America and Africa across the Atlantic Ocean. As far back as 1620, **Sir Francis Bacon** wrote about how the continents appeared to fit together like a jigsaw puzzle. The idea of the continents fitting together lay dormant until 1912, when Wegener used the shapes of matching shorelines on different continents as a supporting piece of evidence for his theory of continental drift. However, the fit of shorelines was not as close as initially expected: There were considerable areas of crustal overlap and large gaps.

In the early 1960s, **Sir Edward Bullard** and two associates constructed a computer fit of all the continents. Instead of using the shorelines of the continents, he achieved the best fit with minimal overlaps or gaps (Figure 2–7) using a depth of 2000 meters (6560 feet) below sea level. This depth corresponds to a depth halfway between the shoreline to the ocean basin and is the true edge of the continents.

Matching Sequences of Rocks and Mountain Chains

To test the fit of the continents, geologists began comparing the rocks along their margins to identify those of the same type, age, and **structural style** (the type and degree of deformation). Identification is not always easy in some areas. During the millions of years since continental separation, younger **sedimentary** (*sedimentum* = settling) **rocks** may have been deposited, covering those that might hold the key to the past history of the continents. However, there are many areas where such rocks are available for observation, and the ages of these rock can be compared by using **fossils** (the remains of ancient organisms preserved in rocks) and/or **radiometric age dating** techniques (discussed in Chapter 1).

What was found from these studies was that certain rock sequences on continents tended to match up with the same rock sequences across ocean basins on other continents. In addition, mountain ranges that terminated abruptly at the edges of continents were continued across ocean basins on other continents with identical rock sequences, ages, and structural styles. Figure 2–8 shows how similar rocks from the Appalachian Mountains across North America match up with identical rocks from the British Isles and the Caledonian Mountains in Europe. Evidently, mountains like these formed during the collision of plates about 300 million years ago, when Pangaea was formed. A similar match occurs with mountains extending from South America through Antarctica and across Australia. Later, when the plates split apart, mountain ranges were often separated onto different continents.

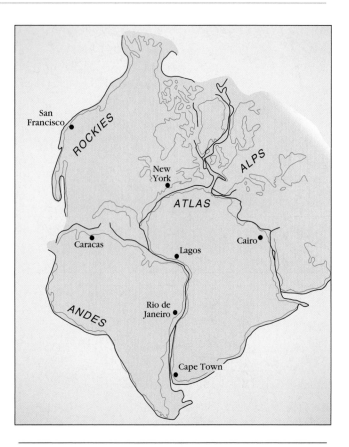

Figure 2–7 Computer fit of the continents.
In 1965, Sir Edward Bullard used a depth of 2000 meters (6560 feet) for a best-fit match of the continents, which shows few gaps and minimal overlap.

Glacial Evidence

Past glacial activity in areas that are now tropical also provides a supporting piece of evidence for drifting continents. Currently, the only places in the world where large continental **ice sheets** occur [where glacial ice is several kilometers (a few miles) thick] are in the polar regions of Greenland and Antarctica. However, 300 million-year-old glacial features and glacial deposits found in South America, Africa, India, and Australia indicate one of two possibilities: (1) There was a worldwide **ice age** and even tropical areas were covered by thick ice, or (2) some continents which are now in tropical areas were once located much closer to a polar region.

The possibility that the entire world was covered by ice 300 million years ago is implausible, since at that time there were vast semitropical swamps in North America and Europe, which were depositing material that would later become large coal deposits. This suggests that some of the continents were in a more polar position than they are today (Figure 2–9). Additional support for this idea exists in grooves caused by glaciation. These grooves indicate the flow direction of the glaciers as they moved and abraded the underlying rocks. Putting the continents back together as the supercontinent Pangaea 300 million

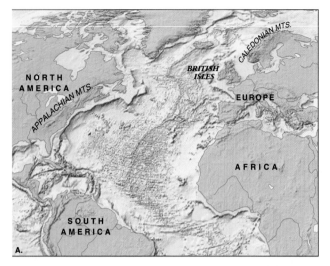

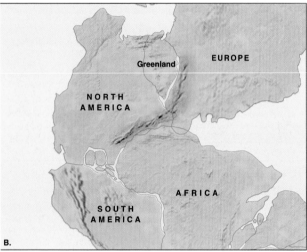

Figure 2–8 Matching mountain ranges across the North Atlantic Ocean.
A. Present-day positions of continents and mountain ranges.
B. Positions of the continents about 300 million years ago, showing how mountain ranges with similar age, type, and structure form one continuous belt.

years ago, the grooves (blue arrows in Figure 2–9A) show how the glaciers flowed away from the South Pole. These help to explain the seemingly impossible pattern of glacial flow directions (indicated by the blue arrows in Figure 2–9B) found on many continents today. Therefore, glacial evidence indicates that many of the continents have moved from more polar regions to their present geographic positions in the last 300 million years.

Fossil and Climate Evidence

The fossil record of plants and animals contained in sedimentary rocks can tell us much about the environments of Earth's past. For instance, it can be determined whether an organism lived in the ocean or on land by looking at its body structure and adaptations. It can also

be determined where the organism lived by examining the characteristics of the sedimentary rock itself.

Certain plants and animals need specific environmental conditions in which to live. For example, corals (small organisms related to jellyfish) cannot live in seawater that is too cold. However, scientists have discovered fossil assemblages and rocks that could not have formed under the climatic conditions that exist there today (for example, fossil corals found in arctic Alaska). These anomalous assemblages and rocks can be explained in one of two ways: (1) Worldwide climate has changed dramatically; or (2) the rocks have moved from their original geographic position.

The dominant factor controlling climatic distribution is latitudinal position (distance north or south of the Equator) on Earth. Since there is no evidence to indicate that Earth's axis of rotation has changed significantly throughout its history, it may reasonably be concluded that a given latitudinal belt has possessed similar climatic characteristics to what it has today. Also, it is reasonable to conclude that these characteristics have not changed greatly during the evolution of life on Earth. This leaves us with the only conclusion that fits all the data: These assemblages and rock types must have moved to their present position through the movement of tectonic plates.

Also, there is the intriguing problem of identical fossils on widely separated landmasses. A case in point is the fossil remains of ***Mesosaurus***, an extinct, presumably aquatic reptile that lived about 250 million years ago. The known distribution of *Mesosaurus* fossils are located only in eastern South America and western Africa (Figure 2–10). If *Mesosaurus* had been a good enough swimmer to cross an ocean, why wouldn't its remains be more widely distributed? This would be a perplexing problem indeed if there wasn't another explanation: Perhaps the continents were closer together in the geologic past, so *Mesosaurus* didn't have to be a good swimmer to leave remains on two different continents. Later, the continents moved to their present-day positions, and a large ocean now separates the once-connected landmasses.

Continental Magnetism

In the previous sections on evidence for plate tectonics, we examined data that Wegener used to support his theory of continental drift. It wasn't until other researchers using new technology to analyze the way rocks retained the signature of Earth's **magnetic field** that the idea of moving continents was reexamined. The first clues to such movements came from the study of the magnetism of continental rocks.

Earth's Magnetic Field   Earth's magnetic field resembles the magnetic field produced by a large bar magnet (Figure 2–11A). Bar magnets have oppositely charges ends (either labeled + and − or N for north and S for south) which cause magnetic objects to align parallel to

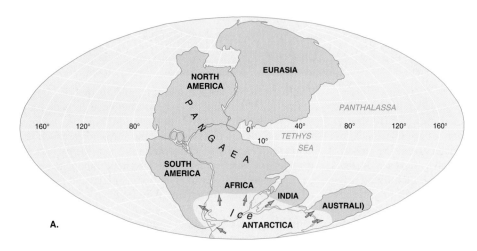

Figure 2–9 Ice age on Pangaea.

A. The supercontinent Pangaea showing the area covered by glacial ice about 300 million years ago. Blue arrows indicate the direction of ice flow. **B.** The positions of the continents today.

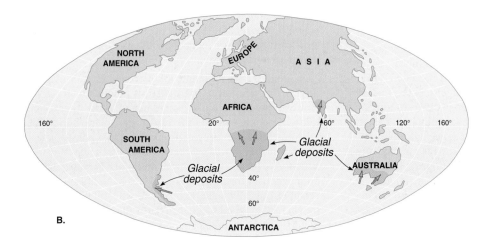

Figure 2–10 Fossils of *Mesosaurus.*

Mesosaurus fossils on South America and Africa appear to link these two continents.

Box 2–1
Do Sea Turtles (and Other Animals) Use Earth's Magnetic Field for Navigation?

It is well known that sea turtles travel great distances across the open ocean to arrive at a distant island so that they can lay their eggs where they themselves were hatched. However, how do they know an island's location, a veritable speck in a huge ocean several thousand kilometers from the closest landmass? Furthermore, how do they navigate at sea during their long voyage? These questions have long puzzled scientists. Studies have indicated that during their migration, green sea turtles (*Chelonia mydas*; Figure 2C) often travel in an essentially straight-line path to reach their destination. One hypothesis suggests that, like the Polynesian navigators, the sea turtles use wave direction to help them steer. However, sea turtles have been radio-tagged and tracked by satellites, which reveals that they continue along their straight-line path independent of wave direction.

Research in *magnetoreception*, the study of an animal's ability to sense magnetic fields, suggests that sea turtles may use Earth's magnetic field for navigation. It has been demonstrated that hatchling turtles can distinguish between different magnetic inclination angles, which in effect would allow the turtles to sense latitude. Recent research indicates that sea turtles can also distinguish magnetic field intensity, a rough indication of longitude. By sensing the magnetic field intensity along with the magnetic inclination, a sea turtle can construct a magnetic map with grid coordinates that can be used for navigation. Thus, a migrating turtle can use this magnetic map for determining its position at sea and can relocate a tiny island hundreds or thousands of kilometers away. Like any good navigator, sea turtles may also use other tools, such as olfactory (scent) clues, sun angles, local landmarks, and oceanographic phenomena.

Perhaps turtles are not the only animals that use Earth's magnetic field to navigate. It has long been suspected that some whales and dolphins detect and follow the magnetic stripes on the sea floor during their movements, which may help to explain why whales sometimes beach themselves. The discovery of bacteria that use the magnetic mineral magnetite to align themselves parallel to the magnetic field precipitated an enormous search for magnetite in other animals. Scientists reported finding magnetite in many other organisms that have a "homing" ability, including tuna, salmon, honeybees, pigeons, turtles, and even humans. What has been unclear is how these animals detect—and potentially use—Earth's magnetic field. Recent findings by a research team studying rainbow trout (close relatives of salmon) have traced magnetically receptive fibers of nerves back to the brain, more closely linking a magnetic sense with an organism's particular sensory system.

Do humans have an innate ability to use Earth's magnetic field for navigation? That question remains unanswered. However, studies conducted on humans indicate that the majority of people can identify north after being blindfolded and disoriented. Interestingly, many people point *south* instead of north, but this direction is along the lines of magnetic force. In a similar fashion, migratory animals that rely on magnetism for navigation will not be confused by a reversal in Earth's magnetic field and will still be able to get to where they need to go. Certainly, the detection of a directional sense in animals seems likely to remain an intriguing and elusive mystery in animal behavior.

Figure 2C Green sea turtle.

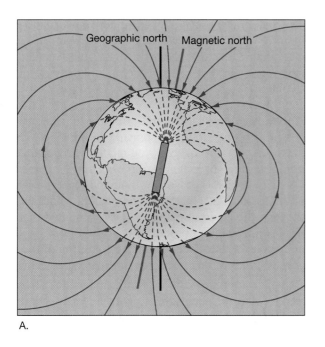

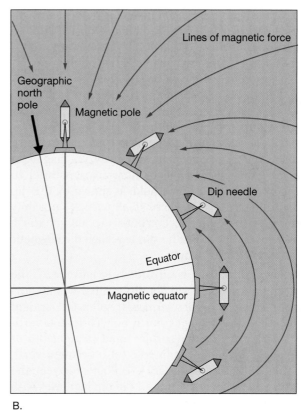

Figure 2–11 Earth's magnetic field.
A. Earth's magnetic field resembles that produced by a large bar magnet. Invisible lines of magnetic force are shown. Note that magnetic north and true north are not in the same exact location.
B. Earth's magnetic field causes a dip needle to align parallel to line of magnetic force. The dip needles change orientation relative to the surface with increasing latitude. Consequently, the latitude can be determined based on the dip angle.

that magnetic field.[6] Although Earth does not really have a bar magnet inside, it behaves as if it does. Invisible lines of magnetic force that originate within Earth travel through Earth and out into space. Notice in Figure 2–11A how Earth's geographic north pole (the rotational axis) and Earth's magnetic north pole (magnetic north) do not coincide.

Rocks Affected by Earth's Magnetic Field
Igneous (*igne* = fire) **rocks** are rocks that solidify from molten **magma** (*magma* = a mass), either underground or after volcanic eruptions at the surface. Nearly all igneous rocks contain some particles of **magnetite**, a naturally magnetic iron mineral. Magnetite particles are influenced by Earth's magnetic field and align themselves with

Earth's magnetic field at the time of the rocks' formation. They can do so because magma is fluid, allowing the magnetite particles to achieve an orientation parallel to Earth's magnetic field before the magma cools. Magma which occurs at Earth's surface is called **lava** (*lava* = to wash). Volcanic eruptions produce lavas such as basalt which are high in magnetite content and solidify from molten material in excess of 1000 degrees centigrade (1800 degrees Fahrenheit). As the lava cools below 600 degrees centigrade (1100 degrees Fahrenheit), the magnetite particles are frozen into position parallel to the direction of Earth's magnetic field, *permanently recording the angle of Earth's magnetic field relative to the rock location.* In essence, grains of magnetite serve as tiny three-dimensional compass needles, permanently frozen in place, which record the strength and alignment of Earth's magnetic field. Unless the rock is again melted, these magnetite grains can be interpreted later regardless of where the rock moves.

Magnetite is also deposited in sediments. While the deposit is in the form of sediment surrounded by water, these magnetite particles also have an opportunity to

---

[6]The properties of a magnetic field can be explored easily enough with a bar magnet and some iron particles. Place the iron particles on a table and place a bar magnet nearby. Depending on the strength of the magnet, you should get a pa tern resembling Figure 2–11A.

align themselves with Earth's magnetic field. This alignment is preserved when the sediment is buried and becomes solidified into sedimentary rock. Although a number of rock types may be used to examine Earth's ancient magnetic field, basaltic lavas and other igneous rocks that are high in magnetite content are best.

Paleomagnetism The study of Earth's ancient magnetic field is called **paleomagnetism** (*paleo* = ancient). Paleomagnetism is concerned with analyzing magnetite particles which have aligned themselves parallel to Earth's prevailing magnetic field, as shown by the dip needle in Figure 2–11B. These small particles not only point in a north–south direction but also point into Earth at an angle relative to Earth's surface called the **magnetic dip**, or **magnetic inclination**.

The magnetic dip is related directly to latitude. At the Equator, the "needle" will not dip at all; it will lie horizontal *relative to Earth's surface*. It will point straight *into* Earth at the magnetic north pole (oriented vertically relative to Earth's surface) and straight *out of* Earth at the magnetic south pole (also oriented vertically relative to Earth's surface). At points between the Equator and the pole, the amount of dip increases with increasing latitude from 0 degrees at the Equator to 90 degrees at the magnetic poles (Figure 2–11B). It is this dip that is retained in magnetically oriented rocks. By measuring the dip angle, the latitude at which the rock initially formed can be determined. Done with

care, paleomagnetism is an extremely powerful tool for helping interpret where rocks first formed. Based on paleomagnetic studies, convincing arguments could finally be made that the continents had drifted relative to one another.

Apparent Polar Wandering Because the present magnetic north pole does not coincide with the geographic north pole (Figure 2–11A), you might expect that determining latitude by magnetic dip would give incorrect results. However, for the last few thousand years, the *average* positions of the magnetic poles have nearly coincided with those of the geographic poles. If it is assumed that this has been true in the past, average positions of the magnetic poles which closely match the position of the geographic poles during a specific time can be determined.

As these average positions for the magnetic poles were determined for rocks on the continents, it was found that their positions changed with time. It appeared that the magnetic poles were wandering. For example, the **polar wandering curve** for North America shows an interesting relationship with that determined for Europe (Figure 2–12A). Both curves have a similar shape, but for all rocks older than about 70 million years the pole determined from North American rocks lies to the west of that determined by the study of European rocks. This difference implies that North America and Europe have changed position relative to the pole and relative to each other.

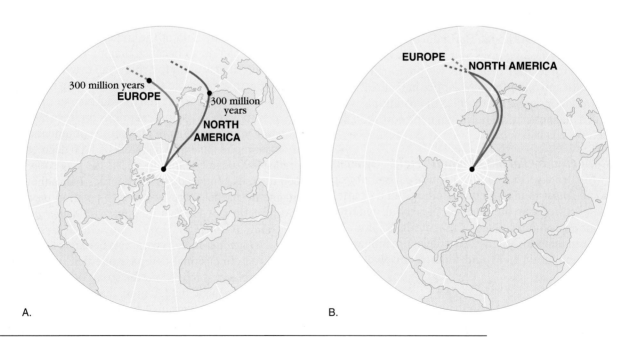

A.   B.

Figure 2–12 Apparent polar wandering paths.
A. Simplified apparent polar wandering paths for North America and Eurasia. A dilemma existed because they were not in alignment. **B.** The positions of the polar wandering paths when the landmasses are assembled.

If there can be only one north magnetic pole at any given time and its position must be at or near the north geographic pole, the wandering curves can be resolved only by moving the continents. By rotating the continents into the positions they occupied when they were part of Pangaea, the two wandering curves can be made to coincide (Figure 2–12B). The fact that moving the continents was the only solution to this problem was strong evidence in support of the movement of continents throughout geologic time. In other words, it was discovered that the continents were wandering, not the poles! Hence the name, *apparent* polar wandering. This discovery added additional evidence for the movement of plates through time.

Magnetic Polarity Reversals Magnetic compasses on Earth today follow lines of magnetic force and point toward magnetic north. However, compasses have not always pointed north on Earth. The **polarity**, or the directional orientation of the magnetic field, has *reversed* itself periodically throughout geologic time that is, the north magnetic pole has become the south magnetic pole and vice versa. A compass needle that would point to magnetic north today would, during a period of reversal, *point south*. Acting like tape recorders, rocks have faithfully recorded this dramatic switching of Earth's magnetic polarity through time (Figure 2–13). It is not known why these reversals occur, but for the last 76 million years they have occurred once or twice each million years on an irregular basis.

The period of time during which a change in polarity occurs lasts for a few thousand years. It is identified in magnetic properties of rocks by a gradual decrease in the intensity of the magnetic field of one polarity until it disappears, followed by the gradual increase in the intensity of the magnetic field with the opposite polarity. The time during which a particular paleomagnetic condition existed ("normal" or "reversed") can be determined by the radiometric dating of the igneous rock from which the paleomagnetic measurements were taken. Earth's present magnetic field has been weakening during the last 150 years, and some scientists believe that the present polarity will diminish to the point of being very weak or nonexistent in another 2000 years.

Evidence from the Ocean Floor

We have so far considered only data obtained from the *continents* in considering the theory of moving plates. As compelling as this evidence was, it did not convince all geologists that the continents had moved. For many years no other data were available because extensive sampling of the deep-ocean floor did not become technologically feasible until the 1960s. As we will see, convincing support for the theory of plate tectonics came from the ocean floor, not from the continents.

Paleomagnetism Detailed paleomagnetism studies of different areas of the ocean floor were conducted by several oceanographic institutions. Based on these studies, narrow strips were identified where the magnetic properties of oceanic crust differed from the properties of ocean crust currently forming. These areas of differently magnetized sea floor are called **magnetic anomalies**

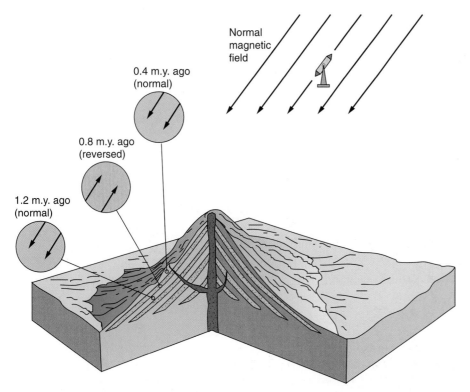

Normal magnetic field

0.4 m.y. ago (normal)

0.8 m.y. ago (reversed)

1.2 m.y. ago (normal)

Figure 2–13 **Paleomagnetism preserved in rocks.** The switching of Earth's magnetic field through time is preserved in rocks such as these lava flows.

(*a* = without, nomo = law; an anomaly is a departure from normal conditions). These magnetic anomalies are oriented parallel to the mid-ocean ridge and occur as wide bands or magnetic stripes on the sea floor. Each anomaly represents a period when the magnetic polarity of Earth's magnetic field was "reversed" compared with today's. These "reversed" anomalies are separated by bands of oceanic crust that display present-day "normal" polarity (Figure 2–14).

Researchers also observed that the sequence of polarity changes on one side of an oceanic ridge is identical to the sequence of reversals on the opposite side of the ridge—one is a *mirror image* of the other. How could such a pattern of normal and reversed polarity form? Rock samples from the ocean floor indicated that new oceanic crust is being formed along segments of the mid-ocean ridge and that the new crust was moving away in opposite directions from the axis of the mid-ocean ridge (arrows in Figure 2–14). Thus, the process of sea floor

spreading could account for the pattern of normal and reverse stripes on the ocean floor. The discovery that the mid-ocean ridge segments are indeed separating and the plates are moving away from the crest of the mid-ocean ridge provided confirmation of a mechanism to move continents.

Age of the Ocean Floor Much of the history of the ocean basins can be learned by examining the pattern of age distribution for the ocean crust. Figure 2–15 shows a map of the age of the ocean floor beneath deep-sea deposits based on thousands of samples that have been radiometrically age dated. This map shows how symmetric the pattern of age dates are with respect to the mid-ocean ridge, which are colored bright red in the figure. Radiometric dating of ocean floor rocks near the crest of mid-ocean ridge segments revealed that the rocks became *older* with *increasing* distance *in either direction* away from the ridge axis. As sea floor spreading occurs,

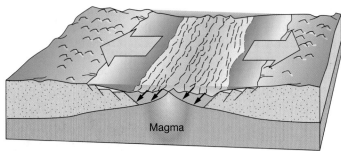

A. Period of normal magnetism

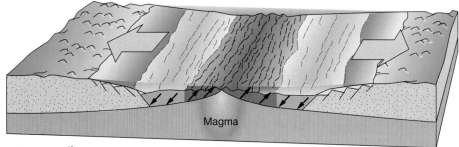

B. Period of reverse magnetism

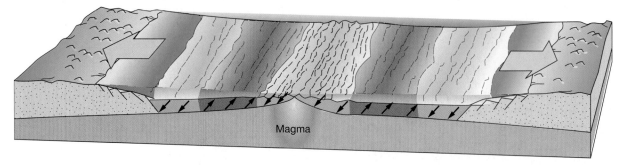

C. Period of normal magnetism

Figure 2–14 Magnetic evidence of sea floor spreading.
As new basalt is added to the ocean floor at mid-ocean ridges, it is magnetized according to Earth's existing magnetic field.

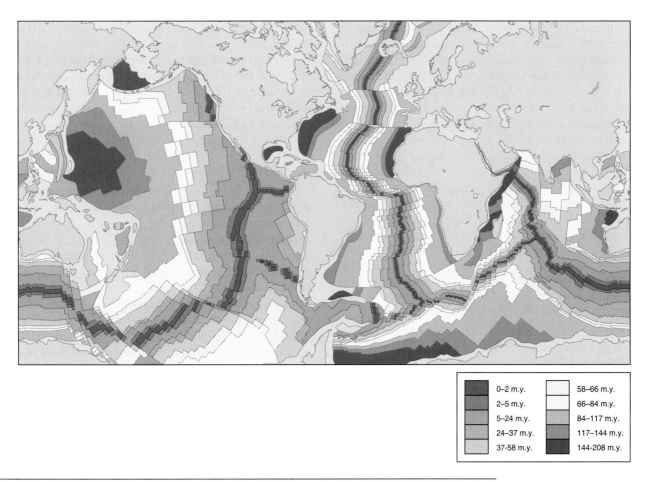

0–2 m.y.	58–66 m.y.
2–5 m.y.	66–84 m.y.
5–24 m.y.	84–117 m.y.
24–37 m.y.	117–144 m.y.
37-58 m.y.	144-208 m.y.

Figure 2–15 Age of the ocean crust beneath deep-sea deposits.

The youngest rocks (*bright red areas*) are found along the mid-ocean ridge. Farther away from the mid-ocean ridge, the rocks increase linearly in age in either direction.

new oceanic crust is continually being created. Like dual conveyer belts moving in opposite directions, new sea floor moves away as even younger sea floor is created along the axis of the mid-ocean ridge. Notice how the Atlantic Ocean has the simplest and most symmetric pattern of age distribution. This was created as the newly formed Mid-Atlantic Ridge rifted Pangaea apart. The Pacific Ocean has the least symmetric pattern because of the large number of subduction zones that surround it. The ocean floor east of the East Pacific Rise that is older than 40 million years old has already been subducted, whereas ocean floor in the northwestern Pacific is about 180 million years old. A portion of the East Pacific Rise has even disappeared under North America. The wider age bands on the Pacific Ocean floor are clear evidence that its spreading rate has been greater than in the Atlantic or Indian Oceans. The process of sea floor spreading was confirmed by the symmetric pattern of ocean floor ages.

Interestingly, there is quite a disparity between the age of the ocean floor and the age of the oceans themselves. As we learned in Chapter 1, the ocean is at least 4 billion years old (and may be even older). Yet, the *oldest* ocean floor is only 180 million years old (or 0.180 billion years old), and the majority of the ocean floor is not even half that old. Indeed, the ocean floor is surprisingly young! These observations suggest an apparent dilemma: *How could the ocean floor be so incredibly young, while the oceans themselves are so phenomenally old?* In other words, how could it be that the oceans have been on Earth for over 85 percent of Earth history, yet the oldest ocean floor which underlies the oceans has been there for only the last 4 percent of Earth history?[7]

An answer to this apparent dilemma is suggested by plate tectonic processes. New ocean floor is created at the mid-ocean ridge by sea floor spreading and moves off the ridge to eventually be subducted and remelted

---

[7] This is similar to having a neighbor who builds a swimming pool and says, "The water has been there for six weeks, but the sides of the pool were just put in two days ago." How could the water in the pool have been there before the boundaries of the pool were built? This perplexing inconsistency is similar to the apparent dilemma about the age of the oceans relative to the age of the ocean floor.

in the mantle. In this way, the ocean floor keeps regenerating itself. Thus, the ocean floor that exists beneath the oceans today is *not* the same ocean floor that existed beneath the oceans 4 billion years ago. What happened to that old ocean floor? According to the plate tectonic model, it has been subducted and remelted as sea floor spreading continued operating over geologic time. The fate will be the same for the present-day sea floor: It too will eventually be subducted and remelted in the mantle. That is why the sea floor has always been (and will continue to be) so incredibly young. Without sea floor spreading constantly creating new ocean floor and old ocean floor being subducted, the disparity in age between the oceans and the ocean floor would indeed be difficult to explain.

Features of the Ocean Floor

The morphology and function of ocean floor features also help support the theory of plate tectonics.

The mid-ocean ridge, which is a spreading center where new sea floor is created, has been described as Earth's zipper. Indeed, Earth is becoming unzipped as the plates move away from the axis of the mid-ocean ridge and new ocean crust is created along the crest to replace the material that slowly moves away from the mid-ocean ridge.

As Earth unzips, is the planet increasing in size (and might it someday expand so far that is explodes)? Fortunately, no, because Earth has a natural recycler of ocean crust that works at the same rate as new sea floor is created. These are the trenches, which mark subduction zones, where sea floor is destroyed

Heat Flow

Earth's interior is hot, and some of this heat is released to the surface. Measuring Earth's **heat flow** can tell scientists a great deal about the internal structure of Earth. Current models indicate that this heat moves to the surface with magma in convective motion and that most of the heat is carried to regions of the mid-ocean ridge spreading centers. This also suggests that cooler portions of the mantle descend in other places to complete each circular-moving convection cell. What is the source of this heat? It was once believed that most of the heat driving global plate tectonics was generated primarily by radioactivity in the upper mantle. However, now there is evidence that a significant amount of heat is "original" heat remaining from Earth's formation, which is slowly being released from the core.

Heat flow measurements taken throughout Earth's crust show that the quantity of heat flowing to the surface along the mid-ocean ridge can be up to eight times greater than the average for Earth's crust. In addition, areas where ocean lithosphere subducts at trenches have heat flow as little as one-tenth the average. This pattern of increased heat flow at the mid-ocean ridge (where lithosphere is thin) and decreased heat flow at subduction zones (where there is a double thickness of lithosphere) is what would be expected based on the plate tectonic model. Table 2–1 summarizes relationships concerning events occurring at the mid-ocean ridge.

Worldwide Earthquakes

Earthquakes are sudden releases of energy caused by fault movement or by volcanic eruptions. All earthquakes occur at or near Earth's surface. The distribution of earthquakes worldwide was something of a mystery until the advancement of plate tectonic theory. For instance, why did some areas of the world have so many earthquakes, and others did not? Looking at an earthquake map of the world (Figure 2–16A), it is clear that most large earthquakes occur along certain features, especially sea floor features. The majority of large earthquakes are along ocean trenches and reflect the energy released during subduction. Other large earthquakes occur along the mid-ocean ridge, reflecting the energy released during sea floor spreading. Still others are along

Table 2–1   Relationships relative to increasing distance in either direction from the axis of a mid-ocean ridge.

	Relationship	Reason
1.	Volcanic activity decreases	Magma chambers are concentrated only along the crest of the mid-ocean ridge.
2.	The age of ocean crust increases	New ocean floor is created at the mid-ocean ridge.
3.	The thickness of sea floor deposits increases	Older ocean floor has more time to accumulate a thicker deposit of debris due to settling.
4.	The thickness of the lithospheric plate increases	As plates move away from the spreading center, they are cooler and gain thickness as material beneath the lithosphere attaches to the bottom of the plate.
5.	Heat flow decreases	As the thickness of the lithospheric plate increases (see relationship 3), less heat can escape.
6.	Water depth increases	The mid-ocean ridge is a topographically high feature, and as plates move away from there, they move into deeper water because plates undergo thermal contraction and isostatic adjustment downward.

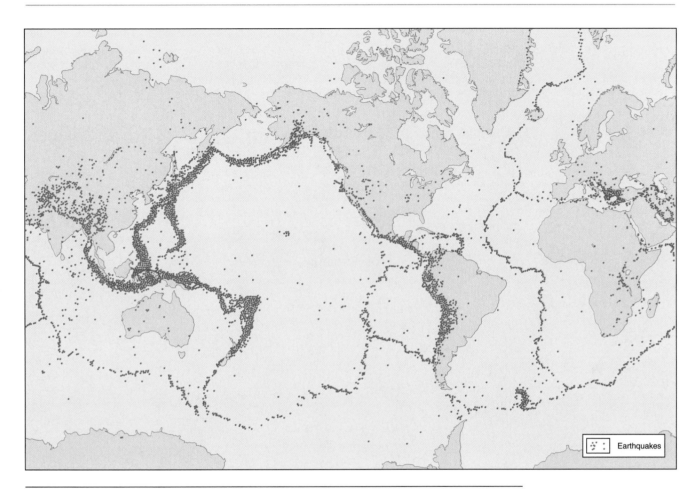

Figure 2–16A Earthquakes and lithospheric plates.
A. Distribution of earthquakes with magnitudes equal to or greater than $M_w = 5.0$ for the period 1980–1990.

major faults in the sea floor and on land, which reflect energy being released in these areas.

As predicted by plate tectonic theory, energy will be released in the form of large earthquakes when moving lithospheric plates contact other plates along their edges. In Figure 2–16, the edges of plates are highlighted by the locations of large earthquakes. In fact, to outline the edges of plates on a map, the only information that one would need is a map of large earthquakes that have occurred over a period of time. From this, it would be a simple exercise of connecting regions with many earthquakes to outline the edges or boundaries of plates. Today, it is clear why certain areas of the world have so many earthquakes: These areas are located close to plate boundaries.

Plate Boundaries

Plate boundaries are areas where plates interact with each other. These areas are associated with a great deal of tectonic activity such as mountain building, volcanic activity, and earthquakes. In fact, the first clues to the locations of plate boundaries were the dramatic tectonic events that commonly occur in these areas. Earth's surface is composed of seven major plates and many smaller plates (Figure 2–16B). Most plates contain both oceanic and continental crust, and the boundaries of plates do not always follow coastlines.

Closer examination of these active regions identified three types of plate boundaries. Plate boundaries where new lithosphere is being added along oceanic ridges are called **divergent** (*di* = apart *vergere* = to incline) **boundaries**. Boundaries where plates are moving together and one plate subducts beneath the other are called **convergent** (*con* = together, *vergere* = to incline) **boundaries**. Lithospheric plates also slowly grind past one another along **transform** (*trans* = across, *form* = shape) **boundaries** (Figure 2–17). Table 2–2 summarizes characteristics, processes, and examples of these plate boundaries.

Divergent Boundaries

Divergent boundaries occur where two plates move apart, resulting in new material from the mantle rising to the surface to create new sea floor (Figure 2–18). The physiographic ocean floor feature associated with a divergent plate boundary is a mid-ocean ridge, where the process of sea floor spreading occurs.

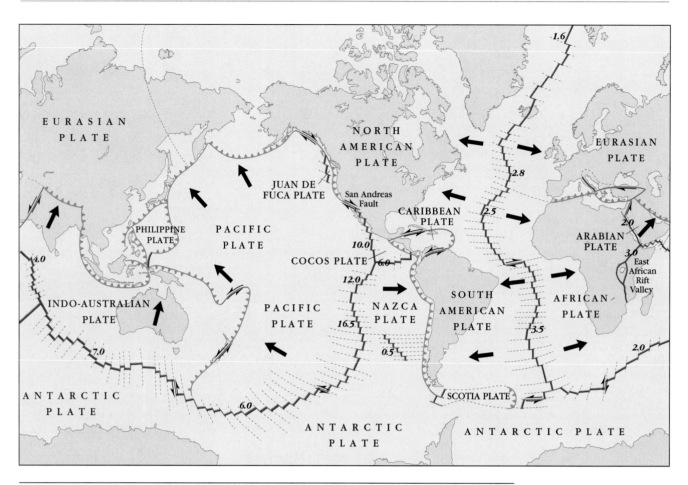

Figure 2–16B *Continued*

B. Plate boundaries define the major lithospheric plates, with arrows indicating the direction of plate movement. Notice how closely the pattern of major earthquakes follows plate boundaries.

Volcanic features and volcanic activity are common along the mid-ocean ridge, as is a central downdropped linear valley called a **rift valley** (Figure 2–19). The existence of pull-apart faults along the central rift valley supports the model that the plates are being *continuously pulled apart* rather than being pushed apart by the upwelling of material beneath the mid-ocean ridge. It is likely that the upwelling of magma beneath the mid-ocean ridge is simply filling in the void left by the separating plates of lithosphere. The volume of magma emplaced along the mid-ocean ridge is impressive: The process of sea floor spreading produces about 20 cubic kilometers (4.8 cubic miles) of new ocean crust worldwide each year.

The stages in the development of a mid-ocean ridge indicate how ocean basins are created. Initially, molten material rises to the surface in an area, causing upwarping and thinning of the crust (Figure 2–20). Volcanic activity produces vast quantities of high-density basaltic rock. As the plates begin to move apart, a linear rift valley is formed and volcanism continues. Further **rifting** of the land and more spreading cause the area to be downdropped below sea level. Once this occurs, the rift val-

ley will eventually be flooded with seawater and a young linear sea is formed. After millions of years of sea floor spreading, a full-fledged ocean basin is created with a mid-ocean ridge in the middle of the two landmasses. With this in mind, it is easy to see how large distances across an ocean can separate features on once-connected landmasses. All it requires are the forces involved in moving plates apart over long periods of time.

The East African rift valleys are being actively rifted apart and are in the rift valley stage of formation (Figure 2–21). To the north of these rift valleys, the Red Sea is in the linear sea stage of development: It has rifted apart so far that the land surface has downdropped below sea level. Like the Gulf of California in Mexico, where similar rifting processes are occuring, the Red Sea is considered to be one of the youngest seas in the world, having been created only a few million years ago. Both of these seas are early in their evolution. If plate motions continue rifting the plates apart, they will eventually become large oceans by their final stage of their development.

Spreading rates (the total widening rate of an ocean basin resulting from the motion of *both* plates away from

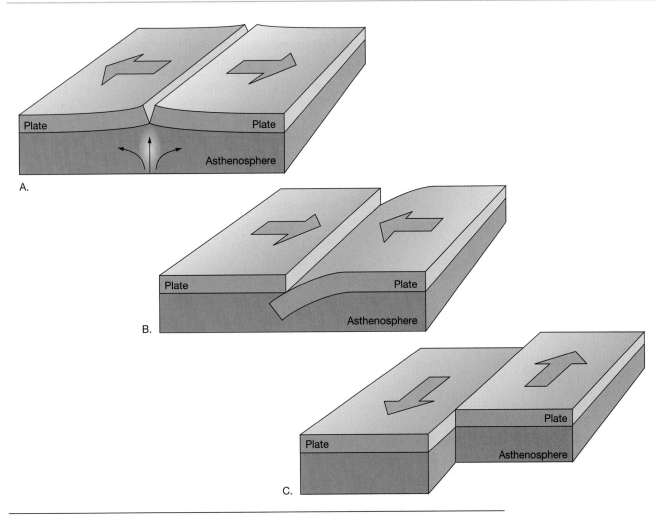

Figure 2–17 The three types of plate boundaries.
A. Divergent, where plates move away from each other. **B.** Convergent, where plates approach each other. **C.** Transform, where plates slide past each other.

a spreading center) vary along the mid-ocean ridge and profoundly affect the steepness of the mountain range. The faster the spreading rate, the broader the mountain range associated with the spreading. The fast-spreading and gently sloping parts of the mid-ocean ridge are called **oceanic rises.** An example of an oceanic rise is the broad **East Pacific Rise** in the eastern Pacific Ocean between the Pacific Plate and the Nazca Plate, where spreading rates are as high as 16.5 centimeters (6.5 inches) per year. The slower-spreading and steeper-sloped areas of the mid-ocean ridge are called **oceanic ridges**. An example of an oceanic ridge is the steep **Mid-Atlantic Ridge** in the central Atlantic Ocean between the South American Plate and the African Plate. The Mid-Atlantic Ridge has spreading rates in the range of 2 to 3 centimeters (0.8 to 1.2 inches) per year and displays the steeper slopes characteristic of slow spreading. Both oceanic ridges and oceanic rises are part of the global mid-ocean ridge system.

Features are significantly different at the axes of fast- and slow-spreading divergent plate boundaries. Fig-

ure 2–22 compares the sizes of axial rift valleys, which are much larger on the slow-spreading ridges. This difference is probably due to a much greater magma supply being available at fast-spreading oceanic rises than at slow-spreading ones.

The magnitude of energy released by earthquakes along the divergent plate boundaries is closely related to the spreading rate. Earthquake intensity is usually measured on a scale called the **seismic moment magnitude**, which reflects the energy released to create very long-period seismic waves. The moment magnitude scale is replacing the well-known Richter scale for measuring earthquakes and is represented by the symbol M_w. Earthquakes in the rift valley of the slow-spreading Mid-Atlantic Ridge reach a maximum magnitude of about $M_w = 6.0$, whereas those occurring along the axis of the fast-spreading East Pacific Rise seldom exceed $M_w = 4.5$.[8]

---

[8] Note that each one-unit increase of magnitude represents an increase of energy release of about 30 times.

Table 2–2 Characteristics, tectonic processes, and examples of plate boundaries

Plate boundary	Plate movement	Crust type(s)	Sea floor created or destroyed?	Tectonic process	Geographic examples
Divergent plate boundaries	Apart ← →	Ocean-ocean	New sea floor created	Sea floor spreading	Mid-Atlantic Ridge, East Pacific Rise
		Continent-continent	As continent splits apart, new sea floor created	Continental rifting	East Africa Rift Valleys, Red Sea, Gulf of California
Convergent plate boundaries	Together → ←	Ocean-continent	Old sea floor destroyed	Subduction	Andes Mountains, Cascade Mountains
		Ocean-ocean	Old sea floor destroyed	Subduction	Aleutian Islands, Mariana Islands
		Continent-continent	N/A	Collision	Himalaya Mountains, Alps
Transform plate boundaries	Past each other → ←	Oceanic	Sea floor neither created nor destroyed	Transform faulting	Mendocino fault, Eltanin fault (between mid-ocean ridges)
		Continental	Sea floor neither created nor destroyed	Transform faulting	San Andreas fault, Alpine fault (New Zealand)

Convergent Boundaries

Convergent boundaries are where two plates move together and collide, resulting in the destruction of ocean crust as one plate plunges below the other to be remelted in the mantle. The physiographic ocean floor feature associated with a convergent plate boundary is a deep-ocean trench, where the process of subduction occurs. Trenches are deep linear scars on the ocean floor where a lithospheric plate begins to plunge beneath the surface. Near each trench are highly active and explosive-erupting volcanoes arranged in an arc-shaped row, called a volcanic arc.

Three subtypes of convergent boundaries result from interactions between the two different types of crust (oceanic and continental) present on lithospheric plates (Figure 2–23). The first subtype of a convergent boundary involves the collision of *oceanic* lithosphere with *continental* lithosphere. The second subtype involves the collision of *oceanic* with *oceanic* lithosphere, and the third subtype involves the collision of *continental* with *continental* lithosphere.

Oceanic-Continental Convergence

When an oceanic and a continental plate converge, the denser oceanic plate will be subducted (Figure 2–23A). As the oceanic lithosphere is subducted into the asthenosphere, it becomes heated. Some of the basaltic rock of the oceanic plate is melted, which allows it to become mobile. This molten rock mixes with superheated gases (mostly water) and begins to rise to the surface through the overriding continental plate. Consequently, the rising basalt-rich magma passes through and mixes with the granite of the continental crust. This produces lava in volcanic eruptions at the surface that is intermediate in composition between basalt and granite. One type of volcanic rock with this composition is called **andesite**, which, as its name suggests, is a common rock type of the Andes Mountains of South America. Because of the high gas content associated with andesitic volcanic eruptions, these eruptions are usually quite explosive and have historically been very destructive. The result of this volcanic activity is an arc-shaped row of volcanoes that parallel the trench and are located on the continent above the subduction zone. This type of volcanic arc is called a **continental arc**. Along with andesitic volcanoes, continental arcs are created by folding and uplifting during the collision of the two plates.

If the spreading center producing the subducting plate is far enough from the subduction zone, an oceanic trench becomes well developed along the margin of the continent. The Peru-Chile Trench is an example, and the Andes Mountains are the associated continental arc produced by melting of the subducting plate. However, if the spreading center producing the subducting plate is close to the subduction zone, the trench is not nearly as well developed. This is the case where the Juan de Fuca Plate subducts beneath the North American Plate off the coasts of Washington and Oregon to produce the Cascade Mountains continental arc (Figure 2–24). Here, the Juan de Fuca Ridge is so near the North American Plate that the subducting lithosphere is less than 10 million years old and has not cooled enough to produce deep-ocean depths. In addition, the large amount of sediment carried

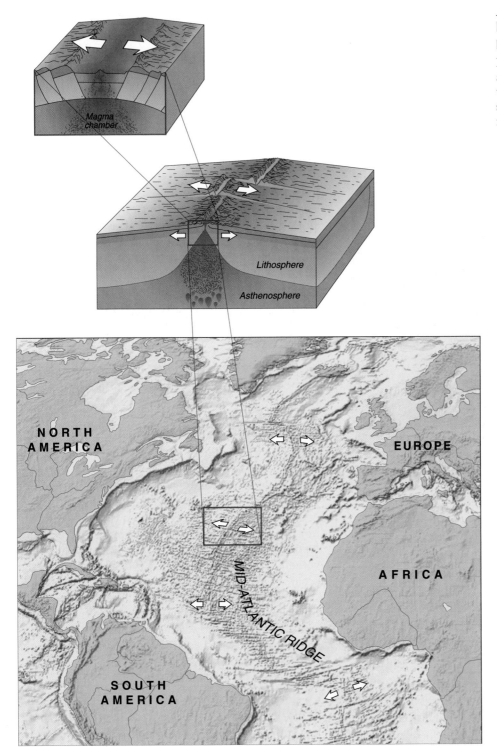

Figure 2–18 Divergent boundary.

Most divergent plate boundaries occur along the crest of the mid-ocean ridge, where sea floor spreading creates new oceanic lithosphere.

to the ocean by the Columbia River has filled most of the trench with sediment. Many of the Cascade volcanoes of this continental arc have been active within the last 100 years. Most recently, Mount St. Helens erupted in May 1980, killing 62 people.

Oceanic-Oceanic Convergence   When two oceanic plates converge, the denser oceanic plate will be subducted (Figure 2–23B). Typically, the older oceanic plate will be denser because it has had more time to cool and contract. This type of convergence produces the deepest trenches in the world, such as the Mariana Trench in the western Pacific Ocean. As the downgoing oceanic plate is subducted into the asthenosphere, it becomes heated and melts. Similar to oceanic-continental convergence, molten material mixes with superheated gases and rises to the surface to produce volcanoes. Here, however, the composition of the molten material is mostly

Figure 2–19 Rift valley.

Rift valley in Iceland, which sits atop the Mid-Atlantic Ridge.

basaltic because there is no mixing with granitic rocks from the continents, and the eruptions are not quite so destructive. The result of this volcanic activity is a volcanic arc in the form of a row of islands that parallel the trench. This type of volcanic arc is called an **island arc**. Well-known examples of island arc/trench systems are the West Indies' Leeward and Windward Islands/Puerto Rico Trench in the Caribbean Sea and the Aleutian Islands/Aleutian Trench in the North Pacific Ocean.

Continental-Continental Convergence When two continental plates converge, which one will be subducted? It may seem reasonable that the older of the two (which is probably the denser one) will subduct. However, continental lithosphere does not form in the same way that oceanic lithosphere does, and old continental lithosphere is no denser than young continental lithosphere. So which one subducts? You may be surprised to find that *neither* subducts. Since both are composed of low-density continental material, they are too low in density to be pulled very far down into the mantle. Instead, a tall uplifted mountain range forms (Figure 2–23C). These mountains are composed of folded and deformed sedimentary rocks. For the most part, these sedimentary rocks were deposited on the sea floor

that previously separated the two continental plates. The relatively low density of the sedimentary deposits prevents them from being subducted. Instead, they have been uplifted and deformed during the plate collision. The oceanic crust itself may, however, subduct beneath such mountains. The classic example of this continental-continental lithosphere convergence is the Himalaya Mountains, created when India began to collide with Asia about 45 million years ago (Figure 2–25). The impressive Himalaya Mountains are the tallest mountains in the world, with all 10 of the world's highest peaks among them.

Earthquakes Associated with Convergent Boundaries
Both spreading centers and trench systems are characterized by earthquakes, but in different ways. Spreading centers have shallow earthquakes, usually less than 10 kilometers (6 miles) in depth. However, earthquakes in the trenches vary from shallow depths near the surface down to 670 kilometers (415 miles) deep, which are the deepest earthquakes in the world. These earthquakes are clustered in a band about 20 kilometers (12.5 miles) thick that closely corresponds to the location of the subduction zone. In fact, the downgoing plate in a convergent plate boundary can be traced below the surface by examining the pattern of successively deeper earthquakes extending from the trench.

Many contributing factors combine to produce large earthquakes at convergent boundaries. The forces involved in convergent-plate boundary collisions are enormous. Huge lithospheric slabs of rock are relentlessly pushing against each other, and the subducting lithosphere must actually bend as it dives below the surface. In addition, thick crust associated with convergent boundaries tends to store more energy than the thinner crust at divergent boundaries. As well, mineral structure changes occur at the higher pressures encountered deep below the surface, which are thought to produce volume changes. Consequently, convergent boundaries are associated with some of the most powerful earthquakes in the world. Currently, the largest earthquake ever recorded was the Chilean earthquake of 1960, which occurred near the Peru-Chile Trench and had a magnitude of $M_w = 9.5$!

Transform Boundaries

Spreading rates vary along different segments of the mid-ocean ridge. To accommodate different rates of spreading along closely spaced segments of the mid-ocean ridge, the segments are **offset** or displaced from one another. Looking at a sea floor map showing the mid-ocean ridge (see inside front cover), it is clear that the once-continuous ridge is offset by these features which are oriented at right angles (90 degrees) from the crest of the ridge. These offsets, called **transform faults**, give the mid-ocean ridge a zigzag appearance. There are thousands of these transform faults, some large and some small, which dissect the global mid-ocean ridge.

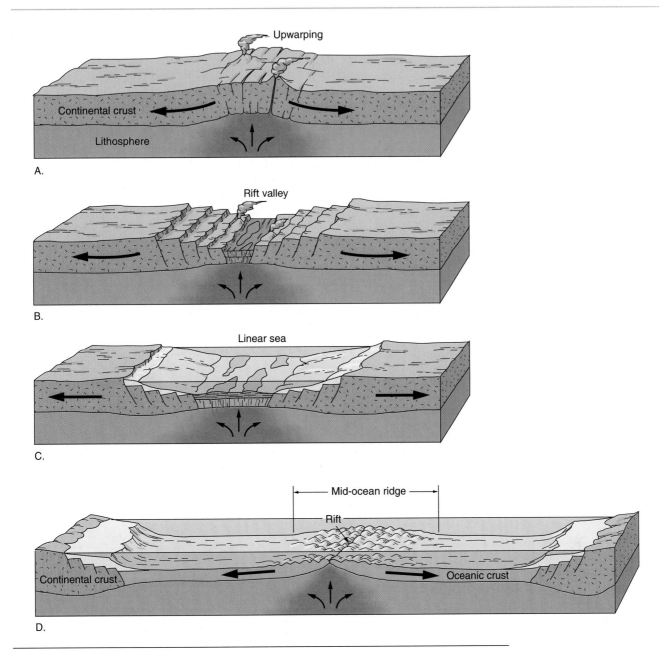

Figure 2–20 Formation of an ocean basin by sea floor spreading.

There are two types of transform faults. One type occurs wholly on the ocean floor, and is called an **oceanic transform fault**. The other type cuts across a continent, and is called a **continental transform fault**. Most of the transform faults in the world are of the first type. Note that for either type, transform faults *always offset two segments of a mid-ocean ridge* (Figure 2–26).

Along these transform faults, one plate slowly grinds past another by a process called **transform faulting**. This movement produces shallow but often strong earthquakes in the lithosphere. Magnitudes of $M_w = 7.0$ have been recorded along some oceanic transform faults. One of the most studied faults in the world is the **San Andreas Fault**, a continental transform fault which runs across California from the Gulf of California through the San

Francisco region to northern California. Because the San Andreas Fault cuts through continental crust, which is much thicker than oceanic crust, earthquakes are considerably larger than those produced by oceanic transform faults. Earthquakes along this fault have had magnitudes up to $M_w = 8.5$. Along the San Andreas Fault, the Pacific Plate continues to move to the northwest past the North American Plate at a rate of about 5 centimeters (2 inches) a year. At this rate of movement, Los Angeles on the Pacific Plate will be adjacent to San Francisco on the North American Plate in about 18.5 million years. In human terms, 18.5 million years represents almost *one million* generations of people. It shouldn't be a concern that California will "fall off into the ocean" anytime soon. However, there certainly should be much

Figure 2–21 East African rift valleys and associated features.

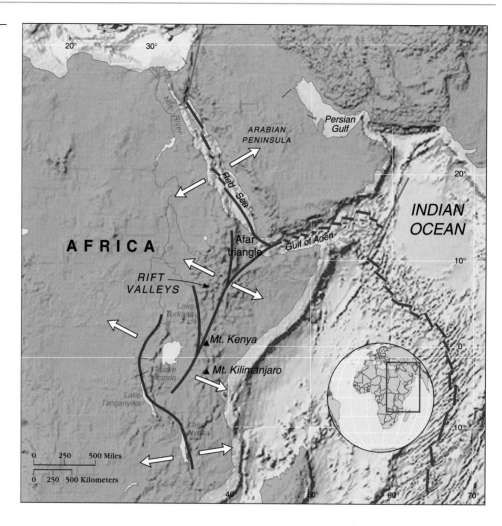

Figure 2–22 Comparing oceanic ridges and rises.
A. Cross-sectional view of the fast-spreading Mid-Atlantic Ridge. **B.** Cross-sectional view of the fast-spreading East Pacific Rise. On both cross sections, vertical exaggeration is 50 times normal.

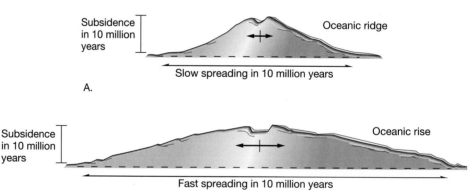

more of a concern about large earthquakes that occur periodically along heavily populated regions of this fault.

Case Studies: Some Applications of Plate Tectonics

One of the strengths of the theory of plate tectonics is how it ties together so many seemingly separate events and relates them all in a consistent model. Once the ideas of plate tectonics began to be accepted, researchers were able to use the model to help explain the origin of many other features that, up until that time, were difficult to explain. Let's look at a few examples to illustrate how plate tectonic processes can be used to explain the origin of these features.

Seamounts and Tablemounts

Highly characteristic of the Pacific Plate are tall volcanic peaks called **seamounts** if conical on top, or **tablemounts** (also called **guyots**, after the Swiss scientist Arnold Guyot[9]) if flat on top. Until recently, it was unclear ex-

---

[9] Guyot is pronounced "GEE-oh" with a hard g as in "give."

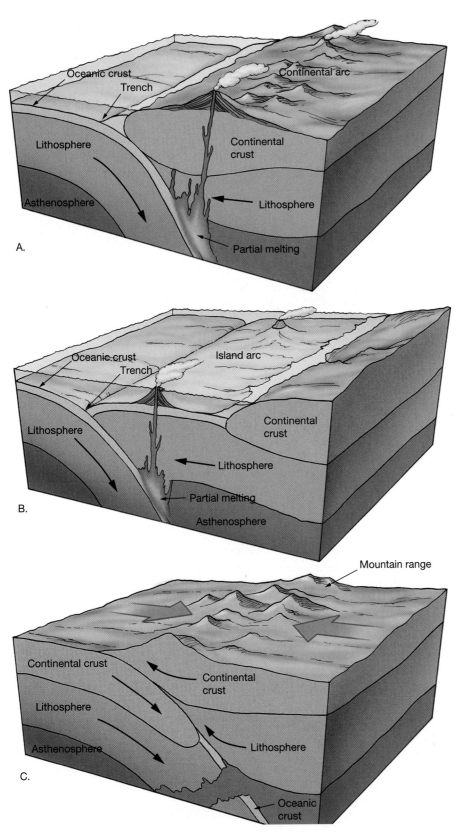

Figure 2–23 Three sub-types of convergent plate boundaries.

A. Oceanic-continental convergence. **B.** Oceanic-oceanic convergence. **C.** Continental-continental convergence.

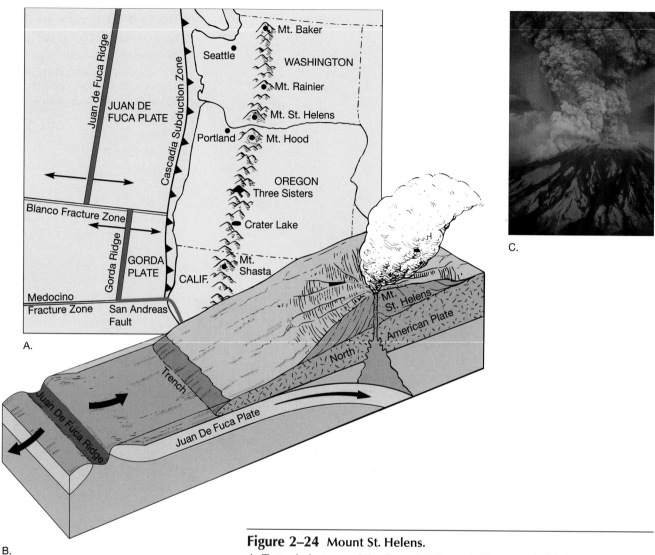

Figure 2–24 **Mount St. Helens.**
A. Tectonic features of the Cascade Mountain Range and vicinity. **B.** The volcanoes of the Cascade Mountains are created by the subduction of the Juan de Fuca Plate beneath the North American Plate. **C.** The eruption of Mount St. Helens in 1980.

actly why tablemounts were flat on top and why some tablemounts had evidence of shallow-water deposits despite being presently located in very deep water. Plate tectonics helped scientists determine how seamounts and tablemounts originated.

The origin of seamounts and tablemounts is related to processes occurring at the mid-ocean ridge (Figure 2–27). Because of sea floor spreading, active volcanoes are characteristic of the crest of the ridge. Once these seamounts are constructed by volcanic activity, they are carried off the ridge as volcanic activity diminishes and the plates move away from the ridge axis. A seamount on the mid-ocean ridge may be built so large that it rises above sea level and becomes an island. Once the volcano's top is above sea level, it is affected by wave attack at the surface. Waves are very powerful and can flatten off the top of a seamount in just a few million years once

the seamount lacks volcanic activity and is no longer receiving magma from below. The flattened seamount continues to be carried down the flank of the oceanic ridge, and after millions of years it is ultimately submerged as the plate carries the extinct flattened volcano into deeper water. The flat-topped structure (now called a tablemount) often carries with it evidence of the shallow-water environment in which it formed.

Coral Reef Development

On his famous voyage aboard the HMS *Beagle*, **Charles Darwin** noticed that there was a progression of stages in **coral reef** development. He hypothesized an origin for **coral reefs** that depended on the subsidence (sinking) of volcanic islands (Figure 2–28) and published the concept in *The Structure and Distribution of Coral Reefs* in

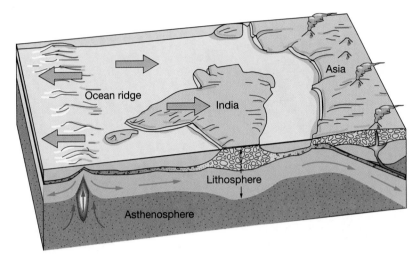

North

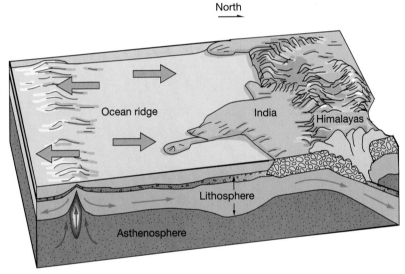

Figure 2–25 The collision of India with Asia.

The ongoing collision of India with Asia began about 45 million years ago and has uplifted the Himalaya Mountains.

1842. What Darwin's hypothesis lacked was a mechanism for how volcanic islands subside. Much later, advances in plate tectonic theory and samples of the deep structure of coral reefs provided evidence to help support Darwin's hypothesis.

Reef-building corals are colonial animals. They secrete a hard skeleton of limestone around their bodies and live in the sunlit surface waters on the flanks of volcanic islands and continents in warm, tropical seawater. Once the corals have colonized all of the sunlit waters available to them, their primary direction of growth is upward. Thus, corals grow on top of each other and each new generation is attached to the skeletons of its predecessors.

There are three stages of development in coral reefs: fringing, barrier, and atoll. **Fringing reefs** (Figure 2–28A) initially develop along the margin of any landmass where the temperature, salinity, and turbidity (cloudiness) of the water are suitable for reef-building corals. Often, fringing reefs are associated with active volcanoes and are not very thick or well developed. Also, because of the close proximity of the landmass to the reef, runoff from

the landmass can carry so much sediment that the reef is buried. The amount of living coral in a fringing reef at any given time is relatively small, with the greatest concentration in areas that are protected from sediment and salinity changes. If sea level does not rise or the land does not subside during the existence of a fringing reef, the process stops here, and the other two stages, barrier reef and atoll, do not develop.

The next stage of development is the **barrier reef** stage. Barrier reefs are linear or circular reefs separated from the landmass[10] by a well-developed lagoon (Figure 2–28B). As the landmass subsides, the reef maintains its position close to sea level by growing upward. Studies of reef growth rates indicate most have grown 3 to 5 meters (10 to 16 feet) per 1000 years during the recent geologic past. There is evidence, however, that some fast-growing reefs in the Caribbean have grown at rates of over 10 meters (33 feet) per 1000 years. Note

---

[10] In this case, the landmass may be an island or the margin of a continent.

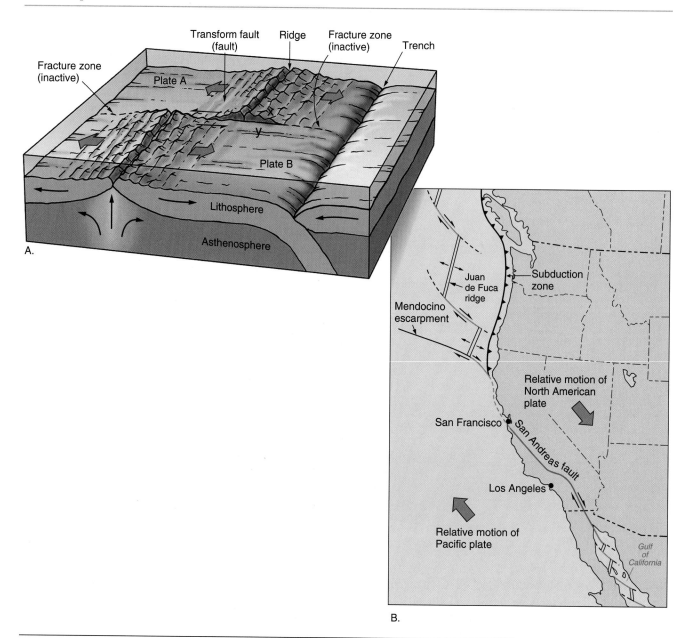

Figure 2–26 The role of transform faults.

The Juan de Fuca Ridge is offset by several oceanic transform faults. Also shown is the San Andreas Fault, a continental transform fault connecting the Juan de Fuca Ridge and the spreading center in the Gulf of California.

that if the landmass subsides at a rate faster than coral can grow upward, the coral reef will be submerged in water too deep for it to live.

The longest reef in the world is Austalia's Great Barrier Reef, a series of over 3000 individual reefs which collectively are in the barrier reef stage of development. The Great Barrier Reef is 150 kilometers (90 miles) wide and extends for more than 2000 kilometers (1200 miles) along Australia's northeastern coast (Figure 2–29). The effects of plate movement are clearly visible in the structure of the Great Barrier Reef. It is oldest (around 25 million years old) and thickest at its northern end. The reef thins to the south, where it is also younger. This relationship

is exactly what plate tectonics would predict as the Indian-Australian Plate moved north toward the Equator from colder Antarctic water. The northern part of Australia would reach water warm enough to grow coral first, and thus the coral in the northern region would be older and thicker than in the south. Smaller barrier reefs are found around many of the tropical islands throughout the Pacific, Indian, and Atlantic Oceans.

The third stage in coral reef development is the **atoll** stage. Atolls form either on a subsiding continental shelf or around the margin of a sinking volcanic island in the open ocean. Most atolls are of the oceanic variety, and the greatest number occur in the equatorial Pacific

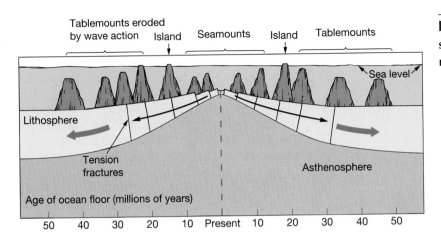

Figure 2–27 Formation of seamounts and tablemounts.

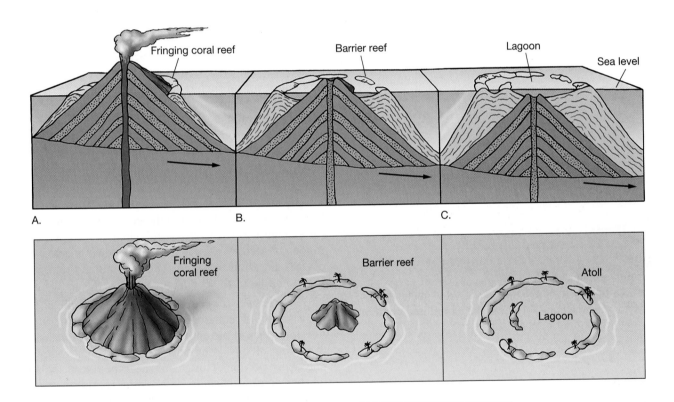

Figure 2–28 Stages of development in coral reefs.
Cross-sectional view (*above*) and map view (*below*) of: **A.** Fringing reef. **B.** Barrier reef. **C.** Atoll.

Ocean. As a barrier reef around a volcano continues to subside, coral builds up toward the surface. After millions of years, the volcano becomes completely submerged, but the coral reef continues to grow. If the rate of subsidence is slow enough for the coral to keep up, a circular reef called an atoll is formed (Figure 2–28C). Enclosed by the atoll is a lagoon that is usually not more than 30 to 50 meters (100 to 165 feet) deep. The reef generally has many channels that allow circulation between the lagoon and the open ocean. Buildups of crushed-coral debris can form narrow, picturesque islands large enough for people to inhabit (Figure 2–30).

The stages of coral reef development as proposed by Darwin can be explained using plate tectonic processes. When a volcanic island (a seamount) forms in a tropical area, it is colonized by coral and a fringing reef is built. As the plate carries the volcano into deeper water and the volcano subsides further, coral keeps building up toward sea level and a barrier reef is created. Finally, with more plate movement and more subsidence, the volcano is completely submerged. The coral continues to build up (by now, it can be several hundred meters thick) and an atoll forms. Darwin's idea about coral reef development was correct, but he did not have knowledge

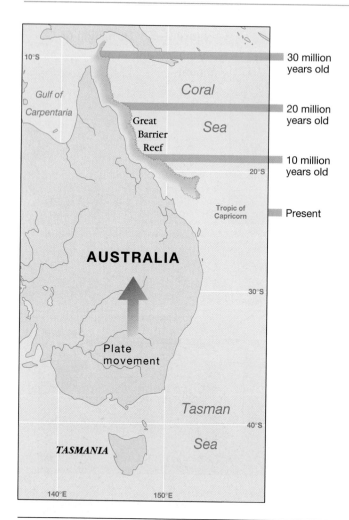

Figure 2–29 Australia's Great Barrier Reef records plate movement.

The Great Barrier Reef first began to develop about 30 million years ago, when northern Australia moved into tropical waters which allowed coral growth.

of plate tectonic processes that are known today to help support his hypothesis.

To take Darwin's hypothesis of coral reef development one step further, plate tectonics can be used to explain how an atoll turns into a tablemount. With further subsidence, the atoll will sink and carry with it the coral reef deposits that were near its summit. In this way, coral reef deposits can be transported several thousand meters below sea level. Analysis of tablemounts reveals that many have ancient coral reef deposits along their flattened tops.

Mantle Plumes and Hotspots

As the theory of plate tectonics was being developed, certain features did not seem to fit well within the model. Plate tectonics helped explain the origin of features near plate boundaries, but what about those features far from any plate boundary (often called **intraplate features**)?

Volcanic islands in the *middle* of a plate are an example of one such feature.

Studies of Earth's interior indicate that plumes (columnar areas) of hot molten rock arise from deep within the mantle, called **mantle plumes**. The areas where mantle plumes come to the surface are marked by an abundance of volcanic activity, called a **hotspot**. The volcanism in places such as Yellowstone National Park and Hawaii are attributed to the fact that these areas are hotspots. In general, the positions of hotspots do *not* coincide with plate boundaries[11] (Figure 2–31). However, there are a few notable exceptions where hotspots are associated with divergent boundaries because these are places where the lithosphere is thin. The Galapagos Islands and Iceland, which are on divergent boundaries, owe at least part of their origin to hotspot activity. For instance, Iceland straddles the Mid-Atlantic Ridge, a divergent plate boundary. It is also directly over a 150-kilometer (93-mile)-wide mantle plume, which accounts for the dramatic amount of volcanic activity there. In fact, the amount of volcanism has been so great that it has caused Iceland to be one of the few areas of the global midocean ridge that is above sea level. The positions of hotspots are probably related to the positions of convection cells in the mantle.

Throughout the Pacific Plate, many island chains trend in a northwestward–southeastward orientation. The most intensely studied of these is the **Hawaiian Islands-Emperor Seamount chain** in the northern Pacific Ocean (inset, Figure 2–32). How was this chain of intraplate volcanoes created? What caused the prominent bend in the overall direction of islands and seamounts that occurs in the middle of the chain?

To help answer these questions, let's look at the ages of the islands and seamounts in this chain. There are no active volcanoes in the entire chain northwest of the island of Hawaii; all have long since become extinct. Today, the only active portion of the chain is near the volcano Kilauea on the island of Hawaii, which is the southeasternmost island of the chain. Interestingly, the age of islands and seamounts progressively *increases* northwestward from Hawaii. You can see this pattern by examining the age dates for the Hawaiian Islands in Figure 2–32. To the northwest, the seamounts increase in age to Suiko Seamount (63 million years old) and beyond.

These age relationships suggest that the Pacific Plate has moved steadily northwestward over a relatively stationary mantle plume. The resulting Hawaiian hotspot created each of the islands and seamounts in the chain. As the plate moved, it carried the active volcano off the hotspot and a new volcano began forming, younger in

---

[11] Note that hotspots are different than either a volcanic arc or a mid-ocean ridge, even though all are marked by a high degree of volcanic activity.

Figure 2–30 Atolls.
View from space of a group of atolls in the Pacific Ocean.

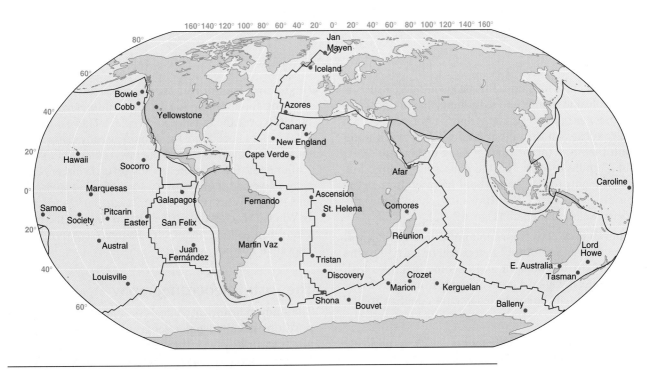

Figure 2–31 Global distribution of major hotspots.

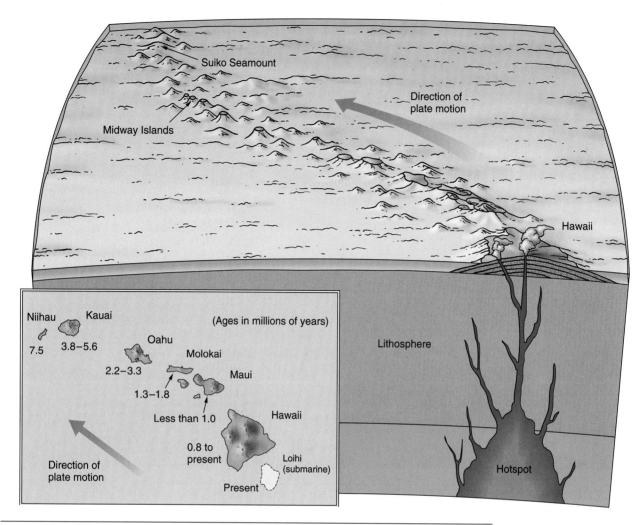

Figure 2–32 Hawaiian Islands—Emperor Seamount Chain.
The chain of islands and seamounts that extends from Hawaii to the Aleutian Trench results from
the movement of the Pacific Plate over the relatively stationary Hawaiian hotspot. Numbers indicate radiometric age dates in millions of years before present.

age than the previous one. This has generated a succession of volcanic islands and seamounts that are progressively older as one travels away from the hotspot. A chain
of volcanoes with this age relationship is called a **nematath** (*nema* = thread, *tath* = dung or manure[12]). At
about 43 million years ago, the plate shifted direction,
which accounts for the bend about halfway through the
chain that separates the Hawaiian chain from the
Emperor Seamount chain (Figure 2–32).

If Hawaii is directly above the hotspot now, what will
become of it in the future? It, too, will be carried to the
northwest off the hotspot and become inactive. With
enough time, it will be a great distance from the hotspot
and will eventually be subducted into the Aleutian Trench

(if you own land in Hawaii, it's never too soon to start
thinking about selling it!). In turn, another volcano will
build up over the hotspot and replace Hawaii. In fact, a
3500-meter (11,500-foot) seamount already exists 32 kilometers (20 miles) southeast of Hawaii, named Loihi. Still
1 kilometer (0.6 mile) below sea level, Loihi is volcani-
cally active and should surface sometime between 50,000
and 100,000 years from now at its current rate of activity. When it reaches the surface, it will become the newest
island in the long chain of volcanoes created by the
Hawaiian hotspot.

The Past: Paleoceanography

The study of historical changes of *continental* shapes and
positions is called **paleogeography** (*paleo* = ancient,
geo = earth, *graphy* = the name of a descriptive science).
Paleoceanography (*paleo* = ancient, *ocean* = the marine

---

[12] The highly descriptive scientific term *nematath* was obviously
coined by someone with a sense of humor!

environment, *graphy* = the name of a descriptive science) is the study of changes in the physical shape, composition, and character of the *oceans* brought about by paleogeographical changes.

Figure 2–33 shows a series of world maps showing the paleogeographical reconstruction of Earth at 60-million-year intervals (like a snapshot of Earth every 60 million years going back in time). At 540 million years ago, many of the present-day continents were unrecognizable. North America was on the Equator and rotated 90 degrees clockwise. Also on the Equator was Antarctica, which was connected to many other continents.

Between 540 and 300 million years ago, the continents began to come together to form Pangaea. Notice that Alaska had not formed yet. Continents are thought to add material through the process of **continental accretion** (*a* = not, *crete* = separated). Like adding layers onto a snowball, bits and pieces of continents, islands, and volcanoes are added to the edges of continents and create larger landmasses.

From 180 million years ago to the present, Pangaea separated and the continents moved toward their present-day positions. North America and South America rifted away from Europe and Africa to produce the Atlantic Ocean. In the Southern Hemisphere the early stages of Pangaea's breakup showed South America and a continent composed of India, Australia, and Antarctica separating from Africa.

By 120 million years ago, there was a clear separation between South America and Africa, and India had moved northward, away from the Australia-Antarctica mass, which began moving toward the South Pole. As the Atlantic Ocean continued to open, India moved rapidly northward and collided with Asia about 45 million years ago. Australia has also begun its rapid journey to the north since its separation from Antarctica.

One major outcome of global plate tectonic events over the past 180 million years has been the creation of the Atlantic Ocean, which continues to grow due to sea floor spreading along the Mid-Atlantic Ridge. At the same time, the Pacific Ocean continues to shrink as oceanic plates produced within it subduct into the many trenches that surround it and continent-bearing plates bear in from both the east and west.

The Future: Some Bold Predictions

Figure 2–34 shows what the world may look like 50 million years in the future (assuming that the rate of plate motion and the direction of plate motion remain the same[13]). There are many notable differences. The east African rift valleys may enlarge to form a new linear sea. The Red Sea may be greatly enlarged due to rifting that occurs there. India may continue to plow into Asia, further uplifting the Himalaya Mountains. As Australia moves north toward Asia, it may use New Guinea like a snow plow and accrete various islands to grow larger. North America and South America may continue to move west, enlarging the Atlantic Ocean but decreasing the size of the Pacific Ocean. The land bridge of Central America may no longer connect North and South America, which would have dramatic effects on ocean circulation. Lastly, the thin sliver of land that lies west of the San Andreas Fault may become an island in the North Pacific, soon to be accreted onto southern Alaska. In 50 million years from now, our world will look very different!

---

[13] These assumptions are probably not entirely valid, but they provide some general guidelines to allow a prediction of the positions of the plates in the future.

Students Sometimes Ask…

Plate tectonics is interesting, but it is still just a theory. Are there other scientific ideas for which there is better proof?

Don't discount the concept of plate tectonics because it is a theory! Scientific theories are rigorously tested and serve as good working models of various observable phenomena. In fact, new data (using recently developed technology) very strongly supports the theory of plate tectonics. Satellite positioning of locations on Earth allows us the ability to track the movement of plates very precisely. This tracking indicates that fixed positions on Earth's surface are indeed moving relative to one another, in good agreement with the directions and rates predicted by plate tectonics.

Certainly, plate tectonic theory is among the most significant scientific findings in human history. In a book entitled *The Five Biggest Ideas in Science* (1997), Charles Wynn and Arthur Wiggins suggest which models have been the most important ones generated in the sciences. Their list includes: (1) physics' model of the atom; (2) chemistry's periodic law; (3) astronomy's Big Bang theory; (4) geology's plate tectonics theory; and (5) biology's theory of evolution. It's nice to know that the theory of plate tectonics is considered important enough to be listed among these great scientific discoveries.

How long has plate tectonics been operating on Earth?

This is difficult to say with much certainty because our planet has been so dynamic over the past several billion years. Evidence to support any conclusion about plate tectonics prior to 180 million years ago has been largely destroyed by ongoing subduction of the sea floor. However, very old marine rock sequences uplifted onto continents show certain characteristics of plate motion and suggest that plate tectonics has been operating for at least the last 2 billion years of Earth history.

Will plate tectonics eventually "turn off" and cease to operate on Earth?

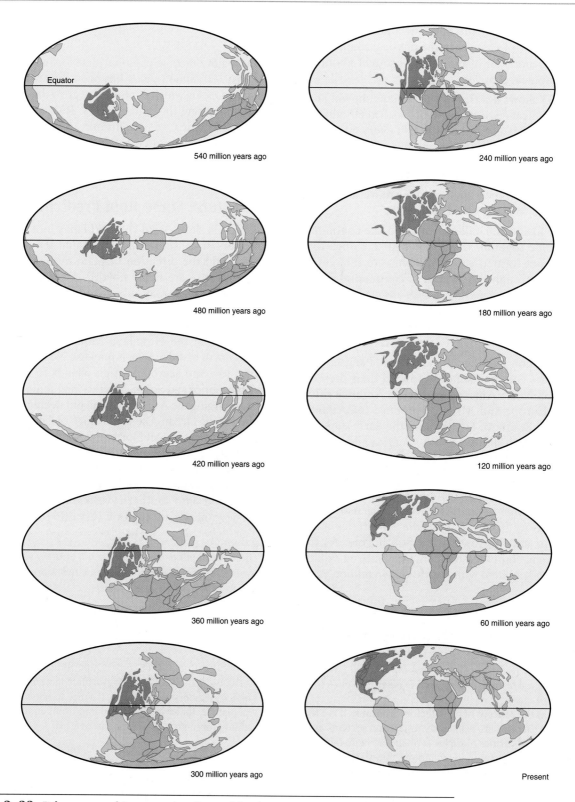

Figure 2–33 Paleogeographic reconstructions of Earth.
The positions of the continents at 60-million-year intervals.

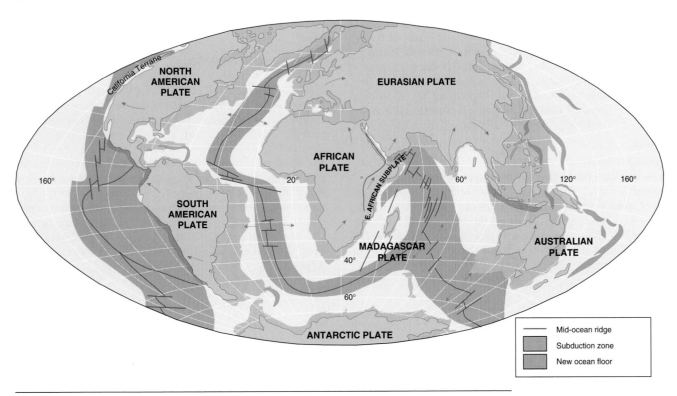

Figure 2–34 The world as it may look in the future.

Dark blue shadows indicate the present-day positions of continents; tan shading indicates positions of the continents in 50 million years. Arrows indicate direction of plate motion.

Because plate tectonic processes are thought to be powered by heat release from within Earth (which is of a finite amount), the forces someday will slow to the point that the plates will cease to move. What a different world it will be—an Earth with no earthquakes, no volcanoes, and no mountains. Flatness will prevail!

You mentioned that the plates move at speeds of 2 to 12 centimeters (1 to 5 inches) per year. Have plates always moved at the same rate?

No, they have actually moved faster in the past. This brings up an even more interesting question: How can scientists determine what the rate of plate motion was millions of years ago? Plate motions in the geologic past can be determined by analyzing the *time* it takes to produce a certain *width* of rock on the ocean floor. As sea floor spreading occurs, new basaltic crust is created. The *faster* the rate of sea floor spreading (and, therefore, the faster the plate motion), the *wider* the stripe of rock which is created per unit of time. It's like producing a recording on a standard VCR tape: On fast record speed, you'll use a lot of tape and the images will be of better quality (although you can only record a two-hour program). With a slower record speed, you'll use a lot less tape (with most tapes, you'll fit six hours of programs but with lower image quality). Just as the VCR tape records a program, the rocks near the mid-ocean ridge faithfully record the spreading rate on Earth: Like the fast record speed using more tape, fast sea floor spreading produces more rock. In fact, this pattern shows up nicely in Figure 2–15. By looking at that figure, does the Pacific Ocean or the Atlantic Ocean have a faster spreading rate?

Recent studies using the same technique indicate that about 50 million years ago, India attained a speed of 19 centimeters

(7.5 inches) per year. Other research indicates that about 530 million years ago, plate motions may have been as high as 30 centimeters (1 foot) per year! What causes these rapid bursts of plate motion? Scientists are not sure why plates have moved more rapidly at times, but increased heat release from Earth's interior may be the mechanism.

What ever happened to Panthalassa, the large ocean that existed 200 million years ago,?

Panthalassa, the ocean that surrounded the supercontinent of Pangaea, is the ancestral Pacific Ocean, which has been decreasing in size since the breakup of Pangaea about 180 million years ago.

If the continents move around on Earth's surface, do other features, such as segments of the mid-ocean ridge, also move?

That's a good observation, and, yes, they do! It is interesting to note that nothing is really fixed in place on Earth's surface. When we talk about movement of features on Earth, we must consider the question, "Moving relative to what?" Certainly, the mid-ocean ridge does move relative to the continents (which sometimes causes segments of the mid-ocean ridges to be subducted beneath the continents!). In addition, the mid-ocean ridge is moving relative to a fixed location outside Earth. This means that an observer orbiting above Earth would notice, after only a few million years, that all continental and sea floor features, as well as plate boundaries, are indeed moving.

What causes the Earth's magnetic field?

Scientists who study Earth's magnetic field aren't exactly certain why Earth has a magnetic field. Certainly, most other planets have magnetic fields. The sun has a strong magnetic field, which reverses its polarity about every 22 years (and is closely tied to the sun's 11-year sunspot cycle). Theories have been proposed which suggest that convective movement of fluids in Earth's iron-nickel liquid outer core is the cause of the magnetic field. The most widely accepted view is that the core behaves like a self-sustaining *dynamo*, a device that converts the energy in convective motion into magnetic energy. The study of Earth's magnetic field (called magnetodynamics) is a topic of current research.

Will the continents come back together and form a single landmass anytime soon?
Yes, it is very likely that the continents will come back together. Since all of the continents are on the same planetary body, there is only so far a continent can travel before it collides with other continents. Based on current plate motions, it appears that the continents may meet up again in the Pacific Ocean. Recent research suggests that the continents may form a supercontinent once every 500 million years or so. That means we only have about 300 million years to establish world peace!

Can earthquakes be predicted?
Earthquake prediction—in particular short-term prediction—is a notoriously difficult task! Despite extensive efforts in several countries during the last several decades, scientists are no better at accurately predicting earthquakes. However, plate tectonics has given scientists a workable model to use in understanding why earthquakes occur in certain areas of the world but only rarely in other areas. By carefully analyzing the past seismic record, seismologists can then make a *forecast*, which is based on probability. A forecast is stated as the likelihood that a certain-magnitude earthquake will occur in a particular area over a certain period of time (for example, there is a 70 percent probability that a $M_w = 7.0$ earthquake will occur along the San Andreas Fault sometime over the next 30 years).

Summary

The theory of plate tectonics suggests that the outermost portion of Earth is composed of a patchwork of thin, rigid plates that move horizontally with respect to one another. The idea began as a hypothesis called continental drift, proposed by Alfred Wegener at the start of the twentieth century. However, many geologists and geophysicists resisted this hypothesis throughout the first half of the twentieth century.

Studies of Earth structure indicate that Earth is a layered sphere, with the brittle plates of the lithosphere riding on a plastic, high-viscosity asthenosphere below. Near the surface, the lithosphere is composed of two different types of crust: low-density, light-colored, thin granite and higher-density, dark-colored, thicker basalt. Both types of crust float isostatically on the denser mantle below.

About 200 million years ago, all the continents were combined into one large continent, called Pangaea, which was surrounded by a single large ocean, called Panthalassa. Continents can drift at speeds that make it possible for a continent to move from a polar position to the Equator in tens of millions of years. The mid-ocean ridge is where new crust is created by the process of sea floor spreading. Old crust is being destroyed at an equivalent rate by subduction at deep-ocean trenches. The driving mechanism of plate tectonics may be gravity.

Evidence from the continents that supports plate tectonics includes the jigsaw-like fit of continents, matching sequences of rocks and mountain chains across oceans, glacial evidence, fossil and climate evidence, and continental magnetism. Tiny magnetic grains in rocks faithfully record the periodic switching of Earth's magnetic field, acting as miniature three-dimensional compass needles. By studying the paleomagnetism of these rocks, an apparent polar wander has occurred, which is interpreted to represent the land moving, not the poles.

Evidence from the ocean floor that supports plate tectonics includes its paleomagnetism, age, and features. The paleomagnetism of the ocean floor is permanently recorded in oceanic crust, indicating that as new material is added along the crest of the mid-ocean ridge, it moves down the slopes of the mid-ocean ridge to make room for new material surfacing at the crest. Thus, rocks beneath the oceans increase in age from the axis of the mid-ocean ridge into the deep-ocean basins. Other evidence from the ocean includes heat flow measurements and the pattern of worldwide earthquakes.

As new crust is added to the lithosphere at the mid-ocean ridge (divergent boundaries), the opposite ends of the plates are subducted into the mantle at ocean trenches or beneath continental mountain ranges such as the Himalayas (convergent boundaries). Additionally, oceanic ridges and rises are offset and plates slide past one another along transform faults (transform boundaries).

Many features and occurrences confirm the existence of plate tectonics. Some case studies of how plate tectonics can help to explain various phenomena include the origin of tablemounts, the developmental stages of coral reefs, mantle plumes and their associated hotspots, and paleocenographic changes. In the future, the geographic positions of the continents will look very different than today.

Key Terms

Questions and Exercises

1. Describe three different types of submersibles and the differences between them.

2. If you could travel back in time with three figures from this chapter to help Alfred Wegener convince the scientists of his day that continental drift does exist, what would they be and why?

3. Discuss how the chemical composition of Earth's interior differs from its physical properties. Include specific examples.

4. What are differences between the lithosphere and the asthenosphere?

5. Describe differences between granite and basalt. Which property is responsible for making the granitic crust "float" higher in the mantle?

6. How does isostatic adjustment help support the idea that continents can move *horizontally*?

7. When did the supercontinent of Pangaea exist? What was the ocean called that surrounded the supercontinent?

8. Assuming that a continent moves at a rate of 10 centimeters per year, how long would it take the continent to travel from your present location to a nearby large city?

9. Describe a possible mechanism responsible for moving the continents over Earth's surface, including a discussion of the lithosphere and asthenosphere.

10. Describe the evidence from the *continents* that helps support plate tectonics.

11. Describe Earth's magnetic field, including how it has changed through time.

12. How does the magnetic dip of magnetite particles found in igneous rocks tell us the latitude at which they formed?

13. How does continental movement account for the two apparent wandering curves of the north magnetic pole as determined from Europe and North America?

14. Describe the evidence from the *ocean floor* that helps support plate tectonics.

15. Describe how sea turtles use Earth's magnetic field for navigation.

16. Describe the general relationships that exist among distance from the spreading centers, heat flow, age of the ocean crustal rock, and ocean depth.

17. Why does a map of worldwide earthquakes closely match the locations of worldwide plate boundaries?

18. List and describe the three types of plate boundaries. Include in your discussion any sea floor features that are related to these plate boundaries, and include a real-world example of each. Construct a map view and cross section showing each of the three boundary types and direction of plate movement.

19. Explain why most lithospheric plates contain both oceanic- and continental-type crust.

20. Describe the differences between oceanic ridges and oceanic rises. Include in your answer why these differences exist.

21. Convergent boundaries can be divided into three types based on the type of crust contained on the two colliding plates. Compare and contrast the different types of convergent boundaries that result from these collisions.

22. Will California suddenly fall off into the ocean? Why or why not?

23. Describe the difference in earthquake magnitudes that occur between the three types of plate boundaries, and include why these differences occur.

24. How can plate tectonics be used to help explain the difference between a seamount and a tablemount?

25. Discuss the possible relationship of coral reef development to sea floor spreading. Include the effects of both vertical and horizontal motions.

26. How is the age distribution pattern of the Emperor Seamount and Hawaiian Island chains explained by the position of the Hawaiian hotspot? What could have caused the curious bend in between the two chains of seamounts/islands?

27. Describe the differences in origin between the Aleutian Islands and the Hawaiian Islands. Provide evidence to support your explanation.

28. What are differences between a mid-ocean ridge and a hotspot?

29. When does evidence for the following events *first* show up in the geologic record? (see Figure 2–33).

   a. North America lies on the Equator.

   b. The continents come together as Pangaea.

   c. The North Atlantic Ocean opens.

   d. India separates from Antarctica.

30. In your own words, explain how it can be that the floor of the ocean is so incredibly young and the ocean itself is so phenomenally old.

References

Allmendinger, E. 1982. Submersibles: Past-present-future. *Oceanus* 25:1, 18–35.

Apperson, K. D. 1991. Stress fields of the overriding plate at convergent margins and beneath active volcanic arcs. *Science* 254:5032, 670–678.

Bercovici, D., Schubert, G., and Glatzmaier, G. A. 1989. Three-dimensional spherical models of convection in the earth's mantle. *Science* 244:4907, 950–954.

Bolt, B. 1993. *Earthqukes.* New York: W. H. Freeman and Co.

Bullard, Sir Edward. 1969. The origin of the oceans. *Scientific American* 221:66–75.

Burnett, M. S., Caress, D. W., and Orcutt, J. A. 1989. Tomographic image of the magma chamber at 12°50'N on the East Pacific Rise. *Nature* 339:206–208.

Carrigan, C. R., and Gubbins, D. 1979. The source of the earth's magnetic field. *Scientific American* 240:2, 118–133.

Davies, P. J., Symonds, P. A., Feary, D. A., and Pigram, C. J. 1987. Horizontal plate motion: A key allocyclic factor in the evolution of the Great Barrier Reef. *Science* 238:1697–1700.

Dietz, R. S., and Holden, J. C. 1970. Reconstruction of Pangaea: Breakup and dispersion of continents, Permian to present. *Journal of Geophysical Research* 75:26, 4939–4956.

———. 1970. The breakup of Pangaea. *Scientific American* 223:4 30–41.

Hamilton, W. B. 1988. Plate tectonics and island arcs. *Bulletin of the Geological Society of America* 100:1503–1526.

Hess, H. H. 1962. History of ocean basins. In Engel, A. E. J., Lames, H. L., and Leonard, B. F., eds., *Petrologic studies: A volume to honor A. F. Buddington,* 559–620. New York: Geological Society of America.

Kious, W. J., and Tilling, R. I. 1996. *This dynamic Earth: The story of plate tectonics.* Washington D.C.: U.S. Geological Survey.

Liu, M., Yuen, D. A., Zhao, W., and Honda, S. 1991. Development of diapiric structures in the upper mantle due to phase transitions. *Science* 252:1836–1839.

Lohmann, K. J., and Lohmann, C. M. F. 1996. Detection of magnetic field intensity by sea turtles. *Nature* 380:59–61.

Marx, R. F. 1978. An illustrated history of submersibles: From Alexander the Great to *Alvin.* Oceans 11:6, 22–29.

Menard, H. W. 1986. *The ocean of truth: A personal history of global tectonics.* Princeton, NJ: Princeton University Press.

Olson, P., Silver, P. G., and Carlson, R. W. 1990. The large-scale structure of convection in the Earth's mantle. *Nature* 344:209–214.

Schmidt, V. A. 1986. *Planet Earth and the new geoscience.* DuBuque, IA: Kendall/Hunt.

Sclater, J. G., Parsons, B., and Jaupart, C. 1981. Oceans and continents: Similarities and differences in the mechanisms of heat loss. *Journal of Geophysical Research* 86:11, 535–552.

Scotese, C. R. 1997. *Phanerozoic plate tectonic recontructions,* PALEOMAP Progress Report #36 (Edition 7), Dept. of Geology. Arlington: University of Texas.

Seachrist, L. 1996. Sea turtles master migration with magnetic memories. In Pirie, R. G., ed., *Oceanography: Contemporary Readings in Ocean Sciences,* 3rd ed. New York: Oxford University Press.

Shepard, F. P. 1977. *Geological oceanography.* New York: Crane, Russak.

Smith, D. K., and Cann, J. R. 1993. Building the crust at the Mid-Atlantic Ridge. *Nature* 365:6448, 707–715.

Tarbuck, E. J., and Lutgens, F. K. 1996. *The earth: An introduction to physical geology,* 5th ed. Upper Saddle River, NJ: Prentice-Hall.

———. 1997. *Earth science,* 8th ed. Upper Saddle River, NJ: Prentice-Hall.

The Open University Course Team. 1989. *The ocean basins: Their structure and evolution.* Oxford: Pergamon Press.

van der Hilst, R., Engdahl, R., Spakman, W., and Nolet, G. 1991. Tomographic imaging of subducted lithosphere below northwest Pacific island arcs. *Nature* 353:37–42.

Various authors. 1992. Special issue featuring several articles focusing on mid-ocean ridges. *Oceanus* 34:4, 1–111.

Various authors. 1993. Special issue featuring several articles on marine geology and geophysics. *Oceanus* 35:4, 1–104.

Walker, M. M., et al., 1997. Structure and function of the vertebrate magnetic sense. *Nature* 390:6658, 371–376.

Wessel, P., and Kroenke, L. 1997. A geometric technique for relocating hotspots and refining absolute plate motions. *Nature* 387:365–369.

Wiltschko, R., and Wiltschko, W. 1995. *Magnetic orientation in animals*. Berlin: Springer-Verlag.

Wolfe, C. J., Bjarnason, I. T., VanDecar, J. C., and Solomon, S. C. 1997. Seismic structure of the Iceland mantle plume. *Nature* 385:245–247.

Wynn, C. M., and Wiggins, A. W. 1997. *The five biggest ideas in science*. New York: Wiley.

Suggested Reading

Earth

Broad, W. J. 1997. The hot dive. 6:4, 26–33. An award-wining journalist experiences a dive in the *Alvin* to the hydrothermal vents along the Juan de Fuca ridge.

Davidson, K. 1994. Predicting earthquakes. 3:3, 56–63. An examination of the techniques seismologists use to predict earthquakes.

———. 1994. Learning from Los Angeles. 3:5, 40–49. An analysis of the tectonic forces that produced the January 1994 Northridge earthquake, including pictures of the damage.

Dawson, J. 1993. CAT scanning the Earth. 2:3, 36–41. Supercomputers are enabling geophysicists to visualize Earth's interior and gain new insights into mechanisms operating within the deep Earth.

Flanagan, R. 1994. Journeys in Iceland. 3:4, 24–33. A discussion of the tectonic processes that are responsible for the continuing evolution of Iceland, along with how people have learned to live in a place where the crust is literally ripping apart.

Frank, A. 1998. Forging the core. 7:1, 24–27. A look at current research aimed at understanding mechanisms operating in Earth's core.

Grossman, D., and Shulman, S. 1995. Earthly attraction. 4:1, 34–41. Examines how migrating birds, sea turtles, and other animals may be able to sense Earth's magnetic field and use it like a map to navigate.

Hopkins, R. L. 1997. Land torn apart. 6:1, 36–41. A look at the plate tectonic forces responsible for shaping Baja California and the interesting landforms and distribution of life that has resulted.

Lewis, G. B. 1995. Island of fire. 4:5, 32–39. A photojournalist's account of the eruption of Kilauea, Hawaii's most active volcano.

Moores, E. 1996. The story of Earth. 5:6, 30–33. A brief review of plate tectonics and how it has changed the way scientists view Earth.

Parks, N. 1994. Exploring Loihi: The next Hawaiian Island. 3:5, 56–63. Using submersibles, researchers study the features and origin of the active volcano Loihi, the newest volcano in the Emperor-Hawaiian chain.

———. 1997. Loihi rumbles to life. 6:2, 42–49. Using a submersible, scientists visit the underwater volcano Loihi after a period of intense earthquake activity in 1996.

Vogel, S. 1994. Inner Earth revealed. 4:4, 42–49. While scanning Earth's deep interior, seismologists have discovered towering peaks, open plains, and molten-iron oceans.

Waters, T. 1993. Roof of the world. 2:4, 26–35. Explores the relationship between the uplift of the Himalaya Mountains due to the collision of India with Asia and the initiation of cyclic ice ages.

Wysession, M. E. 1996. Journey to the center of the Earth. 5:6, 46–49. A review of the techniques seismologists use to study

deep Earth processes and how those processes drive plate tectonics.

Sea Frontiers

Burton, R. 1974. Instant islands. 19:3, 130–136. A description of the 1973 eruption on the island of Heimaey south of Iceland and its relationship to plate tectonics processes.

Dietz, R. S. 1971. Those shifty continents. 17:4, 204–212. A readable and informative presentation of the crustal features and the possible mechanism of plate tectonics.

———. 1976. Iceland: Where the mid-ocean ridge bares its back. 22:1, 9–15. A description of the rift zone of Iceland and its relationship to the Mid-Atlantic Ridge and sea floor spreading.

———. 1977. San Andreas: An oceanic fault that came ashore. 23:5, 258–266. A discussion of the San Andreas Fault and its relationship to global plate tectonics.

Emiliani, C. 1972. A magnificent revolution. 18:6, 357–372. A discussion of advances in studying Earth science from the sea, including climate cycles, plate tectonics, and the economic potential of resources lying within marine sediments.

Mark, K. 1972. Ocean fossils on land. 18:2, 95–106. A discussion of the significance of marine fossils found in rocks now many miles inland is centered on the work of an early American paleontologist, James Hall.

———. 1974. Earthquakes in Alaska. 20:5, 274–283. A discussion of the origin, nature, and effects of earthquakes in Alaska.

———. 1976. Coral reefs, seamounts, and guyots. 22:3, 143–149. A discussion of the role of global plate tectonics in explaining the distribution of seamounts, guyots, and the evolution of coral reefs.

Van Dover, C. 1988. Dive 2000. 34:6, 326–333. Events related to the 2000th dive of the *Alvin* on the East Pacific Rise are recounted.

Scientific American

Bloxham, I., and Bubbins, D. 1989. The evolution of the earth's magnetic field. 261:6, 68–75. The theory of how Earth's magnetic field is associated with motions in Earth's liquid outer core.

Bonatti, E. 1994. The Earth's mantle below the oceans. 270:3, 44–51. An analysis of how sea floor spreading processes affect such diverse phenomena as the shape of the mid-ocean ridge, the composition of rocks along the mid-ocean ridge, the migratory behavior of sea turtles, and even the spin axis of Earth.

Bonatti, E., and Crane, K. 1984. Oceanic fracture zones. 250:5, 40–51. The role of oceanic fracture zones in the plate tectonics process is related to the spherical shape of Earth.

Dietz, R. S., and Holden, J. C. 1970. The breakup of Pangaea. 223:4 30–41. A discussion of how Pangaea broke up into the separate continents that we know today; includes maps of the positions of the continents through time.

Hawkes, G. S. 1997. Microsubs go to sea. 277:4, 132–135. Written by the co-inventor of *Deep Flight I*, the author explains the reasoning and engineering behind developing a single-passenger, maneuverable microsub.

Herbert, S. 1986. Darwin as a geologist. 254:5, 116–123. Before publication of *The Origin of Species*, Charles Darwin considered himself to be primarily a geologist. This article outlines his contributions in this field, with special attention given to his theory of coral reef formation.

Jeanloz, R., and Lay, T. 1993. The core-mantle boundary. 261:5, 48–55. The boundary between the liquid core that generates Earth's magnetic field and the overlying solid mantle is surprisingly active. Core fluids may rise into the mantle hundreds of meters above the average position of the core-mantle boundary.

Macdonald, K. C., and Fox, P. J. 1990. The mid-ocean ridge. 262:6, 72–95. An interesting and comprehensive overview of present knowledge of the oceanic ridges and rises is presented.

Pinter, N, and Brandon, M. T. 1997. How erosion builds mountains. 276:4, 74–79. A new look at how erosion can help construct mountains, which includes a good description of isostatic adjustment.

Schneider, D. 1997. Hot-spotting. 276:4 22–24. A new method is presented that may help accurately locate hotspots beneath the ocean floor.

Schneider, D., and Gibbs, W. W. 1997. Flight of fancy. 276:1, 22–24. A critical analysis of whether the new microsub *Deep Flight I* would be a real asset to the oceanographic research community.

White, R. S., and McKenzie, D. 1989. Volcanism at rifts. 261:1. 62–71. An analysis of the different types of volcanic activity at rift valleys along the mid-ocean ridge.

Oceanography on the Web

www.prenhall.com/thurman

Visit the *Essentials of Oceanography* home page for on-line resources for this chapter. There you will find an on-line study guide with review exercises, and links to oceanography sites to further your exploration of the topics in this chapter. *Essentials of Oceanography* is at: **http://www.prenhall.com /thurman** (click on the Table of Contents menu and select this chapter).

CHAPTER 3
MARINE PROVINCES

Experiments in Underwater Living

The shallow areas next to the continents represent an enormous untapped area of the world. The shallow sea floor of the continental shelves which adjoin the continents cover about 26 million square kilometers (10 million square miles) of Earth's surface, an area almost as large as the entire African continent. Because of an abundance of food, minerals, and other resources, the shallow areas of the continental shelves have often been considered an ideal place to establish permanent underwater living habitats.

Many attempts have been made to advance humankind's settlement of the continental shelves. One of the first trials in underwater living took place in 1962, when Albert Falco and Claude Wesly lived underwater at a depth of 26 meters (85 feet) for one week in **Jacques-Yves Cousteau's** Continental Shelf Station offshore of Marseilles, France. Cousteau's Continental Shelf Project (called Conshelf) culminated with five "aquanauts" staying a month at 11 meters (36 feet) below the surface in the Red Sea. In 1964, the **Sealab** project was initiated by the U.S. Navy to study the feasibility of living underwater for long periods. A U.S. Navy team lead by George F. Bond remained 11 days at a depth of 59 meters (193 feet) off Bermuda in Sealab I. The aquanauts were watched closely for any signs of pressure-induced problems. Sealab II in 1965 provided a habitat for a 10-person team to stay 30 days at a depth of 62 meters (205 feet) off La Jolla, California (Figure 3A). This project generated much public interest because it included M. Scott Carpenter, the second U.S. astronaut to orbit Earth, and a trained Navy dolphin named Tuffy, who, by responding to signals, was useful in swimming errands and reaching lost divers.

In all of these habitats, the aquanauts made daily dives from their habitats without incident because the habitats were at the same pressure as the outside water depth.

Figure 3A Sealab II.

How could these aquanauts stay at these high pressures for so long? Experiments using high-pressure chambers in the late 1950s revealed that different mixtures of gases increased the length of time divers could stay at depth. Once divers became saturated with whatever they were breathing, they could stay at that pressure almost indefinitely. The discovery of **saturation diving** opened the door for underwater habitats. In the Sealab project, the deleterious effects of nitrogen were avoided by replacing it with helium gas within the habitat. The helium-rich atmosphere did distort the aquanauts' voices, making them high-pitched and cartoon-like. This provided for some interesting moments, such as the time that President Lyndon Johnson called to congratulate the aquanauts in Sealab II. Unfortunately, the Sealab project came to a halt when an aquanaut died during the testing of Sealab III in 1969 off San Clemente Island, California, at a depth of 183 meters (600 feet).

Today, only three underwater habitats remain functional, all near Key Largo, Florida: the Jules Verne Undersea Lodge is a small hotel at a depth of 9 meters (30 feet); Scott Carpenter's Man in the Sea Program oversees an underwater habitat as an educational lab; and Aquarius 2000 is an underwater habitat devoted to sciences. Aquarius 2000 accommodates a crew of six at a depth of 15 meters (49 feet) on 10-day missions to study coral reef communities.

Over a century ago, most scientists dismissed the bottom of the sea as scientifically uninteresting. It was suspected that the ocean was deepest somewhere in the middle of the ocean basins. The ocean floor was assumed to be flat and carpeted with a thick layer of muddy sediment that contained little of scientific interest. However, as more and more vessels crisscrossed the seas to map the ocean floor and to lay transoceanic cables, scientists found that the sea floor was a highly varied terrain: There were many interesting features, including deep troughs, ancient volcanoes, submarine canyons, and great mountain chains. It was unlike anything on land, and, as it turned out, the deepest parts of the oceans were actually close to land!

As marine geologists and oceanographers began to analyze the features of the ocean floor, they realized that certain features had profound implications not only for the history of the ocean floor, but also for the history of Earth. How could all these remarkable features have formed, and how can their origin be explained? Through the passage of time, the shape of the ocean basins has changed as continents have ponderously migrated across Earth's surface in response to forces within Earth's interior. The ocean basins as they presently exist reflect the processes of plate tectonics at work for long periods of Earth history. Plate tectonics—the topic of the previous chapter—helps us explain the origin of most sea floor features.

Bathymetry

Bathymetry (*bathos* = depth, *metry* = measurement) is the study of ocean depths. Because bathymetry examines the vertical distance between the ocean surface and the mountains, valleys, and plains of the sea floor, bathymetry is analogous to topography of the dry land surface. Instead of measuring height above sea level, as on land, *bathymetry is the measurement of depth below sea level*. As investigations into the depth of the ocean have proceeded, it has become apparent that a broad shelf like feature is located around the continents. At varying distances from shore, it steepens and slopes off into deep basins. Quite commonly, linear mountain ranges run through these deep basins which, as we saw in the last chapter, are formed by volcanic eruptions at divergent plate boundaries. In all oceans, but especially around the margin of the Pacific Ocean, linear deep scar-like features called **trenches** separate the shallow areas from the deep-ocean basin. We saw in the last chapter that these are the areas where subduction occurs along convergent plate boundaries.

Bathymetric Techniques

The first recorded attempt to measure the ocean's depth was conducted in the Mediterranean Sea in about 85 B.C. by a Greek named Posidonius. His mission was to answer an age-old question: How deep is the ocean? Posidonius' crew let out nearly 2 kilometers (1.2 miles) of line before the heavy weight on the end of the line touched bottom. How surprised they must have been to let out such a long length of line, but little did they realize that the oceans are much deeper in other places. This technique of using a **sounding** line (a long line with a heavy weight attached to the end) was used for the next 2000 years by voyagers the world over to probe the ocean's depths. The standard unit of measurement used to record depth is the **fathom** (*fathme* = outstretched arms[1]) and is equal to 1.8 meters (6 feet).

The first systematic bathymetry of the oceans was made in 1872, when the HMS *Challenger* undertook its historic three-and-one-half-year voyage. Every few thousand kilometers, *Challenger's* crew stopped and measured the

---

[1]This term is derived from how depth sounding lines were brought back on board a vessel: By hauling in the line and counting the number of arm-lengths collected (and by measuring the length of the person's outstretched arms), the amount of line taken in could be calculated. Later, the fathom was standardized to equal exactly 6 feet.

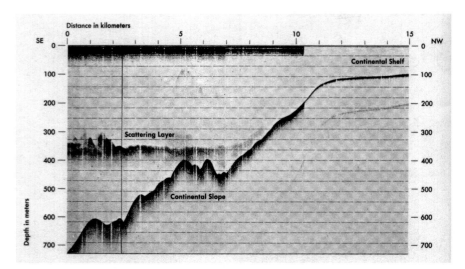

Figure 3–1 An echosounder record.

An echosounder record from the east coast of the United States shows the provinces of the sea floor. Vertical exaggeration (amount of vertical distortion) is 12 times normal. The scattering layer probably represents a concentration of marine organisms.

depth, along with many other ocean properties. These measurements indicated that the deep ocean floor was not flat but had significant topographic relief, just as dry land does. However, determining bathymetry by making soundings rarely gives a complete picture of the ocean floor.[2]

It wasn't until 1925 that the German vessel *Meteor*, equipped with an **echosounder**, identified a mountain range running through the center of the South Atlantic Ocean. An echosounder sends a sound signal (called a **ping**) into the ocean that bounces off any density difference, such as organisms within the water column, or the ocean floor (Figure 3–1). Similar to how radar is used on land, an echosounder records echoes that have bounced off objects. The time it takes for the echoes to return[3] is used to determine the depth and shape of the ocean floor. Until recently, most of our knowledge of ocean bathymetry was provided by the echosounder. The idea of using sound in the ocean instead of sight is not new: Certain marine mammals have been "seeing" by using sound waves for millions of years.

Echosounding lacks detail because, from a ship 4000 meters (13,100 feet) above the ocean floor, the sound beam emitted from the ship widens to a diameter of about 4600 meters (15,000 feet) at the bottom. Consequently, the first echoes to return from the bottom will likely be from the closest (highest) peak within this broad area. This often results in an inaccurate view of the relief (the variations in elevation) of the deep sea floor.

Oceanographers who wish to know more than just ocean depth often use strong low-frequency sounds produced by air guns, that penetrate beneath the ocean floor and reflect off the boundaries between different layers of sediment. This provides **seismic reflection profiles** (Figure 3–2), which show subbottom ocean structure.

The **precision depth recorder (PDR)** was developed in the 1950s. With a more focused high-frequency sound beam, it could provide depths to a resolution of about 1 meter (3.3 feet). Throughout the 1960s, PDRs were used extensively and provided a reasonably good representation of the ocean floor. From thousands of research vessel tracks, the first reliable global maps of sea floor bathymetry were produced. These maps played an important role in the acceptance of the ideas of sea floor spreading and plate tectonics.

Today, multibeam echosounders[4] and side-scan **sonar** (an acronym for <u>so</u>und <u>na</u>vigation <u>a</u>nd <u>r</u>anging) give oceanographers a more precise picture of the ocean floor. The first multibeam echosounder, **Seabeam**, made it possible for a survey ship to map the features of the ocean floor along a strip up to 60 kilometers (37 miles) wide. The system uses sound emitters directed away from both sides of the ship, with receivers permanently mounted on the hull of the ship. With this data, a map of the sea floor can be developed with the aid of computers. Another type of side-scan sonar system is towed behind a survey ship. These systems include **Sea MARC** (<u>Sea</u> <u>M</u>apping and <u>R</u>emote <u>C</u>haracterization) and **GLORIA** (<u>G</u>eological <u>Long-R</u>ange <u>I</u>nclined <u>A</u>coustical instrument), which is shown in Figure 3–3.

To obtain a more detailed picture of the ocean floor, a side-scan instrument must be towed behind a ship on a

[2]Imagine trying to determine what the surface features on Earth look like by flying over the surface in a blimp at an elevation of several kilometers, on a foggy night, with only a long rope with a bucket on the end. Certainly, sampling done by this method would hardly be representative of Earth's surface!

[3]The speed of sound in seawater is about 1450 meters (4750 feet) per second.

[4] As their name implies, multibeam echosounders use multiple beams of sound, sending out and receiving back many different sound frequencies at the same time.

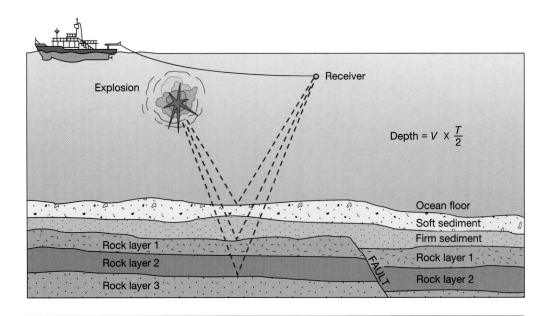

Figure 3–2 Seismic profiling.

Low-frequency sound that can penetrate bottom sediments is emitted by an air-gun explosion. It reflects off the boundaries between rock layers and returns to the receiver.

Figure 3–3 GLORIA system for side-scanning sonar surveys.

The side-scan sonar system GLORIA is towed behind a survey ship. It can map a swath of ocean floor up to 60 kilometers (37 miles) wide.

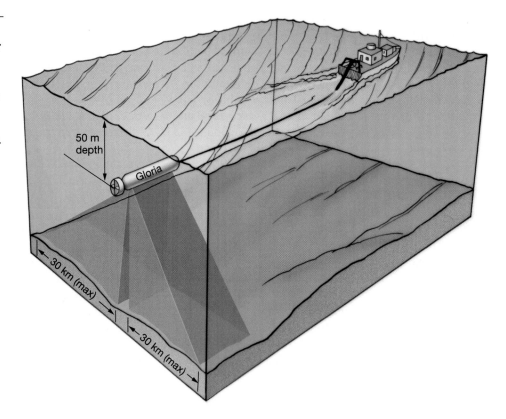

cable so that it "flies" just above the ocean floor. The most recent technology in these deep-tow systems combines a side-scan sonar instrument with a sub-bottom imaging package (Figure 3–4). This allows a simultaneous view of the surface of the ocean floor and a cross section of the sediment below at water depths to 6500 meters (21,325 feet).

Provinces of the Ocean Floor

Study of ocean bathymetry reveals that the ocean floor can be divided into three major provinces: **continental margins** (shallow-water areas close to continents), **deep-ocean basins** (deep-water areas farther from land), and the **mid-ocean ridge** (shallower areas near the middle

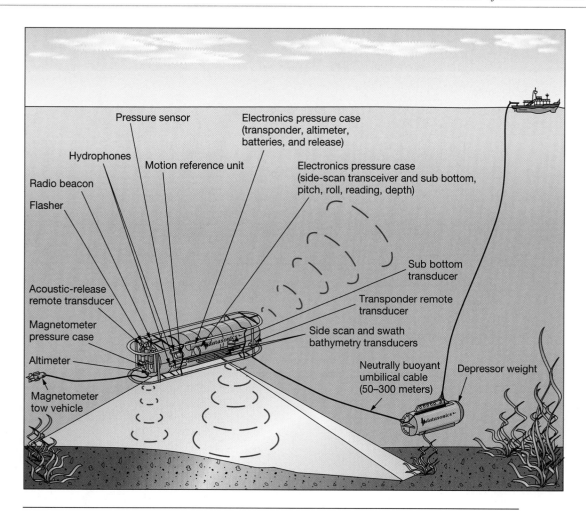

Figure 3–4 Deep-tow sea floor imaging system.
Deep-tow side-scan sonar systems are towed close to the ocean floor. These sea floor imaging systems provide detailed sonar maps of the ocean floor as well as a profile view into sediments on the ocean floor.

of an ocean). Plate tectonic processes are integral to the formation of these provinces. Through the process of sea floor spreading, mid-ocean ridges and deep-ocean basins are created. Elsewhere, as a continent is split apart, new continental margins are formed. A schematic view of the major provinces of the North Atlantic Ocean is shown in Figure 3–5.

Features of Continental Margins

Active and Passive Continental Margins Continental margins can be classified as either active or passive, depending on their proximity to plate boundaries (Figure 3–6). **Active margins** are continental margins that are associated with lithospheric plate boundaries, and are marked by a high degree of tectonic activity. Two types of active margins exist. **Transform active margins** are associated with transform plate boundaries where transform or strike-slip faulting occurs. An example of a transform active margin is the western United States

along the San Andreas Fault. **Convergent active margins** are associated with oceanic-continental convergent plate boundaries. Here, a continental arc and an offshore trench delineate the plate boundary. An example of a convergent active margin is along western South America, where the Nazca Plate is being subducted beneath the South American Plate.

Passive margins are continental margins that are imbedded within the interior of lithospheric plates and are therefore not in close proximity to any plate boundary. Rifting of continental landmasses and continued sea floor spreading produces passive margins. An example of a passive margin is the east coast of the United States, where there is no plate boundary. Passive margins usually lack major tectonic activity (large earthquakes, eruptive volcanoes, and mountain building episodes). A passive continental margin includes the continental shelf and shelf break, the continental slope, and the continental rise that extends toward the deep-ocean basins

Figure 3–5 Major regions of the North Atlantic Ocean floor.

Map view above and profile view below.

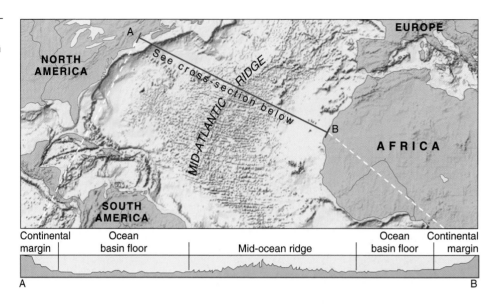

Figure 3–6 Active and passive continental margins.

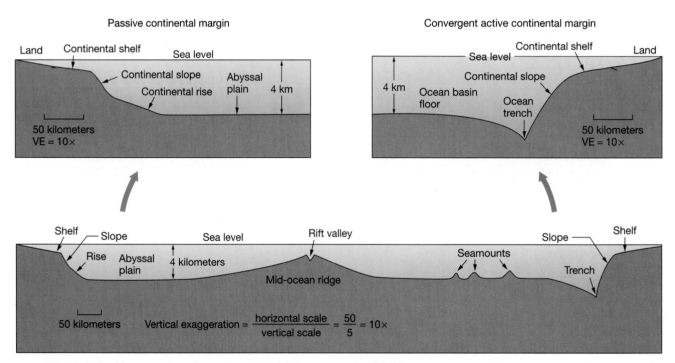

(Figure 3–6). Let's examine the features along a typical passive continental margin (Figures 3–6 and 3–7).

Continental Shelf Extending from the shoreline is a gently sloping, low-relief feature called the **continental shelf**, which is geologically part of the continent (meaning that it is underlain by granitic continental crust) and is often covered by marine sediments. During the geologic past, much of it was exposed above the shoreline when colder climates prevailed during the Ice Age (a time when more of Earth's water was frozen as glaciers on land) which lowered sea level. Over time, as sea level has fluctuated, the shoreline has migrated back and forth across the shelf. Thus, a shelf's general bathymetry can

usually be predicted by examining the topography of the adjacent coastal region. With few exceptions, this coastal topography extends beyond the shore and onto the continental shelf.

The continental shelf is defined as a shelf-like zone extending from the shore beneath the ocean surface to a point at which a marked increase in slope angle occurs. This point is referred to as the **shelf break**, and the steeper portion beyond the shelf break is known as the **continental slope** (Figures 3–6 and 3–7).

Examining a map of the sea floor (see the inside front cover of this book) reveals many aspects of continental shelves. Shelves typically slope gently toward the deep

Box 3–1
Sea Floor Mapping from Space

Mapping the sea floor by ship is a time-consuming process. Multibeam and side-scan sonar produce fairly detailed maps, but a research vessel must travel back and forth (called, appropriately enough, "mowing the lawn") throughout an area to produce an accurate map of bathymetric features. This is because as a ship travels, it collects a swath of data immediately below and a little bit to either side of the ship's track. A satellite, by contrast, offers the advantage that it can observe large areas of the ocean at one time. How does a satellite, which can only view the ocean's *surface,* obtain a picture of the ocean floor?

Sea floor features directly influence Earth's gravitational field. Deep areas such as trenches correspond to a lower gravitational attraction, and large undersea objects such as seamounts exert an extra gravitational pull. These differences affect sea level, which cause the ocean surface to bulge outward and sink inward, mimicking the relief of the ocean floor. A 2000-meter (6500-foot)-high seamount, for example, exerts a small but measurable gravitational pull on the water around it, creating a bulge 2 meters (7 feet) high on the ocean surface that is easily detectable by a satellite. Satellites bounce microwaves off the ocean surface to measure its shape, and the resulting pattern of lumps and bulges reflects what is underneath (Figure 3B). In this way, satellites remotely sense the shape of the ocean surface due to gravity. After corrections are made for waves, tides, currents, and atmospheric effects, this data is used to produce

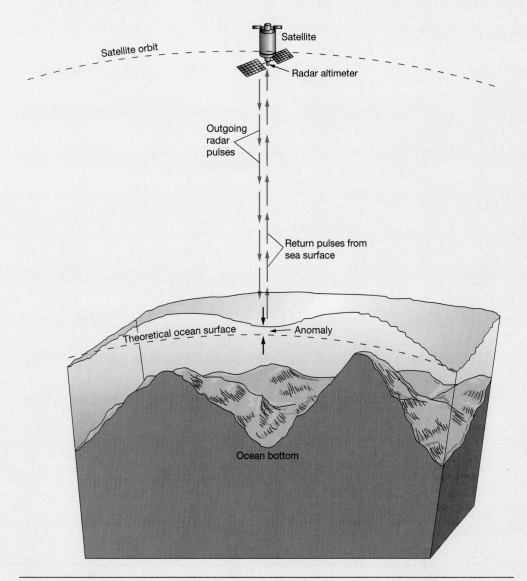

Figure 3B   Satellite measurements of the ocean surface.
Due to gravitational attraction, the shape of the sea floor is mimicked in the relief of the ocean surface.

(continued)

(continued)

maps of the ocean surface, which represent ocean floor bathymetry. The maps in Figure 3C show how much more resolution can be obtained by using these new maps.

New data from the European Space Agency's ERS-1 satellite and from Geosat, a U.S. Navy satellite, were collected during the 1980s. When the data was recently declassified, Walter Smith of the National Oceanic and Atmospheric Administration and David Sandwell of Scripps Institution of Oceanography began producing sea floor maps based on gravity. What is unique about these researchers' maps is that they allow a view of Earth similar to being able to drain the oceans and view the ocean floor clearly. Their latest map of ocean surface gravity (Figure 3D) uses depth soundings to calibrate the gravity measurements. Although gravity is not exactly bathymetry, this new map of the ocean floor clearly delineates many ocean floor features, such as the mid-ocean

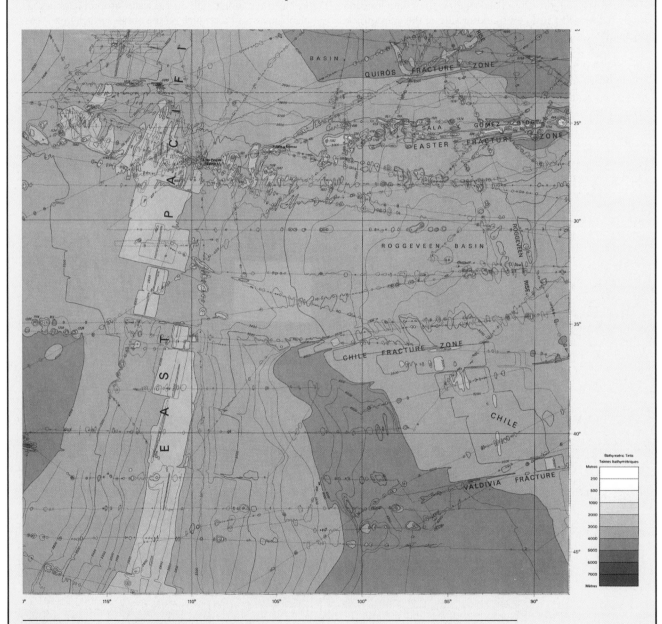

Figure 3C Maps of the sea floor.
Both maps show the same portion of the east Pacific Ocean floor. **A.** A conventional map made using an echosounder.

ocean, and contain various features such as coastal islands, reefs, and banks (shallowly submerged areas). The width of continental shelves varies from a few tens of meters to a maximum of 1500 kilometers (930 miles). The broadest shelves occur off the northern coasts of Siberia and North America in the Arctic Ocean and in the North and western Pacific Ocean from Alaska to Australia. Note that the east coast of North America has a broader shelf than the west coast, and South America has a similar relationship between east and west coasts. This

(continued)

ridge, trenches, seamounts, and nemataths (island chains). Maps like this have many uses: for navigators, to ensure safe travel for vessels across the oceans; for commercial fishers, to assist in the hunt for new fishing grounds; for geophysi- cists, to aid in refining their understanding of sea-floor spread- ing processes; and for ocean miners, to help define explo- ration targets. These new ocean floor maps may even help countries assess the importance of nearshore features with- in their jurisdiction.

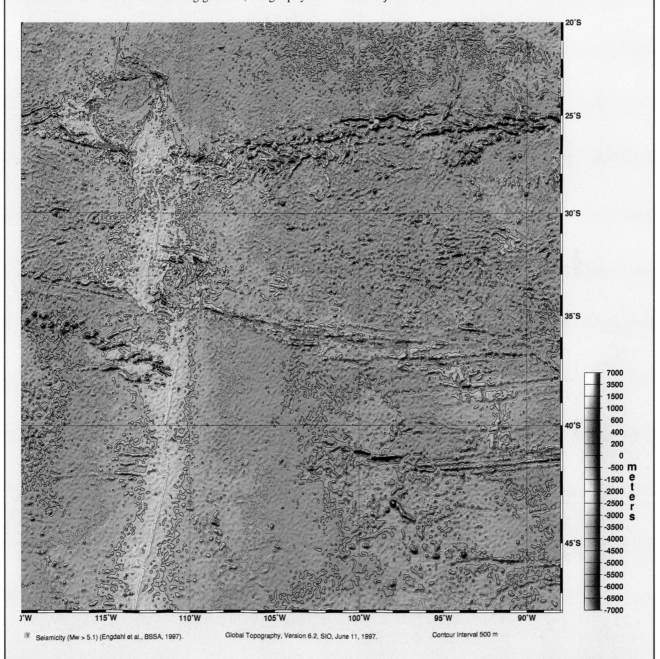

Seismicity (Mw > 5.1) (Engdahl et al., BSSA, 1997).   Global Topography, Version 6.2, SIO, June 11, 1997.   Contour Interval 500 m

Figure 3C *Continued*
B. A map from Geosat satellite data made using measurements of the ocean surface. Red dots are earthquake locations.

(continued)

is because the eastern coasts of these continents are pas- sive margins, which usually tend to have wider shelves.

The average width of the continental shelf is about 70 kilometers (43 miles), and the average depth at which the shelf break occurs is about 135 meters (443 feet). Around the continent of Antarctica, however, the shelf break occurs much deeper, at 350 meters (2200 feet). The average slope of the continental shelf is quite gentle, only about a tenth of a degree, which is about the same slope that a large parking lot has for drainage purposes.

(continued)

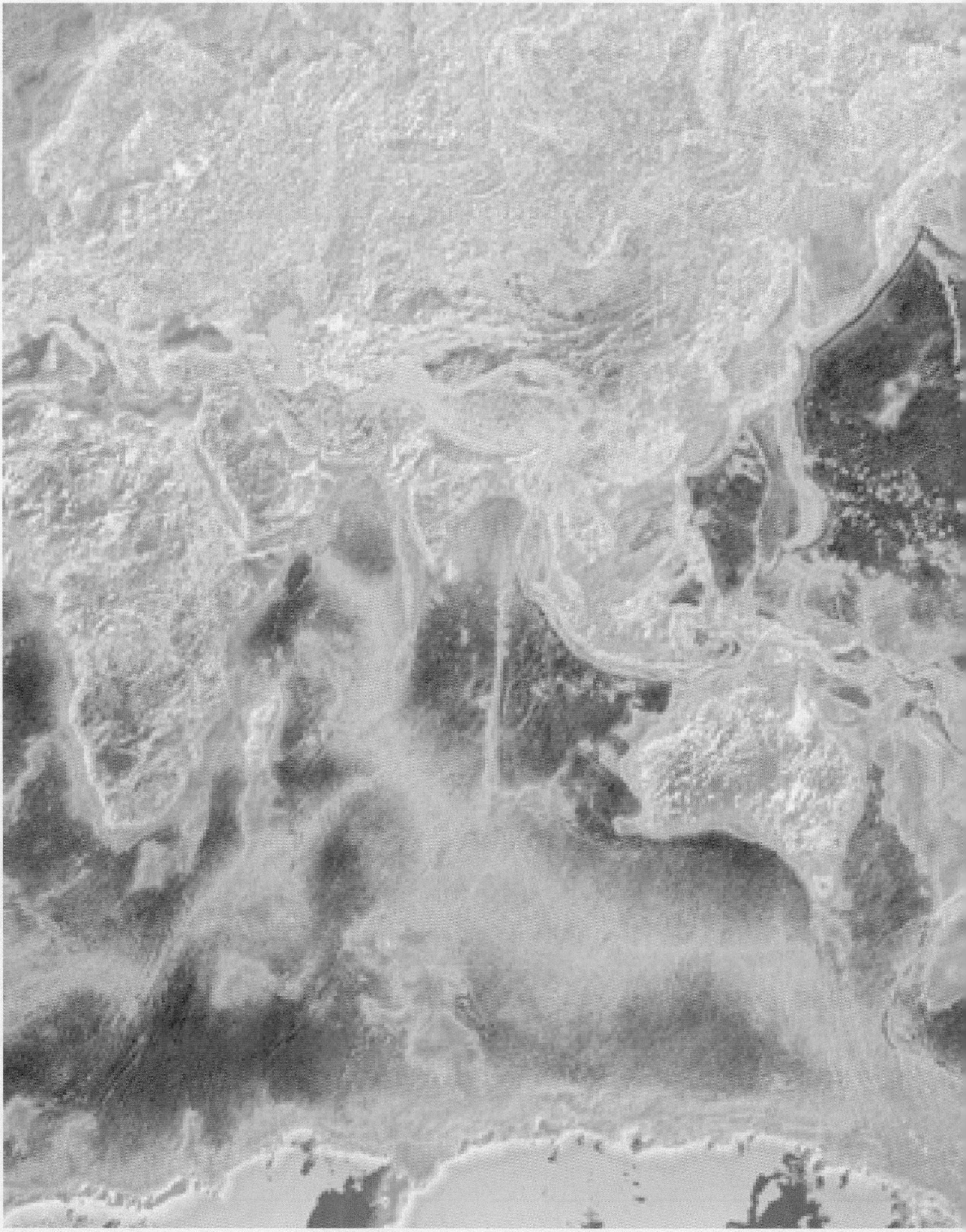

Figure 3D Global sea surface elevation map.

Map showing the global gravity field, which, when adjusted using measured depths, corresponds closely to ocean depth. Light green color indicates shallowest water; purple indicates deep water.

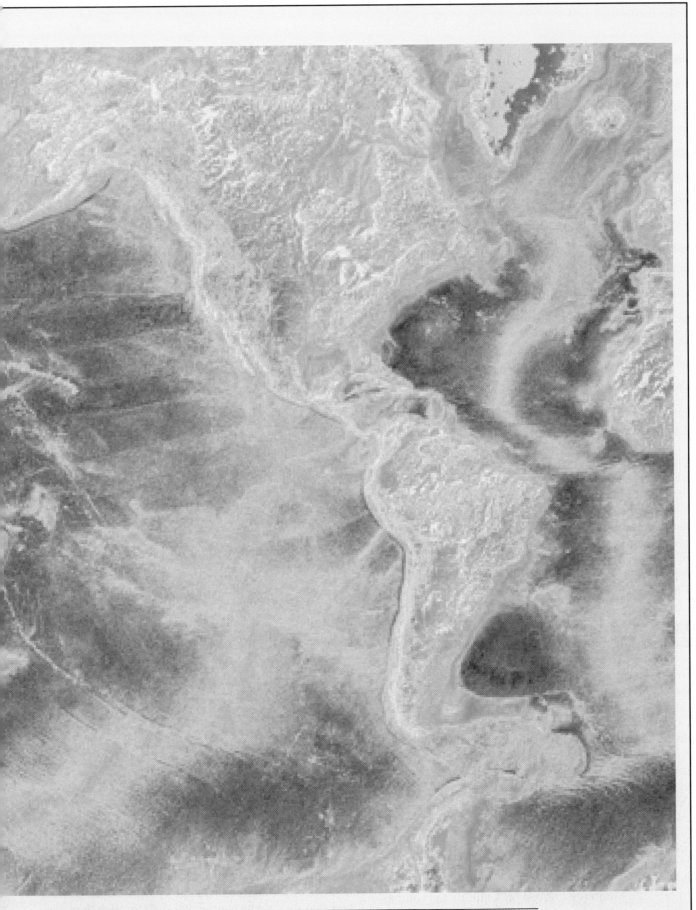

Figure 3D *Continued*
Map also shows land surface elevations, with yellow color indicating low areas and light pink color indicating high areas.

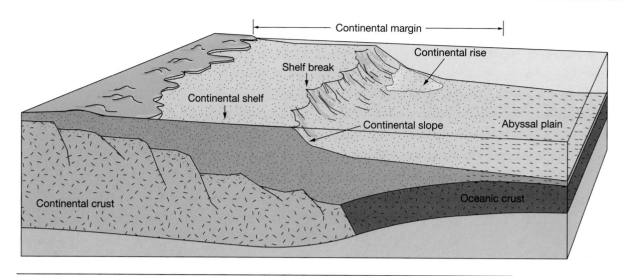

Figure 3–7 Continental margin features.
Schematic view shows the main features of a passive continental margin.

Sediment is transported across the continental shelf by current flow and underwater landslides, which are often generated by earthquakes.

Continental Slope The continental slope, which lies beyond the shelf break, is where the deep-ocean basins begin. Total relief in this region is similar to that found in mountain ranges on the continents. The break at the top of the slope may be from 1 to 5 kilometers (0.6 to 3 miles) above the deep-ocean basin at its base. Along active margins where the slope descends into submarine trenches, even greater vertical relief is measured. Off the west coast of South America, for instance, the total relief from the top of the Andes Mountains to the bottom of the Peru-Chile Trench is about 15 kilometers (9.3 miles).

Worldwide, continental slopes vary in steepness from 1 to 25 degrees[5] and average about 4 degrees, which is the same slope as a rather steep portion of a multi-lane interstate highway. In the United States, the average of five different continental slopes is just over 2 degrees. Around the margin of the Pacific Ocean, the continental slopes are steeper than in the Atlantic and Indian Oceans, because these are active margins. Here, the slope drops directly into a deep offshore trench. Consequently, continental slopes in the Pacific Ocean average more than 5 degrees. The Atlantic and Indian Oceans contain many passive margins, which lack plate boundaries. Thus, the amount of relief is lower and slopes in these oceans average about 3 degrees.

Submarine Canyons and Turbidity Currents The continental slope is incised with large **submarine canyons** that resemble the largest of canyons cut on land by rivers. In fact, the Monterey Canyon off California is compa-

rable in depth, steepness, and length to the Grand Canyon of Arizona. To a lesser extent, the continental shelf exhibits some of these canyons as well (Figure 3–8). Similar to canyons cut by rivers on the continents, submarine canyons have branches or tributaries with steep to overhanging walls. Exposed in the walls of the canyons are rocks of widely varied age and type.

How are submarine canyons formed? Initially, it was thought that their formation was related to *land* river systems, because some canyons are directly offshore from where rivers enter the sea. Presumably, sediment derived from these rivers provided the erosive power to carve submarine canyons as it moved through the canyon when sea level was lower. The majority of submarine canyons, however, do not correspond so closely with land drainage systems; many are confined exclusively to the continental slope. The main objection to explaining submarine canyons as ancient river valleys (created by the erosive work of rivers on land when sea level was lower) is that they continue to the base of the continental slope. Typically, the base of the continental slope averages some 3500 meters (11,500 feet) below sea level, and there is no other evidence that sea level was ever that much lower. Because rivers lose their ability to erode shortly after reaching the ocean, how could submarine canyons have been carved?

Side-scan sonar surveys along the Atlantic coast indicate that the continental slope is dominated by submarine canyons from Hudson Canyon near New York City to Baltimore Canyon in Maryland. Canyons confined to the continental slope are straighter and have steeper canyon floor gradients than those that extend into the continental shelf. These characteristics suggest that the canyons are initiated on the continental slope by some marine process and enlarge into the continental shelf through time.

The most widely supported hypothesis to explain the origin of submarine canyons involves the erosive power of **turbidity currents**. These currents are avalanches of

---

[5] For comparison, the windshield of an aerodynamically designed car has a slope of about 25 degrees.

Figure 3–8 Submarine canyons.

Turbidity currents move downslope, eroding the continental margin to enlarge submarine canyons. Deep-sea fans are created by turbidite deposits, which consist of sequences of graded bedding.

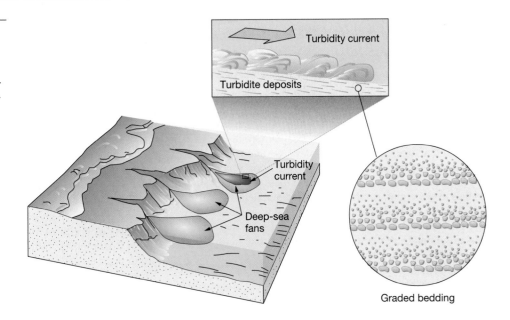

muddy water mixed with rock material and other debris. Turbidity currents in many canyons apparently are initiated after sediment moves across the continental shelf into the head of the canyon and accumulates there. Initiation of the turbidity current may result from a number of events, such as shaking by an earthquake, the oversteepening of sediment that accumulates on the shelf, hurricanes passing over the area, or the rapid input of sediment from flood waters. Once set into motion, the mass moves down the slope under the force of gravity. Turbidity currents occur at irregular intervals, depending on the supply of material from the continent. Every so often, an erosive turbidity current speeds down through a submarine canyon and carves the canyon as it moves downhill, analogous to a flash-flood event on land.

Continental Rise

The **continental rise** is a huge submerged pile of debris located at the base of the continental slope, a transition zone between the continental margin and the deep-ocean floor. Where did all this debris come from, and how did it get there?

Turbidity currents are thought to be responsible for the creation of continental rises. Once the turbidity current moves through and erodes the submarine canyon, it exits through the mouth of the canyon. Here, the slope angle decreases, and the turbidity current slows. Material that was suspended in the turbidity current settles out in a distinct pattern: The larger pieces often settle first, then progressively smaller pieces settle, and eventually even very fine pieces settle out, which may take weeks or months. This results in a characteristic type of layering called **graded bedding**. Graded bedding is named for the fact that the bedding is *graded* in size, meaning that it gets progressively finer upwards within a sequence (Figure 3–8). An individual turbidity current deposits one graded bedding sequence. The next turbidity current may partially erode the previous deposit, and will deposit another graded bedding sequence on top of the previous one. After some time, a thick deposit of superimposed graded bedding sequences will develop, one on top of another, creating a thickness of thousands of meters of sediment. Since it is so indicative of turbidity current deposition, stacks of graded bedding are called **turbidite deposits**. The continental rise is composed of turbidite deposits.

As viewed from above, the deposits at the mouths of submarine canyons are fan-, lobate-, or apron-shaped. Consequently, these deposits are called **deep-sea fans** or **submarine fans** (Figure 3–8). These deep-sea fans merge together along the base of the continental slope, creating the continental rise.

Features on Active Margins

Features on active margins are similar to those features on passive margins described above. However, the plate boundary associated with active margins modifies the shape and relief of the margin as compared to passive margins.

For transform active margins, the continental shelf can be marked with considerable relief due to faults that are oriented parallel to the main transform plate boundary. Thus, many more submarine canyons may exist on the shelf. The shelf break can be found at a great distance from shore, but may be difficult to identify because the shelf is not generally flat. At the shelf break, the continental slope leads down to the continental rise and the deep-ocean basin.

For convergent active margins, the continental shelf is usually quite narrow and the shelf break is close to shore. From there, the slope leads directly into a deep-ocean trench, where sediment from the continent is deposited. Thus, on convergent active margins, there is no continental rise.

Box 3–2
A Grand "Break": Evidence for Turbidity Currents

Information used to help resolve the mystery of how turbidity currents move across the ocean floor and carve submarine canyons comes from instances such as a well-documented earthquake in the North Atlantic Ocean at Grand Banks in 1929. At that time, closely monitored trans-Atlantic telephone and telegraph cables lay across the sea floor. During the Grand Banks earthquake, a number of these communication cables in the region south of Newfoundland near the earthquake were severed (Figure 3E). At first, it was assumed that sea floor movement caused all these breaks. Further analysis of the data revealed that these cables broke in an interesting pattern. The cables closest to the earthquake broke simultaneously with the occurrence of the earthquake. Cables that crossed the slope and deeper ocean floor at greater distances from the earthquake were broken progressively later in time. It seemed unusual that certain cables were affected by the failure of the slope due to ground shaking, but others were broken several minutes later. At the time, it was unclear how the pattern of cable breaks could have been produced. Reanalysis of the pattern several years later suggested that a turbidity current moving down the slope could account for the pattern of cable breaks. Based on the sequence of breaks, the turbidity current must have reached speeds approaching 80 kilometers (50 miles) per hour on the steep portions of the continental slope, and about 24 kilometers (15 miles) per hour on the more gently sloping continental rise. The documentation of turbidity currents moving at these speeds can certainly help to explain how powerfully erosive turbidity currents must be as they move through submarine canyons.

Further evidence of the existence of turdidity currents comes from reports of similar damage done to objects on the ocean floor as well as direct measurement. In one study in Rupert Inlet in British Columbia, Canada, turbidity currents moving through a channel were documented with sonar devices. Studies such as these make it easier to understand how submarine canyons are carved by turbidity currents over long periods of time, just as canyons on land are carved by the work of running water.

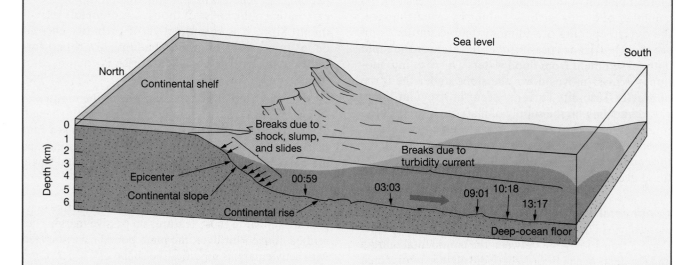

Figure 3E Grand Banks earthquake.
Diagrammatic view of the sea floor showing the sequence of events for the 1929 Grand Banks earthquake. The epicenter is the point on Earth's surface directly above the earthquake.

Features of the Deep-Ocean Basin

Traveling from shore into deeper water, one passes the continental margin province (the shelf, slope, and the rise) into the deep-ocean floor, where there are many interesting features.

Abyssal Plains   Extending from the base of the continental rise into the deep-ocean basins are flat depositional surfaces with slopes of less than a fraction of a degree that cover extensive portions of the deep-ocean basins. These **abyssal** (*a* = without, *byssus* = bottom) **plains** average between 4500 meters (15,000 feet) and 6000 meters (20,000 feet) deep. They are not literally bottomless, of course, but it seemed so to people in the past. However, they are some of the deepest and flattest regions on Earth.

Abyssal plains are formed by fine particles of sediment slowly drifting down onto the crust of the deep-ocean floor and, over millions of years, producing thick sedimentary deposits analogous to marine dust. This process is called **suspension settling** because the fine sediment is constantly settling out from being suspended in water. With enough time, most irregularities of the deep ocean are covered by these deposits (Figure 3–9). In addition, sediment traveling directly from the continent along the ocean floor adds to the sediment load.

Few abyssal plains are located in the Pacific Ocean; most abyssal plains occur in the Atlantic and Indian Oceans. Not surprisingly, the type of continental margin plays an essential role in the distribution of abyssal plains. On convergent active margins, such as the ones found along the margins of the Pacific Ocean, the deep-ocean trenches prohibit the movement of sediment past the continental slope. In essence, the trenches act like a gutter and serve as a sediment trap. On passive margins, sediment derived from land-based sources travels directly down the continental margin to be deposited on the abyssal plains. That helps explain why abyssal plains are so well developed in the Atlantic and Indian Oceans, while they are almost absent in the Pacific Ocean.

Volcanic Peaks of the Abyssal Plains Poking through the sediment cover of the abyssal plains are a variety of volcanic peaks, which extend to various elevations above the ocean floor. Those that extend above sea level are, of course, called islands. Those that are below sea level but rise more than 1 kilometer (0.6 mile) above the deep-ocean floor are called seamounts (Figure 3–10). If they have flattened tops, they are called tablemounts, or guyots. The origin of seamounts and tablemounts was discussed as a piece of supporting evidence for plate tectonics in Chapter 2.

Volcanic features on the ocean floor that are not as tall as seamounts are called **abyssal hills** or **seaknolls**.

Abyssal hills are small volcanic peaks, many of which are gently rounded in shape. They are one of the most abundant features on the planet, totaling several hundred thousand in number, and cover a large percentage of the entire ocean basin floor. They are all less than 1000 meters (0.6 mile) tall and have an average height of about 200 meters (650 feet). Many such features are found buried beneath the sediments of abyssal plains of the Atlantic and Indian Oceans. In the Pacific Ocean, the abundance of active margins has caused a lower rate of sediment deposition. Consequently, extensive regions dominated by abyssal hills have resulted, called **abyssal hill provinces**. The evidence of volcanic activity on the bottom of the Pacific Ocean is particularly widespread—more than 20,000 volcanic peaks exist there.

Ocean Trenches Along passive margins, the continental rise commonly occurs at the base of the continental slope and merges smoothly into the abyssal plain. In convergent active margins, however, the slope descends into a long, narrow, steep-sided **ocean trench**. Ocean trenches are deep linear scars in the ocean floor, caused by the collision of two plates along convergent plate margins (as discussed in the previous chapter). The landward side of the trench rises as a **volcanic arc** that may produce islands (such as the islands of Japan, an **island arc**) or a volcanic mountain range along the margin of a continent (such as the Andes Mountains, a **continental arc**).

The deepest portions of the world's oceans are found in these trenches. It should come as no surprise that the deepest point on Earth's surface is found in a trench: the Challenger Deep area of the Mariana Trench, with a depth of 11,022 meters (36,163 feet). Such features are characteristic of the margins of the Pacific Ocean along

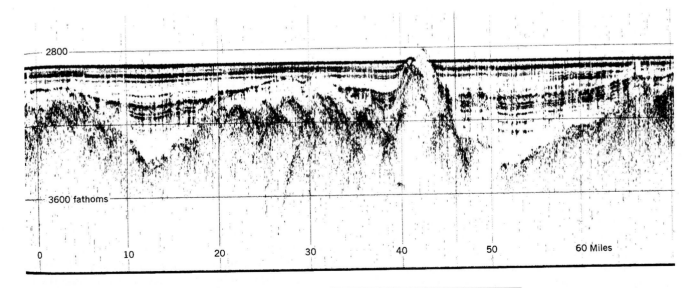

Figure 3–9 Abyssal plain.
Seismic cross section across part of the Madiera Abyssal Plain in the eastern Atlantic Ocean, showing irregular volcanic terrain buried by flat-lying sediments.

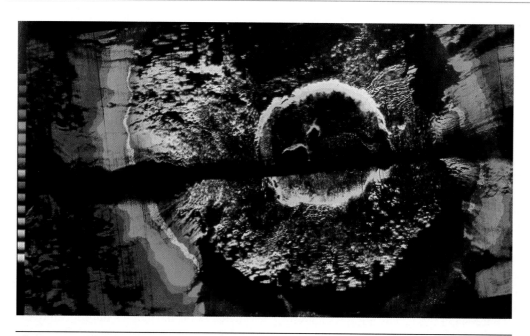

Figure 3–10 Seamount.

Side-scan sonar image of a seamount on the East Pacific Rise at about 10 degrees north latitude. The summit crater is about 2 kilometers (1.2 miles) in diameter.

the coast of South America, Central America, the Aleutian Islands, Japan, and southeast Asia (Figure 3–11). Table 3–1 presents the dimensions of selected trenches.

The tectonic activity associated with these convergent plate boundaries around the Pacific Ocean has lead to the Pacific Rim being called the **Pacific Ring of Fire**. Because the majority of Earth's active volcanoes and large earthquakes occur in this region, the area certainly deserves the name. A part of the Pacific Ring of Fire is South America's western coast, including the Andes Mountains and the associated Peru-Chile Trench. Figure 3–12 is a cross-sectional view across this area and shows the large amount of relief characteristic of these convergent boundaries.

Features of the Mid-Ocean Ridge

The global mid-ocean ridge is a continuous, fractured-looking mountain ridge that extends through all the ocean basins (Figure 3–13). As discussed in the previous chapter, the origin of the mid-ocean ridge is associated with sea floor spreading processes along divergent plate boundaries. The mid-ocean ridge forms Earth's longest mountain chain, extending across some 65,000 kilometers (40,400 miles) of the deep-ocean basin. The width of the mid-ocean ridge varies along its length, but averages about 1000 kilometers (620 miles). The mid-ocean ridge is a topographically high feature, extending an average of 2.5 kilometers (1.5 miles) above the abyssal plains or abyssal hill provinces. In some areas such as in Iceland, the mid-ocean ridge even extends above sea level. In terms of total area, the mid-ocean ridge covers a surprising 23 percent of Earth's surface.

The mid-ocean ridge is entirely volcanic and is composed of lavas with a basaltic composition characteristic of the oceanic crust. Along its crest is a central downdropped rift valley (Figure 3–14) created by seafloor spreading processes where two plates are diverging. Consequently, earthquake activity is common along the central rift valley.

Volcanic features include volcanoes (seamounts), and recent underwater lava flows. Hot lava flows spill onto the sea floor and come into contact with cold seawater, creating a distinct smooth and rounded margin where lava met the seawater and suddenly chilled. This makes the lava look like a stack of bed pillows, called **pillow lava**, which produces a distinctive rock called **pillow basalt**[6] (Figure 3–15). Volcanic activity along the mid-ocean ridge occurs regularly. Bathymetric studies along the Juan de Fuca Ridge off Washington and Oregon revealed that sometime between 1981 and 1987 there was a release of magma equaling 50 million cubic meters (1800 million cubic feet) of new lava. In 1993, a research vessel was actually on-site during the underwater eruption. Later, surveys of the area indicated many changes along the mid-ocean ridge, including new volcanic features, recent lava flows, and depth changes of up to 37 meters (121 feet).

Other features in the rift valley include hot springs called **hydrothermal** (*hydro* = water, *thermo* = heat) **vents**. Seawater seeps along fractures in the ocean crust and is heated when it comes in contact with underlying

---

[6] Since most pillow lavas are basaltic in composition, the rocks they produce are called pillow basalts.

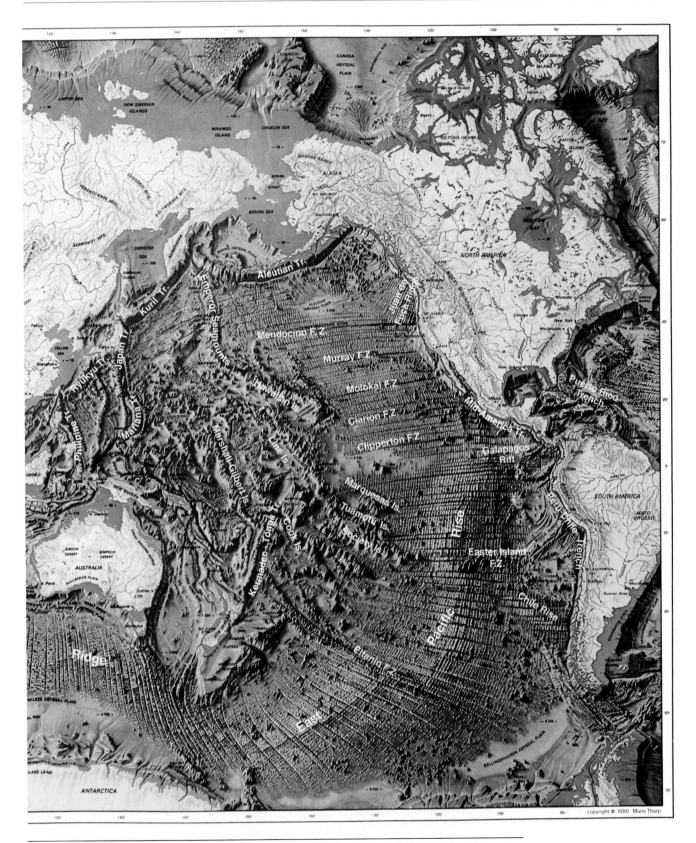

Figure 3–11 The majority of the world's ocean trenches occur along the margins of The Pacific Ocean.

Table 3-1 Dimensions of selected trenches.

Trench	Depth (meters)	Average width (kilometers)	Length (kilometers)
Peru-Chile	8,055	100	5,900
Aleutian	7,700	50	3,700
Middle America	6,700	40	2,800
Mariana	11,022	70	2,550
Kuril	10,500	120	2,200
Kermadec	10,000	40	1,500
Tonga	10,000	55	1,400
Philippine	10,500	60	1,400
Japan	8,400	100	800

Figure 3–12 Profile across the Peru-Chile Trench and the Andes Mountains.
Over a distance of 200 kilometers (125 miles), a change in elevation occurs of more than 15,000 meters (49,200 feet). Vertical scale is exaggerated 10 times.

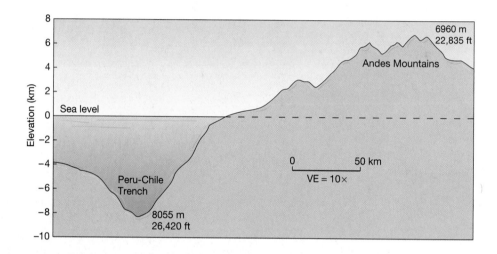

Figure 3–13 Floor of the North Atlantic Ocean.
The global mid-ocean ridge traverses the center of the Atlantic Ocean.

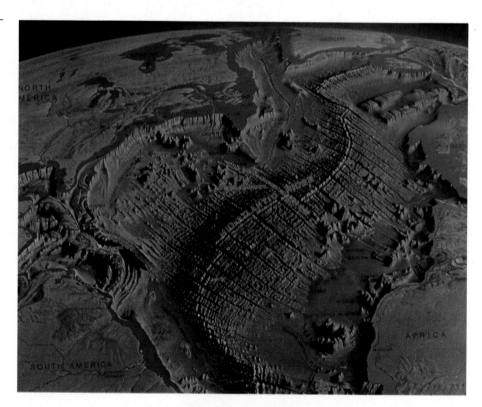

Figure 3–14 Rift valley fissures.
A. A fissure in the rift valley of the Mid-Atlantic Ridge photographed by the submersible *Alvin*.
B. A more readily observable fissure in the rift valley of Iceland.

magma (Figure 3–16). It then rises back toward the surface and exits through the sea floor. This rising water may rush into the ocean as **warm-water vents** at temperatures below 30 degrees centigrade (86 degrees Fahrenheit), as **white smokers**[7] at temperatures between 30 and 350 degrees centigrade (86 to 662 degrees Fahrenheit), or as **black smokers**[8] at temperatures above 350 degrees

centigrade (662 degrees Fahrenheit). Many of these black smokers spew out of chimney-like structures up to 20 meters (66 feet) high (Figure 3–16). The dissolved metal particles of hydrothermal vents often come out of solution, or **precipitate**,[9] when the hot water mixes with cold seawater. This causes nearby rocks to be coated with deposits of the previously dissolved materials. Chemical analyses of these deposits reveal that they are composed

[7] The white color of white smokers is caused by particles of light-colored barium sulfate.

[8] The black color of black smokers is caused by particles of dark-colored metal sulfides.

[9] A chemical precipitate is formed whenever dissolved materials change from existing in the dissolved state to existing in the solid state.

Figure 3–15 Pillow lava.
Pillow lava on the sea floor
near the East Pacific Rise.
Photo also shows ripple marks
in the sea floor sediment.

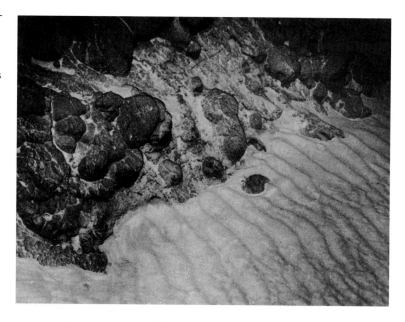

of **metal sulfides**, including iron, nickel, copper, zinc, and silver (Figure 3–16). In addition, the hydrothermal vents support unusual biological communities, including large clams, mussels, tubeworms, and many other animals— some of which are still unidentified. The reason these organisms are able to survive in the absence of sunlight is that the vents discharge hydrogen sulfide gas. Archaeons[10] oxidize the hydrogen sulfide gas and provide a food source for the community. The interesting associations of these organisms will be discussed in Chapter 15, "Animals of the Benthic Environment."

Segments of the mid-ocean ridge are called **oceanic ridges** where they are mountainous with steep and irregular slopes and **oceanic rises** where the slopes are more gentle. As explained in Chapter 2, the differences in overall shape between ridges and rises are due to differences in spreading rates: Oceanic ridges have slower spreading rates than oceanic rises. The best explored of these features are the Mid-Atlantic Ridge and the East Pacific Rise. The Mid-Atlantic Ridge was a focal point during the development of the theory of global plate tectonics because of its position, dividing the Atlantic Ocean into equal halves. Recently, the East Pacific Rise has been studied intensely because of the discovery of hydrothermal vents and associated biological communities along its central rift valley.

Fracture Zones and Transform Faults   As described in Chapter 2, the nearly continuous mid-ocean ridge is cut by a number of **transform faults**, which offset the spreading zones. Oriented perpendicular to the spreading zones, transform faults give the mid-ocean ridge a zigzag appearance (Figure 3–13). This is because the segments of the mid-ocean ridge spread apart at different

rates. Hence, transform faults occur to accommodate the different rate of plate motion between adjoining segments of the mid-ocean ridge.

In the Pacific Ocean, where the scars are less rapidly covered by sediment than in other ocean basins, transform faults appear to extend for thousands of kilometers away from the mid-ocean ridge and have widths of up to 200 kilometers (120 miles). However, these extensions beyond the mid-ocean ridge area are not transform faults. Continuing along in the same direction of a transform fault *away from the mid-ocean ridge,* these features are called **fracture zones**.

What is the difference between a transform fault and a fracture zone? Both are along the same long linear zone of weakness in Earth's crust. However, a transform fault is a seismically active area that offsets the axis of a mid-ocean ridge; a fracture zone is a seismically inactive area that shows evidence of past transform-fault activity. A helpful way to visualize the difference is that transform faults occur *between* the ends of offset segments of mid-ocean ridges, while fracture zones occur *beyond* the end of segments of mid-ocean ridges. Note that by following the same zone of weakness from one end to the other, it changes from a fracture zone to a transform fault and back again to a fracture zone (Figure 3–17).

The relative direction of plate motion across transform faults and fracture zones further differentiates these two features. Across a transform fault, the lithospheric plates are moving in *opposite* directions. Across a fracture zone, there is no relative motion because the lithospheric plates are moving in the *same* direction (Figure 3–17). It is interesting to note that transform faults are actual plate boundaries, whereas the fracture zones are not. Fracture zones are ancient fault scars embedded in a plate.

Earthquake activity is also influenced by the different type of relative plate motion in transform faults and fracture zones. Because of plate motion in opposite directions

---

[10] Archaeons are microscopic bacteria-like organisms—a newly discovered domain of life.

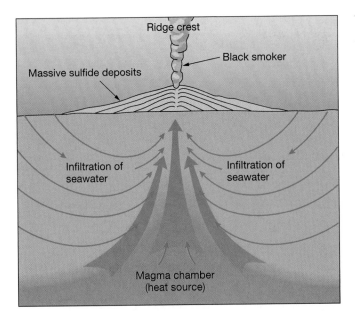

A.

Figure 3–16 **Hydrothermal vents.**
Diagram shows hydrothermal circulation along the mid-ocean ridge and the creation of black smokers. Photo shows a close-up view of a black smoker along the East Pacific Rise.

B.

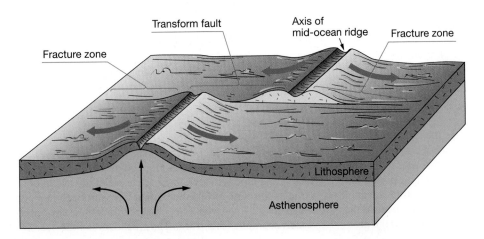

Figure 3–17 **Transform faults and fracture zones.**

Table 3–2   Characteristics of transform faults and fracture zones.

	Transform faults	Fracture zones
Plate boundary?	Yes—transform plate boundary	No—intraplate feature
Relative movement across feature	Movement in opposite directions	Movement in same direction
Earthquakes	Many	Few
Relationship to mid-ocean ridge	Occur *between* mid-ocean ridge segments	Occur *outside of* mid-ocean ridge segments
Examples	San Andreas Fault, Alpine Fault, Dead Sea Fault	Mendocino Fracture Zone, Molokai Fracture Zone

along transform faults, earthquakes shallower than 10 kilometers (6 miles) are common. Along fracture zones, where plate motion is in the same direction, seismic activity is almost completely absent. Table 3–2 lists some of the characteristics of transform faults and fracture zones that help differentiate the two.

Students Sometimes Ask ...

I didn't know that Jacques Cousteau was involved in developing underwater habitats. Whatever happened to him?

Ocean explorer Jacques Cousteau has done much to popularize oceanography. His widely acclaimed television show—*The Undersea World of Jacques Cousteau*—and documentaries on ocean life, many of them filmed while at sea aboard his ship *Calypso,* have inspired a generation of oceanographers and environmentalists. Always an advocate of human presence in the ocean, he experimented with many types of scuba equipment, underwater habitats, and submersibles (the most notable of which was his "diving saucer"). Cousteau died in 1997, at the age of 87.

Has anyone ever seen pillow lava forming?

Amazingly, yes! In the 1960s, an underwater film crew ventured to Hawaii during an eruption of the volcano Kilauea where lava spilled into the sea. They braved high water temperatures and risked being burned on the red-hot lava, but filmed some incredible footage. Underwater, the formation of pillow lava occurs where a tube emits molten lava directly into the ocean. Here, it comes into contact with seawater and forms the characteristic smooth and rounded margins of pillow basalt. The divers also experimented with a hammer on newly formed pillows and were able to initiate new lava outpourings. The creation of pillow lava is vaguely reminiscent of the display produced by a type of firework called "snakes."

What effect does all this volcanic activity along the mid-ocean ridge have at the ocean's surface?

That's an interesting question. Sometimes, the underwater volcanic eruption is large enough to create what is called a "megaplume" of warm, mineral-rich water that rises to the surface. A few lucky research vessels have actually reported experiencing a megaplume directly above an erupting volcano! Researchers on board describe bubbles of gas and steam at the surface, a marked increase in water temperature, and the presence of so much volcanic material that the water turns cloudy. In terms of warming the ocean, the heat released at mid-ocean ridges is not very significant, mostly because the ocean is good at absorbing and redistributing heat. However, there is some evidence that the heat being released at mid-ocean ridges may have significant effects on other surface phenomena. A recent study suggests that major oceanic and atmospheric disturbances like the unusually warm current called El Niño may be initiated by volcanic activity along the East Pacific Rise (the topic of El Niño is discussed in Chapter 7, "Ocean Circulation").

Do the metal sulfides associated with hydrothermal vents create any economically interesting deposits?

Yes, these metal sulfides are prime exploration targets for mining the ocean floor in the future. However, not all sea floor deposits always remain on the sea floor: The deposition of metallic sulfides around sea floor hydrothermal vents is also a major source of *continental* sulfide ore deposits. Initially formed at the mid-ocean ridges and transported by sea floor spreading, they may be *obducted* (uplifted) onto continents during subduction of an oceanic plate. In fact, the copper deposits in Cyprus that have been so influential in Mediterranean cultures throughout history were created by hydrothermal vents on the sea floor. In addition, rifting that begins on land but becomes inactive can also create metal sulfide deposits underneath the ground surface. This is the origin for the large metal sulfide deposits in the states of Missouri and New Jersey.

All this sea floor mapping is interesting, but haven't we found everything on the sea floor already?

Certainly not! Despite all the advances in technology that are used to map the ocean floor, it is surprising that only *5 percent* of the ocean floor has been mapped as precisely as the surface of the moon. In fact, there are areas of the ocean floor as large as the state of Oklahoma that have not been mapped in much detail! Even well-mapped ocean floor areas are based on widely separated ship tracks. The great depth of the oceans and the intervening seawater have certainly been a hindrance to accurately mapping sea floor features, which leaves room in the future for major discoveries (such as new sea floor features, shipwrecks, and mineral deposits).

If black smokers are so hot, why isn't there steam coming out of them instead of hot water?

Indeed, black smokers emit water that can be up to three-and-a-half times the boiling point of water at the surface. However, the water depth where black smokers are found results in much higher pressure than at the surface. At these higher pressures, water has a higher boiling point. Thus, water from hydrothermal vents remains in the liquid state instead of turning into water vapor (steam).

Summary

The varied bathymetry of the ocean floor was first determined using a sounding line to measure water depth. Later, the development of the echosounder gave ocean scientists a more detailed representation of the sea floor. Today, much of our knowledge of the ocean floor has been obtained using seismic reflection profiles (for examining subbottom ocean structure) and various multibeam echosounders or side-scan sonar instruments (used to make strip maps of ocean floor bathymetry).

The sea floor is divided into the continental margin and deep-ocean basin. Continental margins can be characterized as either active (associated with transform or convergent lithospheric plate boundaries) or passive (not associated with any plate boundaries). Extending from the shoreline of passive continental margins are gently sloping, low-relief continental shelves. Shelves typically slope gently toward the deep ocean, and contain various features such as coastal islands, reefs, and banks. Sediment is transported across the continental shelves by currents and underwater landslides. The boundary between the continental slope and the continental shelf is marked by an increase in slope that occurs at the shelf break, which averages 135 meters in depth. Cutting deep into the slopes are submarine canyons, which resemble canyons on land but are created by erosive turbidity currents. Turbidity currents deposit their sediment load at the base of the continental slope, creating deep-sea fans that merge to produce a gently sloping continental rise. The deposits from turbidity currents have characteristic sequences of graded bedding and are called turbidite deposits. Active margins have similar features, although they are modified by their associated plate boundary.

The continental rises gradually become flat, extensive, deep-ocean abyssal plains, which form by suspension settling of fine sediment. Poking through the sediment cover of the abyssal plains are numerous volcanic peaks, including volcanic islands, seamounts, tablemounts, and abyssal hills. In the Pacific Ocean, where sedimentation rates are low, abyssal plains are not extensively developed, and abyssal hill provinces cover broad expanses of ocean floor. Along the margins of many continents are deep linear scars called ocean trenches.

The mid-ocean ridge is a continuous mountain range that winds through all ocean basins and is entirely volcanic in origin. Features that are characteristic of the mid-ocean ridge include seamounts, pillow basalts, hydrothermal vents, deposits of metal sulfides, and unusual life forms. Segments of the mid-ocean ridge can be divided into oceanic ridges and oceanic rises depending on the geometry of the segment, which is influenced by the rate of sea floor spreading.

Long linear zones of weakness—fracture zones and transform faults—cut across vast distances of ocean floor and offset the axes of the mid-ocean ridge. Fracture zones and transform faults are differentiated from one another based on the direction of movement across the feature. Fracture zones have movement in the same direction, while transform faults show movement in opposite directions.

Key Terms

Abyssal hill (p. 89)	Echosounder (p. 77)	Pacific Ring of Fire (p. 90)
Abyssal hill province (p. 89)	Fathom (p. 76)	Passive margin (p. 79)
Abyssal plain (p. 88)	Fracture zone (p. 94)	Precipitate (p. 98)
Active margin (p. 79)	GLORIA (p. 77)	Precision depth recorder (p. 77)
Bathymetry (p. 76)	Graded bedding (p. 87)	Saturation diving (p. 76)
Black smoker (p. 93)	Hydrothermal vent (p. 90)	Seabeam (p. 77)
Continental arc (p. 89)	Island arc (p. 89)	Sealab (p. 75)
Continental margin (p. 78)	Metal sulfides (p. 94)	Seaknoll (p. 89)
Continental rise (p. 87)	Mid-ocean ridge (p. 78)	Sea MARC (p. 77)
Continental shelf (p. 80)	Ocean trench (p. 89)	Seismic reflection profile (p. 77)
Continental slope (p. 80)	Oceanic ridge (p. 94)	Shelf break (p. 80)
Convergent active margin (p. 79)	Oceanic rise (p. 94)	Sonar (p. 77)
Cousteau, Jacques-Yves (p. 75)	Pillow basalt (p. 90)	Sounding (p. 76)
Deep-ocean basin (p. 78)	Pillow lava (p. 90)	Submarine canyon (p. 86)
Deep-sea fan (p. 87)	Ping (p. 77)	Submarine fan (p. 87)

Suspension settling (p. 89) Trench (p. 76) Volcanic arc (p. 89)

Transform active margin (p. 79) Turbidite deposit (p. 87) Warm-water vent (p. 93)

Transform fault (p. 94) Turbidity current (p. 86) White smoker (p. 93)

Questions And Exercises

1. What are some limitations in living in an underwater habitat?

2. How are bathymetry and topography different? How are they similar?

3. Discuss the development of bathymetric techniques, indicating significant advancements in technology.

4. Describe the differences between active and passive continental margins. Be sure to include how these features relate to plate tectonics, and include an example of each type of margin.

5. Describe the major features of a passive continental margin: continental shelf, continental slope, continental rise, submarine canyon, and deep-sea fans.

6. Explain how submarine canyons are thought to be created.

7. What are differences between a submarine canyon and an ocean trench?

8. Explain what graded bedding is and how it forms.

9. Describe the process by which abyssal plains are created.

10. Discuss the origin of the various volcanic peaks of the abyssal plains: seamounts, tablemounts, and abyssal hills.

11. In which ocean basin are most ocean trenches found? Explain why this occurs.

12. Describe characteristics and features of the mid-ocean ridge, including the difference between oceanic ridges and oceanic rises.

13. List and describe the different types of hydrothermal vents.

14. What kinds of unusual life can be found associated with hydrothermal vents? How do these organisms survive?

15. Use pictures and words to describe the differences between a fracture zone and a transform fault.

References

Anderson, R. N. 1986. *Marine-geology: A planet earth perspective.* New York: Wiley.

Carlowicz, M. 1996. New map of seafloor mirrors surface. *EOS, Transactions of the AGU* 76:44, 441–442.

Damuth, J. E., Kolla, V., Flood, R. D., Kowsmann, R. O., Monteiro, M. C., Gorini, M. A., Palma, J. J., and Belderson, R. H. 1983. Distributary channel meandering and bifurcation patterns on the Amazon deep-sea fan as revealed by long-range side-scan sonar (GLORIA). *Geology* 11:2, 94–98.

Field, M. E., Gardner, J. V., Jennings, A. E., and Edwards, D. E. 1982. Earthquake-induced sediment failures on a 0.25° slope. Klamath River delta, California. *Geology* 10:10, 542–545.

Hay, A. E., Burling, R. W., and Murray, J. W. 1982. Remote acoustic detection of a turbidity current surge. *Science* 217:4562, 833–835.

Heezen, B. C., and Ewing, M. 1952. Turbidity currents and submarine slumps, and the 1929 Grand Banks earthquake. *American Journal of Science* 250, 849–873.

Heezen, B. C., and Hollister, C. D. 1971. *The face of the deep.* New York: Oxford University Press.

Pratson, L. F., and Haxby, W. F. 1996. What is the slope of the U.S. continental slope? *Geology* 24:1, 3–6.

Small, C., and Sandwell, D. T. 1996. Sights unseen. *Natural History* 105:3, 28–33.

Smith, W. H. F., and Sandwell, D. T. 1997. Global sea floor topography from satellite altimety and ship depth soundings. *Science* 277:5334, 1956–1962.

Tarbuck, E. J., and Lutgens, F. K. 1996. *The earth: An introduction to physical geology,* 5th ed. Upper Saddle River, NJ: Prentice-Hall.

———. 1997. *Earth science,* 8th ed. Upper Saddle River, NJ: Prentice-Hall.

The Open University Course Team. 1989. *The ocean basins: Their structure and evolution.* Oxford: Pergamon Press.

Twichell, D. C., and Roberts, D. G. 1982. Morphology, distribution and development of submarine canyons on the United States Atlantic continental slope between Hudson and Baltimore canyons. *Geology* 10:8, 408–412.

Various authors. 1986. Special issue on mapping the sea floor (13 articles). *Journal of Geophysical Research,* 91:B3.

Various authors. 1995. Special section: The 1993 volcanic eruption on the CoAxial segment, Juan de Fuca Ridge (14 articles). *Geophysical Research Letters,* 22:2.

Weirich, F. H. 1984. Turbidity currents: Monitoring their occurrence and movement with a three-dimensional sensor network. *Science* 224:4647, 384–387.

Suggested Reading

Earth

Dombrowski, P. 1992. Cruise an ancient rift valley. 1:4, 54–61. A geologic road guide to the Connecticut turnpike reveals features of an ancient rift valley about 500 million years old.

McCommons, J. H. 1993. America's North Coast. 2:2, 56–63. Evidence of an ancient mid-continent rift includes ancient lava flows and copper deposits along the shores of Lake Superior.

Waters, T. 1995. The other Grand Canyon. 4:6, 44–51. With the assistance of a tethered robot submersible, scientists are studying Monterey Canyon, a submarine canyon comparable in size to Arizona's Grand Canyon.

Yulsman, T. 1993. Charting Earth's final frontier. 2:4, 36–41. An examination of how sonar techniques are being used to map the ocean floor.

———. 1996. The seafloor laid bare. 5:3, 42–51. Data recently declassified by the U.S. Navy has enabled scientists to view the seafloor almost as if the oceans had been completely drained away, revealing many interesting features.

Sea Frontiers

Mark, K. 1976. Coral reefs, seamounts, and guyots. 22:3, 143–149. A discussion of the role of global plate tectonics in explaining the distribution of seamounts, guyots, and the evolution of coral reefs.

Rice, A. L. 1991. Finding bottom. 37:2, 28–33. Depth-sounding devices developed by ingenious early navigators are described.

Rona, P. 1984. Perpetual sea floor metal factory. 30:3, 132–141. A discussion of how metallic mineral deposits form in association with hydrothermal vents located on oceanic ridges and rises.

Schafer, C., and Carter, L. 1986. Ocean-bottom mapping in the 1980s. 32:2, 122–130. Sea MARC, a side-scan sonar device, is used in mapping the continental margin off the coast of Labrador.

Scientific American

Brimhall, G. 1991. The genesis of ores. 264:5, 84–91. After being emplaced into Earth's crust at oceanic ridges, metal deposits undergo a number of processes before they become mineable ores embedded in continental rocks.

Edmonds, P. J., and Burton, D. 1996. Ten days under the sea. 275:4, 88–95. Living underwater in the world's only habitat devoted to science, six aquanauts studied juvenile corals and fought off "the funk."

Emery, K. O. 1969. The continental shelves. 221:3, 106–125. The nature of the continental shelves and the effect of the advance and retreat of the shoreline across them as a result of glaciation are discussed.

Heezen, B. C. 1956. The origin of submarine canyons. 195:2, 36–41. Theories explaining the origin of submarine canyons are presented along with data on the location and nature of such features.

Menard, H. W. 1969. The deep-ocean floor. 221:3, 126–145. A summary of the dynamic effects of sea floor spreading is presented with a description of related sea floor features.

Pratson, L. F., and Haxby, W. F. 1997. Panoramas of the seafloor. 276:6, 82–87. A description of how modern sea floor images are produced and used to map features of continental margins.

Wilson, A. C. 1985. The molecular basis of evolution. 253:4, 164–175. Mutations within the genes of organisms play an important role in evolution at the organismal level.

Oceanography on the Web

Visit the *Essentials of Oceanography* home page for on-line resources for this chapter. There you will find an on-line study guide with review exercises, and links to oceanography sites to further your exploration of the topics in this chapter. *Essentials of Oceanography* is at: **http://www.prenhall.com/thurman** (click on the Table of Contents menu and select this chapter).

CHAPTER 4
MARINE SEDIMENTS

Collecting the Historical Record of the Deep-Ocean Floor

During the early exploration of the oceans, the technique involved in sampling the material of the deep-ocean floor was to use a bucket-like device called a **dredge** to scoop up sediment for analysis. This technique was limited in that it could only gather samples from the *surface* of the ocean floor. Later, to enable scientists to penetrate the ocean floor, a device called a **gravity corer** was invented. This device was a hollow steel tube with a heavy weight attached that was dropped onto the ocean floor and retrieved. Similar to a straw thrust into a thick milkshake, the gravity corer was able to collect the first **cores** (cylinders of sediment and rock) from the ocean floor. However, it too was limited to collecting just the uppermost portion of what lay on the ocean floor.

The need for obtaining deep-ocean cores to help understand the history of the deep-ocean floor was recognized in 1963 when the National Science Foundation funded a program for drilling cores from a ship. This type of drilling, called **rotary drilling**, had the advantage that long sections of core could be recovered from deep below the ocean floor's surface. Lead by the Scripps Institution of Oceanography, the program united three other leading oceanographic institutions—the Rosenstiel School of Atmospheric and Oceanic Studies at the University of Miami, Florida; the Lamont-Doherty Earth Observatory of Columbia University; and the Woods Hole Oceanographic Institution in Massachusetts—to form the **Joint Oceanographic Institutions for Deep Earth Sampling (JOIDES)**. They were later joined by oceanography departments of the University of Washington, Texas A&M University, the University of Hawaii, Oregon State University, and the University of Rhode Island.

To implement this program, a ship had to be designed that was capable of drilling into the ocean bottom while floating as much as 6000 meters (3.7 miles) above at the surface. The *Glomar Challenger* was constructed and launched in 1968, commencing the first leg of the **Deep Sea Drilling Project (DSDP)**. In essence, the odd-looking ship was built around a tall drilling rig resembling a steel tower. To facilitate drilling, the crew used technology borrowed from the offshore oil industry to obtain cores of ocean-floor sediment and rock. From these cores, scientists were able to document that: (1) the age of the ocean floor increased progressively with distance from the mid-ocean ridge; (2) sediment thickness increased progressively with distance from the mid-ocean ridge; and (3) Earth's magnetic field polarity reversals were indeed recorded in ocean floor rocks. In essence, the ship made possible a major advancement in the field of Earth science: confirmation of sea floor spreading.

Initially, the oceanographic research program was financed by the U.S. government, but in 1975, it became international. Financial and scientific support was received from West Germany, France, Japan, the United Kingdom, and the Soviet Union. In 1983 the Deep Sea Drilling Project became the **Ocean Drilling Program (ODP)**, with the broader objective of drilling the thick sediment layers near the continental margins, supervised by Texas A&M University. Accompanying this change was the decommissioning (retirement) of *Glomar Challenger* after 15 highly successful years.

Continuing the important work of sampling the world's ocean floor, the *JOIDES Resolution* (Figure 4A) conducted its first scientific cruise in 1985. Like the *Challenger* before it, the *Resolution* contains a tall metal drilling rig that is used to lower sections of drill pipe through the middle of the ship. The pipe is in individual sections of 9.5 meters (31 feet) but can be screwed together to make a single string of pipe up to a length of 8200 meters (27,000 feet) (Figure 4B). The pipe string has on its end a drill bit that rotates as it is pressed against rock at the bottom, allowing the ship to drill deeper and deeper into the ocean floor. Like a twirling soda straw drilling into a layer cake, the drilling operation crushes the rock around the outside and retains a cylinder of rock (a core sample) on the inside of the pipe. The drill crew can then lower core-recovering equipment through the string of drill pipe. Once cores are retrieved from inside the pipe, they are collected and processed by scientists. Every so often, the laborious task of changing the drill

Figure 4A The *JOIDES Resolution*.

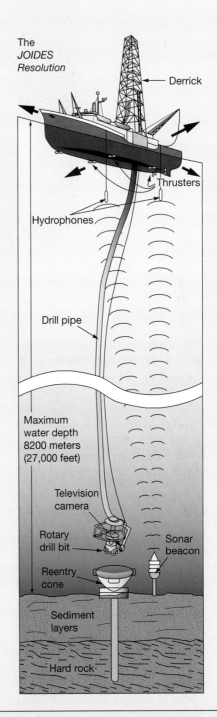

Figure 4B Dynamic positioning of the *JOIDES Resolution*.

bit must be accomplished by retrieving and unscrewing the pipe string section by section. A reentry cone is placed on the ocean floor to aid in reentry of the hole so that drilling can continue (Figure 4B). In addition, sonar beacons and a television camera near the drill bit help drillers reenter the hole. During drilling, the ship must remain in the same place at the surface. In deep water, anchors that could stabilize the ship are impracticable. Instead, the ship uses a series of 12 powerful thrusters—small propellers positioned around the ship's hull—to dynamically position itself and thus hovers above the drill site.

Capable of remaining at sea for 60 days, the *Resolution* includes state-of-the-art scientific laboratory equipment to conduct on-site research. Cores from the deep-ocean floor have helped scientists worldwide interpret the history of Earth from sea floor sediments.

Why are **sediments** (*sedimentum* = settling[1]) interesting to oceanographers? After all, sediments are little more than eroded particles and fragments of dirt, dust, and other debris. On the ocean floor (Figure 4–1), sediment is constantly settling through the water onto the ocean floor and, thus, is often thought of as "marine snow." What is interesting to oceanographers about sediments is that they reveal much about Earth history. Acting like a library of Earth history, the deep-ocean floor is a repository of sediment that dates

Figure 4–1 Oceanic Sediment.

View of the deep-ocean floor from a submersible. Most of the deep-ocean floor is covered with particles of material that have settled out through the water.

back over 170 million years. Clues to past climates, movements of the ocean floor, ocean circulation patterns, and nutrient supplies for marine organisms are embedded in the sedimentary deposits throughout the ocean basins. By examining cores of sediment retrieved from ocean drilling (Figure 4–2) and interpreting them, oceanographers can ascertain the timing of major extinctions, global climate change, and the movement of plates. In fact, most of what is known of Earth's past geology, climate, and biology has been learned through study of these ancient marine sediments—they are the information highway into Earth's ancient past. Author and environmentalist Rachel Carson perhaps put it best:

> When I think of the floor of the deep sea, the single, overwhelming fact that possesses my imagination is the accumulation of sediments. I see always the steady, unremitting, downward drift of materials from above, flake upon flake, layer upon layer—a drift that has continued for hundreds of millions of years, that will go on as long as there are seas and continents. … For the sediments are the materials of the most stupendous snowfall the Earth has ever seen.
>
> —Rachael Carson, *The Sea Around Us* (1961)

Over time, sediments can become **lithified** (*lith* = rock)—turned to rock—and thus form **sedimentary rock**. More than half of the rocks exposed on the continents are sedimentary rocks that were deposited in ancient ocean environments and uplifted onto land by plate tectonic processes. Even in the tallest mountains on the continents, far from any ocean, telltale marine fossils indicate that these rocks originated on the ocean floor in the geologic past.

Figure 4–2 Examination of deep-ocean sediment cores.

Long cylinders of sediment and rock called cores are cut in half and examined, revealing interesting aspects of Earth history.

Classification of Sediment

Not only do particles of sediment come from worn pieces of other rocks, they also are derived from living organisms, from minerals dissolved in water, and even from

---

[1]The term *sedimentum* is highly descriptive, indicating that many of these deposits are formed by material that has settled after being transported.

outer space (Table 4–1). Clues to the origin of sediment are found in its mineral composition and its **texture** (the size and shape of its particles).

Lithogenous Sediment

Lithogenous (*lithos* = stone, *generare* = to produce) **sediment** is derived from preexisting rock material. Since most lithogenous sediment comes from the landmasses, it is also called **terrigenous** (*terri* = land, *generare* = to produce) **sediment**. However, volcanic islands in the open ocean are also important sources of lithogenous sediment. Lithogenous sediment in the ocean is ubiquitous: At least a small percentage of lithogenous sediment is found nearly everywhere on the ocean floor.

Origin Lithogenous sediment begins as rock material on continents or islands. Here, agents of **weathering** (the disintegration and decomposition of rock material) break rock material into smaller pieces (Figure 4–3). Once rocks are in smaller pieces, they can be more easily **eroded** (picked up) and transported because smaller pieces require less energy to move. This eroded material is the basic component from which all lithogenous sediment is composed.

Eroded material from the continents is carried to the oceans by a variety of **transporting media**. These transporting media include streams, wind, glaciers, and gravity (Figure 4–4). The sediment can be deposited in bays or lagoons near the ocean, it can be transported farther by waves along the beach to produce beach deposits

Table 4–1   Classification of marine sediments.

Type	Composition	Source	Main locations
Lithogenous — *Continental margin*	Rock fragments, Quartz sand	Rivers, Coastal erosion, Landslides	Continental shelf
	Quartz silt	Glaciers	Continental shelf in high latitudes
	Clay	Turbidity currents	Continental slope and rise, Ocean basin margins
Lithogenous — *Oceanic*	Clay, Quartz silt, Volcanic ash	Wind-blown dust, Volcanic eruptions, Rivers	Deep-ocean basins
Biogenous — *Calcium carbonate* ($CaCO_3$)	Calcareous ooze (microscopic)	*Warm surface water:* Coccolithophores (algae), Foraminifers (protozoans)	Low-latitude regions, Sea floor above CCD, Along mid-ocean ridges and the tops of volcanic peaks
	Shell/coral fragments (macroscopic)	Macroscopic shell-producing organisms	Continental shelf, Beaches
		Coral reefs	Shallow low-latitude regions
Biogenous — *Silica* ($SiO_2 \cdot nH_2O$)	Siliceous ooze	*Cold surface water:* Diatoms (algae), Radiolarians (protozoans)	High-latitude regions, Sea floor below CCD, Surface current divergence near the Equator
Hydrogenous	Manganese nodules (manganese, iron, copper, nickel, cobalt)		Abyssal plain
	Phosphorite (phosphorous)	Precipitation of dissolved materials directly from seawater due to chemical reactions	Continental shelf
	Oolites ($CaCO_3$)		Shallow shelf in low-latitude regions
	Metal sulfides (iron, nickel, copper, zinc, silver)		Hydrothermal vents at mid-ocean ridges
	Evaporites (gypsum, halite, other salts)		Shallow restricted basins where evaporation is high in low-latitude regions
Cosmogenous	Iron-nickel spherules, Tektites (silica glass)	Space dust	In very small proportions mixed with all types of sediment in all marine environments
	Iron-nickel meteorites, Silicate chondrites	Meteors	Localized near meteor impact structures

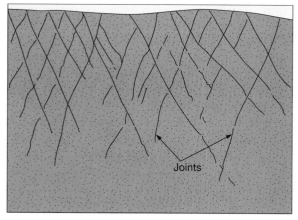

Joints

A.

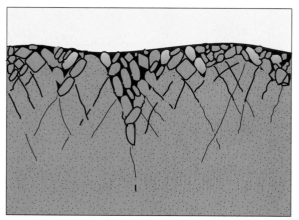

B.

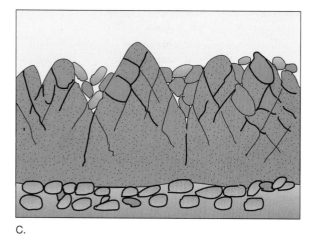

C.

Figure 4–3 Weathering.
Weathering often occurs along weaknesses in the rock (such as joints), breaking rock material into smaller fragments.

along the shoreline, or it may be spread across the continental margin by currents. It can also be carried beyond the continental margin to the deep-ocean basin by turbidity currents, as was discussed in the last chapter.

By far the greatest quantity of lithogenous material is found around the margins of the continents, where it is constantly on the move by high-energy currents along the shoreline and in deeper turbidity currents. However,

there are no parts of the ocean basins where traces of lithogenous sediment are completely absent. Much lower-energy currents distribute the finer components of the sediment to the deep-ocean basins. Microscopic particles can even be carried far out over the open ocean by prevailing winds. Such particles either settle out as the velocity of the wind decreases or they serve as nuclei around which raindrops and snowflakes form, and thus fall to Earth. Some particles that reach high altitudes can be carried very rapidly by the jet stream that exists in the mid-latitudes.

Composition   To a large degree, lithogenous sediment reflects the composition of the material from which it was derived. All rocks are composed of discrete crystals of naturally occurring compounds called **minerals**. One of the most abundant and chemically stable minerals on Earth's crust is **quartz**, composed of silicon and oxygen in the form of SiO_2—the same composition as ordinary glass. Quartz is the major component of nearly all rocks. Because quartz is resistant to abrasion, it can be transported large distances and be deposited far from its source area. Therefore, it is no surprise that the majority of lithogenous deposits are composed primarily of quartz. For example, most lithogenous beach sands throughout the world are composed of quartz (Figure 4–5).

A large percentage of the lithogenous particles that find their way into the deep-ocean sediments far from the continents are transported by prevailing winds that remove small particles from the subtropical desert regions of the continents. Figure 4–6 shows where microscopic fragments of quartz are present in the surface sediments of the ocean floor. It reveals a close relationship between the location of strong prevailing winds and high concentrations of microscopic quartz in ocean sediment. Evidently, desert areas of Africa, Asia, and Australia are important source areas for wind-blown lithogenous sediment.

Sediment Texture   One of the most important properties of sediment is its texture. An important aspect of lithogenous sediment texture is its **grain[2] size**. The Wentworth scale of grain size presented in Table 4–2 classifies particles in sizes ranging from boulders through cobbles, pebbles, granules, sand, silt, and clay. Sediment size indicates the energy condition under which a deposit is laid down. Deposits that are laid down in areas where wave action is strong (areas of high energy) may be composed primarily of larger particles—cobbles and boulders. In general, the deposition of fine-grained particles occurs in areas where the energy level is low and the current speed is minimal. However, once clay-sized particles—which are flat—are deposited, they tend to stick together by cohesive forces. Consequently, higher-energy condi-

[2]Sediment grains are also known as particles, fragments, or clasts.

A.

B.

Figure 4–4 Sediment-transporting media.

Transporting media include: **A.** Streams. **B.** Wind. **C.** Glaciers. **D.** Gravity, which creates landslides.

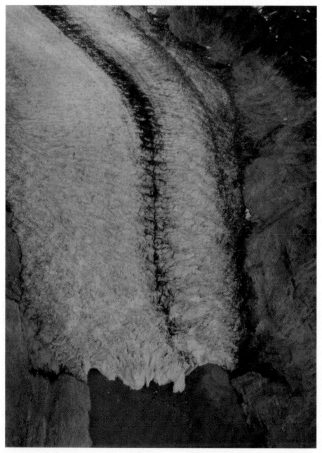

C.

D.

Figure 4–5 Lithogenous beach sand.
Lithogenous beach sand is composed mostly of particles of the mineral quartz (white), plus small amounts of other minerals. This sand is from North Beach, Hampton, New Hampshire and is magnified approximately 107 times.

tions than what would be expected based on grain size alone are required to cause clays to be eroded and initiate their transport.

Another important aspect of lithogenous sediment texture is its **sorting**. Sorting is a measure of how uniform the grain sizes are. For example, sediments composed of particles that are primarily the same size are considered well sorted. Beach sand is usually a well-sorted deposit. Poorly sorted deposits may contain particles ranging in size from clay- to boulder-sized particles. An example of poorly sorted sediment is sediment carried by glaciers that dropped out as the glacier melted.

Maturity is also an important aspect of lithogenous sediment texture. As particles are carried from the source to their point of deposition, they increase in maturity. This results from their association with moving water, which has the capacity to carry particles of a certain size away from a deposit and leave behind other, larger particles. As the time in the transporting medium increases, particles of sand size or larger become more rounded

through chemical weathering and abrasion. Increasing sediment maturity is indicated by (1) decreasing clay content, (2) increased sorting, (3) decrease of nonquartz minerals, and (4) increased **rounding** of the grains within the deposit.

Poorly sorted glacial deposits, which contain relatively large quantities of clay-sized and larger particles that have not been well rounded, are examples of an immature sedimentary deposit. A well-sorted beach sand contains very little clay-sized material and is usually composed of well-rounded particles that have undergone considerable transportation. Beach sand with these characteristics is an example of a mature sedimentary deposit. Figure 4–7 illustrates the nature of sediments of various degrees of maturity based on these criteria.

Biogenous Sediment

Biogenous (bio = life, *generare* = to produce) **sediment** is derived from the remains of hard parts of organisms. Under appropriate environmental conditions that allow marine organisms to live, biogenous material slowly settles onto the ocean floor.

Origin   Biogenous sediment begins as the hard remains of living organisms. In the ocean, numerous organisms—from minute algae and protozoans to large animals—produce hard parts—shells, bones, and teeth. These hard parts are constructed directly from the dissolved materials in seawater or from the material supplied by feeding. Once these organisms die, their hard parts settle on the ocean floor and biogenous sediment begins to accumulate.

Biogenous sediment can be divided into two primary types. **Macroscopic biogenous sediment** is large enough to be seen without the aid of a microscope and includes shells, bones, and teeth of large organisms. Except in certain tropical beach localities, this type of sediment is volumetrically rare in the marine environment, especially in the deep ocean where few large organisms live. Much more abundant is **microscopic biogenous sediment**, which contains particles so small that a microscope is needed to see these hard remains. Microscopic organisms produce tiny shells called **tests**, which begin to sink after the organisms die and rain down on the ocean floor by the uncountable billions each day. These microscopic tests can accumulate on the deep-ocean floor and form deposits called **ooze**. As its name implies, ooze resembles very fine-grained mushy material.[3] Technically, biogenous ooze must contain at least 30 percent

_____

[3] Ooze has the consistency of toothpaste mixed about half and half with water. To help you remember this term, imagine walking barefoot across the deep-ocean floor and how the sediment would *ooze* between your toes.

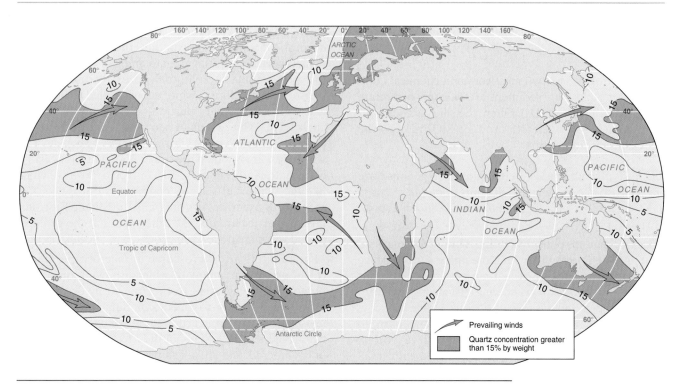

Figure 4–6 Lithogenous quartz in surface sediments of the world's oceans.

Most clay-sized quartz in deep-sea sediments is believed to have been transported by wind. Blue arrows indicate prevailing wind.

Table 4–2 Wentworth scale of grain size for sediments.

Size range (millimeters)	Particle name	Grain size	Example	Energy conditions
Above 256	Boulder	Coarse-grained	Coarse material found in stream beds near the source areas of rivers	High energy
64 to 256	Cobble			
4 to 64	Pebble	↕		
2 to 4	Granule			↕
$^1/_{16}$ to 2	Sand		Beach sand	
$^1/_{256}$ to $^1/_{16}$	Silt		Feels gritty in teeth	
$^1/_{4096}$ to $^1/_{256}$	Clay	Fine-grained	Microscopic; feels sticky	Low energy

(Gravel spans Boulder, Cobble, Pebble, Granule.)

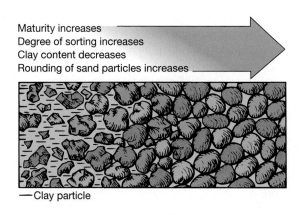

Maturity increases
Degree of sorting increases
Clay content decreases
Rounding of sand particles increases

— Clay particle

Figure 4–7 Sediment maturity.

As sediment maturity increases (*left to right*), the degree of sorting and rounding of particles increases, whereas clay content decreases.

biogenous test material by weight. This implies that biogenous ooze contains up to 70 percent nonbiogenous material. What comprises this other material? Commonly, it is fine-grained lithogenous clay that also settles onto the deep-ocean floor at the same time as microscopic organisms. Volumetrically, microscopic biogenous sediment (ooze) is much more significant on the ocean floor than the macroscopic type.

The organisms that contribute to biogenous sediment are chiefly **algae** and **protozoans** (*proto* = first, *zoa* = animal). Algae are primarily aquatic, eukaryotic,[4] photosynthetic organisms, ranging in size from single-celled forms to the giant kelp. A protozoan is any of a large group

---

[4] Eukaryotic (*eu* = good, *karyo* = the nucleus) cells contain a distinct membrane-bound nucleus.

of single-celled, eukaryotic, usually microscopic organisms that are generally not photosynthetic.

Composition   The two most common chemical compounds in biogenous sediment are **calcium carbonate** ($CaCO_3$) and **silica** (SiO_2). Often, the silica is chemically combined with water to produce $SiO_2 \cdot nH_2O$—the same composition as opal.

Silica   Contributing most of the silica in biogenous ooze are microscopic algae called **diatoms** (*diatoma* = cut in half) and protozoans called **radiolarians** (*radio* = a spoke or ray).

Because diatoms photosynthesize, they need strong sunlight and are found only within the upper sunlit surface waters of the ocean. Most diatoms are free-floating or **planktonic** (*planktos* = wandering). The living organism builds a glass greenhouse out of silica as a protective covering and lives inside. Most species have two parts to their test that fit together like a petri dish or pillbox (Figure 4–8A). The tiny tests are perforated with small holes in intricate patterns to allow for the passage of nutrients and waste products. Where diatoms are abundant at the ocean surface, thick deposits of diatom-rich ooze can accumulate below on the ocean floor. Once this ooze lithifies, it becomes **diatomaceous earth**,[5] a lightweight white rock composed of diatoms and clay.

Radiolarians are microscopic single-celled protozoans, most of which are also planktonic. True to their name, they often have long spikes or rays of silica protruding from their siliceous shell (Figure 4–8B). They do not photosynthesize but rely on external food sources such as bacteria and other plankton. Radiolarians typically display well-developed symmetry. Because of this, they have been described as the living snowflakes of the sea.

The accumulation of siliceous tests of diatoms, radiolarians, and other silica-secreting organisms produces **siliceous ooze** (Figure 4–8C).

Calcium Carbonate   The **foraminifers** (*foramen* = an opening), close relatives of radiolarians, are a source for calcium carbonate biogenous ooze. Another significant contributor are microscopic algae called **coccolithophores** (*coccus* = berry; *lithos* = stone; *phorid* = carrying).

Coccolithophores are single-celled supermicroscopic algae, most of which are planktonic. Coccolithophores produce thin plates or shields made of calcium carbonate, 20 or 30 of which are used in an overlapping fashion to produce a spherical test (Figure 4–9A). Like diatoms, coccolithophores photosynthesize in the presence of sunlight. However, their size is about 10 to 100 times smaller than most diatoms (Figure 4–9B). Because of their small size, coccolithophores are often

called **nanoplankton** (*nano* = dwarf, *planktos* = wandering). Once the organism dies, the individual plates (called **coccoliths**) disaggregate and can accumulate on the ocean floor as coccolith-rich ooze. This ooze can lithify over time and form a white deposit called **chalk**, which is used for a variety of purposes (including chalkboard chalk). The White Cliffs of southern England are composed of hardened coccolith-rich calcium carbonate ooze, which was deposited on the ocean floor and has been uplifted onto land (Figure 4–10). In fact, deposits of chalk like the White Cliffs are so common in Europe, as well as in North America, Australia, and the Middle East, during this particular geologic time span that the geologic period was named the Cretaceous (*creta* = chalk) Period.

Foraminifers are microscopic to macroscopic single-celled protozoans, many of which are planktonic. They do not photosynthesize, so they must rely on ingesting other organisms for food. Foraminifers produce a hard test made of calcium carbonate in which the organism lives (Figure 4–9C). Many of their tests look segmented or chambered, and all have a prominent opening in one end. Although very small in size, the tests of foraminifers resemble larger shells that one might find at a beach.

Deposits comprised primarily of tests of foraminifers, coccoliths, and other calcareous-secreting organisms are called **calcareous ooze** (Figure 4–9D).

Hydrogenous Sediment

Hydrogenous (*hydro* = water, *generare* = to produce) **sediment** is derived from the dissolved material in water.

Origin   Seawater contains many dissolved materials. Chemical reactions within seawater cause certain minerals to come out of solution or precipitate. This is usually caused by a change in conditions, such as a change in temperature or pressure or the addition of chemically active fluids. A good example of a precipitate is rock candy. To make rock candy, a pan of water is heated and sugar is added. Once the water is hot and the sugar is dissolved (stirring the water helps), the pan is removed from the stove to let the water cool. This *change in the conditions* (the drop in temperature) causes the sugar to become supersaturated and it will begin to precipitate. The sugar will precipitate on anything that is put in the pan, including kitchen utensils or pieces of string.

Composition   Although hydrogenous sediments are volumetrically not very significant in the ocean, they have many different compositions and environments of deposition.

Manganese Nodules   **Manganese nodules** are rounded, hard lumps of manganese, iron, and other metals up to 30 centimeters (12 inches) long. When they are cut in half, they often reveal that they have formed by

---

[5] Diatomaceous earth is also called diatomite, tripolite, or kieselguhr.

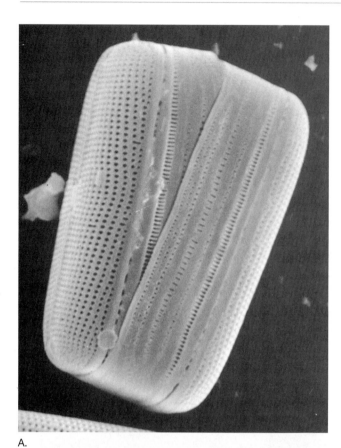

A.

Figure 4–8 Microscopic siliceous tests.

Scanning electron micrographs: **A.** Diatom (length = 30 micrometers, equal to 30 millionths of a meter), showing how the two parts of the diatom's test fit together.
B. Radiolarian (length = 100 micrometers).
C. Siliceous ooze, showing mostly fragments of diatom tests (magnified 250 times).

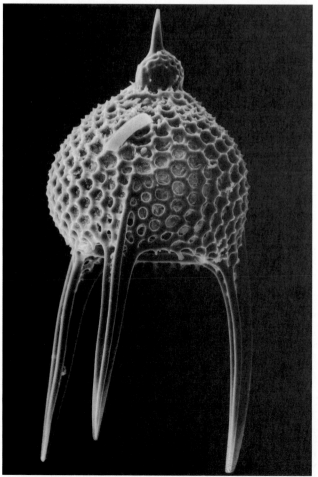

B.

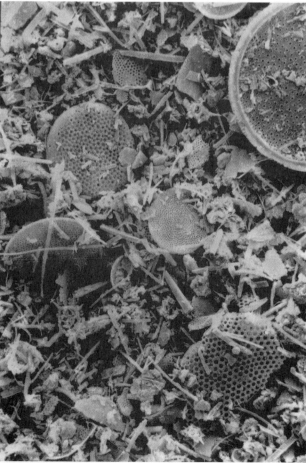

C.

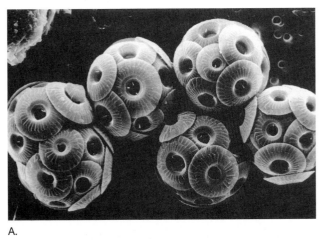

A.

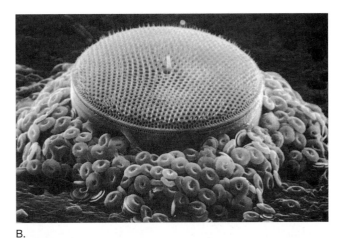

B.

C.

D.

Figure 4–9 Microscopic calcareous tests.

Scanning electron micrographs: **A.** Coccolithophores (diameter of individual coccolithophores = 20 micrometers, equal to 20 millionths of a meter). **B.** Diatom (siliceous) and coccoliths (diameter of diatom = 70 micrometers). **C.** Foraminifers (most species 400 micrometers in diameter). **D.** Calcareous ooze, which also includes some siliceous radiolarian tests (magnified 160 times).

Figure 4–10 The White Cliffs of southern England.

The White Cliffs of southern England are composed of hardened coccolith-rich calcareous ooze (chalk).

Box 4–1
Diatoms: The Most Important Things You Have (Probably) Never Heard Of

Diatoms are amazing creatures. Capable of photosynthesis, a single-celled organism lives inside its protective test that is constructed from dissolved silica in water. The test contains two halves that fit together like a shoebox with its lid. Fist described with the aid of a microscope in 1702, they are exquisitely ornamented with holes, ribs, and spines radiating in spectacular geometry across their minute silica test surface (Figure 4C). Differences in the patterns on their tests have led to the description of more than 10,000 different species. Diatoms can occur individually or linked together into long communities. They are found floating in the oceans, but also floating in lagoons, bays, and freshwater lakes. Diatoms live on the bottom of shallow estuaries (accounting for the brown film on tidal mudflats), on the undersides of polar ice, and even on the skins of whales. On land, they live under moist conditions within soil, attached to moss, in thermal springs, or on trees and brick walls. The fossil record indicates that diatoms have been on Earth for at least 150 million years. Marine diatoms are the most abundant life on Earth, exceeding in mass all other living organisms in the ocean and on land. In fact, they are one of the most important groups of aquatic organisms on which all other aquatic life to some degree depends.

In terms of their life history, diatoms live for a few days to as much as a week. Since they photosynthesize, marine species must remain in the upper, sunlit surface water. When conditions allow for the right mix of sunlight and nutrients, they can reproduce a few times a day. They can reproduce asexually by dividing into two cells, with each cell taking half

of its original test. When this happens, diatoms construct another half to complete their test, albeit a smaller half. After a few generations of reproduction in this fashion, diatoms attain such a small size that they must reproduce sexually by releasing reproductive gametes to return to their original size. Once a diatom dies, its test can accumulate on the sea floor or the bottom of a lake as siliceous ooze. Lithified deposits of siliceous ooze, called diatomaceous earth, can be as much as 900 meters (3000 feet) thick.

Because diatomaceous earth consists of billions of minute silica tests, the material has many unusual properties. It is lightweight, has an inert chemical composition, is resistant to high temperatures, and has excellent filtering properties. Thus, diatoms are used in a variety of common products (Figure 4D). The main uses of diatoms include filters (for refining sugar, straining yeast from beer, and filtering swimming pool water), mild abrasives (in toothpaste, facial scrubs, matches, and in household cleaning and polishing compounds), an absorbent (for chemical spills and as pest control), and as a chemical carrier (in paint and dynamite). Dynamite, which was invented by Alfred Nobel in 1867, uses diatoms in an interesting way. Nobel's brother was killed by an explosion of nitroglycerine, an oily liquid that can suddenly explode if subjected to even the slightest shock. Nobel invented a way to make a much safer explosive by saturating diatoms with nitroglycerine and thus effectively spreading the explosive throughout the deposit. Although intended of peaceful purposes such as construction, it inevitably was used in military ammunition. When Nobel died in 1896, his will stipulated that his estate was to be used for the permanent establishment of cash prizes to reward "those who, during the preceding year, shall have conferred the greatest benefit on mankind." In part, he established what became known as the Nobel prizes because his invention of dynamite was used for warfare instead of for the benefit of humankind.

Other products from diatoms include optical-quality glass (because of the pure silica content of diatoms), space shuttle tiles (because they are lightweight and provide good insulation), an additive in concrete, a filler in tires, an anti-caking agent, and even building stone for construction of houses. Because diatoms are algae that photosynthesize, they produce the vast majority of oxygen that we breathe. In addition, diatoms have created large offshore oil deposits. Each living diatom contains a tiny droplet of oil. Once diatoms die, their tests containing droplets of oil accumulate by the uncountable trillions on the sea floor. These deposits are the beginnings of offshore petroleum source areas, and much of the offshore oil reserves in California are thought to be the oily remains of ancient diatoms.

The life history of single-celled microscopic diatoms is truly amazing and the elegant geometry of the tests they build is an engineering accomplishment. In addition, they have many practical uses in a variety of everyday products. It is difficult to imagine how different our lives would be without diatoms!

Figure 4C Diatom *Thalassiosira eccentrica,* diameter = 40 micrometers.

(continued)

(continued)

Figure 4D Products containing diatoms.

precipitating in layers around a central nucleation object (Figure 4–11A). Sometimes, a piece of lithogenous sediment, coral, or volcanic rock is the central nucleation object; other times, a fish bone or a shark's tooth is at the center. In some areas, manganese nodules cover the deep-ocean floor in great abundance (Figure 4–11B), resembling a pumpkin patch of softball-sized pumpkins. Their formation requires extremely low rates of lithogenous or biogenous input so that the nodules are not buried by another sediment.

The major components of these nodules are manganese dioxide (around 30 percent by weight) and iron oxide (around 20 percent). Also occurring in the nodules are copper, nickel, and cobalt. The element manganese is important for making high-strength steel alloys. Copper is valuable in electrical wiring, pipe, and in making brass and bronze. Cobalt is alloyed with iron to make strong magnets and steel tools. Nickel is used in making stainless steel. Although their concentrations are usually less than 1 percent, they can exceed 2 percent by weight, and it may be economically feasible to mine them in the future. Technologically, mining the deep-ocean floor for

nodules such as these is possible. However, the political issue of determining international mining rights at great distances from land has hindered the mining of this resource. As well, environmental concerns about mining the deep-ocean floor have not been fully addressed. In addition, recent evidence suggests that it takes a very long time (millions to tens of millions of years) for these nodules to form, and that their formation depends on a particular set of physical and chemical conditions that probably do not last long at any location. Once they are mined, they will not be replaced for many lifetimes.

Phosphates Phosphorus-bearing compounds (**phosphates**) occur abundantly as coatings on rocks and as nodules on the continental shelf and on banks at depths shallower than 1000 meters (3300 feet). Concentrations of phosphates in such deposits commonly reach 30 percent by weight and indicate abundant biological activity in surface water above where they accumulate. Phosphates are valuable as fertilizers, and ancient marine deposits on land are extensively mined to supply agricultural needs.

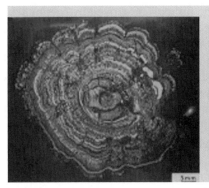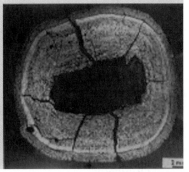

A.

Figure 4–11 Manganese nodules.

A. Manganese nodules cut in half, revealing their central nucleation object and layered internal structure. **B.** Manganese nodules on the floor of the South Pacific Ocean.

B.

Carbonates The two most important **carbonates** (a mineral containing CO_3 in its chemical formula) in marine sediment are the minerals **aragonite** and **calcite**. Both aragonite and calcite are composed of calcium carbonate ($CaCO_3$), but they have different crystalline structures. Aragonite is less stable and over time changes into calcite. The most common forms of precipitated (nonbiogenous) carbonate are short aragonite crystals (less than 2 millimeters, or 0.08 inch) and oolites. **Oolites** are small onion-like spheres found in some shallow tropical waters where concentrations of $CaCO_3$ are high. Requiring wave action to form, they precipitate around a nucleus and reach diameters of less than 2 millimeters (0.08 inch).

Ocean water is essentially saturated with calcium ions. The only factor preventing precipitation of more calcium carbonate is the absence of carbonate ions. Carbon dioxide (CO_2) is important in carbonate chemistry, for it is the source of "carbonate" in calcium carbonate.

The presence of carbon dioxide and the carbonate ion in ocean water is mutually exclusive, meaning that where there is a high concentration of CO_2, there will be few or no carbonate ions in the water. Where photosynthesis removes significant carbon dioxide from the water, more carbonate ions are available for carbonate precipitation. In addition, warm water contains less carbon dioxide. Both factors are at work in places such as the Bahama Banks near Florida, where water is heated and algae photosynthesis is high. These factors reduce the carbon dioxide content of the water and enhance precipitation of carbonates.

Ancient marine carbonate deposits constitute 2 percent of Earth's crust and 25 percent of all sedimentary rocks on Earth. These deposits, which include the rock **limestone**, form the bedrock underlying Florida and many midwestern states from Kentucky to Michigan and from Pennsylvania to Colorado. Percolation of groundwater through these deposits has dissolved the limestone to produce dramatic phenomena, such as formation of sudden sinkholes and spectacular cavern features.

Metal Sulfides Associated with hydrothermal vents and black smokers along the mid-ocean ridge are deposits of **metal sulfides**. The composition of these metal sulfides includes iron, nickel, copper, zinc, silver and other metals in varying proportions. Transported away from the mid-ocean ridge by sea floor spreading, these deposits can be found throughout the ocean floor and can even be uplifted onto continents.

Evaporites **Evaporite minerals** form where there is restricted open ocean circulation and where evaporation rates are high. As evaporation removes water from these areas, the remaining seawater becomes saturated with certain dissolved elements and evaporite minerals begin to precipitate. Heavier than seawater, they sink to the bottom or form around the margins of seas. Typically, they create a white crust of evaporite minerals around the edges of these areas (Figure 4–12). Collectively termed "salts," some evaporite minerals, such as **halite** (table salt, NaCl), taste salty and some, such as **gypsum** ($CaSO_4 \cdot 2H_2O$) do not. Another important evaporite mineral is calcite.

Cosmogenous Sediment

Cosmogenous (*cosmos* = universe, *generare* = to produce) **sediment** is derived from extraterrestrial sources. Volumetrically insignificant on the ocean floor, cosmogenous sediment consists of two main types: microscopic **spherules**, and macroscopic **meteor** debris.

Spherules are microscopic globular masses composed of silicate rock material, called **tektites** (*tektos* = molten), or composed mostly of iron and nickel (Figure 4–13). They had long been thought to have formed as ejecta from meteors entering Earth's atmosphere. Recent evidence suggests that they form in the asteroid belt between the orbits of Mars and Jupiter and are produced when asteroids collide. This material constantly rains down on Earth as a general component of microscopic **space dust** that floats harmlessly down through the atmosphere. Recent studies indicate that about 90 percent of micrometeorite influx is destroyed by frictional heating as it enters the atmosphere. Still, it has been estimated that as much as 300,000 metric tons (331,000 short tons) of space dust fall on Earth's surface each year. The iron-rich space dust that lands in the oceans often dissolves in seawater. Glassy tektites, however, do not dissolve as easily and sometimes constitute minute proportions of various marine sediments.

Meteor debris is rare on Earth and is associated with the more than 110 meteor impact sites identified worldwide. Meteors collide with Earth at great speeds and some larger ones release energy equivalent to the explosion of a large nuclear bomb (Box 4–2). The debris from meteors, called **meteorite** material, is primarily one of two compositions: silicate rock material (called **chondrites**), or iron and nickel (called **irons**). As a result of these cosmic cannonball collisions with Earth, altered rock material is ejected into the atmosphere and settles out around the impact site.

Mixtures

Rarely do lithogenous and bigenous sediment occur as an absolutely pure deposit that does not contain other types of sediment. Most calcareous oozes contain some siliceous material, and vice versa. The abundance of clay-sized lithogenous particles throughout the world combined with the fact that they are easily transported by winds and currents means that these particles are incorporated into every sediment type. The composition of biogenous ooze includes up to 70 percent fine-grained lithogenous clays. Conversely, most lithogenous sediment contains small percentages of biogenous particles. Hydrogenous sediment is rarely pure, and tiny amounts of cosmogenous sediment are mixed in with all other sediment types. Thus, each sediment is a mixture of different sediment types. Typically, however, one type of

Figure 4–12 Evaporite salts.
Due to high evaporation rate, salts (white material) precipitate from water.

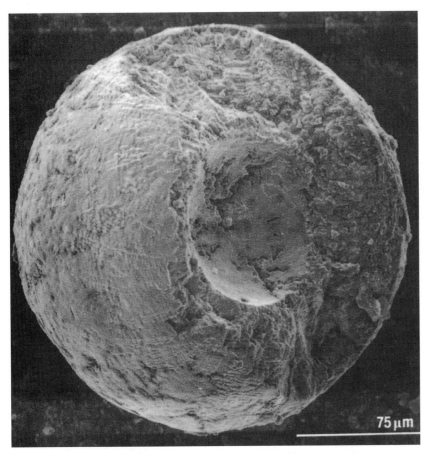

Figure 4–13 Cosmogenous spherule.

Scanning electron micrograph of an iron-rich spherule of cosmic dust. Bar scale of 75 micrometers is equal to 75 millionths of a meter.

75 μm

sediment dominates, which allows it to be classified into one of the sediment types.

Neritic and Pelagic Deposits

The relative amounts of lithogenous, biogenous, hydrogenous, and cosmogenous particles found in marine sedimentary deposits vary considerably. The two largest categories into which marine sedimentary deposits can be divided are **neritic** (*neritos* = of the coast) **deposits**, found near the continents and islands, and **pelagic** (*pelagios* = of the sea) **deposits**, characteristic of the deep-ocean basins.

Lithogenous sediment dominates most neritic deposits. This is not surprising, since lithogenous material is derived from rocks on nearby landmasses. It consists of coarse-grained deposits and accumulates rapidly on the continental shelf, slope, and rise. Neritic deposits also contain biogenous, hydrogenous, and cosmogenous particles, but these constitute only a minor percentage of the total sediment mass. The amount of lithogenous sediment being deposited in neritic environments completely overwhelms the component of other sediment types. For example, both microscopic and macroscopic biogenous material may be incorporated into lithogenous sediment in neritic environments, but it will not reach concentrations high enough to be considered biogenous ooze. Neritic deposits are distributed along the continental shelf by surface currents and carried down the submarine canyons by turbidity currents. Here the sediment is further distributed by deep currents.

Pelagic deposits are composed of fine-grained material that accumulates at a slow rate on the deep-ocean floor. Biogenous ooze is the most common type of pelagic deposit. The reason that microscopic biogenous sediment is so common on the deep-ocean floor is because there is so little lithogenous sediment deposited at great distances from the continents. Curiously, in areas deeper than about 4.5 kilometers (3 miles), there is very little biogenous sediment, and fine-grained lithogenous sediment dominates the deeper regions of the ocean floor. In shallower pelagic areas, biogenous and hydrogenous components are abundant.

Continental Margin Sediments (Neritic Deposits)

Neritic deposits include **relict** (*relict* = left behind) **sediments**, turbidite deposits, glacial deposits, and carbonates.

Relict Sediments At the end of the last Ice Age, around 18,000 years ago, glaciers melted and sea level rose. As a result, many rivers of the world today deposit their sediment in drowned river mouths rather than carry it onto the continental shelf as they did during the geologic past. In many areas the sediments that cover the continental

Box 4–2
When the Dinosaurs Died: The Cretaceous-Tertiary (K–T) Event

A long-standing debate in Earth sciences concerns the demise of the giant land-dwelling reptiles called dinosaurs. Although birds appear to be the direct descendants of dinosaurs, a major extinction of dinosaurs and 75 percent of all plant and animal species on Earth (including many marine species) occurred 65 million years ago. The extinction of these species make it possible for mammals of that day eventually to rise to the position of dominance they hold on Earth today. This extinction marks the boundary between the Cretaceous (K) and Tertiary (T) Periods of geologic time, and is often referred to as the "**K–T event**." Did slow climate change lead to the extinction of these organisms, or was it a catastrophic event? Was their demise related to disease, diet, predication, or volcanic activity? Earth scientists have long sought clues to this mystery.

A report in 1980 of trace elements in sediments by geologist Water Alvarez and his father, physicist Luis Alvarez, revealed some startling results. In deposits collected in northern Italy from the K–T boundary, there is a clay layer containing high proportions of the metallic element iridium (Ir). Iridium is rare in rocks from Earth, but occurs in greater concentration in meteorites. Therefore, layers of sediment that contain unusually high concentrations of iridium suggest that the material may be of extraterrestrial origin. An additional component of the clay layer was the presence of shocked quartz grains. These grains indicate that an event occurred with enough force to fracture and partially melt pieces of quartz. A worldwide search for other deposits from the K–T boundary reveals similar features. These deposits suggested that a large extraterrestrial body had impacted Earth around the time the dinosaurs died. However, this was circumstantial evidence and the scientific community was skeptical. One problem with the impact hypothesis was that, theoretically, volcanic eruptions on Earth could create similar clay deposits enriched in iridium and containing shocked quartz. In fact, large outpourings of basaltic volcanic rock in India (called the Deccan Traps) and other locations occurred at the same time as the dinosaur extinction. Another problem involves

the location of the "smoking gun": If there truly was a catastrophic meteor impact, where was the crater?

Evidence for the impact crater came in the early 1990s with the recognition that the **Chicxulub** (pronounced "SCHICK-sue-lube") **Crater** off the Yucatàn coast in the Gulf of Mexico was an impact crater. The largest impact crater on Earth, it wasn't recognized as such because of its huge size: The crater is at least 190 kilometers (120 miles) in diameter. The shape and age of the crater suggest that a 16-kilometer (10-mile)-wide meteoroid composed of rock and/or ice (a *bolide*) impacted Earth about 65 million years ago, exploding with a force equivalent to a 100-million-megaton nuclear bomb blast (Figure 4E). The meteoroid was so large that when the front hit the ocean floor, the back side of it was still above the atmosphere! The meteoroid that caused this collision was traveling at speeds up to 72,000 kilometers (45,000 miles) per hour. This collision almost penetrated the crust and left an impact crater 30 kilometers (20 miles) deep, which is now partially buried by marine sediment. The impact probably bared the sea floor in the area, and created huge waves [estimated to be up to 914 meters (3000 feet) high], which traveled throughout the ocean basins. This impact probably kicked up so much dust that it blocked sunlight, chilled Earth's surface, and brought about the extinction of dinosaurs and other species. In addition, acid rains and global fires may have added to the environmental disaster. Such a large impact on Earth could easily cause major volcanic activity, so it seems likely that the basalt outpourings could be related to the impact.

Supporting evidence for the meteor impact hypothesis was provided in 1997 by the Ocean Drilling Program (ODP). Previous drilling close to the impact site did not reveal any K–T deposits. Evidently, the impact and resulting huge waves had stripped the ocean floor of its sediment. However, at 1600 kilometers (1000 miles) from the impact site, some of the telltale sediments were preserved on the sea floor. Drilling into the continental margin off the coast of Florida into an underwater peninsula called the Blake Nose, the ODP scien-

Figure 4E K–T meteor impact.

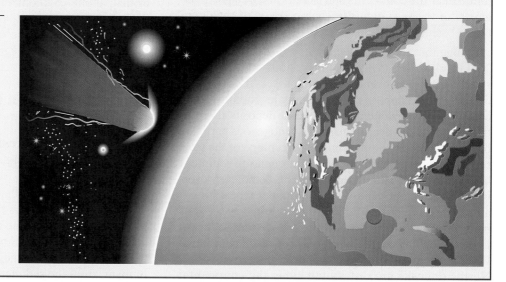

(continued)

tific party recovered cores from the K-T boundary that contain the most complete record of the impact known to date (Figure 4F).

The cores reveal that before the impact, the layers of Cretaceous age sediment are filled with abundant ancient planktonic marine microfossils, mostly coccoliths and foraminifers. These deposits show signs of underwater landslide activity—perhaps the effect of an impact-triggered earthquake. Above this calcareous ooze is a 20-centimeter (8-inch) thick layer of rubble containing spherules, tektites, shocked quartz from hard-hit terrestrial rock, and even a piece of reef rock from the Yucatàn peninsula 2 centimeters (1 inch) across. This layer is rich in iridium, which is characteristic of the K–T boundary. Atop this layer is a thick gray clay deposit containing meteor debris and only two species of coccolithophores and two of foraminifers. These species gave rise to all the coccolithophores and foraminifers alive in the world today. They had a hard time of it right after the impact (these species no doubt struggled for a while in an inhospitable ocean) and gave rise to a few new species before going extinct. Life in the ocean apparently recovered slowly, taking at least 5,000 years before sediment teeming with new, Tertiary-age microorganisms began to be deposited. Similar deposits have also been discovered in New Jersey and in other locations near the impact site. Truly, evidence for the "smoking gun" had been discovered.

This raises another concern: How often does an impact the size of the K–T event occur on Earth? Certainly, the solar system is littered with various-sized debris, including space dust, meteors, asteroids, and comets. Some of this material zips through space at supersonic speeds and can strike Earth with tremendous force. To date, over 110 meteor impact sites, called astroblemes, have been detected, with more being added as they are discovered. Large meteor impacts have only recently been recognized as important events that have shaped Earth history. Based on current evidence, a meteor impact the size of the K–T event should occur about every 100 million years. The time is drawing near to seriously consider how to prevent large meteoroids from striking Earth.

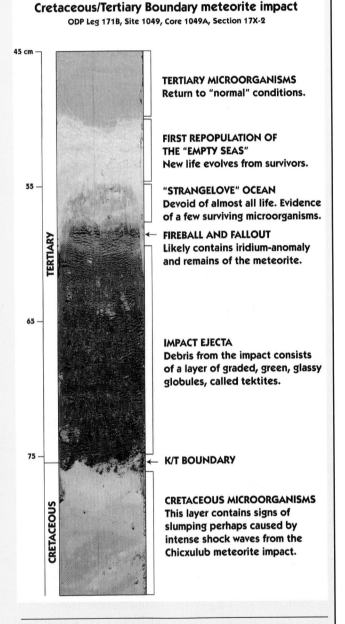

Figure 4F K–T boundary meteor impact core.

shelf, called relict sediments, were deposited from 3000 to 7000 years ago and have not yet been covered by deposits that are more recent. Such sediments presently cover about 70 percent of the world's continental shelves.

Turbidite Deposits Although it seems unlikely that wave action and ocean current systems could carry coarse material beyond the continental shelf into the deep-ocean basin, there is evidence that much neritic material has been deposited at the base of the continental slope forming the continental rise. These accumulations thin gradually toward the abyssal plains. Such deposits are called **turbidite deposits** and are thought to have been deposited by **turbidity currents** that periodically move down the continental slopes through the submarine canyons, carrying loads of neritic material that spreads out across the continental rise as discussed in the preceeding chapter. Turbidite deposits are composed of characteristic layering called graded bedding.[6]

---

[6] For a discussion on turbidite deposits and graded bedding, see Chapter 3 and Figure 3–8.

Glacial Deposits Poorly sorted deposits containing particles of all sizes, from boulders to clays, may be found in the high-latitude[7] portions of the continental shelf. These **glacial deposits** were laid down by melting glaciers that covered the continental shelf area during the Ice Age, when glaciers were more widespread than today and sea level was lower. Glacial deposits are still forming around the continent of Antarctica and around the island of Greenland by **ice rafting**. In this process, rock particles trapped in glacial ice are carried out to sea by icebergs that break away from glaciers as they push into the coastal ocean. As the icebergs melt, lithogenous particles of many sizes are released and settle to the ocean floor.

Carbonate Deposits During the geologic past, deposits of limestone ($CaCO_3$) in the marine environment appear to have been widespread and indicate shallow, warm water conditions. Some limestones contain evidence of a biogenous origin in the form of fossil marine shells. Others, such as ones containing oolites, appear to have formed as chemical precipitates. There appear to be very few places where nonbiogenous carbonates are presently forming. All these deposits are found in low-latitude shallow waters of continental margins or adjacent to islands.

The Bahama Banks is the largest region where significant carbonate deposition is presently occurring, although deposits are also forming on Australia's Great Barrier Reef, in the Persian Gulf, and in other areas in the low latitudes. Coral reefs are another example of biogenous deposits of calcium carbonate found extensively on the continental margins of low-latitude continents and around oceanic islands.

Deep-Ocean Sediments (Pelagic Deposits)

The continental rise is a transitional feature between the continental margin and the deep-ocean basin. Consequently, some neritic material is deposited in significant quantities in the deep-ocean basin as part of the deep-sea fan deposits making up the continental rise. However, in the total area of the deep-ocean basin, it is more common that no neritic material exists. Two types of pelagic deposits cover most of the deep-ocean floor: biogenous oozes and abyssal clay.

Biogenous Oozes At somewhat shallower depths than those where abyssal clay forms, biogenous ooze is a significant portion of the sediment. It consists of at least 30 percent biogenous material from the hard tests of microscopic organisms. The rate of accumulation of biogenous material on the ocean floor depends on three fundamental processes—productivity, destruction, and dilution. *Productivity* is the amount of organisms present in the surface water above the ocean floor. Surface

waters with high biologic productivity contain many living and reproducing organisms. If biologic productivity is low in surface waters, biogenous oozes will never be deposited on the ocean floor, because of the lack of test material being produced in the waters above. *Destruction* of skeletal remains occurs by dissolving in seawater at depth. *Dilution* refers to the inability of oozes to form where other sediments keep the amount of biogenous test material below 30 percent of the sediment. Mostly, dilution is due to the abundance of coarse-grained lithogenous material in neritic environments. Thus, biogenous oozes are not usually found along continental margins.

Siliceous Ooze Siliceous ooze contains at least 30 percent of the hard remains of silica-secreting organisms. If accumulations consist mostly of diatoms, **diatomaceous ooze** is produced. If radiolarians are particularly abundant in an ooze, it is called a **radiolarian ooze**. Another type of siliceous ooze is **silicoflagellate ooze**, produced by an abundance of biogenous test material from single-celled silicoflagellates, another type of algae.

The ocean is undersaturated with silica at all depths, so seawater continually but slowly dissolves silica. How can siliceous ooze accumulate on the ocean floor if it is being dissolved? One way is to accumulate the siliceous tests faster than seawater dissolves them. For instance, many tests sinking at the same time will effectively create a deposit.[8] Once buried beneath other siliceous tests, they no longer dissolve. Consequently, siliceous ooze is commonly found in areas below surface waters with high biologic productivity of silica-secreting organisms (Figure 4–14).

Calcareous ooze Calcareous ooze contains at least 30 percent of the hard remains of calcareous-secreting organisms. If accumulations consist mostly of coccolithophores, **coccolith ooze** is formed. If foraminifers abound, it is called a **foraminifer ooze**. One of the most common types of foraminifer oozes is **Globigerina ooze**, named for a type of foraminifer that is especially widespread in the Atlantic and South Pacific Oceans. Other types of calcareous oozes include **pteropod oozes** (where pteropods dominate) and **ostracod oozes** (oozes full of ostracods).

In the case of calcium carbonate (calcite), destruction (solubility) varies with depth. At the warmer surface, the water is generally saturated with calcium carbonate and calcite does not dissolve. At greater depths, the colder water contains more carbon dioxide (which forms carbonic acid), causing calcareous fragments to be readily

---

[7] High-latitude regions are those far from the Equator (either north or south); low latitudes are areas close to the Equator.

[8] An analogy to this is trying to get a layer of sugar to form on the bottom of a cup of hot coffee. If a few grains of sugar are slowly dropped into the cup, a layer of sugar won't accumulate. However, if a whole bowl full of sugar is dumped into the coffee, a thick layer of sugar will form on the bottom of the coffee cup.

dissolved. In addition, the higher pressure at depth helps speed the dissolution of calcium carbonate. Below 4500 meters (15,000 feet), carbon dioxide increases until calcium carbonate dissolves quite readily.

The depth at which calcium carbonate is dissolved as fast as it falls from above is the **calcite compensation depth**[9] **(CCD)**. The average depth of the CCD is 4500 meters (15,000 feet) below sea level, but may be as deep as 6000 meters (20,000 feet) in portions of the Atlantic Ocean, and as shallow as 3500 meters (11,500 feet) in regions of low biological productivity in the Pacific Ocean. Below the CCD, calcite dissolves rapidly in seawater. Studies have indicated that even thick-shelled foraminifers dissolve within a day or two below the CCD.

The presence of the CCD throughout all oceans of the world helps explain why modern carbonate oozes are generally rare below 5000 meters (16,400 feet). Still, buried ancient deposits of calcareous ooze are found beneath the CCD. How can calcareous ooze exist below the CCD if the CCD dissolves calcium carbonate so quickly? The conditions that allow calcareous ooze to be found *below* the CCD are shown in Figure 4–15. Not surprisingly, plate tectonic processes help to explain how this can occur. Because the mid-ocean ridge is a topographically high feature that rises an average of 2.5 kilometers (1.5 miles) above the sea floor, it often sticks up above the CCD while the surrounding deep-ocean floor is

---

[9] Because the mineral calcite is composed of calcium carbonate, the calcite compensation depth is also known as the calcium carbonate compensation depth, or the carbonate compensation depth. All go by the handy abbreviation of CCD.

below the CCD. Thus, calcareous ooze can be deposited on top of the mid-ocean ridge and will not be dissolved. In essence, the crest of the mid-ocean ridge is above the calcareous ooze "snow line" in the ocean. Sea floor spreading causes the newly created sea floor and the calcareous sediment on top of it to move away from the ridges into greater water depths. Over millions of years, the sea floor is transported below the CCD. Calcareous ooze dissolves below the CCD, but if the ooze is covered by a deposit that is unaffected by the CCD (such as siliceous ooze or abyssal clay), then it is preserved. Consequently, the calcareous ooze is transported along with the plate to depths below the CCD.

The percentage by weight of calcium carbonate accumulation in the modern surface sediments of ocean basins is shown in Figure 4–16. It shows that high concentrations of calcareous ooze (sometimes exceeding 80 percent) are found along segments of the mid-ocean ridge, whereas little is found in deep-ocean basins below the CCD. The deeper the ocean floor, the less calcium carbonate occurs in the sediment, as is seen in the deep North Pacific Ocean. Calcium carbonate is also rare in sediments accumulating beneath cold, high-latitude waters where calcareous-secreting organisms do not live in great abundance.

A comparison of environmental differences that can be inferred from siliceous and calcareous oozes is listed in Table 4–3.

Abyssal Clay **Abyssal clay** is composed of at least 70 percent by weight clay-sized particles derived from the continents and carried by winds or ocean currents. These particles, along with space dust and fine-grained

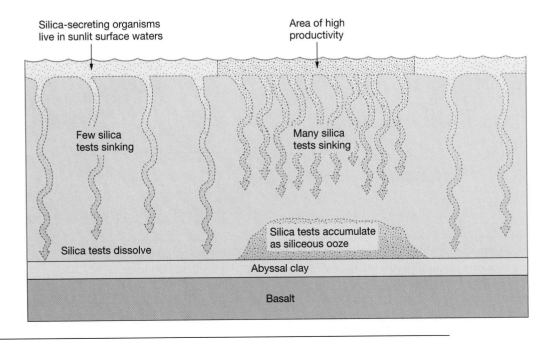

Figure 4–14 Accumulation of siliceous ooze.
Siliceous ooze accumulates on the ocean floor beneath areas of high productivity, where the rate of accumulation of siliceous tests is greater than the rate at which silica is being dissolved.

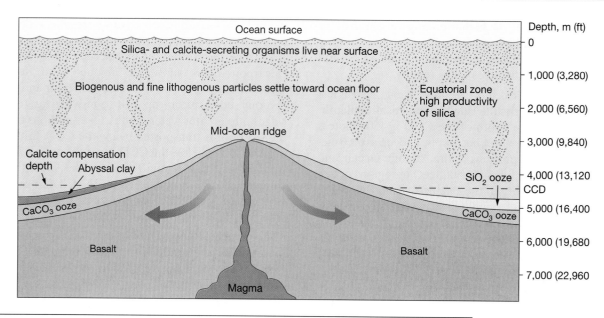

Figure 4–15 Sea floor spreading and sediment accumulation.

Relationships among carbonate compensation depth, the mid-ocean ridge, sea floor spreading, productivity, and destruction that allow calcareous ooze to be found below the CCD.

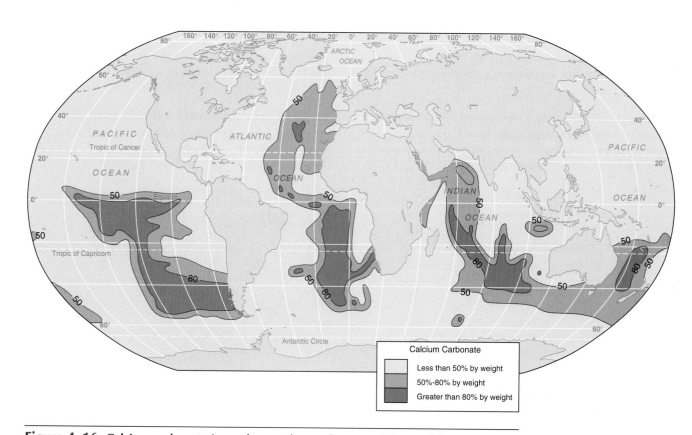

Figure 4–16 Calcium carbonate in modern surface sediments of the world's oceans.

High percentages of calcareous ooze closely follows the regions of relatively shallow mid-ocean ridge segments throughout the oceans.

Table 4–3   A comparison of environments interpreted from siliceous ooze and calcareous ooze.

	Siliceous ooze	Calcareous ooze
Surface water temperature above sea floor deposits	Cool	Warm
Main location found	Sea floor beneath cool surface water in high latitudes	Sea floor beneath warm surface water in low latitudes
Other factors	Upwelling brings deep, cold, nutrient-rich water to the surface	Calcareous ooze dissolves below the CCD
Other locations found	Sea floor beneath areas of upwelling, including along the Equator	Sea floor beneath warm surface water in low latitudes along the mid-ocean ridge

Table 4–4   Average rates of deposition of selected marine sediments.

Type of sediment/environment	Average rate of deposition (per 1000 years)	Thickness of deposit after 1000 years equivalent to...
Coarse lithogenous sediment, neritic environment	1 meter (3.3 feet)	A meter stick
Biogenous ooze, pelagic environment	1 centimeter (0.4 inch)	The diameter of a dime
Abyssal clay, pelagic environment	1 millimeter (0.04 inch)	The thickness of a dime
Manganese nodule, pelagic environment	0.001 millimeter (0.00004 inch)	A microscopic dust particle

volcanic material transported great distances through the atmosphere, settle slowly onto the ocean floor (Table 4–4). Abyssal clay covers most of the deep abyssal plain regions. Abyssal clay deposits are commonly red-brown or buff in color because they contain oxidized iron, and are sometimes referred to as **red clays**. It is interesting to note that the predominance of abyssal clay on abyssal plains is due to an *absence* of other material which elsewhere dilutes the fine-grained lithogenous component, rather than to an increased amount of clay particles arriving on the ocean floor. In the deep-ocean basins, very little other sediment is being deposited except for fine abyssal clays.

Distribution of Neritic and Pelagic Deposits

The distribution of neritic and pelagic deposits is illustrated in Figure 4–17. Coarse-grained neritic deposits dominate continental margin areas (dark brown shading), while abyssal clays (yellow shading) dominate the deep-ocean basins. Calcareous oozes (blue shading) are abundant on the relatively shallow deep-ocean areas along the mid-ocean ridge. Siliceous oozes are found beneath areas of unusually high biologic productivity such as the Antarctic (light green shading, where diatomaceous ooze occurs) and equatorial Pacific Ocean (dark green shading, where ra-

diolarian ooze occurs). Figure 4–18 shows the distribution of sediment across a passive continental margin.

While neritic deposits cover about one-quarter of the ocean floor, the other three-quarters are covered by pelagic deposits. Figure 4–19 shows the proportion of each ocean floor that is covered by pelagic calcareous ooze, siliceous ooze, and abyssal clay. Calcareous oozes are clearly the dominant pelagic deposits, covering almost 48 percent of the world's deep-ocean floor. Abyssal clay covers 38 percent and siliceous oozes 14 percent of the world ocean floor area. The graph shows that less ocean basin floor is covered by calcareous ooze in deeper basins. This is because deeper ocean basins have more area that lies beneath the CCD. The dominant oceanic sediment in the deepest basin, the Pacific, is abyssal clay (see also Figure 4–17), whereas calcareous ooze is the most widely deposited oceanic sediment in the shallower Atlantic and Indian Oceans. Siliceous oozes cover a smaller percentage of the ocean bottom in all the oceans. This is because regions of high productivity of radiolarians and diatoms, the major components of these deposits, are restricted to the equatorial region (for radiolarians) and in the Antarctic (for diatoms).

Microscopic biogenous tests typically take from 10 to 50 years to sink from the ocean surface where organisms live to abyssal depths where biogenous sediment accumulates. During this time interval, a horizontal

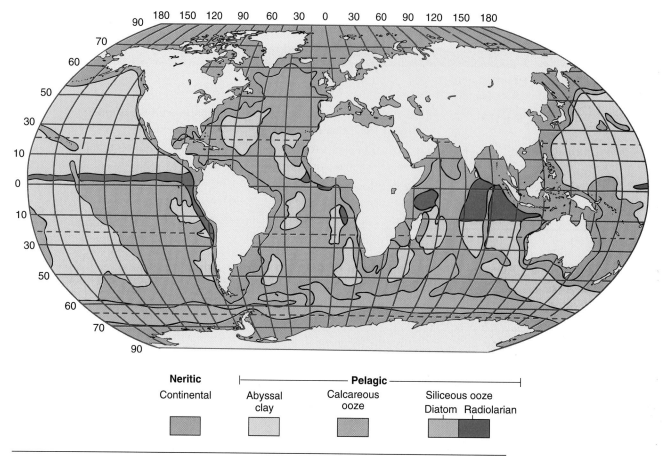

Figure 4–17 Distribution of neritic and pelagic sediments.

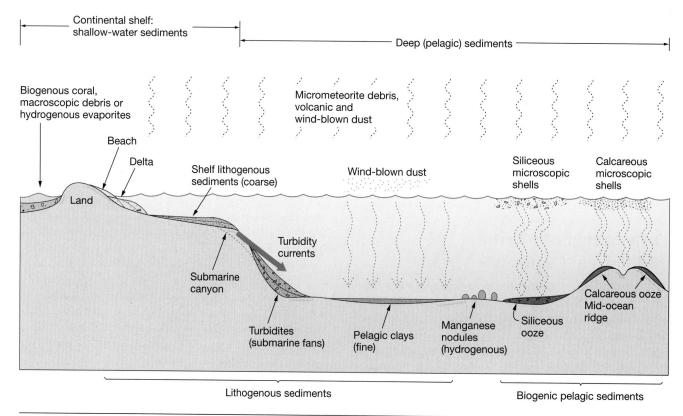

Figure 4–18 Distribution of sediment across a passive continental margin.

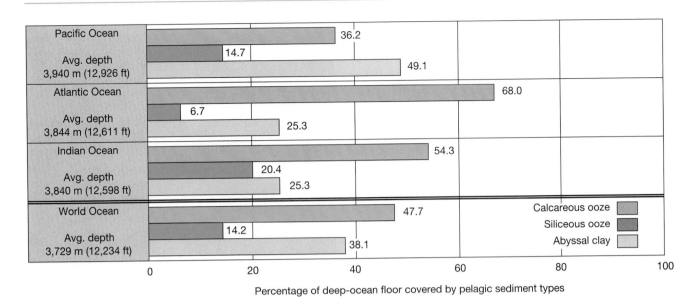

Figure 4–19 Distribution of pelagic sediment.

Distribution of pelagic calcareous ooze, siliceous ooze, and abyssal clay in each ocean, and for the world ocean. The average depths shown exclude shallow adjacent seas, where little pelagic sediment accumulates.

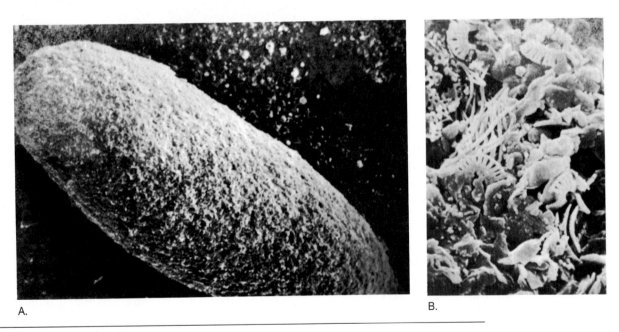

A.

B.

Figure 4–20 Fecal pellets.

A. A 200-micrometer (0.008 inch)-long fecal pellet, large enough to sink rapidly from the surface to the ocean floor. **B.** Close-up of the surface of a fecal pellet showing the remains of coccoliths and other debris.

ocean current of only 1 centimeter per second (about 0.02 mile per hour) could carry the test as much as 15,000 kilometers (9300 miles) before they settled onto the deep-ocean floor. Why do biogenous tests on the deep-ocean floor closely reflect the population of organisms living in the surface water directly above? It appears that the tests are settling at a much faster rate, but how could they do so? Closer examination of the way sediments are deposited reveals that 99 percent

of particles that fall to the ocean floor do so as part of **fecal pellets** produced by tiny animals that feed on algae and protozoans living in the water column above. These pellets (Figure 4–20), though still small, are large enough to allow the particulate matter to reach the abyssal ocean floor in only 10 to 15 days. The occurrence of fecal pellets helps explain the remarkable similarity in particle composition of the surface waters and the sediment on the sea floor immediately below.

Ocean Sediments as a Resource

The sea floor is rich in potential mineral and organic resources. Because of its inaccessibility (and therefore, the high cost involved to reach these areas), the deep-ocean floor is unlikely to be widely exploited in the near future. However, some notable resources make appealing exploration targets.

Petroleum

The remains of microscopic organisms, buried within marine sediments before they could decompose, are the source of our most valuable marine resource—**petroleum** (oil and natural gas). Of the nonliving resources extracted from the oceans, more than 95 percent of the value is in petroleum products.

The term *petroleum* encompasses both oil and natural gas and provides most of the energy on which the world economy depends. At current rates of consumption, experts have estimated that global petroleum production will begin to decline before the year 2010. Beyond that date, major changes in the energy supply will be required. Societies will then need to convert to nonconventional sources such as extra-heavy oil, tar sands, and other high-cost sources. Converting to natural gas would extend the fuel supply only a few years.

The percentage of world oil production coming from offshore has increased from a trace amount in 1930 to over 30 percent in the 1990s. Most of this increase was due to technological advancements employed by offshore drilling platforms (Figure 4–21). Major offshore reserves exist in the Persian Gulf, in the Gulf of Mexico, off southern California, in the North Sea, and in the East Indies. Additional reserves are probably located off the north coast of Alaska and in the Canadian Arctic, Asian seas, Africa, and Brazil, but there is little likelihood that they will provide enough to reverse the existing trend of dwindling reserves. With almost no hope of finding major undiscovered reserves on land, and only a slightly increased likelihood of finding them offshore, future oil exploration will still be intense in offshore waters. A major drawback to offshore petroleum exploration is the possibility of offshore oil spills caused by inadvertent leaks or blowouts during the drilling process.

Sand and Gravel

The offshore sand and gravel industry is second in value only to that of petroleum. This resource, which includes rock fragments and shells of marine organisms, is mined by offshore barges using a suction dredge. The resource is used primarily for resupplying recreational beaches, filling in areas on which to build, and for use as an aggregate in concrete.

Offshore deposits are a major source of sand and gravel in New England, in New York, and throughout the Gulf

Figure 4–21 Offshore drilling rig.
Constructed on tall stilts, rigs like this one in the Gulf of Mexico off Texas are important for extracting petroleum reserves from beneath the continental shelves.

Coast. Many European countries, Iceland, Israel, and Lebanon also depend heavily on such deposits.

Some gravel deposits are rich in valuable minerals. For example, diamonds are recovered from gravel deposits offshore of South Africa and Australia. Other sediments rich in tin have been mined for years from Thailand to Indonesia. Platinum and gold have been found in deposits in gold mining areas throughout the world, and some Florida beach sands are rich in titanium.

A major concern in sand and gravel mining is that it not be conducted too near shore. Nearshore mining operations may damage beaches, because beach material is washed by wave action and currents offshore into the mined areas.

Evaporative Salts

When seawater evaporates, water is removed and the salts remain. Once the salts increase in concentration, they can no longer be held in solution. Consequently, the salts precipitate out of solution and form salt deposits. The most economically notable salts are gypsum and halite. Gypsum is used in plaster of Paris to make casts and molds, and is the main component in wallboard. Halite, or table salt, is widely used for seasoning, curing, and preserving foods. Its major use is in the production of chlorine, sodium, and sodium hydroxide. Manufacture and use of salt is one of the oldest chemical industries.[10]

A large-scale example of evaporite deposition occurred in the Mediterranean Sea about 6 million years ago. Based on evidence obtained from deep-sea cores, the Mediterranean Sea was cut off from the Atlantic Ocean at the narrow Strait of Gibraltar—in effect, creating a dam. Because of the warm and dry climate in this region, the Mediterranean Sea nearly evaporated in a few thousand years, leaving a thick deposit of gypsum on the hot, dry basin floor. A half-million years later, erosion, further tectonic activity, or a rise in worldwide sea level caused the dam to breach at Gibraltar. The world's largest waterfall spilled into the Mediterranean and filled the sea with seawater once again. Other seas with restricted ocean circulation and high evaporation rates, such as the Gulf of Mexico, also have thick layers of evaporite minerals buried within their sea floor sediments.

Phosphorite (Phosphate Minerals)

Phosphorite is a sedimentary rock made of various phosphate minerals, which contain the element phosphorus, an important plant nutrient. Although no commercial phosphorite mining from the oceans is presently occurring, the marine reserve is estimated to exceed 45 billion metric tons (50 billion short tons). Phosphorite occurs at depths less than 300 meters (1000 feet) on the continental shelf and slope. Deposits can be used to produce phosphate fertilizer if economic conditions make them recoverable.

Some shallow sand and mud deposits contain up to 18 percent phosphate. Some phosphorite deposits occur as nodules, with a hard crust formed around a nucleus. The nodules may be as small as a sand grain or up to a meter in diameter and may contain over 25 percent phosphate. Most land sources have had their phosphate concentration enriched to more than 31 percent by groundwater leaching.

Manganese Nodules and Crusts

Manganese nodules contain significant concentrations of manganese and iron and smaller concentrations of copper, nickel, and cobalt. In the 1960s, mining companies began to assess the feasibility of mining the vast expanses of manganese nodules found on the deep-ocean floor (Figure 4–22). Exploration efforts revealed that vast areas of the sea floor contain manganese nod-

ules (Figure 4–23), with the richest metallic content in the eastern Pacific Ocean between Hawaii and Mexico.

Of the five metals, cobalt is the only metal deemed "strategic" to the United States, which means that it is essential to national security. It is required to produce dense, strong alloys with other metals for use in high-speed cutting tools, powerful permanent magnets, and jet engine parts. Essentially, the United States must import all its supply of cobalt. Major enrichments of cobalt in Earth's crust are confined to central southern Africa and deep-ocean nodules and **crusts** (hard coatings on other rocks). Because of the unstable political situation in Africa, the United States has been looking to the ocean floor as a more dependable source.

In the 1980s cobalt-rich manganese crusts were discovered on the upper slopes of islands and seamounts that lie within 200 miles from the shorelines of the United States and its territories. The cobalt concentrations in these crusts are half again as rich as in the best African ores and at least twice as rich as in the deep-sea manganese nodules. However, interest in mining these deposits has faded in the face of lower metal prices.

Figure 4–22 Mining manganese nodules.
Manganese nodules can be collected by dredging the ocean floor. The dredge is shown unloading nodules onto the deck of a ship.

---

[10] An interesting historical note about salt is that part of a Roman soldier's pay was in salt. That portion was called the *salarium*, from which the word *salary* is derived. If a soldier did not earn it, he was not worth his salt.

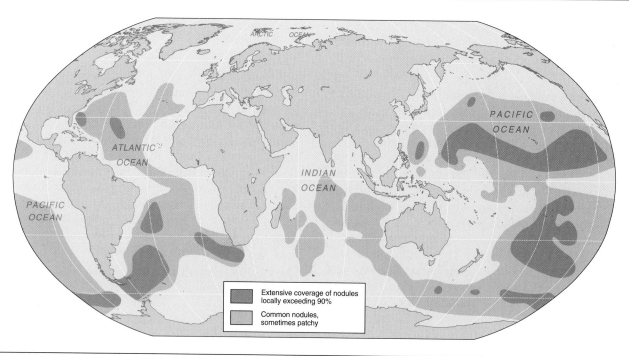

Figure 4–23 Distribution of manganese nodules.
Generalized map showing the distribution of manganese nodules on the sea floor.

Students Sometimes Ask...

Are there any areas of the ocean floor where no sediment is being deposited?
Essentially, all areas of the ocean floor receive various types of sediment that rain down through the water and accumulate to form sedimentary deposits. Even the deep-ocean floor far from land receive small amounts of wind-blown dust, microscopic biogenous particles, and cosmic dust. However, the mid-ocean ridge is one of the few areas where very little sediment exists. The reason for this has to do with the *age* of the ocean floor at the mid-ocean ridge. Because of sea floor spreading, new sea floor is created along the crest of the mid-ocean ridge. The sea floor is so young and the rates of sediment accumulation so slow in the middle of ocean basins that most segments of the mid-ocean ridge have only a thin dusting of sediment.

How effective is wind as a transporting agent?
Surprisingly, wind can transport huge volumes of silt- and clay-sized particles great distances into the ocean. A recent study indicated that each year an average of 11.4 million metric tons (12.6 million short tons) of dust from the Sahara Desert is carried downwind up to 6500 kilometers (4000 miles) across the Atlantic Ocean. Ships traveling across the Atlantic Ocean downwind from the Sahara Desert often arrive at their destinations quite dusty!

I've been to Hawaii and seen a black volcanic sand beach. Since it forms by lava flowing into the ocean and being broken up by waves, is it hydrogenous sediment?
No. Many active volcanoes in the world create black sand beaches. The material that produces black sand is derived from a continent or an island, and is consequently considered lithogenous sediment. A test for whether a deposit is hydrogenous is this: Was the material ever *dissolved* in water? Even though

lava flows enter the water, the lava is never dissolved in water. All hydrogenous sediment is derived from material that was once dissolved in water.

I've heard that a flammable substance called methane hydrate has been found on the deep-ocean floor. What is it?
Methane hydrate is the most common type of gas hydrate. **Gas hydrates**, also known as clathrates (*clathr* = a lattice), are a combination of two common substances: water and natural gas, usually methane. In deep-ocean sediments, where pressures are high and temperatures are low, these substances combine to form an unusually compact chemical structure, in which the gas is trapped inside a lattice-like cage of water molecules. Cores drilled into gas hydrates on the ocean floor have retrieved mud mixed with golf-ball-sized chunks of gas hydrates, which fizzle and evaporate quickly once they are removed from the conditions where they form. Gas hydrates resemble chunks of ice, but ignite when lit by a flame because methane and other flammable gases are released as gas hydrates melt.

There has been widespread interest in exploring the extent of gas hydrates in ocean sediment. Because of their methane content, gas hydrates may potentially be the world's largest source of usable energy. Also, there is interest in knowing what role they play in sea floor instability and as contributors to global warming. In addition, methane seeps support a rich community of organisms, many of which are species new to science.

Most gas hydrates are created when bacteria break down organic matter trapped in sea floor sediments, producing methane gas with minor amounts of ethane and propane. These gases can be incorporated into gas hydrates under high pressure and low temperature conditions. Most ocean floor areas below 525 meters (1720 feet) provide these conditions, but gas hydrates seem to be confined to continental margin areas, where high productivity surface waters enrich ocean floor sediments below with organic matter.

Since their discovery in ocean drill cores in 1976, studies of the deep-ocean floor reveal that at least 50 sites worldwide may contain extensive deposits of gas hydrates trapped in sediments. In fact, some estimates indicate that as much as 20 quadrillion cubic meters (700 quadrillion cubic feet) of methane are locked up in sediments containing gas hydrates. In terms of fossil fuel content, this amount is equivalent to about twice as much as Earth's coal, oil, and conventional gas reserves *combined*. One of the major drawbacks in exploiting these reserves is that gas hydrates rapidly decompose at surface temperatures and pressures. In the future, however, we may see the day when these vast sea floor reserves are used to power modern society.

How do manganese nodules form?

The answer to that question has puzzled oceanographers since manganese nodules were first discovered during the world's first great oceanographic voyage, begun in 1872 by the British research ship HMS *Challenger*. If manganese nodules are truly hydrogenous and precipitate from seawater, then how can they have such high concentrations of manganese (which occurs in seawater at concentrations often too small to measure accurately)? Furthermore, why are they on *top* of ocean floor sediment? They should be buried by the constant rain of particles falling down on the ocean floor. Unfortunately, nobody has definitive answers to these questions. Perhaps the creation of manganese nodules is the result of one of the slowest chemical reactions known. Recent research suggests that the formation of manganese nodules may be aided by bacteria or some other as-yet-unidentified marine organism. Other studies reveal that the nodules don't form continuously over time but in spurts that are related to specific conditions such as a low sedimentation rate of lithogenous clay and strong deep-water currents. The mystery of manganese nodules is awaiting someone to investigate what is considered the most interesting unresolved problem in marine chemistry. If you do solve this mystery, you'll be famous throughout the field of oceanography!

I've heard that some countries want to bury their nuclear waste in sediments at sea. Is this true?

Not only is this true, but the United States is one of the counties that has seriously considered high-level nuclear waste disposal at sea. Nuclear waste that remains from the production of nuclear weapons and from commercial power generation must be safely disposed of until it is no longer dangerous, typically on the order of several hundred thousand years.

The major emphasis for disposal of nuclear waste has been focused on land sites because the materials remain accessible. However, from 1976 until 1986, research was conducted into disposal of high-level nuclear waste within the sea floor of the deep sea. Given the requirements that the waste must be kept away from human activities, kept safe from exposure by natural erosion, and kept away from seismically active regions, the centers of oceanic lithospheric plates are attractive disposal sites. In addition, the fine-grained abyssal clay that exists in the deep-ocean serves as an ideal medium in which to place the high-level nuclear waste, presumably trapping any containment leaks. Proposals specify that a drill-hole in the deep-ocean floor (similar to the holes drilled by ODP's drill ship *Resolution* to recover deep-sea sediment cores) could be used for the disposal of canisters containing nuclear waste hundreds of meters beneath the sea floor. The plan calls for stacking the canisters within the drill hole, which would later be filled with mud.

In 1986, the United States halted research into seabed disposal of nuclear waste because of budget constraints. After a lengthy evaluation of potential sites nationwide, Yucca Mountain in Nevada was selected as the future repository site for U.S. nuclear waste. However, concerns about the safety of the Yucca Mountain site have recently been raised. In addition, it appears that the site is not large enough to hold all the nuclear waste that will be generated by the year 2020.

Given the political realities that may be faced as work proceeds toward land-based disposal, deep-sea disposal will most likely be considered again. The seabed disposal program did produce encouraging results, and, in fact, Japan is currently investigating disposing of its nuclear waste in the Mariana Trench. Studies of areas near where nuclear weapons and nuclear submarines have accidentally been lost at sea indicate that clays on the deep-ocean floor tend to hold fast to certain radioactive elements, effectively isolating them from the environment. Thus, it would seem that careful ocean disposal, if conducted safely, might be a disposal strategy with merit. However, environmental consequences must be carefully considered before such a form of disposal is put into operation.

Summary

The sea floor spreading hypothesis was confirmed when the *Glomar Challenger* began the Deep Sea Drilling Project to sample ocean sediments and crust. Today, the *JOIDES Resolution* continues the important work of retrieving sediments from the deep-ocean floor. Analysis and interpretation of these sediments revels that the ocean floor has had an interesting and complex history.

Sediments that accumulate on the ocean floor are classified by origin as lithogenous (derived from rock), biogenous (derived from organisms), hydrogenous (derived from water), or cosmogenous (derived from outer space).

Lithogenous sediment is composed of rock fragments that have a composition that reflects the composition of the material from which it is derived. Its texture, determined in part by the size, sorting, and rounding of particles, is affected greatly by the type of transporting media (water, wind, ice, or gravity) and the energy conditions under which it was deposited.

Biogenous sediment is composed of either silica (SiO_2) from organisms such as diatoms and radiolarians or calcium carbonate ($CaCO_3$) from the remains of organisms such as foraminifers and coccolithophores. Accumulations of these microscopic shells (tests) of organisms must comprise at least 30 percent of the deposit for it to be classified as biogenic ooze.

Hydrogenous sediment includes a wide variety of materials that precipitate directly from water or are formed by the interaction of substances dissolved in water with materials on the ocean floor. Manganese nodules, phosphates, carbonates, metal sulfides, and evaporites are examples.

Cosmogenous sediment is composed of either macroscopic meteor debris (such as that produced during the K-T event) or microscopic iron-nickel and silicate spherules that may result from asteroid collisions. Most ocean sediment is a mixture of the various types of sediment.

Neritic deposits accumulate rapidly along the margins of continents and are dominated by sediment of lithogenous origin. Neritic deposits include relict sediment deposited in drowned river valleys since the Ice Age, turbidite deposits at the base of continental slopes, poorly sorted glacial deposits from ice rafting, and carbonate deposits indicating deposition in low-latitude shallow waters.

Pelagic deposits accumulate at low rates on the floor of the deep-ocean basins, far from the continents. In the deeper basins, where most biogenous sediment is dissolved before reaching bottom, abyssal clay deposits predominate. Biogenous oozes are found in shallower basins on ridges and rises. The rate of biological productivity measured against the rates of destruction and dilution of biogenous sediment determines whether abyssal clay or oozes will form on the ocean floor. At the carbonate compensation depth, ocean water dissolves calcium carbonate at a rate equal to the rate it settles from above, thus preventing accumulation on the ocean floor below that depth.

The distribution of neritic and pelagic sediment is influenced by many factors, including proximity to sources of lithogenous sediment, productivity of microscopic marine organisms, depth of the ocean floor, and the distribution of various sea floor features. Fecal pellets are important in transporting particles to the deep-ocean floor and helps explain the similarity in particle composition of the surface waters and sediment on the ocean floor immediately below them.

The most valuable nonliving resource from the ocean today is petroleum, which is recovered from below the continental shelves and used as a source of energy. Other important resources include sand and gravel, evaporative salts, phosphorite, and manganese nodules and crusts.

Key Terms

Abyssal clay (p. 119)

Algae (p. 107)

Aragonite (p. 113)

Biogenous sediment (p. 106)

Bolide (p. 116)

Calcareous ooze (p. 108)

Calcite (p. 113)

Calcite compensation depth (CCD) (p. 119)

Calcium carbonate (p. 108)

Carbonate (p. 113)

Chalk (p. 108)

Chicxulub Crater (p. 116)

Chondrite (p. 114)

Coccolith (p. 108)

Coccolith ooze (p. 118)

Coccolithophore (p. 108)

Core (p. 100)

Cosmogenous sediment (p. 114)

Crusts (p. 125)

Deep Sea Drilling Project (DSDP) (p. 100)

Diatom (p. 108)

Diatomaceous earth (p. 108)

Diatomaceous ooze (p. 118)

Dredge (p. 100)

Eroded (p. 103)

Evaporite mineral (p. 114)

Fecal pellet (p. 123)

Foraminifer (p. 108)

Foraminifer ooze (p. 118)

Gas hydrate (p. 126)

Glacial deposit (p. 118)

Globigerina ooze (p. 118)

Glomar Challenger (p. 100)

Gravity corer (p. 100)

Grain size (p. 104)

Gypsum (p. 114)

Halite (p. 114)

Hydrogenous sediment (p. 108)

Ice rafting (p. 118)

Irons (p. 114)

JOIDES Resolution (p. 100)

Joint Oceanographic Institutions for Deep Earth Sampling (JOIDES) (p. 100)

K–T event (p. 116)

Limestone (p. 113)

Lithified (p. 102)

Lithogenous sediment (p. 103)

Macroscopic biogenous sediment (p. 106)

Manganese nodule (p. 108)

Maturity (p. 106)

Metal sulfide (p. 113)

Meteor (p. 114)

Meteorite (p. 114)

Microscopic biogenous sediment (p. 106)

Mineral (p. 104)

Nanoplankton (p. 108)

Neritic deposit (p. 115)

Ocean Drilling Program (ODP) (p. 100)

Oolite (p. 113)

Ooze (p. 106)

Ostracod ooze (p. 118)

Pelagic deposit (p. 115)

Petroleum (p. 124)

Phosphate (p. 112)

Phosphorite (p. 125)

Planktonic (p. 108)

Protozoan (p. 107)

Pteropod ooze (p. 118)

Quartz (p. 104)

Radiolarian (p. 108)

Radiolarian ooze (p. 118)

Red clay (p. 121)

Relict sediment (p. 115)

Rotary drilling (p. 100)

Rounding (p. 106)

Sediment (p. 102)

Sedimentary rock (p. 102)

Silica (p. 108)

Siliceous ooze (p. 108)

Silicoflagellate ooze (p. 118)

Sorting (p. 106)

Space dust (p. 114)

Spherules (p. 114)

Tektites (p. 114)

Terrigenous sediment (p. 103)

Test (p. 106)

Texture (p. 103)

Transporting media (p. 103)

Turbidite deposits (p. 117)

Turbidity currents (p. 117)

Weathering (p. 103)

Questions And Exercises

1. Describe the process of how a drilling ship like the *JOIDES Resolution* obtains core samples from the deep-ocean floor.

2. What kind of information can be obtained by examining and analyzing core samples?

3. List and describe the characteristics of the four basic types of marine sediment.

4. How does lithogenous sediment originate?

5. Why is most lithogenous sediment composed of quartz grains? What is the chemical composition of quartz?

6. If a deposit has a coarse grain size, what does this indicate about the energy of the transporting medium? Give several examples of various transporting media that would produce such a deposit.

7. What characteristics of marine sediment indicate increasing maturity?

8. List the two major chemical compounds of which most biogenous sediment is composed and the organisms that produce them. Sketch these organisms.

9. What are several reasons that diatoms are so remarkable? List uses of diatoms in products.

10. Describe manganese nodules, including what is currently known about how they form.

11. Lava flows often enter the ocean, where crashing waves produce black volcanic beach sand. Is this sand lithogenous or hydrogenous sediment? Explain.

12. Describe the most common types of cosmogenous sediment and give the probable source of these particles.

13. Why is lithogenous sediment the most common neritic deposit? Why are biogenous oozes the most common pelagic deposits?

14. How do oozes differ from abyssal clay? Discuss how productivity, destruction, and dilution combine to determine whether an ooze or abyssal clay will form on the deep-ocean floor.

15. Describe the environmental conditions at the surface that influence the distribution of siliceous and calcareous ooze.

16. If siliceous ooze is slowly but constantly dissolving in seawater, how can deposits of siliceous ooze accumulate on the ocean floor?

17. Explain the stages of progression that results in calcareous ooze existing below the CCD.

18. How do fecal pellets help explain why the particles found in the ocean surface waters are closely reflected in the particle composition of the sediment directly beneath? Why would one not expect this?

19. Discuss the present importance and the future prospects for the production of petroleum; sand and gravel; phosphorite; and manganese nodules and crusts.

20. What are gas hydrates, where are they found, and why are they important?

21. Describe the K–T event, including its effect on the environment.

References

Alvarez, L. W., Alvarez, W., Asaro, F., and Michel, H. V. 1980. Extraterrestrial cause for the Cretaceous-Tertiary extinction. *Science* 208:4484, 1095–1108.

———. 1982. Current status of the impact theory for the terminal Cretaceous extinction. In Silver, L. T., and Schultz, P. H., eds. *Geological implications of impacts of large asteroids and comets on the Earth*, Geological Society of America Special Paper 190. Boulder, CO: Geological Society of America.

Alvarez, W. 1997. *T. rex and the crater of doom*. Princeton, NJ: Princeton University Press.

Berger, W. H., Adelseck, C. G., Jr. and Mayer, L. A. 1976. Distribution of carbonate in surface sediments of the Pacific Ocean. *Journal of Geophysical Research* 81:15, 2617–2629.

Biscaye, P. E., Kolla, V., and Turedian, K. K. 1976. Distribution of calcium carbonate in surface sediments of the Atlantic Ocean. *Journal of Geophysical Research* 81:15, 2592–2602.

Brasier, M. D. 1980. *Microfossils*. London: George Allen & Unwin.

Brewer, P. G., et al. 1997. Deep-ocean field test of methane hydrate formation from a remotely operated vehicle. *Geology* 25:5, 407–410.

Broadus, J. M. 1987. Seabed minerals. *Science* 235:853–860.

Burnett, J. L. 1991. Diatoms—The forage of the sea. *California Geology* 44:4, 75–81.

Carson, R. L. 1961. *The sea around us*, revised edition. New York: Oxford University Press.

Emery, K. O., and Uchupi, E. 1984. *The geology of the Atlantic Ocean*. New York: Springer-Verlag.

Frank, L. A., and Huyghe, P. 1990. *The big splash*. New York: Birch Lane Press.

Glassby. G. P., ed. 1977. *Marine manganese deposits*. New York: Elsevier.

Halbach, P., Friedrich, G., and von Stackelberg, U., eds. 1988. The manganese nodule belt of the Pacific Ocean: Geological environment, nodule formation, and mining aspects. Stuttgart, Germany: Ferdinand Enke Verlag.

Hallegraeff, G. M. 1988. *Plankton: A microscopic world*. New York: E. J. Brill.

Hildebrand, A. R., and Boynton, W. V. 1990. Proximal Cretaceous-Tertiary boundary impact deposits in the Caribbean. *Science* 248:4957, 843–846.

Honjo, S. 1992. From the Gobi to the bottom of the North Pacific. *Oceanus* 35:4, 45–53.

Howell, D. G., and Murray, R. W. 1986. A budget for continental growth and denudation. *Science* 233:4762, 446–449.

Hsü, K. 1983. *The Mediterranean was a desert: A voyage of the Glomar Challenger*. Princeton, NJ: Princeton University Press.

Inter-University Program of Research on Ferromanganese Deposits of the Ocean Floor. *Phase 1 Report.* Unpublished.

Kennett, J. P. 1982. *Marine geology.* Englewood Cliffs, NJ: Prentice-Hall.

Kolla, V., and Biscaye, P. E. 1976. Distribution of calcium carbonate in surface sediments of the Atlantic Ocean. *Journal of Geophysical Research* 81:15, 2602–2616.

Krajick, K. 1997. The crystal fuel. *Natural History* 106:4, 26–31.

Kvenvolden, K. A. 1996. Gas hydrates-geological perspective and climate change, *in* Pirie, R. G., ed., *Oceanography: Contemporary Readings in Ocean Sciences,* 3rd ed., New York: Oxford University Press.

Kyte, F. T., Zhou, L., and Wasson, J. T. 1988. New evidence on the size and possible effects of a late Pliocene oceanic asteroid impact. *Science* 241:4861, 63–65.

Leinen, M., Cwienk, D., Heath, G. R., Biscaye, P. E., Kolla, V., Thiede, J., and Dauphin, J. P. 1986. Distribution of biogenic silica and quartz in recent deep-sea sediments. *Geology* 14:3, 199–203.

Masters, C. D., Root, D. H., and Attanasi, E. D. 1991. Resource constraints in petroleum production potential. *Science* 253:146–152.

Olsson, R. K., Miller, K. G., Browning, J. V., Habib, D., and Sugarman, P. J. 1997. Ejecta layer at the Cretaceous-Tertiary boundary, Bass River, New Jersey (Ocean Drilling Program Leg 174AX). *Geology* 25:8, 759–762.

Power to the people: A survey of energy. 1994. *The Economist* 331:7868 (18-page insert following p. 60).

Ryder, G., Fastovsky, D., and Gartner, S., eds. 1996. *The Cretaceous-Tertiary event and other catastrophes in Earth history.* Geological Society of America Special Paper 307 (a compendium of 37 papers). Boulder, CO: Geological Society of America.

Sharpton, V. L., et al. 1992. New links between the Chicxlub impact structure and the Cretaceous/Tertiary boundary. *Nature* 359:6398, 819–821.

Stanley, D. J. 1990. Med desert theory is drying up. *Oceanus* 33:1, 14–23.

Sverdrup, H. U., Johnson, M. W., and Fleming, R. H. 1942. Renewal 1970. *The oceans: Their physics, chemistry, and general biology.* Englewood Cliffs, NJ: Prentice-Hall.

The Open University Course Team. 1989. *Ocean chemistry and deep-sea sediments.* Oxford: Pergamon Press.

Udden, J. A. 1898. Mechanical composition of wind deposits. *Augustana Library Pub. 1.*

Various authors. 1994. Special issue featuring several articles focusing on 25 years of ocean drilling. *Oceanus.* 36:4, 1–136.

Weaver, P. P. E., and Thomson, J., eds. 1987. *Geology and geochemistry of abyssal plains.* Palo Alto, CA: Blackwell Scientific.

Wentworth, C. K. 1922. A scale of grade and class terms for clastic sediments. *Journal of Geology* 30, 377–392.

Suggested Reading

Earth

Barnes-Svarney, P. 1992. Salt of the Earth. 1:1, 52–57. A look at the importance of salt, and how deposits of salt form.

Desonie, D. 1996. The threat from space. 5:4, 24–31. Examines the possibility of an extraterrestrial body colliding with Earth and what could be done about it.

Elliott, R. R. 1993. Paradise of form and color. 2:1, 64–69. Close-up photography exposes the beauty and variety of beach sands.

Grossman, D., and Shulman, S. 1994. Verdict at Yucca Mountain. 3:2, 54–63. An analysis of the safety of using Yucca Mountain, Nevada as a long-term repository site for high-level nuclear waste.

Harben, P. 1992. Strategic minerals. 1:4, 36–45. Examines the politics and importance of four important minerals, including manganese.

Sea Frontiers

Dietz, R. S. 1978. IFOs (Identified Flying Objects). 24:6, 341–346. The source of Australasian tektites and microtektites is discussed.

Dudley, W. 1982. The secret of the chalk. 28:6, 344–349. An informative discussion of the formation of marine chalk deposits.

Dugolinsky, B. K. 1979. Mystery of manganese nodules. 25:6, 364–369. The problems of origin, growth, and the environmental implications of manganese nodule mining.

Feazel, C. T. 1986. Asteroid impacts, sea-floor sediments, and extinction of the dinosaurs. 32:3, 169–178. A discussion of the evidence in marine sediments that may support the theory that the demise of the dinosaurs was caused by the impact of a large meteor.

———. 1989. Inner space: Porosity of seafloor sediments. 35:1, 49–52. A petroleum geologist discusses how porosity forms in marine sediments and the significance of the pores.

Mark, K. 1988. Ancient ocean rocks: High in the Sangre de Cristo Mountains. 34:1, 22–29. The history of an ancient ocean is revealed in the sediments of the mountains of Colorado and New Mexico.

Prager, E. J. 1988. Curious nodules on Florida's outer shelf. 34:6, 354–358. The origin of calcareous nodules found near Molasses Reef south of Key Largo is discussed.

Shinn, E. A. 1987. Sand castles from the past: Bahamian stromatolites discovered. 33:5, 334–343. The formation of stromatolites, sand domes trapped by algal growth, in the Bahamas is discussed.

Victory, J. J. 1973. Metals from the deep sea. 19:1, 28–33. The formation and economic potential of mining manganese nodules are discussed.

Wood, J. 1987. Shark Bay. 33:5, 324–333. The history of Shark Bay, Australia, since its discovery by Dirk Hartog in 1616 is summarized. The stromatolites and "tame" dolphins that attract tourists are also discussed.

Scientific American

Cambell, C. J., and Laherrère, J. H. 1998. The end of cheap oil. 278:3, 78–83. An updated look at the status of global petroleum reserves, including a forecast of when the world will run out of cheap oil.

Gehrelt, T. T. 1996. Collisions with comets and asteroids. 274:3, 54–61. The nature of asteroids and comets is reviewed. The age and size of some objects involved in major Earth-jarring hits by extraterrestrial bodies and the odds of hits by different size ranges of objects are discussed.

Hollister, C. D., and Nadis, S. 1998. Burial of radioactive waste under the seabed. 278:1, 60–65. Reexamines the merits of the disposal of nuclear waste by burying it within the sea floor.

Mack, W. N., and Leistikow, E. A. 1996. Sands of the world. 275:2, 62–67. A photoessay of sands of the world, showing a wide variety of beach sand material.

Nelson H., and Johnson, K. R. 1987. Whales and walruses as tillers of the sea floor. 256:2, 112–118. The 200,000 walruses and 16,000 gray whales that feed by scooping sediment from the Bering Sea continental shelf suspend large amounts of sediment that is transported by bottom currents.

Schneider, D. 1997. Not in my back yard: Could ocean mud trap nuclear waste from old Russian subs? 276:3, 20–22. A critical look at using the deep-ocean floor as a nuclear waste repository, with examples from accidental Russian submarine sinkings.

Oceanography on the Web

Visit the *Essentials of Oceanography* home page for on-line resources for this chapter. There you will find an on-line study guide with review exercises, and links to oceanography sites to further your exploration of the topics in this chapter. *Essentials of Oceanography* is at: **http://www.prenhall.com/thurman** (click on the Table of Contents menu and select this chapter).

CHAPTER 5
WATER AND SEAWATER

The HMS *Challenger* Expedition: Birth of Oceanography

The existence of oceanography as a scientific discipline is commonly considered to date from the HMS *Challenger* Expedition, which began in 1872. At that time, there were many unanswered questions about the oceans (as there are today). For instance, did life forms live in the deep-ocean? If so, what were the physical and chemical conditions there? What was the nature of deep-ocean deposits? These were the questions that helped stimulate interest in the first large-scale voyage with the express purpose of studying the ocean. Because knowledge is power, there was also interest in exploring the deep ocean for military and commercial interests.

In 1871, the Royal Society (the most prestigious scientific society of its day) recommended that funds be raised from the British government for an expedition to investigate the following subjects:

1. The physical conditions of the deep sea in the great ocean basins

2. The chemical composition of seawater at all depths in the ocean

3. The physical and chemical characteristics of the deposits of the sea floor and the nature of their origin

4. The distribution of organic life at all depths in the sea, as well as on the sea floor

Because the general public's enthusiasm for these types of scientific journeys was at an all-time high, the government agreed to sponsor the expedition. In 1872 the HMS *Challenger* (Figure 5A), a reserve warship, was refitted to explore the world's deep oceans and conduct scientific investigations.

A staff of six scientists under the direction of C. Wyville Thomson left England with great fanfare in December 1872 aboard the vessel HMS *Challenger*. During the course of the voyage, the ship would sail to a specific location in the ocean and begin its scientific work. John Young Buchanan, the expedition's chemist, held that oceanography started on the day of *Challenger's* Station I,

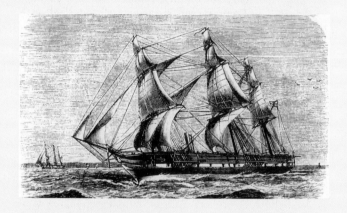

Figure 5A HMS *Challenger*.

when they first retrieved samples from the depths of the Atlantic Ocean south of the Canary Islands. At each station, the scientific party would:

- Measure the exact depth of the ocean using a sounding line
- Collect a sample of the bottom sediment
- Collect a sample of the bottom water
- Measure the temperature of the bottom (with newly developed thermometers that could withstand the high pressure)
- Record the atmospheric and meteorologic conditions

In addition, at most stations, the bottom was trawled for life with a net, organisms at the surface were collected, temperature measurements were made at various depths, and samples of seawater were collected at various depths. At a few stations, surface and deep current measurements were attempted.

Challenger returned in May 1876 after circumnavigating the world (Figure 5B). During the nearly three-and-a-half years that the expedition lasted, *Challenger* covered 127,500 kilometers (79,200 miles). Spending

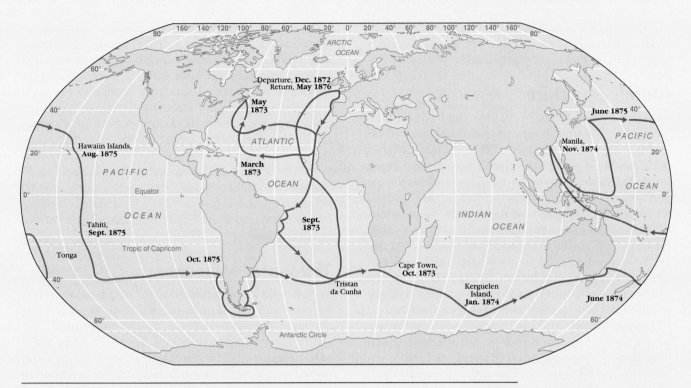

Figure 5B Route of the HMS *Challenger*.

most of their time in the Atlantic and Pacific Oceans, the scientists collected 492 deep-sea soundings, 133 bottom dredges, 151 open-water trawls, 263 water temperature observations, and collected water samples from as deep as 1830 meters (6000 feet). As with most oceanographic expeditions, the real work was just beginning. In fact, it took nearly 20 years to compile the results of the expedition into a total of 50 volumes.

The more outstanding aspects of the voyage were the netting and classification of 4717 new species of marine life, the measurement of a record water depth of 8185 meters (26,850 feet) from the Mariana Trench in the western North Pacific Ocean, the realization that the ocean floor was not flat but had significant relief, and the discovery of manganese nodules. The biologic objectives of the cruise were certainly met: The existence of life at all depths was confirmed. Truly, the expedition changed humankind's view of the ocean to one that included a third dimension: depth. In addition, pioneering work on the chemistry of the oceans was completed. Seventy-seven samples of ocean water collected during the *Challenger* expedition were analyzed in 1884 by chemist William Dittmar. The analysis of these samples revealed that the oceans were remarkably consistent in their chemical composition, even down to minor dissolved substances, including dissolved gases. Not only were the ratios between various salts virtually constant across the *surface* from ocean to ocean, but they were also distinctively constant at *depth*, establishing the "consistency of seawater" principle. This first refined analysis of the dissolved components in seawater established the *Principle of Constant Proportions* and contributed greatly to our understanding of ocean salinity.

At first you might not consider water to be an unusual substance. After all, it is one of the most common substances on Earth. Water most certainly is abundant, but it has unique properties that make it a very *unusual* substance. Generally, these properties include the arrangement of its atoms, how its molecules stick together, its ability to dissolve almost everything, and its heat storage capacity. Truly, it is one of the most remarkable substances on Earth.

Water is extremely important to all forms of life on Earth. The chemical properties of water are absolutely essential for the continuance of life, and all forms of life need water to survive. In fact, the primary component of all living organisms is water, with the water content of organisms ranging between about 65 and 95 percent (humans consist of about 65 percent water by weight). Water is the ideal medium to have within our bodies because it facilitates chemical reactions. Our blood, so important for transporting nutrients and removing wastes within our bodies, is 83 percent water. Life as we know it simply would not exist without water. As well, water controls the distribution of heat over Earth's surface, and thus controls Earth's climate. The very presence of water on our planet makes life possible, and the unusual properties of water make our planet livable.

Atomic Structure

Atoms (*a* = not, *tomos* = cut) are the basic building blocks of all matter. Every physical substance in our world is composed of atoms. For instance, a chair, a table, this book, your pencil, the air we breathe—everything, including people—is composed of atoms. An atom resembles a microscopic sphere (Figure 5–1) and was originally thought to be the smallest form of matter (hence the name, implying that it could be cut no further). Con-tinued study of atoms revealed that they are composed of even smaller particles, called subatomic particles.[1] As shown in Figure 5–1, the **nucleus** (*nucleos* = a little nut) of an atom is composed of **protons** (*protos* = first) and **neutrons** (*neutr* = neutral), which are bound together by strong forces. Protons have a positive electrical charge, and neutrons, true to their name, have no electrical charge. Both protons and neutrons have a mass of about one atomic mass unit, which is a very small mass. Orbiting at a distance from the nucleus are very small particles called **electrons** (*electr* = electric). Electrons have a negative electrical charge and about $1/2000$ the mass of either protons or neutrons. Held in their orbits by electrical attraction, electrons move around the nucleus at speeds approaching one-tenth the speed of light. Electrons are arranged in layers or shells around the nucleus.

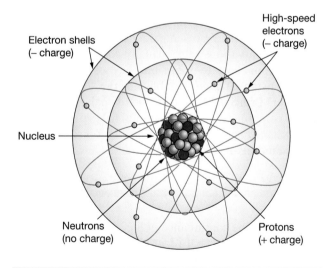

Figure 5–1 Simplified model of an atom.
An atom consists of a central nucleus composed of protons and neutrons that is encircled by electrons.

The overall electrical charge of atoms is balanced because each atom contains an equal number of protons and neutrons. For example, an oxygen atom has eight protons and eight electrons, with the electrical charges balancing each other. The oxygen atom also has eight neutrons, which do not affect its electrical charge because they are electrically neutral. Atoms are differentiated from one another by the number of protons (and the corresponding number of electrons) that they contain. For instance, a hydrogen atom has one proton, as compared to eight protons in the oxygen atom. In some cases, an atom will lose or gain one or more electrons and thus have an overall electrical charge. These atoms are called **ions**.

The Water Molecule

Most people know the chemical formula for water: H_2O. In fact, some people often order a drink of water by its chemical formula—and are widely understood. The chemical formula for water indicates that a water molecule is composed of two hydrogen atoms combined with one oxygen atom. A **molecule** is a group of two or more atoms that stick together by mutually shared electrons.[2] A molecule is the smallest form of a substance that can exist yet still retain the original properties of that substance. In a single drop of water, for instance, there are billions of water molecules.

Geometry

Atoms are represented as balls of various sizes corresponding to the size of atoms, with atoms containing more electrons having a larger size. In a water molecule, the oxygen atom is about twice the size of a hydrogen atom. The atoms in a water molecule join in such a manner that the hydrogen atoms are *not* on opposite sides of the oxygen atom (180 degrees apart). In essence, *both hydrogen atoms are on the same side of the oxygen atom* (Figure 5–2A). The two hydrogen atoms are separated by an angle of 105 degrees, which is related to the bonding sites available on the oxygen atom.

Sometimes, a water molecule is represented in a more compact form to show how tightly bonded the oxygen atom is to its two hydrogen atoms (Figure 5–2B). When a water molecule forms, **covalent bonds** are created between each hydrogen atom and the oxygen atom. The covalent bonds are created by the atoms sharing their electrons. These covalent bonds are relatively strong chemical bonds that require much energy to break. In other instances, the water molecule is represented with letter symbols, with the O representing the oxygen atom and each H representing a hydrogen atom (Figure 5–2C).

---

[1] It has recently been discovered that subatomic particles themselves are composed of even smaller particles, called quarks.

[2] When atoms combine with other atoms to form molecules, they share or trade electrons and establish chemical bonds.

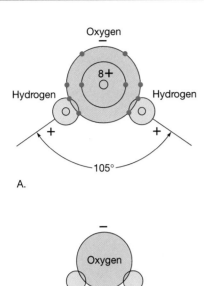

Figure 5–2 The water molecule.

A. Geometry of a water molecule. The oxygen end of the molecule is negatively charged, and the hydrogen regions exhibit a positive charge. Covalent bonds occur between the oxygen and the two hydrogen atoms **B.** A more compact representation of the water molecule. **C.** The water molecule represented by letters (H = hydrogen, O = oxygen).

Note that in all cases, the geometry of the water molecule is maintained, with the two hydrogen atoms on the same side of the oxygen atom.

Polarity

As the electrons are shared within the water molecule, the greater positive electrical charge of the eight protons contained in the nucleus of the oxygen atom provides a greater attraction for the negatively charged electrons. Conceptually, one might think of the oxygen atom as being "greedy" for the shared electrons. Thus, the shared electrons spend more of their time orbiting around the oxygen portion of the molecule. This results in the formation of a slight overall *negative charge* on the side of the oxygen atom *away from* the hydrogen atoms (Figure 5–2A). Conversely, a slight *positive charge* from the unshielded proton in each hydrogen atom nucleus is located on the side of the water molecule *containing* the two hydrogen atoms. This situation gives the entire molecule an electrical **polarity**, which is the property of having separated charges.

Since each water molecule has two electrical poles, water molecules are considered **dipolar** (*di* = two, *polar* = poles). Other common objects that are dipolar (often represented as having positive and negative ends) are a flashlight battery, a car battery, and a bar magnet. Although the electrical charges are weak, water molecules behave as if they contain a tiny bar magnet. The fact that water molecules have polarity is a direct result of the unusual geometry of water molecules. Many of water's unique properties are caused by the polarity of the water molecule.

Interconnections of Molecules

Imagine placing some small bar magnets together in a box. Because bar magnets have polarity, the bar magnets would begin to orient themselves relative to one another. The arrangement would be in such a way that the positive end of one bar magnet would be attracted to the negative end of another. Its positive end would be attracted to another's negative end, and so on. The end result would be that the bar magnets would all be interconnected, with positive and negative charges attracting each other.

The same is true for water molecules. Owing to their polar nature, water molecules are attracted to one another and form *intermolecular* bonds between adjacent water molecules. The positively charged hydrogen areas of each water molecule attract the negatively charged oxygen ends of neighboring water molecules. These bonds are referred to as **hydrogen bonds** (Figure 5–3). The molecules become bonded together by **electrostatic forces**, similar to the forces that hold clothes together when they come out of a dryer if no fabric softener was added. Hydrogen bonds that form *between* water molecules are

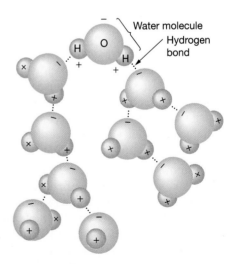

Figure 5–3 Hydrogen bonds.

Dashed lines indicate locations of hydrogen bonds, which occur between water molecules.

much weaker than the covalent bonds that occur *within* a water molecule. Thus, water molecules tend to stick to one another, exhibiting **cohesion**.

The cohesive properties of water cause water to "bead up" on a waxed surface, such as a freshly waxed car. Another example of water's cohesive properties can be observed when a glass is just a little *too* full with water. The water can be just above the brim of the glass without spilling it, demonstrating water's **surface tension**. Surface tension is the phenomenon in which a substance's surface molecules cling together, behaving like a weak membrane. In water, surface tension is formed as hydrogen bonds create a strong attraction between the outermost layer of water molecules and the underlying molecules. Without surface tension, water poured into a container would overflow as soon as the container's top was reached. Water is unique in that it has the highest surface tension of any liquid except the element mercury, which is the only metal that is a liquid at normal surface temperatures.

Water: The Universal Solvent

Not only does water tend to stick to itself, it also tends to stick to other chemical compounds that have polarity. When polar water molecules are bound together in groups (as in a cup of water), their ability to reduce the intensity of an electrical field acting on the water is greatly increased. Therefore, the electrostatic attraction between ions of opposite charges introduced into water is greatly reduced. In fact, this electrostatic attraction can be reduced to $1/80$ its value out of water.

Consider ordinary table salt, or sodium chloride (NaCl), a compound held together by **ionic bonds**.[3] An ionic bond is the electrostatic attraction that exists between ions that have opposite charges (Figure 5–4A). Simply by placing this solid substance in water, the electrostatic attraction (ionic bonding) between the sodium and chloride ions is reduced by 80 times. This sharply reduced attraction makes it much easier for the sodium ions and chloride ions to dissociate, or spread apart, which is the reason salt dissolves so easily in water. Once these ions dissociate, the positively charged sodium ions become attracted to the negative ends of the water molecules. The negatively charged chloride ions become attracted to the positive ends of the water molecules (Figure 5–4B).

As more and more ions of sodium and chloride are freed by the weakening of the electrostatic attraction that binds them together, they become surrounded by the polar molecules of water. This condition of being surrounded by water molecules is called **hydration**. The sodium ions are surrounded by water molecules whose negative ends are toward the positive sodium ion. The

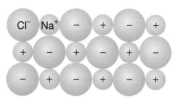

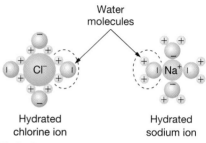

A. Sodium chloride, solid crystal structure

Water molecules

Hydrated chlorine ion

Hydrated sodium ion

B. Sodium chloride, in solution

Figure 5–4 Water as a solvent.
A. Table salt, composed of sodium chloride (Na$^+$ = sodium ion, Cl$^-$ = chlorine ion). **B.** As sodium chloride is dissolved, the positively charged end of the water molecule is attracted to the negatively charged Cl$^-$ ion, while the negatively charged end is attracted to the positively charged Na$^+$ ion.

chloride ions are surrounded by water molecules whose positively charged ends are toward the negative chloride ion.

The fact that water molecules are cohesive to other water molecules as well as to other polar molecules makes water able to dissolve nearly everything.[4] Given enough time, water can dissolve more substances and in greater quantity than any known substance. This has led to water being called "the universal solvent." This is why the ocean contains so much dissolved material and tastes "salty." In fact, dissolved substances comprise about $3^1/_2$ percent of the ocean's mass—an estimated 50 quadrillion tons of material.

Water's Thermal Properties

Water exists on Earth in many forms and is unusual in its capacity to store and release great amounts of heat energy. Water's unusual thermal properties are in part responsible for the development of tropical cyclones. As well, the thermal properties of water influence the heat budget of the world, the development of worldwide wind belts, and even ocean surface currents.

---

[3] Sodium is represented by the letters Na because the Latin term for sodium is *natrium*.

[4] If water is such as good solvent, then why doesn't oil mix with water? As you might have guessed, the chemical structure of oil is remarkably nonpolar. With no positive or negative ends to attract the polar water molecule, oil will not dissolve in water.

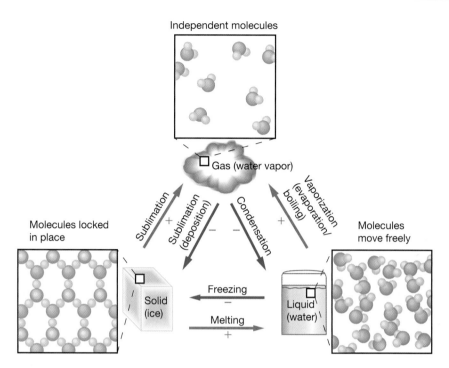

Figure 5–5 Water in the three states of matter.
Red arrows (+) indicate heat absorbed by water (which cools the environment); blue arrows (−) indicate heat released by water (which warms the environment).

Heat, Temperature, and Changes of State

Normally, matter can exist in three states: solid, liquid, and gas.[5] To change the state of a compound such as water, what must happen? Among all molecules of any compound, there exists a relatively weak attraction (intermolecular bond) called the **van der Waals force**, named for a Dutch scientist who first described the phenomenon. It becomes significant only when molecules are very close together, as in the solid and liquid states (but not the gaseous state).

The van der Waals force, which causes molecules of a substance to be attracted to one another, must be overcome if the state of the substance is to be changed from solid to liquid or from liquid to gas. To break this attraction, energy must be given to the molecules so that they can move faster to overcome this force. As energy is added to molecules, their motion increases.

What form of energy changes the state of matter? Very simply, it is *heat,* which is what causes ice cubes to melt or water to boil. Before proceeding, let's define some important terms: heat, calorie, and temperature.

- **Heat** *is the energy of moving molecules.* It is proportional to the energy level of molecules, and thus is the total kinetic energy[6] of a substance. Water is a solid, liquid, or gas depending on the number of calories of heat added. Heat may be generated by

combustion (a chemical reaction commonly called "burning"), through other chemical reactions, by friction, or from radioactivity.

- A **calorie** *is the amount of heat required to raise the temperature of 1 gram of water[7] by 1 degree centigrade.* To measure the amount of heat energy that is being added to or removed from molecules, calories are used. Note that these calories are different from food calories (such as those featured on nutritional labels), which are equal to 1000 of these calories.

- **Temperature** *is the direct measure of the average kinetic energy of the molecules that make up a substance.* Temperature change is a substance's response to an input or removal of heat energy. Since kinetic energy is energy of motion, the higher the temperature, the faster the molecules of the substance move.

Figure 5–5 shows the relationships of water molecules to one another in the solid, liquid, and gaseous states. In the *solid state* (ice), water has a rigid form and structure and does not flow. Bonds are constantly being broken and reformed. Nevertheless, the prevailing relation between molecules is one of rather firm attachment, produced by the nearness of the molecules, which causes the van der Waals force to be greater. In the solid state, the molecules vibrate with energy but remain in relatively fixed positions.

In the *liquid state* (water), water molecules are unstructured and can flow to take the shape of their container.

---

[5] Plasma, an electrically neutral, highly ionized gas composed of ions, electrons, and neutral particles, is a phase of matter distinct from solids, liquids, and normal gases.

[6] **Kinetic** (*kineto* = movement) **energy** is energy of motion.

[7] One gram of water is equal to about 10 drops.

When heat is added to melt ice, water molecules have gained enough energy to overcome many of the van der Waals forces that bound them together in the solid state. The molecules are free to move relative to one another, but are still attracted by one another. Bonds are being formed and broken at a much greater rate than in the solid state.

In the *gaseous state* (water vapor), molecules are unattached to one another and flow very freely, filling the volume of whatever container they are placed in. Molecules now are moving at random, and there exists no significant attraction among individual molecules. The only effect that one molecule may have on another results from collision during random movement.

Water's Freezing and Boiling Points

If enough heat energy is added continuously to a solid, it converts to a liquid (it melts). This happens at a temperature called the substance's **melting point**. Also, it will change from a liquid back to a solid at its **freezing point**, which is the same temperature (Figure 5–5). For water, this melting/freezing point is 0 degrees centigrade (32 degrees Fahrenheit).

If enough heat energy is added continuously to a liquid, it converts to a gas. This happens at a temperature called its **boiling point**. Also, it will change from a gas back to a liquid at its **condensation point**, which is the same temperature (Figure 5–5). For water, this boiling/condensation point is 100 degrees centigrade[8] (212 degrees Fahrenheit).

The fact that water freezes at 0 degrees centigrade (32 degrees Fahrenheit) and boils at 100 degrees centigrade (212 degrees Fahrenheit) demonstrates the great significance of the polarity of the water molecule and the hydrogen bond that this structure produces. The high freezing and boiling points of water are manifestations of the additional heat energy required to overcome two types of bonds to achieve a change of state: both the van der Waals forces and the hydrogen bonds.

It is interesting to compare how different the properties of water would be if water molecules were not polar. If *nonpolar* water followed the pattern of similar chemical compounds, it would freeze at −90 degrees centigrade (−130 degrees Fahrenheit) and would boil at −68 degrees centigrade (−90 degrees Fahrenheit). Note how much lower the temperature would be for water to remain in the liquid state, and how much more reduced the *range* of temperature under which liquid water would exist. If water molecules were nonpolar, all water on Earth would be in the gaseous state, severely impacting every life form on Earth.

Water's Heat Capacity

Another direct result of the cohesion between water molecules is the high heat capacity of water. Because of the great strength of the hydrogen bonds between water molecules, more heat energy must be added to accelerate the molecules and raise the water temperature than is required for substances in which the dominant intermolecular bond is the weaker van der Waals force. As noted, 1 calorie is the amount of heat required to raise the temperature of 1 gram of water by 1 degree centigrade.

Heat capacity *is the amount of heat that is required to raise the temperature of 1 gram of any substance by 1 degree centigrade.*[9] A substance that has a high heat capacity is one that has the ability to absorb (or lose) large quantities of heat without much change in temperature. A substance that heats up quickly when heat is applied has a low heat capacity (for instance, oil and all types of metals). The heat capacity is exactly 1 calorie for water and less than 1 calorie for other substances.

The heat capacity of water, compared with that of most other substances, is very high. In fact, water has the highest heat capacity of all common substances. This means that water gains or loses much more heat than other common substances while undergoing an equal temperature change. It also means that water resists any change in temperature. A practical example of this occurs when a large pot of water is boiled. A heat source is applied to the pot, which has low heat capacity and heats up quickly when heat is applied. However, the water inside the pot takes a long time to heat up (hence, the tale that a watched pot never boils but an *unwatched* pot boils over). Changing water to steam requires that the hydrogen bonds be broken, which accounts for the very large amount of heat needed. This exceptional capacity of water to absorb heat is used in cooking, home heating, industrial cooling, and in automobile cooling systems.

The difference in the heat capacities of ocean water and the rocks that make up the continents is strikingly illustrated in Figure 5–6. It shows the average difference between day and night temperatures. Throughout most of the ocean area, day and night temperatures vary by less than 1 degree centigrade (white areas), because the high heat capacity of water easily accommodates the daily heat gains from the sun and losses at night. By contrast, the much lower heat capacity of continental rocks, soil, vegetation, and air can cause day and night temperatures to fluctuate by as much as 30 degrees centigrade (86 degrees Fahrenheit). It is the high heat capacity of water that is responsible for the moderate climate of the coastal regions of the continents.

---

[8] Note that the temperature scale centigrade (*centi* = a hundred, *grad* = step) is based on 100 even divisions between the melting and boiling points of pure water.

[9] Note that the heat capacity of water is used as the unit of heat quantity, the calorie. Thus, water is the standard against which the heat capacities of other substances are compared.

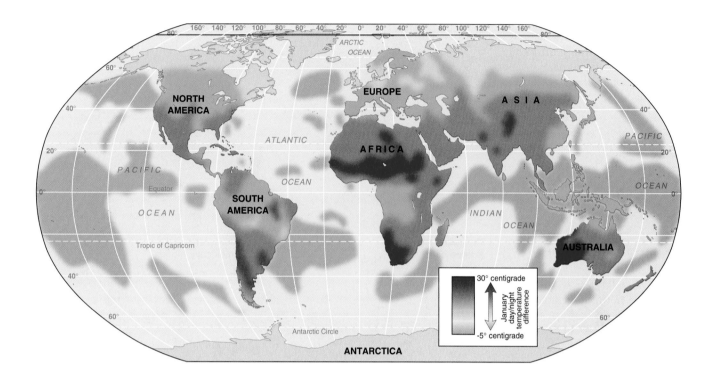

Figure 5–6 Day and night temperature differences.

This map shows the *difference* between day and night temperatures, averaged for the month of January 1979, and was constructed by subtracting 2:00 A.M. temperatures from 2:00 P.M. temperatures. The high heat capacity of water limits the day/night difference in the oceans to a minimal amount. Continental regions, with their much lower heat capacity, have much greater differences.

Water's Latent Heat of Melting

Closely related to water's unusually high heat capacity are its high latent (*latent* = hidden) heats of melting and vaporization. The term *latent* refers to the fact that heat is stored *invisibly* and *dormantly* in water. The only time the effect of this latent heat can be observed is when a change of state occurs—ice to water or vice versa, and water to water vapor or vice versa. During any of these changes of state, a remarkable amount of heat is either absorbed or released. For instance, as water evaporates from your skin, it cools your body by absorbing heat (this is the reason why sweating cools your body). Conversely, if you ever have been scalded by water vapor—steam—you know how much latent heat steam can release.

Figure 5–7 shows the amount of energy expended as water increases in temperature and changes state, and demonstrates how latent heat is involved during this process. Beginning with 1 gram of ice (lower left), the addition of 20 calories of heat raises its temperature by 40 degrees, from −40 degrees centigrade to 0 degrees

centigrade (point *a* on the graph). Once the temperature has been raised to 0 degrees centigrade (32 degrees Fahrenheit), it remains at that temperature even though more heat is continually being added. This is represented by the plateau on the graph between points *a* and *b*. The addition of more heat does not raise the temperature of the water (thus the heat is "hidden") until 80 more calories of heat energy have been added. There is no temperature increase during the addition of these 80 calories because all the heat energy goes to break the intermolecular bonds that structure the water molecules into ice crystals. The temperature remains unchanged until most of the bonds are broken and the mixture of ice and water has changed completely to 1 gram of water.

The amount of heat that must be added to 1 gram of a substance, at its melting point, to break the required bonds to complete the change of state from solid to liquid is the **latent heat of melting**. For water, the amount of heat required to convert 1 gram of ice to 1 gram of

Figure 5–7 Latent heats and changes of state of water.

The latent heat of melting (80 calories) is much less than the latent heat of vaporization (540 calories).

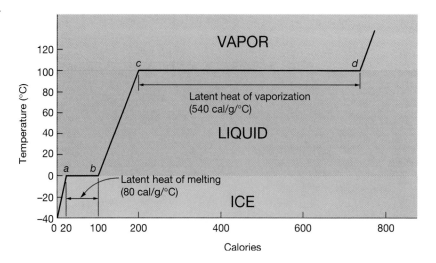

Figure 5–8 Hydrogen bonds in H$_2$O.

A. In the solid state, water exists as ice, in which there are hydrogen bonds between all water molecules. **B.** In the liquid state, there are some hydrogen bonds. **C.** In the gaseous state, there are no hydrogen bonds and the water molecules are moving rapidly and independently.

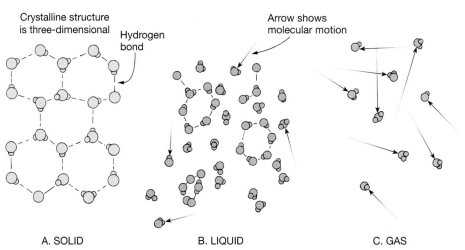

A. SOLID   B. LIQUID   C. GAS

water is 80 calories.[10] This latent heat of melting is greater for water than for any other common substance.

After the change from ice to liquid water has occurred at 0 degrees centigrade, the addition of heat to the water causes the temperature to rise again (between points *b* and *c* in Figure 5–7). As it does, note that it requires 1 calorie of heat to raise the temperature of water 1 degree centigrade (or 1.8 degrees Fahrenheit). Therefore, another 100 calories must be added before the gram of water reaches the boiling point of 100 degrees centigrade (212 degrees Fahrenheit). So far, a total of 200 calories have been added (point *c*).

Water's Latent Heats of Vaporization and Evaporation

At 100 degrees centigrade, another plateau is seen between points *c* and *d* in Figure 5–7. This plateau is far more prominent on the graph than the plateau for the latent heat of melting, which was 80 calories per gram. This 100-degree centigrade plateau represents the addi-

tion of *540 calories* to the gram of water. The amount of heat that must be added to 1 gram of a substance, at its boiling point, to break the required bonds to complete the change of state from liquid to vapor (gas) is the **latent heat of vaporization**. For water, the latent heat of vaporization is 540 calories.[11] This indicates that 540 calories of energy must be added before complete conversion to the vapor state occurs.

Why is so much more heat energy required to convert 1 gram of water to water vapor (540 calories) than was required to convert 1 gram of ice to water (80 calories)? Refer to Figure 5–8, recalling that in the gaseous state molecules are free from the influence of other molecules and therefore move randomly. To make the conversion from ice to water, not all the hydrogen bonds have to be broken—just enough to allow movement among the ice clusters and individual molecules. But to convert water to water vapor, every molecule must be freed from the attraction of all neighboring water molecules. Therefore, every hydrogen bond must be broken (Figure 5–8). This requires a much greater amount

---

[10] In terms of heat energy, 80 calories is $^1/_{1000}$ the food energy in 1 egg.

[11] In terms of heat energy, 540 calories is about $^1/_{1000}$ the food energy in a hot fudge sundae.

of heat energy: 540 calories per gram of water instead of 80 calories.

This example illustrates that large amounts of latent heat energy are stored in water, and that this heat energy is absorbed or released by water as it changes state—particularly if the change is from liquid to vapor or vice versa. It also demonstrates that water not only resists any change in temperature, it also resists any change of state.

Of course, boiling temperatures of 100 degrees centigrade do not occur at the surface of the ocean, where conversion of water to vapor occurs in nature. Sea-surface temperatures average 20 degrees centigrade (68 degrees Fahrenheit) or less. The conversion of a liquid to a gas below the boiling point is called **evaporation**. At ocean-surface temperatures, individual molecules that are being converted from the liquid to the gaseous state have a lower amount of energy than do the molecules of water at 100 degrees centigrade. Therefore, to gain the additional energy necessary to break free of the surrounding ocean water molecules, an individual molecule must capture heat energy from its neighbors.

This phenomenon explains the cooling effect of evaporation. The molecules left behind have lost heat energy to those that escape. To produce 1 gram of water vapor from the ocean surface at temperatures less than 100 degrees centigrade requires more than the 540 calories of heat that are required to make this conversion at the boil-

ing point. For instance, the **latent heat of evaporation** is 585 calories per gram at 20 degrees centigrade (68 degrees Fahrenheit). This higher value is due to the fact that more hydrogen bonds must be broken at this lower temperature for a gram of water molecules to enter the gaseous state. At higher temperatures, liquid water has fewer hydrogen bonds because the molecules are vibrating and jostling about more.

The evaporation–condensation cycle (and the great amount of latent heat energy exchanged) is quite important for life on Earth. The sun radiates energy to Earth, where some of it becomes stored in the oceans. Evaporation removes this heat energy from the oceans and carries it high into the atmosphere as the water vapor rises. In the cooler environment of the upper atmosphere, the water vapor condenses into clouds, releasing its latent heat energy. *It is water's latent heat of evaporation that accounts for the removal of great quantities of heat energy from the low latitude oceans by evaporation.* It is later released in the heat-deficient higher latitudes after the vapor is transported through the atmosphere and condenses there as precipitation, mostly as rain and snow (Figure 5–9). For every gram of water that condenses in cooler latitudes, the same amount of heat is released to warm these regions as was removed from the tropical ocean when that gram of water was evaporated to become water vapor. Water's unusual thermal properties

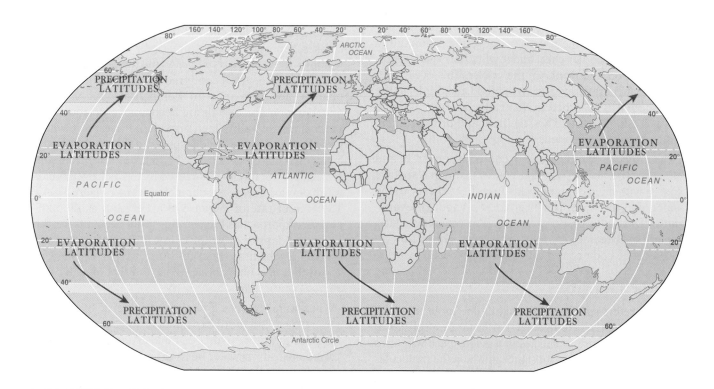

Figure 5–9 Atmospheric transport of surplus heat from low latitudes into heat-deficient high latitudes.

The heat removed from the tropical ocean (*evaporation latitudes*) is carried toward the poles and is released at higher latitudes through precipitation (*precipitation latitudes*).

have prevented large variations in Earth's temperature, and thus have served to moderate Earth's climate. Because rapid change is the enemy of all life, Earth's moderated climate is one of the main reasons that life exists on Earth.

Release of Heat by Water

We have looked at what happens when heat energy is added and temperature increases. Now let us examine what happens when heat energy is removed, and temperature falls. When water vapor is cooled sufficiently, it returns to a liquid state (an example of this can be seen when drops of water form on the *outside* of a glass containing a cold beverage). Thus, the water vapor **condenses** and heat is released into the surrounding air. On a small scale, the heat released is enough to cook food, as in a "steamer." On a large scale, this amount of heat is tremendous, and is sufficient to power such large-scale examples as thunderstorms and hurricanes.

Release of heat also occurs with the freezing of water, which is the change of state from liquid to solid. In Earth's polar regions, heat released by formation of sea ice is absorbed by the heat-deficient atmosphere, thus moderating the polar climate. Note that the *amount* of heat involved in the latent heat of vaporization and condensation is identical: Vaporization calories *must equal* condensation calories. Also, the *amount* of heat involved in the latent heat of melting and freezing is identical: Melting calories *must equal* freezing calories.

A practical application of these principles of heat transfer can be seen in the use of ice for refrigeration. Put a bag of ice in a cooler chest and it will lower the temperature of food and drinks in the cooler. This occurs because heat energy travels from the food and drinks to the ice, where it is absorbed to change the ice from its solid state to a liquid state. Thus, the food and drinks remain cold until all the ice melts.

The principle of using water to cool air is similar. In arid climates, the hot dry air that passes through a surface coated with liquid water quickly loses heat energy to the water, and the water is converted to a vapor (it evaporates). Thus, after passing across a water-covered surface, the hot air becomes considerably cooler. This is how an evaporative cooler (also called a swamp cooler) works.

Water Density

Density refers simply to how tightly the molecules of a substance are packed together. For instance, a substance that contains more molecules in a given space will have a higher density than a substance in which the molecules are not as tightly packed. As mentioned in Chapter 2, *density* is defined as mass per unit of volume, or how heavy something is for its size. Units of density are typically expressed in grams per cubic centimeter (g/cm^3), with pure water having a density of $1.0 \ g/cm^3$. Density of water is affected by its temperature, salinity, and pressure.

With most substances, it can be observed that "colder is denser." For example, cold air is dense and sinks. This density increase happens because the same number of molecules occupy less space as they cool down, lose energy, and molecular motion slows. This condition, called **thermal contraction**, also occurs in water, but only to a certain point. As water cools, it gets denser—as long as the temperature decrease occurs above 4 degrees centigrade (39 degrees Fahrenheit). As the temperature of water is lowered from 4 degrees to 0 degrees centigrade (32 degrees Fahrenheit), its density actually *decreases*. In other words, the water stops contracting and actually *expands*. This is highly unusual among Earth's many substances. The result is that ice takes up more volume than liquid water, causing ice to be less dense. Thus, *ice floats on water.* This may not seem like a revelation to most people, but it is just the opposite of most other substances: The solid state of most substances is denser than the liquid state, and thus the solid sinks. Why is ice less dense than water?

This anomalous behavior of frozen water can be explained only by considering water's molecular structure and its hydrogen bonds. Figure 5–10 shows the progression of molecular packing as water approaches its freezing point. From points *a* to *c* in the figure, the temperature of water is lowered from 20 degrees centigrade (68 degrees Fahrenheit) to 4 degrees centigrade (39 degrees Fahrenheit). This reduces the amount of thermal motion of the water molecules, resulting in the unbonded water molecules occupying less volume because of their decreased energy. Thus, the water contracts and density increases, represented by more water molecules in the windows shown in parts *a* to *c*. As the temperature is lowered below 4 degrees centigrade, this reduction in volume is not sufficient to compensate for another phenomenon: Ice crystals are becoming more abundant. Ice crystals are bulky, open, six-sided structures in which water molecules are widely spaced. Their characteristic hexagonal shape (Figure 5–11) mimics the hexagonal molecular structure that is caused by the hydrogen bonds which develop between water molecules (see Figure 5–8A). The greater rate of increase of ice crystals as the temperature approaches the freezing point accounts for the *decreased* density of water below 4 degrees centigrade (part *d*). By the time water fully freezes (part *e*), the density of the ice is only about 90 percent that of water at 4 degrees centigrade.

When water freezes, it becomes less dense by forming a more rigid and more expanded arrangement of molecules due to water's hydrogen bonds. Associated with this lowered density as water turns to ice is an increase in volume of about 9 percent. Anyone who has put a beverage in a freezer for "just a few minutes" to cool it down and inadvertently forgotten about it has experienced the volume increase associated with water's expansion as it freezes—usually resulting in a burst beverage container. A more serious problem occurs in colder climates when

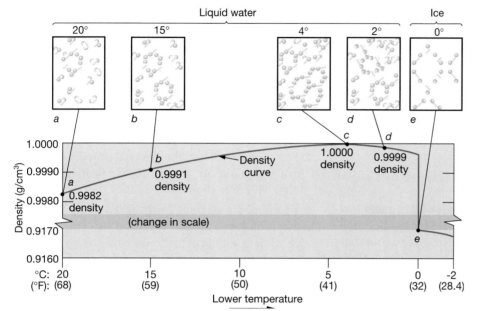

Figure 5–10 The formation of ice.

The formation of ice clusters in fresh water, showing how water reaches its maximum density at 4°C. Below 4°C, water becomes less dense as ice begins to form. At 0°C, ice forms, the crystal structure expands, and density decreases, causing ice to float.

Figure 5–11 Snowflakes.
Hexagonal snowflakes indicate the internal structure of water molecules held together by hydrogen bonds.

water pipes freeze, causing them to crack. The force of ice expansion is so powerful that it can break apart rocks and split pavement on roads and sidewalks.

As can be seen on Figure 5–10 at point *c*, water attains its maximum density at 4 degrees centigrade (39.2 degrees Fahrenheit). Water at this temperature is denser than either warmer or colder water and thus will sink, creating vertical currents in the ocean. Conversely, water at either higher or lower temperatures will float, also creating vertical currents. Density is of great importance in considering the movement of water in the ocean because water of greater density sinks, whereas water of lower density rises.

The temperature of maximum density for fresh water, 4 degrees centigrade (39.2 degrees Fahrenheit), can be lowered by two means. Increasing the pressure on water will lower the temperature of maximum density. Adding dissolved substances will also lower the temperature of maximum density, because the salts inhibit the formation

of hydrogen bonds. In both cases, the formation of bulky ice crystals is inhibited. Thus, to produce crystals equal in volume to those that could be produced at 4 degrees centigrade (39.2 degrees Fahrenheit) in fresh water, more energy must be removed, causing a reduction in the temperature of maximum density.

This interference with the formation of ice crystals also produces a progressively lower freezing point for water as more dissolved solids are added. This is why most seawater never freezes, except near Earth's frigid poles (and even then, only at the surface). In cold climates on land, a similar strategy is employed by spreading salt on roads and sidewalks during the winter. The addition of salt lowers the freezing point of water, thus creating ice-free passageways at temperatures that are several degrees below freezing.

Salinity of Ocean Water

So far in this chapter, we have looked at the unusual properties of water but have not yet considered seawater. What is the difference between pure water and seawater? One of the most obvious differences is that seawater is salty. Even though the dissolved substances in seawater constitute only a small proportion, one can certainly taste the distinct saltiness of seawater. However, these dissolved substances are not simply sodium chloride (table salt)—they include various salts,[12] metals, and

---

[12] Chemically, a salt is a compound formed by the replacement of the hydrogen ion (H^+) of an acid with some positively charged chemical unit, usually a metal, such as sodium. For example, most of the "salt" in the ocean results from replacing the hydrogen ion in hydrochloric acid (HCl) with sodium ions (Na^+), to form NaCl, or sodium chloride.

even dissolved gases. The salt content of seawater is unfortunate: We can't drink it, it is unsuitable for irrigating most crops, and it is highly corrosive to many synthetic materials. Some scientists estimate that the oceans contain as much as 50 quadrillion tons (50 million billion tons) of salt—enough to cover the entire planet with a layer of salt more than 150 meters (500 feet) thick, about the height of a 50-story skyscraper.

Salinity

Salinity is the total amount of solid material that is dissolved in water. Note that the material must have a *solid* state, and thus includes dissolved gases, which, under low enough temperatures, do exist as solids (for example, the solid state of CO_2 is known as dry ice). Also, note that the material must be *dissolved* in water, as opposed to fine particles being held in suspension or solid material in contact with water. Lastly, note that salinity applies to *all water,* not just seawater. One way to think about salinity is that salinity is the *ratio of the mass of dissolved substances to the mass of the water sample.*

Seawater typically contains 3.5 percent salinity, which is about 220 times saltier than fresh lake water. A salinity of 3.5 percent also indicates that seawater typically contains 96.5 percent pure water molecules (Figure 5–12). The high concentration of pure water in seawater is the controlling factor that influences the physical properties of seawater. Thus, the physical properties of seawater are remarkably similar to the properties of pure water, with only slight variations.

Six elements account for over 99 percent of the dissolved solids in seawater (Figure 5–12). These are chlorine, sodium, sulfur (in the form of the sulfate ion, SO_4^{-2}), magnesium, calcium, and potassium. At least 89 chemical elements have been identified in seawater, most in extremely small amounts. Probably all of Earth's natu-

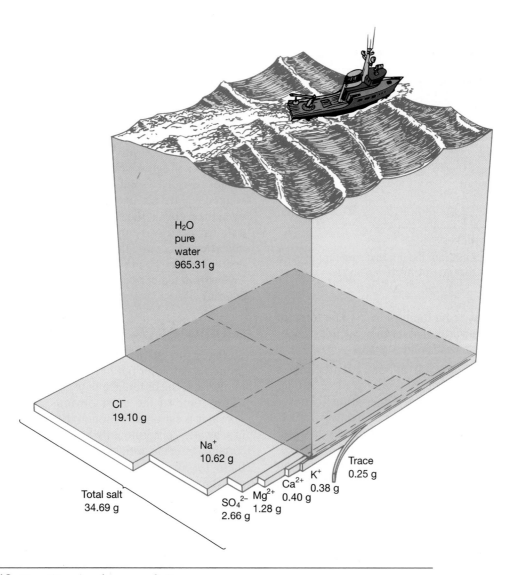

Figure 5–12 Constituents of ocean salinity.
The composition of 1 kilogram of ocean water in grams (g), showing the concentrations of various dissolved substances. Average seawater salinity is approximately 35‰ or 3.5%.

Table 5–1   Selected dissolved materials in 35‰ seawater

Major constituents (in parts per thousand, ‰)		
Constituent	Concentration (‰)	Ratio of constituent/total salts (%)
Chloride (Cl^-)	19.3	55.04
Sodium (Na^+)	10.7	30.61
Sulfate (SO_4^{2-})	2.7	7.68
Magnesium (Mg^{2+})	1.3	3.69
Calcium (Ca^{2+})	0.41	1.16
Potassium (K^+)	0.38	1.10
Total	34.79‰	99.28%

Minor constituents (in parts per million, ppm[a])	
Constituent	Concentration (ppm)
Gases	
Carbon dioxide (CO_2)	90
Nitrogen (N_2)	14
Oxygen (O_2)	5
Nutrients	
Silicon (Si)	3.0
Nitrogen (N)	0.5
Phosphorous (P)	0.07
Iron (Fe)	0.002
Others	
Bromine (Br)	65.0
Carbon (C)	28.0
Strontium (Sr)	8.0
Boron (B)	4.6

Trace elements (in parts per billion, ppb[b])	
Constituent	Concentration (ppb)
Iodine (I)	60
Manganese (Mn)	2
Lead (Pb)	0.03
Mercury (Hg)	0.03
Gold (Au)	0.005

[a] Note that 1 ppm equals 1000‰.
[b] Note that 1 ppb equals 1000 ppm.

rally occurring elements exist in the sea. The concentrations of major constituents plus selected minor constituents and trace elements are listed in Table 5–1.

Salinity is often expressed using the quantity **parts per thousand (‰)**. Just as 1 percent (1%) is one part in 100, one part per thousand (1‰) is one part in 1000. When converting from percent to parts per thousand, the decimal is simply moved over one place to the right. For instance, typical seawater salinity of 3.5 percent is equivalent to 35‰. Instead of reporting seawater salinities in percent, salinity is usually reported in parts per thousand, which often has the advantage of avoiding decimals.

Box 5–1
How to Avoid Goiters

Many who have read the nutritional label on salt containers have noticed that the label often proudly proclaims "this product contains iodine, a necessary nutrient." Why is it necessary to have iodine in our diets? The answer is that if a person does not receive a sufficient amount of iodine in his or her diet, a potentially life-threatening disease called **goiters** may result.

Although not a problem in many developed nations like the United States, goiters are a serious health hazard in many underdeveloped nations, especially those far from the sea. Our bodies need trace amounts of certain elements to enable our metabolism to function properly, and one of those trace elements is iodine. Iodine is used by the thyroid gland, which is a butterfly-shaped organ located in the neck in front of and on either side of the trachea (windpipe). The thyroid gland manufactures hormones that regulate cellular metabolism that is essential for mental development and physical growth. If people lack iodine in their diet, their thyroid glands cannot function properly. Often, this results in the enlargement or swelling of the thyroid gland. Severe symptoms include dry skin, loss of hair, puffy face, weakness of muscles, weight

increase, diminished vigor, mental sluggishness, and a large nodular growth on the neck called a goiter (Figure 5C). If proper steps are not taken to correct this disease, it can lead to cancer. Iodine ingested regularly often begins to reverse the effects. In advanced stages, surgery to remove the goiter or exposure to radioactivity is the only course of action.

How can you avoid goiters? Fortunately, goiters are preventable by a diet with just *trace amounts* of iodine. Where can you get iodine in your diet? All products from the sea contain trace amounts of iodine because iodine is one of the many elements that is dissolved in seawater. Sea salt, seafood, seaweed, and other sea products contain enough iodine to help prevent goiters. In the United States, we get plenty of iodized salt in our diets, so goiters are quite uncommon. In fact, the reverse situation is much more prevalent in the United States: too much iodine in our diets, creating the overproduction of hormones by the thyroid gland. That's why most stores that sell iodized salt also carry noniodized salt for those people who have a hyperthyroid condition and must restrict their intake of iodine.

Figure 5C Goiters.

Salinity Variations

From place to place, salinity varies within the oceans. Salinity in the open ocean (far from land) varies between 33 and 38‰. In some coastal areas, salinity variations can be extreme. For example, in the Baltic Sea, salinity averages 10‰. Here, physical conditions create low-salinity water called **brackish** water. Brackish water is produced in areas where fresh water (from rivers and high rainfall) and seawater mix.

In the Red Sea, salinity averages 42‰, which is caused by physical conditions that produce **hypersaline** water. Hypersaline water is typical of seas and inland bodies of water that experience high evaporation rates and limited open-ocean circulation. Some of the most hypersaline water in the world is found in inland lakes, where their high salinity results in them being called seas. The Great Salt Lake in Utah has a salinity of 280‰, and the Dead Sea on the border of Israel and Jordan has a salinity of 330‰. The water in the Dead Sea contains 33 percent dissolved solids and is almost *10 times saltier than seawater*. This produces some interesting properties: The hypersaline conditions causes the water to be so dense that one can easily float in the water—with arms and legs sticking up above water level! As you might imagine, the water also tastes much saltier than seawater.

Salinity of seawater in coastal areas also varies seasonally. For example, the salinity of seawater off Miami Beach, Florida, varies from about 34.8‰ in October to 36.4‰ in May and June when evaporation is high. Offshore of Astoria, Oregon, seawater salinity is always relatively low because of the vast fresh water input from the Columbia River. Here, seawater salinity varies from 0.3‰ in April and May (when the Columbia river is at its maximum flow rate) to 2.6‰ in October (the dry season).

Other types of water have much lower salinity. Tap water contains dissolved substances, but at concentrations 50–100 times lower than seawater. Typical tap water has salinity somewhere below 0.8‰, and good-tasting tap water is below 0.6‰. Salinity of premium bottled water is on the order of 0.3‰, with the salinity often displayed prominently on its label (usually as total dissolved solids (TDS) in units of parts per million (ppm), where 1000 ppm is equal to 1‰).

Determining Salinity

One of the first ways devised to determine the salinity of water is to evaporate a carefully weighed amount of seawater completely and weigh the precipitate. However, the accuracy of this method is limited and it is a time-consuming process.

Another way to measure salinity is to take advantage of the fact that the ratio of major elements that make up the dissolved components is remarkably constant worldwide. This *Principle of Constant Proportions* was firmly established by William Dittmar in his analysis of the *Challenger* Expedition water samples. The **Principle of Constant Proportions** states that *the major dissolved constituents that comprise the salinity of seawater occur nearly everywhere in the ocean in the exact same proportions, independent of salinity.* Essentially, this implies that the ocean is well mixed, and that when salinity changes, the salts don't leave (or enter) the ocean but water molecules do. Because of seawater's constancy of composition, the concentration of *a single* major constituent can be measured to determine the *total salinity* of a given water sample. The constituent that occurs in the greatest abundance and is the easiest to measure accurately is the chloride ion, Cl^-. The weight of this ion in a water sample is called **chlorinity**.

In any sample of ocean water worldwide, the chloride ion always accounts for 55.04 percent of the total proportion of dissolved solids (Table 5–1). Therefore, by measuring only the chloride ion concentration, the total

salinity of a seawater sample can be determined by the following relationship:

$$\text{Salinity (‰)} = 1.80655 \times \text{chlorinity (‰)}$$

For example, given the average chlorinity of the ocean, which is 19.2‰, the ocean's average salinity[13] is 1.80655 × 19.2‰, which rounds to 34.7‰. In other words, on average there are 34.7 parts dissolved material in every 1000 parts of seawater.

Standard seawater consists of ocean water analyzed for chloride ion content to the nearest ten-thousandth of a part per thousand by the Institute of Oceanographic Services in Wormly, England. It is then sealed in small glass vials called ampules and sent to laboratories throughout the world for use as a reference standard in calibrating analytical equipment.

The chloride content in seawater can be measured very accurately, but this requires much time and care. Fortunately, with advanced oceanographic instruments, this task has been greatly simplified. Measuring the **electrical conductivity** (the ability of a substance to transmit electric current) of water is now the most common method used to measure seawater salinity. Oceanographers know that the electrical conductivity of ocean water increases with increased salinity. The more dissolved substances there are in water, the better the water transmits electric current (Figure 5–13). Thus, salinity can be determined by measuring the conductivity of ocean water by passing a weak electric current through the water. An instrument capable of making such a measurement is called a **salinometer**. Very accurate salinity determinations can be made with salinometers to resolutions of better than 0.003‰.

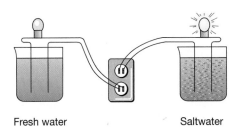

Fresh water   Saltwater

Figure 5–13 Salinity affects water conductivity.
Increasing the amount of dissolved substances increases the conductivity of the water, allowing the water to transmit electrical current more easily. This can be demonstrated by using a light bulb with bare electrodes dipped in waters of different salinities: The higher the salinity, the more electricity is transmitted, and more brightly lit the bulb will be.

---

[13] The number 1.80655 comes from dividing 1 by 0.5044 (the chloride ion's proportion in seawater of 55.04 percent). However, if you actually divide this, you will get 1.81686. Oceanographers have agreed to use 1.80655 because the constancy of composition has been found to be an approximation.

Dissolved Components Added and Removed from Seawater

Salts do not remain in the ocean forever but are cycled into and out of seawater by various processes (Figure 5–14). The majority of dissolved components that contribute to the salinity of seawater are continually added to the ocean from two main sources. One of these sources is streams that dissolve the ions from continental rocks and carry them to the sea. The other source is volcanic eruptions, both on the land and on the sea floor. Other sources include the atmosphere (which contributes gases that are dissolved in seawater) and biologic interactions.

The primary method in which dissolved substances are added to the oceans is from stream runoff. Table 5–2 shows the major dissolved components contained in stream runoff. The low concentrations of dissolved substances is why "fresh" water does not taste salty. Comparing Table 5–1 with Table 5–2, it is clear that streams have far lower salinity, and that rivers have a vastly different composition of dissolved substances than seawater. For example, although calcium (Ca^{2+}) is a major constituent in streams, its concentration is slight

Table 5–2   Major constituents of dissolved materials in streams

Constituent	Concentration (parts per million)
Carbonate ion (HCO_3^-)	58.4
Calcium ion (Ca^{2+})	15.0
Silicate (SiO_2)	13.1
Sulfate ion (SO_4^{2-})	11.2
Chloride ion (Cl^-)	7.8
Sodium ion (Na^+)	6.3
Magnesium ion (Mg^{2+})	4.1
Potassium ion (K^+)	2.3
Total	119.2[a]

[a] Note that 119.2 ppm equals 0.1192‰.

compared with seawater, where its concentration is almost 30 times greater.

Salts do remain in the ocean for considerable lengths of time, called their **residence time**, which is the average length of time that an atom of an element resides in the ocean.

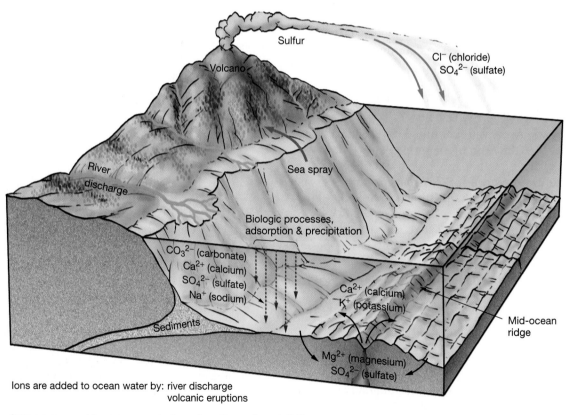

Ions are added to ocean water by: river discharge
volcanic eruptions

Ions are removed from ocean water by: adsorption and precipitation
sea spray
biologic processes
mid-ocean ridge infiltration

Figure 5–14 The cycling of dissolved components in seawater.
The dissolved components of seawater are cycled into and out of seawater by various processes.

Long residence times lead to higher concentrations of dissolved material. For many of the dissolved components, their residence time is on the order of several thousand to several million years. For instance, the ion sodium (Na^+) has a residence time in the ocean of 260 million years.

Even though various salts have long residence times in the ocean, the ocean is not becoming saltier through time because the rate at which an element is added to the ocean is equal to the rate at which it is being removed. Salts are removed by several processes that cycle dissolved substances out of seawater. When waves break at sea, salt spray releases many tiny salt particles into the atmosphere, where they may be blown over land before being washed back to Earth by precipitation. Along mid-ocean ridges, the infiltration of seawater near hydrothermal vents causes magnesium and sulfate ions to become incorporated into sea-floor mineral deposits. These hydrothermal vents directly and actively influence the chemical composition of seawater. It has been estimated that *the entire volume of ocean water* is recycled through this hydrothermal circulation system at the mid-ocean ridge every 3 million years. Therefore, the chemical exchange between ocean water and the basaltic crust has a major influence on the nature of ocean water.

Dissolved substances are also removed from seawater in other ways. Calcium, sulfate, sodium, and silicon are deposited in ocean sediments within the shells of dead microscopic organisms and animal feces. Vast amounts of dissolved substances can be removed when inland arms of the sea dry up, leaving salt deposits. In addition, ions dissolved in ocean water are also removed by adsorption (physical attachment) to the surfaces of sinking clay and biologic particles.

Processes Affecting Seawater Salinity

One way to affect seawater salinity is to change the amount of dissolved substances within the water. Another way to affect seawater salinity is to *change the amount of water molecules,* since salinity is the ratio between the mass of dissolved components to the mass of the water sample. For instance, adding more water molecules to seawater dilutes the dissolved components, thus lowering the salinity of the sample. Conversely, removing water molecules from seawater increases salinity. Note that changing the salinity in these ways does not affect the *amount* or the *composition* of the dissolved components, which remain in constant proportions to one another.

Processes That Decrease Seawater Salinity

Several processes *decrease* seawater salinity by adding additional water molecules to seawater (Table 5–3). **Precipitation** is the process by which atmospheric water

Table 5–3 Processes affecting seawater salinity.

Process	How accomplished	Adds or removes	Effect on salt in seawater	Effect on H_2O in seawater	Salinity increase or decrease?	Source of fresh water from the sea?
Precipitation	Rain, sleet, hail, or snow falls directly on the ocean	Adds very fresh water	None	More H_2O	Decrease	—
Runoff	Streams carry water to the ocean	Adds mostly fresh water	Negligible addition of salt	More H_2O	Decrease	—
Icebergs melting	Glacial ice calves into the ocean and melts	Adds very fresh water	None	More H_2O	Decrease	Yes, icebergs from Antarctic have been towed to South America
Sea ice melting	Sea ice melts in the ocean	Adds mostly fresh water and some salt	Adds a small amount of salt	More H_2O	Decrease	Yes, sea ice can be melted and is better than drinking seawater
Sea ice forming	Seawater freezes in cold ocean areas	Removes mostly fresh water	30% of salts in seawater are retained in ice	Less H_2O	Increase	Yes, through multiple freezings, called *freeze separation*
Evaporation	Seawater evaporates in hot climates	Removes very pure water	None (essentially all salts are left behind)	Less H_2O	Increase	Yes, through evaporation of seawater and condensation of water vapor, called *distillation*

returns to Earth as rain, snow, sleet, and hail. Of all precipitation, the majority (76 percent) falls directly back into the oceans and 24 percent falls onto land. Precipitation falling directly into the oceans adds water molecules to seawater, thus reducing seawater salinity.

Most of the precipitation that falls on land returns to the oceans directly as **runoff** in streams. Even though this water dissolves minerals on land, the runoff is composed of relatively pure water (see Table 5–2). In a similar fashion to precipitation, more water molecules are added to seawater and thus seawater salinity is decreased.

Glaciers are bodies of ice that originate on land and flow downhill. Some glaciers flow so far that they enter an ocean or marginal sea. When these glaciers melt, large chunks of ice break off in spectacular fashion by the process of **calving**. The calving of glacial ice produces **icebergs**. Once in the ocean, glacial ice melts. Since glacial ice was produced by snowfall in high mountain accumulation areas, glacial ice is composed of fresh water. Consequently, the melting of icebergs in the ocean also adds water molecules to seawater and thus seawater salinity is decreased.

Sea ice forms when ocean water freezes in high-latitude regions. As will be explained shortly, sea ice is composed primarily of fresh water. When warmer summer temperatures return to high-latitude regions, sea ice melts in the ocean. Sea ice melting adds mostly fresh water with a small amount of salt to the ocean. This results in adding more water molecules (and a slight amount of salt) to seawater. Thus, seawater salinity is decreased.

Processes That Increase Seawater Salinity

The formation of sea ice and evaporation are processes that *increase* seawater salinity by removing water molecules from seawater (Table 5–3). When seawater freezes, sea ice forms. The salinity of seawater influences the temperature of sea ice formation and how the sea ice is created. Depending on the salinity of seawater and the rate of ice formation, about 30 percent of the dissolved components in seawater are retained in sea ice. For example, 35‰ seawater will create sea ice with about 10‰ salinity (30 percent of 35‰ is 10‰). Consequently, the formation of sea ice removes mostly fresh water from seawater, increasing the salinity of the remaining unfrozen water. This high-salinity water also has a high density, which causes the water to sink below the surface.

Evaporation is the transfer of water molecules from the liquid state to the vapor state at temperatures below the boiling point. Water is removed from the liquid reservoirs on Earth (the ocean, streams, and lakes) by evaporation. Evaporation occurs most rapidly in warm climates. During evaporation of seawater, water molecules are removed but the dissolved substances remain behind, causing the salinity of seawater to increase.

The Hydrologic Cycle

The processes that affect seawater salinity are related by the **hydrologic cycle** (Figure 5–15), which is also known as the **water cycle**. These processes cause water to be recycled among the ocean, the atmosphere, and the continents. Earth's water supply exists in these proportions:

- 97 percent in the world ocean
- 2 percent frozen in glaciers and ice caps
- 0.6 percent in groundwater and soil moisture
- 0.02 percent in streams and lakes
- 0.001 percent as water vapor in the atmosphere

Water passes freely between the different locations where it is found, known as *reservoirs*. Thus, the hydrologic cycle implies that water is in continual motion between the various reservoirs.

Surface and Depth Salinity Variation

Average seawater salinity is 35‰, but it does vary significantly from place to place at the surface and also with depth in the ocean. Let's first examine surface salinity variation, and then turn our attention to how salinity varies with depth.

Surface Salinity Variation

Figure 5–16 shows how salinity varies at the surface, based on measurement of surface salinity by ships. The red curve shows temperature, which is shaped as might be expected: Temperatures are coldest at high latitudes and warmest near the Equator. The green curve shows salinity, which indicates that the lowest salinity areas are in high-latitude regions. Conversely, the highest salinity occurs in low-latitude regions near the Tropic of Cancer and the Tropic of Capricorn, respectively.[14] At the Equator, there is an interesting dip in the curve, representing that surface salinity is slightly *lower* than just to the north or south.

Why does surface salinity have this pattern? Processes affecting seawater salinity hold the key to understanding why the curve looks the way it does. At high latitudes, surface salinity is affected by abundant precipitation and runoff, which decreases salinity. The melting of freshwater icebergs further decreases salinity. In addition, the cool temperatures experienced there limit the amount of evaporation that takes place (which would increase salinity). Interestingly, the formation and melting of sea ice balance each other out in the course of a year and so are not a factor.

At lower latitudes, near the Tropic of Cancer and the Tropic of Capricorn, surface salinity is at its highest. This

---

[14] The region between the Tropic of Cancer and the Tropic of Capricorn is known appropriately enough as the tropics.

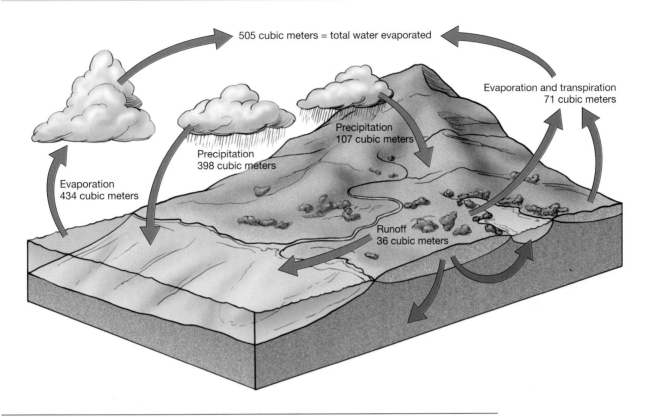

Figure 5–15 The hydrologic cycle.

All water is in continual motion between various reservoirs (ice not shown). Volumes shown are average yearly amounts.

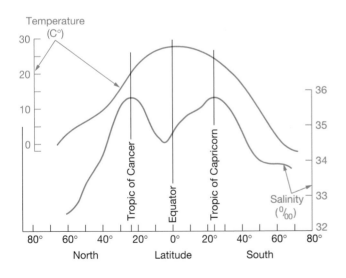

Figure 5–16 Surface salinity variation.

Surface seawater temperature (red curve) is lowest at the poles and highest at the Equator. Surface seawater salinity (green curve) is lowest at the poles, peaks at the Tropics of Cancer and Capricorn, and dips at the Equator.

is because warm dry air descends in these areas (as will be explained in Chapter 6, "Air-Sea Interaction"). This results in high evaporation rates that cause increased salinity. Additionally, there is little precipitation and runoff in these arid regions. Truly, the regions near the Tropic of Cancer and the Tropic of Capricorn are the continental *and* maritime deserts of the world.

Near the Equator, there is an unexpected dip in the surface salinity curve. After all, one would expect that the warm temperatures found at the tropics would also be found at the Equator. This is true, so the rate of evaporation remains high. Partially offsetting this, however, is an increase in precipitation (and, where near land, runoff). Daily thunderstorms are common along the Equator, thus adding water to seawater and lowering its salinity.

The pattern of surface salinity variation worldwide is shown in Figure 5–17, where the salinity pattern illustrated in Figure 5–16 can be seen.

Depth Salinity Variation

Figure 5–18 shows how seawater salinity varies with depth, based on lowering a salinometer probe from a ship on the surface into deep water. The graph shows two curves: one for low-latitude regions, and one for high-latitude regions.

For low-latitude regions, the curve begins at the surface with relatively high salinity (as was discussed in the preceding section; even at the Equator, surface salinity is still relatively high). With increasing depth, however, the curve swings toward an intermediate value of salinity, between high and low salinity.

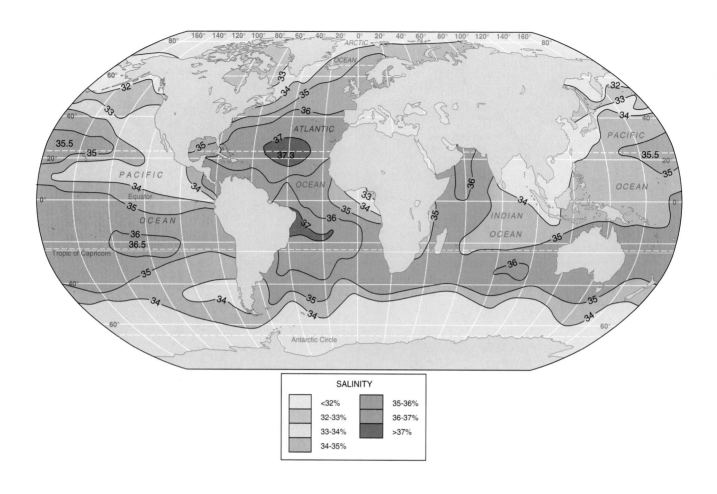

Figure 5–17 Surface salinity of the oceans.

Average seawater surface salinity shown for August. The highest surface salinities occur in the tropics, where descending dry air results in a high rate of evaporation and little precipitation. Values in parts per thousand (‰).

For high-latitude regions, the curve begins at the surface with relatively low salinity (again, see the discussion in the preceding section). With increasing depth, the curve also swings toward an intermediate value of salinity, approaching the value of the low-latitude salinity curve at depth.

These two curves, which together resemble the outline of a wine glass, show that at the surface there is a wide range of salinity variation. For instance, surface seawater salinities fall anywhere between the low-latitude and high-latitude values used in the example. At depth, however, the curves nearly merge, indicating that there is a much narrower range of salinity values in the deep ocean. Why is this so? Again, consider the processes affecting seawater salinity. The processes that affect seawater salinity (such as precipitation, runoff, icebergs or sea ice melting, sea ice forming and evaporation) are all processes that occur *at the surface*. Consequently, deep water below is not affected by these processes. This helps explain why there is such a wide variation of sea-

water salinity at the surface and such little variation in salinity at depth.

Halocline Another interesting aspect of Figure 5–18 involves the region of the curves between the depths of about 300 meters (980 feet) and 1000 meters (3300 feet). Between these depths, both curves on the graph show a rapid change in salinity. Note that for the low-latitude curve, this change is represented as a *decrease* in salinity. Conversely, for the high-latitude curve, this change is represented as an *increase* in salinity. In both cases, this *zone of rapidly changing salinity* is called a **halocline** (*halo* = salt, *cline* = slope). Haloclines are common within the ocean and represent different salinity conditions above and below the halocline.

Seawater Density

The density of pure water is 1.000 gram per cubic centimeter (g/cm³) at 4 degrees centigrade (39 degrees Fahrenheit), which serves as a standard against which

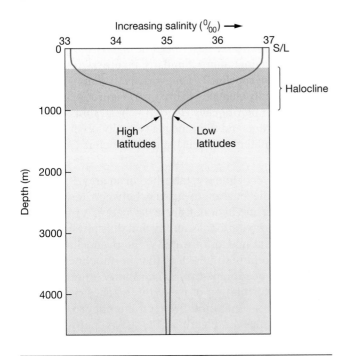

Figure 5–18 **Salinity variation with depth.**
Graph showing salinity variation (horizontal scale, in ‰) with depth (vertical scale, in km) for high- and low-latitude regions. Sea level is at the top; the vertical scale is often labeled as the z axis. The zone of rapidly changing salinity is known as the halocline.

the density of all other substances can be measured. Seawater contains various dissolved substances within it that increase its density as compared to pure water. In the open ocean, seawater density averages between 1.022 and 1.030 g/cm^3 (depending on its salinity). Thus, the density of seawater is 2 to 3 percent greater than pure water. Unlike fresh water, seawater continues to increase in density toward its freezing point, and reaches its maximum density at −1.3 degrees centigrade (29.6 degrees Fahrenheit).

Density is an important property of ocean water because density differences make water masses sink or float, thus determining the vertical position and movement of ocean water masses. For example, if seawater with a density of 1.030 g/cm^3 were added to fresh water with a density of 1.000 g/cm^3, the denser seawater would sink below the fresh water.

Ocean water is layered based on density. Since denser water sinks, low-density water exists near the surface and higher-density water occurs below.[15] Except for some shallow inland seas where there is a high rate of evaporation, the highest-density water is found at the deepest ocean depths.

Three factors influence seawater density: temperature, salinity, and pressure, in the following ways.

- As temperature increases, density *decreases*[16] (due to thermal expansion).
- As salinity increases, density *increases* (due to the addition of more dissolved material).
- As pressure increases, density *increases* (due to the compressive effects of pressure).

Of the three factors, pressure has the smallest influence on density because water is nearly *incompressible*. In fact, water is considered an incompressible fluid for engineering purposes. Unlike air, which can be compressed and put in a tank for scuba diving, water cannot be compressed. Pressure is a factor only when very high pressures are encountered, such as in deep-ocean trenches, where the density of seawater is only about 5 percent greater than at the ocean surface. Thus, pressure is a minor factor that can largely be ignored.

The two other factors, salinity and temperature, are much more important. Of these two factors, *temperature is the most important factor influencing seawater density*. Only in the extreme polar areas of the ocean, where temperatures remain relatively constant, does salinity significantly affect density. Thus, cold water that also has high salinity is some of the highest-density water in the world. The density of seawater—the result of its salinity and temperature—is an important factor that influences currents in the deep ocean because high-density water sinks below less-dense water.

The effects of temperature and salinity changes on density are shown in Figure 5–19. The figure shows that temperature change affects density (blue curving lines) much more in warm, low-latitude regions than in high-latitude regions. Points *a, b, c,* and *d* show various densities of water at a constant salinity of 35‰. These points demonstrate that the density change is much greater over a 5-degree temperature span at a higher temperature (line *a–b*) than at a lower temperature (line *c–d*). Note that the lines of constant density more nearly parallel the temperature axis at the bottom (low temperatures) than at the top (high temperatures). For example, the density change across a temperature range from 20 to 25 degrees centigrade (68 to 77 degrees Fahrenheit) is 0.0012 g/cm^3, compared with a change of only 0.0004 g/cm^3 across an equal range at the lower temperatures of 0 to 5 degrees centigrade (32 to 41 degrees Fahrenheit) (Figure 5–19). Thus, a change in the temperature of warm, low-latitude water has about three times the effect on density of an equal change in temperature occurring in colder, high-latitude waters.

---

[15] Note that this also occurs within Earth, where the densest material is in the core, with progressively less-dense material toward the surface.

[16] A relationship where one variable *decreases* as a result of another variable's *increase* is known as an inverse relationship, in which the two variables are *inversely proportional.*

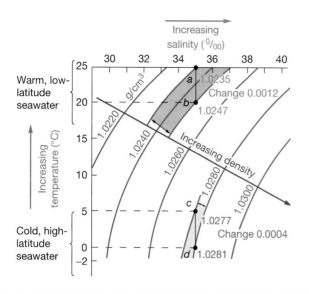

Figure 5–19 **Seawater density varies with temperature and salinity.**

Blue curves show density, which increases with increasing salinity and decreases with increasing temperature. In high-latitude areas of cold water, temperature has less effect on density than in high-temperature, low-latitude areas.

Figure 5–20 shows how seawater density and seawater temperature vary with depth. Figure 5–20A shows the variation of density with depth, while Figure 5–20B shows the variation of temperature with depth. Each of these two graphs shows two curves: one for low-latitude regions, and one for high-latitude regions.

Let's examine the density curve for low-latitude regions (where sea surface temperatures are warm) in Figure 5–20A. At the surface, there is relatively low-density water because of high temperatures (remember that temperature is the primary factor influencing density and that temperature and density are inversely proportional). Below the surface, density initially remains constant until a depth of about 300 meters (980 feet), where the density increases rapidly until a depth of about 1000 meters (3300 feet), from which point density again remains constant down to the ocean floor.

The density curve for high-latitude regions (where sea surface temperatures are cold) shows a very different pattern. At the surface, there is relatively high-density water because of low temperatures. Below the surface, there is also low-temperature water and correspondingly high-density water. In essence, there is high-density water at the surface and high-density water below. This results in the density curve for high-latitude regions resembling a straight line.

Because temperature is the most important factor influencing density, it is no surprise that the temperature graph (Figure 5–20B) strongly resembles the density graph (Figure 5–20A). The only difference is that the scales of the two graphs are inverted, owing to the fact that temperature and density are inversely proportional.

Pycnocline and Thermocline

We have previously described the halocline, a zone of rapidly changing salinity. Similarly, Figure 5–20A shows a prominent **pycnocline** (*pycno* = density, *cline* = slope), a zone of rapidly changing density, and Figure 5–20B shows a prominent **thermocline** (*thermo* = heat, *cline* = slope). The pycnocline and thermocline typically occur between about 300 meters (980 feet) and 1000 meters (3300 feet) depth.

When a pycnocline is established in an area, it presents an incredible barrier to mixing between low-density water above and high-density water below. A pycnocline has a high gravitational stability and thus physically isolates adjacent masses of water.[17] The pycnocline results from the combined effect of the thermocline and the halocline, since temperature and salinity influence density. The interrelation of these three zones determines the degree of separation between the upper-water and deep-water masses. Figure 5–20C shows that the **mixed surface layer** occurs above a strong permanent thermocline (and corresponding pycnocline), where uniform water characteristics result from wave and current mixing. The zone containing the thermocline and pycnocline is a relatively low-density zone called **upper water**, which is well developed throughout the low and middle latitudes. It is underlain by the denser, cold **deep-water** mass that extends from below the thermocline/pycnocline to the deep-ocean floor.

At depths shallower than the main thermocline, divers often experience lesser thermoclines (and corresponding pycnoclines) while descending into the ocean. They are distinct boundaries that divers can certainly feel, where the temperature of the ocean dramatically decreases. Another example of the development of a thermocline occurs in a swimming pool (and also in ponds and lakes). During the spring and fall, when nights are cool but days can still be quite warm, solar heating heats the surface water of the pool yet the water below the surface can be quite cold. If the pool has not been mixed, the temperature at the surface can be misleading because the warm surface layer is effectively isolated from the deeper cold water by a well-developed thermocline. The cold water below the thermocline can be quite a surprise for anyone who dives into the pool!

Notice that a pycnocline and thermocline are absent in high-latitude regions. In general, the temperature of the surface water remains cold at these latitudes year round. There is very little difference between the temperature at the surface and in deep water below. Thus, a thermocline and corresponding pycnocline rarely develop in high-latitude regions. Only during the short sum-

---

[17] This is similar to a temperature inversion in the atmosphere, which traps cold (high-density) air underneath warm (low-density) air.

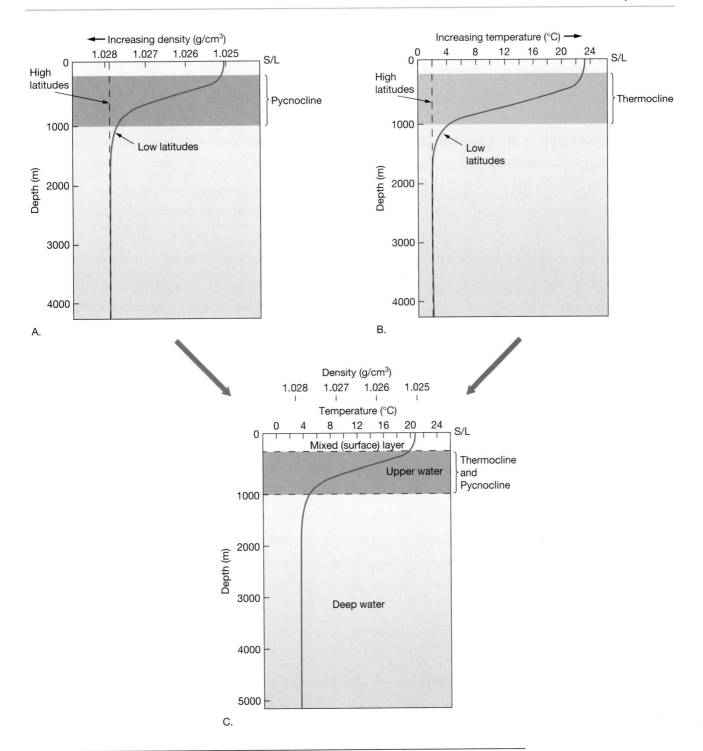

Figure 5–20 Density and temperature variations with depth.
A. Density variation with depth for high-latitude and low-latitude regions. The zone of rapidly changing density is the pycnocline. Note that the density scale is shown increasing to the left, which was done to make the density and temperature curves easier to compare. **B.** Temperature variation with depth for high-latitude and low-latitude regions. The zone of rapidly changing temperature is the thermocline. **C.** A typical ocean profile showing the variation of temperature and density with depth. The thermocline and pycnocline coincide. Note that the slight increase in density below 2 kilometers (1.2 miles) depth is caused by an increase in pressure.

mer, when the days are long, does solar heating begin to heat the surface water. Even then, the water does not heat up very much. Thus, there is typically more mixing between surface water and deeper water in high-latitude regions because of the lack of a pycnocline.

Light Transmission in Ocean Water

As Figure 5–21 reveals, most solar energy is in the range of wavelengths called **light**. It is essential to understand this radiant energy from the sun, because it powerfully affects three major components of the oceans.

1. The major wind belts of the world, which produce ocean currents and wind-driven ocean waves, ultimately derive their energy from solar radiation. Wind belts and ocean currents strongly influence world climates.

2. A thin layer of warm water at the ocean surface overlies the great mass of cold water that fills most of the ocean basins. This is the "life layer" where most sea life exists, created by solar heating.

3. Phytoplankton (floating photosynthetic cells such as diatoms and coccolithophores), which are the base of the oceanic food chain. They produce their own food through photosynthesis. Photosynthesis can occur only where sunlight penetrates the ocean water, so phytoplankton and most animals that eat them must live where the light is, in the relatively thin layer of sunlit surface water.

The Electromagnetic Spectrum

To understand color, it is important to understand something about the electromagnetic spectrum and the narrow portion of it that is visible as light. The many wavelengths of electromagnetic energy radiated by the sun are displayed in Figure 5–21 (*upper portion*), the **electromagnetic spectrum**. The very narrow segment of this spectrum designated as *visible light* can be divided by wavelength into violet, blue, green, yellow, orange, and red energy levels. Combined, these different wavelengths of light produce white light. The shorter wavelengths of energy to the left of visible light are highly dangerous and include X rays and gamma rays. To the right of the visible segment of the electromagnetic spectrum are longer wavelengths of energy. Our technology uses these wavelengths for heat transfer and communication.

We will further consider the electromagnetic spectrum when we examine the role of solar energy in setting the ocean masses in motion and in the greenhouse effect. For now, we will consider only that portion of the spectrum that includes visible light. We call it "visible" light because our electromagnetic sensors—our eyes—are adapted to detect only certain wavelengths. In essence, our eyes "tune into" the visible light wavelengths, just as a radio "tunes into" specific radio waves.

The Color of Objects

Our basic source of light is the sun, and its radiation includes all the visible colors. Most of the light we see is reflected from objects. All objects are selective reflectors, reflecting different wavelengths of light, and each wavelength represents a color in the visible spectrum. For example, vegetation absorbs most wavelengths except green and yellow, which they reflect, so most plants look green. A red jacket absorbs all other wavelengths of color except red, which is reflected.

Why are some areas of the ocean blue, whereas others appear greener? Indeed, the color of ocean water ranges from a deep indigo blue to a yellow-green. The indigo blue color is typical of tropical and equatorial regions, where there is little primary biologic productivity (plant growth that provides food for animals). The yellow-green color occurs in the coastal waters of high-latitude areas, where primary biologic productivity occurs seasonally at a very high rate. In addition, the lack of particulate matter in tropical waters minimizes the molecular scattering of solar radiation, which causes the water to appear blue.[18] Greater concentrations of particulate matter, especially where phytoplankton are abundant, result in greater scattering and absorption of light. This decreases the transmission of solar radiation, producing the greenish color characteristic of such waters.

If a certain wavelength of light is missing from the light that falls upon an object, that color cannot be seen. The ocean is a selective absorber of visible light, and its absorption is greater for the longer-wavelength colors (red, orange, and yellow). Thus, the shorter-wavelength portion of the visible spectrum (violet and blue light) is all that can be transmitted to greater depths (Figure 5–21, *lower portion*).

When one submerges only 100 meters (330 feet) into the ocean, why are there no other colors besides blue and green? This is because red wavelengths are absorbed within the upper 10 meters (33 feet), and yellow disappears before a depth of 100 meters (330 feet). Only the blue and some green wavelengths extend beyond these depths, and their intensity becomes low. Only in the surface waters can the true colors of objects be observed in natural light, since only in the surface waters can all wavelengths of the visible spectrum be found. In the open ocean no sunlight penetrates below a depth of 1000 meters (3300 feet).

To measure the depth of transmission of visible light in the ocean, a **Secchi disk** (Figure 5–22) is used. The Secchi (pronounced "SECK-ee") disk is named after its inventor, Angelo Secchi, an Italian astronomer who first used the device in 1865. Its straightforward construction consists of a round flat disk about 30 centimeters

---

[18] In fact, molecular scattering is also responsible for the blue color of the sky.

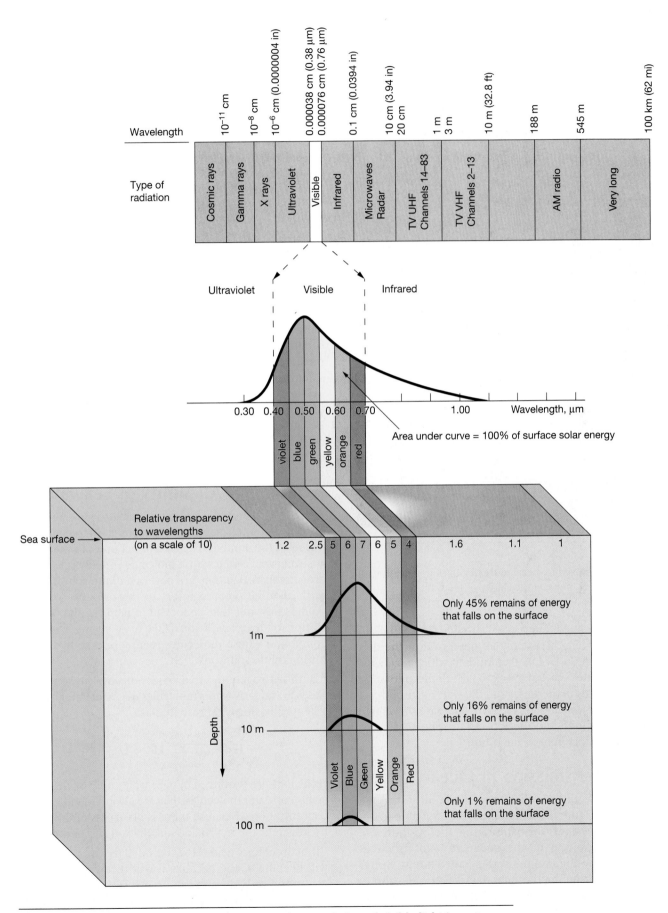

Figure 5–21 The Electromagnetic spectrum and transmission of visible light in water. The spectrum (*top*) runs from extremely short cosmic rays (*left*) with progressively increasing wavelength shown toward the right. The narrow portion of the spectrum that we see as visible light is shown passing through seawater (*bottom*), which absorbs the longer wavelengths (red, orange, and yellow).

Figure 5–22 Secchi disk.
A Secchi disk is used to measure the depth of penetration of sunlight, and thus indicate the clarity of the water.

(12 inches) in diameter attached to a line that is marked off at regular lengths, usually in meters. After the disk is lowered into the ocean, the depth at which it can last be seen indicates the water's clarity. Both the amount of microorganisms in the water and the water's turbidity—the amount of suspended material—increase the degree of light absorption, thus decreasing the depth of transmission of visible light in the ocean.

Sound Transmission in Ocean Water

Sound is transmitted much more efficiently through water than through air. This has made it possible to develop sound systems for determining the position and distance of objects in the ocean. As described in Chapter 3, the technique used is called **sonar**. Sonar relies on determining the time that it takes an echo to bounce off an underwater object and return to the source of the sound.

Average sound velocity in the ocean is 1450 meters (4750 feet) per second. This is over four times faster than

sound velocity in the air, which is 334 meters (1100 feet) per second in dry atmosphere at 20 degrees centigrade. Further, sound travels faster in the ocean with increases in temperature, salinity, density, and pressure. Because these factors vary with the seasons, the velocity of sound through water in a particular area varies seasonally.

At a depth of around 1000 meters (3300 feet), there exists a layer of ocean water where these three variables—temperature, salinity, and pressure—combine to create a relatively low-velocity zone for sound that transmits sound very efficiently (Figure 5–23). Sound originating above and below this low-velocity layer travels to the layer and there becomes refracted, or bent, into the layer. In effect, this layer acts as a wave guide (or lens) that traps and transmits sound within the layer. Theoretically, sound would be able to be transmitted within this sound channel over large distances. For instance, if a light bulb is submerged in the ocean to a depth that makes it implode, the sound of this implosion can be heard for over 1000 kilometers (600 miles) because of transmission through this sound channel. This channel is called the **SOFAR channel** (an acronym for <u>SO</u>und <u>F</u>ixing <u>A</u>nd <u>R</u>anging) because of its practical application in distance determination. Because sound travels so efficiently in the SOFAR channel, it may be used by marine mammals, especially whales, in communicating across entire ocean basins.

Acidity and Alkalinity of Seawater

An **acid** is a compound containing hydrogen that, when dissolved in water, sets hydrogen ions (H^+) free and thus increases the number of such ions per unit of water mass. A strong acid is one whose molecules readily release hydrogen ions when they are part of a dilute solution. The term **alkaline** or **base** refers to the presence of compounds in water that dissociate to increase the concentration of hydroxide ion (OH^-). A strong base is a compound whose molecules readily release hydroxide ions in dilute solutions.

Both hydrogen and hydroxide ions are present at all times in water as water molecules dissociate and reform. Chemically, this is represented as

$$H_2O \underset{\text{reform}}{\overset{\text{dissociate}}{\rightleftharpoons}} H^+ + OH^-$$

If the ions are produced in pure water only by the dissociation of water molecules, they are always found in equal concentrations. Because there is no excess of either the hydrogen or the hydroxide ion, the water is **neutral**.

If hydrochloric acid (HCl) is added to water, the water will become acidic because it will receive a high concentration of hydrogen ions as a result of the dissociation of the HCl molecules. Conversely, if a base such as baking soda (sodium bicarbonate, $NaHCO_3$) is added to water, an excess of hydroxide ions (OH^-) results as the $NaHCO_3$ molecules break up.

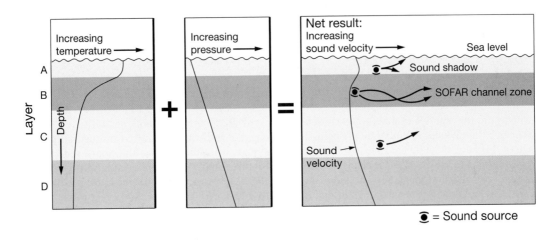

= Sound source

Figure 5–23 Transmission of sound in the ocean.
In water, sound travels faster with increasing temperature and pressure, as shown in the left two panels of the figure. These variables combine to produce a *shadow zone* where a layer of maximum speed exists (*layer A*) and a channel that traps sound energy, the SOFAR channel (*layer B*). The sound waves traveling to the right from the sound source in layer C will be bent upward into the SOFAR channel. Once within the SOFAR channel, the sound remains within the channel, effectively trapping sound and transmitting it across ocean basins.

The pH Scale

The **pH** (potential of hydrogen) **scale** is a measure of the acidity or alkalinity of a solution.[19] The pH is 7.0 for neutral solutions, increasing with increasing alkalinity and decreasing with increasing acidity. Values for pH range from 0 (strongly acid) to 14 (strongly alkaline or basic).

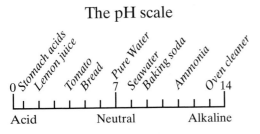

Pure water has a pH of 7.0, which is neutral. This might seem surprising in light of water's tremendous ability to dissolve substances; it seem intuitive that water should be acidic and thus have a low pH. However, pH measures the activity of the hydrogen ion in solution, not the ability of a substance to dissolve by forming hydrogen bonds (as water does).

In the ocean, water combines with carbon dioxide to form a weak acid, called carbonic acid (H_2CO_3) which readily dissociates and sets hydrogen ions (H^+) free.

$$H_2O + CO_2 \rightarrow H_2CO_3 \rightarrow H^+ + HCO_3^-$$

---

[19] The value of pH is equal to the negative log of the hydrogen ion concentration (pH = $-\log[H^+]$).

It would appear that this reaction would make the ocean slightly acidic. However, carbonic acid has an important role in keeping the ocean slightly alkaline, through the process of buffering.

The Carbonate Buffering System

The introduction of carbon dioxide into the oceans serves to maintain the slightly alkaline conditions of the oceans (Figure 5–24). When a carbon dioxide molecule combines with a water molecule, two of the electrons in the oxygen atom of the water molecule are shared with two of the electrons of the carbon atom from the carbon dioxide molecule. This situation leaves the hydrogen nuclei weakly attached to the water molecule as it becomes part of the H_2CO_3 molecule. One of these hydrogen nuclei may be lost and become a hydrogen ion (H^+), transforming the carbonic acid molecule into a negatively charged **bicarbonate ion (HCO_3^-)**.

Should the bicarbonate ion lose its hydrogen ion, it then is transformed into a double-charged negative **carbonate ion (CO_3^{-2})**, some of which combines with calcium ions to form **calcium carbonate ($CaCO_3$)**. Some of the calcium carbonate is dissolved at depth and some is deposited on the ocean floor. The arrows in Figure 5–24 indicate that the reaction can proceed in *both* directions. If for any reason the pH of the ocean rises, the reaction proceeds downward, releasing H^+ and lowering the pH. Should the pH of the ocean drop, the reaction proceeds upward and raises the pH by removing H^+. The process of minimizing changes in pH as described above is termed **buffering**. Thus, buffering is responsible for maintaining the ocean's pH at an average of 8.1 with little variation through time.

Box 5–2
The ATOC Experiment: SOFAR So Good?

The SOFAR channel efficiently traps and transmits sounds long distances through the ocean. It has long been suspected that certain whales use the SOFAR channel to send sounds across entire ocean basins. Whales are not the only beings to take advantage of this natural sound channel in the oceans: Humans may also be able to use the SOFAR channel to help determine the amount of global warming. Several studies have been conducted that send sound transmissions through the SOFAR channel to investigate whether the oceans are experiencing any warming due to the greenhouse effect.

An increase in global warming is thought to occur by the addition of greenhouse gas concentrations as a result of human activities. The fact that certain greenhouse gases have increased in the atmosphere due to human activities is well established. However, the amount of global temperature change over the last century is more difficult to determine. Although there are records of temperatures on land going back at least that far, they may be biased by their proximity to heat-absorbing structures. Because of water's unusual thermal properties, any global climate change would be represented by a corresponding change in seawater temperatures.

Dr. Walter Munk of the Scripps Institution of Oceanography (Figure 5D) proposed an experiment to determine if sending sound through the ocean might aid in detecting global warming of the oceans. Munk's experiment, called **Acoustic Thermometry of Ocean Climate (ATOC)**, uses low-frequency sound signals sent worldwide through the SOFAR channel. To test the feasibility of this idea, his group selected Heard Island in the southern Indian Ocean as the ideal test site because sound could reach many different receiving sites along straight-line paths (Figure 5E). In 1991, researchers used an underwater array deployed from a ship to set off low-frequency acoustical signals for six days at a depth of 150 meters (492 feet). The sound was refracted into the SOFAR channel and was transmitted throughout the oceans. These signals were received at shipboard recording stations up to three-and-a-half hours later after traveling as far as 19,010 kilometers (11,801 miles), where the precise time of their arrival was noted. Since the speed of sound in seawater increases with an increase in temperature, if the oceans are experiencing warming, then the sound will take slightly less time to travel the same distance in the future. The study proposes that these same sound signals will be sent out periodically during the next decade to compare the travel times with those measurements taken previously. If ATOC shows that the oceans are indeed warming, it could provide some of the most compelling evidence in support of global warming.

The success of the first trial has led to further development of ATOC. ATOC will differ from the Heard Island experiment in two ways: The sources and the receivers will be at fixed locations (instead of aboard ships), and it has been proposed that the signals run for at least one year. The most recent study was initiated in 1995, when sound signals near Pioneer Seamount off California were sent to an array of fixed receivers throughout the Pacific Ocean. Even though many precautions were taken to avoid any unwanted effects on marine mammals, three humpback whales in the area were found dead a few days after the sound transmissions were begun. The sounds were halted and the U.S. National Marine Fisheries Service (NMFS) conducted studies to determine if the sounds constitute a threat to marine mammals. The concern was that the sounds affected the hearing of nearby whales and somehow caused their deaths. Other concerns are that the sounds may interfere with marine mammal communication through the SOFAR channel. However, preliminary research indicates that marine mammals are unaffected by the transmissions and the dead whales were an unfortunate coincidence. The long-term effect on marine mammals is poorly understood, but is also likely to be minimal. Recently, the NMFS has approved the continuance of the project, and future transmissions from offshore California and Hawaii are planned.

A similar smaller-scale operation, the **Transarctic Acoustic Propagation Experiment (TAP)**, was successfully undertaken in 1994. Scientists sent sound signals through Arctic waters from near Spitzbergen in the north Atlantic Ocean to receivers off the north coast of Baffin Island and in the Beaufort Sea. Because of the importance of heat exchange mechanisms in high-latitudes, it is thought that any global change in temperature would be experienced first in high-latitude waters. When compared with 10-year-old temperature measurements, it was discovered that the sound traveled faster than predicted, indicating that Arctic waters have warmed. Further research indicated that the warming has resulted from an influx of water from the North Atlantic. What is unclear is *why* this influx has occurred. Perhaps it is part of a natural cycle that the Atlantic Ocean experiences. Another interpretation is that the higher seawater temperatures are an early manifestation of global warming.

Figure 5D Walter Munk.

(continued)

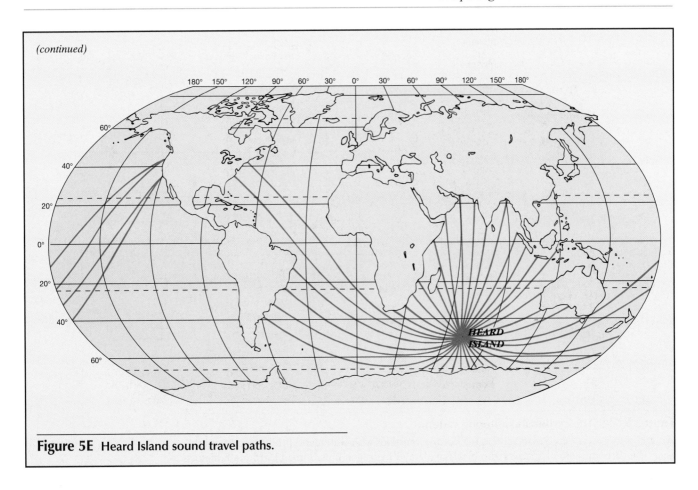

Figure 5E Heard Island sound travel paths.

Deep-ocean water contains more carbon dioxide than at the surface because at colder temperatures water can hold more of a dissolved gas. This carbon dioxide combines with water to form carbonic acid, thus decreasing the pH of the deep ocean, which has the potential for making the deep ocean acidic. However, microscopic marine organisms that make their shells out of calcium carbonate (calcite) have an important role in helping the deep ocean buffer itself. When these microscopic organisms die, their shells sink. The input of calcium carbonate into the deep ocean neutralizes the acid through buffering (Figure 5–24). In essence, these organisms act as an "antacid" for the deep ocean in a way similar to the way commercial antacids work to neutralize stomach acid with calcium carbonate. As explained in Chapter 4, the depth at which these shells dissolve is the calcite (calcium carbonate) compensation depth (CCD).

Comparing Pure Water and Seawater

Included are two tables that summarize much of what has been discussed in this chapter. Table 5–4 summarizes the physical and biological significance of the properties of water. Table 5–5 compares the properties of pure water and seawater.

The physical properties listed in Table 5–5 demonstrate that since seawater is 96.5 percent water molecules, sea-water has properties remarkably similar to those of pure water with a few notable exceptions. For instance, the color of pure water and seawater is identical, yet the dissolved substances in seawater give it a distinct odor and taste.

Another difference between pure water and seawater is that the density of 35‰ seawater is 3 percent higher than pure water because of its increased amount of dissolved substances. Remember that if salinity of seawater is increased, its density also increases.

We have also learned that dissolved substances interfere with water changing state, as seen by comparing the freezing and boiling points of water and seawater. Seawater freezes at a temperature 1.9 degrees centigrade (3.4 degrees Fahrenheit) *lower* than pure water. Thus, seawater needs to be a little below the freezing point of pure water to freeze. At the other extreme, seawater boils at a temperature 0.6 degree centigrade (1.1 degrees Fahrenheit) *higher* than pure water. Thus, seawater needs to be a little hotter than the boiling point of pure water to boil. Moreover, the salts in seawater *extend the range of temperatures* in which water is a liquid. The same principle applies to antifreeze/summer coolant used in automobile radiators. Adding impurities (antifreeze) to radiator water extends the range of temperatures that the water will remain in the liquid state. That is why it is unnecessary to change antifreeze in the summer: It also acts as a summer coolant.

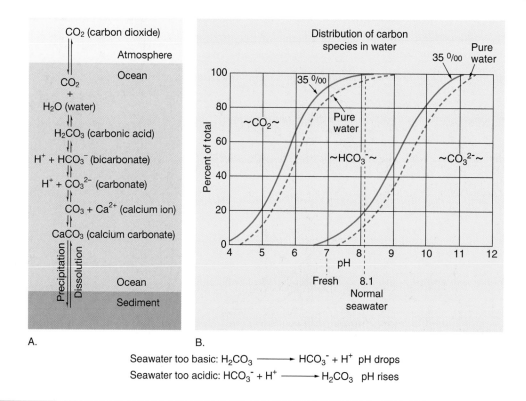

Seawater too basic: $H_2CO_3 \longrightarrow HCO_3^- + H^+$ pH drops
Seawater too acidic: $HCO_3^- + H^+ \longrightarrow H_2CO_3$ pH rises

Figure 5–24 The carbonate buffering system.
A. Atmospheric carbon dioxide enters the ocean, combines with water, and produces carbonic acid. The carbonic acid dissociates into bicarbonate and hydrogen ions. Some bicarbonate dissociates into carbonate ions, which combine with calcium ions to form calcium carbonate. **B.** More than 80 percent of the carbon in ocean water is in the form of bicarbonate (HCO_3^-). If pH rises, the reactions shown in part A move downward. This releases more H^+ into the water and lowers pH. If the pH drops, the reactions move upward, removing H^+ from the water and causing pH to rise. Thus, the pH of ocean water varies little from a value of 8.1.

Desalination

Earth's growing population is consuming fresh water supplies in greater volumes each year. As fresh water becomes a scarcer commodity, several countries have begun to use the ocean as a source of water. **Desalination**, or salt removal from seawater, has been proposed to take advantage of the greatest water supply on Earth—the oceans. Desalination can provide fresh water for domestic, home, and agricultural use. Not counting those desalination units on board ships, there are more than 1500 desalination plants in existence worldwide, although most of them are small scale. The majority of these plants are located in the Middle East. The United States produces only about 10 percent of the world's desalted water, with Florida and California leading the way. More than half of the world's desalination plants use distillation to purify water, with the remaining plants using various membrane processes.

Distillation involves boiling saltwater, capturing the water vapor, and leaving the salt behind (Figure 5–25). The water vapor is passed through a cooling condenser, where it condenses and is collected as fresh water. This simple procedure is very efficient: From 35‰ seawater,

water with a salinity of only 0.03‰ is produced, which is about 10 times fresher than bottled water! Often, distillation plants mix distilled water with less pure water to make it taste better. However, distillation is expensive because it requires large amounts of heat energy to boil the saltwater (remember that water's high latent heat of vaporization requires an enormous amount of heat— 540 calories—to convert *1 gram*—about 10 drops—of water at the boiling point to the vapor state[20]). Increased efficiency is required to make distillation practical on a large scale. Using the waste heat from a power plant is one way to increase efficiency.

Solar humidification or **solar distillation** does not require supplemental heating and has been used successfully in small-scale agricultural experiments in arid regions such as Israel, West Africa, and Peru. It is similar to distillation in that it involves the evaporation of saltwater in a covered container, but the water is heated by direct sunlight (Figure 5–25). Saltwater in the con-

---

[20]Assuming 100 percent efficiency, it takes a whopping *270,000 calories* of heat energy to make one half-liter bottle of distilled water.

Table 5–4   Summary of the properties of water and their significance.

Property	Comparison with other substances	Physical and/or biological significance
Physical states: gas (*water vapor*), liquid (*water*), solid (*ice*).	Water is the only substance that occurs as a gas, liquid, and solid within the range of normal surface temperatures on Earth.	*Water vapor* is an important component of the atmosphere, because water vapor transfers great quantities of heat from warm, low latitudes to cold, high latitudes. *Liquid water* runs across land, dissolving minerals from the rocks and carrying them to the oceans. Most organisms are composed primarily of water. Serves as a transport medium and facilitates chemical reactions. *Ice* formation at high latitudes increases surface water salinity, increasing the density of the water and thus making it sink. When ice forms at the surface, it protects the water and life below it from freezing.
Solvent property: the ability to dissolve substances.	Water can dissolve more substances than any other liquid, hence it is often called "the universal solvent."	*Wide-ranging implications* in both physical and biological phenomena. Explains why seawater is "salty." In addition, seawater carries dissolved within it the nutrients required by marine algae and the oxygen needed by animals.
Surface tension: cohesive attraction of hydrogen bonds causes a molecule-thick "skin" to form on water surfaces.	Water has the greatest surface tension of all common liquids.	*Causes water to form drops*, to "bead up," and to overfill a glass without spilling. Water striders use this "skin" as a walking surface, while other organisms hang from its undersurface. Also causes *capillarity*, in which water is drawn upward in small tubes.
Heat capacity: the quantity of heat required to change the temperature of 1 gram of a substance by 1°C. The heat capacity of water is used as the unit of heat quantity, the *calorie*.	Water has the highest heat capacity of all common liquids and gains or loses much more heat than other common substances while undergoing an equal temperature change.	*A major factor* in moderating Earth's climate. It helps explain the narrow range of temperature change occurring near the oceans as compared to interior regions of the continents.
Latent heat of melting: the quantity of heat gained or lost per gram by a substance changing from a solid to a liquid, or from a liquid to a solid, without a temperature change.	For water, it is 80 cal at 0°C (32°F), the highest of any common substance.	*Heat energy lost when ice forms* is mostly absorbed by the heat-deficient atmosphere at high latitudes. *Heat energy gained by water when ice melts* is manifested as molecular energy of the liquid water. This prevents the high-latitude ocean from becoming much warmer or colder than the freezing temperature of ocean water.
Latent heat of vaporization: the quantity of heat gained or lost per gram by a substance changing from a liquid to a gas, or from a gas to a liquid, without a temperature change.	It is greater for water, 540 cal at 100°C (212°F), than for any other common substance. At 20°C (68°F), the temperature at which much of the evaporation from the ocean surface occurs, it is 585 cal.	*Extremely important* in global heat and water transfer in the atmosphere. Evaporation from the low-latitude ocean removes a great amount of excess heat energy that is released through precipitation at heat-deficient higher latitudes. This greatly moderates temperatures at the poles and Equator, which otherwise would be far more extreme.
Density: mass per unit volume (g/cm³).	Water's density increases as water cools, but it *decreases* below 4°C (39°F), which is highly unusual. Ocean water density increases at lower temperatures and as salinity and pressure increase.	*Plankton* that stay near the surface through buoyancy and frictional resistance to sinking are greatly influenced by the effect of temperature on density. In low-density warm water, plankton must be smaller or more ornate to obtain the increased ratio of surface area to body mass necessary to remain afloat Also produces layering of ocean water.

continued...

Table 5–4 *(continued)*

Property	Comparison with other substances	Physical and/or biological significance
Thermal expansion: the expansion of a substance when it is heated, and the contraction of a substance when it is cooled.	As water is cooled, it contracts until 4°C (39°F), at which point it expands due to ice formation. This is highly unusual for a substance. Water expands by 9% when frozen.	*Ice floats.* In the ocean at high latitudes where sea ice forms, ice provides a protective layer below which seawater does not freeze.
Light transmission: the ability of a substance to transmit light.	Water has high transparency, so is a good transmitter of light. In the ocean, no sunlight penetrates below a depth of 1000 m (3300 ft).	*Wind-driven ocean waves and surface currents* are driven by solar energy. The sun heats the ocean surface, producing a warm surface layer with cold deep water below. The energy for life in the oceans, most of which is supported by photosynthesis, is dependent on solar radiation.
Sound transmission: the ability of a substance to transmit sound.	Water's dense molecular packing makes water a good transmitter of sound. Sound transmits 4 times better in water than in air.	*A zone of rapid temperature decrease* beneath the warm surface layer of ocean water creates a low-velocity sound channel called the SOFAR channel. This channel traps sound and can conduct it over great distances. The SOFAR channel may be used by marine mammals to communicate over long distances.

Table 5–5 **Comparison of selected properties of pure water and seawater.**

Property	Pure water	35‰ seawater
Color (light transmission)	High transparency—in small quantities, water is transparent; in large quantities, water looks blue-green because water scatters blue and green wavelengths best	Same as for pure water
Odor	Odorless	Distinctly marine
Taste	Tasteless	Distinctly salty
pH	7.0	8.1 (slightly alkaline)
Density at 4°C (39°F)	1.000 g/cm^3	1.030 g/cm^3
Freezing point	0°C (32°F)	−1.9°C (28.6°F)
Boiling point	100°C (212°F)	100.6°C (213°F)

tainer evaporates, and the water vapor that condenses on the cover runs into collection trays. The major difficulty lies in effectively concentrating the energy of sunlight into a small area to speed evaporation.

A process demanding large amounts of energy is **electrolysis**. In this method, two volumes of fresh water—one containing a positive electrode and the other a negative electrode—are placed on either side of a volume of seawater. The seawater is separated from each of the fresh water reservoirs by semipermeable membranes. These membranes are permeable to salt ions but not to water molecules. When an electrical current is applied, positive ions such as sodium ions are attracted to the negative electrode, and negative ions such as chloride ions are attracted to the positive electrode. In time, enough ions are removed through the membranes to convert the seawater to fresh water.

Another application that is limited to small-scale use is **freeze separation**. Because seawater selectively excludes dissolved substances as it freezes, the salinity of sea ice (once it is melted) is typically 70 percent lower than seawater. To make this technique effective, the water must be frozen and thawed multiple times, with the salts washed from the ice between each thawing. The process of freeze separation requires plentiful energy, so it may not be practical on a large scale.

Yet another way is to melt naturally formed ice. Imaginative thinkers have proposed using tugboats to tow large icebergs to coastal waters off countries that need fresh water. There, the fresh water that is produced as the

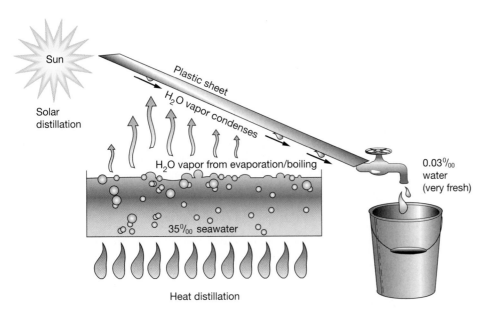

Figure 5–25 Distillation.
The process of distillation requires boiling saltwater (heat distillation) or using the sun's energy to evaporate seawater (solar distillation). In either case, the water vapor is captured and condensed, which produces very pure water.

icebergs melt could be captured and pumped ashore. Studies of the resource potential of towing large Antarctic icebergs to arid regions for drinking water reveal that ample technology exists and that for certain Southern Hemisphere locations, the economics are favorable.

A method that has much potential for large-scale projects is **reverse osmosis**. In natural osmosis, water molecules pass through a water-permeable membrane from a fresh water solution into a saltwater solution. In reverse osmosis, pressure is applied to the saltwater solution, forcing it to flow "backwards" through the water-permeable membrane into the fresh water solution (Figure 5–26). This is the reverse of natural osmosis, giving the method its name. A significant problem is that the membranes must be replaced frequently. At least 30 countries located in arid climates are operating reverse osmosis units. Santa Barbara, California, operates a reverse osmosis plant that produces up to 34 million liters (9 million gallons) daily, which supplies that city with up to 60 percent of its water needs. This method is used in many household water purification units and aquariums.

Other novel approaches to desalination include crystallization, solvent demineralization, and even salt-eating bacteria!

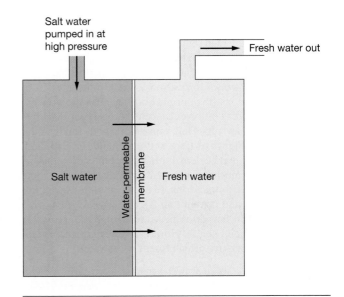

Figure 5–26 Reverse osmosis.
This process requires that pressure be applied to a reservoir of saltwater to force it through a water-permeable membrane, removing the salts and producing fresh water.

Students Sometimes Ask...

How can it be that water—a liquid at room temperature—can be created by combining hydrogen and oxygen—two gases at room temperature?
That's what amazes most people about chemistry. It is true that combining these two gases together in the proportion of two parts hydrogen to one part oxygen produces liquid water. This can be accomplished as a chemistry demonstration, although care should be taken because much energy is released during the reaction (don't try this at home!). Often times, when combining two elements together, the product has very different properties than the pure substances. For instance, if you were to combine elemental sodium (Na), a highly reactive metal, with pure chlorine (Cl_2), a toxic nerve gas, cubes of harmless table salt (NaCl) can be produced. Even a glimpse into the chemical properties of substances can certainly be astounding!

I've seen the warning labels on electric cords warning about using electrical appliances close to water. Are these warnings because water's polarity allows electricity to be transmitted through it?
Yes and no. Because water molecules have polarity, one would assume that water is a good conductor of electricity. However,

pure water is not. Pure water is a very poor conductor because the molecules will not move toward either the negatively charged pole or the positively charged pole in an electrical system. If an electrical appliance is dropped into a tub of absolutely pure water, no electricity will be transmitted by the water molecules. The water molecules will simply rotate in place until they become oriented with their positively charged hydrogen ends toward the negative pole of the appliance and their negatively charged oxygen ends toward its positive pole. This orientation of water molecules in an electrical field tends to neutralize the field. Interestingly, it is the *dissolved substances* that transmit electrical current through water. Even slight amounts of dissolved substances, such as those in tap water, allow electricity to be transmitted. That's why there are warning labels on the electric cords of household appliances that are commonly used in the bathroom, such as blow-dryers, electric razors, and heaters. Also, that's why it is recommended to stay out of any water—including a bathtub or shower—during a lightning storm!

What would happen to a person if he or she drank seawater?
It depends on the quantity. Drinking large quantities of seawater is definitely not recommended because the salinity of seawater is about four times greater than that of your body fluids. In your body, the high salinity of seawater would cause your internal membranes to lose water through the process of *osmosis* (water molecules passing through a semipermeable membrane). Osmosis works to transport water molecules from higher concentrations (the normal body chemistry of your internal fluids) to areas of lower concentrations (your digestive tract containing seawater). Thus, your natural body fluids would move into your digestive tract and eventually be expelled. Essentially, you would be dehydrating yourself from the *inside*.

Don't worry too much if you've inadvertently swallowed some seawater. As a nutritional drink, seawater provides seven important nutrients and contains no fat, cholesterol, or calories. Some people claim that drinking a small amount of seawater daily gives them good health! However, beware of microscopic contaminants that are often present in seawater.

You mentioned that when seawater freezes, it produces ice with about 10‰ salinity. Once that ice melts, can a person drink it with no ill effects?
Early Arctic explorers found out the answer to your question by necessity. Some explorers who traveled by ship in high-latitude regions became inadvertently or purposely entrapped by sea ice. Lacking other water sources, they melted sea ice as an emergency form of relatively fresh water. Although newly formed sea ice contains little salt, it does trap a significant quantity of brine (drops of salty water) during the freezing process. Depending on the rate of freezing, newly formed ice may have a total salinity from 4 to 15‰. The more rapidly it forms, the more brine will be captured and the higher the salinity will be. Certainly, melted sea ice with salinity this high doesn't taste very good, and it will still cause dehydration, but not as quickly as drinking 35‰ seawater does. Over time, however, the brine will trickle down through the coarse structure of the sea ice, and its salinity will decrease. By the time it is a year old, sea ice normally has become relatively pure. Thus, drinking melted sea ice that had been purified by this method enabled the early explorers to survive.

Would it be dangerous to drink chemically purified water?
For short periods of time, no. However, drinking absolutely pure water for extended periods of time is not advisable. Because water is so good at dissolving other substances, if you were to consume pure water regularly, it would begin to steal some of the important trace elements that your body needs to survive. It wouldn't taste very good, either, since pure water is tasteless. In fact, some companies actually *add* dissolved substances to their water in carefully controlled amounts to enhance the taste of their bottled water. It's best to stick with tap water or good bottled water.

If seawater is so good at dissolving things, why aren't marine organisms being dissolved?
They are! Consider what life is like for both large and small marine organisms: They live in a medium that is actively working to dissolve them. Fortunately, marine organisms have special defenses to prevent them from being dissolved. One of the more common defenses is *mucus.* Composed of an oily organic substance, mucus is used as a protective coating by many marine organisms since oil does not dissolve in water. That's why fish feel slimy when you hold them.

What is the strategy behind adding salt to a pot of water when making pasta? Does it make the water boil faster?
Contrary to popular belief, adding salt to water will not make the water boil faster. Beyond the obvious reason that it might make the pasta taste better, there is a sound chemical reason for putting salt in your pasta water. Consider how salt affects the boiling point of water. As shown in Table 5–5, when impurities are added to water, they raise its boiling point. Thus, adding salt to water makes the water boil at a *higher temperature,* thus creating a *hotter boil.* The pasta will cook in less time. But add the salt *after* the water has come to a boil—otherwise, it takes longer to reach a boil. This is a wonderful use of chemical principles—helping you to cook better!

If the ocean is such a good transmitter of sound, can marine organisms hear noises?
Yes, many marine organisms can. Indeed, the ocean is a very noisy place! The constant sound of waves breaking at the shoreline, fish making cracking noises, whales clicking and singing, crabs snapping their claws, boats motoring around on the surface—all of these sounds can be heard, sometimes for hundreds or thousands of kilometers in all directions. Sound travels as a compressive wave, so sounds in the ocean are felt by an organism's entire body. The effect is similar to being able to "feel" the music from a loud rock concert even if you have earplugs in your ears. If you ever go whale watching, don't think for a minute that you are sneaking up on whales. They have heard you coming toward them since your boat left the harbor! However, seeing whales in their natural environment can be enhanced by knowing that the whales are aware of your presence and that they are *allowing* you to observe them at close range.

Summary

Water's unusual properties are extremely important to all living organisms on Earth. These properties include the arrangement of its atoms, how its molecules stick together, its ability to dissolve almost everything, and its heat storage capacity.

The water molecule is composed of one atom of oxygen and two atoms of hydrogen (H_2O). The geometry of the water molecule is such that the two hydrogen atoms are attached to the same side of the oxygen atom by covalent bonds. This unusual geometry results in the water molecule having polarity, which causes water to form hydrogen bonds between molecules or with other substances and gives water its remarkable properties. For example, water is "the universal solvent" because its bipolar molecules can attach themselves to charged particles—ions—that make up many substances. This hydrates them and places them into solution.

Water is one of the few substances that exists naturally on Earth in all three states (solid, liquid, gas), and it has the capacity to store great amounts of heat energy. The hydrogen bond also accounts for the unusual thermal properties of water, such as its high freezing point (0 degrees centigrade or 32 degrees Fahrenheit) and boiling point (100 degrees centigrade or 212 degrees Fahrenheit), high heat capacity (1 calorie), high latent heat of melting (80 calories per gram), and high latent heat of vaporization (540 calories per gram).

The density of water increases as temperature drops, as does the density of most substances, but only to 4 degrees centigrade (39.2 degrees Fahrenheit), its temperature of maximum density. Below 4 degrees centigrade, water density decreases with temperature, due to the formation of bulky ice crystals. By the time water freezes fully, the density of the ice is only about 90 percent that of water at 4 degrees centigrade, causing ice to float on water.

Salinity is the amount of dissolved solids in ocean water, averaging about 35 grams of dissolved solids per kilogram of ocean water [35 parts per thousand (‰)], but ranges from brackish to hypersaline. Over 99 percent of the dissolved solids in ocean water are accounted for by six ions—chloride, sodium, sulfate, magnesium, calcium, and potassium. In any seawater sample, these ions always occur in a constant proportion to one another, so salinity can be determined by measuring the concentration of only one, usually the chloride ion.

The dissolved components in seawater are added and removed by a variety of processes. The residence time of various elements indicates how long they stay in the ocean.

Precipitation, runoff, and the melting of icebergs and sea ice are processes that add fresh water to seawater and decrease seawater salinity. The formation of sea ice and evaporation are processes that remove fresh water from seawater and increase salinity. These processes are related by the hydrologic cycle, which indicates that the oceans contain 97 percent of Earth's water.

The salinity of surface water varies considerably due to surface processes, with the maximum salinity found near the Tropic of Cancer and the Tropic of Capricorn. As compared with surface salinity variation, the salinity of deep water is very consistent. A halocline develops where there is a rapidly changing zone of salinity.

Seawater density increases with decreasing temperature and increasing salinity, with temperature controlling density more strongly than salinity (the influence of pressure is negligible). Density and temperature vary considerably with depth in low-latitude regions, creating a pycnocline and corresponding thermocline.

In ocean water of low biologic productivity, the molecule-sized particles scatter the short wavelengths of visible light, producing a blue color. Greater amounts of dissolved organic matter and photosynthetic algae in more-productive ocean water scatter more green light wavelengths, which produces a green color. With increasing depth in the ocean, the colors of the visible spectrum are absorbed by ocean water in a way that removes red and yellow at relatively shallow depths. The blue and green wavelengths of light are the last to be removed.

Velocity of sound transmission in the ocean increases with temperature, salinity, and pressure. A low-velocity SOFAR sound channel can conduct sound over great distances in the oceans. This sound channel is being used in an experiment to document global warming.

A natural buffering system exists in the ocean, created by the reaction of carbon dioxide in water. This buffering system regulates any changes in pH, and has resulted in a stable ocean environment. The physical properties of pure water and seawater are remarkably similar, with a few notable exceptions.

Desalination of ocean water to provide fresh water for domestic, home, and agricultural use is of growing interest. It may be achieved by solar distillation (solar humidification), electrolysis, freeze separation, reverse osmosis and other techniques for processing seawater.

Key Terms

Acid (p. 158)

Acoustic Thermometry of Ocean Climate (ATOC) (p. 160)

Alkaline (p. 158)

Atom (p. 134)

Base (p. 158)

Bicarbonate ion (HCO_3^-) (p. 159)

Boiling point (p. 138)

Brackish (p. 146)

Buffering (p. 159)

Calcium carbonate ($CaCO_3$) (p. 159)

Calorie (p. 137)

Calving (p. 150)

Carbonate ion (CO_3^{2-}) (p. 159)

Challenger, HMS (p. 132)

Chlorinity (p. 147)

Cohesion (p. 136)

Condensation point (p. 138)

Condense (p. 142)

Covalent bond (p. 134)

Deep-water (p. 154)

Desalination (p. 162)

Dipolar (p. 135)

Distillation (p. 162)

Questions and Exercises

1. List some major achievements of the voyage of HMS *Challenger.*

2. Sketch a model of an atom, showing the positions of the subatomic particles protons, neutrons, and electrons.

3. Describe what condition exists in water molecules to make them dipolar.

4. Sketch several water molecules, showing all covalent and hydrogen bonds. Be sure to indicate the polarity of each water molecule.

5. How does hydrogen bonding produce the surface tension phenomenon of water?

6. Discuss how the dipolar nature of the water molecule makes it such an effective solvent for ionic compounds.

7. Describe the differences between the three states of matter, using the arrangement of molecules in your explanation.

8. There is a fundamental difference between the intermolecular bonds that result from the dipolar nature of the water molecule (hydrogen bond) and the van der Waals force as compared with the chemical bonds. What is it?

9. Why are the freezing and boiling points of water higher than would be expected for a compound of its molecular makeup?

10. How does the heat capacity of water compare with that of other substances? Describe the effect this has on climate.

11. The heat energy added as latent heat of melting and latent heat of vaporization does not increase water temperature. Explain why this occurs, and where the energy is used.

12. Why is the latent heat of vaporization so much greater than the latent heat of melting?

13. Describe how excess heat energy absorbed by Earth's low-latitude regions is transferred to heat-deficient higher latitudes through a process that uses water's latent heat of evaporation.

14. As water cools, two distinct changes take place in the behavior of molecules: Their slower movement tends to increase density, whereas the formation of bulky ice crystals decreases density. Describe how the relative rates of their occurrence cause pure water to have a temperature of maximum density at 4 degrees centigrade (39.2 degrees Fahrenheit) and make ice less dense than liquid water.

15. What is your state sales tax, in parts per thousand?

16. What are goiters? How can one avoid goiters?

17. What physical conditions create brackish water in the Baltic Sea and hypersaline water in the Red Sea?

18. What condition of salinity makes it possible to determine the total salinity of ocean water by measuring the concentration of only one constituent, the chloride ion?

19. Describe the ways that dissolved components are added and removed from seawater.

20. List the reservoirs of water on Earth and the percentage of Earth's water that each holds. Describe the processes by which water moves among these reservoirs.

21. Explain why there is such a wide variation of surface salinity, but such a narrow range of salinity at depth.

22. Why is there such a close resemblance between (a) the curve showing seawater density variation with ocean depth and (b) the curve showing seawater temperature variation with ocean depth?

23. How is the color of the ocean surface water related to biological productivity? Why does everything in the ocean at depths below the shallowest surface water take on a blue-green appearance?

24. How does the decrease in temperature at the base of the warm surface water produce a SOFAR, or sound channel, below the ocean's surface? How is the SOFAR channel being used to determine if global warming has begun?

25. Explain the difference between an acid and an alkali (base) substance. How does the ocean's buffering system work?

26. Compare and contrast the following seawater desalination methods: distillation, solar humidification, and reverse osmosis.

References

Bailey, H. S., Jr. 1953. The voyage of the *Challenger. Scientific American* 188:5, 88–94.

Behrman, A. S. 1968. *Water is everybody's business: The chemistry of water purification.* Garden City, NY: Doubleday.

Charnock, H. 1996. H.M.S. *Challenger* and the development of marine science. In Pirie, R. G., ed., *Oceanography: Contemporary readings in ocean sciences,* 3rd ed. New York: Oxford University Press.

Davis, K. S., and Day, J. S. 1961. *Water: The mirror of science.* Garden City, NY: Doubleday.

Friederici, P. 1996. Across a crowded sea: How (and what) whales hear. *Pacific Discovery* 49:1, 28–31.

Hammond, A. L. 1977. Oceanography: Geochemical tracers offer new insight. *Science* 195:164–166.

Harvey, H. W. 1960. *The chemistry and fertility of sea waters.* New York: Cambridge University Press.

Kuenen, P. H. 1963. *Realms of water.* New York: Science Editions.

MacIntyre, F. 1970. Why the sea is salt. *Scientific American* 223:5, 104–115.

Munk, W. 1993. The sound of oceans warming. *The Sciences* 33:5, 20–26.

Packard, G. L. 1975. *Descriptive physical oceanography,* 2nd ed. New York: Pergamon Press.

Revelle, R. 1963. Water. *Scientific American* 209:3, 92–108.

Swenson, H. 1993. *Why is the ocean salty?* Washington D.C.: U.S. Geological Survey pamphlet.

The Open University Course Team. 1989. *Seawater: Its composition, properties and behaviour.* Oxford: Pergamon Press.

Wadhams, P. 1996. The resource potential of Antarctic icebergs. In Pirie, R. G., ed., *Oceanography: Contemporary readings in ocean sciences,* 3rd ed. New York: Oxford University Press.

Yokohama, Y., Guichard, F., Reyss, J-L., and Van, N. H. 1978. Oceanic residence times of dissolved beryllium and aluminum deduced from cosmogenic tracers ^{10}Be and ^{26}Al. *Science* 201:1016–1017.

Suggested Reading

Earth

Barnes-Svarney, P. 1992. Salt of the Earth. 1:1, 52–57. A look at the importance of salt and how deposits of salt form.

Sea Frontiers

Alper, J. 1991. Munk's hypothesis. 37:3, 38–43. Walter Munk's proposal to monitor global warming by transmitting sound signals across vast ocean reaches in the SOFAR channel is discussed.

Friedman, R. 1990. Salt-free water from the sea. 36:3, 48–54. The economics and nature of the various processes used in desalination are discussed.

Gabianelli, V. J. 1970. Water: The fluid of life. 16:5, 258–270. The unique properties of water are lucidly described and explained. Topics covered include hydrogen bond, capillarity, heat capacity, ice, and solvent properties.

Rice, A. L. 1972. HMS *Challenger:* Midwife to oceanography. 18:5, 291–305. An interesting account of the achievements recorded during the first major oceanographic expedition.

Smith, F., and Charlier, R. 1981. Saltwater fuel. 27:6, 342–349. The potential for using the salinity difference between river water and coastal marine water to generate electricity is considered.

Scientific American

Bailey, H. S., Jr. 1953. The voyage of the *Challenger.* 188:5, 88–94. A summary of the accomplishments of the English oceanographic expedition of 1872.

Baker, J. A., and Henderson, D. 1981. The fluid phase of matter. 245:5, 130–139. The structure of gases and liquids is modeled using hard spheres.

Buckingham, M. J., Potter, J. R., and Epiphany, C. L. 1996. Seeing underwater with background noise. 274:2, 86–91. A technique called acoustic-daylight imaging uses ambient sound in the ocean to form images just as daylight is used for that purpose in the atmosphere. The resolution may be a bit less than that used for seeing in daylight, but it has tremendous potential for use in the ocean.

Gerstein, M., and Leavitt, M. 1998. Simulating water and the molecules of life. 279:5, 100–105. Computer modeling reveals how the unusual properties of water affect biological molecules such as proteins and DNA.

Liss, T. M., and Tipton, P. L. 1997. The discovery of the top quark. 277:3, 54–59. A review of the quest to find the top (6th) quark, which involved the world's most energetic collisions and a cast of thousands.

MacIntyre, F. 1970. Why the sea is salt. 223:5, 104–115. A summary of what is known of the processes that add and remove the elements dissolved in the ocean.

Oceanography on the Web

Visit the *Essentials of Oceanography* home page for on-line resources for this chapter. There you will find an on-line study guide with review exercises, and links to oceanography sites to further your exploration of the topics in this chapter. *Essentials of Oceanography* is at: **http://www.prenhall.com /thurman** (click on the Table of Contents menu and select this chapter).

CHAPTER 6
AIR–SEA INTERACTION

RMS *Titanic:* Lost (1912) and Found (1985)

The early twentieth century was a time of rapidly advancing technology. Many people put their faith in this technology, and there was a widely held belief that science could find cures for diseases, contribute to a higher standard of living, and generally improve everyone's lives. It was during this time and with great acclaim that the world's largest passenger ship—and the world's largest movable object—of its day was constructed: the RMS *Titanic* (Figure 6A). The luxury ship was 269 meters (882 feet) in length, almost as long as three football fields. She was designed to be unsinkable, with multiple compartments that would keep her afloat even if some of the compartments filled with water. Because of her unsinkable design, certain safety precautions were ignored. For instance, the *Titanic* had only enough room on its lifeboats for about half of its passengers. Operated by the White Star Line, she departed on her maiden voyage in April 1912 to travel from England to the United States across the North Atlantic Ocean.

Icebergs in the North Atlantic Ocean have been a constant navigational hazard to ships because ocean currents carry the icebergs into shipping lanes. The icebergs originate from glaciers on land at Greenland and Ellesmere Island. Some of these icebergs are gigantic, and take years to melt in the ocean. Consequently, they can sometimes be carried as far south as 40 degrees north latitude, the same latitude as Philadelphia.

On the evening of April 14, the *Titanic* was traveling under a clear sky and calm seas at an excessive speed of 41 kilometers (25.5 miles) per hour through an area where icebergs had been reported in the vicinity of the Grand Banks south of Newfoundland. She was running ahead of schedule and was hoping to make a good impression by arriving in New York early. Even though the ship had received repeated warnings of iceberg hazard, the crew was unconcerned: After all, the ship was thought to be unsinkable. About a half-hour before midnight, she sideswiped a huge iceberg and sustained a 100-meter

Figure 6A RMS *Titanic,* circa 1912.

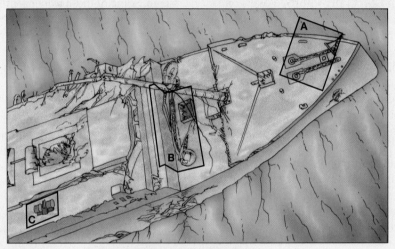

A.

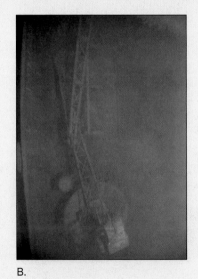

B.

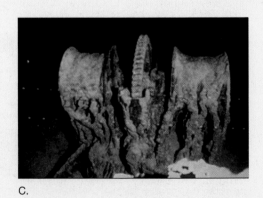

C.

Figure 6B RMS *Titanic* found.

(330-foot) slash in her hull. The *Titanic's* compartments began to fill with water, causing the ship to tip toward the front and resulting in more compartments filling with water. Within two-and-a-half hours after impacting the iceberg, the ship broke in half and sank. Because of the cold temperatures, the nighttime conditions, the lack of aid from other ships in the area, and the shortage of lifeboats, 1513 of the 2224 passengers perished. It was the world's greatest maritime disaster.

The tragedy brought about more stringent safety rules for ships and the formation of an iceberg patrol. The ice patrol regularly surveys an area larger than the state of Pennsylvania to prevent further loss of life from such accidents. The U.S. Navy began this patrol immediately following the *Titanic* disaster; it became international in 1914. Today, the International Ice Patrol is maintained by the U.S. Coast Guard.

There had been much interest in locating the wreck of the *Titanic*, but only recently did the technology become available to find it in such deep water. As well, the exact location of the collision was somewhat of a mystery, and it was unclear whether she sank straight down onto the sea floor or if she drifted on the way to the bottom. After several unsuccessful attempts, an expedition led by oceanographer Dr. Robert Ballard of the Woods Hole Oceanographic Institution discovered the wreck in September 1985 (Figure 6B). The expedition used the unmanned, tethered submersible *Argo* to locate the wreckage of the *Titanic* on the ocean floor.

The *Titanic* lay in two pieces at a depth of 3844 meters (12,612 feet), with debris scattered all around the wreckage. In 1986, the manned research submersible *Alvin* touched down on the deck of the *Titanic*, carrying the first humans actually to observe the *Titanic* in almost 75 years. Dr. Ballard believes that the wreck should be consecrated as a burial ground, and that retrieval of any items would constitute grave robbing. However, since Ballard's discovery of the *Titanic*, several other submersibles have visited the site and retrieved personal items and other belongings.

The atmosphere and the ocean act together as one interdependent system. What happens in one causes changes in the other, and the two are linked by complex feedback loops. For instance, the currents at the surface in the oceans are a direct result of the atmospheric wind belts of the world. Conversely, physical processes called weather occur in the atmosphere that are manifested in the oceans. If either the oceans or the atmosphere is truly to be understood, then their interactions and relationships must be considered.

Most of the currents and waves that exist in the surface ocean are created directly by atmospheric winds. The winds themselves are created by heating, which is provided by solar energy. Thus, the ultimate source of energy that drives the motion in the atmosphere and in the ocean is radiant energy from the sun. The atmosphere and ocean constantly exchange this energy. Together, the atmosphere and ocean are a powerful team that shape Earth's global weather patterns.

In addition to the day-to-day influence of the oceans on global weather, periodic extremes of atmospheric weather, such as droughts and profuse precipitation, are related to periodic changes in oceanic conditions. The well-publicized "El Niño" events of the equatorial Pacific Ocean appear to have such a relation. In fact, El Niño was one of the first major phenomena to spur study of air–sea interaction, because in the 1920s it was recognized that El Niño—an ocean event—was tied to catastrophic weather events worldwide. Certainly, El Niño illustrates the interdependence of the ocean and the atmosphere. However, it is not yet clear if changes in the ocean produce changes in the atmosphere that lead to the El Niño phenomenon—or vice versa. The El Niño–Southern Oscillation phenomenon is discussed in Chapter 7, "Ocean Circulation."

Another example of air–sea interaction is apparent in the greenhouse effect. In the 1980s, studies confirmed that Earth's average temperature had risen over the past century. This led to much investigation of the cause of the global warming phenomenon called the greenhouse effect. This warming of our planet's atmosphere may be due to human-caused increases in carbon dioxide and other gases that absorb heat. However, atmospheric carbon dioxide has increased only half as much as predicted from human activities, so where did the rest of the carbon dioxide go? If the oceans are responsible for removing carbon dioxide from the atmosphere, how does carbon dioxide enter the oceans and where does it go?

Answers to these questions are produced slowly as new research is completed and more sophisticated computer models are developed. In this chapter, we will examine the redistribution of solar heat by the atmosphere and its influence on oceanic conditions. First, large-scale phenomena that influence this interaction will be studied, and then smaller-scale phenomena will be examined.

Earth's Seasons

Earth revolves around the sun along an invisible ellipse in space (Figure 6–1). Imagine a flat surface in space that includes this ellipse: This flat surface is called the **ecliptic**. Earth's axis of rotation is not perpendicular ("upright") on the ecliptic; it leans on its side at an angle of 23.5 degrees. Figure 6–1 shows that Earth's tilted axis is parallel to the position of the axis at different places along Earth's orbital path. Consequently, the

Figure 6–1 Earth's seasons.
As Earth orbits the sun during one year, its axis of rotation constantly tilts 23.5 degrees from perpendicular (relative to the plane of the ecliptic) and causes the Earth to have seasons.

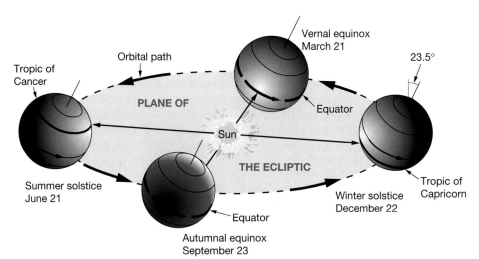

axis *always points in the same direction* throughout the yearly cycle.[1]

This tilt of Earth's rotational axis causes the seasons: spring, summer, fall, and winter. The yearly cycle of the seasons occurs as follows:

- At the **vernal equinox** (*vernus* = spring; *equi* = equal, *noct* = night), which occurs about March 21, the sun is directly overhead along the Equator. During this time, all places in the world experience equal lengths of night and day (hence the name *equinox*). In the Northern Hemisphere, the vernal equinox is also known as the spring equinox.

- On about June 21, the **summer solstice** (*sol* = the sun, *stitium* = a stoppage) occurs. At this time the sun reaches its most northerly point in the sky, directly overhead along the **Tropic of Cancer**, which is at 23.5 degrees north latitude.

- The sun then moves southward in the sky each day, and on about September 23 it is directly overhead along the Equator again, producing the **autumnal** (*autumnus* = fall) **equinox**. In the Northern Hemisphere, the autumnal equinox is also known as the fall equinox.

- During the next three months the sun is more southerly in the sky, until the **winter solstice** on about December 22, when the sun is directly overhead along the **Tropic of Capricorn**,[2] at 23.5 degrees south latitude.

As Earth orbits the sun during one year, its axis of rotation constantly tilts 23.5 degrees from perpendicular

relative to the plane of the ecliptic and causes the Earth to have seasons. Without Earth's 23.5 degrees tilt, seasonal differences would disappear. Because of the tilt, the sun's **declination** (angular distance from the equatorial plane) varies between 23.5 degrees north and 23.5 degrees south of the Equator on a yearly cycle.

These seasonal changes have a profound influence on Earth's climate by affecting the sun angle and the length of day experienced at any location on Earth throughout the year. For example, the longest day in the Northern Hemisphere occurs on the summer solstice and the shortest day in the Northern Hemisphere is on the winter solstice. In addition to these seasonal changes, daily heating of Earth influences climate.

Uneven Solar Heating on Earth

Essentially all of the energy available to Earth comes from the sun. The side of Earth facing the sun (the daytime side) is exposed to intense solar radiation and receives a tremendous amount of incoming solar energy. It is this energy that drives the global ocean–atmosphere engine, creating pressure and density differences that stir currents and waves in both air and sea.

Distribution of Solar Energy

Image for a moment that Earth is not spherical, but a flat plate in space, with a flat side directly facing the sun (that the plate is perpendicular to the sun). If this were the case, sunlight would fall equally on all parts of Earth. However, Earth is spherical, so solar radiation strikes its surface at 90 degrees (perpendicular) only at the center of the sphere (Figure 6–2, point A). Anywhere else on the sphere, sunlight strikes at angles less than 90 degrees (for example, at point B), down to 0 degrees at the edge of the sphere. The significance of this is that *solar radiation is most intense at the perpendicular point but becomes progressively less intense toward the poles* because sunlight is spread over a larger area at high latitudes.

---

[1] Earth's axis points toward a distant star called Polaris, the North Star.

[2] The tropics are named for their constellations: directly over the Tropic of Cancer is the constellation Cancer (the crab) and over the Tropic of Capricorn is Capricorn (the goat).

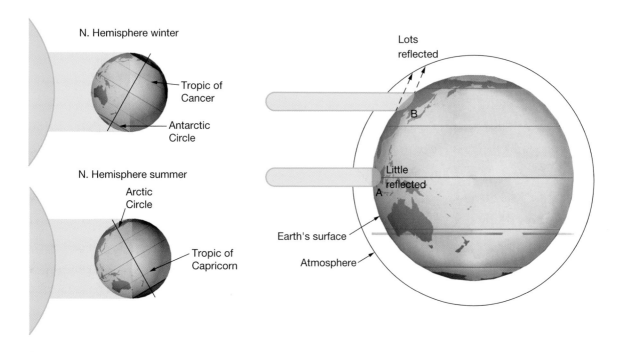

Figure 6–2 Solar radiation received on Earth.

The amount of solar energy received at higher latitudes is much less than that at lower latitudes (*right*). During the Northern Hemisphere winter (*upper left*), the Southern Hemisphere is tilted toward the sun and the Northern Hemisphere receives reduced radiation. During the Northern Hemisphere summer (*lower left*), the situation is reversed.

As an example, consider point A in Figure 6–2, where a beam of solar radiation is striking Earth. The beam of sunlight covers exactly 1 square kilometer (0.4 square mile). By contrast, consider an identical beam of sunlight falling on a polar region of Earth's surface (point B), where it does not strike at a right angle. Here the square kilometer of radiation spreads out over considerably more than a square kilometer of Earth's surface.

Three other factors also decrease the amount of radiation received at high latitudes:

- Less radiation strikes Earth at high polar latitudes than at low latitudes, because sunlight passes through a *greater apparent thickness* of atmosphere at high latitudes.
- More radiation is reflected at high latitudes because of the high reflectivity of ice at high latitudes as compared to other Earth surfaces.
- More radiation is reflected at high latitudes due to the low angle at which it strikes Earth's surface. The angle at which direct sunlight strikes the ocean sur-

face is important in determining how much of the solar energy is absorbed and how much is reflected. If the sun shines down on a smooth sea from directly overhead, only 2 percent of the radiation is reflected, but if the sun is at a very low angle, only 5 degrees above the horizon, 40 percent of the radiation is reflected back into the atmosphere (Table 6–1).

Because of all these reasons, the intensity of radiation at high latitudes is greatly decreased compared with that falling in the equatorial region. Figure 6–2 also shows that throughout the year, the sun is directly overhead between the Tropic of Cancer and the Tropic of Capricorn, affording intense, closer-to-vertical radiation to the hot equatorial area. The region between these two tropics (called, appropriately enough, the **tropics**) receives much greater annual radiation per square kilometer than polar areas.

Calculating the amount of radiation received at any point on Earth is complicated by two important Earth movements: its daily rotation, and its tilt combined with its annual revolution around the sun. Lets examine both of these movements:

Table 6–1 Reflection and absorption of solar energy relative to the angle of incidence on a flat sea

Elevation of sun above horizon	90°	60°	30°	15°	5°
Reflected radiation (%)	2	3	6	20	40
Absorbed radiation (%)	98	97	94	80	60

Figure 6–3 Heat gain and heat loss from oceans.

Heat gained by the oceans at equatorial latitudes (*red*) equals heat lost at polar latitudes (*blue*). This creates a balanced heat budget for Earth's oceans.

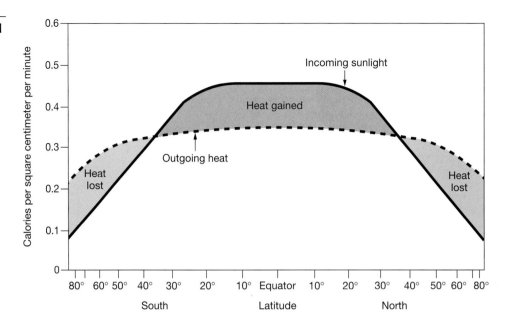

- Earth's rotation on its axis ensures that the sunlight received varies *daily*, simply due to the rotation through daylight and darkness each day.

- Earth's axis of rotation is tilted 23.5 degrees with respect to the ecliptic. As Earth revolves on its annual trip around the sun, the Northern and Southern Hemispheres take turns "leaning toward" the sun every six months. Consequently, the amount of solar radiation received on Earth varies *annually*, causing the change of seasons.

During the Northern Hemisphere winter, a significant portion of Earth's surface north of the **Arctic Circle** (66.5 degrees north latitude) receives no direct solar radiation at all, and the Northern Hemisphere receives reduced radiation because of the low sun angles. This causes areas north of the Arctic Circle to experience up to six months of darkness, while areas south of the **Antarctic Circle** (66.5 degrees south latitude) receive continuous radiation ("midnight sun"). Half a year later, during the Northern Hemisphere summer (the Southern Hemisphere winter), the situation is reversed.

Oceanic Heat Flow

Because of the low angle at which solar radiation strikes Earth's surface toward the poles, and the very reflective ice cover in polar latitudes, more energy is reflected back into space than is absorbed. In contrast, because of the high angle at which sunlight strikes Earth between 35 degrees north latitude and 40 degrees south latitude, more energy is absorbed between these latitudes than is radiated back into space.[3] Figure 6–3 shows how this phe-

[3] The reason this latitudinal range extends farther into the Southern Hemisphere is because the Southern Hemisphere has more ocean surface area in the mid-latitudes than the Northern Hemisphere does.

nomenon is manifested in the average daily heat flow in oceans by latitude.

It might be expected that, as a direct result of this condition, the equatorial zone would grow warmer as the years pass and the polar regions would become progressively cooler. This is not the case; the polar regions always are considerably colder than the equatorial zone, but the temperature *difference* is not increasing. Clearly, excess heat at equatorial latitudes is being transferred toward the poles. How is this accomplished? The mechanism involves heat being transferred by both the oceans and the atmosphere during circulation.

The Atmosphere: Physical Properties

The atmosphere is important in transferring heat and water vapor from place to place on Earth. Within the atmosphere, there are complex relationships among air composition, temperature, density, water vapor content, and pressure. Before we apply these relationships, let's examine the atmosphere's composition and some of its physical properties.

Composition   The composition of dry air (with no water vapor) is shown in Table 6–2. The atmosphere is composed mostly of nitrogen and oxygen gases. Argon is an inert gas, and all other gases are present in very

Table 6–2   Composition of dry air.

Gas	Concentration (%)
Nitrogen (N_2)	78.1
Oxygen (O_2)	20.9
Argon (Ar)	0.9
Carbon dioxide (CO_2)	0.035
All others	Trace

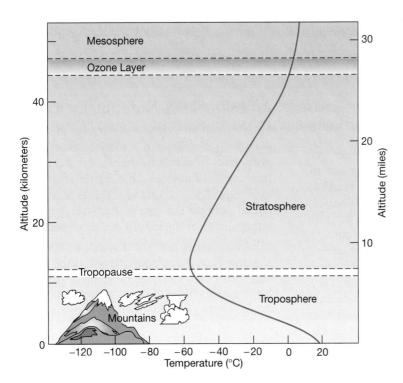

Figure 6–4 Temperature profile of the atmosphere.

small amounts. However, these small amounts can be significant for trapping heat within the atmosphere.

Temperature Figure 6–4 shows a temperature profile through the atmosphere. The lowermost portion of the atmosphere extends from the surface to 12 kilometers (7 miles) and is called the **troposphere** (*tropo* = turn, *sphere* = a ball). There is much atmospheric mixing in the troposphere, which gives the troposphere its name. The troposphere is where all weather is produced. Within the troposphere, it gets cooler with increasing distance above Earth. At high altitudes, the air temperatures are often below freezing. For instance, if you have ever flown in a jet airplane, you may have noticed that any water on the wings or inside your window will freeze during the flight.

Density It may seem surprising that air has density, but since air is composed of molecules, it certainly does. Temperature affects density dramatically, because at higher temperatures the air molecules are moving more quickly and take up more space. Thus, the general relationship between density and temperature is as follows:

- Warm air is less dense, so it rises.
- Cool air is more dense, so it sinks.

It is commonly known that "heat rises," which can be experienced in a room with a heater (Figure 6–5). The heater warms the nearby air, thus making it less dense and causing it to rise. Conversely, a cool window will cause nearby air to cool, causing it to become more dense and sink (Figure 6–5). In this scenario, a **convection cell** will form, composed of the rising and sinking air moving in a circular fashion. Similar to convection in Earth's

mantle discussed in Chapter 2, this convection cell is a circular loop of air: The air warms and rises, then cools and sinks, thus completing the loop.

Water Vapor Content One property related to temperature and density is the air's water vapor content. Warm air can hold more water vapor, so it is typically moist.[4] This is because the air molecules are moving more quickly and come into contact with more water vapor. Conversely, cool air is typically dry. A practical application of this principle is that the best day to hang laundry outside to dry is a warm, breezy day, which speeds evaporation.

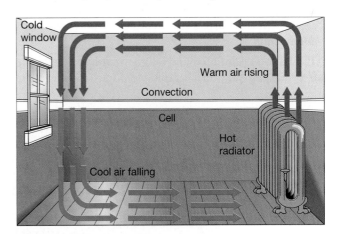

Figure 6–5 Convection in a room.
A circular-moving loop of air (convection cell) is caused by warm air rising and cool air sinking.

---

[4] The addition of water vapor decreases the density of air because water vapor has a lower density than air.

Pressure   Atmospheric pressure is 1.0 atmosphere (14.7 pounds per square inch) at sea level and decreases with increasing altitude. Atmospheric pressure in *non-moving (static) air* is related to the weight of the column of air above. A thicker air column above a location will produce higher atmospheric pressure then at a location where there is a thinner column of air. An analogy to this is water pressure in a swimming pool: the thicker the column of water above, the higher the pressure. Thus, the highest pressure in a pool is at the bottom of the deep end. Similarly, air pressure is higher where there is a thicker column of air—at sea level. At higher elevations, the air pressure decreases. You may have experienced this when sealed bags of potato chips or pretzels are taken to a high elevation. The pressure at high elevations is much lower than where they were sealed, causing the bags to become so inflated that they sometimes burst! You may have also experienced this change in pressure yourself when your ears "pop" during the takeoff or landing of an airplane or while driving on mountain roads.

Moving (dynamic) air also causes changes in atmospheric pressure, caused by changes in the molecular density of the air. The general relationship is as follows (Figure 6–6):

- Sinking cool air (movement *toward* surface, compression) causes high pressure at the surface.
- Rising warm air (movement *away from* the surface, expansion) causes low pressure at the surface.

In addition, sinking air warms due to compression. Conversely, rising air cools because of expansion.

Movement   Although there are complex relationships among air composition, temperature, density, water vapor content, and pressure, the movement of air is straightforward. Air *always* moves from high- to low-pressure regions. On the Earth's surface, this movement of air is called *wind*.

If a balloon is inflated and let go, what happens to the air inside the balloon? It rapidly escapes, moving from a high-pressure region inside the balloon (caused by the balloon pushing on the air inside) to the lower-pressure region outside the balloon. This is one science experiment that you can certainly try at home!

An Example: A Nonspinning Earth

Let's look at an example of some of the concepts of uneven heat distribution on Earth and how the physical properties of the atmosphere affect heat distribution. Imagine for a moment that Earth is not spinning on its axis (Figure 6–7). Although this is fictional, it is a simplified example that can be used to help understand some important relationships in the atmosphere.

Because more solar radiation is received along the Equator than at the poles, the air at the Equator in contact with Earth's surface is warmed. This warm air, typically moist, rises and cools. What kind of air pressure is caused by air rising and moving away from the Equator? Rising air creates low pressure at the surface. This rising air cools (see Figure 6–4) and releases its moisture in the form of rain. So, along the Equator of our fictional non-spinning Earth, there is a zone of low pressure and lots of precipitation.

As the air along the Equator rises upward, it comes to the top of the troposphere and begins to move horizontally. It will move laterally in the upper atmosphere toward the poles. Since the temperature is much cooler at high altitudes, the air will cool and its density will increase. This high-density cool air will have a tendency to sink, and sink it does at the poles. What kind of air pressure is produced by sinking air moving toward the surface at the poles? Sinking air creates high pressure at the surface. This sinking air will be quite dry because cool air cannot hold much water vapor. Thus, at the poles, there will be high pressure and fair, dry, clear weather.

Which way will surface winds blow? Remembering that air always moves from high to low pressure, the air will travel from the high pressure at the poles toward the low pressure at the Equator. Thus, there will be strong

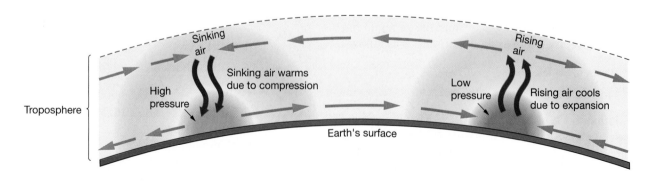

Figure 6–6   High and low atmospheric pressure zones.
Sinking cool air warms due to compression and causes high pressure at the surface (*left*). Rising warm air cools due to expansion and causes low pressure at the surface (*right*).

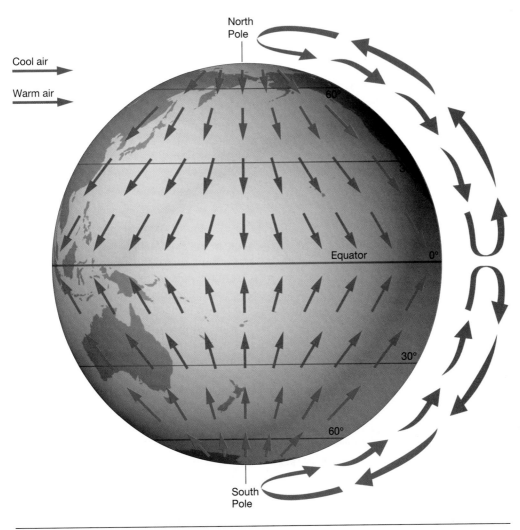

Figure 6–7 A non-spinning Earth.
A fictional non-spinning Earth, showing the pattern of winds that would develop due to uneven solar heating on Earth.

northerly winds in the Northern Hemisphere and strong southerly winds in the Southern Hemisphere.[5] Along the way toward the Equator, the air warms. Finally, it arrives back where it started: warm air rising at the Equator, cooling and moving poleward in the upper troposphere, sinking at the poles, and moving along the surface to complete the cycle. In essence, the air travels in a loop (called a convection cell), driven by uneven heating on Earth.

Is this fictional case of a nonspinning Earth a good analogy to what is really happening on Earth? Actually, it is not, even though the *principles* that drive the physical movement of air remain the same whether Earth is spinning or not. Let's examine how the important factor of Earth's spin influences atmospheric circulation.

The Coriolis Effect

The **Coriolis effect** is named after a French engineer, Gaspard Gustave de Coriolis, who first calculated this effect in 1835. It is often incorrectly called the Coriolis *force*. However, it is not a true force in that it does not provide an *acceleration* of a body, and hence does not influence a moving object's speed. The Coriolis effect changes only the *intended path of travel* of a moving body and thus is truly an effect.

The Coriolis effect causes moving objects on Earth to follow curved paths. In the Northern Hemisphere, an object will follow a path to the *right* of its intended direction; in the Southern Hemisphere, an object will follow a path to the *left* of its intended direction. Note that the directions right and left are the *viewer's perspective looking in the direction in which the object is traveling*. For example, the Coriolis effect very slightly influences the movement of a ball thrown between two people. In the

[5] Notice that winds are named based on the direction *from which they are moving*. The same is true for ocean currents.

Figure 6–8 A merry-go-round.

See text for description of paths A, B, and C.

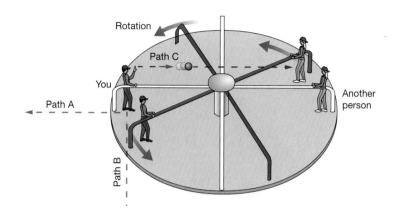

Northern Hemisphere, the ball will veer slightly to its right *from the thrower's perspective.*

The Coriolis effect acts on all moving objects. However, it is much more pronounced on objects traveling long distances, especially north or south. Thus, the Coriolis effect has a dramatic effect on atmospheric circulation and ocean current movement, as we shall see.

The Coriolis effect is caused by the fact that the Earth is rotating toward the east. More specifically, it is the difference in the speed of Earth's rotation at different latitudes that causes the Coriolis effect. In reality, objects travel along straight-line paths.[6] It is the Earth that is actually rotating underneath and causing objects to appear to curve. Let's look at two examples to help clarify this.

Example 1: Perspectives and Frames of Reference on a Merry-Go-Round

A playground merry-go-round is a wonderful experimental apparatus with which to test some of the concepts of the Coriolis effect. A merry-go-round is a large circular wheel that rotates around its center. It has bars that people hang onto while the merry-go-round spins. If you have ever been on a merry-go-round, you know how this works.

Imagine that you are on a merry-go-round that is spinning counterclockwise as viewed from above (Figure 6–8). Let's suppose further that as you are spinning around on the merry-go-round, you let go of the bar. What will happen to you? If you guessed that you would fly off along a straight line path perpendicular to the merry-go-round (Figure 6–8, path A), that's not quite right. Your angular momentum would propel you *in a straight line* tangent to your circular path on the merry-go-round at the point where you let go (Figure 6–8, path B). The law of inertia states that a moving object will follow a straight-line path until it is compelled to change that path by other forces. Thus, your trajectory takes you along a straight-

line path (*path B*) until you collide with some object such as other playground equipment or the ground. Note that from the perspective of a person on the merry-go-round, your departure along path B would *appear* to be curved to the right because of the merry-go-round's rotation.

Let's suppose that you survived the collision and returned to spinning around on the merry-go-round in a counterclockwise direction. Let's say that you are now joined by another person who is facing you directly but on the opposite side of the merry-go-round. If you were to toss a ball to the other person, what would the path of the ball look like? Even though you threw the ball straight at the other person, from *your perspective* the ball's path would appear to curve to its right. That's because the frame of reference (in this example, the merry-go-round) has moved during the time that it took the ball to arrive there (Figure 6–8). Being on the rotating frame of reference—the merry-go-round—causes the ball to appear to curve to its right for those on the merry-go-round. A person with the perspective of viewing the merry-go-round from directly overhead would observe that the ball did indeed travel along a straight-line path (Figure 6–8, path C), just as your path was straight when you let go of the merry-go-round bar. Similarly, the perspective of being on the rotating Earth causes objects to appear to travel along curved paths. This is the Coriolis effect. The analogy of the merry-go-round spinning in a counterclockwise direction is similar to the Northern Hemisphere: As viewed from above the North Pole, Earth is spinning counterclockwise. Thus, moving objects appear to follow curved paths to the *right* of their intended direction in the Northern Hemisphere.

Notice that if the other person on the merry-go-round threw a ball toward you, it would also appear to curve. From the perspective of the other person, the ball curved to its right, just as the ball you threw curved. However, from your perspective, the curvature of the ball thrown toward you would be to its *left*. The perspective to keep in mind when considering the Coriolis effect is the perspective of *looking in the same direction that the object is moving.* For example, the person who threw the ball toward you did indeed see that the ball curved to its right *from his or her perspective.*

---

[6] Newton's first law of motion (the law of inertia) states that every body persists in its state of rest or of uniform motion in a *straight line* unless it is compelled to change that state by forces imposed on it.

To simulate the Southern Hemisphere, the merry-go-round would need to be turned in a *clockwise* orientation, which is the rotational direction of Earth when viewed from above the South Pole. Thus, moving objects appear to follow curved paths to the *left* of their intended direction in the Southern Hemisphere.

Example 2: A Tale of Two Missiles

Now that we have established how the Coriolis effect influences moving objects, let's look at a "real-world" example. Figure 6–9 shows that as Earth rotates on its axis, points at different latitudes rotate at different velocities. The velocity changes with latitude and ranges from 0 kilometers per hour at the poles to over 1600 kilometers (1000 miles) per hour at the Equator. *This difference in velocity with latitude is the true cause of the Coriolis effect.* Let's look at another example to illustrate this difference in velocity with latitude.

This example deals with sending missiles from one point to another on Earth and following their trajectories. Let's imagine that we have two missiles that fly only in straight-line paths toward their destinations. For simplicity, let's assume that the flight of each missile takes one hour regardless of the distance flown. Let's travel to the North Pole to launch the first missile. We'll aim this missile toward a point on the 30 degree north latitude line—say, to New Orleans in Louisiana (Figure 6–10). Does the missile land in New Orleans? Actually, no. Since Earth is rotating eastward at 1400 kilometers (870 miles) per hour along the 30 degrees latitude line (see Figure 6–9), the missile will land somewhere near El Paso in west Texas, 1400 kilometers west of its target. From your perspective watching the missile, you would notice that the path of the missile appeared to curve *to its right* in accordance with the Coriolis effect. In reality, New Orleans has rotated out of the line of fire due to Earth's rotation.

Now let's do something different. Let's go to the Galápagos Islands, which are directly south of New Orleans along the Equator, and fire the second missile toward New Orleans (Figure 6–10). From its position on the Equator, the Galápagos Islands are moving east at 1600 kilometers (1000 miles) per hour, 200 kilometers (124 miles) per hour faster then New Orleans (see Figure 6–9). This means that at take-off, the missile is also moving toward the east 200 kilometers per hour faster than New Orleans. Thus, when the missile returns to Earth one hour later at the latitude of New Orleans, it will land offshore of Alabama, 200 kilometers east of New Orleans. Again, from your perspective, the missile appears to curve *to its right*. Note that the curvature to the right in these examples ignores friction, which would greatly reduce the magnitude of the missles' deflection to the right of their intended course.

Changes in the Coriolis Effect with Latitude

The missiles in the example above illustrate that the Coriolis effect changes with latitude. Why did the first missile, shot from the North Pole, miss the target by more

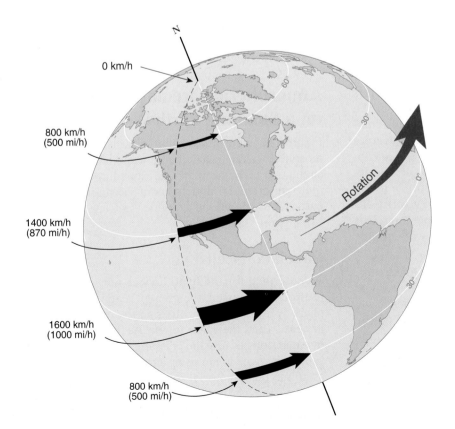

Figure 6–9 The Coriolis effect.

Earth spins at a steady rate, but the actual velocity of any point varies with latitude. Earth's speed at the Equator is about 1600 kilometers (1000 miles) per hour and diminishes poleward to a speed of 0 kilometers per hour at either pole.

0 km/h

800 km/h (500 mi/h)

1400 km/h (870 mi/h)

1600 km/h (1000 mi/h)

800 km/h (500 mi/h)

Figure 6–10 Missile trajectories.

The trajectories of two missiles shot toward New Orleans; one from the North Pole, the other from the Galápagos Islands on the Equator. Dashed lines indicate intended trajectories; solid lines indicate trajectories that the missiles would travel if viewed from Earth's surface.

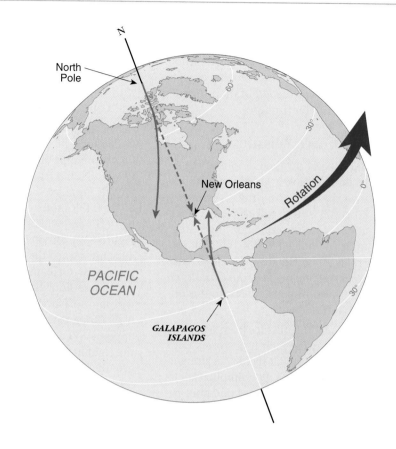

than the second missile did?[7] The answer is that a difference exists in the rotational velocity of points on Earth. The rotational velocity of points on Earth ranges from 0 kilometers per hour at the poles to more than 1600 kilometers (1000 miles) per hour at the Equator. However, the *rate of change* with each degree of latitude is not constant. As the pole is approached from the Equator, the rate of change of rotational velocity (per degree of latitude) increases.

For example, there is a difference of 200 kilometers (124 miles) per hour in velocity of rotation from the Equator (0 degrees) to 30 degrees north latitude. In comparison, from the equivalent distance of 30 to 60 degrees north latitude, there is a difference in velocity of 600 kilometers (372 miles) per hour. If we consider the next 30 degrees of latitude, from 60 degrees north latitude to the North Pole (where the velocity is zero), the velocity difference is over 800 kilometers (500 miles) per hour. This explains why the Coriolis effect increases with latitude toward the poles in a nonlinear fashion.

Thus, the maximum Coriolis effect is at the poles, causing a large deflection of moving objects. Alternatively, at the Equator, there is no Coriolis effect and no deflection of moving objects. However, the magnitude of the Coriolis effect depends much more on the length of time

that an object (such as an air mass or an ocean current) is in motion. Even at low latitudes, where the Coriolis effect is small, a large Coriolis deflection is possible if an object is in motion for a long time. In addition, note that because the Coriolis effect is caused by the *difference* in velocity of different latitudes on Earth, there is no Coriolis effect for those objects moving due east or due west along the equator.

Atmospheric Circulation Cells on a Spinning Earth

Figure 6–11 shows how heat is transferred in the atmosphere, taking into account a spinning Earth. These circulation cells and their corresponding wind belts are quite important for understanding ocean surface current patterns, which will be examined in Chapter 7, "Ocean Circulation."

Circulation Cells

As described for the hypothetical, nonspinning Earth previously discussed, the greater heating of the atmosphere over the Equator causes air to expand, to decrease in density, and therefore to rise. As the air rises, it cools by expansion because the pressure is lower, and the water vapor contained in the rising air mass condenses and falls as rain in the equatorial zone. After losing its moisture, this dry air mass travels some distance north or south of the Equator. When the air becomes drier and cooler, it

----

[7] Notice that the first missile missed its target by 1600 kilometers (1000 miles), versus only 200 kilometers (124 miles) for the second missile.

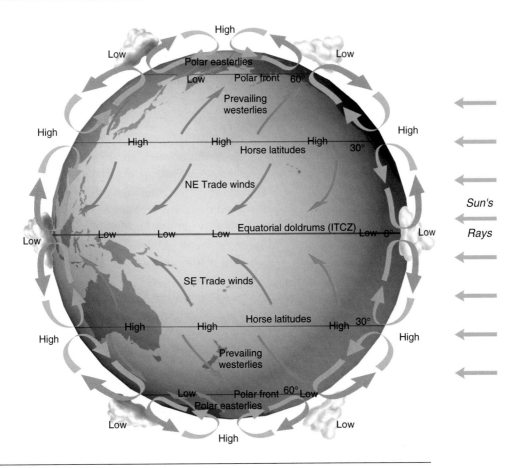

Figure 6–11 Atmospheric circulation and wind belts of the world.
The wind belts of the world are formed by atmospheric circulation cells. The general pattern
shown here is modified by seasonal changes and the distribution of continents.

becomes denser than the surrounding air and begins to descend in the subtropical regions, around 30 degrees latitude in both the Northern and Southern Hemispheres, to complete the loop. Thus, a circulation cell exists in each hemisphere between the Equator and about 30 degrees latitude (purple arrows in Figure 6–11). These circulation cells are called **Hadley cells**, named after a noted meteorologist.

There are two more circulation cells in each hemisphere: one between 30 and 60 degrees latitude, and one between 60 and 90 degrees latitude. These are the **Ferrel cells** (named after another famous meteorologist) and the **polar cells**, respectively. Notice that the Ferrel cell is not driven solely by differences in solar heating; if it were, air within it would circulate in the opposite direction. It is driven mostly by its relationship to the other two cells. Similar to the movement of interlocking gears in a row, the Ferrel cell moves in the direction that coincides with the movement of the two adjoining circulation cells.

Pressure

As we have learned, air moving *toward* the surface creates high pressure. Thus, the descending, dense air at about 30 degrees north and south latitudes and at the

poles creates high-pressure belts in those areas. The high-pressure zones at about 30 degrees north or south latitude are called the **subtropical highs**; the high-pressure regions at the poles are called the **polar highs**.

What kind of weather is experienced in these high-pressure areas? Since the descending air is quite dry, and because air tends to warm under its own weight, these areas typically experience dry, clear, fair conditions. Notice that the conditions are not necessarily warm (such as at the poles)—just dry and associated with clear skies.

Conversely, over the Equator and at about 60 degrees north or south latitude, the low density of the rising air creates a band of low pressure that encircles the globe in these areas. The low-pressure zone at the Equator is called the **equatorial low**, while the low at about 60 degrees north or south latitude is called the **subpolar low**. As you might expect, the weather is dominated by cloudy conditions with lots of precipitation, since the rising air cools and cannot hold its water vapor.

Wind Belts

The lowermost portion of the circulation cells—that is, the part that is closest to the surface—generates the major wind belts of the world. The masses of air that move across

Earth's surface from the subtropical high-pressure belts toward the equatorial low-pressure belt constitute the **trade winds**. These steady winds are named from the term *to blow trade*, which means to blow in a regular course. If Earth did not rotate, these winds would follow a simple north-south movement. However, in the Northern Hemisphere the **northeast trade winds** do not blow directly north to south but from the *northeast* to the *southwest*. This is the result of the Coriolis effect. In the Northern Hemisphere, the Coriolis effect causes a curvature to the *right* of its intended direction. Conversely, in the Southern Hemisphere, where the trade winds blow from south to north and the Coriolis curvature is to the *left* of its intended direction, the **southeast trade winds** occur.

Some of the air that descends in the subtropical regions moves along Earth's surface to higher latitudes as the **prevailing westerly wind belts**. These air masses rise over the dense, cold air moving away from the polar high-pressure caps at the subpolar low-pressure belts located near 60 degrees north and 60 degrees south latitude. Again, these wind belts are influenced by the Coriolis effect. Thus, in the Northern Hemisphere, the prevailing westerlies blow from southwest to northeast; in the Southern Hemisphere, they blow from northwest to southeast. The prevailing westerlies are named because they flow generally from the west, but specifically from the southwest (Northern Hemisphere) or from the northwest (Southern Hemisphere).

The air moving away from the poles produces the **polar easterly wind belts**. Located between one of the poles and about 60 degrees north or south latitude, these winds are deflected by the strong Coriolis curvature that high latitudes experience. Thus, in the Northern Hemisphere the polar easterlies blow from the northeast, and in the Southern Hemisphere they blow from the southeast.

Boundaries

The boundaries between the wind belts have interesting histories and correspondingly unusual names. The boundary between the two trade wind belts along the Equator is known as the **doldrums** (*doldrum* = dull). Long ago, sailors referred to this region as the doldrums because their sailing ships were becalmed by the lack of winds. Sometimes stranded for days or weeks, the situation was unfortunate but not life-threatening: Daily rainshowers supplied sailors with plenty of fresh water. Today, meteorologists refer to this region as the **Intertropical Convergence Zone (ITCZ)**, because it is the region between the tropics where the trade winds converge (Figure 6–11).

The boundary between the trade winds and the prevailing westerlies centered at 30 degrees north or south latitude is known as the **horse latitudes**. High pressure and clear, dry, fair conditions dominate these regions, where the air is sinking and surface winds are light and variable. For sailors on a ship powered only by sails, becoming stuck in the horse latitudes was a potentially serious problem. Here, sailors could run out of fresh water. Legend has it that some ships tried to conserve water by throwing some of their livestock over the side of the vessel. Hence, the name "horse latitudes."

The boundary between the prevailing westerlies and the polar easterlies at 60 degrees north or south latitude is known as the **polar front**. This is a battleground for different air masses, so cloudy conditions and lots of precipitation are common here.

As previously discussed, directly at the poles is a high-pressure area known as the polar high. Clear, dry, fair conditions are associated with the polar high pressure, and precipitation is minimal. In fact, the poles are often classified as cold deserts based on the lack of precipitation that is experienced in these areas.

Circulation Cells: Idealized or Real?

These latitudinal patterns do exist on Earth, but are presented here in an idealized form to help clarify atmospheric circulation. In reality, the patterns are significantly altered by the uneven distribution of land and ocean over Earth's surface. The general effects of this idealized system are, however, clearly visible on a broad scale.

The idealized pressure belts and resulting global wind systems shown in Figure 6–11 are significantly modified by two factors:

1. Earth's tilt on its axis of rotation, which produces seasons

2. The fact that the air over continents gets colder in winter and warmer in summer than the air over adjacent oceans

As a result of these two factors, and because continental rocks have a lower heat capacity than seawater,[8] continents become hotter in the summer and colder in the winter than the adjacent oceans. Thus, during winter the continents usually develop atmospheric high-pressure cells due to the weight of cold air centered over them, and during the summer they develop low-pressure cells (Figure 6–12). Otherwise, the pattern of atmospheric high- and low-pressure zones shown in Figure 6–12 corresponds closely to those shown in Figure 6–11. A summary of the characteristics of global wind belts and boundaries is provided in Table 6–3.

The Oceans, Weather, and Climate

Weather describes the conditions of the atmosphere at a given time and place. **Climate** is the long-term average of weather. If we observe the weather conditions over a long period, we can begin to say some things about the climate. For instance, if the weather in an area over

---

[8] An object that has low heat capacity heats up quickly when heat energy is applied. Water has high heat capacity.

Box 6–1
Why Christopher Columbus Never Set Foot on North America

The Italian navigator and explorer **Christopher Columbus** is widely credited for discovering America in the year 1492. However, America was already populated with many natives, and the Vikings predated his voyage to North America by about 500 years. In Columbus's four voyages, he never did set foot on continental North America. This was due largely to the pattern of the major wind belts of the world, which dictated the travels of his sailing ships.

The idea for a voyage such as the one Christopher Columbus undertook was initiated by an astronomer in Florence, Italy, named Toscanelli. He wrote to the king of Portugal, suggesting that a course be charted to the west in an attempt to reach the East Indies (today the country of Indonesia) by crossing the Atlantic Ocean. Columbus later contacted Toscanelli and was given a copy of the information, indicating that India could be reached at a certain distance. Today, we know that this distance would have carried him just west of the continent of North America.

Rather than sailing to the east as had been done traditionally, Columbus was determined to reach the East Indies by sailing *west*. After years of difficulties in initiating the voyage, Columbus received the backing of the Spanish monarchs Ferdinand V and Isabella I. He finally set sail with 88 men and three ships (the *Niña*, the *Pinta*, and the *Santa María*; Figure 6C) on August 3, 1492, from the Canary Islands off Africa. The Canary Islands are located at 28 degrees north latitude and are within the northeast trade winds, which are steady winds that blow from the northeast. Instead of following the line of latitude directly across the Atlantic to North America (28 degrees north latitude on North America is about the latitude of central Florida), Columbus's sailing ships

traveled along a more southerly route (aided by the northeast trade winds, which blow to the southwest). Thus, Columbus "sailed the parallel" that was farther south than North America (Figure 6D).

During the morning of October 12, 1492, the first land was sighted, which is generally believed to have been Watling Island in the Bahama Islands southeast of Florida. Based on the inaccurate information he had been given, Columbus was convinced that he had arrived in the East Indies and was somewhere near India. Consequently, he called the inhabitants "Indians," and the area is known today as the West Indies. Later on this voyage, he landed on Cuba and Hispaniola.

On his return journey, his ships sailed to the northeast and picked up the prevailing westerlies, which transported him away from North America and toward Spain. Upon his return to Spain and the announcement of his discovery, additional voyages were planned. Columbus made three more trips across the Atlantic Ocean, following similar paths through the Atlantic where his ship's paths were controlled by the trade winds on the outbound voyage and by the prevailing westerlies on the return trip. His next voyage was in 1493 with 17 ships, during which he explored Puerto Rico and the Leeward Islands and established a colony on Hispaniola. In 1498, he explored Venezuela and landed on South America, unaware that it was a continent that was new to Europeans. On his last voyage in 1502, he reached Central America. Although he is today considered a master mariner, he died in neglect in 1506, still convinced that he had explored islands near India. Even though he never set foot on North America, his journeys inspired other Spanish and Portuguese navigators to explore the "New World" including the coasts of North and South America.

Figure 6C A modern-day replica of the *Niña*.

(continued)

(continued)

Figure 6D Route of Christopher Columbus's first voyage.

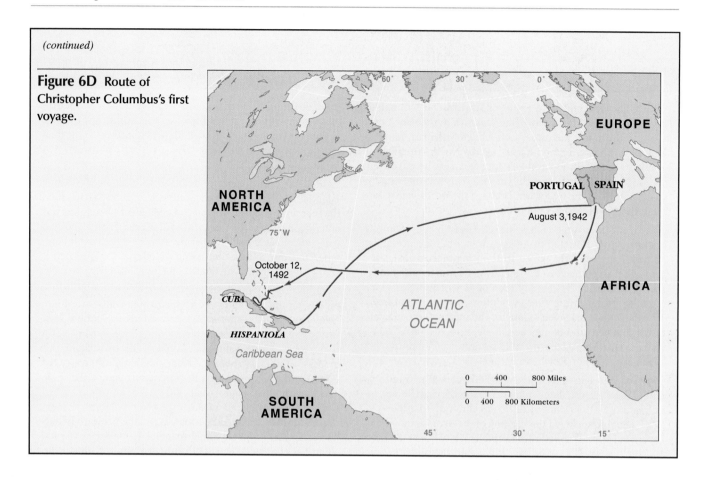

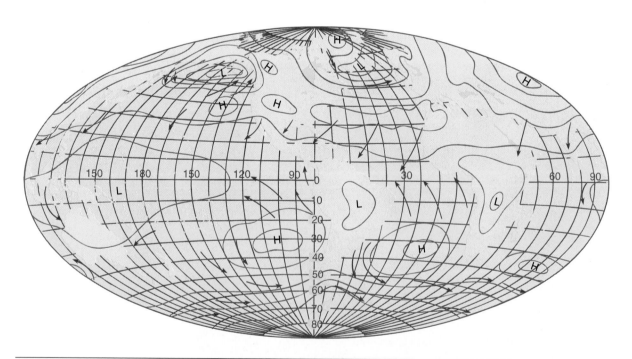

Figure 6–12 January sea-level atmospheric pressures and winds.

Average atmospheric pressure pattern for January. High (H) and low (L) atmospheric pressure zones correspond closely to those shown in Figure 6–11, but are modified by the change of seasons and the distribution of continents. Arrows show direction of winds.

Table 6–3   Characteristics of wind belts and boundaries

Region (north or south latitude)	Wind belt or boundary name	Atmospheric pressure	Characteristics
Equator (0 degrees)	Doldrums	Low	Light, variable winds. Abundant cloudiness and much precipitation. Breeding ground for hurricanes.
0–30 degrees	Trade winds	—	Strong, steady winds generally from the east.
30 degrees	Horse latitudes	High	Light, variable winds. Dry, clear, fair weather with little precipitation. Major deserts of the world.
30–60 degrees	Prevailing westerlies	—	Winds generally from the west. Brings storms that influence weather across U.S.
60 degrees	Polar front	Low	Variable winds. Stormy, cloudy weather year-round.
60–90 degrees	Polar easterlies	—	Cold, dry winds generally from the east.
Poles (90 degrees)	Polar high pressure	High	Variable winds. Clear, dry, fair conditions, cold temperatures, and minimal precipitation. Cold deserts.

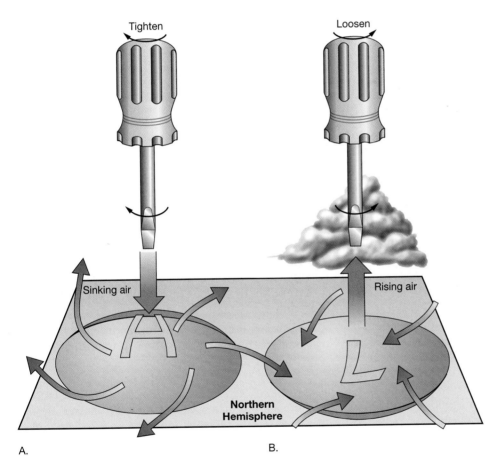

A.   B.

Figure 6–13 High-pressure and low-pressure cells and air flow.

As air moves away from a high-pressure cell (**A**) toward a low-pressure cell (**B**), the Coriolis effect causes the air to curve to the right in the Northern Hemisphere. This results in clockwise winds around high-pressure cells (anticyclonic flow) and counterclockwise winds around low-pressure cells (cyclonic flow).

many years is characterized by a lack of precipitation, we can say that the area has an arid climate.

Winds

The movement of air is called **wind**. Air always moves from high to low pressure. As air moves away from high-pressure cells and toward low-pressure cells, the Coriolis effect modifies the direction of air movement. In the Northern Hemisphere, this results in a counterclockwise[9] flow of air around low-pressure cells [called **cyclonic**

(*cyclo* = a circle) **flow**] and a clockwise flow of air around high-pressure cells (called **anticyclonic flow**). This can be remembered by how a screwdriver works: To tighten a screw (equivalent to high pressure), the screwdriver is turned clockwise; to loosen a screw (low pressure), the screwdriver is turned counterclockwise. Figure 6–13 shows the pattern of wind circulation around high- and

---

[9] These directions are for the Northern Hemisphere; they are reversed in the Southern Hemisphere.

Figure 6–14 Sea and land breezes.

A. Sea breezes occur when air warmed by the land rises and is replaced by cool air from the ocean. **B.** Land breezes occur when the land has cooled, causing dense air to sink and flow toward the warmer ocean.

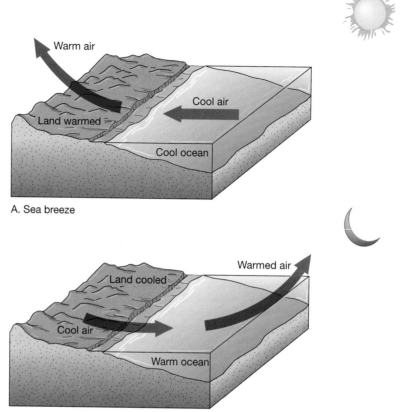

A. Sea breeze

B. Land breeze

low-pressure regions. Because winter high-pressure cells are replaced by summer low-pressure cells over the continents, wind patterns associated with continents often reverse themselves seasonally.

Other factors that influence regional winds are **sea breezes** and **land breezes** (Figure 6–14). These are caused by differential heating of land and ocean surfaces and are common in coastal areas. When an equal amount of solar energy is applied to both land and ocean, the land heats up about five times more due to its lower heat capacity. The land heats the air around it, and, during the afternoon, the warm, low-density air over the land rises. This rising air creates a low-pressure region over the land and causes the cooler air over the ocean to be pulled toward land, creating what is known as a sea breeze. At night, the land surface cools about five times more rapidly than the ocean and cools the air around it. This cool, high-density air sinks, creates a high-pressure region, and causes the wind to blow from the land. This is known as a land breeze, which is commonly well developed in the late evening and early morning hours. Alternating sea and land breezes are common in the Florida peninsula.

Storms

At high and low latitudes, day-to-day weather may change little. Polar regions are usually cold and dry regardless of the season. Near the Equator, day-to-day weather also is the same year-round.[10] In equatorial regions, the air is warm, damp, and typically calm, because the dominant direction of air movement in the doldrums is upward. Midday rains are common, even during the supposedly "dry" season. It is within the mid-latitudes that weather becomes interesting because storms are common.

Storms are atmospheric disturbances characterized by strong winds accompanied by precipitation and often by thunder and lightning. Due to the seasonal change of pressure systems over continents, air masses from the high and low latitudes may move into the mid-latitudes, meet, and produce severe storms. **Air masses** are large areas of air that have a definite area of origin and distinctive characteristics. The United States is influenced by several air masses, including polar air masses and tropical air masses (Figure 6–15). Note that some air masses originate over land (c = continental) and therefore are dryer, but most originate over the sea (m = maritime) and are moist. Some are colder (P = polar; A = Arctic) and some are warm (T = tropical). Typically, the United States is influenced more by polar air masses during the winter and more by tropical air masses during the summer.

---

[10] In fact, in equatorial Indonesia, the vocabulary of Indonesians doesn't include the word *seasons*.

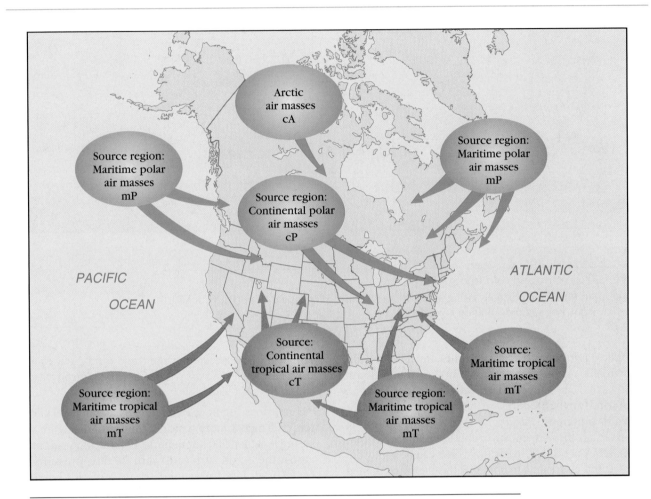

Figure 6–15 Air masses that affect U.S. weather.
Polar air masses are shown in blue, and tropical air masses are shown in red. Air masses are classified based on their source region: The designation continental (c) or maritime (m) indicates moisture content, whereas polar (P), Arctic (A), and tropical (T) indicate temperature conditions.

As polar and tropical air masses move into the mid-latitudes, they also move gradually in an easterly direction. A **warm front** is the contact between a warm air mass moving into an area that is occupied by cold air. A **cold front** is the contact between a cold air mass moving into an area that is occupied by warm air (Figure 6–16).

These confrontations are brought about by the movement of the **jet stream**, a narrow, fast-moving, easterly-flowing air mass. It exists above the mid-latitudes just below the top of the troposphere, centered at about 10 kilometers (6 miles) altitude. It usually follows a wavy path and may cause unusual weather by steering a polar air mass far to the south or a tropical air mass far to the north.

In either case where a warm or cold front is produced, the less-dense, warmer air always rises above the denser cold air. The rising air cools, and the water vapor in it condenses as precipitation. A cold front is usually steeper, and the temperature difference across it is greater than a warm front. Therefore, rainfall along a cold front usually is heavier and briefer than rainfall along a warm front.

Tropical Cyclones

Tropical cyclones (*kyklon* = moving in a circle) are huge rotating masses of low pressure characterized by strong winds and torrential rain. They are the largest storm systems on Earth. However, tropical cyclones are not associated with any fronts.

Origin Tropical cyclones begin as low-pressure cells that break away from the equatorial low-pressure belt and grow as they pick up heat energy from the warm ocean. By means of the large heat capacity and latent heat of evaporation of the ocean's water, they transport great amounts of heat energy into heat-deficient higher-latitude regions. When wind velocity within the storms reaches 120 kilometers (74 miles) per hour, tropical storms are classified as tropical cyclones.[11] Sometimes, the wind in tropical cyclones attains speeds of 400 kilometers

---

[11] A *tropical depression* has winds greater than 61 kilometers (38 miles) per hour. A *tropical storm* has winds between 61 and 120 kilometers (38 and 74 miles) per hour.

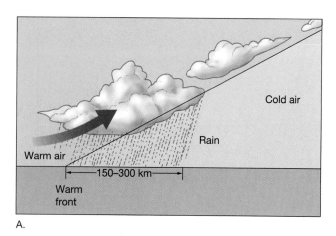
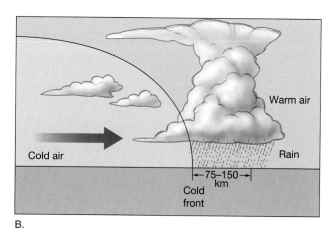

Figure 6–16 Warm and cold fronts.

Cross sections through a gradually rising warm front (**A**) and a steeper cold front (**B**). With both fronts, warm air rises and precipitation is produced.

(250 miles) per hour! Tropical cyclones are called by different names in different regions: In North and South America, they are known as **hurricanes** (*Huracan* = Taino god of wind); in the western North Pacific Ocean, they are known as **typhoons** (*tai-fung* = great wind); and in the Indian Ocean, they are known simply as **cyclones**. No matter what they are called, tropical cyclones can be highly destructive. It is interesting to note that the energy contained in one hurricane is greater than that generated by all energy sources over the last 20 years in the United States.

Worldwide, about 100 storms grow to hurricane status each year. The conditions needed for a hurricane's creation are:

- Warm ocean water greater than 25 degrees centigrade (77 degrees Fahrenheit), which supplies the heat energy to the atmosphere by evaporation.
- Warm moist air, which supplies additional heat as the air's water vapor condenses.
- The Coriolis effect, which causes the hurricane to spin (that's why there are no hurricanes directly on the Equator, where the Coriolis effect is minimized). Thus, the air flowing toward the low-pressure center of a hurricane veers to its right in the Northern Hemisphere and to its left in the Southern Hemisphere, producing counterclockwise and clockwise rotation, respectively.

These conditions are found during the late summer and early fall, when the tropical and subtropical oceans are at their maximum temperature.

Movement Once hurricanes are initiated, they typically remain in the tropics. Rarely, they can pass through the tropics and affect the mid-latitudes (Figure 6–17). If a hurricane moves over land, the hurricane's source of energy is cut off and the storm will eventually dissipate.

Hurricanes are affected by the trade winds, so they move from east to west across the oceans, and typically last five to ten days.

Hurricanes can have diameters exceeding 800 kilometers (500 miles); more typically, they have diameters measuring less than 200 kilometers (124 miles). As air moves across the ocean surface toward the low-pressure center, it is drawn up around the **eye of the hurricane** (Figure 6–18). Once in the vicinity of the eye, the air begins to spiral upward and horizontal wind speeds may be negligible. Thus, the eye of the storm is usually calm. The eye of the hurricane travels across Earth's surface at speeds up to 15 kilometers (25 miles) per hour. Hurricanes are composed of spiral rain bands where intense rainfall caused by severe thunderstorms can drop tens of centimeters (several inches) of rainfall in an hour.

Types of Destruction Destruction from hurricanes is caused by high wind speeds and because of flooding due to intense rainfall. However, the majority of a hurricane's destruction is caused by a phenomenon called **storm surge**. In fact, 90 percent of people killed by hurricanes are killed by storm surge. When a hurricane develops over the ocean, its low-pressure center produces a low "hill" of water (Figure 6–19). As the hurricane migrates across the open ocean, this hill moves with it. As the hurricane approaches shallow water near shore, the portion of the hill over which the wind is blowing shoreward produces a mass of elevated, wind-driven water. This mass of water—the storm surge—is represented by a dramatic increase in sea leavel at the shore, often accompanied by large storm waves.

A storm surge can be extremely destructive to low-lying coastal areas, especially when it occurs at high tide. The coincidence of a storm surge with high tide in areas that normally have particularly high tides frequently produces major catastrophes, with great loss of life and property

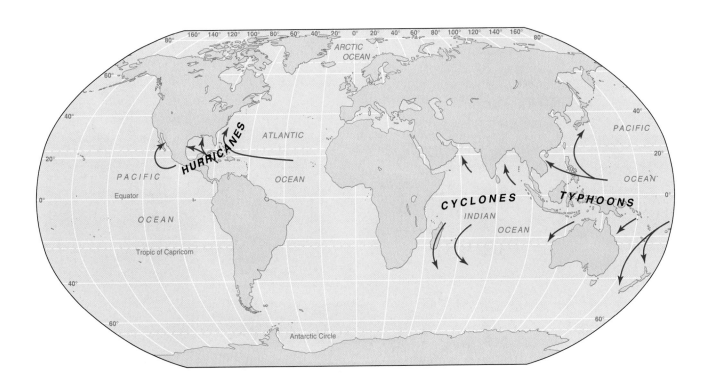

Figure 6–17 Paths of tropical cyclones.

Cyclones originate in tropical areas where ocean surface temperatures are warm and migrate across the tropics, sometimes reaching the mid-latitudes. Depending on the area, cyclones may also be called typhoons or hurricanes.

damage. In severe hurricanes, storm surges can be up to 12 meters (40 feet) in height.

More than 30 percent of hurricanes are formed in the waters north of the Equator in the western Pacific Ocean. These cyclones, called typhoons, do much damage to coastal areas and islands in Southeast Asia. Another 18 percent form in the waters of the eastern Pacific, where the storms usually die out in the vast expanse of cooler water between North America and the Hawaiian Islands. About once every 10 years, an eastern Pacific hurricane will reach Hawaii. About 12 percent of hurricanes form in the Atlantic Ocean—mostly off the coast of Africa. These hurricanes travel across the Atlantic Ocean and frequently make landfall along Caribbean Islands and the North American coastline, where substantial damage can be done.

Historic Destruction Periodic destruction from hurricanes occurs along the East Coast and the Gulf Coast regions of the United States. The worst natural disaster in

U.S. history was from a hurricane. On September 8, 1900, a hurricane with wind speeds of 135 kilometers (84 miles) per hour inundated the low-lying barrier island city of Galveston, Texas. Most of the resulting 6000 deaths were caused by a 6-meter (20-foot)-high storm surge [Galveston averages only 1.4 meters (4.5 feet) above sea level].

In August 1992, Hurricane Andrew came ashore in southern Florida and did more than $20 billion in damage before crossing the Gulf of Mexico and inflicting another $2 billion in damage along the Gulf Coast (Figure 6–20). In regard to property damage, Andrew was by far the most destructive hurricane in U.S. history. Winds reached 258 kilometers (160 miles) per hour as the hurricane crossed the Everglades, ripping down every tree in its path. Over 250,000 people were left homeless. Fortunately, many people heeded the warnings to evacuate, and only 54 deaths resulted from the hurricane.

Other areas of the world experience tropical cyclones on a regular basis. Bangladesh borders the Indian Ocean and is particularly vulnerable since it is a highly populated

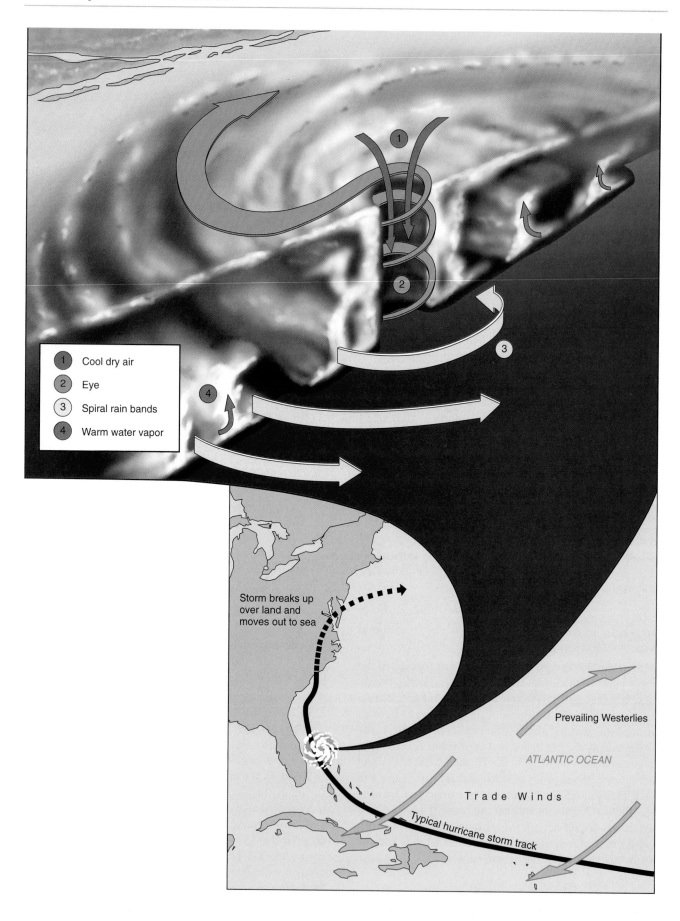

1	Cool dry air
2	Eye
3	Spiral rain bands
4	Warm water vapor

Storm breaks up over land and moves out to sea

Prevailing Westerlies

ATLANTIC OCEAN

Trade Winds

Typical hurricane storm track

Figure 6–18 Structure of a hurricane.

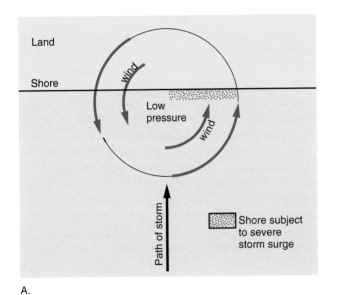

A.

B.

Figure 6–19 Storm surge.

A. As a cyclonic storm in the Northern Hemisphere moves ashore, the low-pressure cell around which the storm winds blow and the onshore winds of the storm produce a high-water storm surge. **B.** Storm waves generated by Hurricane Felix in August 1995 surge ashore in New Jersey.

A.

B.

Figure 6–20 Hurricane Andrew.

A. A satellite view of the hurricane, showing the prominent central eye. **B.** Destruction from Hurricane Andrew.

and low-lying country, much of it only 3 meters (10 feet) above sea level. In 1970, a 12-meter (40-foot)-high storm surge from a hurricane killed an estimated 1 million people. In 1991, Hurricane Gorky's 145-mile-per-hour winds caused extensive damage and killed 200,000 people.

Even islands near the centers of ocean basins may be struck by hurricanes. The Hawaiian Islands were hit hard by Hurricane Dot in August 1959 and by Hurricane Iwa

in November 1982. Hurricane Iwa hit very late in the hurricane season and produced winds up to 130 kilometers (81 miles) per hour. Over $100 million of damage was experienced on the islands of Kauai and Oahu. Niihau, a small island occupied by only 230 native Hawaiians, was directly in the path of the storm and suffered severe damage, but none of the population received serious injuries. Hurricane Iniki roared across the islands of

Kauai and Niihau in September 1992, with 210-kilometer (130-mile)-per-hour winds. It was the most powerful hurricane to hit the Hawaiian Islands this century, and property damage approached $1 billion.

Hurricanes will always be a threat to life and property. However, as a result of accurate forecasts and prompt evacuation, the loss of life during hurricanes has been decreasing. Still, with more and more construction along the coasts because of increasing coastal populations, property damage has been *increasing*. Inhabitants of areas subject to a hurricane's destructive force need to be aware of the danger and destruction caused by this powerful natural phenomenon so that they can plan accordingly.

Climate Patterns in the Oceans

Not only does climate apply to land areas, the term applies to regions of the oceans as well. In fact, the open ocean is divided into climatic regions that run generally east–west (parallel to lines of latitude) and have relatively stable boundaries that are somewhat modified by ocean surface currents (Figure 6–21).

In the **equatorial** region (the region spanning the Equator), the major air movement is upward because air

is heated and rises. Consequently, surface winds are weak and variable, as is typical in the doldrums. Surface waters are warm, and the air is saturated with water vapor. Daily rainshowers are common; this precipitation keeps surface salinity relatively low.

Tropical regions (those areas north or south of the equatorial region and roughly between the Tropic of Cancer and the Tropic of Capricorn) are characterized by strong trade winds. These trade winds blow from the northeast in the Northern Hemisphere and from the southeast in the Southern Hemisphere. These winds push the equatorial currents and create moderately rough seas. Relatively little precipitation falls at higher latitudes within tropical regions, but precipitation increases toward the Equator. The tropical regions are the breeding grounds for tropical cyclones, which carry large quantities of heat into higher latitudes.

Belts of high pressure are centered in the **subtropical** regions. The dry air descending on the subtropics results in little precipitation and a high rate of evaporation, producing the highest surface salinities in the open ocean (see Figure 5–17). Here, winds are weak and currents are sluggish, as is typical of the horse latitudes. However, strong boundary currents (along the boundaries of con-

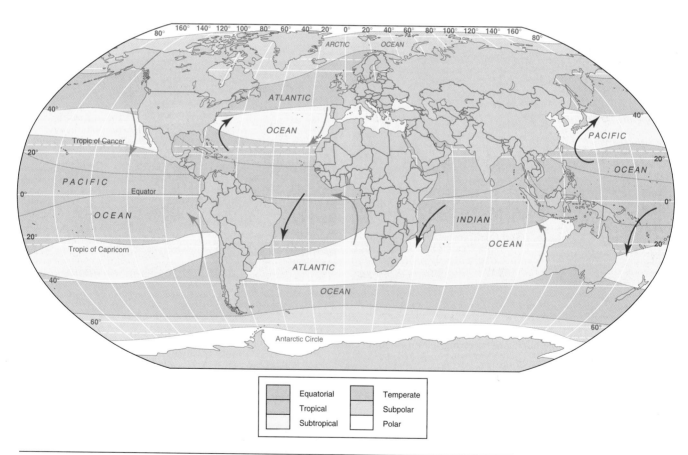

Figure 6–21 Climatic regions of the ocean.
The ocean's climatic regions are relatively stable and are defined by oceanic characteristics, wind belts, and latitude. Red arrows indicate warm surface currents; blue arrows indicate cool surface currents.

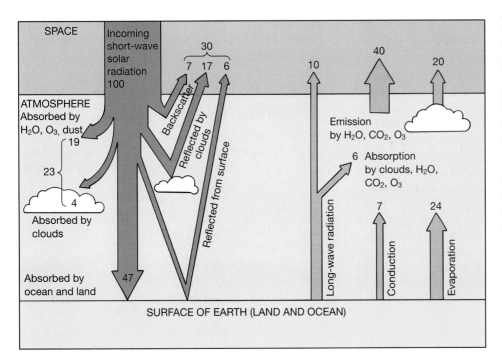

Figure 6–22 Earth's heat budget.

One hundred units of short-wave solar radiation from the sun (mostly visible light) is reflected, scattered, and absorbed by various components of the Earth-atmosphere system. The absorbed energy is radiated back into space from Earth as long-wave infrared radiation (heat). If this infrared radiation does not leave Earth, global warming will occur. The values shown are estimates.

tinents) flow north and south, particularly along the western margins of the subtropical oceans.

The **temperate** regions (also called the mid-latitudes) are characterized by strong westerly winds (the prevailing westerlies) blowing from the southwest in the Northern Hemisphere and from the northwest in the Southern Hemisphere (see Figure 6–11). Severe storms are common, especially during winter, and precipitation is heavy. In fact, the North Atlantic is noted for fierce storms in this zone. These storms have claimed many ships and numerous lives over the centuries.

The **subpolar** region is associated with high precipitation characteristic of the subpolar low. The subpolar ocean is covered in winter by sea ice. The sea ice melts away, for the most part, in summer. Icebergs are common, and the surface temperature seldom exceeds 5 degrees centigrade (41 degrees Fahrenheit) in the summer months.

Surface temperatures remain at or near freezing in the **polar** regions, which are covered with ice throughout most of the year. The polar high pressure dominates the area. In these areas, which include the Arctic Ocean and the ocean adjacent to Antarctica, there is no sunlight during the winter and constant daylight during the summer.

The Atmosphere's Greenhouse Effect

The worldwide average temperature of Earth and the troposphere is about 15 degrees centigrade (59 degrees Fahrenheit). However, if the atmosphere contained no water vapor, carbon dioxide, methane, or other trace gases, the average worldwide temperature would be −18 degrees centigrade (−4 degrees Fahrenheit). At this temperature,

all of the water on Earth would be frozen and it would be too cold to support the present distribution of life. The more pleasant temperatures on Earth to which life is accustomed would not be possible without the greenhouse effect of these gases.

The **greenhouse effect** is analogous to the operation of a greenhouse for raising plants. Greenhouses are built with a glass or thin plastic covering and are designed to trap heat. Energy radiated by the sun covers the full electromagnetic spectrum, but most of the energy that reaches Earth's surface is of short wavelengths, in and near the visible portion of the spectrum. In a greenhouse, short-wave sunlight passes through the transparent glass or plastic. Striking the plants, the floor, and other objects inside, it is converted to infrared radiation (heat). Heat has longer wavelengths than visible light, and when heat energy tries to escape the greenhouse, the glass or plastic around the greenhouse traps the heat.[12] The trapped heat energy causes the temperature inside the greenhouse to increase. This same greenhouse effect causes a car parked in the sun to experience a dramatic increase in interior temperature.

Figure 6–22 diagrams the roles of the various components of Earth's **heat budget**. In the upper atmosphere, most solar radiation within the visible spectrum penetrates the atmosphere to Earth's surface, like sunlight coming through greenhouse glass. After scattering by atmospheric molecules and reflection off clouds, about 47 percent of the solar radiation that is directed toward Earth is absorbed by the oceans and continents. About

---

[12] The glass or plastic of a greenhouse is opaque to, or impenetrable by, long-wavelength infrared radiation (heat).

23 percent is absorbed by the atmosphere and clouds, and about 30 percent is reflected into space by atmospheric backscatter, clouds, and Earth's surface.

Evidence suggests that Earth has maintained a constant average temperature over long periods of time, so Earth must be reradiating energy back to space at the same rate at which it absorbs solar energy. Similar to the heat energy radiated from objects inside a greenhouse, the energy reradiating from Earth falls within the infrared range. Figure 6–23 shows that most of the energy coming to Earth from the sun is within the visible spectrum and peaks at 0.48 micrometers[13] (0.0002 inch). The atmosphere is transparent to much of this radiation, but it is absorbed by materials such as water and rocks at Earth's surface. When these materials reradiate this energy back toward space, they do so as longer-wavelength infrared (heat) radiation, with a peak at a wavelength of 10 micrometers (0.004 inch). Carbon dioxide and water vapor present in the atmosphere absorb some of this infrared radiation and reradiate it to produce the greenhouse effect. *It is this change of wavelengths that is the key to how the greenhouse effect works.*

Although Earth's atmosphere and clouds absorb only 23 percent of the shorter-wavelength incoming solar radiation, the outgoing longer-wavelength infrared radiation from Earth is absorbed in much greater quantities by water vapor, carbon dioxide, and other gases in the atmosphere. This is the same as air in a greenhouse being warmed by trapped heat energy.

Some of the infrared energy absorbed in the atmosphere becomes reabsorbed by Earth to continue the process (the rest of the energy is lost to space). Therefore, the solar radiation received is retained for a time within our atmosphere. It moderates temperature fluctuations between night and day and also between seasons.

Which Gases Contribute to the Greenhouse Effect?

Water vapor is the largest contributor to the greenhouse effect. However, it occurs naturally in the atmosphere, and human activities are not thought to affect the amount of water vapor in the atmosphere. Research has revealed which greenhouse gases are increasing in the atmosphere due to human activity. They are listed in Table 6–4 in order of their potential contribution to increasing the greenhouse effect.

Clearly, carbon dioxide has the greatest relative contribution to the greenhouse effect. As a result of human activities, the atmospheric concentration of carbon dioxide has increased by 25 percent over the past 125 years, to a level of 350 parts per million. The concentration increases by 1.2 parts per million each year.

The other trace gases—methane, nitrous oxide, tropospheric ozone, and chlorofluorocarbons—are present in far lower concentrations. However, they are important because, per molecule, they absorb far more infrared radiation than does carbon dioxide (Table 6–5). Their smaller overall contribution to increasing the greenhouse effect in the atmosphere (Table 6–4) is due to their low concentrations in the atmosphere compared with carbon dioxide. It is clear that the contribution of all of these gases must be considered when calculating the amount of greenhouse warming.

What Should We Do About the Increasing Greenhouse Gases?

Earth's average surface temperature has risen by 0.5 degree centigrade (0.9 degree Fahrenheit) in the last 100 years. But there is no clear, simple proof that this has resulted from the 25 percent increase in atmospheric carbon dioxide or the increase in other greenhouse gases that began with the Industrial Revolution in the late 1700s (Figure 6–24). Many scientists are convinced that it has; others are not so certain. In addition, some scientists who develop sophisticated computer models of climate believe that greenhouse warming will evaporate more seawater and thus produce more cloud cover, which will block the sun's rays and significantly reduce the warming effect.

One novel way that has been proposed to remove carbon dioxide from the atmosphere is to stimulate productivity in the ocean. Through photosynthesis, microscopic marine algae take up dissolved carbon dioxide from the ocean and use the carbon while converting the oxygen to oxygen gas. By removing carbon dioxide from the ocean, the ocean in turn can absorb more carbon dioxide from the atmosphere.

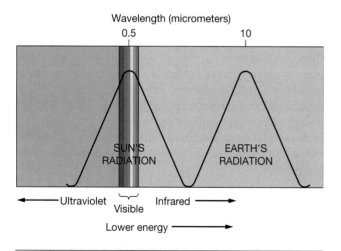

Figure 6–23 **Energy radiated by the sun and Earth.** The intensity of energy radiated by the sun peaks at 0.48 micrometer (0.0002 inch) in the visible spectrum, but that reradiation from Earth peaks at 10 micrometers (0.004 inch) in the infrared range.

---

[13] A micrometer (μm) is one-millionth of a meter.

Table 6–4 Concentration, rate of increase, and relative contribution of greenhouse gases to increasing the greenhouse effect.

Gas	Concentration (ppbv[a])	Rate of increase (% per year)	Relative contribution to increasing the greenhouse effect (%)
Carbon dioxide (CO_2)	353,000	0.5	60
Methane (CH_4)	1,700	1	15
Nitrous oxide (N_2O)	310	0.2	5
Tropospheric ozone (O_3)	10–50	0.5	8
Chlorofluorocarbon (CFC-11)	0.28	4	4
Chlorofluorocarbon (CFC-12)	0.48	4	8
Total	—	—	100

[a]ppbv = parts per billion by volume (not by weight).

Table 6–5 Efficiency of various greenhouse gases in absorbing infrared radiation.

Gas	Infrared radiation absorption per molecule (number of times greater than CO_2)
Carbon dioxide (CO_2)	1
Methane (CH_4)	25
Nitrous oxide (N_2O)	200
Tropospheric ozone (O_3)	2,000
Chlorofluorocarbon (CFC-11)	12,000
Chlorofluorocarbon (CFC-12)	15,000

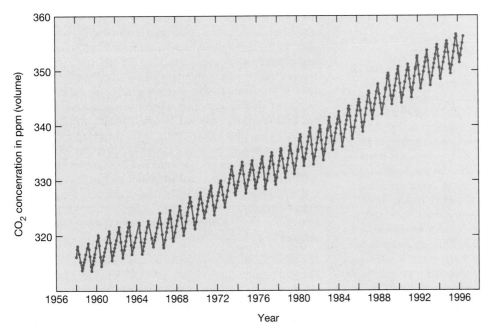

Figure 6–24 Atmospheric carbon dioxide increase since 1958.

Monthly average concentration of atmospheric carbon dioxide at Mauna Loa Observatory in Hawaii, expressed in parts per million by volume of dry air. The annual cycle reflects the seasonal cycles of photosynthesis by plants (which consumes CO_2) and respiration by animals (which releases CO_2).

Areas of the ocean that have relatively low productivity, such as the tropical oceans, would be a good place to stimulate productivity and thus increase the amount of carbon dioxide removed from the atmosphere. In 1987, oceanographer John Martin realized that the factor limiting productivity in tropical oceans is the absence of iron, and thus conceived the **"iron hypothesis."** The iron hypothesis states that an effective way of increasing productivity in the ocean is to fertilize the ocean by adding the only nutrient that appears to be lacking—iron. Martin's associates experimented in 1993 by adding finely ground iron to a test area of the ocean near the Galápagos Islands in the Pacific Ocean. Their results and the results of another voyage in 1995 (collectively called the "Iron Ex" experiments) showed that by adding iron to the ocean, productivity could indeed be increased by a factor of up to 30 times normal.

The results also revealed that there was a problem keeping the iron in suspension for long periods of time. One technique that helped minimize this problem was to grind the iron to a very fine size. In spite of this problem (and other problems with dispersal), this technique holds promise of reducing the amount of carbon dioxide in the atmosphere if done on a larger scale. However, the long-term global environmental effect of adding additional iron (and a correspondingly large amount of carbon dioxide) to the ocean is unknown.

What is very clear, however, is that our agricultural and industrial activities are producing changes in the environment. Although we cannot completely eliminate human-induced modifications of the environment, we can at least reduce human impact on Earth. Steps in the right direction include reducing our combustion of fossil fuels and reducing the widespread deforestation that accompanies the spread of agriculture. In order to halt further damage, we must also preserve Earth's plant communities and replace much of what has been removed. Additionally, we must modify activities that affect the environment most severely, while continuing to pursue research that improves our understanding of how Earth's systems work.

? Students Sometimes Ask…

How effective has the ice patrol been in preventing accidents like the Titanic *disaster?*
The tracking of ice by the Ice Patrol has been extremely effective in helping to prevent disasters similar to the *Titanic* in the busy shipping lanes of the North Atlantic Ocean. In fact, since the International Ice Patrol was established in 1914, not one ship has been lost to ice within the patrolled area (except during wartime). However, accidents still do occur outside the patrolled area. For example, the Soviet cruise liner *Maxim Gorky* rammed an iceberg in the Arctic Ocean well north of the Arctic Circle between Greenland and Spitzbergen in 1989. Fortunately, quick action by the crew and the Norwegian Coast Guard prevented any loss of life among the more than 1300 people on board.

I've heard that gigantic icebergs the sizes of small U.S. states are found near Antarctica. How are they formed?
In Antarctica, the edges of glaciers form large floating sheets of ice called shelf ice. Shelf ice extends into surrounding seas and produces vast plate-like icebergs when the edges of the shelf ice calve. Some of these icebergs have lengths exceeding 150 kilometers (93 miles). Their flat tops may stand up to 200 meters (650 feet) above the ocean surface, although most rise less than 100 meters (330 feet) above sea level. The underwater portion of an iceburg is even more impressive: Depending on the amount of rock material an iceberg may be carrying, as much as 90 percent of its mass will be below waterline. In August 1991, for example, an iceberg the size of Connecticut (13,000 square kilometers or 5000 square miles) broke loose from the ice shelf in the Weddell Sea. Once they are created, ocean currents driven by strong winds carry the icebergs to the north, where they disintegrate by melting. Because this region is not a major shipping route, the icebergs pose no real navigation hazard. Ships sighting these gigantic bergs, which can extend across the horizon, have often mistaken them for land!

Is the Coriolis effect what makes a curve ball curve in baseball?
No, and the Coriolis effect is *not* to blame for why your jump shot in basketball doesn't always go through the hoop either. Basketballs and baseballs travel too quickly and over too short a distance to be influenced significantly by the Coriolis effect. What makes a curve ball curve in baseball is the spin put on the ball, which creates high- and low-pressure regions around the ball that influence its trajectory. The Coriolis effect has a much greater influence over objects that travel long distances, such as ocean currents and air masses. That's why oceanographers and meteorologists (as well as missile launchers and airline pilots) must take into account the Coriolis effect.

Can the Coriolis effect be seen in the motion of objects that are not as large-scale as air masses and ocean currents?
The Coriolis effect has often been credited with being responsible for the creation of the swirling motion that is seen in sinks and bathtubs when they drain. Looking down on a sink, the rotation of the vortex (commonly called an eddy or a whirlpool) that forms would follow a counterclockwise rotation in the Northern Hemisphere. This is because as the water moves toward the drain, it curves to the *right* of the drain before it empties down the drain. Conversely, in the Southern Hemisphere, the vortex would move in a clockwise direction. Many scientists dispute that this phenomenon could occur over such a short distance. Perhaps other characteristics, such as the shape of the sink and local slopes, are influential in creating the phenomenon. This is easy to test at home by observing which way your sink drains. In the Northern Hemisphere, it should drain counterclockwise.

Is a water spout the same thing as a hurricane?
No, water spouts more closely resemble tornadoes, but are usually smaller in scale and can form anywhere there are strong enough winds over water. **Water spouts** are funnel-shaped vortices of water associated with extremely low atmospheric

Box 6–2
Hot and Cold About OTEC Systems

The sun imparts a large amount of radiant energy to Earth, where it is stored in the ocean and the atmosphere. Can this solar-based energy be harnessed for use by humans? The potential for extracting this energy from the movement and heat distribution patterns of the atmosphere and ocean is attractive for several reasons:

- It can be achieved without significant pollution of air or water.
- The amount of energy available at any time is far greater than that in fossil fuels (coal, oil, natural gas) and nuclear fuel (uranium).
- The energy is renewable as long as the sun continues to radiate energy to Earth.
- It can be used 24 hours a day because it does not depend on the sun shining; instead, it relies on solar heat stored in the oceans and atmosphere over time.

The sources of renewable energy are:

1. Heat stored in the oceans
2. Kinetic energy of the winds
3. Potential and kinetic energy of waves
4. Potential and kinetic energy of tides and currents

Of these four sources, the renewable energy source with the greatest store of potential energy is the heat stored in the warm surface layers of the tropical oceans.

Of Earth's surface between the Tropic of Cancer and Tropic of Capricorn, 90 percent is ocean. What makes this warm tropical surface water such an important source of energy is

the presence of much colder water beneath the thermocline. The temperature difference between the surface and from below the thermocline is 20 degrees centigrade (36 degrees Fahrenheit) or more. With a temperature difference as small as 17 degrees centigrade (31 degrees Fahrenheit), this heat difference can be converted to more useable forms by **ocean thermal energy conversion (OTEC)** systems (Figure 6E).

In an OTEC unit, warm surface water heats a fluid, such as liquefied propane gas or ammonia, which is under pressure in evaporating tubes (Figure 6F). Heating vaporizes the fluid, and the vapor pushes against the blades of a turbine, which turns an electrical generator. After passing through the turbine, the fluid is condensed by cold water that has been pumped up from the deep ocean. It is again ready for heating by warm surface water that causes it to vaporize and pass through the turbine. Thus, heat energy stored in the ocean is converted to useful electrical energy. The system works in the opposite way from a typical refrigeration system and on a much larger scale.

By the late 1980s the Japanese had advanced the furthest with OTEC technology and had built some small electricity-generating plants. Along the continental U.S. coast, the only region with good potential for ocean thermal energy conversion is a strip about 30 kilometers (19 miles) wide and 1000 kilometers (600 miles) long along southern Florida. Here, the western margin of the warm surface current called the Gulf Stream provides an adequate temperature difference with cold deeper water.

Another promising U.S. location for the development of OTEC systems is in Hawaii. The Pacific International Center

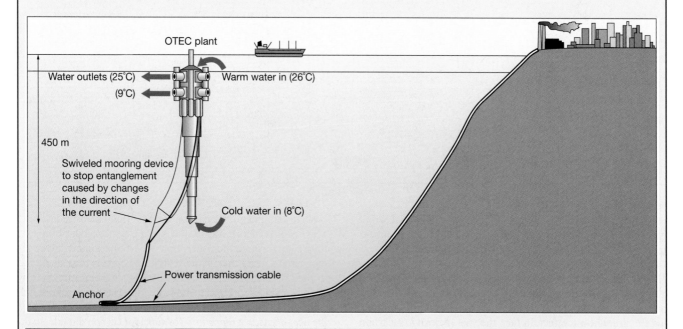

Figure 6E An OTEC system.

(continued)

(continued)

for High Technology Research on the island of Hawaii operates the world's only presently operating OTEC plant. In addition to producing up to 210 kilowatts of electricity, the cold water brought up from below the thermocline is used to condense atmospheric water, and thus the plant also produces a supply of fresh water. The plant is also experimenting with

using the coldness of seawater to chill roots and stimulate growth in crops by channeling the cold seawater through pipelines below the soil. Another benefit is the enhancement of fishing grounds due to artificial upwelling, which results from cold, deep, nutrient-rich water being brought to the surface. Although OTEC requires a large initial investment, the Hawaii site has demonstrated that OTEC power generation can be commercially successful.

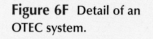

Figure 6F Detail of an OTEC system.

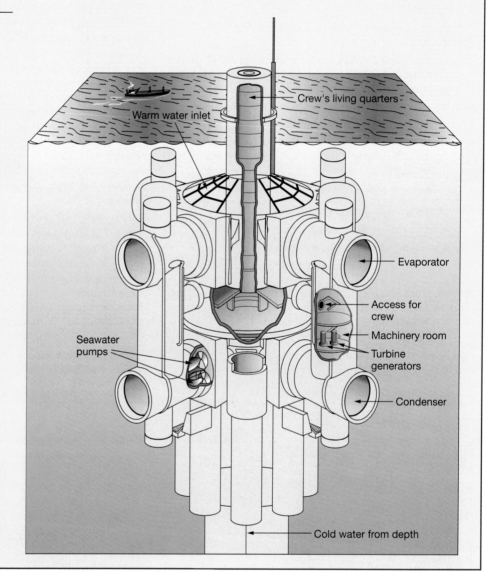

pressure. The low pressure causes air and water to spiral rapidly inward and upward, producing a spectacular funnel that extends from the ocean surface to the clouds (Figure 6G).

How is the ocean important in reducing the greenhouse effect? The ocean is quite important for reducing the amount of atmospheric carbon dioxide in the atmosphere and thus reducing the greenhouse effect. In fact, the vast majority of carbon dioxide in the ocean–atmosphere system is found in the ocean because carbon dioxide is approximately 30 times more soluble in water than are other common gases.

What happens to the carbon dioxide that enters the ocean? Most of it is incorporated into organisms through photosynthesis and through their secretion of carbonate shells. Over geologic time, more than 99 percent of the carbon dioxide added to the atmosphere by volcanic activity has been removed by the ocean and deposited in marine sediments as biogenous calcium carbonate and fossil fuels (oil and natural gas). Essentially, the ocean acts as a *repository* (or *sink*) for carbon dioxide, soaking up the gas and removing it from the environment as sea floor deposits. There have even been recent investigations into the effect of removing carbon dioxide from the atmosphere

Figure 6G A water spout near the Florida Keys.

What kinds of changes will there be if there is an increased greenhouse effect that results in global warming?

Researchers believe there will be many changes, but not all agree as to what those changes will be or how severe they may be. Some of the predicted changes include higher than normal sea surface temperatures, more frequent and more intense tropical storms, longer and more intense heat waves, more precipitation in certain areas and droughts in other areas. In addition, greenhouse warming may dramatically alter the ocean's deep-water circulation pattern, which can cause rapid changes in Earth's climate. Certainly, any kind of global climate change will affect the distribution of plant and animal communities. Some studies suggest that the principal grain-growing regions (such as the U. S. midwest region) will be most affected. Perhaps the best-known possible effect of global warming is the melting of polar ice caps. If this occurs, the water produced from the melting ice would result in a dramatic rise in sea level, flooding low-lying coastal areas.

However, if global temperatures increase, one possible outcome may be that the ice caps will actually *enlarge*. How can global warming cause ice caps to enlarge? If the atmosphere is warmer, then evaporation will occur more quickly (which, incidentally, will probably make tropical storms more intense). Increased evaporation will cause more water vapor to be present in the atmosphere. Much of this water vapor will fall as precipitation on land, thus potentially increasing the amount of snowfall for the formation of ice caps. An increase in the amount of water stored in ice caps (instead of in the ocean) will cause sea level to drop. Certainly, the global climate system is complex and contains many poorly understood feedback loops, including the role of clouds and atmospheric water. Truly, the successful modeling of Earth's climate is one of the biggest scientific challenges today, even with some of the world's most powerful computers.

Since scientists don't know for sure that the greenhouse effect is actually increasing global temperature, why should we do anything about it?

The kinds of measures that prevent the greenhouse effect—being more fuel-efficient, protecting plant communities, eliminating chlorofluorocarbons, reducing carbon dioxide emission—are all sound practices for preserving the environment regardless of reducing the greenhouse effect. Besides, if we find out in the next few decades that there really has been increased global warming due to the addition of human-caused gases, we'll wish that we had started making changes much earlier.

by pumping it into the deep ocean (thereby potentially slowing global warming).

In addition, the thermal properties of the ocean make it ideal for minimizing changes in temperature that may be brought about by global warming. Hence, the oceans also act as a "thermal sponge," absorbing heat but not increasing much in temperature.

Summary

The atmosphere and the ocean act together as one interdependent system, liked by complex feedback loops. There is a close association between many atmospheric and oceanic phenomena.

Seasons on Earth are caused by the tilt of Earth's rotational axis, which is 23.5 degrees from vertical. The change of the seasons as well as daily heating and cooling create a pattern of uneven solar heating on Earth. The uneven distribution of solar energy on Earth is responsible for creating temperature, density, and pressure differences that produce atmospheric and oceanic movement.

The Coriolis effect influences the paths of objects moving in a north or south direction on Earth and is caused by Earth's rotation. Because Earth's surface rotates at different velocities at different latitudes, objects in motion tend to veer to the right in the Northern Hemisphere and to the left in the Southern Hemisphere. The Coriolis effect is nonexistent at the Equator, but increases with latitude and is maximized at the poles.

The atmosphere is unevenly heated and set in motion because more energy is received than is radiated back into space at low latitudes as compared to high latitudes. On the spinning Earth, this creates three circulation cells in each hemisphere: a Hadley cell between 0 and 30 degrees latitude, a Ferrel cell between 30 and 60 degrees latitude, and a polar cell between 60 and 90 degrees latitude. High-pressure regions, where dense air descends, are located at about 30 degrees north or south latitude and at the poles. Belts of low pressure, where air rises, are generally found at the Equator and at about 60 degrees latitude.

The movement of air within the circulation cells produces the major wind belts of the world. The air at Earth's surface that is moving away from the subtropical highs produces trade winds moving toward the Equator and prevailing westerlies moving toward higher latitudes. The air moving along Earth's surface from the polar high to the subpolar low creates the polar easterlies.

Boundaries between the major wind belts of the world are characterized by calm winds. The boundary between the two trade wind belts is referred to as the doldrums, which coincides with the Intertropical Convergence Zone (ITCZ). The boundary between the trade winds and the prevailing westerlies is called the horse latitudes. The boundary between the prevailing westerlies and the polar easterlies is called the polar front.

Earth's tilt on its axis of rotation and the distribution of continents modify the idealized wind and pressure belts. Despite the modification of the idealized picture, atmospheric circulation on Earth closely matches this model.

Weather describes the conditions of the atmosphere at a given place and time, while climate is the long-term average of weather. Atmospheric motion (wind) always moves from high- to low-pressure regions. Consequently, in the Northern Hemisphere, there is a counterclockwise cyclonic movement of air around the low-pressure cells and a clockwise anticyclonic movement around high-pressure cells. Coastal regions commonly experience sea and land breezes, caused by alternating daily heating and cooling.

Many storms are related to the movement of air masses. In the mid-latitudes, cold air masses from higher latitudes meet warm air masses from lower latitudes and create cold and warm fronts that move from west to east across Earth's surface. Tropical cyclones (hurricanes) are large powerful storms that mostly affect tropical regions of the world. Destruction caused by hurricanes is caused mainly by storm surge but also by high winds and intense rainfall.

Climate patterns in the ocean are closely related to the distribution of solar energy and the wind belts of the world. The oceanic climate patterns are modified somewhat by ocean surface currents.

Radiant energy reaching Earth from the sun is mostly in the ultraviolet and visible light range, whereas that radiated back to space from Earth is primarily in the infrared (heat) part of the spectrum. Water vapor, carbon dioxide, and other trace gases absorb the infrared radiation and heat the atmosphere. This phenomenon is called the greenhouse effect. Because human activities are increasing the concentrations of trace gases that enhance the greenhouse effect, there is concern that Earth's life zone may be warming. Ocean thermal energy conversion (OTEC) is a process developed to use the difference in temperature between warm tropical surface water and the cold water below the thermocline to produce electricity.

Key Terms

Air mass (p. 188)

Antarctic Circle (p. 176)

Anticyclonic flow (p. 187)

Arctic Circle (p. 178)

Autumnal equinox (p. 174)

Climate (p. 184)

Cold front (p. 189)

Columbus, Christopher (p. 185)

Convection cell (p. 177)

Coriolis effect (p. 179)

Cyclone (p. 190)

Cyclonic flow (p. 187)

Declination (p. 174)

Doldrums (p. 184)

Ecliptic (p. 173)

Equatorial (p. 194)

Equatorial low (p. 183)

Eye of the hurricane (p. 190)

Ferrel cell (p. 183)

Greenhouse effect (p. 195)

Hadley cell (p. 183)

Heat budget (p. 195)

Horse latitudes (p. 184)

Hurricane (p. 190)

Intertropical Convergence Zone (ITCZ) (p. 184)

Iron hypothesis, the (p. 198)

Jet stream (p. 189)

Land breeze (p. 188)

Northeast trade winds (p. 184)

Ocean thermal energy conversion (OTEC) (p. 199)

Polar (p. 195)

Polar cell (p. 183)

Polar easterly wind belt (p. 184)

Polar front (p. 184)

Polar high (p. 183)

Prevailing westerly wind belt (p. 184)

Sea breeze (p. 188)

Southeast trade winds (p. 184)

Storm surge (p. 190)

Storm (p. 188)

Subpolar (p. 195)

Subpolar low (p. 183)

Subtropical (p. 194)

Subtropical high (p. 183)

Summer solstice (p. 174)

Temperate (p. 195)

Titanic, RMS (p. 171)

Trade winds (p. 184)

Tropic of Cancer (p. 174)

Tropic of Capricorn (p. 174)

Tropical (p. 194)

Tropical cyclone (p. 189)

Tropics (p. 175)

Troposphere (p. 177)

Typhoon (p. 190)

Vernal equinox (p. 174)

Warm front (p. 189)

Water spout (p. 198)

Weather (p. 184)

Wind (p. 187)

Winter solstice (p. 174)

Questions And Exercises

1. What caused the sinking of RMS *Titanic*? What international organization was formed after the sinking of the *Titanic* to prevent other such disasters?

2. Describe the effect on Earth as a result of Earth's axis of rotation being angled 23.5 degrees to the ecliptic. What would happen if Earth were not tilted on its axis?

3. Since there is a net annual heat loss at high latitudes and a net annual heat gain at low latitudes, why does the temperature difference between these regions not increase?

4. Along the Arctic Circle, how would the sun appear during the summer solstice? During the winter solstice?

5. Describe the physical properties of the atmosphere, including its composition, temperature, density, water vapor content, pressure, and movement.

6. Describe the Coriolis effect in the Northern and Southern Hemispheres and include a discussion of why the effect increases with increased latitude.

7. Sketch the pattern of surface wind belts on Earth, showing atmospheric circulation cells, zones of high and low pressure, the names of the wind belts, and the names of the boundaries between wind belts.

8. Why are there high-pressure caps at each pole and a low-pressure belt in the equatorial region?

9. Discuss why the idealized belts of high and low atmospheric pressure shown in Figure 6–11 are modified (see Figure 6–12).

10. What is the difference between weather and climate? If it rains in a particular area during a day, does that mean that the area has a wet climate? Explain.

11. Describe the difference between cyclonic and anticyclonic flow, and show how the Coriolis effect is important in producing both a clockwise and a counterclockwise flow pattern.

12. How do sea breezes and land breezes form? During a hot summer day, which one would be most common and why?

13. Name the polar and tropical air masses that affect U.S. weather. Describe the pattern of movement across the continent and patterns of precipitation associated with warm and cold fronts.

14. What are the conditions needed for the formation of a tropical cyclone? Why do most mid-latitude areas only rarely experience a hurricane? Why are there no hurricanes at the Equator?

15. Describe the types of destruction caused by hurricanes. Of those, which one causes the majority of fatalities and destruction?

16. How are the ocean's climate belts (Figure 6–21) related to the broad patterns of air circulation described in Figure 6–11? What are some areas where the two are not closely related?

17. Describe the fundamental difference between solar radiation absorbed at Earth's surface and the radiation that is primarily responsible for heating Earth's atmosphere.

18. Discuss the greenhouse gases in terms of their relative concentrations and relative contributions to any increased greenhouse effect.

19. Describe the iron hypothesis, and discuss the relative merits and dangers of undertaking a project that could cause dramatic changes in the global environment.

20. Construct your own diagram of how an ocean thermal energy conversion unit might generate electricity, or make a flow diagram presenting the steps of the process.

References

Changing climate and the oceans. 1987. *Oceanus* 29:4, 1–93.

Charlson, R. J., et al. 1992. Climate forcing by anthropogenic aerosols. *Science* 255:5043, 423–430.

Coale, K. H., et al. 1996. A massive phytoplankton bloom induced by an ecosystem-scale iron fertilization experiment in the equatorial Pacific Ocean. *Nature* 383:6493, 495–501.

de la Mare, W. 1997. Abrupt mid-twentieth-century decline in Antarctic sea-ice extent from whaling records. *Nature* 389:6646, 57–60.

Dopyera, C. 1996. The iron hypothesis. *Earth* 5:5, 26–33.

Golden, J. H. 1969. The Dinner Key "tornadic waterspout" of June 7, 1968. *Mariner's Weather Log* 13:4, 139–147.

Keeling, C. D., Whorf, T. P., Wahlen, M., and van der Plicht, J. 1995. Interannual extremes in the rate of rise of atmospheric carbon dioxide since 1980. *Nature* 375:6533, 666–670.

Martin, J. H., Gordon, R. M., and Fitzwater, S. E. 1990. Iron in Antarctic waters. *Nature* 345:3271, 156–158.

Martin, J. H., et al. 1994. Testing the iron hypothesis in ecosystems of the equatorial Pacific Ocean. *Nature* 371:6493, 123–129.

Ocean energy. 1979. *Oceanus* 22:4, 1–68.

Oceans and climate. 1978. *Oceanus* 21:4, 1–70.

Philander, S. G. 1998. *Is the temperature rising? The uncertain science of global warming.* Princeton NJ: Princeton University Press.

Pickard, G. L. 1975. *Descriptive physical oceanography,* 2nd ed. New York: Pergamon Press.

Rodhe, H. 1990. A comparison of the contribution of various gases to the greenhouse effect. *Science* 248:4960, 1217–1219.

Ryan, P. R. 1985. The Titanic: Lost and found, 1985. *Oceanus* 28:4, 1–112.

———. 1986. The Titanic revisited. *Oceanus* 29:3, 2–17.

The Open University Course Team. 1989. *Ocean circulation.* Oxford: Pergamon Press.

Thieler, E. R., and Bush, D. M., 1991. Hurricanes Gilbert and Hugo send powerful messages for coastal development. *Journal of Geological Education* 39:4, 291–298.

Various authors. 1989. The oceans and global warming. *Oceanus.* 32:2, 1–75.

Various authors. 1994. The enigma of weather—A collection of works exploring the dynamics of meteorological phenomena. *Scientific American* (Special Publication).

Suggested Reading

Earth

Dopyera, C. 1996. The iron hypothesis. 5:5, 26–33. John Martin's idea of fertilizing the ocean with iron to reduce atmospheric carbon dioxide—thereby reducing the greenhouse effect—is put to the test.

Flanagan, R. 1996. Engineering a cooler planet. 5:5, 34–39. A critical look at how global engineering (such as adding iron to the ocean to reduce the greenhouse effect) can affect the planet.

Flanagan, R., and Yulsman, T. 1996. On thin ice. 5:2, 44–51. Scientists venture to Antarctica to investigate the role of sea ice in shaping Earth's climate.

Mallinson, D. 1992. 20,000 leagues under the Keys. 1:6, 46–53. Deep exploration off the Florida coast may help scientists determine whether Earth has built-in protection from the greenhouse effect.

Parks, N. 1996. Bounty from the sea. 5:4, 50–55. A look at how OTEC systems can be used to create electricity as well as a host of other unexpected benefits.

Pendick, D. 1996. Hurricane mean season. 5:3, 24–32. Presents evidence to support the idea that the large number of hurricanes that made landfall along the Atlantic coast in 1995 might be a preview of hurricane activity in the future.

Vogel, S. 1995. Has global warming begun? 4:6, 24–35. Examines a variety of evidence that suggests that human-made climate change has begun.

Wigley, T. M. L. 1996. A millennium of climate. 5:6, 38–41. The atmosphere is viewed as part of a vast climate machine that humans may be altering on a global scale.

Sea Frontiers

Boling, G. R. 1971. Ice and the breakers. 17:6, 363–371. An interesting history of people in icy waters and the development and improvement of icebreakers.

Charlier, R. 1981. Ocean-fired power plants. 27:1, 36–43. The potential of ocean thermal energy conversion is discussed.

Heidorn, K. C. 1975. Land and sea breezes. 21:6, 340–343. A discussion of the causes of land and sea breezes.

Kretschmer, J. 1990. When the winds blow: A mythology of gust and squall. 36:3, 40–43. A review of the wind belts and boundaries of the world, as well as a history of various winds and how they received the names that they did.

Land, T. 1976. Europe to harness the power of the sea. 22:6, 346–349. An article emphasizing the clean, safe, permanent nature of ocean tides and waves as a source of power.

Mayor, A. 1988. Marine mirages. 34:1, 8–15. The nature of marine mirages and the research efforts that lead to our understanding them are considered.

Rush, B., and Lebelson, H. 1984. Hurricane! The enigma of a meteorological monster. 30:4, 233–239. An overview of the nature of hurricanes and the problem of predicting where they will go.

Scheina, R. L. 1987. The *Titanic*'s legacy to safety. 33:3, 200–209. A brief summary of the sinking of the *Titanic* and an overview of the ice patrol and iceberg collision history subsequent to the sinking of the "unsinkable" luxury liner.

Smith, F. G. W. 1974. Planet's powerhouse. 20:4, 195–203. A description of how Earth's "heat engine" works, with an emphasis on the nature of tropical cyclonic storms.

———. 1974. Power from the oceans. 20:2, 87–99. A survey of the many tried and untried proposals for extracting energy from the oceans.

Sobey, E. 1979. Ocean ice. 25:2, 66–73. The formation of sea ice and icebergs as well as their climatic and economic effects are discussed.

———. 1980. The ocean-climate connection. 26:1, 25–30. Our increasing knowledge of the effect of the oceans on Earth's climate may be used to predict climatic trends of the future.

Scientific American

Alley, R. B., and Bender, M. L. 1998. Greenland ice cores: Frozen in time. 278:2, 80–85. An analysis of the techniques that scientists use to study ice cores from Greenland and interpret clues to Earth's past—and future—climate.

Gregg, M. 1973. Microstructure of the ocean. 228:2, 64–77. A discussion of the methods of studying the detailed movements of ocean water by observing temperature and salinity changes over distances of 1 centimeter (0.4 inch) and the motions they reveal.

Houghton, R. A., and Woodwell, G. M. 1989. Global climate change. 260:4, 36–44. Presents evidence that global warming has begun and looks at what the future might hold in a warmer climate.

Jones, P. D., and Wigley, T. M. 1990. Global warming trends. 263:2, 84–91. Data relating to evidence of global warming over the past 100 years are presented.

Karl, T. P., Nicholls, N., and Gregory, J., 1997. The coming climate. 276:5, 78–83. An analysis of meteorological records and computer models indicate how a warmer climate will affect plant and animal life in the future.

MacIntyre, F. 1974. The top millimeter of the ocean. 230:5, 62–77. Processes that are confined to a thin film at the surface of ocean water and their role in the overall nature of the oceans are discussed.

Nadis, S. 1998. Fertilizing the sea. 278:4, 33. A news item about a U. S. company that is planning to stimulate fishery production in the Marshall Islands by using John Martin's iron hypothesis.

Penney, T. R., and Bharathan, D. 1987. Power from the sea. 256:1, 86–93. A prediction that generating electricity by ocean thermal energy conversion will be competitive with fossil fuel plants as the price of oil rises.

Pollack, H. N., and Chapman, D. S. 1993. Underground records of changing climate. 268:6, 44–53. Direct measurement of temperature shows that Earth has warmed over the past 150 years. Data from boreholes reveal ancient temperatures that may give us a more complete picture of the history of global climate.

Revelle, R. 1982. Carbon dioxide and world climate. 247:2, 35–43. Some of the possible effects of increasing atmospheric temperature due to carbon dioxide accumulation are considered.

Stanley, S. M. 1984. Mass extinctions in the oceans. 250:6, 64–83. Geological evidence suggests that most major periods of species extinction over the last 700 million years occurred during the brief intervals of ocean cooling.

Stolarski, R. S. 1988. The Antarctic ozone holes. 258:1, 30–37. The discovery of the Antarctic ozone hole and its possible significance are discussed.

White, R. M. 1990. The great climate debate. 263:1, 36–45. The controversy over the degree of global warming that can be expected in the future is discussed.

Oceanography on the Web

Visit the *Essentials of Oceanography* home page for on-line resources for this chapter. There you will find an on-line study guide with review exercises, and links to oceanography sites to further your exploration of the topics in this chapter. *Essentials of Oceanography* is at: **http://www.prenhall.com/thurman** (click on the Table of Contents menu and select this chapter).

CHAPTER 7
OCEAN CIRCULATION

Benjamin Franklin: The World's Most Famous Physical Oceanographer

People often think of **Benjamin Franklin** as an inventor, a statesman, and one of the founding fathers of our country. He was also deputy postmaster general of the colonies from 1753 to 1774. What many people don't know about him is that he is perhaps the most famous and well-known physical oceanographer—an oceanographer who studies physical phenomena in the ocean, including waves, tides, and currents. Franklin earned this distinction because he contributed greatly to the understanding of the Gulf Stream, a North Atlantic Ocean surface current that influenced mail routes between the new colonies and England.

Franklin became interested in the North Atlantic Ocean circulation patterns when he was perplexed by the fact that it took mail ships coming from Europe two weeks *longer* to reach New England by a northerly route than it did ships that came by a more southerly route. In about 1769 or 1770, Franklin happened to mention this dilemma to his cousin, a Nantucket sea captain named Timothy Folger. Folger told Franklin that there was a strong current with which the mail ships were not familiar that was impeding their journey because it flowed *against* their direction of movement. The whaling ships were familiar with the current because they often hunted whales along the boundaries of the current. The whalers often met the mail ships within the current and told the crew aboard the mail ships that they would make better progress if they were to avoid the current. However, the British captains of the mail ships thought themselves too wise to be counseled by simple American fishers. The mail ships were not just making slow progress within the current; if the winds were light, the ships were actually carried *backwards*!

Folger sketched the current on a map for Franklin, including directions for avoiding it when sailing from Europe to North America. With this knowledge, Franklin then asked other ship captains for information concerning the movement of surface waters in the North Atlantic Ocean. Franklin inferred that there was a significant current moving northward along the eastern coast of the United States. After passing along the coast, the current moved out in a easterly path across the North Atlantic. He concluded that this east-flowing current, which the ships traveling a northerly route from Europe had to combat, was responsible for increasing the time of their voyage. He subsequently published a map of the current in 1777 based on these observations (Figure 7A) and distributed it to the captains of the mail ships (who initially ignored the map). This strong current was named the "Gulf Stream" because it carried warm water from the Gulf of Mexico, and because it was narrow and well defined—similar to a stream, but in the ocean.

Figure 7A Benjamin Franklin's chart of the Gulf Stream (1777).

206

More recently, a study of the Gulf Stream was conduced in 1969 by a vessel that was carried for a month while it was submerged within the Gulf Stream. During the vessel's 2640-kilometer (1650-mile) journey as a floating object within the Gulf Stream, the six scientists aboard observed and measured the properties of and cataloged life forms within the Gulf Stream. Appropriately enough, the vessel was named the *Ben Franklin*.

Ocean currents are masses of oceanic water that flow from one place to another. This movement can involve large or small masses of water. It can occur at the surface or deep below the surface. The flow phenomenon that accomplishes the movement can be quite complex or it can be relatively simple. Simply put, currents are water masses in motion.

The surfaces of the major oceans are dominated by huge current systems. These currents serve to transfer heat from warmer to cooler areas on Earth, just as the major wind belts of the world do. The major wind belts of the world are responsible for transferring about two-thirds of the total amount of heat from the tropics to the poles; the surface currents of the ocean transfer the other third. Ultimately, these surface currents are driven by radiant energy from the sun and closely follow the pattern of major wind belts of the world.

More locally, surface currents also affect the climates of coastal continental regions. Cold currents flowing toward the Equator on the western sides of continents produce arid conditions, whereas warm currents flowing poleward on the eastern sides of continents produce warm, humid conditions. Additionally, ocean currents contribute to the mild climate of northern Europe and Iceland, whereas conditions at similar latitudes along the Atlantic coast of North America (such as Labrador) are much colder.

The effect of ocean currents on life is profound. Life in the deep sea is made possible only by a continuing supply of oxygen. This oxygen is carried there by cold, dense water that sinks in polar regions and spreads across the deep-ocean floor. Ocean currents greatly influence the distribution of life in surface waters by affecting the growth of algae, which is the basis of most oceanic food chains. Ocean currents also powerfully affect the distribution of humans on land: Currents have aided the travel of prehistoric peoples from Europe and Africa to the New World and on to the Pacific Ocean islands.

Measuring Ocean Currents

Ocean currents can be categorized as being either *wind-driven* or *density-driven*. Wind-driven currents are set in motion by moving air masses, most notably the major wind belts of the world. This motion is parallel to the surface (horizontal) and occurs primarily in the ocean's surface waters. Therefore, currents produced in this way are called **surface currents**. Density-driven circulation, on the other hand, has a significant vertical component and accounts for the thorough mixing of the deep masses of ocean water. Density-driven circulation is initiated at the ocean surface by temperature and salinity conditions that produce a high-density water mass, which sinks and spreads slowly beneath the surface waters. Thus, these currents are referred to as **deep currents**.

Obtaining useful averages of flow rates for surface currents is difficult due to their constantly shifting nature. Rarely do surface currents flow in the same direction and at the same rate for very long. However, there is some consistency to the *overall* surface current pattern worldwide. Surface currents can be measured directly or indirectly.

Direct current measuring is done by two main methods. One method involves releasing a device that is transported by the current and is tracked through time. Typically, radio-transmitting float bottles or other devices are used (Figure 7–1A). The other method is done from a fixed position, such as from a pier, where a current-measuring device capable of measuring flow rates (usually with a propeller) is lowered into the water (Figure 7–1B). These propeller devices can also be towed behind ships, and the ship's speed is then subtracted to determine a current's true flow rate.

Indirect current measurements can be completed by using one of three methods. One method is to determine the internal distribution of density and the corresponding pressure gradient across an area of the ocean, since water flows parallel to a pressure gradient. A second method is to use radar altimeters mounted on satellites to determine the shape of the ocean surface (dynamic topography), from which current flow can be inferred One such radar altimeter was launched with the TOPEX/Poseidon satellite in 1992 as a joint mission of the United States and French space agencies. The satellite accurately measures the bulges and depressions on the ocean surface caused by surface current movement to within 4 centimeters (1.5 inches). From this data, dynamic topography maps can be produced that show the speed and direction of surface currents (Figure 7–2). A third method uses a *Doppler flow meter* to transmit low-frequency sound signals through the water. The instrument measures the shift in frequency between the sound waves emitted and those backscattered by particles in the water to determine current movement.

Measuring deep currents is even more difficult because they exist deep below the surface. Often, they are mapped using various drift devices or based on the presence of a telltale chemical tracer. Some tracers are naturally absorbed into seawater, while others are intentionally added. Some useful tracers that have inadvertently been added to seawater include tritium (a radioactive isotope of hydrogen produced by nuclear bomb tests in the 1950s and early 1960s) and man-made chlorofluorocarbons (freons and other gases now thought to be depleting the ozone layer). Other techniques used to identify deep currents include measuring the distinctive temperature and salinity characteristics of a deep-water mass.

Surface Currents

Surface currents in the ocean develop from friction between the ocean and the wind that blows across its surface. Only about 2 percent of the energy of the wind is transferred to the ocean surface. For instance, a 50-knot[1] wind will create a 1-knot current. On a tiny scale, you can simulate this simply by blowing gently and steadily across a cup of coffee, not hard enough to cause waves, but enough to start the surface moving as a current.

If there were no continents on Earth, the pattern of surface currents worldwide would be quite simple: The currents would follow the major wind belts of the world; that is, in each hemisphere, there would be a current flowing within 0 to 30 degrees latitude that would be affected by the trade winds, another current between 30 and 60 degrees latitude that would follow the prevailing westerlies, and another current from 60 to 90 degrees latitude that would be a result of the polar easterlies. However, the distribution of continents on Earth influences the nature and flow directions of surface currents. Other factors that influence surface current movement include gravity, friction, and the Coriolis effect.

Surface currents occur within and above the pycnocline (the zone of rapidly changing density) to a depth of about 1 kilometer (0.6 mile). Thus, surface currents affect only about 10 percent of the world's ocean water.

Equatorial Currents, Boundary Currents, and Gyres

Trade winds blowing from the southeast in the Southern Hemisphere and from the northeast in the Northern Hemisphere are the principal drivers of the system of ocean surface currents. The trade winds thus set in motion the water masses between the tropics. Figure 7–3 shows how the world's major wind belts influence the circulation pattern for the Atlantic Ocean. The trade winds develop the **equatorial currents**, which travel along the Equator worldwide (Figure 7–4). They are called

A.

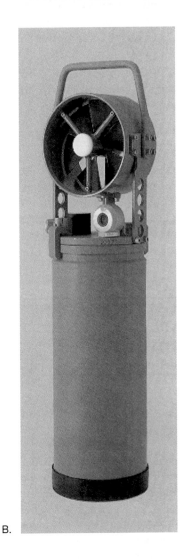

B.

Figure 7–1 Current-measuring devices.
A. Drift current meter. Depth of metal vanes is 1 meter (3.3 feet). **B.** Propeller-type flow meter. Length of instrument is 0.6 meter (2 feet).

[1]A knot is one nautical mile per hour. A nautical mile is defined as the distance of one minute of latitude, and is equivalent to 1.15 statute (land) miles or 1.85 kilometers.

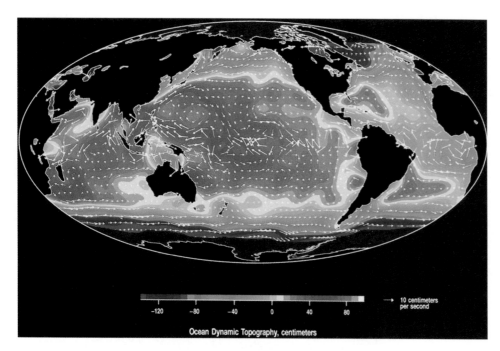

Figure 7–2 Satellite view of dynamic topography. TOPEX/POSEIDON radar altimeter data from September 1992 to September 1993. Colors represent the dynamic topography of the ocean in centimeters, measured relative to an arbitrary datum. White arrows indicate the flow direction of currents, with longer arrows indicating faster flow rates.

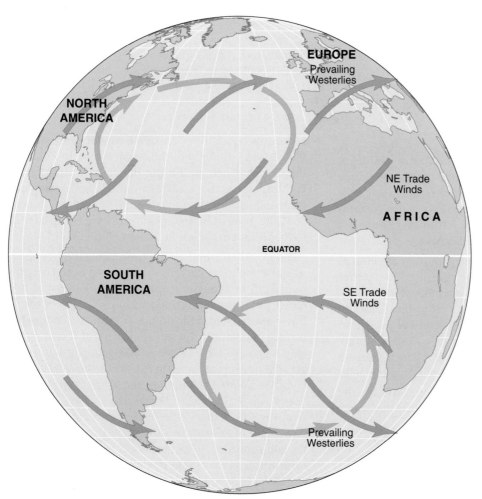

Figure 7–3 Atlantic Ocean surface circulation pattern. The trade winds in conjunction with the prevailing westerlies create surface current flow within the subtropical gyres. The gyres are also affected by the positions of the continents, gravity, and the Coriolis effect.

north or south equatorial currents, depending on their position north or south of the Equator.

These currents move westward parallel to the Equator. When the water reaches the western portion of the ocean basin, it must turn since it cannot cross land. Deflection caused by the Coriolis effect directs these currents away from the Equator as **western boundary currents**. The name means that they are currents traveling along the

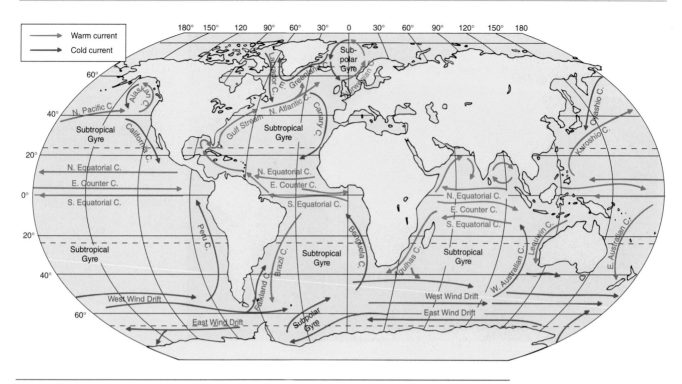

Figure 7–4 Wind-driven surface currents.
Major wind-driven surface currents of the world's oceans in February and March. The subtropical gyres that dominate the five major ocean basins are the North and South Pacific Ocean Gyres, the North and South Atlantic Ocean Gyres, and the Indian Ocean Gyre. The smaller subpolar gyres rotate in the reverse direction of the adjacent subtropical gyres.

western boundary of their ocean basin.[2] In Figure 7–4, the Gulf Stream and the Brazil Current are examples. Notice that since these western boundary currents come from equatorial regions, where water temperatures are warm, the western boundary currents are warm currents that carry warm water to higher latitudes. Figure 7–4 shows these warm currents as red arrows.

At the same time, between 30 degrees and 60 degrees latitude in both hemispheres, the prevailing westerlies blow surface water in an easterly direction (Figure 7–3). These winds blow from the northwest in the Southern Hemisphere and from the southwest in the Northern Hemisphere. This directs the water in a westerly direction across the ocean basin. The North Atlantic Current and the West Wind Drift[3] are examples (Figure 7–3).

Once the current flows back across the ocean basin, the Coriolis effect and continental barriers turn this water

toward the Equator as **eastern boundary currents** along the eastern boundary of the ocean basins. In Figure 7–3, the Canary Current and the Benguela Current are examples.[4] Since these eastern boundary currents come from high-latitude regions where water temperatures are cool, the eastern boundary currents are cool-water currents. These are shown in blue in Figure 7–4.

The flow described above from equatorial current to western boundary current to prevailing westerly current to eastern boundary current creates a pattern of water moving in a circular fashion within an ocean basin. This flow pattern creates the dominant feature of ocean basin surface circulation, the **subtropical gyres** (*gyros* = a circle). A **gyre** is a large circular-moving loop of water. There are five subtropical gyres in the world: the *North Atlantic Gyre*, the *South Atlantic Gyre*, the *North Pacific Gyre*, the *South Pacific Gyre*, and the *Indian Ocean Gyre* (which is mostly within the Southern Hemisphere) (Figure 7–4). The center of each of the subtropical gyres coincides with the subtropics at 30 degrees north or south latitude. The subtropical gyres rotate in a clockwise direction in the Northern Hemisphere (see the North Atlantic Ocean in

---

[2]Notice that the western boundary currents are off the *eastern* coasts of continents. This sounds confusing, but is a result of the fact that we have a land-based perspective. From an oceanic perspective, the *western* side of the ocean basin is where the western boundary current resides.

[3] Currents, like winds, are often named based on the direction from which they flow. Thus, the West Wind Drift is initiated by the westerlies and flows *from* the west.

---

[4]Currents are sometimes named for a prominent geographic location near where they pass. For instance, the Canary Current passes the Canary Islands; the Benguela Current is named for the Benguela Province in Angola, Africa.

Figure 7–4) and in a counterclockwise direction in the Southern Hemisphere (see the South Atlantic Ocean in Figure 7–4). Generally, each subtropical gyre is composed of four main currents that flow progressively into one another (Table 7–1). For instance, the North Atlantic Gyre is composed of the North Equatorial Current, the Gulf Stream, the North Atlantic Current, and the Canary Current (Figure 7–3).

At subpolar latitudes, the polar easterlies drive the surface currents in a westerly direction and, in combination with the Coriolis effect, produce **subpolar gyres** that rotate in a pattern opposite that of the adjacent sub-

tropical gyres. These subpolar gyres are best developed in the Atlantic Ocean between Greenland and Europe and the Weddell Sea off Antarctica (Figure 7–4). The *Antarctic Circumpolar Gyre* is unusual in that it passes through the South Pacific, South Atlantic, and South Indian Oceans as it completely encircles Antarctica.

Ekman Spiral and Ekman Transport

During the voyage of the *Fram* (Box 7–2), Fridtjof Nansen observed that Arctic Ocean ice moved 20 to 40 degrees to the *right* of the wind blowing across its surface (Figure 7–5). Not only is this true for ice, it is also

Table 7–1   Gyres and surface currents.

Pacific Ocean	Atlantic Ocean	Indian Ocean	Antarctic Circulation
North Pacific Gyre	North Atlantic Gyre	Indian Ocean Gyre	Antarctic Circumpolar Gyre
North Pacific Current	North Atlantic Current	South Equatorial Current	West Wind Drift
California Current[a]	Canary Current[a]	Agulhas Current[b]	Other Major Currents
North Equatorial Current	North Equatorial Current	West Wind Drift	East Wind Drift
Kuroshio (Japan) Current[b]	Gulf Stream[b]	West Australian Current[a]	
South Pacific Gyre	South Atlantic Gyre	Other Major Currents	
South Equatorial Current	South Equatorial Current	Equatorial Countercurrent	
East Australian Current[b]	Brazil Current[b]	North Equatorial Current	
West Wind Drift	West Wind Drift	Leeuwin Current	
Peru (Humboldt) Current[a]	Benguela Current[a]		
Other Major Currents	Other Major Currents		
Equatorial Countercurrent	Equatorial Countercurrent		
Alaskan Current	Florida Current		
Oyashio Current	East Greenland Current		
	Labrador Current		
	Falkland Current		

[a] Denotes an eastern boundary current of a gyre, which is relatively *slow*, *wide*, and *shallow* (and is also a *cold water* current).
[b] Denotes a western boundary current of a gyre, which is relatively *fast*, *narrow*, and *deep* (and is also a *warm water* current).

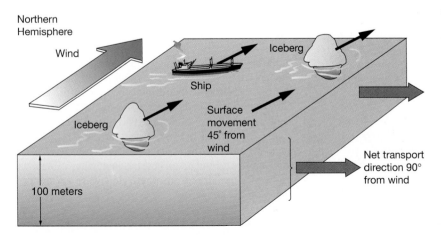

Northern Hemisphere
Wind
Iceberg
Ship
Iceberg
Surface movement 45° from wind
100 meters
Net transport direction 90° from wind

Figure 7–5 Transport of floating objects.

Nansen first noticed that floating objects, such as icebergs and ships, were moved to the right of the wind direction in the Northern Hemisphere.

Box 7-1
Running Shoes as Drift Meters: Just Do It

Essentially, any floating object can serve as a makeshift drift meter, as long as two locations are known: (1) where the object entered the ocean; and (2) where it was retrieved. From this information, the path of the object can be inferred, providing information about surface current flow directions and speeds. Oceanographers have long used **drift bottles** (a floating "message in a bottle" or a radio-transmitting device set adrift in the ocean) as a method to track the movement of currents.

Interestingly, many objects have inadvertently become drift meters when ships carrying containers full of cargo lose some (or all) of their cargo at sea. In this way, Nike athletic shoes and colorful floating bathtub toys (Figure 7B) have become drift meters that have advanced the understanding of current movement in the North Pacific Ocean.

In May 1990, the container vessel *Hansa Carrier* was en route to Seattle from Korea when it encountered a severe North Pacific storm. The ship was transporting 12.2-meter (40-foot)-long metal containers that resemble large rectangular boxes, many of which were staked onto the deck and lashed to the ship for the voyage. The ship was tossed around by the storm waves, causing 21 large deck containers to be lost overboard. Five of these containers held Nike brand athletic shoes, releasing up to 39,466 pairs of shoes into the ocean. Since the shoes floated, those that were released from their containers were carried to the east by the North Pacific Current. Within six months of the spill, the shoes had traveled 2400 kilometers (1500 miles) eastward, where thousands began to wash up along the beaches of British Columbia, Washington, and Oregon (Figure 7C). A few shoes were found on beaches in northern California, and over two years later shoes from the spill were even recovered from the north end of the Big Island of Hawaii!

Even though the shoes had spent considerable time drifting in the ocean, they were in good shape and wearable (once barnacles and oil were removed). Because the shoes were not tied together, many beachcombers found individual shoes or pairs that did not match. People who had found shoes were interested in finding matching pairs, since many of the shoes

retailed for around $100. Matching pairs of shoes were obtained through newspaper want ads and by exchange at local swapmeets.

With help from collectors who found shoes at the beach, locations and numbers of shoes collected were compiled during the months following the spill. Serial numbers inside the shoes could be traced to individual containers. The serial numbers indicated that only four of the five containers had released their shoes; evidently, one entire container holding shoes sank without opening. Thus, a maximum of 30,910 pairs of shoes (61,820 individual shoes) were released. The almost instantaneous release of such a large number of drift items provided oceanographers with a way to study the current patterns of the North Pacific Ocean. Previously, the largest number of drift bottles purposefully released at one time in the North Pacific was about half the number of shoes released during the spill.

Analysis of the data revealed that shoe drift rates and locations where the shoes were found agreed very well with models of North Pacific circulation. The models also revealed that the shoes would have drifted to the north or south had they been released at a different time of the year or in different years. In terms of the number of shoes reported found, only 2.6 percent of the shoes were recovered. Even though this seems like an unusually small number, it compares favorably with the 2.4 percent recovery rate of drift bottles released by oceanographers conducting research. Based on this data, it seems likely that many of the shoes did indeed escape the containers in which they were being carried.

The models were put to the test again in January 1992 when more drift items were released in the North Pacific from another container spill. This time, 12 containers were lost from a vessel in the vicinity where the shoes had spilled. One of these containers held 29,000 small, floatable, colorful plastic bathtub toys in the shapes of blue turtles, yellow ducks, red beavers, and green frogs (Figure 7B). Even though the toys were housed in plastic packaging glued to a cardboard backing, studies showed that after 24 hours in seawater, the glue deteriorated and the toys were released. These drift items began to come ashore in southeast Alaska 10 months later.

Again, the help of the beachcombing public (as well as lighthouse operators) was essential in recording the locations and numbers of bathtub toys that arrived at the beach. Computer modeling of the spill indicated that the toys would follow a more northerly route than the shoes. Since none of the toys was found south of Sitka, Alaska, it appears that the models were correct. The models indicate that many of the bathtub toys will continue to float within the Alaska Current, eventually dispersing throughout the North Pacific Ocean. Some may find their way into the Arctic Ocean, where they could spend time within the Arctic Ocean ice pack. From there, the toys may drift into the North Atlantic, eventually washing up on beaches in northern Europe, thousands of kilometers from where they were accidentally released in the ocean.

Figure 7B Some inadvertent float meters.

(continued)

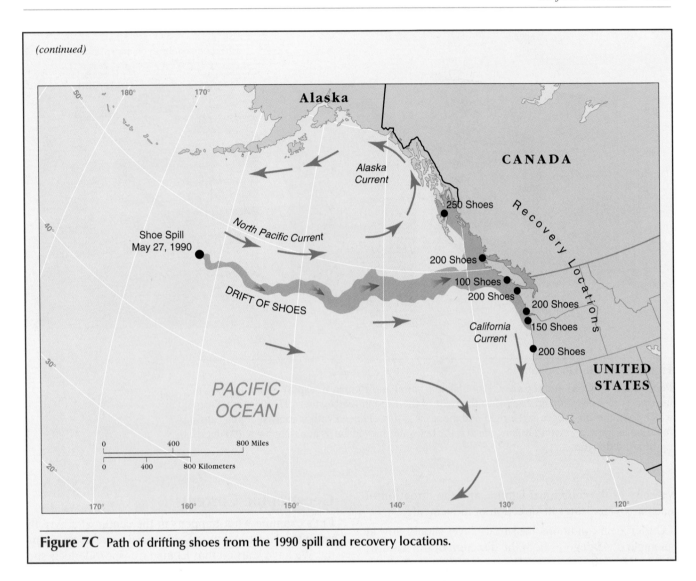

Figure 7C Path of drifting shoes from the 1990 spill and recovery locations.

true for surface water at different depths in the Northern Hemisphere. In the Southern Hemisphere, surface currents move to the left of the wind direction. Why does surface water move as it does relative to wind direction? Nansen passed this information on to **V. Walfrid Ekman**, a physicist who developed the mathematical relationships that explain Nansen's observations in accordance with the Coriolis effect.

Ekman developed a circulation model that is called the **Ekman spiral** (Figure 7–6). The Ekman spiral describes the speed and direction of flow of surface waters at various depths. It is caused by wind blowing across the surface and is modified by the Coriolis effect. The model assumes that a uniform water column is being set in motion by wind blowing across its surface. Owing to the Coriolis effect, the immediate surface water moves in a direction 45 degrees to the right of the wind (in the Northern Hemisphere). The water moves as a thin "layer," called a *lamina*. As the surface "layer" moves, it also sets in motion another "layer" beneath it.

It may be helpful to think of this in another way. As the wind energy put into the ocean surface is transmitted downward, it is consumed by water motion until none is left. Thus, with increasing depth, current speed progressively decreases, and the Coriolis effect causes an increased curvature to the right (like a spiral). An analogy to this is a deck of cards smeared between two hands; the motion is transferred successively to each card, analogous to each layer of water.

The energy of the wind is passed downward through the water column. Each successive "layer" of water is set in motion at a progressively slower velocity, and in a direction progressively to the right of the one above that set it in motion. At some depth, a lamina of water may move in a direction *exactly opposite to the wind direction that initiated it!* If the water is deep enough, the energy imparted by the wind will be consumed by friction and no motion will occur below that depth. Although it depends on wind speed and latitude, this stillness normally occurs at a depth of about 100 meters (330 feet).

Figure 7–6 shows the spiral nature of this movement with increasing depth from the ocean's surface. The length of each arrow in the figure is proportional to the

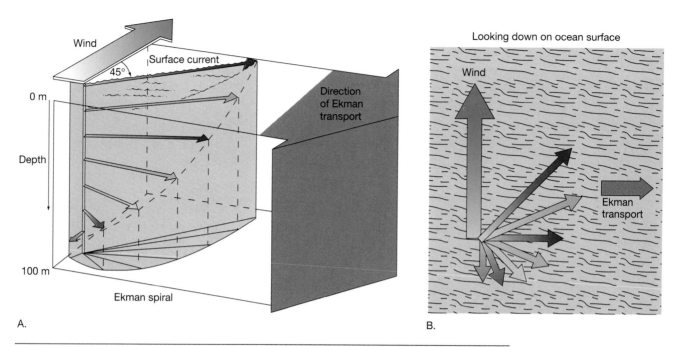

Figure 7–6 Ekman Spiral.

Perspective view (**A**) and top view (**B**) of Ekman spiral and Ekman transport. Wind drives surface water in a direction 45 degrees to the right of the wind in the Northern Hemisphere. Deeper water continues to deflect to the right and moves at a slower speed with increased depth, causing the Ekman spiral. Ekman transport, which is the net water movement, is at a right angle (90 degrees) to the wind direction.

velocity of the individual lamina, and the direction of each arrow indicates its direction of movement.[5]

Under ideal conditions, the surface layer should flow at an angle of 45 degrees from the direction of the wind, as shown. However, the overall average transport of all the laminae is a net water movement at an angle of 90 degrees to the right of the wind direction. This average movement of all the laminae is called **Ekman transport**. In the Northern Hemisphere, Ekman transport is 90 degrees to the right of the wind direction.

Of course, "ideal" conditions rarely exist in the ocean, so movements actually resulting from wind friction on the ocean surface often deviate from this idealized picture. Generally, experiments conducted in the ocean indicate that the surface current will move at an angle somewhat less than 45 degrees from the direction of the wind. In addition, experiments show that Ekman transport will be at an angle somewhat less than 90 degrees from the direction of the wind. Typical Ekman transport in the open ocean is on the order of about 70 degrees from the wind direction. In shallow coastal waters, Ekman transport may be in a direction very nearly the same as that of the wind.

---

[5] The name Ekman *spiral* refers to the spiral observed by connecting the tips of the arrows shown in Figure 7–6.

Geostrophic Currents

Let's examine what happens in the center of a gyre by using the North Atlantic Subtropical Gyre as an example. We have learned that Ekman transport deflects surface water to the right in the Northern Hemisphere. As a result, water constantly turns to the right in the Northern Hemisphere within an ocean basin. Thus, a clockwise rotation develops and produces a **subtropical convergence** of water in the middle of the gyre, causing water literally to pile up in the center of the subtropical gyre. We find within all such ocean gyres a hill of water that rises as much as 2 meters (6.6 feet) above the water level at the margins of the gyres.

Water that has built up into a "hill" structure has a tendency to flow downhill due to gravity. However, the Coriolis effect deflects the water flowing down the hill to the right in a curved path (Figure 7–7A). As the Coriolis effect continually curves the path of the water and tends to push the water into the hill, gravity also acts to move water down the slope and away from the hill. In essence, the water is affected by two opposing factors: the Coriolis effect (which moves the water *into the hill* and toward the center of the gyre) and gravity (which moves the water *down the hill*). The water piles up on these hills until the down-slope component of gravity acting on individual particles of water balances the Coriolis effect. When these two factors balance, the net effect is a

Box 7–2
The Voyage of the *Fram*

One of the most remarkable voyages in oceanography was initiated by a Norwegian, **Fridtjof Nansen** (1861-1930). Nansen developed a great interest in exploring the North Atlantic and the Arctic area. A previous expedition of particular interest to Nansen was the ill-fated American voyage of George Washington DeLong and his crew on the *Jeanette* in 1879. This expedition had attempted to sail through the Bering Strait between North America (Alaska) and Asia (Russia) to Wrangell Island, from which it planned to go overland to the North Pole. Scientists believed that Wrangell Island was possibly the southern tip of a peninsula extending southward from the great, but yet undiscovered, Arctic continent that lay beneath the Arctic Ocean permanent ice sheet.

The *Jeanette* became stuck in the polar ice pack on September 6, 1879, and drifted north of Wrangell Island. This proved that the island was not a peninsula of an Arctic continent, for if it was, the ship could not have drifted past it in the ice. After two years of drifting, the *Jeanette* was crushed while still embedded within the ice off the New Siberian Islands. Unfortunately, DeLong and many of his crew perished in the Lena Delta region of eastern Siberia.

In 1884, five years later, a number of articles that could be traced to the *Jeanette* were found frozen in the pack ice off the southwestern coast of Greenland, over 4800 kilometers (3000 miles) from where she sank. Based on this unusual evidence, Nansen reasoned that the articles must have drifted with Arctic ice from the wreck of the *Jeanette* to the observed location by following a path that would have taken them near the North Pole. If Arctic ice could be proven to drift that far, then this would help prove that an Arctic continent did not exist. It could also provide a means of transporting an expedition to the North Pole, which had never been visited before. He envisioned that a ship, once locked in the ice, might be able to follow a path similar to the transported articles, carrying the vessel near the North Pole.

Nansen began to plan the expedition and eventually raised funds for building the **Fram**, a 39-meter (128-foot)-long wooden ship (Figure 7D). The *Fram* was designed with extra reinforcing in its hull so that the expanding ice would not crush it but rather would force it up to the surface, free of the grip of the growing ice.

Provisions for 13 men for five years were stored on the small ship, and the crew set sail from Oslo, Norway, on June 24, 1893. On September 21, after an arduous voyage along the northern coast of Siberia, they had not yet sighted the New Siberian Islands when the ice captured them at 78.5 degrees north latitude. At that point, they were 1100 kilometers (683 miles) from the North Pole. They had now begun what would be a long and lonely endeavor, drifting slowly with the pack ice as it was moved by winds and currents (Figure 7E). During

Figure 7D The *Fram*.

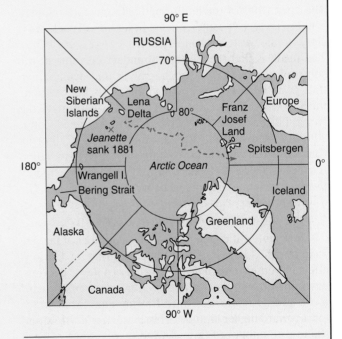

Figure 7E Route of the Fram, 1893–1896.
The course followed by the Fram after becoming frozen in ice near the New Siberian Islands. The route passes close to the North Pole and near Spitsbergen, where the vessel was released by the ice.

(continued)

(continued)

their time locked in the Arctic ice, the explorers were able to collect depth soundings, seawater samples, and ice measurements.

On November 15, 1895, the *Fram* reached its northernmost point, only 394 kilometers (244 miles) from the pole. On August 13, 1896, after almost three years of being locked in ice, the *Fram* broke free of the ice and was again floating in the open sea. She had drifted a total of 1658 kilometers (1028 miles) during an amazingly successful journey.

The drift of the *Fram* proved that no continent existed under the Arctic Ocean. It also showed that the ice that covered the polar area throughout the year was not of glacial origin but was a freely moving accumulation of ice that was formed by the freezing of seawater. During the voyage, the crew had found that the depth of this ocean exceeded 3000 meters

(9840 feet). They also discovered a rather surprising body of relatively warm water with temperatures as high as 1.5 degrees centigrade (35 degrees Fahrenheit) between the depths of 150 and 900 meters (500 and 3000 feet). Nansen correctly described this water as being a mass of Atlantic Ocean water that had sunk below the less saline Arctic water.

Nansen's realization of the need for more accurate measurements of water salinity and temperature led him to develop the Nansen bottle, a device widely used for collecting water samples at depth. His observations of the direction of ice drift relative to the wind direction helped V. Walfrid Ekman, a Norwegian physicist, develop the mathematical explanation of this phenomenon, known today as the Ekman spiral. Nansen (who was awarded the Nobel Peace Prize in 1922), Ekman, and other Scandinavian scientists led the way in developing the discipline of physical oceanography in the early twentieth century.

---

geostrophic (*geo* = earth, *strophio* = turn) **current** that moves in a circular path around the hill.[6] This path is labeled in Figure 7–7A as the path of ideal geostrophic flow.

What has just been described is accurate but idealized. Due to friction between water molecules, the water does converge and build up, but it gradually moves down the slope of the hill as it flows around it. This is the path of actual geostrophic flow indicated in Figure 7–7A.

Western Intensification

The apex (top) of the hill formed within a rotating gyre is not in the gyre's center. The highest point of each hill is closer to the western boundary of the gyre (Figure 7–7A). The result is that the western boundary currents of the subtropical gyres are faster, narrower, and deeper than their eastern boundary current counterparts. For example, the Kuroshio Current (a western boundary current) of the North Pacific Subtropical Gyre is up to 15 times faster, 20 times narrower, and 5 times as deep as the California Current (an eastern boundary current). This phenomenon is called **western intensification**, and currents affected by this phenomenon are said to be western intensified. *The western boundary currents of all subtropical gyres are western intensified, even in the Southern Hemisphere.*

Western intensification is caused by a number of factors, including the Coriolis effect. Remember that the Coriolis effect increases toward higher latitudes (poleward). Thus, eastward-flowing high-latitude water turns equatorward more strongly than westward-flowing equatorial water turns toward higher latitudes. This causes a wide, slow, and shallow equatorward flow of water across most of each subtropical gyre, leaving only a narrow band through which the poleward flow can occur along the western margin of the ocean basin. In Figure 7–7B, if there is a

constant volume of water rotating around the apex of the hill, then the velocity of the water along the western margin will be much faster than the velocity around the eastern side.[7] This is indicated by the spacing of lines in the figure: The closer the lines are together, the faster the flow. Indeed, this is what is observed in the ocean. Its main manifestation is a high-speed western boundary current that flows along the hill's steeper westward slope. The eastern slope has a slow drift of water toward the Equator. Table 7–2 summarizes the differences between western and eastern boundary currents of subtropical gyres.

Directly related to this difference in speed is the steepness of the hill's slope. The slope of the hill on the side of the slow-moving eastern boundary current is quite gentle; the western margin has a comparatively steep slope corresponding to the high velocity of the western boundary current.

Equatorial Countercurrents

A large volume of water is driven westward as flow within the north and south equatorial currents. Because the Coriolis effect is minimal near the Equator, much of the water is not turned toward higher latitudes. Instead, it piles up at the western margin of an ocean basin. This causes average sea level on the western side of the basin to be as much as 2 meters (6.6 feet) higher than the eastern side. This elevated water flows downhill under the influence of gravity, creating narrow **equatorial countercurrents** that flow to the east *counter to* and *between* the equatorial currents.

This pattern is particularly apparent in the western Pacific Ocean (see Figure 7–4), where a dome of equa-

---

[6]The term *geostrophic* for these currents is appropriate, since the currents behave as they do beause of Earth's rotation.

[7]A good analogy for this phenomenon is a funnel: In the narrow end of a funnel, the flow rates are speeded up (such as in western intensified currents); in the wide end, the flow rates are sluggish (such as in eastern boundary currents).

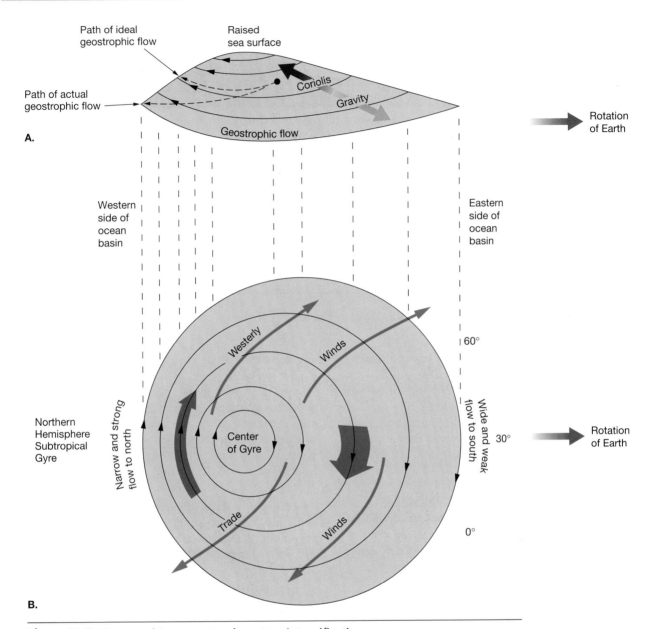

Figure 7–7 Geostrophic current and western intensification.
A. A cross-sectional view of a subtropical gyre showing how water literally piles up in the center, forming a hill up to 2 meters (6.6 feet) high. Gravity and the Coriolis effect balance to create an ideal geostrophic current that flows in equilibrium around the hill. However, friction makes the current gradually run downslope (actual geostrophic flow). **B.** A map view of the same subtropical gyre, showing that the flow pattern is restricted (lines are closer together) on the western side of the gyre, resulting in western intensification.

torial water is trapped in the island-filled embayment between Australia and Asia. This dome of water with very weak current flow displays the highest year-round surface temperature found anywhere in the world ocean, as can be seen in Figure 7–8. Continued influx of water carried by the equatorial currents builds the dome and creates an eastward countercurrent of water that continues across the Pacific toward South America.

Referring back to Figure 7–2, which shows a satellite image of average sea surface elevation using different colors, the hills of water within the subtropical gyres of

the Atlantic Ocean are clearly visible. In the Pacific Ocean, the hill in the North Pacific is easily seen, but the low equatorial elevations one might expect to see between the northern and southern subtropical gyres is missing. These data were recorded during a moderate El Niño event (which will be explained shortly). Thus, the high stand of equatorial water is likely due to the fact that the equatorial countercurrent is well developed.

Also contributing to the lack of definition between the two Pacific Ocean gyres is the fact that the South Pacific subtropical gyre is less intense than other gyres are. This

Table 7–2   Characteristics of western and eastern boundary currents of subtropical gyres.

Current type (examples)	Width	Depth	Speed	Transport volume (millions of cubic meters per second[a])	Comments
Western boundary currents (Gulf Stream, Brazil Current, Kuroshio Current)	Narrow, usually less than 100 kilometers (60 miles)	Deep, to depths of 2 kilometers (1.2 miles)	Fast, hundreds of kilometers per day	Large, as much as 50 sv[a]	Waters derived from low latitudes and are warm; little or no upwelling
Eastern boundary currents (Canary Current, Benguela Current, California Current)	Wide, up to 1000 kilometers (600 miles)	Shallow, to depths of 0.5 kilometer (0.3 mile)	Slow, tens of kilometers per day	Small, typically 10 to 15 sv[a]	Waters derived from mid-latitudes and are cool; coastal upwelling common

[a] One million cubic meters per second is a flow rate equal to one sverdrup (sv).

is due to the great area it covers, its lack of confinement by continental barriers along its western margin, and the presence of numerous islands (really the tops of tall sea floor mountains). The South Indian Ocean hill is rather well developed, although its northeastern boundary stands high because of the influx of warm Pacific Ocean water through the East Indies islands.

Upwelling and Downwelling

Upwelling and downwelling are important processes that affect both surface and deep currents. These vertical movements are essential in stirring the ocean, delivering oxygen to its depths, bringing nutrients to the surface, and distributing heat energy.

Upwelling is the movement of cold, deep, nutrient-rich water to the surface; **downwelling** is the movement of surface water to deeper depths. Since upwelling brings water to the surface that is enriched in dissolved oxygen and nutrients, it stimulates productivity (algal growth), which in turn supports incredible amounts of marine life. Thus, upwelling is a more intensely studied process than downwelling. Upwelling and downwelling occur in a variety of ways in the marine environment.

Diverging Surface Water

Current divergence[8] is caused when surface waters move *away from* an area on the ocean's surface. For instance, the trade winds drive equatorial currents toward the west of each ocean basin on either side of the Equator (see Figure 7–4). Once the currents move toward the margin of the ocean basin, Ekman transport moves the water

___
[8]Motion in current divergence is similar to a divergent plate boundary, where two plates are moving *away from* each other.

on the north side of the Equator to the right (toward a higher northern latitude), while in the Southern Hemisphere the water moves to the left (toward a higher southern latitude). In each ocean basin, this movement causes a divergence of surface currents. The net effect is a water deficiency at the surface between the two currents. Water from below rises to the surface to fill the void (Figure 7–9). This type of upwelling commonly occurs along the Equator and is called *equatorial upwelling*.

Because upwelling creates areas of high productivity, these areas of equatorial upwelling are associated with regions of rich marine life. Indeed, some of the most prolific fishing grounds in the world are located in these areas.

Converging Surface Water

Conversely, current convergence is caused when surface waters move *toward* each other. For instance, in the North Atlantic Ocean, the Gulf Stream, the Labrador Current, and the East Greenland Current all come together in the same vicinity. Once the currents converge, the water stacks up and has no place to go but downward. The surface water slowly sinks down in a process called, appropriately enough, downwelling. (Figure 7–10).

Unlike upwelling, areas of downwelling are not associated with prolific marine life. This is because the nutrients get used up and are not continuously resupplied from the cold, deep, nutrient-rich water below the surface. Consequently, downwelling areas have low productivity.

Coastal Upwelling and Downwelling

Let's look at examples of how coastal winds can cause upwelling or downwelling due to Ekman transport. Figure 7–11 shows a coastal region in the Southern Hemisphere with winds moving parallel to the coast. If the winds are from the south (Figure 7–11A), Ekman

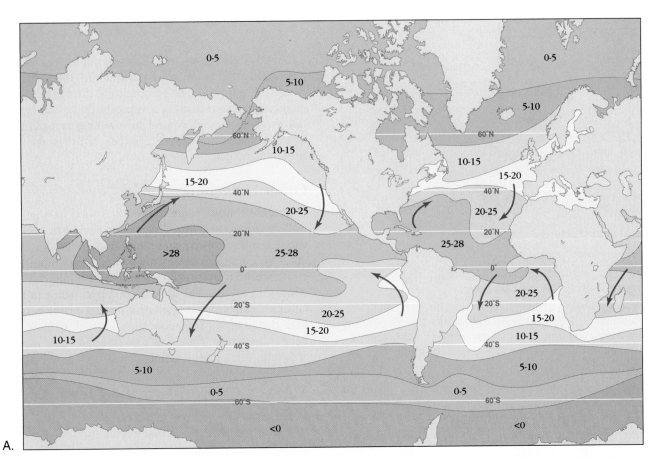

A.

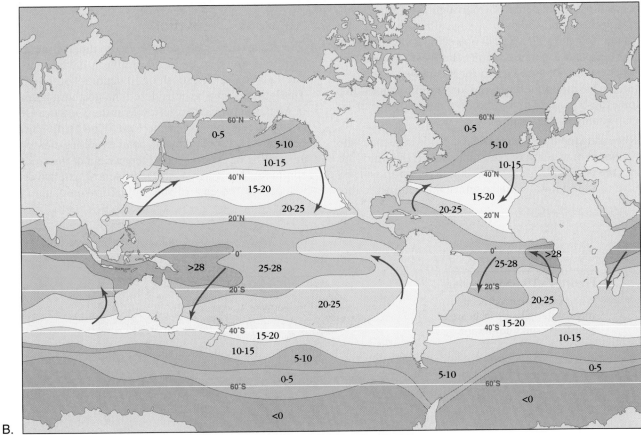

B.

Figure 7–8 Surface temperatures of the world ocean.

Average sea surface temperature distribution in degrees centigrade for August (**A**) and for February (**B**). Note that temperatures migrate north–south with the seasons. Red arrows indicate warm surface currents; blue arrows indicate cool surface currents.

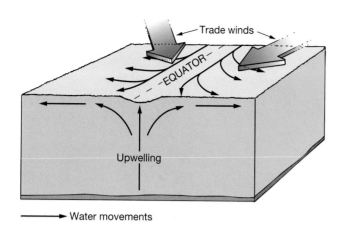

Water movements

Figure 7–9 Equatorial upwelling.
Westward-flowing equatorial currents are driven by trade winds, and partly steered by the Coriolis effect, which pulls surface water away from the equatorial region (surface currents diverge). This water is replaced by upwelling subsurface water.

transport will move the coastal water to the left of the wind direction, causing the water to flow *away from* the shoreline. The replacement of this water comes from the lower portions of the upper water. Such areas are characterized by low surface temperatures and high concentrations of nutrients, both brought up from depth, making these areas high in biological productivity and rich in marine life. This type of upwelling is called **coastal upwelling** and is common along the West Coast of the United States, an area known for its low water temperatures, even in summertime. The low water temperatures along these coastal areas provide a natural form of air conditioning in the summer.

Reversing the direction of coastal winds influences the movement of water in the coastal region. If the winds are from the north (Figure 7–11B), Ekman transport will still move the coastal water to the left of the wind direction,

but in this case, the water will flow *toward* the shoreline. This causes the water to stack up along the shoreline, where it has nowhere to go but down. It sinks below the surface in these regions of low productivity. This type of downwelling is called **coastal downwelling**, and can occur in areas that typically experience coastal upwelling when the winds reverse.

Other Upwelling

Figure 7–12 shows some of the other ways in which upwelling can occur. Upwelling can be caused by offshore winds, sea floor obstructions, or a sharp bend in a coastline. Upwelling also commonly occurs in high-latitude regions, where a pycnocline (a zone of rapidly changing density) is absent. The absence of a pycnocline in high-latitude regions allows significant vertical mixing between high-density cold surface water and high-density cold deep water below. Thus, both upwelling and downwelling are common in high-latitudes.

Surface Currents of the Oceans

Now that we have discussed ocean currents in general, let's tour the world's oceans and examine the surface circulation pattern in each.

Antarctic Circulation

The **Antarctic Circumpolar Gyre** dominates the movement of water masses in the southern Atlantic, Indian, and Pacific Oceans, south of 50 degrees south latitude. This latitude may be considered the northern boundary of what is unofficially referred to as the Southern Ocean. A surface current called the **East Wind Drift**, so called because it is propelled by the polar easterlies, moves from an easterly direction around the margin of the Antarctic continent. The East Wind Drift is most extensively developed to the east of the Antarctic Peninsula in the

Figure 7–10 Downwelling.
When surface currents converge, water piles up in an area and slowly sinks downward.

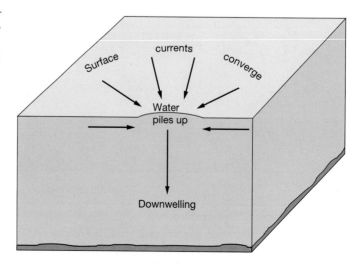

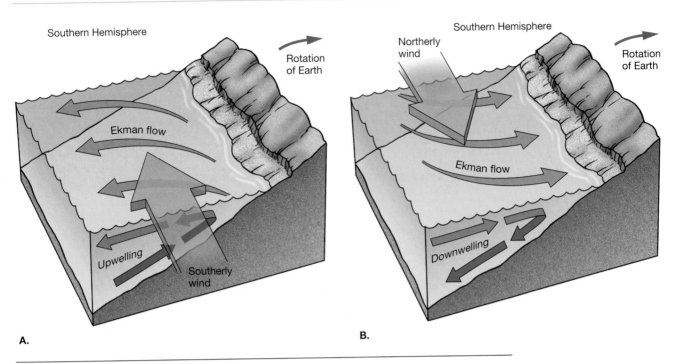

Figure 7–11 Coastal upwelling and downwelling.
A. Where southerly coastal winds blow parallel to a coastline in the Southern Hemisphere, Ekman transport carries surface water away from the continent. An upwelling of deeper water replaces the surface water that has moved away from the coast. **B.** A reversal of the direction of the winds that cause upwelling causes water to pile up against the shore and forces downwelling.

Weddell Sea region and in the area of the Ross Sea (Figure 7–13).

The main current in Antarctic waters is the **West Wind Drift**, which encircles Antarctica at a position centered at approximately 50 degrees south latitude but varies between 40 and 65 degrees south latitude, giving the area its nickname, "The Roaring Forties." This mass is driven by the westerly winds, which are very strong throughout much of the year. The West Wind Drift is the only current to circumscribe Earth completely. It meets its greatest restriction as it passes through the Drake Passage (named for explorer Sir Francis Drake) between the Antarctic Peninsula and the southern islands of South America. The passage is about 1000 kilometers (600 miles) wide. Although the current is not speedy [its maximum surface velocity is about 2.75 kilometers (1.65 miles) per hour], it does transport more water than any other surface current with an average of about 130 million cubic meters per second.[9]

Thus, there are two currents flowing around Antarctica in opposite directions. Recall that the Coriolis effect deflects moving masses to the left in the Southern Hemisphere, so the East Wind Drift is deflected toward the continent and the West Wind Drift is deflected away from it. This creates a zone of divergence called the **Antarctic Divergence**, which occurs between the two currents. The Antarctic Divergence is an area of abundant marine life in the Southern Hemisphere summer because of good nutrient supply caused by the mixing of these two currents.

Atlantic Ocean Circulation

Atlantic Ocean surface currents are shown in Figure 7–14. As mentioned previously, the basic surface circulation pattern in the Atlantic Ocean is that of two large subtropical gyres.

The North and South Atlantic Gyres
The **North Atlantic Gyre** rotates clockwise and the South Atlantic Gyre rotates counterclockwise, in response to the Coriolis effect. As stated earlier, both rotations are driven by the trade winds and the prevailing westerlies. As shown in Figure 7–14, each gyre consists of a poleward-moving warm current (*red*) and an equatorward-moving cold "return" current (*blue*). The two gyres are partially separated by the **Atlantic Equatorial Countercurrent**.

The **South Atlantic Gyre** includes the **South Equatorial Current**, which reaches its greatest strength just below the Equator and is split in two by topographic interference from the eastern prominence of Brazil. Part of the South Equatorial Current moves off along the northeastern coast of South America toward the Caribbean Sea and the North Atlantic. The rest is turned southward as the **Brazil Current,** which ultimately merges with the West Wind Drift and moves eastward across the South

---

[9] One million cubic meters per second is a useful flow rate for describing ocean currents, so it has become a standard unit, named the **sverdrup (sv)** after Norwegian explorer Otto Sverdrup.

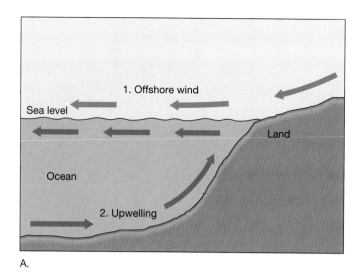

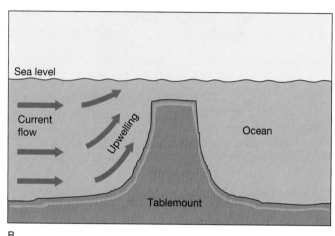

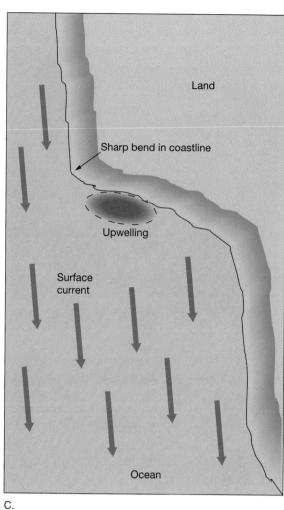

Figure 7–12 Other types of upwelling.
Upwelling can be caused by: **A.** Offshore winds; **B.** A sea floor obstruction, in this case, a tablemount; **C.** A sharp bend in coastal geometry.

Atlantic. The Brazil Current is much smaller than its Northern Hemisphere counterpart, the Gulf Stream, due to the splitting of the South Equatorial Current. The gyre is completed by a slow-drifting movement of cold water, the **Benguela Current**, which flows equatorward along Africa's western coast.

Outside the gyre, a significant northbound flow of cold water also moves along the western margin of the South Atlantic. This current, called the **Falkland Current** (Figure 7–14), is an important cold current that moves along the coast of Argentina as far north as 25 to 30 degrees south latitude, wedging its way between the continent and the southbound Brazil Current.

The Gulf Stream
The **Gulf Stream** moves northward along the U.S. East Coast, warming coastal states and moderating winters in these and northern European regions. It is interesting to trace the origins and behavior of this important current, the most famous and best studied of all ocean currents.

Figure 7–15 shows the network of currents involved in the flow of the Gulf Stream within the North Atlantic Ocean. The **North Equatorial Current** moves parallel to the Equator in the Northern Hemisphere, where it is joined by that portion of the South Equatorial Current that is shunted northward along the South American coast. This flow then splits into two masses: the **Antilles Current**, which flows along the Atlantic side of the West Indies; and the **Caribbean Current,** which passes through the Yucatán Channel into the Gulf of Mexico. These masses reconverge as the **Florida Current**.

The Florida Current flows close to shore over the continental shelf, carrying a volume that at times exceeds 35 sverdrups. As it moves off North Carolina's Cape Hatteras and flows across the deep ocean in a northeasterly direction, it is called the Gulf Stream. The Gulf Stream is a western boundary current, and as such it experiences western intensification: It is quite narrow, 50 to 75 kilometers (31 to 47 miles) wide, reaches depths as deep as 1.5 kilometers (1 mile), and has speeds from

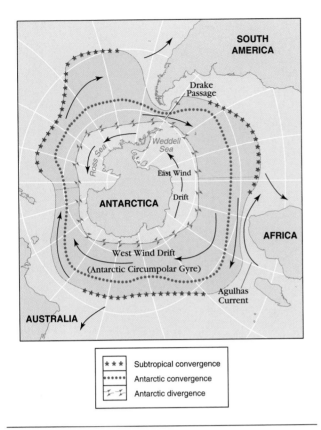

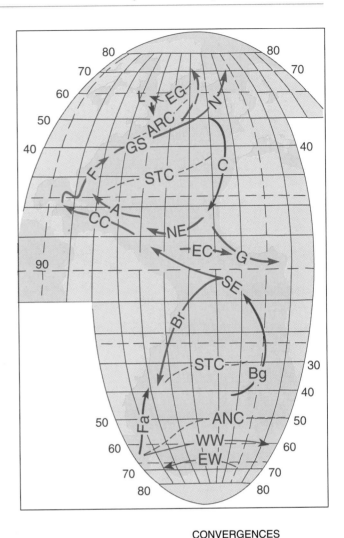

Figure 7–13 Antarctic surface circulation.

The East Wind Drift flows around Antarctica westward, driven by the polar easterly wind system. In the opposite direction, and farther out from the continent, the West Wind Drift circles Antarctica with an easterly flow, caused by the prevailing westerlies. Antarctic Convergence and Divergence is caused by interactions at the boundaries of these currents.

CONVERGENCES
ARC–Arctic
STC–Subtropical
ANC–Antarctic

CURRENTS

A–Antilles	G–Guinea
Bg–Benguela	GS–Gulf Stream
Br–Brazil	I–Irminger
C–Canary	L–Labrador
CC–Caribbean	NE–North Equatorial
EG–East Greenland	N–Norwegian
EW–East Wind Drift	SE–South Equatorial
EC–Equatorial Counter	WW–West Wind Drift
Fa–Falkland	
F–Florida	

Figure 7–14 Atlantic Ocean surface currents.

Atlantic Ocean circulation is composed primarily of two subtropical gyres.

3 to 10 kilometers (2 to 6 miles) per hour, making it the fastest current in the world ocean.

The western margin of the Gulf Stream can frequently be defined as a rather abrupt boundary that periodically migrates closer to and farther away from the shore. Its eastern boundary becomes very difficult to identify because it is usually masked by meandering water masses that change their position continuously.

The Gulf Stream gradually merges eastward with the water of the **Sargasso Sea**. The Sargasso Sea is the water that circulates around the rotation center for the North Atlantic gyre. Essentially, the Sargasso Sea is the stagnant eddy of the North Atlantic Gyre. Its name is derived from a type of floating marine alga called *Sargassum* (*sargassum* = grapes) that abound on its surface. Evidently, sailors thought that the small floats on this alga were reminiscent of bunches of grapes.

A transport volume of over 90 sverdrups[10] off Chesapeake Bay indicates that a large volume of Sargasso Sea

water has been added to the flow provided by the Florida Current. However, by the time the Gulf Stream nears Newfoundland, the Gulf Stream's volume has been reduced to 40 sverdrups. This indicates that most of the Sargasso Sea water that joined the Florida Current to

---

[10] The Gulf Stream's flow of 90 sverdrups is more than 100 times greater than the combined flow of all the rivers in the world!

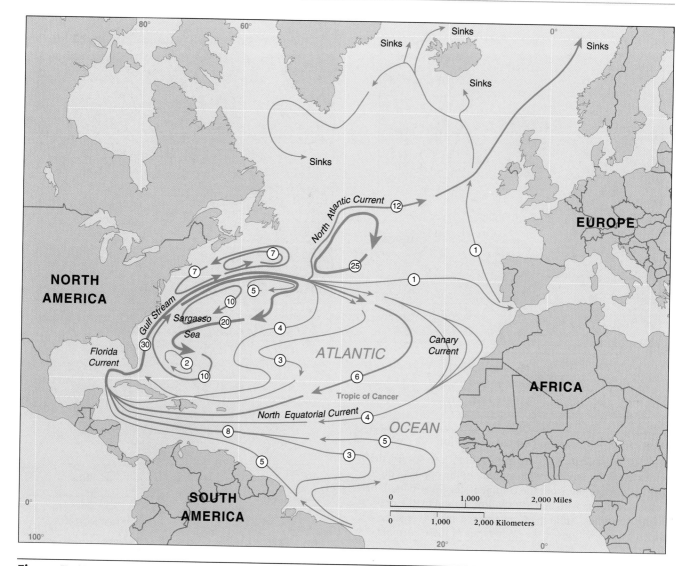

Figure 7–15 The Gulf Stream and North Atlantic Ocean circulation.
The Gulf Stream consists of water recirculated through the North Atlantic Gyre from the North Equatorial Current system via the Gulf of Mexico by the Florida Current. Its northern extension is called the North Atlantic Current. Numbers indicate flow volume in sverdrups (millions of cubic meters per second).

make up the Gulf Stream has returned to the diffuse flow of the Sargasso Sea.

Just how this loss of water occurs has yet to be determined. However, much of it may be caused by *meanders* (snake-like bends in the current) that often completely disconnect from the Gulf Stream and form large rotating masses called *rings* or *eddies* (vortexes). Figure 7–16 shows several of these rings, which are noticeable near the center of each image. The figure shows how meanders along the north boundary of the Gulf Stream pinch off and trap warm Sargasso Sea water in eddies that rotate clockwise, creating **warm core rings** (*yellow*) surrounded by cooler (*blue and green*) water. These warm rings contain shallow, bowl-shaped masses of warm water about 1 kilometer (0.6 mile) deep, with diameters of about 100 kilometers

(60 miles). As warm core rings disconnect from the Gulf Stream, they remove large volumes of water from it.

Cold near-shore water spins off to the south of the Gulf Stream as counterclockwise-rotating **cold core rings** (*green*) surrounded by warmer (*yellow and red-orange*) water (Figure 7–16). The cold rings include spinning cone-shaped masses of cold water that extend over 3.5 kilometers (2.2 miles) in depth. These rings may exceed 500 kilometers (310 miles) in diameter at the surface. The diameter of the cone increases with depth, and sometimes reaches all the way to the sea floor, where they have a tremendous impact on sea-floor sediment features. These cold rings move southwest at speeds from 3 to 7 kilometers (2 to 4 miles) per day toward Cape Hatteras, where they often rejoin the Gulf Stream.

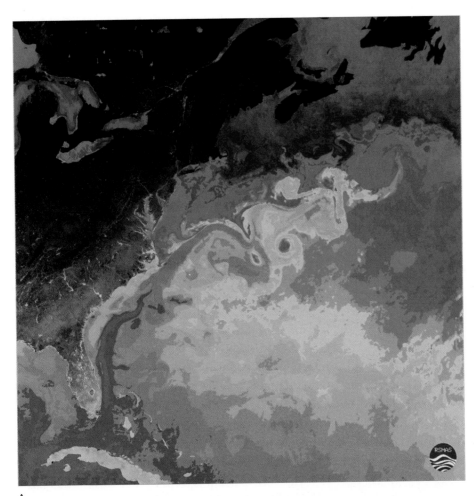

A.

Figure 7–16 The Gulf
Stream and Sea surface
temperatures.
A. NOAA satellite false-color
image of sea surface temperature. The warm waters of the
Gulf Stream are shown in red
and orange. Colder waters are
shown in green, blue, and purple. As the Gulf Stream meanders northward, some of its
meanders close, forming rings,
which trap warm or cold water
within them. **B.** Schematic diagram of the same view as in
A, showing how rings are
formed by the pinching off of
a meander loop.

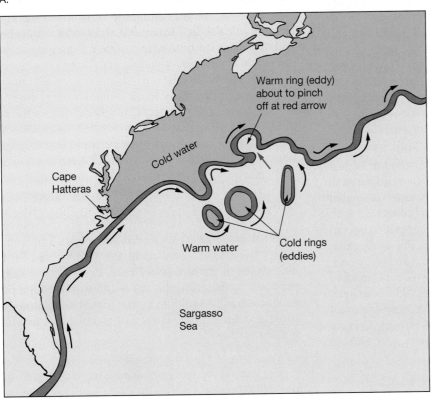

B.

Southeast of Newfoundland, the Gulf Stream continues in an easterly direction across the North Atlantic (Figure 7–15). Here the Gulf Stream breaks into numerous branches. Two branches combine water through the mixing of the cold **Labrador Current** and the warm Gulf Stream, producing abundant fog in the North Atlantic. These are the **Irminger Current**, which flows along Iceland's west coast, and the **Norwegian Current**, which moves northward along Norway's coast. The other major branch crosses the North Atlantic as the **North Atlantic Current**[11] and turns southward to become the cool **Canary Current**, which passes between the Azores Islands and Spain. This broad, diffuse southward flow eventually joins the North Equatorial Current, completing the gyre.

Climatic Effects of North Atlantic Currents

The warming effects of the Gulf Stream are far ranging. Not only does the Gulf Stream moderate temperatures along the U.S. East Coast, the Gulf Stream (in conjunction with heat transferred by the atmosphere) moderates European climate as well. In fact, it has been estimated that northern Europe is as much as 9 degrees centigrade (20 degrees Fahrenheit) warmer because of the Gulf Stream. Certainly, the Gulf Stream contributes to a climate in Northern Europe that is far warmer than would otherwise be expected and helps keep high-latitude Baltic ports ice free throughout the year.

The warming effects of western boundary current flow in the North Atlantic Ocean can be seen on average sea-surface temperature maps such as the one for February (see Figure 7–8B). For example, off the coast of North America from latitudes 20 degrees north (the latitude of Cuba) to 40 degrees north (the latitude of Philadelphia), there is a 20-degree centigrade (36-degree Fahrenheit) temperature difference in sea surface temperatures. By contrast, on the eastern side of the North Atlantic, only a 5-degree centigrade (9-degree Fahrenheit) range in temperature can be observed between the same latitudes, indicating the moderating effect of the Gulf Stream.

The average sea-surface temperature map for August (see Figure 7–8A) also shows clearly how northwestern Europe is warmed by the North Atlantic and Norwegian Currents (branches of the Gulf Stream) compared with the same latitudes along the North American coast. On the western side of the North Atlantic, the southward-flowing Labrador Current—which is cold and often contains icebergs from western Greenland—keeps Canadian coastal waters much cooler. During the Northern Hemisphere winter (Figure 7–8B), North Africa's coastal waters are cooled by the southward-flowing Canary Current and are much cooler than waters near Florida and the Gulf of Mexico.

---

[11] The North Atlantic Current is often called the North Atlantic Drift, emphasizing its sluggish nature.

Pacific Ocean Circulation

Dominating the circulation pattern in the Pacific Ocean are two large subtropical gyres, resulting in similar surface water movement and climatic affects to those found in the Atlantic. However, the Equatorial Countercurrent is much better developed in the Pacific Ocean than in the Atlantic (Figure 7–17), because of the larger size and unobstructed geometry of the Pacific Ocean basin.

Normal Conditions

Figure 7–17 shows that the **North Pacific Gyre** is composed of the North Equatorial Current, which flows into the western intensified **Kuroshio Current** near Asia. Because of its proximity to Japan, the Kuroshio Current is also called the *Japan Current*. Its warm waters make Japan's climate warmer than would be expected from its latitude. This current flows into the **North Pacific Current**, which connects to the cool-water **California Current**, which travels from northerly latitudes along the coast of California to complete the loop. Some North Pacific Current water also flows to the north and merges into the **Alaskan Current** in the Gulf of Alaska.

The **South Pacific Gyre** is composed of the South Equatorial Current, which flows into the western intensified **East Australian Current**.[12] From there, it joins the West Wind Drift, and completes the gyre through flow within the **Peru Current** (also called the *Humboldt Current*, after German naturalist Friedrich Heinrich Alexander von Humboldt) (Figure 7–17).

The Peru Current is a cool-water current that has historically provided one of Earth's richest fishing grounds. Figure 7–18A is a diagrammatic view across the equatorial South Pacific Ocean that shows the atmospheric and oceanic conditions responsible for creating such an incredible abundance of marine life. Along the west coast of South America, coastal winds cause Ekman transport that moves water away from shore, causing upwelling of cool, nutrient-rich water. This upwelling fuels the productivity of phytoplankton, which results in an abundance of larger marine life. Near Peru and Ecuador, small silver-colored fish called *anchovettas* (anchovies) become particularly plentiful. Anchovies are an essential link in the region's food web because they feed by swimming through the water with their mouths open, straining plankton from the water. They in turn are the source of food for many other marine species. Peru's fishing industry became established in the 1950s based on the abundance of anchovies. By 1970, Peru was the largest producer of fish from the sea in the world, with a peak production of 13.1 million metric tons (14.4 million short tons), accounting for nearly 22 percent of *all fish* from the sea worldwide.

---

[12] Note that the western intensified East Australian Current was named because it lies off the *East Coast* of Australia, even though it occupies a position along the *western* margin of the Pacific Ocean basin.

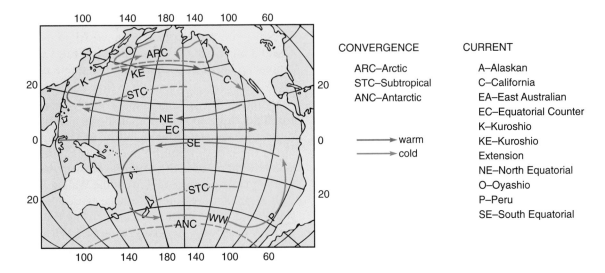

Figure 7–17 Pacific Ocean surface currents.

Similar to the Atlantic Ocean, the Pacific contains two large subtropical gyres. However, the equatorial countercurrent is more strongly developed than in smaller ocean basins.

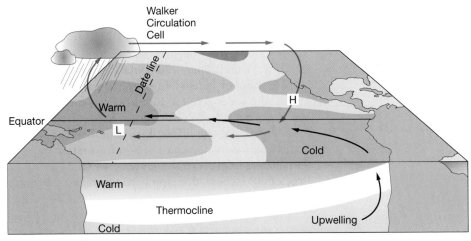

A. Normal conditions

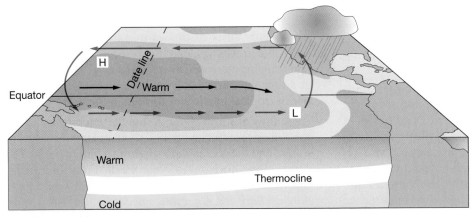

B. ENSO conditions

Figure 7–18 Normal and ENSO conditions.

A. Normal oceanic and atmospheric conditions in the equatorial Pacific.
B. El Niño–Southern Oscillation (ENSO) conditions.

Figure 7–18A shows that high pressure (caused by sinking air within the South Pacific High) dominates the coastal region of South America, resulting in clear, fair, and dry weather conditions. Across the Pacific, rising air within the Indonesian Low is responsible for creating cloudy conditions with plentiful precipitation in Indonesia, New Guinea, and northern Australia. This pressure difference causes the strong southeast trade winds to move from high to low pressure across the equatorial South Pacific (see Figure 6–12). The circulation cell in the equatorial South Pacific Ocean created by the movement of air between high- and low-pressure regions is named the **Walker Circulation Cell**, after G. T. Walker, a British meteorologist who first described the effect in the 1920s.

The southeast trade winds set ocean water in motion, which also moves across the Pacific towards the west. The water warms as it flows in the equatorial region and creates a wedge of warm water on the western side of the Pacific Ocean, called the **Pacific Warm Pool** (this wedge of water can be clearly seen in Figure 7–8). The Pacific Warm Pool also extends vertically into the ocean as a result of convergence, causing a greater thickness of warm water along the western side of the Pacific than along the eastern side. The thermocline beneath the warm pool occurs at a depth exceeding 100 meters (about 330 feet); in contrast, the thermocline develops within 30 meters (100 feet) of the surface in areas of the eastern equatorial Pacific. This can be seen by the sloping boundary between the warm surface water and the cold deep water in Figure 7–18A. As remarkable as these conditions might seem, these are normal conditions for this region. Let's look at how an El Niño event changes these conditions.

El Niño–Southern Oscillation (ENSO) Conditions

Historically, Peru's residents knew that a current of warm water every few years reduced the population of anchovies in coastal waters. The decrease in anchovies not only caused a dramatic decline in the fishing industry, it also caused the decline of marine life such as marine mammals and sea birds that depend on anchovies as a source of food. This warm current also brought about changes in the weather—usually intense rainfall—and even brought such interesting items as floating coconuts from tropical islands from near the Equator. At first, these events were called *años de abundancia* (years of abundance) because of the increase in plant growth on land due to the increased rainfall in normally arid regions. What was once thought of as a joyous event, however, soon became associated with the ecological and economic disaster that is now a well-known consequence of the phenomenon.

This warm-water phenomenon usually occurred around Christmas and thus was given the name **El Niño**, Spanish for "the child," in reference to baby Jesus. Because an El Niño event is associated with the switching of atmospheric pressure regions called the Southern Oscillation, El Niños are more correctly called **El Niño–Southern Oscillation (ENSO)** events, which indicates the interrelationship between ocean and atmosphere during such an event.

Figure 7–18B is a diagrammatic view across the equatorial South Pacific Ocean that shows how normal conditions change as a result of an ENSO. The high pressure along the coast of South America weakens, causing a smaller difference in atmospheric pressure between the high- and low-pressure regions of the Walker Circulation Cell. A smaller pressure difference causes the southeast trade winds to diminish, or in very strong ENSO events, actually to *blow in the reverse direction*. When the trade winds fail, this is a sure sign of an impending ENSO.

Without the trade winds, the Pacific warm pool that has built up on the western side of the Pacific begins to flow back across the ocean toward South America (Figure 7–18B). This is accomplished by an increase in the flow of the Equatorial Countercurrent. Thus, the temperature of surface waters across the Pacific increase as the water flows to the east across the ocean basin toward Peru. This creates a band of warm water that stretches across the equatorial Pacific Ocean (Figure 7–19). The warm water usually begins to move in September of an ENSO year, and comes into contact with South America by December or January. In strong to very strong ENSO events, the water temperature off Peru can increase by 8 degrees centigrade (14 degrees Fahrenheit) as compared to normal conditions. In addition, the presence of warm water along the coast of South America can cause an increase in average sea level by as much as 20 centimeters (8 inches), due simply to thermal expansion of the warm water.

As the warm water moves across the Pacific, sea surface temperatures increase in the equatorial region. This causes a devastating effect on corals because the water temperatures are simply too high for corals to survive. Thus, corals are decimated in Tahiti, the Galápagos, and other tropical Pacific islands. After the warm water contacts South America, the warm water also moves north and south, traveling along the west coast of the Americas and bringing increased average sea level along with it. In addition, the warm water causes an increase in the number of tropical hurricanes formed in the eastern Pacific.

The flow of warm water across the Pacific also causes the sloped thermocline boundary between warm surface waters and the cooler waters below to flatten out and become more horizontal. Near Peru, upwelling brings warmer, nutrient-depleted water to the surface. In fact, *downwelling* can sometimes occur as the warm water stacks up along coastal South America. Thus, productivity diminishes and most types of marine life in the area are dramatically reduced.

As the warm water moves to the east across the Pacific, the low-pressure zone also migrates. In a strong

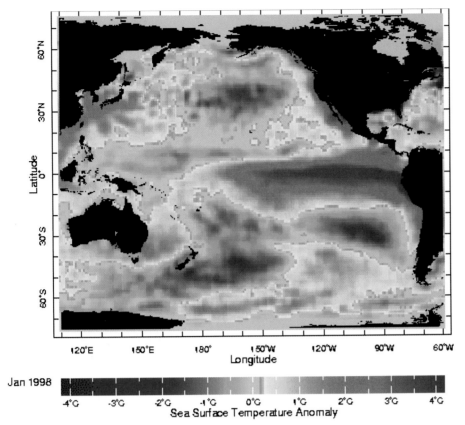

Figure 7–19 Sea-surface temperature anomaly.

Sea-surface temperature anomaly map for the ENSO event of 1997–1998. Anomalies represent departures from normal conditions compared to the average. Red colors represent warmer than normal temperatures and blue colors represent cooler than normal temperatures.

to very strong ENSO event, the low pressure can move entirely across the Pacific and remain over South America. The low pressure causes much precipitation along coastal South America, an area that normally is influenced by high pressure and receives very little precipitation. Conversely, the Indonesian low-pressure region is replaced by a high-pressure region, bringing dry conditions or, in strong to very strong ENSO events, drought conditions to Indonesia and northern Australia. ENSO events usually end 12 to 18 months after they start, with a gradual return to near normal conditions that begins in the southeastern tropical Pacific and spreads westward.

Effects of ENSOs   Mild ENSO events influence only the equatorial South Pacific Ocean. Strong to very strong ENSO events not only influence the weather in the equatorial South Pacific, but often can be felt worldwide. Typically, an ENSO will alter the atmospheric jet steam and produce weather *different from normal conditions* in most parts of the globe. Sometimes this is manifested as drier than normal conditions; at other times, conditions may be wetter than normal. In addition, the weather in a region may be warmer or cooler than normal. It still remains a difficult task to predict exactly how a particular ENSO will affect any region's weather.

Figure 7–20 shows some of the typical effects felt worldwide as a result of very strong ENSO events. Flooding, erosion, droughts, fires, tropical storms, and ef-

fects on marine life are all associated with ENSO events. These weather perturbations also affect the production of certain commodities such as corn, cotton, and coffee. ENSO events are clearly related to other periodic climatic fluctuations observed worldwide, and the relations among these occurrences are being intensely studied to gain further insight into understanding the world's climate.

More locally, the warm water from an ENSO event can work its way along the west coast of the United States all the way to British Columbia, Canada. Figure 7–21 shows satellite images of sea surface temperatures off southern California during a normal and an ENSO year.

Examples from Recent ENSOs   During this century, there have been a number of ENSO events of varying severity (Table 7–3). The average frequency of moderate, strong, and very strong ENSO events is one every 2 to 10 years, but in a highly irregular pattern. For instance, in some decades there has been an ENSO event every few years, while in other decades there may have been only one ENSO event. Let's examine some of the effects of some of the more recent ENSOs.

In the winter of 1976, a moderate ENSO event coincided with California's worst drought of this century. Thus, it is clear that ENSO events don't just bring torrential rains to the western United States. During this same winter, the eastern United States experienced record cold conditions.

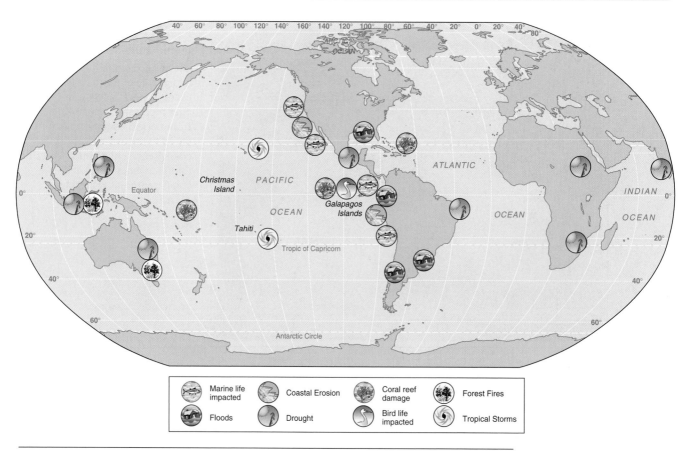

Figure 7–20 Effects of ENSO.

Flooding, erosion, droughts, fires, tropical storms, and effects on marine life are all associated with very strong ENSO events.

The 1982–1983 ENSO stands as the strongest ENSO event ever recorded, causing far-ranging effects across the globe (Figure 7–20). In the United States, the jet stream swung much farther south than normal and brought a series of powerful storms. This caused up to three times the normal rainfall in the southwestern United States, which resulted in severe flooding and landslides. Higher than normal sea level combined with high surf caused much damage to coastal structures and an increase in coastal erosion. There was much snowfall across the Rocky Mountains. Alaska and western Canada experienced a warm winter, and the eastern United States had its mildest winter in 25 years. Meanwhile, in Europe, there was severe cold weather. Worldwide, there were droughts in Australia, Indonesia, China, India, Africa, and Central America.

In the Pacific Ocean, normally arid Peru was drenched with more than 3 meters (10 feet) of rainfall, causing extreme damage by flooding and landslides. Sea surface temperatures were so high that corals were decimated across the equatorial Pacific. In this region, marine mammals and sea birds that depend on the food source associated with normal high productivity died or went

elsewhere. French Polynesia had not experienced a hurricane in 75 years; that year, it endured six. Kauai also experienced a rare hurricane.

Worldwide, there was an estimated $8 billion in damage ($2 billion in the United States) and over 1000 deaths were attributed to the 1982–1983 ENSO event. However, an ENSO event can be beneficial for some regions. For instance, tropical hurricane formation is generally suppressed in the Atlantic Ocean, some desert regions receive much-needed rain, and organisms adapted to warm-water conditions thrive in the Pacific.

The 1982–1983 ENSO event was not predicted, nor was it recognized until it was near its peak. Because of the extensive damage and worldwide influence on weather phenomena caused by the 1982–1983 ENSO event, a major study of the mechanism of ENSO events was initiated in 1985, called the **Tropical Ocean—Global Atmosphere (TOGA)** program. The goal of the TOGA program was to monitor the equatorial South Pacific Ocean during the time when ENSO events occurred to enable scientists to model and predict future ENSO events. The 10-year program involved studying the ocean from research vessels, analyzing surface and subsurface

A. Normal (Jan. 1982)   B. ENSO (Jan. 1983)

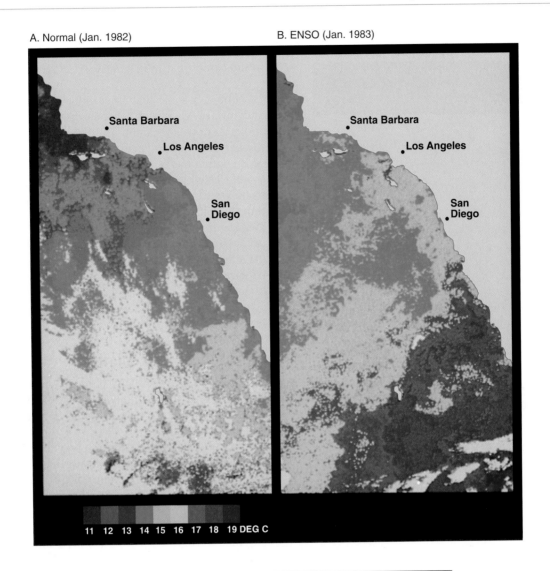

Figure 7–21 Water temperatures off southern California.
Water temperature maps (in degrees centigrade) for southern California from data collected by
the satellite-mounted Advanced Very High Resolution Radiometer. Blue is cold water, red is
warm water. **A.** Water temperatures in January 1982, a non-ENSO year. **B.** Water temperatures a
year later, during an ENSO event.

data from radio-transmitting sensor buoys, monitoring
oceanic phenomena by satellite, and developing computer
models. The study was completed in 1995, offering a wealth
of new data that has enabled scientist to develop more reliable models that link the combined ocean–atmosphere
system. These models have given scientists the ability to
predict ENSO events up to a year in advance, as was done
for the ENSO events of 1987, 1992, and 1998. Since the
completion of TOGA, the **Tropical Atmosphere and
Ocean (TAO)** project has continued to monitor the equatorial Pacific Ocean with a series of 70 moored buoys. This
multinational project is sponsored by the United States,
Canada, Australia, and Japan.

In the fall of 1987, a strong ENSO event started to form
but for some reason lost energy. Eventually, it expressed
itself as a moderate ENSO event.

In 1991–1992, an ENSO event began that produced
global weather modifications similar to those of the
1982–1983 event. For instance, warm water was found
offshore of California, heavy rains in the mid-continent
region of the United States caused flooding of the
Missouri and Mississippi Rivers, and the Pacific Northwest and Canada were very dry. TOGA computer models were able to predict this ENSO event, even though
the causes of ENSO events are still not fully understood.
The 1991–1992 ENSO was unusual because it maintained
itself until the spring of 1995.

The 1997–1998 ENSO event was unusual in that it
began several months earlier than normal and peaked
in January 1998. In terms of the amount of Southern
Oscillation and sea surface warming in the equatorial
Pacific, the 1997–1998 ENSO was initially as strong as

Table 7-3   Moderate, strong, and very strong ENSO events during the last century

ENSO event (during the winter of)	Years since previous ENSO event	Event strength
1899–1900	2	Strong
1902	2	Moderate
1907	5	Moderate
1911–1912	4	Strong
1914	2	Moderate
1917	3	Strong
1923	6	Moderate
1925–1926	2	Very strong
1932	6	Strong
1939	7	Moderate
1940–1941	1	Strong
1943	2	Moderate
1953	10	Moderate
1957–1958	4	Strong
1965	7	Moderate
1972–1973	7	Strong
1976	3	Moderate
1982–1983	6	Very strong
1987	4	Moderate
1991–1994	4	Strong
1997	3	Very strong

the 1982–1983 ENSO, which was the strongest event of the century. However, the 1997–1998 ENSO weakened in the last few months of 1997 before reintensifying in early 1998. The impact of the 1997–1998 ENSO event was felt mostly in the tropical Pacific, where surface water temperatures were up to 8 degrees centigrade (14 degrees Fahrenheit) warmer than normal in eastern Pacific areas such as the Galápagos Islands and off the coast of Peru. In the western Pacific, drought conditions persisted because of the high-atmospheric-pressure system that replaced the area's normal low pressure causing wildfires to burn out of control in Indonesia. Also, there was warmer water than normal along the west coast of Central and North America, which caused an increase in hurricane occurrence off Mexico. In the United States, many unusual weather events were linked to the 1997–1998 ENSO, such as killer tornadoes in the southeast, massive blizzards in the upper Midwest, and flooding of the Ohio River Valley. In California, most of the state received two times the normal rainfall, which caused flooding and landslides in many parts of the state. However, the lower Midwest, the Pacific Northwest, and the eastern seaboard had relatively mild weather. In spite of predictions that the 1997–1998 ENSO would

be the most severe ENSO of the century, it did not affect worldwide weather or cause damage and casualties to the extent that the 1982–1983 ENSO did.

La Niña   In some years, the trade winds are very strong and surface waters in the equatorial South Pacific are cooler than usual. Commonly, below-average temperatures in the eastern Pacific occur following an El Niño. This cooling has been given the name **La Niña** (opposite of El Niño; Spanish for "the female child") and seems to be associated with weather phenomena opposite to those of El Niño. For instance, Indian Ocean monsoons are *drier* than usual in El Niño years but *wetter* than usual in La Niña years. In addition, the exceptionally hot and dry summer of 1988 in North America is thought to have been related to the strong La Niña of that year. Cooling of equatorial Pacific waters in mid to late 1998 indicated that La Niña conditions followed the 1997–1998 ENSO event.

Indian Ocean Circulation

Because the Indian Ocean is mostly in the Southern Hemisphere (extending only to about 20 degrees north latitude), surface circulation varies considerably from that in the Atlantic and Pacific. From November to March, the equatorial circulation is similar to that in the other oceans, with two westward-flowing equatorial currents (North and South Equatorial Currents) separated by an eastward-flowing Equatorial Countercurrent. However, in contrast to the Atlantic and Pacific wind systems, which shift northward of the geographical Equator, the meteorological equator is shifted *southward* in the Indian Ocean.

The Equatorial Countercurrent flows between 2 and 8 degrees south latitude, bounded on the north by the North Equatorial Current (which extends as far as 10 degrees north latitude) and on the south by the South Equatorial Current (which extends to 20 degrees south latitude). The winds of the northern Indian Ocean have a seasonal pattern that are called **monsoon** (*monsoon* = season) winds.

During winter, the typical northeast trade winds are called the *northeast monsoon*. They are strengthened by the rapid cooling of air over the Asian mainland during winter. This creates a high-pressure cell, which forces atmospheric masses off the continent and out over the ocean, where air pressure is less (*green arrows* in Figure 7–22A).

During summer, the Asian mainland warms faster than the adjacent ocean.[13] As a result, a summer low-pressure cell develops over the continent, allowing higher-pressure air to reverse direction and move from the Indian Ocean

---

[13] Recall that this is due to the lower heat capacity of continental rocks and soil compared with water.

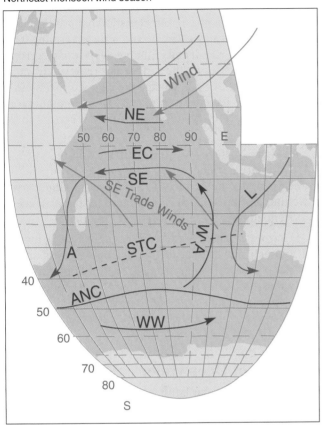

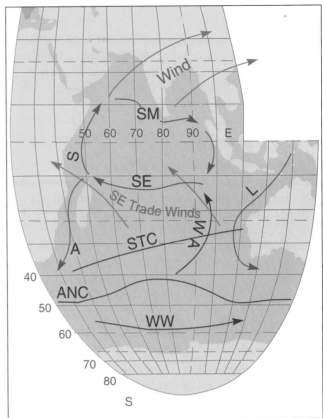

Warm ➝   Cold ➝

CURRENTS A—Agulhas   EC—Equatorial Countercurrent
L—Leeuwin   NE—North Equatorial   S—Somali
SE—South Equatorial   SM—Southwest Monsoon
WA—West Australian   WW—West Wind Drift

CONVERGENCES STC—Subtropical   ANC—Antarctic

Figure 7–22 Indian Ocean surface currents.
Surface currents in the Indian Ocean are influenced by the seasonal monsoons.

onto the Asian landmass. This gives rise to the *southwest monsoon* (*green arrows* in Figure 7–22B), which may be thought of as a continuation of the southeast trade winds across the Equator.

During this season, the North Equatorial Current disappears and is replaced by the *Southwest Monsoon Current*. It flows from west to east across the North Indian Ocean. In September or October, the northeast trade winds are reestablished, and the North Equatorial Current reappears (Figure 7–22A).

Surface circulation in the southern Indian Ocean (the **Indian Ocean Gyre**) is similar to the counterclockwise circulation of gyres observed in other southern oceans. When the northeast trade winds blow, the South Equatorial Current provides water for the Equatorial Countercurrent and the **Agulhas Current**, which flows

southward along Africa's eastern coast and joins the West Wind Drift. Turning northward out of the West Wind Drift is the **West Australian Current**, an eastern boundary current that completes the gyre by merging with the South Equatorial Current.

During the southwest monsoon, a northward flow from the Equator along the coast of Africa, the **Somali Current**, develops with velocities approaching 4 kilometers (2.5 miles) per hour.

The eastern boundary current in the southern Indian Ocean is unique. Other eastern boundary currents of subtropical gyres are cold drifts toward the Equator that produce arid coastal climates, which receive less than 25 centimeters (10 inches) of rain per year. However, in the southern Indian Ocean, the West Australian Current is displaced offshore by a southward-flowing current called the **Leeuwin Current**. This current is driven southward along the Australian coast from the warm-water dome piled up in the East Indies by the Pacific equatorial currents.

The Leeuwin Current produces a mild climate in southwestern Australia, which receives about 125 centimeters (50 inches) of rain per year. This current weakens during ENSO events and contributes to Australian drought conditions associated with these events.

Deep Currents

Deep currents influence the deep zone below the pycnocline, thus influencing about 90 percent of ocean water. The mechanism for movement of deep currents is a result of density differences. Although these density differences are usually small, they are large enough to cause denser waters to sink. Changes that increase water density normally are changes in temperature and salinity, so this circulation is referred to as **thermohaline** (*thermo* = heat, *haline* = salt) **circulation**.

Origin of Thermohaline Circulation

As discussed in Chapter 5, either decreasing temperature or increasing salinity causes an increase in density. Thus, cold, high-salinity water is very dense. Of these two factors, temperature has the greatest influence on density. Salinity appears to have a minimal effect on the movement of water masses in equatorial latitudes. Density changes due to salinity are important only in very high latitudes, where water temperature remains low and relatively constant.

The water involved in deep-ocean current movement (thermohaline circulation) is initially formed in high-latitude regions *at the surface*. In these regions, surface water becomes cold and achieves increased salinity through the process of sea ice formation. Once this surface water acquires the characteristics that produce high-density water, it sinks beneath the surface, carrying those characteristics with it and initiating deep-ocean currents. After the water leaves the surface, it is removed from the physical processes that change the characteristics of the water, and so its temperature and salinity remain largely unchanged for the duration of time it spends in the deep ocean.

As these surface-water masses are sinking in high-latitude areas, deep-water masses are also upwelling and rising to the surface. This vertical mixing (downwelling *and* upwelling) is facilitated by the absence of a pycnocline in these high-latitude regions.

Deep-water currents are much larger in volume and are also much slower than surface currents. Typical speeds of deep currents range from 10 to 20 kilometers (6 to 12 miles) per year. It is interesting to note that it takes a deep current *an entire year* to travel the same distance that a western intensified surface current can move in *one hour*.

Sources of Deep Water

In southern subpolar latitudes, the most significant area where deep-water masses form is beneath sea ice along the margins of the Antarctic continent. Here, rapid winter freezing produces very cold, high-density water that sinks down the continental slope of Antarctica and becomes **Antarctic Bottom Water**, the densest water in the open ocean (Figure 7–23). Antarctic Bottom Water slowly sinks beneath the surface and spreads into all the world's ocean basins. In nearly all parts of the world's

Figure 7–23 Antarctic deep-water masses.

Upwelling and downwelling occur near Antarctica, creating deep-water masses.

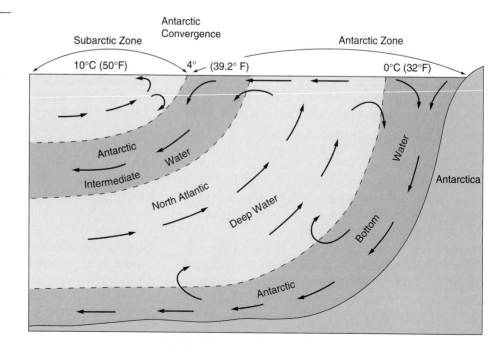

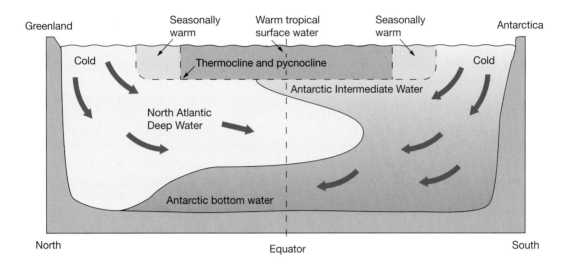

Figure 7–24 Atlantic Ocean subsurface water masses.
Schematic diagram of the various water masses in the Atlantic Ocean. Similar layering based on density occurs in the Pacific and Indian Oceans as well.

oceans,[14] if one samples deep enough water, one will collect water that began its journey perhaps over 1000 years ago as surface water near Antarctica.

In the northern subpolar latitudes, an important sinking of cold surface waters that become deep-water masses occurs in the Atlantic Ocean. In the North Atlantic, the major sinking of surface water is thought to occur in the Norwegian Sea. From there it flows as a subsurface current into the North Atlantic. This flow becomes part of what is called **North Atlantic Deep Water**. Contributing to North Atlantic Deep Water is additional surface water that sinks at the margins of the Irminger Sea off southeastern Greenland and the Labrador Sea as well as water coming from the dense, salty Mediterranean Sea. Not quite as dense as Antarctic Bottom Water, North Atlantic Deep Water also spreads throughout the ocean basins but occupies a position above the denser Antarctic Bottom Water (Figure 7–23).

On a broad scale, some surface-water masses converge, which may cause sinking. This convergence occurs within the subtropical gyres and in the Arctic and Antarctic. Subtropical convergences do not produce deep sinking because warm surface waters have relatively low densities and will not sink into the ocean because of their lower density. However, major sinking does occur along the **Arctic Convergence** and **Antarctic Convergence** (Figure 7–23). The major water mass formed from sinking at the Antarctic convergence is called the **Antarctic Intermediate Water** mass (Figure 7–23). The Antarctic Intermediate Water is a water mass that has been largely ignored and remains a true frontier of knowledge, awaiting further discovery.

Figure 7–24 is a generalized cross-section of water masses in the Atlantic Ocean, showing how the ocean is layered based on density. The highest-density water is found along the ocean bottom, with less-dense water above. In low-latitude regions, the boundary between the warm surface water and the deeper cold water is marked by a prominent thermocline and corresponding pycnocline that prevents vertical mixing. In high-latitude regions, this barrier to vertical mixing is absent, and substantial vertical mixing (upwelling and downwelling) occurs.

This same general pattern of layering based on density is typical of the Pacific and Indian Oceans as well. However, the Pacific and Indian Oceans have no source of Northern Hemisphere deep water as the Atlantic does, so their layering lacks a deep-water mass. In the northern Pacific Ocean, the low salt content of surface waters prevents sinking of water into the deep ocean. In the northern Indian Ocean, surface waters are too warm. Instead, these two oceans contain **Oceanic Common Water** along their bottoms, which is created by mixing of Antarctic Bottom Water and North Atlantic Deep Water.

Worldwide Deep-Water Circulation

For every liter of water that sinks from the surface into the deep ocean, a liter of deep water is displaced and therefore must return to the surface somewhere else. It is difficult to identify specifically *where* this vertical flow to the surface is occurring. It is generally believed that this return vertical flow occurs as a gradual, uniform upwelling throughout the ocean basins. This return to the surface may be somewhat greater in low-latitude regions, where surface temperatures are higher.

Figure 7–25 shows a general deep-water circulation model of the world ocean. Note how the most intense

---

[14] The exception to this is in shallow coastal seas and in the Arctic Ocean.

deep-water flow in each ocean basin is along the western side. This is because of the Coriolis effect influencing the motion of deep currents and that the flow pattern is also affected by bathymetric features along the sea floor (such as the mid-ocean ridge).

Conveyer-Belt Circulation

An integrated model combining deep thermohaline circulation and surface currents is shown in Figure 7–26. The circulation shown in this model resembles a conveyer belt, with surface water carrying heat to high latitudes in the North Atlantic Ocean within the Gulf Stream. During the cold winter months, this heat is transferred to the overlying atmosphere, greatly supplementing that received by solar radiation, and, as previously described, warms northern Europe.

Cooling in the North Atlantic increases the density of this upper ocean water to the point where it sinks to the bottom and flows southward, initiating the lower limb of the "conveyor." Here, a volume of water equal to 100 Amazon rivers begins its long journey into the deep basins of all of the world's oceans. This limb extends all the way to the southern tip of Africa, where it joins deep water that encircles Antarctica. The deep water that encircles Antarctica includes deep water that descends along the margins of the Antarctic continent. This mixture of deep waters is dispensed northward into the deep Pacific and Indian Ocean basins, which eventually surfaces and completes the conveyer belt by being carried northward into the North Atlantic Ocean.

Dissolved Oxygen in Deep Water

Cold water has a greater ability to hold dissolved oxygen than warm water. Thus, worldwide deep-water circulation provides the deep ocean with oxygen-rich water by the sinking of dense, cold, oxygen-enriched surface water. During its time in the deep ocean, deep water becomes enriched in nutrients as well (due to decomposition and the lack of organisms using nutrients in the deep ocean).

The sinking of surface waters in high-latitude oceans is extremely important to life on our planet. Without this sinking and the return flow from the deep sea to the surface, distribution of life in the sea would be considerably different. Because of the lack of oxygen, there would be no life in the deep ocean. Life in surface waters would be significantly reduced and confined to the extreme margins of the oceans, where the only oxygen and nutrient source would be runoff from streams.

Thermohaline Circulation and Climate Change

Evidence from deep-sea sediments confirmed by recently developed computer models suggest that worldwide deep-water circulation is closely tied to global climate change. These computer models indicate that changes in the global deep-water circulation pattern can dramatically and abruptly change climate. For instance, if the sinking of surface waters were to halt, the oceans would become less efficient in absorbing and redistributing heat from solar radiation. This might cause much

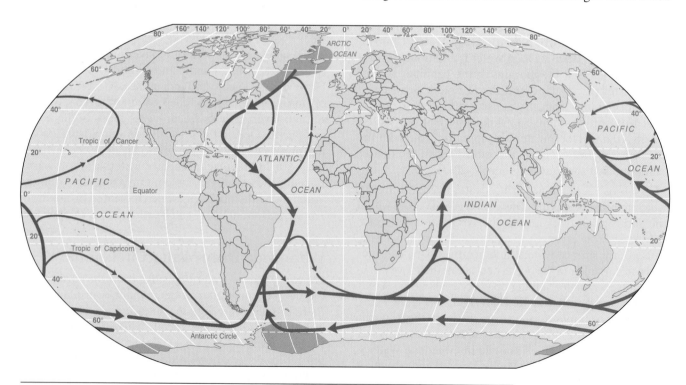

Figure 7–25 Deep-water circulation model.

Schematic model of deep-water circulation first developed by oceanographer Henry Stommel in 1958. Heavy lines mark the major western boundary currents, which result from the same forces that produce western intensified surface currents. The shaded areas indicate source areas for deep water.

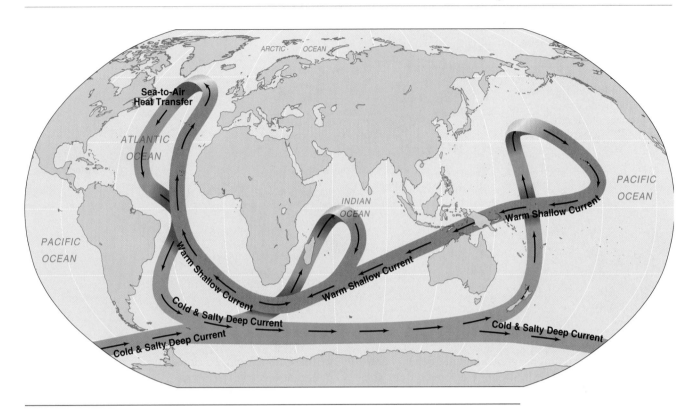

Figure 7–26 Conveyer-belt circulation.

Conveyer-belt circulation is initiated in the North Atlantic Ocean, where warm water cools and sinks below the surface. This water moves southward as a subsurface flow and joins water near Antarctica. This deep water spreads into the Indian and Pacific Oceans, where it slowly rises and completes the conveyer as it travels along the surface into the North Atlantic Ocean.

warmer surface water temperatures and much higher land temperatures than are experienced now.

Alternatively, the buildup of greenhouse gases in the atmosphere may trigger a reorganization of ocean circulation. For example, warmer temperatures may increase the rate of melting of glaciers in Greenland, forming a pool of fresh, low-density surface water in the North Atlantic Ocean. This fresh water could inhibit the downwelling that generates North Atlantic deep water, thus disturbing global deep-water circulation and causing a corresponding change in climate. Because changes such as these are capable of occurring rapidly, it would be difficult for plant and animal life to adapt successfully to the new conditions of the planet.

Power from Winds and Currents

As mentioned previously, Earth's uneven distribution of solar energy drives the winds. The winds, in turn, drive the ocean's surface currents. Steady winds like the prevailing westerlies off the coast of New England are a near-shore wind resource that contains a large amount of energy. There is some optimism that **offshore wind power systems (OWPS)** could generate electricity to meet the needs of large areas of the U.S. North Atlantic coast (Figure 7–27).

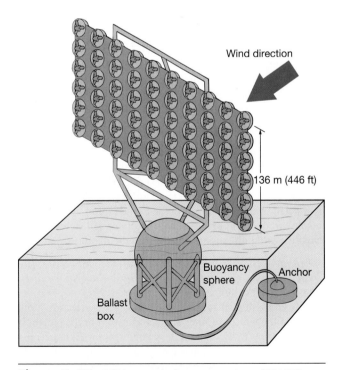

Figure 7–27 Offshore wind power system (OWPS).

A system such as this one could generate electricity to meet the needs of large areas of the U.S. North Atlantic coast.

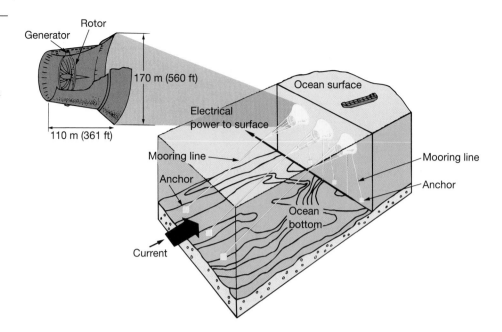

Figure 7–28 Harnessing current power.
Buoyant hydroturbines anchored to the sea floor are powered by an ocean current to produce electricity.

Many have considered the great amount of energy contained in the Florida-Gulf Stream Current System and imagined harnessing it to help supply modern society's energy demands. Devices proposed for extraction range from underwater "windmills" to a water low-velocity energy converter (WLVEC), which is operated by parachutes attached to a continuous belt. Calculations indicate that such systems can be economically competitive.

A novel approach to take advantage of the potential energy source of western intensified current flow off the coast of Florida has been proposed by William J. Moulton of Tulane University. The proposed method uses the flow of the Gulf Stream to power large hydroturbines that generate electricity (Figure 7–28). This method, called the **Coriolis Program**, would employ an array of 242 units covering an area 30 kilometers (18.6 miles) wide and 60 kilometers (37 miles) long. This array could generate up to 10,000 megawatts of electricity per year—the equivalent of 130 million barrels of oil. If Coriolis Program hydroturbines are placed in the Gulf Stream off Florida, they could provide up to 10 percent of the present electricity needs of Florida. Unfortunately, environmental and engineering concerns have delayed further development of this project.

Students Sometimes Ask…

What does an Ekman spiral look like at the surface? Is it strong enough to disturb ships?
The Ekman spiral creates different layers of surface water moving in slightly different directions at slightly different speeds. The Ekman spiral is not strong enough to create an eddy or whirlpool (a vortex) at the surface, and so is of no danger to ships. In fact, the Ekman spiral is unnoticeable at the surface. However, it can be observed by lowering oceanographic equipment over the side of a vessel to various depths. The equipment is affected by different layers of the Ekman spiral, and can often be observed to drift in interesting directions at different angles from the wind direction depending on the depth.

When the westerlies blow offshore from the east coast of the United States, does Ekman transport move water in the direction opposite the Gulf Stream?
One would assume that since the prevailing westerlies are blowing from the west, and since Ekman transport in the Northern Hemisphere is to the right of the wind direction, Ekman transport would move water in a direction opposite to that of the Gulf Stream (to the south). However, Ekman transport is relatively weak as compared to powerful ocean currents such as the western intensified Gulf Stream. Thus, the flow of the Gulf Stream overwhelms any Ekman transport.

I know that when objects are heated, they expand. Does thermal expansion cause any current flow in the ocean?
That's alert thinking, but it is not observed to happen in the oceans. As mentioned previously, the intensity of solar radiation is much greater in the equatorial region than in high-latitude regions. Therefore, you might expect heated equatorial surface water to expand and move away from the Equator toward the poles, spreading over the colder, denser, high-latitude waters. You might further expect some return flow from the high latitudes toward the Equator. But this exchange does not occur. The reason is that the energy imparted to surface waters by winds greatly exceeds the effect of density changes due to heating. Thus, wind overcomes any tendency for such a pattern of circulation to develop.

I've heard that unusual weather—such as severe storms, droughts, flooding, and warm weather—are a result of El Niño conditions. Is this true?
Perhaps. Certainly, strong to very strong ENSO events are the largest single weather disturbance on the planet and cause unusual weather to occur nearly worldwide. It is interesting to note that many people incorrectly attribute *any* kind of unusual weather to an ENSO. Actually, an ENSO event is only responsible for unusual weather conditions some of the time. This is be-

cause many different phenomena influence weather conditions at a particular location and because not every year is an El Niño year. To determine if an El Niño is occurring, one can easily check the sea-surface temperature patterns in the tropical Pacific Ocean and compare any deviation from normal conditions—such as anomalous warming. Unfortunately, ENSO is blamed for many phenomena that are unrelated to actual ENSO conditions.

Do El Niño events occur in other ocean basins?
Yes, the Atlantic and the Indian Oceans both experience events similar to the Pacific's El Niño. However, these events are not nearly as strong nor do they influence worldwide weather phenomena to the same extent as those that occur in the equatorial Pacific Ocean.

In the Atlantic Ocean, this phenomenon is related to the North Atlantic Oscillation (NAO), which is a periodic change in atmospheric pressure between Iceland and Portugal. This pressure difference determines the strength of the prevailing westerlies in the North Atlantic, which in turn affects ocean surface currents there. The Atlantic Ocean experienced an El Niño-type event in 1984, which caused heavy rainfall along the normally arid coast of southwest Africa.

The amount of anchovies produced by Peru is impressive! Besides a topping for pizza, what are some other uses of anchovies?
Anchovies are an ingredient in certain dishes, hors d'oeuvres, sauces, and salad dressing, and they are also used as bait by fishers. However, this hardly accounts for the tons of anchovies that Peru has produced. Historically, most of the *anchovetta* caught in Peruvian waters were exported and used as fishmeal (consisting of ground anchovies), which is in turn used largely in pet food and as a high-protein chicken feed. As unbelievable as it may seem, ENSO events affected the price of eggs! Prior to the collapse of the Peruvian *anchovetta* fishing industry in 1972–1973, ENSO events significantly reduced the availability of anchovetta. This drastically cut the export of anchovies from Peru, causing U.S. farmers to pursue more expensive options for chicken feed. Thus, egg prices typically increased.

The collapse of the *anchovetta* fishing industry in Peru that was triggered by the 1972–1973 ENSO event was due largely to chronic overfishing of anchovies in prior years. It is interesting to note how closely ENSO events are interrelated to global food availability. For example, the shortage of fishmeal after 1972–1973 led to an increase in demand for soyameal, an alternative source of high-quality protein. Increased demand for soyameal products caused an increase in the price of soy commodities, thereby encouraging U.S. farmers to plant more soybeans, which reduced the production of wheat. This caused a major global food crisis to occur, all because of the lack of wheat.

Is the Gulf Stream rich in life?
The Gulf Stream *itself* isn't, but its *boundaries* often are. The oceanic areas that have abundant marine life are typically associated with cool water—either in high-latitude regions, or in any region where upwelling occurs. These areas are constantly resupplied with oxygen- and nutrient-rich water, which results in high productivity. Warm-water areas develop a prominent thermocline that isolates the surface water from colder, nutrient-rich water below. As nutrients are used up in warm waters, the nutrients tend not to be resupplied. Thus, warm surface water results in low productivity (although tropical coral reefs are an exception to this generality). The Gulf Stream, a western intensified current, is a warm-water current that is associated with an absence of marine life. The reason that New England fishers knew about the Gulf Stream (see chapter opener) was because they sought their catch along the *sides* of the current, where mixing and upwelling occur due to the current's movement.

Actually, all western intensified currents are warm-water currents and are associated with low productivities. In fact, the Kuroshio Current in the North Pacific Ocean is named for its conspicuous absence of marine life. In Japanese, the term *Kuroshio* literally means "black current," in reference to its clear, lifeless waters.

If high-salinity water has high density, why doesn't surface water sink in the subtropics, where salinities are the highest?
It is true that the highest-salinity water in the open ocean is found in the subtropical regions. Remember that density is affected by both temperature and salinity, and that temperature is the dominant factor. There is no sinking water in the subtropics because water temperatures are high enough to maintain a low density for the surface-water mass and prevent it from sinking. In such areas, a strong halocline (zone of rapidly changing salinity) develops. It features a relatively thin surface water layer with salinity exceeding 37‰. Salinity decreases rapidly with increasing depth to typical seawater salinity of about 35‰.

Summary

Ocean currents are masses of oceanic water that flow from one place to another, and can be divided into surface currents that are wind driven or deep currents that are density driven. Measuring currents can be accomplished through direct or indirect means.

Surface currents occur within and above the pycnocline and are set in motion by the major wind belts of the world. The surface circulation pattern is composed of circular-moving loops of water called gyres, which are a result of winds but are modified by the Coriolis effect and the positions of the continents. Water is pushed toward the center of clockwise-rotating subtropical gyres in the Northern Hemisphere and counterclockwise-rotating subtropical gyres in the Southern Hemisphere, forming low "hills" in the gyres. A subpolar gyre, the Antarctic Circumpolar Gyre, completely encircles Antarctica.

The Ekman spiral influences shallow surface water and is caused by winds and the Coriolis effect. The average net flow of water affected by the Ekman spiral causes the water to move at 90-degree angles to the wind direction, depending on the hemisphere. As water in the Northern Hemisphere runs down the slope of these hills, the Coriolis effect causes it to turn right and into the clockwise flow pattern. As a result, a geostrophic current flowing parallel to the contours of the hill is maintained.

The apex (top) of the hill is located to the west of the geographical center of the gyre as a result of Earth's spin. This causes the western boundary currents of subtropical gyres to have intensified flow, described as "western intensification."

Upwelling and downwelling are important processes that provide vertical mixing of deep and surface waters. Upwelling—the movement of cold, deep, nutrient-rich water to the surface—is a more-studied process because it stimulates biologic productivity and creates large amount of marine life. Upwelling and downwelling can occur in a variety of ways.

Antarctic circulation is dominated by the largest of all ocean currents, the West Wind Drift, which comprises the Antarctic Circumpolar Gyre. The West Wind Drift flows in a clockwise direction around Antarctica and is driven by the Southern Hemisphere's prevailing westerly winds. A smaller current occurs between the West Wind Drift and the Antarctic continent, called the East Wind Drift, which is powered by the polar easterly winds. In between the two currents is a zone of divergence called the Antarctic Divergence, which contains abundant marine life because of current mixing.

The Atlantic Ocean is dominated by two subtropical gyres, the North Atlantic Gyre and the South Atlantic Gyre, which are separated by a poorly developed equatorial countercurrent. The highest-velocity and best-studied ocean current is the Gulf Stream, which carries warm water along the southeastern U.S. Atlantic coast. Meanders of the Gulf Stream produce warm- and cold-core rings. The warming effects of the Gulf Stream extend along its route and reach as far away as Northern Europe.

Pacific Ocean circulation is also composed of two subtropical gyres, the North Pacific Gyre and the South Pacific Gyre, which are separated by a well-developed equatorial countercurrent. A periodic disruption of normal surface circulation and atmospheric circulation patterns in the Pacific Ocean is called an El Niño–Southern Oscillation (ENSO) event. ENSO events are associated with the eastward movement of the Pacific warm pool, halting or reversal of the trade winds, a rise in sea level along the eastern tropical Pacific, a decrease in ocean productivity along the west coast of South America, and, in very strong ENSOs, worldwide changes in weather. Cooler than normal ocean temperatures in the eastern tropical Pacific are related to La Niña conditions.

The Indian Ocean consists of one gyre, the Indian Ocean Gyre, which exists mostly in the Southern Hemisphere. Indian Ocean circulation is dominated by the monsoon wind system, which changes direction with the seasons. The monsoons blow from the northeast in the winter and from the southwest in the summer.

Deep currents occur below the pycnocline and affect the majority of ocean water. Deep currents are set in motion by slight increases in density caused by temperature and/or salinity changes at the surface and are termed thermohaline circulation. As compared to surface currents, deep currents are much larger in size and move much more slowly.

The deep ocean is layered based on density. Antarctic Bottom Water, the densest deep-water mass in the oceans, forms near Antarctica and sinks along the continental shelf into the South Atlantic Ocean. Farther north, at the Antarctic Convergence, the low-salinity Antarctic Intermediate Water sinks to an intermediate depth dictated by its density. Sandwiched between these two masses is the North Atlantic Deep Water, rich in algal nutrients after hundreds of years in the deep ocean. The layering in the Pacific and Indian oceans is similar to that in the Atlantic, except that there is no source of Northern Hemisphere deep water.

Worldwide circulation models that include both surface and deep currents resemble a conveyer belt. Deep currents carry oxygen into the deep ocean, which is extermely important for life on the planet. Recent investigations indicate that worldwide deep-water circulation is closely linked to global climate change.

Continued investigation of the energy potential of ocean winds and currents may lead to new renewable energy technology that can be exploited by society.

Key Terms

Questions And Exercises

1. Why did Benjamin Franklin want to know about the surface current pattern in the North Atlantic Ocean?

2. Compare the forces that are directly responsible for creating horizontal and deep vertical circulation in the oceans. What is the ultimate source of energy that drives both circulation systems?

3. Describe the different ways in which currents are measured.

4. What would the pattern of ocean surface currents look like if there were no continents on Earth?

5. On a base map of the world, plot and label the major currents involved in the surface circulation gyres of the oceans. Use colors to represent warm versus cool currents and indicate which currents are western intensified. On an overlay, superimpose the major wind belts of the world on the gyres.

6. What atmospheric pressure is associated with the centers of subtropical gyres? With subpolar gyres? Explain why the subtropical gyres in the Northern Hemisphere move in a clockwise fashion while the subpolar gyres rotate in a counterclockwise pattern.

7. Diagram and discuss how the Ekman transport produces the "hill" of water within subtropical gyres that causes geostrophic current flow. As a starting place on the diagram, use the prevailing wind belts, the trade winds, and the westerlies.

8. Describe the voyage of the *Fram* and how it helped prove that there was no continent beneath the Arctic ice pack.

9. What causes the apex of these geostrophic "hills" to be offset to the west of the center of the ocean gyre systems?

10. Draw or describe several different oceanographic conditions that produce upwelling.

11. The largest river in the world—the mighty Amazon River—dumps 200,000 cubic meters of water into the Atlantic Ocean each second. Compare its flow rate with the volume of water transported by the West Wind Drift and the Gulf Stream. How many times larger than the Amazon is each of these two ocean currents?

12. Observing the flow of Atlantic Ocean currents in Figure 7–14, offer an explanation as to why the Brazil Current has a much lower velocity and volume transport than the Gulf Stream.

13. Explain why Gulf Stream eddies that develop northeast of the Gulf Stream rotate clockwise and have warm-water cores, whereas those that develop to the southwest rotate counterclockwise and have cold-water cores.

14. Describe changes in oceanographic phenomena including Walker Circulation, the Pacific Warm Pool, trade winds, equatorial countercurrent flow, upwelling/downwelling, and the abundance of marine life that occur during an El Niño–Southern Oscillation event.

15. Over the last century, has the pattern of moderate, strong, and very strong ENSO events occurred at regular intervals? What are some global effects of ENSOs? How is a La Niña different from an El Niño?

16. Describe the relationship between atmospheric pressure, winds, and surface currents during the monsoons of the Indian Ocean.

17. Discuss the origin of thermohaline vertical circulation. Why do deep currents form only in high-latitude regions?

18. Name the two major deep-water masses and give the locations of their formation at the ocean's surface.

19. The Antarctic Intermediate Water is identifiable throughout much of the South Atlantic on the basis of a temperature minimum, salinity minimum, and dissolved oxygen content. Why should it be colder and less salty and contain more oxygen than the surface-water mass above it and the North Atlantic Deep Water below it?

20. Where is the potential for using ocean wind power systems and ocean current power systems greatest, and why?

References

Behringer, D., et al., 1982. A demonstration of ocean acoustic tomography. *Nature* 299: 121–125.

Boyle, E., and Weaver, A. 1994. Conveying past climates. *Nature* 372:41–42.

Broecker, W. S. 1997. Thermohaline circulation, the Achilles heel of our climate system: Will man-made CO_2 upset the current balance? *Science* 278:5343, 1582–1588.

Ebbesmeyer, C. C., and Ingraham, W. J., 1992. Shoe spill in the North Pacific. *EOS, Trans. AGU* 73:27, 361.

———. 1994. Pacific toy spill fuels ocean current pathways research. *EOS, Trans. AGU* 75:37, 425–427.

Gill, A. 1982. *Atmosphere-ocean dynamics.* Orlando, FL: Academic Press.

Glantz, M. H. 1996. *Currents of change: El Niño's impact on climate and society*. Cambridge: Cambridge University Press.

Goldstein, R., Barnett, T., and Zebker, H. 1989. Remote sensing of ocean currents. *Science* 246:4935, 1282–1286.

Gordon, A. L. 1986. Interocean exchange of thermocline water. *Journal of Geophysical Research* 91:C4, 5037–5046.

Ledwell, J. R., Watson, A. J., and Law, C. S. 1993. Evidence for slow mixing across the pycnocline from an open-ocean tracer-release experiment. *Nature* 364:6439, 701–703.

McCartney, M. 1997. Is the ocean at the helm? *Nature* 388:6642, 521–522.

McCartney, M. S. 1994. Towards a model of Atlantic Ocean circulation: The plumbing of the climate's radiator. *Oceanus* 37:1, 5–8.

McDonald, A. M., and Wunsch, C. 1996. An estimate of global ocean circulation and heat fluxes. *Nature* 382:6590, 436–439.

Montgomery, R. 1940. The present evidence of the importance of lateral mixing processes in the ocean. *American Meteorological Society Bulletin* 21:87–94.

National Research Council. 1996. *Leaning to predict climate variations associated with El Niño and the southern oscillation: Accomplishments and legacies of the TOGA program*. Washington, DC: National Academy Press.

Pedlosky, J. 1990. The dynamics of the oceanic subtropical gyres. *Science* 248:4935, 316–322.

Philander, G. 1996. El Niño and La Niña. In Pirie, R. G., ed., *Oceanography: Contemporary readings in ocean sciences*, 3rd ed., New York: Oxford University Press.

Philander, S. G. 1992. El Niño. *Oceanus* 35:2, 56–61.

Philander, S. G. H. 1983. El Niño Southern Oscillation phenomena. *Nature* 302:5906, 295–301.

Pickard, G. L. 1975. *Descriptive physical oceanography*, 2nd ed. New York: Pergamon Press.

Quinn, W. H., Neal, V. T., and Antunez de Mayolo, S. E. 1987. El Niño occurrences over the past four and a half centuries. *Journal of Geophysical Research* 92:C13, 14449–14461.

Schmitz, W. J., and McCartney, M. S. 1993. On the North Atlantic circulation. *Reviews of Geophysics* 31:1, 29–49.

Stommel, H. 1958. The abyssal circulation. Letters to the editor. *Deep sea research* 5, New York: Pergamon Press.

———. 1987. *A view of the sea*. Princeton, NJ: Princeton University Press.

Street-Perrott, A. F., and Perrott, R. 1990. Abrupt climate fluctuations in the tropics: The influence of Atlantic Ocean circulation. *Nature* 343:6259, 607–612.

Stuiver, M., Quay, P. D., and Ostlund, H. G. 1983. Abyssal water carbon-14 distribution and the age of the world oceans. *Science* 219:4586, 849–851.

Sverdrup, H. U., Johnson, W., and Fleming, R. H. 1942. Renewal 1970. *The oceans*. Englewood Cliffs, NJ: Prentice-Hall.

The Open University Course Team. 1989. *Ocean circulation*. Oxford: Pergamon Press.

University Corporation for Atmospheric Research. 1994. *El Niño and climate prediction*. Reports to the Nation on our Changing Planet, Spring 1994, no. 3.

Wood, J. D. 1985. The world ocean circulation experiment. *Nature* 314:6011, 501–511.

Suggested Reading

Earth

Alper, J. 1994. Virtual Earth: The silicon ocean. 3:2, 33–38. An examination of how supercomputers are used to model ocean currents.

Bell, M. 1994. Is our climate unstable? 3:1, 24–31. Analysis of ice cores from Greenland reveals that Earth's climate is linked to ocean circulation patterns and that climate may not be as stable as was once believed.

Davidson, K. 1995. El Niño strikes again. 4:3, 24–33. A look at prediction of and destruction caused by the lingering 1994 ENSO event.

Flannagan, R. 1998. Sea change. 7:1, 42–47. Examines the effect of the creation of the land bridge of Central America on ocean currents and worldwide climate.

Knox, P. N. 1992. El Niño—A current catastrophe. 1:5, 30–37. Discusses various weather phenomena that are related to ENSO events, and focuses on unusual weather in 1992 that was the result of an ENSO.

Sea Frontiers

Frye, J. 1982. The ring story. 28:5, 258–267. A discussion of the Gulf Stream, its rings, and how the rings are studied.

Miller, J. 1975. Barbados and the island-mass effect. 21:5, 268–272. A discussion of the phenomenon by which waters around tropical islands are much more productive than the surface waters of the open ocean.

Smith, F. G. W. 1972. Measuring ocean movements. 18:3, 166–174. Discussed are some practical problems related to current flow and some methods used to determine current direction, speed, and volume.

Smith, F. G. W., and Charlier, R. 1981. Turbines in the ocean. 27:5, 300–305. A discussion of the potential of extracting energy from ocean currents is presented.

Sobey, E. 1982. What is sea level? 28:3, 136–142. The factors that cause sea level to change are discussed.

Scientific American

Baker, D. J., Jr. 1970. Models of ocean circulation. 222:1, 114–121. This article discusses observations of a model depicting a segment of the surface of Earth over which fluids move and helps explain geostrophic flow within ocean gyres. Some knowledge of basic physics is necessary for full comprehension of the material presented.

Hollister, C. D., and Nowell, A. 1984. The dynamic abyss. 250:3, 42–53. Submarine "storms" are associated with deep, cold currents flowing away from polar regions toward the Equator.

Munk, W. 1955. The circulation of the oceans. 191:96–108. A physical oceanographer who has been highly honored by the scientific community presents his views of ocean circulation.

Spindel, R. C., and Worcester, P. F. 1990. Ocean acoustic tomography. 263:4, 94–99. The theoretical basis and practical application of OAT are clearly presented.

Stewart, R. W. 1969. The atmosphere and the ocean. 221:3, 76–105. The exchange of energy between the atmosphere and ocean and the resulting phenomena, currents, and waves are covered in this readable, comprehensive article.

Stommel, H. 1955. The anatomy of the Atlantic. 190:30–35. Another vintage article by one of the world's preeminent physical oceanographers.

Webster, P. J. 1981. Monsoons. 245:5, 108–119. The mechanism of the monsoons and their role in bringing water to half Earth's population are discussed.

Welbe, P. H. 1982. Rings of the Gulf Stream. 246:3, 60–79. The biological implications of the large cold-water rings are considered.

Whitehead, J. A. 1989. Giant ocean cataracts. 260:2, 50–57. An examination of the effect on ocean circulation of giant ocean cataracts, which are large and fast flows of deep water that play a crucial role in maintaining the chemistry and climate of the deep ocean.

Oceanography on the Web

Visit the *Essentials of Oceanography* home page for on-line resources for this chapter. There you will find an on-line study guide with review exercises, and links to oceanography sites to further your exploration of the topics in this chapter. *Essentials of Oceanography* is at: **http://www.prenhall.com /thurman** (click on the Table of Contents menu and select this chapter).

CHAPTER 8
WAVES AND WATER DYNAMICS

The Biggest Wave in Recorded History: Lituya Bay, Alaska (1958)

Lituya Bay is located in southeast Alaska about 200 kilometers (125 miles) west of Juneau, Alaska's capital. Lituya Bay is an 11-kilometer (7-mile)-long T-shaped bay that is separated from the Pacific Ocean by a sand bar (Figure 8A). The bay was created during the ice age by glaciers carving out a deep valley that is now flooded by the sea. Its glacial origin causes the bay to have steep sides, and the bay itself has water depths between 180 and 240 meters (600 to 800 feet). The largest wave ever authentically recorded was a **splash wave** (a long-wavelength wave produced by an object spashing into water) that occurred in Lituya Bay on July 9, 1958. Remarkably, there were eyewitnesses to the event: Three small fishing trawlers were near the entrance to the bay (Figure 8A) and experienced the great wave.

On the evening of July 9 at about 10:00 P.M., a large earthquake of magnitude $M_w = 7.9$[1] occurred along the Fairweather Fault, which runs along the top of the "T" portion of the bay. The earthquake itself didn't produce the wave, but the earthquake did trigger a giant rockslide. Based on the scar left by the rockslide, it has been estimated that at least 90 million tons of rock—some of it from as high as 914 meters (3000 feet) above sea level—crashed down upon the upper part of the bay. The rockslide created a huge splash wave that began to travel down the bay toward the boats. The wave swept over the ridge facing the rockslide area and uprooted, de-barked, or snapped off trees high up on the ridge. Based on damage to the trees and soil on the ridge, the wave must have reached an astounding height of 530 meters (1740 feet) above the water level of the bay—a full 87 meters (285 feet) *higher* than the world's tallest building, the Sears Tower in Chicago. As the wave traveled down the bay, it snapped off trees

along the shoreline and completely overtopped the island in the middle of the bay.

During the summer in Alaska, it was still light enough at 10:00 P.M. for the people on the boats to see the rockslide occur—and the giant wave moving down the bay toward them at a speed of over 160 kilometers (100 miles) per hour. At this speed, it took only four minutes for the wave to reach the boats. The *Badger*, a 13.4-meter (44-foot) fishing vessel, had its anchor chain snapped and was lifted up bow-first into the oncoming wave. Amazingly, the vessel surfed the wave over the sand bar! The two people on board reported looking down on the tops of the trees on the sand bar from a height of 24 meters (80 feet) above the tops of the tall trees—and trees in the area reach heights of 30 meters (100 feet). The *Badger* plunged into the Pacific Ocean on the other side of the sand bar stern-first, where it foundered and eventually sank. The two people on board were able to launch a small skiff before the *Badger* sank and were rescued a few hours later, shaken but alive.

Two other vessels were in the bay at the time of the giant wave. The *Edrie* was at anchor in the bay when the wave arrived. The vessel's anchor chain snapped, and the vessel (including two people on board) was washed onto land. After the wave passed, the withdrawal of water washed it back into the bay, and the vessel was largely undamaged. In fact, they were able to leave the bay the next day under its own power. The two people on board the *Sunmore* were not nearly so lucky. Their vessel met the wave from the side near the bay's entrance. The wave sank the *Sunmore*, and the two people on board were lost with their vessel.

Damage to the shoreline area of the bay was extensive (Figure 8B). The wave created a trimline of trees that extended around the bay and across the island. The wave also completely bared the sand bar of trees and killed most of the shellfish living near the water's edge. Floating

---

[1] The symbol M_w indicates the *moment magnitude* of an earthquake, as discussed in Chapter 2.

Figure 8A Lituya Bay, Alaska: Before.

Figure 8B Lituya Bay, Alaska: After.

logs from the destruction filled Lituya Bay for many years. The wave spread out into the Pacific, and was even registered at a tide-recording station in Hawaii over six hours later [the wave was only 10 centimeters (4 inches) tall after its journey].

Older downed trees suggest that faulting in Lituya Bay periodically produces rockslides that generate giant splash waves. For instance, there is evidence of a 61-meter (200-foot) wave in 1899; an 1853 wave reached 120 meters (395 feet) in height; and on October 27, 1936, a splash wave rose to 150 meters (490 feet) above the level of the bay. Interestingly, events in other regions have also created giant waves. Some such events include landslides from islands [there is evidence that the island of Hawaii has experienced several large landslides within the past 100,000 years that generated waves as tall as 300 meters (984 feet)] and meteorite impacts in the ocean [such as the large meteorite impact that occurred in the Gulf of Mexico at the end of the Cretaceous Period about 65 mil-

lion years ago,[2] creating a wave estimated to be 914 meters (3000 feet) high].

___

[2] See Box 4–2 in Chapter 4 for a description of the Cretaceous-Tertiary (K–T) impact event.

When one observes the ocean, there is rarely perfectly calm water. The ocean surface seems restless and in continual motion. The most noticeable movement is that of waves traveling across its surface. The majority of these waves are driven by the wind and break in familiar fashion along the shore. Most of the time, they release their energy gently, although ocean storms sometimes push them ashore hard and fast, with devastating effects.

Waves have inspired poets and writers for centuries because of their beauty and awesome power. Mariners in the days of sail always greeted imminent storms with a sense of challenge, because their humdrum daily shipboard routine would be replaced with excitement and—because of the waves—some moments of sheer terror. To surfers, big waves are where one can experience thrills unparalleled by any on land. Whatever your view, waves are *moving energy* traveling along the interface between ocean and atmosphere, often transferring energy from a storm far out at sea over distances of several thousand kilometers.

Waves are among the best-known ocean phenomena, yet it was not until well into the 1800s that some understanding of waves was learned, including their causes and behavior. Let's look first at what causes waves, and then we'll examine the characteristics and properties of waves, paying special attention to oceanic waves.

What Causes Waves?

All waves begin as *disturbances*. If you throw a rock into a still pond (thereby creating a disturbance), it creates waves that radiate out in all directions from where the rock hit the water. Releases of energy, similar to the rock hitting the water, are the cause of all waves.

Most ocean waves are generated by the wind. If the wind is strong enough, it will create waves by blowing across the surface of the ocean. This disturbance generates waves that radiate out in all directions, similar to the rock thrown in the pond but on a much larger scale.

The movement of fluids with different densities can also create waves. These waves travel along the interface (boundary) between the two different fluids. Since both the air and the ocean are fluids, waves can be created along interfaces *between* and *within* these fluids. For example:

- Along an *air–water interface*, the movement of air across the ocean surface creates **ocean waves**, or more simply for our discussion, *waves*.

- Along an *air–air interface*, the movement of different air masses creates **atmospheric waves**, which are often represented by a ripple-like cloud pattern in the sky. Atmospheric waves are especially common when cold fronts (which contain high-density air) invade an area.

- Along a *water–water interface*, the movement of water of different densities creates **internal waves**, so called because they are *internal* to the ocean (Figure 8–1A). Internal waves may be created by tidal movement, turbidity currents, wind stress, or even by the passage of ships at the surface. Because these waves travel along the boundary between waters of different density, they are associated with a pycnocline.[3] Internal waves can be much larger than surface waves, with heights sometimes exceeding 100 meters (330 feet). This can present quite a danger to submarines, which can be carried by internal waves to depths exceeding their designed pressure limits. At the surface, parallel slicks caused by a film of surface debris may indicate the presence of internal waves below (Figure 8–1B). On a smaller scale, internal waves are prominently featured sloshing back and forth in "desktop oceans," which contain two fluids that do not mix.

Mass movement into the ocean also creates waves. Examples of this type of disturbance include coastal landslides that fall into the ocean and glaciers that calve icebergs directly into seawater. These waves are commonly called *splash waves*, a descriptive term indicating how these waves originate.

Underwater sea floor movement is another mechanism that can create very large waves. During underwater sea floor movement, the ocean bottom changes its shape and, potentially, a large amount of energy is released to the entire water column (as opposed to wind-driven waves that affect only the surface water). Examples include underwater avalanches in the form of turbidity currents, un-

___

[3] As discussed previously, a pycnocline is a zone of rapidly changing density.

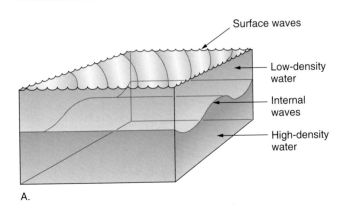

Surface waves
Low-density water
Internal waves
High-density water

A.

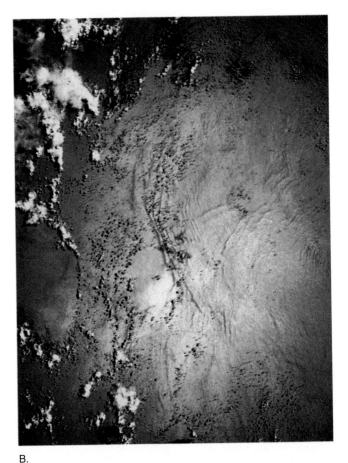

B.

Figure 8–1 Internal wave.
A. An internal wave moving along the density interface (pycnocline) below the ocean surface. **B.** A view from space of internal waves near the Seychelles Islands in the Indian Ocean. The internal waves are the broad fronts running down the middle of the photograph, and are moving from right to left; white spots are clouds.

derwater volcanic eruptions, and underwater fault slippage. These waves are called *seismic sea waves* or *tsunami*. Fortunately, these potentially destructive waves are infrequent visitors to coastal regions, where they may inundate the shore up to 40 meters (131 feet) above normal sea level.

The gravitational pull that the moon and the sun have on Earth also creates waves. These waves are vast, low, highly predictable waves that are generated by the gravitational attraction of the sun and moon on every particle of ocean water. These waves are called *tides*, which will be discussed in the next chapter.

Lastly, human activities in the ocean cause waves. When ships ply through the ocean, they leave behind a wake, which is a wave. In fact, smaller boasts are often carried along in the wake of larger ships, and marine mammals sometimes play in a vessel's wake. The detonation of nuclear devices at or near sea level is a huge release of energy that also creates waves.

In all cases, though, waves are created by some type of energy release. Figure 8–2 shows that most of the ocean's energy is typically concentrated in wind waves, indicating that most ocean waves are wind generated.

How Waves Move

Waves are energy in motion. Wave phenomena involve the transmission of energy by means of cyclic movement through matter. The medium itself (solid, liquid, or gas) does not actually travel in the direction of the energy that is passing through it. The particles in the medium simply oscillate, or cycle, in a back-and-forth, up-and-down, or orbital pattern, transmitting energy from one particle to another. For example, if you thump your fist on a table, the energy travels through the table as waves that can be felt by someone sitting at the other end, but the table itself does not move. Another example is how the wind causes waves to travel across a field of wheat (amber waves of grain): The wheat itself doesn't travel across the field, but the waves do.

Waves move in different ways. Simple *progressive waves* (in which the shape of the wave can be observed to oscillate uniformly and *progress* or travel without breaking) are shown in Figure 8–3A. Progressive waves may be *longitudinal, transverse,* or a combination of the two motions, called *orbital.*

In **longitudinal waves** (also known as push-pull waves), the particles that are in vibratory motion "push and pull" in the same direction that the energy is traveling, like a coil spring whose coils are alternately compressed and expanded (you can observe this with a Slinky™ as well). The shape of the wave (called a *waveform*) travels through the medium by compressing and decompressing as it travels. For instance, sound travels as longitudinal waves. Clapping your hands initiates a percussion that compresses and decompresses the air as the sound moves through a room. Energy can be transmitted through all states of matter—gaseous, liquid, or solid—by this longitudinal movement of particles.

In **transverse waves** (also known as side-to-side waves), energy travels at right angles to the direction of particle vibration. For example, if one end of a rope is tied to a

Figure 8–2 Distribution of energy in ocean waves.

Most of the energy possessed by ocean waves is in wind-generated waves (the tall blue peak). The two lower peaks to the right of the large peak represent tsunami, whereas the two sharp peaks at the far right represent ocean tides.

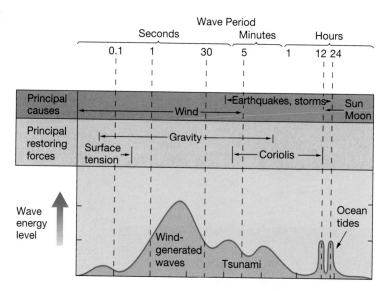

 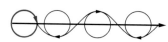

LONGITUDINAL WAVE
Particles (color) move back and forth in direction of energy transmission. These waves transmit energy through all states of matter.

A.

TRANSVERSE WAVE
Particles (color) move back and forth at right angles to direction of energy transmission. These waves transmit energy only through solids.

ORBITAL WAVE
Particles (color) move in orbital path. These waves transmit energy along interface between two fluids of different density (liquids and/or gases).

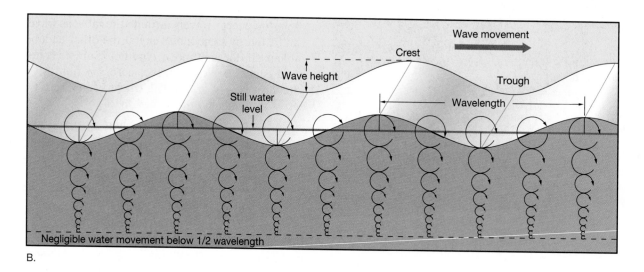

B.

Figure 8–3 Types and characteristics of progressive waves.
A. Types of progressive waves. **B.** A diagrammatic view of an idealized ocean wave showing its characteristics.

doorknob while the other end is moved up and down by hand, a waveform is set up in the rope and progresses along it, and energy is transmitted from the motion of the hand to the doorknob. If you watch any segment of the rope, you can see it move up and down (or side to side) at right angles to a line drawn from the fastened end to the hand that is putting energy into the rope. Generally, this type of wave transmits energy only

through solids, because in solid material, particles are strongly bound to one another and can transmit such a motion.

Longitudinal and transverse waves are called *body waves* because they transfer energy through a body of matter. Conversely, ocean waves transmit energy along an interface between the atmosphere and the ocean. The movement of particles along such an interface involves

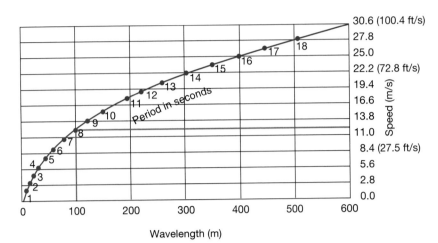

Figure 8–4 Speed of deep-water waves.

Ideal relations among wave speed, wavelength, and period for deep-water waves.

components of *both* longitudinal and transverse waves, causing the particles to move in circular orbits. Thus, waves at the ocean surface are **orbital waves** (also called *interface waves*).

Wave Characteristics

When one observes the ocean surface, waves of various sizes can be seen moving in various directions. This produces a complex wave pattern that constantly changes. Let's examine a simple, uniform, moving waveform (Figure 8–3B). It represents the transmission of energy from a single source and traveling along the ocean–atmosphere interface. Even though such an idealized waveform does not truly exist in nature, it is a starting point for understanding wave characteristics.

As our idealized progressive wave passes a permanent marker, such as a pier piling, there are a succession of high parts of the waves, called **crests**, separated by low parts, called **troughs**. Halfway between the crests and the troughs is the **still water level**, or *zero energy level*. This is the level of the water if there were no waves.

If the water level on the piling when the troughs pass is marked and then the same is done for the crests, the vertical distance between the marks is the **wave height**, designated by the symbol **H**. Thus, the wave height is the vertical distance between a crest and a trough.

The horizontal distance between any two corresponding points on successive waveforms, such as from crest to crest or from trough to trough, is the **wavelength, L**. The ratio of wave height to wavelength, **H/L**, is called the **wave steepness**, and is defined as

$$\text{Wave steepness} = \frac{\text{wave height } (H)}{\text{wavelength } (L)}$$

Note that if the wave steepness is ever greater than $^1/_7$, the wave *breaks* (spills forward) because the wave is literally too steep to support itself. A wave can break anytime the 1:7 ratio is exceeded, either along the shoreline

or out at sea. The 1:7 wave steepness ratio also indicates the maximum height of a wave. For example, a wave 7 meters long can only be 1 meter high.

The time that elapses during the passing of one full wave—one wavelength—past a fixed location like a pier piling is the **wave period**, designated by the symbol T (for time, usually measured in seconds). Typical wave periods range between 6 and 16 seconds. The **frequency** (f) is the inverse of the period, and is defined as the number of wave crests passing a fixed location per unit of time. For instance, waves with a period of 12 seconds have a frequency of $^1/_{12}$ or 0.083 waves per second.[4]

Figure 8–4 shows the relations among wavelength, period, and speed for waves. Because the period is the *time required for the passing of one wavelength*, if either the wavelength or the period of a wave is known, then the other can be calculated. Mathematically, Figure 8–4 can be expressed as

$$\begin{aligned}\text{Wavelength } (L,&\text{ in meters})\\ &= 1.56 \text{ meters/second} \times T^2 (\text{seconds}).\end{aligned}$$

Since **wave speed**[5] (S) is

$$\text{Speed } (S) = \frac{\text{wavelength } (L)}{\text{period } (T)}$$

the example shown by the red lines in Figure 8–4 shows that

$$\text{Speed } (S) = \frac{L}{T} = \frac{100 \text{ meters}}{8 \text{ seconds}} = 12.5 \text{ meters/second.}$$

---

[4] This frequency converts to a value of 5 waves per minute.

[5] Wave speed is more correctly known as *celerity* (C). Celerity is different from the traditional concept of speed in that it is used only in relation to waves where no mass is in motion, just the wave form.

Figure 8–4 also shows that *the longer the wavelength, the faster the wave travels*. Note that a fast wave does not necessarily have a large wave height, since the wave speed depends *only* on wavelength.

Circular Orbital Motion

Waves have been observed to travel great distances across ocean basins. In one study, waves generated near Antarctica were tracked as they traveled through the Pacific Ocean basin. After a journey of more than 10,000 kilometers (over 6000 miles), the waves finally expended their energy a week later along the shoreline of the Aleutian Islands of Alaska. The water itself doesn't travel the entire distance, but the waveform (a transmission of energy) does. What is surprising is that as the wave travels, the water passes the energy along by moving in a circle. This movement is called **circular orbital motion**.

At first glance, an object floating in the waves appears to have an up-and-down movement. Upon closer inspection, however, the object is also moving slightly forward and back with each successive wave (Figure 8–5). A floating object moves up and forward as the crest passes, moves down and back as the trough approaches, and rises and moves forward again as the next crest advances. In essence, the shape of the wave moves forward through the water, *not the water particles themselves*. If the pattern of the rubber duck shown in Figure 8–5 were to be traced as a wave passes, it could be seen that the movement of the rubber duck is in the shape of a circle, returning the rubber duck to near its original position.[6]

The circular orbits followed by floating objects at the surface (such as the rubber duck or water particles) have a diameter equal to the wave height (Figure 8–3B). Circular orbital motion dies out quickly below the surface: With increasing depth, the diameters of particle orbits decrease (Figure 8–6). At some depth below the surface, the circular orbits become so small that movement is negligible. This depth is called the **wave base**, and is equal to a depth of one-half the wavelength (*L/2*) *measured from still water level*. Thus, the depth of the wave base is controlled only by the wavelength, where the longer the wave, the deeper the wave base.

The decrease of orbital motion with depth has many practical applications. For instance, submarines can ride out large ocean waves simply by submerging into the water below the wave base. Even for the largest storm waves, submarines submerged to a depth of only 150 meters (500 feet) will not notice any wave orbital motion. Floating bridges and floating oil rigs are constructed so

---

[6] Actually, the circular orbit would not quite return the floating object to its original position because the half of the orbit accomplished in the trough is slower than the crest half of the orbit. This slight forward movement is called *wave drift*.

Figure 8–5 A rubber duck in water.
The pattern of motion of a floating rubber duck resembles that of a circular orbit as waves pass.

that most of the mass of the structure is below wave base, so that they will be unaffected by wave motion. In the future, there may even be offshore floating airport runways. Another useful application of decreasing orbital motion with depth comes when you enter the ocean from the beach. As you walk into the waves, there comes a time when it is advantageous to dive under an incoming wave rather than jump over it. This is because it is much easier to swim through the smaller orbital motion below the surface than to fight the motion of large orbits at the

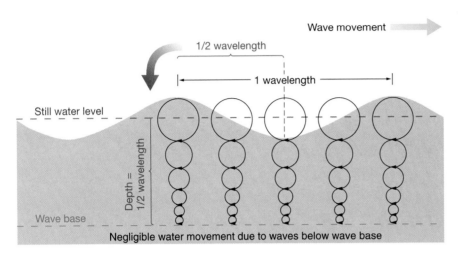

Wave movement

1/2 wavelength

1 wavelength

Still water level

Depth = 1/2 wavelength

Wave base

Negligible water movement due to waves below wave base

Figure 8–6 Orbital motion in waves.

The orbital motion of water particles in waves extends to a depth of one-half the wavelength, measured from still water level, which is the wave base.

surface! Also, seasick scuba divers find relief from motion sickness when they submerge into the calm, motionless water below wave base.

Deep-Water Waves

If the water depth (d) is greater than the wave base ($L/2$), the waves are called **deep-water waves** (Figure 8–7A). Thus, deep-water waves do not have any interference with the ocean bottom. Deep-water waves include all wind-generated waves as they move across the open ocean, where water depths are quite deep relative to wave base.

As is true of all waves, the speed of deep-water waves is related to wavelength (L) and period (T) (Figure 8–4). The easiest of these characteristics to measure is the period (the time required for one wave to pass). Thus, the following equation is most commonly used for computing wave speed, where g is acceleration due to gravity:

$$\text{Speed } (S) = \frac{\text{gravity } (g) \times \text{period } (T)}{2\pi} = 1.56T$$

Shallow-Water Waves

Waves in which depth (d) is less than $1/20$ of the wavelength ($L/20$) are classified as **shallow-water waves**, or *long waves* (Figure 8–7B). In this case, the waves are said to *touch bottom* or *feel bottom* because the orbital motion of the wave is interfered with by the ocean floor.

Included in this category are wind-generated waves that have moved into shallow near-shore areas; *tsunami* (seismic sea waves), generated by earthquakes in the ocean floor; and *tide waves*, generated by the gravitational attraction of the sun and moon. In these waves, the wavelength is very great relative to water depth, and speed is determined by water depth.

Particle motion in shallow-water waves is in a very flat elliptical orbit that approaches horizontal (back-and-forth) oscillation. The vertical component of particle motion decreases with increasing depth, causing the orbits to become flattened.

Transitional Waves

Transitional waves have wavelengths greater than twice the water depth, but less than 20 times. The speed of transitional waves is controlled partially by wavelength and partially by water depth.

Figure 8–8 shows the relation of wave speed to wavelength and/or water depth. It is divided into three zones because *water depth* determines the speed of shallow-water waves (*red*), *wavelength* determines the speed of deep-water waves (*green*), and *both* affect the speed of transitional waves (*purple*). The blue lines are different wavelengths. To determine wave speed for shallow-water waves (*red zone*), first find the water depth on the bottom scale (red example: 50 meters). Second, move up to the "shallow-water waves" line, and then over to the wave-speed scale. In the example, a wave in 50-meter-deep water travels at a speed of about 21 meters per second.

For deep-water waves (*green zone*), water depth is not a factor. To determine wave speed, simply extend the wavelength line (*blue*) to the left until it intersects the wave-speed scale. The green example is for a 50-meter wavelength, showing a wave speed of about 9 meters per second.

For transitional waves (*purple zone*), combine the shallow-water and deep-water procedures. The purple example is for a 1000-meter wavelength and 200-meter water depth. Note that transitional waves have a slower wave speed than they would have if they were deep-water waves.

Wind-Generated Waves

Wind-generated waves have a life history that includes their origin in windy regions of the ocean, their movement across great expanses of open water without subsequent aid of wind, and their termination when they

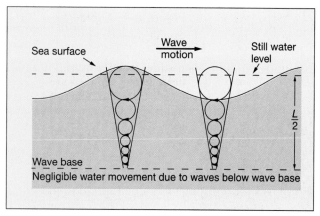

A. DEEP-WATER WAVE

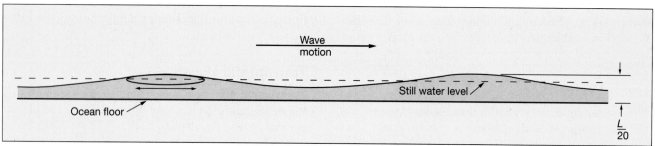

B. SHALLOW-WATER WAVE

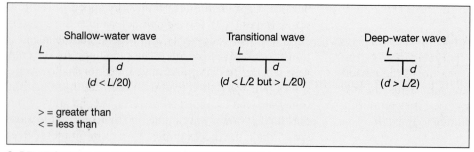

C. RELATIONSHIP OF WAVELENGTH TO WATER DEPTH

Figure 8–7 Deep-water and shallow-water waves.
A. Wave profile and water-particle motions of a deep-water wave, showing the diminishing size of the orbits with increasing depth. **B.** Motions of water particles in shallow-water waves, where water motion extends to ocean floor. **C.** Relationship of wavelength to water depth.

break and release their energy, either in the open ocean or against the shore.

"Sea"

As the wind blows over the ocean surface, it creates pressure and stress. These factors deform the ocean surface into small, rounded waves with extremely short wavelengths, less than 1.74 centimeters (0.7 inch). Commonly they are called *ripples,* and oceanographers call them **capillary waves**. The name comes from *capillarity,* a property that results from the surface tension of water. Capillarity is the dominant **restoring force** that works to destroy these tiny waves, restoring the smooth ocean surface once again. Capillary waves characteristically have rounded crests and V-shaped troughs (Figure 8–9, *left*).

As capillary wave development increases, the sea surface takes on a rougher character. This "catches" more of the wind, allowing the wind and ocean surface to interact more efficiently, transferring more energy to the water. As more energy is transferred to the ocean, **gravity waves** develop. They have wavelengths exceeding 1.74 centimeters (0.7 inch) and are shaped more like a sine curve (the familiar waveform shown in the *middle* of Figure 8–9). Because they reach greater height at this stage, gravity replaces capillarity as the dominant restoring force, giving these waves their name.

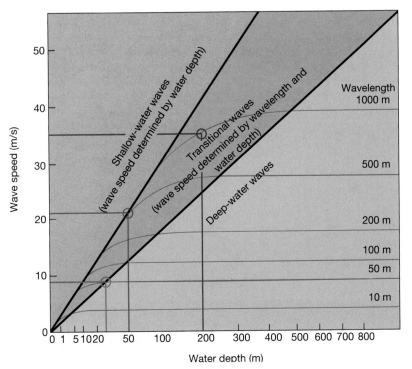

Figure 8–8 Wave speed.

Wave speed is a function of wavelength and/or water depth. The graph is divided into three zones because *water depth* determines the speed of shallow-water waves (*red*), *wavelength* determines the speed of deep-water waves (*green*), and *both* affect the speed of transitional waves (*purple*). The blue horizontal lines represent different wavelengths.

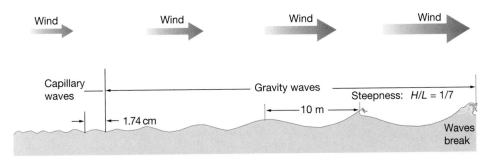

Figure 8–9 Capillary and gravity waves.

As energy is transferred to the ocean surface by wind and the water gains energy, the waves increase in height and length. They begin as *capillary waves* (*left*), and progress to become *gravity waves*. As the steepness reaches the point where the wave steepness (*H/L*) exceeds a 1:7 ratio, the waves become unstable and break. (Not to scale.)

The length of these "young" waves is generally 15 to 35 times their height. As additional energy is gained, wave height increases more rapidly than wavelength. The crests become pointed and the troughs are rounded, resulting in a waveform that is described as *trochoidal* (*trokhos* = wheel) (Figure 8–9, *right*).

Energy imparted by the wind increases the height, length, and speed of the wave. When wave speed reaches that of the wind, neither height nor length can change because there is no net energy exchange, so the wave is at its maximum size.

Figure 8–10 shows important relations between wind and the area where waves are initiated. The area where wind-driven waves are generated is called, simply, **sea**. For clarity in this discussion, we will call it the *sea area*. It is characterized by choppiness and short wavelengths, with waves moving in many directions and having many different periods and lengths. The variety of wave periods and wavelengths is caused by frequently changing wind speed and direction.

Factors that are important in increasing the amount of energy that waves obtain are (1) wind speed, (2) the duration of time during which the wind blows in one direction, and (3) the *fetch*—the distance over which the wind blows in one direction.

Wave height is related directly to the energy in a wave. Wave heights in a sea area are usually less than 2 meters (6.6 feet), although it is not uncommon to observe waves with heights of 10 meters (33 feet) and periods of 12 seconds. As sea waves gain energy, their steepness increases. When steepness reaches a critical value of $1/7$, open ocean breakers—called *whitecaps*—form.

Figure 8–11 is a satellite image of average wave heights worldwide during October 3–12, 1992. Note how large the waves are in the Southern Hemisphere. This is because the westerly wind belt in the Southern Hemisphere

Figure 8–10 The "sea" area and swell.

As wind blows across the *sea area* (outlined in red), wave size increases with wind speed, duration, and fetch. As waves advance beyond the sea area, they continue to advance across the ocean surface out of the area of origination as *swell.*

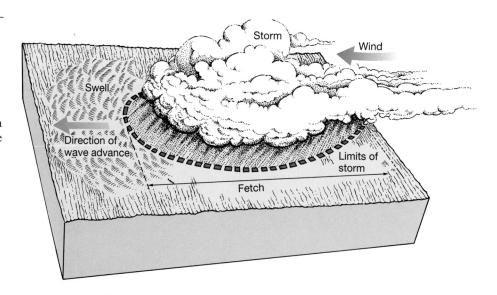

Figure 8–11 TOPEX/POSEIDON wave height, October 3–12, 1992.

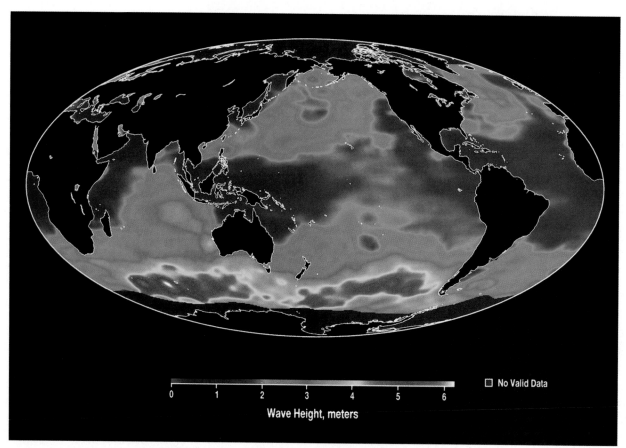

The TOPEX/POSEIDON satellite receives a return of stronger signals from calm seas and weaker signals from seas with large waves. Based on these data, a map of wave height can be produced. The largest average wave heights (*red areas*) are in the prevailing westerly wind belt in the Southern Hemisphere. (Scale in meters.)

south of 40 degrees latitude reaches the highest average wind speed on Earth, producing the largest ocean waves and creating the aptly named "Roaring Forties."

There are many stories of huge waves in the ocean, which can do substantial damage to ships. The largest wind-generated waves authentically measured were generated by a typhoon in the western Pacific Ocean in 1935. While in a 108-kilometer (67-mile)-per-hour windstorm, the 152-meter (500-foot)-long U.S. Navy tanker USS *Ramapo* encountered the waves as she was en route from

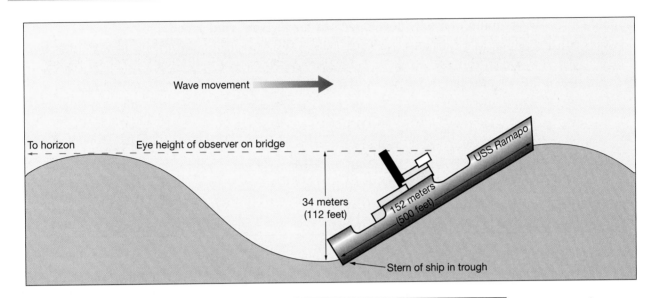

Figure 8–12 USS *Ramapo* in heavy seas.

Bridge officers aboard the *Ramapo* measured the largest authentically recorded wave by sighting from the bridge to the horizon while the ship's stern was directly in the trough of a large wave. A wave height of 34 meters (112 feet) was calculated based on geometric relationships of the vessel and the waves.

Figure 8–13 The aircraft carrier *Bennington*.

The *Bennington* returns from heavy seas encountered in a typhoon off Okinawa in 1945 with part of its flight deck bent down over the bow. Note that the ship did not hit an island; rather, the damage was caused by large waves. The reinforced steel deck is 16.5 meters (54 feet) above still water level.

the Philippines to San Diego (Figure 8–12). The waves were symmetrical, uniform, and had a period of 14.8 seconds. The vessel's officers made very careful measurement of the waves, which included consulting the design drawings of the ship and calculating the length and height of the ship down to the *eye height* of an observer on the ship's bridge. The calculations revealed that the waves were an astounding 34 meters (112 feet) high, taller than an 11-story building. Fortunately, the *Ramapo* was trav-

eling in the same direction that the waves were moving, and so the ship wasn't damaged. Other ships traveling in heavy seas aren't so lucky (Figure 8–13).

For a given wind speed, there is a maximum fetch and duration of wind beyond which the waves cannot grow. When both maximum fetch and duration are attained for a given wind speed, a **fully developed sea** has been created. A fully developed sea is an equilibrium condition and is defined as the maximum size to which waves can

Table 8–1 Description of a fully developed sea for a given wind speed.

Wind speed in km/h (mi/h)	Average height in m (ft)	Average length in m (ft)	Average period in sec	Highest 10% of waves in m (ft)
20 (12)	0.33 (1.0)	10.6 (34.8)	3.2	0.75 (2.5)
30 (19)	0.88 (2.9)	22.2 (72.8)	4.6	2.1 (6.9)
40 (25)	1.8 (5.9)	39.7 (130.2)	6.2	3.9 (12.8)
50 (31)	3.2 (10.5)	61.8 (202.7)	7.7	6.8 (22.3)
60 (37)	5.1 (16.7)	89.2 (292.6)	9.1	10.5 (34.4)
70 (43)	7.4 (24.3)	121.4 (398.2)	10.8	15.3 (50.2)
80 (50)	10.3 (33.8)	158.6 (520.2)	12.4	21.4 (70.2)
90 (56)	13.9 (45.6)	201.6 (661.2)	13.9	28.4 (93.2)

grow under given conditions of wind speed, duration, and fetch (Table 8–1). The reason that waves can grow no further in a fully developed sea is that the waves are losing as much energy breaking as whitecaps under the force of gravity as they are receiving from the wind.

Swell

As waves generated in a sea area move toward its margins, where wind speeds diminish, they eventually move faster than the wind. When this occurs, the wave steepness decreases, and they become long-crested waves called **swell**. Swell describes uniform, symmetrical waves that have traveled out of their area of origination. The swell moves with little loss of energy over large stretches of the ocean surface, transporting energy away from the wind-energy input area of the sea area. Thus, there can be waves at distant shorelines where there is no wind.

Waves with longer wavelengths travel faster, and thus leave the sea area first. They are followed by shorter-length, slower **wave trains**, or groups of waves.

This progression illustrates the principle of **wave dispersion**, which is a sorting of waves by their wavelength. In the generating area, waves of many wavelengths are present. Since wave speed is a function of wavelength in deep water (Figure 8–4), the longer waves "outrun" the shorter ones.

As a group of waves leaves a sea area and becomes a swell wave train, the group moves across the ocean surface at only half the velocity of an individual wave in the group. Progressively, the leading wave disappears. However, the same number of waves always remains in the group. As the leading wave disappears, a new wave replaces it at the back of the group (Figure 8–14).

Interference Patterns Because swells may move away from a number of storm areas in any ocean, it is inevitable that the swell from different storms will run together and the waves will clash, or *interfere* with one another. This gives rise to a special feature of wave motion: **interference patterns**. An interference pattern produced when two or more wave systems collide is the algebraic sum of the disturbance that each wave would have produced individually. The result may be a larger or smaller trough or crest, depending on conditions (Figure 8–15).

When swells from two storm areas collide, the interference pattern may be constructive or destructive, but it is more likely to be mixed. Ideally, **constructive interference** results when wave trains having the same wavelength come together *in phase*, meaning crest to crest and trough to trough. If we add the displacements that would result from each wave individually, we find that the resulting interference pattern is a wave having the same length as the two converging wave systems, but with a wave height equal to the sum of the individual wave heights (Figure 8–15, *left*).

Destructive interference results when wave crests from one swell coincide with the troughs from a second swell (the swells are said to be *out of phase*). If the waves have identical characteristics, the algebraic sum of the crest plus the trough would be zero, and the energy of these waves would cancel each other. This canceling of wave energy results in a lack of waves for a short time (Figure 8–15, *center*).

However, it is more likely that the two swell systems possess waves of various heights and lengths and would come together with a mixture of constructive and destructive interference. Thus, a more complex **mixed interference** pattern would develop (Figure 8–15, *right*). It is such interference that explains the occurrence of a varied sequence of high and lower waves (called **surf beat**) and other irregular wave patterns that are observed when swell approaches the seashore. In the open ocean, several swell systems often interact, creating complex wave patterns (Figure 8–16).

You can experience the effect of wave interference in another fluid—the air—by listening to a steady, high-pitched sound (for example, a steadily blown note on a brass instrument). As you move your head, the sound will grow loud (constructive interference) or virtually disappear (destructive interference), but mostly it will vary (mixed interference).

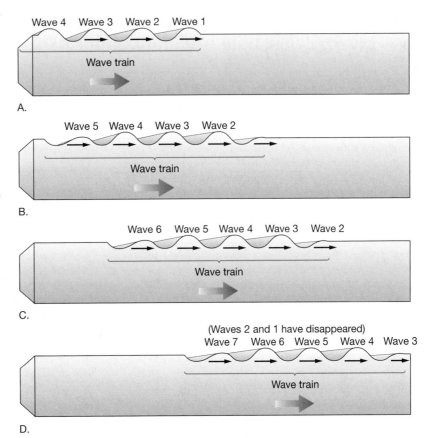

Figure 8–14 Movement of a wave train.

As energy in the leading waves (**A**, waves 1 and 2) is transferred into circular orbital motion, the waves in front die out and are replaced by new waves from behind (**B**). Even though new waves take up the lead (**C** and **D**), the length of the wave train and the total number of waves remain the same.

Rogue Waves

Rogue waves are massive, solitary waves that can reach enormous height and are often experienced at times when normal ocean waves are not unusually large. For example, in a sea of 2-meter (6.5-foot) waves, there might suddenly be a 20-meter (65-foot) rogue wave. The name *rogue* means "unusual," and in this case, they are unusually large waves, which are also called **superwaves**. Because of their unpredictability, these waves have severely damaged a number of large ships and are even responsible for rolling and sinking others.

Rogue waves result from rare coincidence in ordinary wave behavior. On average in the open ocean, one wave in 23 will be over twice the height of the wave average, one in 1175 will be three times as high, and only one in 300,000 will be four times as high. The chances of a truly monstrous wave are one in billions, but it does happen. Of course, this statistical information does not indicate specifically when and where a rogue wave will arise.

Even with satellites that can measure average wave size and forecast storms, several large ships each year are reported missing without a trace. The total number of vessels lost of all sizes may reach 1000 per year; some are the victims of rogue waves. Satellites that have gone into service during the 1990s (such as the TOPEX/POSEIDON satellite) have provided a wealth of data about ocean waves, but they still have not made it possible to predict rogue waves.

The main cause of rogue waves is theorized to be an extraordinary case of constructive wave interference, where multiple waves overlap and produce an extremely large wave. It has also been discovered that rogue waves tend to occur more frequently in locations that are downwind from islands or shoals. In addition, rogue waves can occur when storm-driven waves move against strong ocean currents, causing the waves to steepen, shorten, and become larger. Such a situation exists in the Agulhas Current off the southeastern coast of Africa, where Antarctic storm waves move to the northeast directly into the oncoming current (Figure 8–17). This stretch of water is probably responsible for sinking more ships than any other place on Earth. Consequently, the area has been nicknamed the "Wild Coast."

Surf

Most waves generated in the sea area by storm winds move across the ocean as swell. Generally, they release their energy along the margins of continents in the **surf zone**, which is the zone of breaking waves. Breaking waves are a picture of power and persistence, sometimes moving objects weighing several tons. Breaking waves are also quite interesting to watch: Energy from distant storms travels thousands of kilometers and is finally expended along the shoreline in a few wild moments.

As deep-water waves of swell move toward continental margins over gradually **shoaling** (*shold* = shallow) water, they eventually encounter water depths that are

Box 8–1
The Drilling Ship *Resolution* Almost Lost at Sea

Since 1985, the Ocean Drilling Program's research vessel—the *Resolution*—has conducted research in all the oceans of the world, collecting drill cores for analysis from below the sea floor (see chapter-opening feature in Chapter 4). The ship is equipped with one of the best (and most expensive) research laboratories in the world. In the process of collecting the samples, the ship takes two-month-long cruises, during which it has withstood wind and wave alike. However, in September 1995, it experienced a large storm in the North Atlantic Ocean from which it almost did not return.

The *Resolution* had aboard its usual contingency of 120 scientists, technicians, officers, and crew during a late-season research cruise in the East Greenland Sea. A couple of weeks after leaving Iceland, the seas became very rough. In addition, the vessel had to maneuver around icebergs from Greenland's coastal glaciers. The officers watched with concern during the last few days of September as the barometer plummeted, revealing that a major storm was approaching. In fact, two major low-pressure regions to the north and to the east were converging, building into a powerful storm.

Initially, the officers attempted to ride out the storm by motoring the vessel slowly forward into the wind and waves. As the two storms merged, the winds intensified and gusts from the storm were so intense that they caused the ship's wind-speed indicator to reach its maximum reading of 100 knots [185 kilometers (115 miles), per hour]. The ship was battered with huge storm waves that reached 21 meters (70 feet) in size (Figure 8C). As the ship was rocked by the waves, the propellers were sometimes lifted completely out of water, which caused the ship to lose its ability to be steered. Riding against the storm wasn't working, but turning the ship sideways into the waves might cause the ship to capsize. Clearly, another plan had to be devised.

Even though the 143-meter (470-foot) vessel was constructed to withstand heavy seas, certain elements of the ship's design as a scientific vessel do not stand up well under storm conditions. For instance, the tall drilling rig acts as a giant sail and also makes the ship top-heavy. In addition, the open space in the middle of the ship through which drill pipe is lowered reduces the ship's structural integrity. Fortunately, the ship is equipped with a *dynamic positioning (DP) system* that enables the ship to remain stationary above a drill site for weeks at a time. The DP system uses a series of 12 thrusters mounted around the hull of the ship to overcome currents, winds, and seas. The DP system is normally used only during drilling, but it became the officers' only hope of keeping the ship from turning sideways and catching a wave broadside.

By running the thrusters at 20 percent above their rated capacity, the ship was stabilized and remained pointing into the wind. However, the ship was still being pushed backward at a speed of 3 knots [5.5 kilometers (3.5 miles) per hour], raising fears that the vessel might ram an iceberg. Lookouts were strapped in under the helicopter deck at the rear of the vessel to alert the officers if icebergs were sighted in the heavy seas.

Figure 8C Stormy seas from the JOIDES *Resolution*, 1995.

(continued)

Tension ran high among the people on board, as instructions were given for each person to carry a rubberized survival suit in case they needed to abandon ship. Given the extreme conditions of the storm, the special suits would probably do little to protect them from drowning, and being rescued by another ship in the rough seas was unrealistic. As some of the thrusters gave out, a giant wave suddenly crashed over the bow, blasting out a window and flooding the bridge with water. The window was repaired with plywood and two-by-fours, rescuing the critical computer that regulates the DP system but knocking out radar and satellite communications.

After several more hours of battering, the winds and waves began to dissipate. Damage to the ship's radar, floodlights, and much of the communications gear rendered the ship unable to track icebergs effectively. Eventually, the seas finally calmed enough to allow the ship to limp back to port. Those who have been in heavy seas aboard a ship that is in danger of sinking know what welcome relief it is to come into a protected harbor. Fortunately for the people on board (and for the general scientific community), the ship did not sink, but it was as close a call as most would ever like to experience.

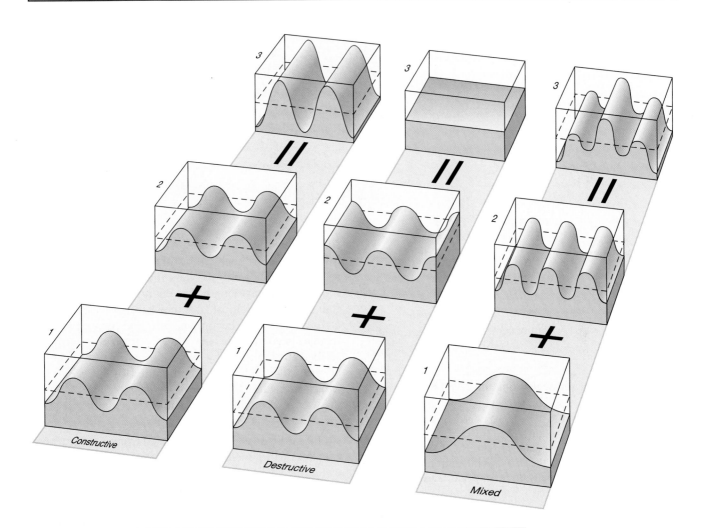

Figure 8–15 Interference patterns.

As different wave trains (1 and 2) come together, three possible interference patterns may result (3). *Constructive interference* (*left*) occurs when waves of the same wavelength come together in phase (crest to crest and trough to trough), producing waves of greater height. *Destructive interference* (*center*) occurs when overlapping waves have identical characteristics but come together out of phase, resulting in a canceling effect with no remaining wave at all. More commonly, waves of different lengths and heights encounter one another and produce a complex, *mixed interference* pattern (*right*).

Figure 8–16 Wave interference.

The observed wave pattern in the ocean (**A**) can be considered to consist of many different overlapping wave sets (**B**).

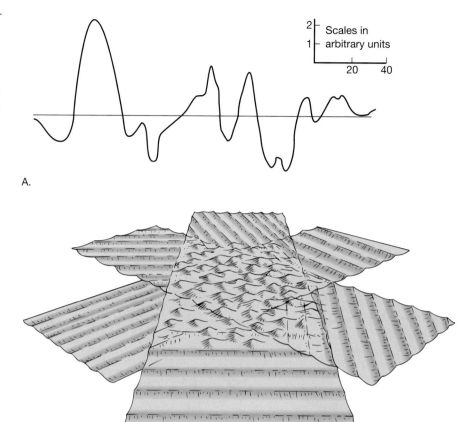

A.

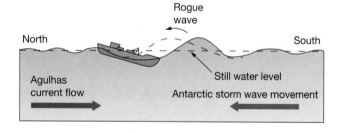

Figure 8–17 Rogue waves along the "Wild Coast."

As a large oil tanker travels south along the "Wild Coast" off the southeastern coast of Africa, it may encounter a rogue wave that results from the northward travel of Antarctic storm waves against the Agulhas current. The rogue wave can crash onto the bow as the ship dips into a trough, overcoming the structural capacity of the ship and causing the ship to sink.

less than one-half of their wavelength (Figure 8–18) and become transitional waves. Actually, any shallowly submerged obstacle (such as a coral reef, sunken wreck, or sand bar) will cause waves to experience physical changes that cause the wave to release some energy. Navigators have long known that breaking waves indicate dangerously shallow water.

There are many physical changes that occur to a wave before the wave breaks. The shoaling depths interfere with water particle movement at the base of the wave, so the *wave speed decreases*. As one wave slows, the following waveform, which is still moving at unrestricted speed, comes closer to the wave that is being slowed, thus reducing the wavelength. The energy in the wave, which remains the same, must go somewhere, so *wave height increases*. The crests become narrow and pointed and the troughs become wide curves, a waveform previously described for high-energy waves in the open sea. The increase in wave height accompanied by a *decrease in wavelength* causes an *increase in the steepness (H/L)* of the waves. When the wave steepness reaches the 1:7 ratio, the waves break as surf (Figure 8–18).

If the surf is swell that has traveled from distant storms, breakers will develop relatively near shore in shallow water, the shoaling of which is primarily responsible for their breaking. By the time waves break, they have become shallow-water waves. The horizontal water motion associated with such waves moves water alternately toward and away from the shore in a sort of oscillation. The surf will be characterized by parallel lines of relatively uniform breakers.

However, if the surf is composed of waves that have been generated by local wind, the waves may not have been sorted into swell. The surf may be more nearly characterized by unstable, deep-water, high-energy waves with steepness already near the 1:7 ratio. They will break shortly after feeling bottom some distance from shore, and the surf will be rough, choppy, and irregular.

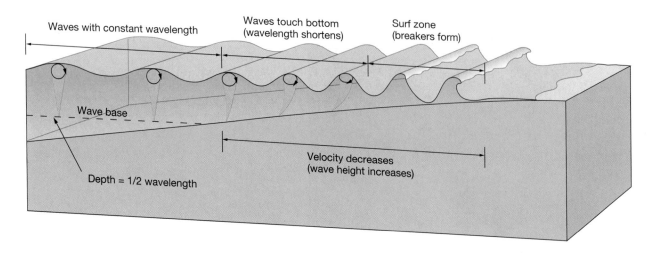

Figure 8–18 Surf zone.
As waves approach the shore and encounter water depths of less than one-half wavelength, the waves "feel bottom." The *wave speed decreases* and waves stack up against the shore, causing the *wavelength to decrease.* This results in an *increase in wave height* to the point where the *wave steepness is increased* beyond the 1:7 ratio, causing the wave to pitch forward in the surf zone.

Ideally, when the water depth is about one and one-third times the breaker height, the crest of the wave breaks, producing surf.[7] When the water depth becomes less than $\frac{1}{20}$ the wavelength, waves in the surf zone begin to behave as shallow-water waves (Figure 8–7). Particle motion is greatly hampered by the bottom, and a significant transport of water toward the shoreline occurs (Figure 8–18).

Waves that break in the surf zone result from severe restriction of particle motion near the wave's bottom. This slows the waveform, but at the surface, individual water particles that are orbiting have not been slowed because they have no contact with the bottom. Thus, the top of the waveform leans shoreward as the bottom of the waveform experiences drag as the wave touches bottom. The entire waveform slows, and water moves faster in the circular orbits near the surface due to increasing wave height. This causes water particles at the ocean/atmosphere interface to move faster toward shore than the waveform itself. This situation is similar to people who lean far forward: If they don't catch themselves, they may also "break" something when they fall.

Breakers and Surfing The most commonly observed breaker is the **spilling breaker** (Figure 8–19A). These result from a relatively gentle slope of the ocean bottom, which extracts energy from the wave more gradually, producing a turbulent mass of air and water that runs down the front slope of the wave instead of producing a spectacular cresting curl. Because of the gradual extraction

of energy of spilling breakers, they have a longer life span and give surfers a long, but somewhat less exciting, ride than do other breakers.

A wave's orbital motion can be seen particularly well in **plunging breakers** (Figure 8–19B), which have a curling crest that moves over an air pocket. This results because the curling particles literally outrun the wave, and there is nothing beneath them to support their motion. Plunging breakers form on moderately steep beach slopes, and are the best waves for surfing.

When the ocean bottom has an abrupt slope, the wave energy is compressed into a shorter distance and the wave will surge forward, creating a **surging breaker** (Figure 8–19C). Because these waves build up and break right at the shoreline, the short distance over which surging breakers occur discourages surfers. However, these waves present the greatest challenge to body surfers.

Surfing is analogous to riding a gravity-operated water sled by balancing the forces of gravity and buoyancy. Recalling the particle motion of ocean waves in Figure 8–3, you can see that, in front of the crest, water particles are moving up into the crest. It is this force, along with the buoyancy of the surfboard, that helps maintain a surfer's position in front of a breaking wave. The trick is to perfectly balance the force of gravity (directed downward) with the buoyant force (directed perpendicular to the wave face) to enable a surfer to be propelled forward by the wave's energy. A skillful surfer, by positioning the board properly on the wave front, can regulate the degree to which the propelling gravitational forces exceed the buoyancy forces, and high speeds—up to 40 kilometers (25 miles) per hour—can be obtained while moving along the face of the breaking wave. When the upward motion

---

[7] This is a handy way of estimating water depth in the surf zone: The depth of the water where waves are breaking is one and one-third times the breaker height.

A.

Figure 8–19 Breakers.
A. Spilling breaker, resulting from a gradual beach slope.
B. Plunging breaker at Oahu, Hawaii, resulting from a steep beach slope. **C.** Surging breaker, resulting from an abrupt beach slope.

B.

C.

of water particles is interrupted by the wave passing over water that is too shallow to allow this movement to continue, the ride is over.

Wave Refraction

As discussed previously, waves begin to bunch up and wavelengths become shorter as swell begins to "feel bottom" upon approaching the shore. However, waves seldom approach a shore at a perfect right angle (90 degrees). Some segment of the wave can be expected to feel bottom first, and therefore it will slow before the rest of the wave. This results in the *bending* or **refraction** of each wave crest (also called a wave front) as the waves approach the shore.

Figure 8–20 shows how waves refract along a straight shoreline. Waves coming toward the shore are refracted and tend to align themselves *nearly* parallel to the shore-

line. This explains why all waves come almost straight in toward a beach, no matter what their original orientation was.

As shown in Figure 8–21, an irregular shoreline causes waves to refract such that they also nearly align with the shoreline. However, the refraction of waves along an irregular shoreline distributes wave energy unevenly along the shoreline. Lines constructed perpendicular to the wave fronts and evenly spaced so that the energy between lines is equal at all times are called **orthogonal** (*ortho* = straight, *gonia* = angle) **lines** (Figure 8–21). Orthogonals are used to help determine how energy is distributed along the shoreline by breaking waves.

Orthogonals indicate the direction that waves travel. Notice in Figure 8–21 that the orthogonals initially have equal spacing far from shore. However, as the orthogo-

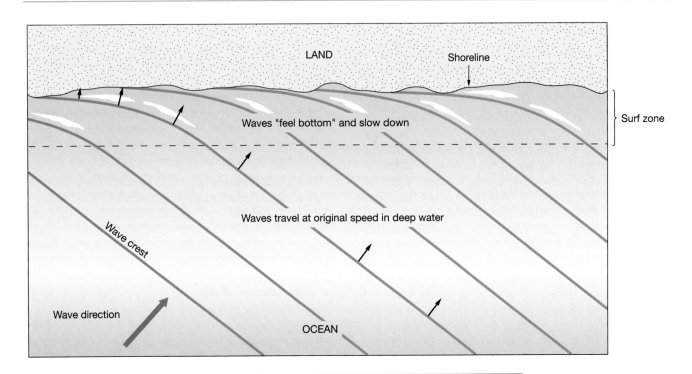

Figure 8–20 Refraction along a straight shoreline.
Waves approaching the shore at an angle first "feel bottom" close to shore. This causes the segment of the wave in shallow water to slow, causing the crest of the wave to *refract* or bend so that the waves arrive at the shore nearly parallel to the shoreline.

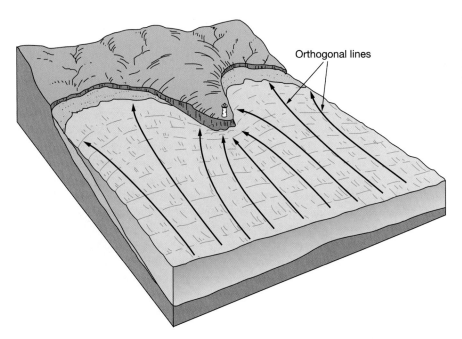

Figure 8–21 Refraction along an irregular shoreline.
As waves first "feel bottom" in the shallows off the headlands, they are slowed, causing the waves to refract and align nearly parallel to the shoreline. Evenly spaced orthogonal lines help to show how wave energy is concentrated on headlands (causing erosion) and dispersed in bays (resulting in deposition).

nals approach the shore, they *converge* on headlands that jut into the ocean, and they *diverge* in bays. This indicates that a concentration of energy is released against the headlands, whereas energy is more dispersed in bays. The result is heavy erosion of headlands, whereas deposition of sediment may occur in bays. The greater energy of waves breaking on headlands is reflected in an increased wave height.[8] Conversely, the smaller waves in bays provide areas for good boat anchorages.

_____

[8] Sailors and have known for years that "the points draw the waves." Surfers also know how wave refraction causes good "point breaks."

Wave Reflection

Not all of the energy of waves is expended as they rush onto the shore. A vertical barrier, such as a seawall or a rock ledge, can reflect swell back into the ocean with little loss of energy. Just as a mirror reflects (bounces) back light, **wave reflection** is the bouncing back of wave energy. If the incoming wave strikes the barrier at a right (90-degree) angle, the wave energy is reflected back parallel to the incoming wave, often interfering with the next incoming wave and creating unusual waveforms. More commonly, waves approach the shore at an angle, causing reflection of wave energy at an angle equal to the angle at which the wave approached the barrier.

An outstanding example of reflection phenomena occurs in an area called "The Wedge," which develops west of the jetty that protects the harbor entrance at Newport Harbor, California (Figure 8–22A). The jetty is a solid man-made object that extends into the ocean 400 meters (1300 feet) and has a vertical western side. As incoming waves strike the vertical side of the jetty at an angle, they are reflected at an equivalent angle. Because the original waves and the reflected waves have the same wavelength, a constructive interference pattern develops. When the crests of incoming waves overlap with crests of reflected waves, the size of the waves doubles in a wedge-shaped area where the two waves combine. This results in plunging breakers that may exceed 8 meters (26 feet)

in height (Figure 8–22B). Too dangerous to be surfed by board surfers, these waves present a fierce challenge to the most experienced body surfers. The Wedge has crippled or even killed many who have come to try it.

Waves that hit a barrier at right angles and have wave energy that is reflected often produce **standing waves**, also called *stationary waves.* Standing waves are the product of two waves of the same wavelength moving in opposite directions, resulting in no net movement.

The standing wave is a special interference pattern. In a standing wave, no net momentum is carried because the two waves move in opposite directions. The water particles continue to move vertically and horizontally, but there is none of the circular motion that is characteristic of a progressive wave.

Standing waves are characterized by lines along which there is no vertical movement. There may be one or more such *nodes* (node = knot), or nodal lines. *Antinodes*, crests that alternately become troughs, represent the points of the greatest vertical movement within a standing wave (Figure 8–23).

There is no particle motion when an antinode is at its greatest vertical displacement, and the maximum particle movement occurs when the water surface is level. At this time, the maximum movement of the water is in a horizontal direction directly beneath the nodal lines. The movement of water particles beneath the antinodes is entirely vertical.

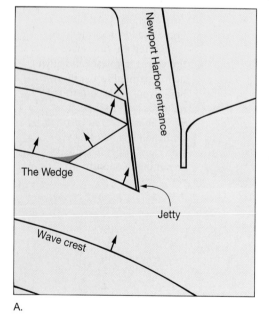

A.

B.

Figure 8–22 Reflection at The Wedge.
A. Map view of The Wedge, showing the jetty at the entrance to Newport Harbor, California. Constructive interference between original waves and the reflected waves causes a wedge-shaped wave (dark blue triangle) that may reach heights exceeding 8 meters (26 feet). The X on shore is the location where photo B was taken. **B.** View of The Wedge from the landward end of the jetty. The three dots in the water in front of the wave are the heads of body surfers waiting to catch the wave.

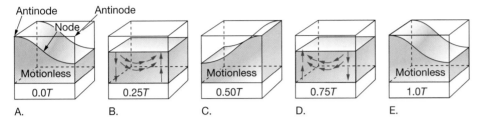

Figure 8–23 Standing waves.

An example of water motion viewed at four points during a wave cycle (*A* and *E* are identical). Water is motionless when *antinodes* reach maximum displacement (*A, C, E*). Water movement is at a maximum (blue arrows) when the water is level (*B* and *D*). Movement is vertical beneath the antinodes, and maximum horizontal movement occurs beneath the nodes.

We will consider standing waves further when we discuss tidal phenomena in the next chapter. Under certain conditions, the development of standing waves significantly affects the tidal character of coastal regions.

Tsunami

The Japanese term for the large, destructive waves that occasionally roll into their harbors is **tsunami** (*tsu* = harbor, *nami* = waves[9]). Tsunami are unusual waves that originate from sudden changes in the topography of the seafloor caused by underwater fault slippage, underwater avalanches, or underwater volcanic eruptions. Many people mistakenly call them "tidal waves," which implies that they are related to the tides, but they are not. Since the mechanisms that trigger tsunami are typically related to seismic events, tsunami are appropriately termed *seismic sea waves.*

Tsunami are usually caused by *fault movement,* or displacement in Earth's crust along a fracture. This causes not only an earthquake but also a sudden change in water level at the ocean surface above (Figure 8–24A). It is interesting to note that only certain types of fault movement generate tsunami. Faults that produce *vertical* displacement (the uplift or downdropping of ocean floor) change the volume of the ocean basin, which affect the entire water column and often generate tsunami. Conversely, faults that produce *horizontal* displacement (such as the lateral movement associated with transform faulting) generally do not generate tsunami because the side-to-side movement of these faults does not cause a change in the volume of the ocean basin. Secondary events, such as underwater avalanches produced by the faulting, or underwater volcanic eruptions, which create the largest waves, may also produce tsunami. All tsunami are the result of releases of energy expended along the sea floor.

Because the wavelength of a typical tsunami exceeds 200 kilometers (125 miles), it is considered a shallow-water wave everywhere in the ocean.[10] Thus, the speed of the tsunami is determined only by water depth. Moving at great speeds in the open ocean—well over 700 kilometers (435 miles) per hour, which is comparable to jet airplane speeds—these waves have heights of only approximately 0.5 meter (1.6 feet) in the open ocean. Thus, tsunami are not readily observable until they reach shore. In shallow water, they slow, and the water begins to pile up.

Coastal Effects

Contrary to popular belief, a tsunami does not express itself as a huge breaking wave at the shoreline. Rather, it is represented as a strong flood or surge of water that causes the ocean to advance (or, in certain cases, retreat) dramatically. In fact, a tsunami resembles a sudden, *extremely* high tide, which is why they are misnamed "tidal waves." It takes several minutes for the tsunami to express itself fully, during which time normal sea level can be up to 40 meters (131 feet) higher than normal, with normal waves superimposed on top of the higher sea level. The strong surge of water can rush into unsuspecting low-lying areas with destructive results (Figures 8–24B to 8–24D).

As the trough of the tsunami arrives at the shore, the water will rapidly drain off the land. In coastal areas, it will appear that a sudden and *extremely* low tide is occurring, where sea level is many meters lower than even the lowest low tide. Since tsunami are typically a series of waves, they are often represented as a series of alternating dramatic surges and withdrawals of water, separated by only a few minutes. The first surge may not always be the largest: In some cases, the third, fourth,

---

[9] Notice that the Japanese term *nami* is plural for waves, indicating that tsunami are typically more than a single wave.

[10] Remember that the depth of the wave base is equal to one-half the wavelength. Thus, tsunami can typically be felt to depths of 100 kilometers (62 miles), which is deeper than even the deepest ocean trenches.

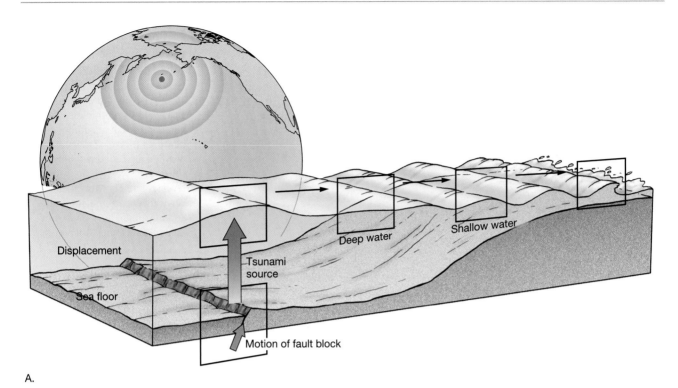

A.

Figure 8–24 Tsunami.
A. Abrupt vertical movement along a fault on the sea floor pushes up or drops the ocean water column above the fault. The energy released is then distributed laterally along the atmosphere–ocean interface in the form of long, fast waves called *tsunami*. **B.** A 1983 tsunami in northern Japan surges toward fleeing spectators in a harbor. The tsunami killed 100 people and injured 104.

or even seventh surge may be the maximum surge event within a tsunami.

In some cases, the trough of a tsunami will arrive at a coastal location first. Eyewitness accounts indicate that it appears as if an extremely low tide suddenly occurs. This often exposes parts of the lowermost shoreline that are rarely seen. For people at the shoreline, the temptation is to explore these newly exposed areas and catch stranded fish. However, the imminent danger is that within a few minutes, a strong surge of water (the crest of the tsunami) will arrive at shore.

The alternating surges and retreats of water by tsunami can severely damage coastal structures. Tsunami can

be deadly as well: The speed of the advance—up to 4 meters (13 feet) per second—is faster than a person can run. Those who are trapped by tsunami are often drowned or crushed by floating debris.

Historic Tsunami

Many small tsunami are created each year, which go largely unnoticed. Historically, a large tsunami occurs somewhere in the world every two to three years. An extremely large and damaging one occurs every 15 to 20 years. The ocean with the highest incidence of tsunami is the tectonically active Pacific, from which about 80 percent of all great waves are generated. This is be-

Figure 8–24 (*continued*)
Tsunami.
C and **D.**

cause the Pacific is ringed by a series of trenches that are the convergent margins of lithospheric plates, along which large-magnitude earthquakes occur. The vast majority of the tsunami in the Pacific are related to earthquake activity associated with the Pacific Rim. Volcanic activity is also common along the Pacific "Ring of Fire," which is aptly named for the abundance of active volcanoes and large earthquakes that occur along its margin, capable of producing extremely large tsunami.

One of the most destructive tsunami ever generated came from the greatest release of energy from Earth's interior observed during historical times. On August 27, 1883, the volcanic island of Krakatau[11]—approximately the same size as a small Hawaiian Island—in what is now Indonesia exploded and was nearly obliterated. The sound of the explosion was heard an incredible 4800 kilometers (2980 miles) away at Rodriguez Island in the western Indian Ocean. Dust from the explosion ascended into the atmosphere and circled Earth on high-altitude winds. This dust produced unusual and beautiful sunsets for nearly a year.

Not many were killed by the outright explosion of the volcano because the island was uninhabited. However, the displacement of water from the energy released during the explosion was enormous. The resulting tsunami exceeded 35 meters (116 feet) in height. It devastated the coastal region of the Sunda Strait between the nearby islands of Sumatra and Java, drowning over 1000 villages and taking more than an estimated 36,000 lives. The energy carried by this wave reached every ocean basin, and

---

[11] The volcanic island Krakatau (which is *west* of Java) is also called Krakatoa.

was detected by tide recording stations as far away as London and San Francisco.

Several ships were along the coast of Java during the eruption of Krakatau and were eyewitnesses to the tsunami and its destruction. N. van Sandick, an engineer aboard the Dutch vessel *Loudon*, gives this account:

> Suddenly we saw a gigantic wave of prodigious height advancing from the sea-shore with considerable speed. Immediately the crew set to under considerable pressure and managed after a fashion to set sail in face of the imminent danger; the ship had just enough time to meet with the wave from the front. After a moment, full of anguish, we were lifted up with a dizzy rapidity. The ship made a formidable leap, and immediately afterwards we felt as though we had plunged into the abyss. But the ship's blade went higher and we were safe. Like a high mountain, the monstrous wave precipitated its journey towards the land. Immediately afterwards another three waves of colossal size appeared. And before our eyes this terrifying upheaval of the sea, in a sweeping transit, consumed in one instant the ruin of the town; the lighthouse fell in one piece, and all the houses of the town were swept away in one blow like a castle of cards. All was finished. There, where a few moments ago lived the town of Telok Betong, was nothing but the open sea.

A strong recession marked the arrival of a large tsunami in the port of Hilo, Hawaii, on April 1, 1946. The tsunami was from a magnitude $M_w = 7.3$ earthquake in the Aleutian Trench off the island of Unimak, Alaska, over 3000 kilometers (1850 miles) away. The bathymetry in horseshoe-shaped Hilo Bay tends to focus a tsunami's energy directly toward town, causing waves to build up to tremendous heights. In this case, the tsunami elevated the water nearly 17 meters (55 feet) above normal high tide, causing more than $25 million in damage and killing 159 people in Hawaii, mostly near Hilo. Even today, it stands as Hawaii's worst natural disaster.

Closer to the source of the earthquake, the tsunami was considerably larger. The tsunami struck Scotch Cap, Alaska, on Unimak Island, where a two-story reinforced concrete lighthouse stood 14 meters (46 feet) above sea level at its base. The lighthouse was destroyed by a wave that is estimated to have attained a height of 36 meters (118 feet), killing all five people inside the lighthouse at the time. Vehicles on a nearby mesa 31 meters (103 feet) above water level were also moved by the onrush of water.

Tsunami Warning System

Until 1948, it was impossible to warn of coming tsunami in time for people to avoid the destructive waves. As a result of a wave that struck Hawaii in 1946, a tsunami warning system was established throughout the Pacific Ocean. It led to what is now the **Pacific Tsunami Warning Center (PTWC)**, a consortium of 25 Pacific Rim countries that is geographically and administratively centered in Honolulu, Hawaii. Since tsunami have small wave heights and are difficult to predict in the open ocean, the tsunami warning system uses seismic waves—some of which travel through Earth at speeds 15 times faster than tsunami—to predict the occurrence of large waves.[12] When a seismic disturbance occurs beneath the ocean surface that is large enough to be considered tsunamigenic (capable of producing a tsunami), a *tsunami watch* is issued. At this point, a tsunami may or may not have occurred, but certainly the potential for one exists.

Since the PTWC is linked to over 50 tide-measuring stations throughout the Pacific, the closest recording station to the earthquake is closely monitored for any indication of unusual wave activity. If unusual wave activity is verified, the tsunami watch is upgraded to a *tsunami warning*. Generally, earthquakes smaller than magnitude $M_w = 6.5$ are not considered tsunamigenic because they lack the duration of ground shaking necessary to initiate a tsunami. As well, transform faults do not usually produce tsunami because the lateral movement of these faults does not offset the ocean floor and impart energy to the water column in the same way that vertical fault movements do.

Once a tsunami is detected, warnings are sent to all the coastal regions that might encounter a destructive wave, along with its estimated time of arrival. This warning, usually just a few hours in advance of a tsunami, allows for evacuation of people from low-lying areas and removal of ships from harbors before the waves arrive. However, if the disturbance is nearby, there is not enough time for a warning to be issued because a tsunami travels so rapidly. Unlike hurricanes, whose high winds and waves threaten ships at sea and send them to the protection of a coastal harbor, a tsunami washes ships from their coastal moorings into the open ocean or onto shore. Thus, the best strategy for ship owners during a tsunami warning is to get their ships out of coastal harbors and into deep water, where tsunami are not easily felt.

Since the PTWC became effective in 1948, it has been tremendously effective in preventing loss of life due to tsunami when people heed the evacuation warnings. Property damage, however, has increased as people construct more buildings close to shore. To combat the threat of tsunami, proposals have been introduced to build protective coastal structures capable of diminishing tsunami energy along a coastline. Perhaps one of the best strategies is not to build structures in low-lying coastal regions where tsunami have frequently struck in the past.

Power from Waves

There is a great amount of energy in moving water, which is why there are a large number of hydroelectric power plants on rivers to generate electricity. Even greater en-

---

[12] Another method that can be used to detect tsunami is to place sensitive pressure sensors on the ocean floor to detect the passage of tsunami in the open ocean.

Box 8–2
A Cascadia Tsunami in 1700 Felt in Japan

Even though the potential for large earthquakes along subduction zones always exists, no large earthquake has been recorded in historic times along the Cascadia subduction zone. Located offshore of the U.S. Pacific Northwest, the Cascadia subduction zone is created by the Juan de Fuca Plate sliding beneath the North America Plate.

Compelling evidence for large local earthquakes along the Cascadia subduction zone has recently been discovered. This evidence is of two types. One type is paleo-earthquake evidence, which includes offshore turbidity current deposits, buried peat deposits and dead forests that indicate sudden coastal downdropping, and sediments that have liquefied on land due to ground shaking. The other type is paleo-tsunami evidence in the form of sand layers deposited in low-lying coastal areas. Presumably, the Cascadia subduction zone is the source of the earthquakes, although other major faults on land (such as the Seattle Fault) might also produce large earthquakes. The evidence suggests that many large earthquakes have occurred in the area during the last 1000 years, but the exact size of the earthquakes is unknown. As well, the date of these earthquakes can be resolved only to within a decade.

A new piece of supporting evidence for large Cascadia earthquakes comes from the historic records of tsunami damage in Japan. The records indicate that a tsunami of unknown origin

hit the coast of Japan on January 27–28 in the year 1700. Based on analysis of the time of arrival and the height of the tsunami in different locations in Japan, other Pacific Rim regions (South America, Alaska, and the Kamchatka Peninsula in Russia) can be eliminated as potential sources. Even a locally generated tsunami does not fit the reported effects, leaving the Cascadia subduction zone as the most likely source (Figure 8D). Because Japan received tsunami heights of 2 to 3 meters (6.5 to 10 feet), the earthquake must have been huge: Research suggests that the earthquake must have been magnitude $M_w = 9.0$. Only two earthquakes of this size have ever been recorded: one along the coast of Chile in 1960, and the other in southern Alaska in 1964. An earthquake this large must have ruptured the entire length of the Cascadia subduction zone, which has profound implications for forecasting the size and frequency of future Cascadia earthquakes.

What is perhaps even more interesting is that the *time* of the earthquake can be determined. Based on the historic records in Japan (which indicate when the tsunami hit different parts of the coast) and knowing how fast tsunami travel, the exact time can be extrapolated. The estimated time of the earthquake is at about 9:00 P.M. Pacific Northwest local time on January 26, 1700. This estimate is consistent with Native American legends, which indicate that a large earthquake occurred at about this date during a winter night.

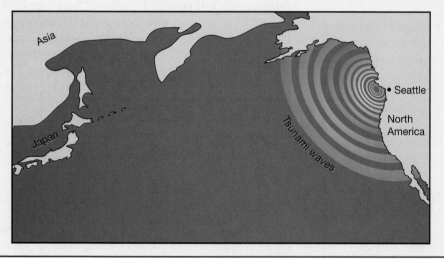

Figure 8D The tsunami of 1700.

ergy exists in ocean waves, if it can be harnessed efficiently. However, significant problems must be overcome. In places where waves refract and converge, as they do around headlands (see Figure 8–21), energy is focused, creating a potential setting for power generation. Such a system might extract up to 10 megawatts of power[13] per kilometer of shoreline.

One of the disadvantages of wave energy is that the system could produce significant power only when large storm waves break against it. Thus, such a system would operate only as a power supplement. In addition, the construction of a series of perhaps a hundred or more such structures along the shore would be required. Structures of this type could have a significant impact on the environment, with negative effects on marine organisms that rely on wave energy for dispersal, transporting food supplies, or removing wastes. Also, natural processes might be altered, which could lead to

---

[13] Ten megawatts of power is comparable to the electricity consumed by 20,000 average U.S. households in one month.

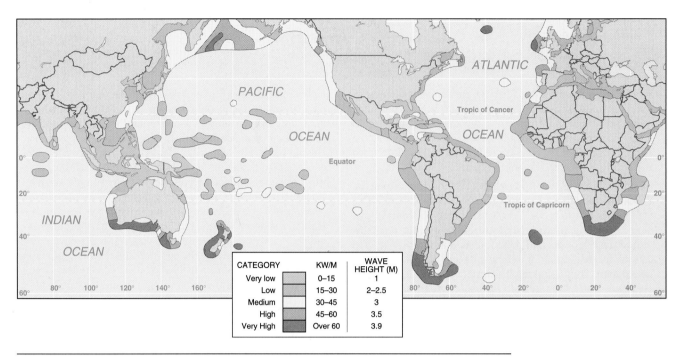

Figure 8–25 Global coastal wave energy resources.
Distribution of coastal wave energy shows that more wave energy is available along western shores of continents, especially in the Southern Hemisphere. KW/M is kilowatts per meter; for example, every meter of "red" shoreline is a potential site for generating over 60 kilowatts of electricity. Average wave height is in meters.

serious coastal erosion problems in areas deprived of sediment.

Internal waves also have the potential for being utilized as a source of energy. Along shores that have favorable ocean floor topography for focusing wave energy, internal waves with their great heights may be effectively concentrated by refraction. This energy could be used to power an energy-conversion device that would produce electricity.

Economic conditions in the future may enhance the appeal of converting wave energy to electrical energy.

The coastal regions with the greatest potential for exploitation are shown in Figure 8–25. The map shows the effect of west-to-east movement of storm systems in the mid-latitudes. The western coasts of continents are struck by larger waves than eastern coasts, meaning that more wave energy is available along western shores. Further, larger waves appear to be associated with the prevailing westerly wind belts between 30 and 60 degrees latitude, especially in the Southern Hemisphere.

 ## Students Sometimes Ask...

Do waves always travel in the same directions as currents?
Not always. It seems logical that since surface currents and most waves are created by winds blowing across the ocean surface, they should move in the same direction. Even though most of the wind evergy that is put into surface waves generates waves that travel in the same direction as the wind blows, waves radiate outward in *all* directions from the disturbance (release of energy) that creates them. In addition, as waves travel away from the sea area where they were generated, they move into areas where other currents exist. Consequently, the direction of wave movement is often unrelated to that of currents. Fundamentally, movement of waves is completely independent of currents, and wave trains can travel at various directions in relation to currents. In fact, waves can even travel in a direction *completely opposite* to that of a current. An example that illustrates this principle is a surface

current of water called a rip current that moves away from shoreline, opposite to the direction of incoming waves.

Can internal waves break?
Internal waves do not break in the classical way that surface waves break in the surf zone. However, when internal waves approach the edges of continents, they do experience similar physical changes as waves do in the surf zone. This causes the waves to build up and expend their energy with much turbulent motion, in essence "breaking" against the continent.

Can a wave break twice?
Certainly. This is commonly seen anywhere offshore shallowly submerged obstacles are located, such as coral reefs, rock reefs, or sand bars. Waves approaching these obstacles will undergo physical changes just as breaking waves do in the surf zone, expending some of the wave's energy by breaking.

However, some of the energy moves over the top of the obstacle, and the wave—now diminished in size—continues to move toward shore. The wave will expend the last of its energy against the shore, effectively breaking twice.

I know that swell is what surfers hope for. Is swell always big?
Not necessarily. Remember that swell is defined as waves that have moved out of their area of origination. Consequently, waves do not have to be a certain wave height to be classified as swell. It is true, however, that the uniform and symmetrical shape of most swell delights surfers.

In your opinion, where is the best surfing?
The answer to that question is highly subjective, with each surfer claiming his or her own favorite surf spot—often revealed to very few others. Certainly, some of the factors that determine an ideal surfing location must include size and regularity of waves, climate, cost, accessibility, and seclusion (not necessarily in that order). Referring back to Figure 8–11, which shows worldwide wave height, it is clear that the Southern Hemisphere has very large waves. Some of the largest waves are found in the south Indian and South Pacific Oceans. Since the Pacific Ocean has the greatest fetch (which allows for the possibility of bigger waves to develop), continents and islands in the South Pacific seem like ideal locations if wave size is the most important factor.

However, surfing can often be enhanced when waves have been sorted into swell, so locations a bit farther from the Southern Hemisphere might have slightly smaller waves but ones that are more regular. In this case, some of the tropical Pacific islands (such as Hawaii, Fiji, and some of the more remote Indonesian islands) might offer better surfing. Besides their location in warm climates, they also have the advantage of being exposed to North Pacific Ocean swell.

If seclusion is the highest priority—and you don't mind donning a full wetsuit—some coastal locations with large waves in Alaska have recently been surfed, and surfing in Antarctica can't be far behind!

Why is surfing so much better along the west coast of the United States than along the east coast?
There are three main reasons why the west coast has better surfing conditions:

- The waves are generally bigger in the Pacific. This is because the Pacific is larger than the Atlantic, so the fetch is larger, allowing bigger waves to develop in the Pacific (see Figure 8–25).

- The beach slopes are generally steeper along the west coast. Along the east coast, the gentle slopes often create spilling breakers, which are not as favorable for surfing. The steeper beach slopes along the west coast cause plunging breakers, which are better for surfing.

- The wind is more favorable. Most of the United States is influenced by the prevailing westerlies, which blow toward shore and enhance waves along the west coast. Along the east coast, the wind blows away from shore.

What is the record height of a tsunami?
Perhaps it is not surprising that Japan holds the record, because Japan's proximity to subduction zones causes that country to endure more tsunamis than any other place on Earth (followed by Chile and Hawaii). The largest documented tsunami on record occurred in the Ryukyu Islands of southern Japan in 1971, when a tsunami raised normal sea level by an astounding 85 meters (278 feet). Remember that the heights of tsunami are reported in *vertical* distances and reflect the amount of water level rise above normal sea level. In low-lying coastal areas, such a vertical rise can extend inland many kilometers, causing flooding and widespread damage. The most deadly tsunami was probably the tsunami in 1703 that hit Aura, Japan, and was responsible for an estimated 100,000 deaths.

Since 5-foot surf is common at the beach where I live, why should I be worried about a 5-foot tsunami?
Remember that a tsunami is quite different from a normal ocean wave, even though both may have the same wave height. Even though a 5-foot (1.5-meter) breaker has a lot of energy that is released at the shore, it pales in comparison to the energy carried by a tsunami. This is because a tsunami influences the *entire* column of water, while an ocean wave is confined to surface waters. A 5-foot tsunami is like a 5-foot rise in normal sea level. Since the tsunami has such a great wavelength, it will reach much farther ashore, too. The biggest storm breakers expend their energy within the surf zone while a tsunami can reach several kilometers inland in low-lying areas. Five-foot surf may be fun to play in, but a 5-foot tsunami can knock down the building you are in, drag you out to sea, and batter you to death.

If there is a tsunami warning issued, what is the best thing to do?
The *smartest* thing to do is to stay out of coastal areas, but people often want to see the tsunami in person. For instance, when an earthquake of magnitude $M_w = 7.7$ occurred offshore of Alaska in May 1986, a tsunami warning was issued for the west coast of the United States. In southern California, people flocked to the beach to observe this natural phenomenon. Fortunately, the tsunami was only a few centimeters in height by the time it reached southern California, so it was unnoticed at the beach.

If you must go to the beach to observe the tsunami, expect crowds, road closings, and general mayhem. It would be a good idea to stay at least 30 meters (100 feet) above sea level. If you happen to be at a remote beach where the water suddenly withdraws, it would be in your best interest to evacuate immediately to higher ground. And if you happen to be at a beach where an earthquake occurs and shakes the ground so hard that you can't stand up, as soon as you *can* stand up, *RUN*—don't walk—for high ground!

After the first surge of the tsunami, it would be a wise to stay out of low-lying coastal areas for several hours because several more surges (and withdrawals) can be expected. There have been many documented cases where curious people have been killed when they are trapped by the third or fourth surge of a tsunami.

Summary

Waves are one of the most noticeable features of the ocean. All ocean waves begin as disturbances and are caused by releases of energy. Such releases of energy include wind, the movement of fluids of different densities (which create internal waves), mass movement into the ocean, underwater sea floor movements, the gravitational pull of the moon and the sun on Earth, and human activities in the ocean.

Once initiated by a disturbance, wave phenomena transmit energy through various states of matter by setting up patterns of oscillatory motion in the particles that make up matter. Progressive waves are longitudinal, transverse, or orbital, depending on the pattern of particle oscillation. Particles in ocean waves move primarily in orbital paths.

Characteristics used to describe waves are wavelength (L), wave height (H), wave steepness (H/L), wave period (T), frequency (f), and wave speed (S). As a wave travels, the water passes the energy along by moving in a circle, called circular orbital motion. This causes the waveform to advance, not the water particles themselves. The depth at which circular motion ceases is called wave base and is at a depth equal to one-half the wavelength measured from still water level.

If water depth is greater than one-half the wavelength, a progressive wave travels as a deep-water wave with a speed that is directly proportional to wavelength. If water depth is less than $1/20$ wavelength ($L/20$), the wave will move as a shallow-water wave, the speed of which increases with increased water depth. Transitional waves have wavelengths between deep- and shallow-water waves, with speeds determined by both wavelength and water depth.

As wind-generated waves form in a sea area, capillary waves with rounded crests and wavelengths less than 1.74 centimeters (0.7 inch) form first. As the energy of the waves increases, gravity waves form, with increased wave speed, wavelength, and wave height. Factors that influence the size of wind-generated waves include wind speed, duration (time), and fetch (distance). An equilibrium condition is often reached of maximum wind speed, duration, and fetch called a fully developed sea.

Energy is transmitted from the sea area across the ocean by uniform, symmetrical waves called swell. Different wave trains of swell can create either constructive, destructive, or mixed interference patterns. Unusually large waves called rogue waves or superwaves can be created by constructive interference.

As a wave approaches shoaling water near shore, it experiences many physical changes. Waves release their energy in the surf zone when their steepness exceeds a 1:7 ratio and break.

If waves break on a relatively flat surface, the result is usually a spilling breaker. Breakers forming on steep slopes have spectacular curling crests called plunging breakers, which are the best breakers for surfing. Abrupt beach slopes create surging breakers.

When swell approaches the shore, segments of the waves that first encounter shallow water are slowed, whereas other parts that have not been affected by shallow water move ahead, causing the wave to refract, or bend. Refraction concentrates wave energy on headlands, whereas low-energy breakers are characteristically found in bays.

Wave reflection is the bouncing back of wave energy. Reflection of waves off seawalls or other barriers can cause an interference pattern called a standing wave. In standing waves, crests do not move laterally as in progressive waves but form alternately with troughs at locations called antinodes. Separating the antinodes are nodes, where there is no vertical movement of the water.

Tsunami, or seismic sea waves, are generated by sudden changes in the elevation of the sea floor, such as fault movement or volcanic eruptions. Such waves often have lengths exceeding 200 kilometers (125 miles) and travel across the open ocean with undetectable heights of about 0.5 meter (1.6 feet) at speeds in excess of 700 kilometers (435 miles) per hour. Upon approaching shore, a tsunami is represented by a series of rapid withdrawals and surges, some of which may increase sea level by over 30 meters (100 feet) in vertical height. Most tsunami occur in the Pacific Ocean and, historically, they have been known to cause millions of dollars of damage and take tens of thousands of lives. An international tsunami warning system within the Pacific Ocean was established in 1948, which has dramatically reduced fatalities by successfully predicting tsunami using real-time seismic information.

Ocean waves can be harnessed to produce hydroelectric power, but significant problems must be overcome to make this a practical source of energy.

Key Terms

Atmospheric wave (p. 246)	Orthogonal line (p. 262)	Surf zone (p. 257)
Capillary wave (p. 252)	Pacific Tsunami Warning Center (PTWC) (p. 268)	Surging breaker (p. 261)
Circular orbital motion (p. 250)		Swell (p. 256)
Constructive interference (p. 256)	Plunging breaker (p. 261)	Transitional wave (p. 251)
Crest (p. 249)	Refraction (p. 262)	Transverse wave (p. 247)
Deep-water wave (p. 251)	Restoring force (p. 252)	Trough (p. 249)
Destructive interference (p. 256)	Rogue wave (p. 257)	Tsunami (p. 265)
Frequency (p. 249)	Sea (p. 253)	Wave base (p. 250)
Fully developed sea (p. 255)	Shallow-water wave (p. 251)	Wave dispersion (p. 256)
Gravity wave (p. 252)	Shoaling (p. 257)	Wave height (p. 249)
Interference pattern (p. 256)	Spilling breaker (p. 261)	Wave period (p. 249)
Internal wave (p. 246)	Splash wave (p. 244)	Wave reflection (p. 264)
Longitudinal wave (p. 247)	Standing wave (p. 264)	Wave speed (p. 249)
Mixed interference (p. 256)	Still water level (p. 249)	Wave steepness (p. 249)
Ocean wave (p. 246)	Superwave (p. 257)	Wave train (p. 256)
Orbital wave (p. 249)	Surf beat (p. 256)	Wavelength (p. 249)

Questions And Exercises

1. How large was the largest wave ever authentically recorded? Where did it occur, and how did it form?

2. Discuss several different ways in which waves form. How are most ocean waves generated?

3. Why is the development of internal waves likely within the pycnocline?

4. Discuss longitudinal, transverse, and orbital wave phenomena, including the states of matter in which each can transmit energy.

5. Draw a diagram of a simple progressive wave. From memory, label the crest, trough, wavelength, wave height, and still water level.

6. Can a wave with a wavelength of 14 meters ever be more than 2 meters high? Why or why not?

7. Calculate the speed (S) in meters per second (m/s) for deep-water waves with the following characteristics:

    a. $L = 351$ meters, $T = 15$ seconds

    b. $T = 12$ seconds

    c. $f = 0.125$ wave/second

8. What physical feature of a wave is related to the depth of the wave base? On the diagram that you drew for Question 5, add the wave base. What is the difference between the wave base and still water level?

9. Explain why the following statements for deep-water waves are either true or false:

    a. The longer the wave, the deeper the wave base.

    b. The greater the wave height, the deeper the wave base.

    c. The longer the wave, the faster the wave travels.

    d. The greater the wave height, the faster the wave travels.

    e. The faster the wave, the greater the wave height.

10. Describe the change in the shape of waves that occur as they progress from capillary waves to increasingly larger gravity waves until they reach a steepness ratio of 1:7. A change in which variable—H, L, S, T, or f—will make gravity the dominant restoring force?

11. Define swell. Does swell necessarily imply a particular wave size? Why or why not?

12. Waves from separate sea areas move away as swell and produce an interference pattern when they come together. If Sea A has wave heights of 1.5 meters (5 feet) and Sea B has wave heights of 3.5 meters (11.5 feet), what would be the height of waves resulting from constructive interference and destructive interference? Illustrate your answer (see Figure 8–15).

13. Describe the physical changes that occur to a wave's wave speed (S), wavelength (L), height (H), and wave steepness (H/L) as a wave moves across shoaling water to break on the shore.

14. Describe the three different types of breakers and indicate the slope of the beach that produces the three types. How is the energy of the wave distributed differently within the surf zone by the three types of breakers?

15. Using examples, explain how wave refraction is different from wave reflection.

16. Using orthogonal lines, illustrate how wave energy is distributed along a shoreline with headlands and bays. Identify areas of high and low energy release.

17. Define the terms *node* and *antinode* as they relate to standing waves.

18. Why is it more likely that a tsunami will be generated by faults beneath the ocean along which vertical rather than horizontal movement has occurred?

19. While shopping in a surf shop, you overhear some surfing enthusiasts mention that they would really like to ride the curling wave of a tidal wave at least once in their life, because it is a single breaking wave of enormous height. What would you say to these surfers?

20. Explain what it would look like at the shoreline if the trough of a tsunami arrives there first. What is the impending danger?

21. What ocean depth would be required for a tsunami with a wavelength of 220 kilometers (136 miles) to travel as a deep-water wave? Is it possible that such a wave could become a deep-water wave any place in the world ocean?

22. Explain how the tsunami warning system in the Pacific Ocean works. Why must the tsunami be verified at the closest tide recording station?

23. Describe the different types of evidence used to support the idea that a large earthquake along the Cascadia subduction zone produced a tsunami in 1700 that was felt in Japan.

24. Discuss some environmental problems that might result from developing facilities for conversion of wave energy to electrical energy.

References

Bascom, W. 1980. *Waves and beaches: The dynamics of the ocean surface* (revised edition). New York: Anchor Books (Doubleday).

Bowditch, N. 1958. *American practical navigator,* rev. ed. H.O. Pub. 9. Washington, DC: U.S. Naval Oceanographic Office.

Brown, J. 1989. Rogue waves. *Discover* 10:4, 47–52.

Folger, T. 1994. Waves of destruction. *Discover* 15:5, 66–73.

Gill, A. E. 1982. *Atmosphere-ocean dynamics.* International Geophysics Series, Vol. 30. Orlando, FL; Academic Press.

Kinsman, B. 1965. *Wind waves: Their generation and propagation on the ocean surface.* Englewood Cliffs, NJ: Prentice-Hall.

Knamori, H., and Kikuchi, M. 1993. The 1992 Nicaragua earthquake: A slow tsunami earthquake associated with subducted sediments. *Nature* 361:6414, 714–716.

McCredie, S. 1994. When nightmare waves appear out of nowhere to smash the land. *Smithsonian* 24:12, 28–39.

Melville, W., and Rapp, R. 1985. Momentum flux in breaking waves. *Nature* 317:6037, 514–516.

Miller, D. J. 1960. *Giant waves in Lituya Bay, Alaska.* U. S. Geological Survey Professional Paper 354-C.

Pickard, G. L. 1975. *Descriptive physical oceanography: An introduction,* 2nd ed. New York: Pergamon Press.

Satake, K., Shimazaki, K., Tsuji, Y., and Ueda, K. 1996. Time and size of a giant earthquake in Cascadia inferred from Japanese tsunami records of January 1700. *Nature* 379:6562, 246–249.

Schneider, D. 1995. Tempest on the high sea: The Ocean Drilling Program narrowly averts catastrophe. *Scientific American* 273:6, 14–16.

Stewart, R. H. 1985. *Methods of satellite oceanography.* Berkeley, CA: University of California Press.

Sverdrup, H. U., Johnson, M. W., and Fleming, R. H. 1942. Renewal 1970. *The oceans: Their physics, chemistry, and general biology.* Englewood Cliffs, NJ: Prentice-Hall.

van Arx, W. S. 1962. *An introduction to physical oceanography.* Reading, MA: Addison-Wesley.

Suggested Reading

Earth

Carey, S., Sigurdson, H., and Mandeville, C. 1992. Fire and water at Krakatau. 1:2, 26–35. Describes the 1883 eruption of the volcano Krakatau (Krakatoa), which created a large and devastating tsunami.

Dvorak, J., and Peek, T. 1993. Swept away. 2:4, 52–59. A look at the power of tsunamis, as well as what can be done to prevent the destruction caused by them. Includes several eye-witness accounts of tsunami.

Parks, N. 1993. The fragile volcano. 2:6, 42–49. Examines evidence that huge chunks of the Hawaiian Islands have broken off and plunged into the Pacific, creating huge waves in the process.

Pendick, D. 1997. Four disasters that shaped the world: Ashes, ashes, all fall down. 6:1, 32–33. Examines how tsunamis may be generated when large portions of the volcanic islands slump into the sea.

———. Four disasters that shaped the world: Waves of destruction. 6:1, 28–29. A description of the tsunami created by the 1883 eruption of Krakatau (Krakatoa).

Wuethrich, B. 1995. Cascadia countdown. 4:5, 24–31. From evidence of past earthquakes, scientists predict that a large and devastating earthquake—and resulting tsunami—is likely to occur in the Pacific Northwest.

Sea Frontiers

Alpner, J. 1993. Waves are just energy moving through the water. 39:1 22–27. Discusses what forms large waves and gives examples, including the great North Atlantic storm of 1991 that occurred during Halloween.

Barnes-Svarney, P. 1988. Tsunami: Following the deadly wave. 34:5, 256–263. The origin of tsunami, a history of major occurrences, and the methods used to detect them are covered.

Changery, M. J. 1987. Coastal wave energy. 33:4, 259–262. The wave energy resources of the world's shores are considered.

Ferrell, N. 1987. The tombstone twins: Lights at the top of the world. 33:5, 344–351. A short history of the two lighthouses on Unimak Island, Alaska.

Land, T. 1975. Freak killer waves. 21:3, 139–141. The British design a buoy that will gather data in areas where 30-meter (98-foot) waves, which may be responsible for the loss of many ships, occur.

Mooney, M. J. 1975. Tragedy at Scotch Cap. 21:2, 84–90. A recounting of the events resulting from an earthquake off the Aleutians on April 1, 1946. The resulting tsunami destroyed the lighthouse at Scotch Cap, Alaska.

Pararas-Carayannis, G. 1977. The International Tsunami Warning System. 23:1, 20–27. A discussion of the history and operations of the International Tsunami Warning System.

Robinson, J. P., Jr. 1976. Newfoundland's disaster of '29. 22:1, 44–51. A description of the destruction caused by a tsunami that struck Newfoundland on November 18, 1929.

———. 1976. Superwaves of southeast Africa. 22:2, 106–116. A discussion of the formation and destruction caused by large waves that strike ships off the southeast coast of South Africa.

Smail, J. 1982. Internal waves: The wake of sea monsters. 28:1, 16–22. An informative discussion of the causes of internal waves and their effect on surface ships and submarines.

Smail, J. R. 1986. The topsy-turvy world of capillary waves. 32:5, 331–337. Capillary waves are clearly described, and their role in transmitting wind energy to the motion of waves and currents is discussed.

Smith, F. G. W. 1970. The simple wave. 16:4, 234–245. This is a very readable explanation of the nature of ocean waves. It deals primarily with the characteristics of deep-water waves.

———. 1971. The real sea. 17:5, 298–311. A comprehensive and readable discussion of wind-generated waves.

———. 1985. Bermuda mystery waves. 31:3, 160–163. A discussion of the possible source of large waves that struck Bermuda on November 12, 1984.

Truby, J. D. 1971. Krakatoa: The killer wave. 17:3, 130–139. The events leading up to the 1883 eruption of Krakatoa and the tsunami that followed are described.

Scientific American

Bascom, W. 1959. Ocean waves. 201:2, 89–97. An informative discussion of the nature of wind-generated waves, tsunami, and tides.

Greenberg, D. A. 1987. Modeling tidal power. 257:5, 44–55. Examines what effect a dam to extract tidal power would have in upper New England.

Hyndman, R. D. 1995. Giant earthquakes of the Pacific Northwest. 273:6, 68–75. A review of the evidence that supports the idea that the Cascadia subduction zone can produce extremely large earthquakes, with special attention to the January 1700 tsunami felt in Japan.

Koehl, M. A. R. 1982. The interaction of moving water and sessile organisms. 274:6, 124–135. A discussion of the adaptations of benthic shore-dwelling animals to the stresses of strong currents and breaking waves.

Oceanography on the Web

Visit the *Essentials of Oceanography* home page for on-line resources for this chapter. There you will find an on-line study guide with review exercises, and links to oceanography sites to further your exploration of the topics in this chapter. *Essentials of Oceanography* is at: **http://www.prenhall.com/thurman** (click on the Table of Contents menu and select this chapter).

CHAPTER 9
TIDES

A Brief History of Some Successful Tidal Power Plants

The history of human efforts to harness tidal energy dates at least to the Middle Ages. It has long been known that water, with its high fluidity and weight, can be used to turn turbines and generate electrical energy (Figure 9A). All that is needed is a way to lift the water to a greater height so that the water runs downhill through the turbines. Tides are wonderfully adapted to do just this, since they raise and lower average sea level in bays and estuaries (*aestus* = tide; an *estuary* is an inland arm of the sea along a river).

One successful tidal power plant is operating in the estuary of La Rance River near the English Channel in northern France. The estuary has a surface area of approximately 23 square kilometers (9 square miles), and the **tidal range** (the vertical difference between high and low tides) at La Rance is 13.4 meters (44 feet). Usable tidal energy increases as the area of the basin increases and also with an increase in the tidal range.

The power-generating barrier was built across the estuary a little over 3 kilometers (2 miles) upstream to protect it from storm waves. The barrier is 760 meters (2500 feet) wide and supports a two-lane road (Figure 9B). The deepest water ranges from just over 12 meters (39 feet) at low tide to more than 25 meters (82 feet) at high tide. To allow water to flow through the barrier when the generating units are shut down, sluices (artificial channels) were built into the barrier.

Water passing through the barrier provides the power for 24 electricity-generating units that operate beneath the power plant. Each unit can generate 10 million watts (10 megawatts) of electricity, enough to serve over 1500 homes for a year.

The plant generates electricity only when sufficient water height exists between the estuary and the ocean—about half of the time. Annual power production of about 540 million kilowatt-hours can be increased to 670 million kilowatt-hours by using the turbine generators as pumps to move water into the estuary at proper times.

Within the Bay of Fundy, which has the largest tidal range in the world, the Canadian province of Nova Scotia has constructed a tidal power plant that has generated up to 40 million kilowatt-hours per year since its completion in 1984. It is the only North American tidal power plant in operation today, and is built on the Annapolis River estuary, an arm of the Bay of Fundy, where maximum tidal range is 8.7 meters (26 feet).

One tidal cycle = 12 hours 25 minutes

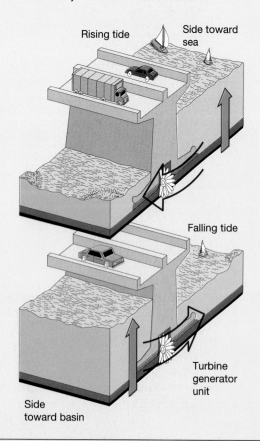

Figure 9A How a tidal power plant works.

276

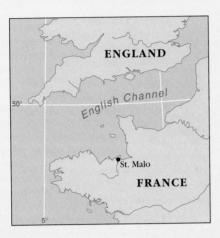

Figure 9B La Rance tidal power plant at St. Malo, France.

Some engineers think that a tidal power plant could be made to generate electricity continually if it were located on the Passamaquoddy Bay near the U.S.–Canadian border at the south end of the Bay of Fundy. Although a tidal power plant across the Bay of Fundy has often been proposed, it has never been built. Potentially, the usable tidal energy seems great compared to the La Rance plant, because the flow volume is about 117 times greater.

I f you have spent time at the seashore, you are probably familiar with the rise and fall of the tides. **Tides** are the periodic raising and lowering of average sea level that occurs throughout the oceans of the world. Even casual visitors notice that the edge of the sea slowly shifts landward and seaward daily as sea level rises and falls, destroying sand castles built during low tide. Tides are so important that accurate records of the tides have been kept at nearly every port for several centuries. Historically, the tides have determined when it is the best time to do certain things—such as sail from port. The term *tide* has even crept into our everyday vocabulary (for instance, "to tide someone over," "to go against the tide," or even to wish someone "good tidings").

People undoubtedly observed the rise and fall of the tides since they first inhabited the coastal regions of continents. However, there is no written record of tides prior to observations of the Mediterranean Sea by Herodotus (circa 450 B.C.). Even the earliest sailors knew that the moon had some connection with the tides because both followed a similar cyclic pattern. But it was not until **Sir Isaac Newton** (1642–1727) developed his universal law of gravitation that the tides could be adequately explained.

Although the study of the tides can be a large and complex subject, tides are fundamentally very long and regular shallow-water waves. The tides possess wavelengths measured in thousands of kilometers and heights ranging to more than 15 meters (50 feet). Ocean tides are generated by the gravitational attraction of the sun and moon, and affect every particle of water from the surface to the deepest parts of the ocean basin.

Generating Tides

Tides are generated through a combination of *gravity* and *motion* among Earth, the moon, and the sun. Let's examine the forces that interact among these three celestial bodies to develop an understanding of the ocean's daily rhythms.

Tide-Generating Forces

Sir Isaac Newton published his *Philosophiae Naturalis Principia Mathematica* ("Philosophy of Natural Mathematical Principles") in 1686; he stated the following in his preface:

> I derive from the celestial phenomena the forces of gravity with which bodies tend to the sun and several planets. Then from these forces, by other propositions which are also mathematical, I deduce the motions of the planets, the comets, the moon, and the sea.

What followed from his thinking was our first understanding of why tides behave as they do. It is well known that gravity tethers the sun, its planets, and their moons

277

Figure 9–1 Earth–moon system rotation.

A. The dashed line through Earth's center is the path followed as it moves around the common center of gravity (the *barycenter*) of the Earth–moon system. **B.** If a ball with a string attached is swung overhead, it stays in a circular orbit because the string exerts a centripetal (center-seeking) force on the rock. If the string breaks, the rock will fly off along a straight path at a tangent to the circle.

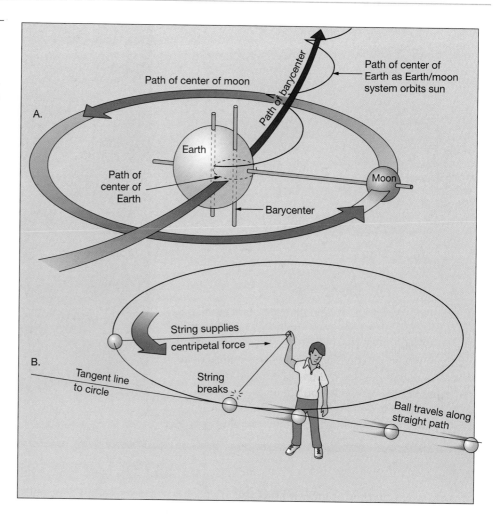

together (Figure 9–1A). Not surprisingly, it is also gravity that tugs every particle of water on Earth toward the moon and the sun, thus creating tides on Earth.

Gravitational and Centripetal Forces in the Earth–Moon System To understand how *tide-generating forces* influence the oceans, let's examine how *gravitational forces* and *centripetal forces* affect objects on Earth within the Earth–moon system (we'll ignore the influence of the sun for the moment).

The **gravitational force** follows Newton's law of gravitation, which states that *every particle of mass in the universe attracts every other particle of mass.* This occurs with a force that is directly proportional to the product of their masses and inversely proportional to the square of the distance between the masses. What this means is that the *greater* the mass of the objects and the *closer* they are together, the greater will be the gravitational attraction. Since gravitational attraction varies with the *square* of distance, even small changes in distance between two objects result in a significant change in the gravitational attraction between them.

The gravitational forces for various points on Earth (caused by the moon) are shown in Figure 9–2. In the figure, note how the differences in the gravitational

forces (which are shown by the varying lengths of the arrows) are due to the varying distances from the moon. For instance, the greatest attraction toward the moon is at the point on Earth directly facing the moon, called the **zenith** (*zenith* = a path over the head). At a point on the opposite side of Earth called the **nadir** (*nadir* = opposite the zenith), there is a much weaker gravitational pull toward the moon. In addition, notice that the direction of the gravitational attraction between most particles and the center of the moon is at an angle relative to a line connecting the center of Earth and the moon (Figure 9–2). This causes the force of gravitational attraction between each particle and the moon to be slightly different.

One important difference between gravitational attraction and the tide-generating force is that distance is a more highly weighted variable in the tide-generating force than it is in gravitational attraction force. Tide-generating forces vary inversely as the *cube* of the distance from each point on Earth to the *center* of the tide-generating object (moon or sun), instead of varying inversely to the *square* of the distance, as does gravitational attraction. Although the tide-generating force is similar to the gravitational force, it is not linearly proportional to gravity.

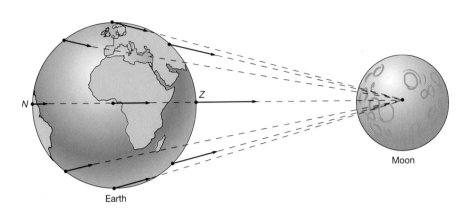

Figure 9–2 Gravitational forces due to the moon on Earth.

The gravitational forces due to the moon on objects located at different places on Earth are shown by arrows. The length and orientation of the arrows indicate the strength and direction of the gravitational force. Notice the length and angular differences of the arrows for different points on Earth. The letter *Z* represents the zenith; *N* represents the nadir.

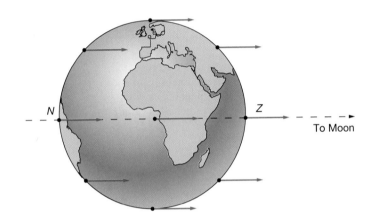

Figure 9–3 Centripetal forces due to the moon on Earth.

The centripetal forces required to keep identical-sized particles in identical circular orbits are shown by arrows. This required force is supplied by gravitational attraction between the particles and the moon. Notice that the arrows are all the same length, and are oriented in the same direction for all points on Earth. *Z* = zenith; *N* = nadir.

The **centripetal** (*centri* = the center, *pet* = seeking) **force**[1] required to keep planets in their orbits is provided by the gravitational attractions between them and the sun. Centripetal force "tethers" an orbiting body to its parent, pulling the object *inward* toward the parent, "seeking the center" of its orbit. For example, if you tie a string to a ball and swing the tethered ball around your head (Figure 9–1B), the string pulls the ball toward your hand. The string provides a *centripetal force* on the ball, forcing the ball to *seek the center* of its orbit. If the string should break, the force is gone and the ball can no longer maintain its circular orbit. When the string breaks, the ball will fly off in a *straight* line,[2] and the path of the ball is oriented tangent (*tangent* = touching) to the circle (Figure 9–1B).

_____

[1] This is not to be confused with the so-called *centrifugal* (*centri* = the center, *fug* = flee) *force*, an apparent force which is oriented outward.

[2] At the moment that the string breaks, the ball will continue along a straight-line path, obeying Newton's first law of motion (the law of inertia), which states that moving objects follow straight-line paths until they are compelled to change that path by other forces.

The same is true of the Earth–moon system. The two bodies are tethered not by strings but by gravity in such a way that the required centripetal force is provided by gravity. If all gravity in the solar system could abruptly be shut off, centripetal force would vanish. If this were to happen, the momentum of the celestial bodies would cause them to fly off into space along straight-line paths, tangent to their orbits.

Resultant and Tide-Generating Forces Particles of identical mass rotate in identical sized paths as a result of the Earth–moon rotation system (Figure 9–3). Thus, each particle requires an identical centripetal force to maintain it in its circular path. This centripetal force is supplied by the gravitational attraction between the particle and the moon, and this *supplied* force is not identical to the *required* force except at the center of Earth. It is this difference that creates tiny **resultant forces**, which are the mathematical difference between the two sets of arrows shown in Figures 9–2 and 9–3. Figure 9–4 combines Figures 9–2 and 9–3 to show the forces involved in producing resultant forces.

These resultant forces are small, averaging about one-millionth of the magnitude of Earth's gravity. If the

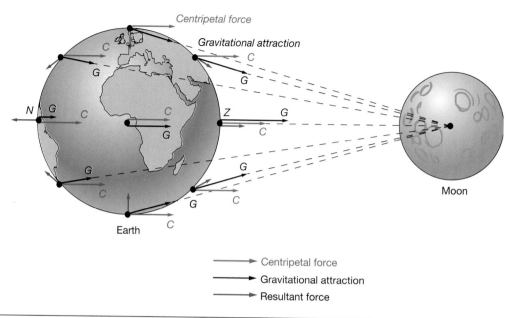

Figure 9–4 Resultant forces.
The length and direction of the arrows represent the magnitude and orientation of the forces acting on each of nine identical points shown on the figure. Red arrows indicate centripetal forces (*C*), which are not equal to the black arrows that indicate gravitational attraction (*G*). The small blue arrows show resultant forces—of which the horizontal component is the tide-generating force. *Z* = zenith; *N* = nadir.

Figure 9–5 Tide-generating forces.

Where the resultant force acts vertically relative to Earth's surface, the tide-generating force is zero. This occurs at the zenith (*Z*) and nadir (*N*), and along an "equator" connecting all points halfway between the zenith and nadir (*black dots*). However, where the resultant force has a significant *horizontal component,* it produces a tide-generating force on Earth. These tide-generating forces reach their maximum value at points on Earth's surface at a "latitude" of 45 degrees relative to the "equator" mentioned here (*blue arrows*).

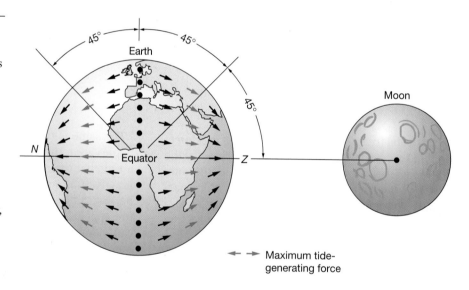

resultant force acts vertically relative to Earth's surface, as it does at the zenith and nadir (oriented *upward*) and along an "equator" connecting all points halfway between the zenith and nadir (oriented *downward*), it has no tide-generating effect (Figure 9–5). However, where the resultant force has a significant *horizontal component*—that is, tangent to Earth's surface—it aids in producing tidal bulges on Earth. Because there are no other large horizontal forces on Earth with which they must compete, these small, ever-present residual forces can push water across Earth's

surface. Thus, only the *horizontal component* of the resultant forces serves to generate the tides, creating what are known as the **tide-generating forces.** These tide-generating forces reach their maximum value at points on Earth's surface at a "latitude" of 45 degrees relative to the "equator" mentioned above (Figure 9–5).

The tide-generating forces cause water to be pushed into the regions located at the zenith and nadir (Figure 9–6). At regions on the side of Earth that face the moon directly, the gravitational force is greater than the required

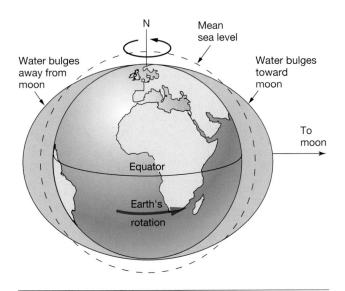

Figure 9–6 Idealized tidal bulges.

Assuming an ocean of uniform depth covering Earth and the moon aligned with the Equator, tide-generating forces produce two bulges in the ocean surface: One extends *toward* the moon and the other *away from* the moon. As Earth rotates, all points on its surface (except the poles) experience two high tides daily, because Earth rotates into and out of the two tidal bulges.

centripetal force and the excess causes a bulging of the ocean's surface toward the moon. Conversely, on the opposite side of Earth, the required centripetal force is greater than the gravitational force provided by the moon, producing a second bulge of water directed away from the moon. The tide-generating forces push the ocean's water into *bulges* of nearly equal size, with one bulge directed *toward* the moon and the other directed *away from* the moon (Figure 9–6).

Tidal Bulges: The Moon's Effect

It is easier to understand the tides if we assume an ideal Earth that has two tidal bulges, one toward the moon and one away from the moon, called the **lunar bulges**. It also simplifies the situation to assume an ideal ocean of uniform depth, with no friction between the seawater and the sea floor. Although this is a simplistic case, it does illustrate the principals involved in how tides on Earth are created. In fact, Sir Isaac Newton made these same simplifications when he first provided an explanation of Earth's tides.

Our ideal, uniformly deep ocean is modified only by the tide-generating forces that causes bulges on opposite sides of Earth—*one bulge facing the moon, and the other bulge oriented on the opposite side of Earth*, as shown in Figure 9–6. Let's assume that a stationary moon is aligned with Earth's Equator so that the maximum bulge will occur on the Equator on opposite sides of Earth. Let's also imagine that you are standing on the Equator.

Earth requires 24 hours for one complete rotation, so on the Equator you would experience two high tides each day. The time that elapses between high tides, the **tidal period**, would be 12 hours. If you moved to any latitude north or south of the Equator, you would experience a similar tidal period, but the high tides would be less high, because you would be at a lower point on the bulge rather than at its apex.

But high tides do not occur every 12 hours on Earth's surface. Instead, in most places on Earth, they occur every 12 hours 25 minutes. Twice this period is equivalent to a **lunar day**. The lunar day is the time that elapses between when the moon is on the meridian of an observer[3] and the next time that the moon is on that meridian. While a **solar day** is equal to 24 hours (which represents the time that elapses between when the sun is on the meridian of an observer and the next time that the sun is on that meridian), a lunar day is equal to 24 hours 50 minutes.[4] Where do the additional 50 minutes come from? During the time that Earth is making a full rotation in 24 hours, the moon has continued moving another 12.2 degrees to the east in its orbit around Earth (Figure 9–7). Thus, Earth must rotate an additional 50 minutes to "catch up" to the moon so that the moon is again on the meridian of our observer.

The difference between a solar day and a lunar day can be seen in some of the natural phenomena related to the tides. For example, alternating high tides are normally 50 minutes *later* each successive day. In addition, if you observe the time at which the moon rises on successive nights, you will see that it rises 50 minutes *later* each night.

Tidal Bulges: The Sun's Effect

Up until now, we have considered only the effect of the moon on Earth in its ability to produce tides. However, the sun also has an effect on the tides. In a similar fashion, the sun produces tidal bulges that exist on opposite sides of Earth, which are oriented *toward* and *away from* the sun. However, these **solar bulges** are about one-half the size of the lunar bulges, due to the fact that the sun is so much farther away from Earth (Figure 9–8). Despite its smaller size, the moon controls tides far more than the sun does, because the moon is so much closer to Earth.

From our perspective on land, the tides appear to cause water to come in toward shore and to go out away from shore (the rising and falling of the tides along the coast are called **flood tide** and **ebb tide**, respectively). Actually, *Earth's rotation is responsible for carrying us into and out of the tidal bulges.* The key to understanding the tides developed in the simplified model presented above is that

---

[3] When the moon is on the meridian of an observer, it is directly overhead from the observer on Earth

[4] A lunar day is exactly 24 hours, 50 minutes, 28 seconds long.

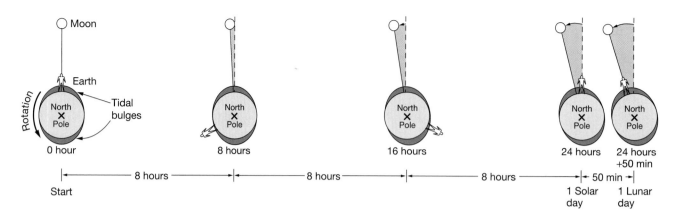

Figure 9–7 The lunar day.

A lunar day is the time that elapses between when the moon is directly overhead and the next time the moon is directly overhead. During one complete rotation of Earth (the 24-hour solar day), the moon moves eastward 12.2 degrees, and Earth must rotate an additional 50 minutes to place the moon in the exact same position overhead. Thus, a lunar day is 24 hours 50 minutes long.

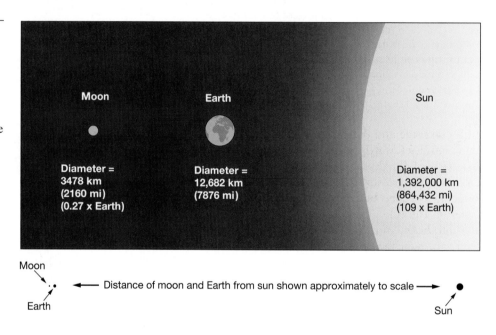

Figure 9–8 Relative sizes and distances of the Earth, moon, and sun.

Top: The diameter of the moon is roughly one-fourth that of Earth, while the diameter of the sun is 109 times the diameter of Earth. *Bottom*: Their relative sizes and distances are shown to scale.

Earth rotates *into* and *out of* these bulges, which are in fixed positions relative to the moon and the sun. In other words, Earth turns constantly inside a fluid envelope of ocean whose watery bulges are supported by the moon and the sun.

The Monthly Tidal Cycle

The monthly tidal cycle (called a *lunar cycle*) is the time it takes for one complete orbit of the moon around Earth. It takes $29\frac{1}{2}$ days to complete a lunar cycle. During this time, the phase of the moon changes dramatically. When the moon is between Earth and the sun, it cannot be seen for a few days, and it is then called the **new moon**. When the moon is on the side of Earth opposite the sun, its entire disk is brightly visible, and it is called a **full moon**. A **quarter moon**—which resembles a half moon when

viewed from Earth—results when the moon is at right angles to the sun relative to Earth.

Figure 9–9 shows the positions of the Earth, moon, and sun at various points during the $29\frac{1}{2}$-day lunar cycle. When the sun and moon are aligned, either with the moon between Earth and the sun (new moon; moon in *conjunction*) or with the moon on the side opposite the sun (full moon; moon in *opposition*), the tide-generating forces of the sun and moon combine (Figure 9–9, *top*). At this time, a large tidal range (the vertical difference between high and low tides) is experienced because of *constructive interference*[5] between the lunar and solar

---

[5] As mentioned in Chapter 8, constructive interference occurs when two waves (or, in this case, two tidal bulges) overlap crest to crest and trough to trough.

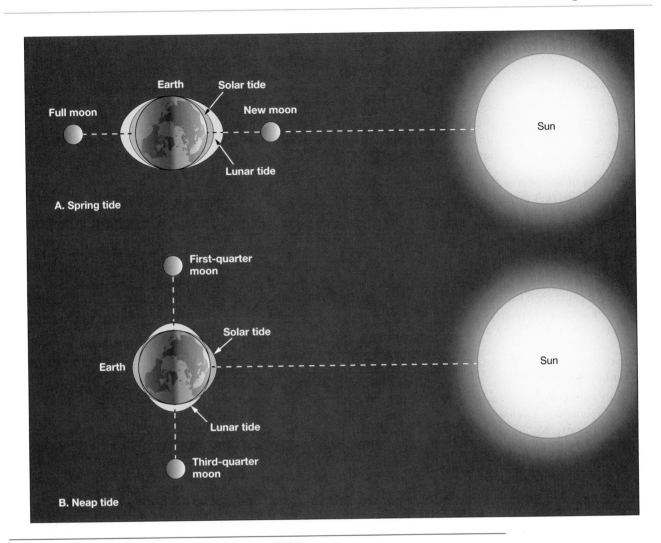

Figure 9–9 Earth–moon–sun positions and the tides.

Top: When the moon is in the new or full position, the tidal bulges created by the sun and moon are aligned, there is a large tidal range on Earth, and spring tides are experienced. *Bottom*: When the moon is in the first- or third-quarter position, the tidal bulges produced by the moon are at right angles to the bulges created by the sun. Tidal ranges are smaller and neap tides are experienced.

tidal bulges. During a large tidal range, there are very *high* high tides, and quite *low* low tides. This maximum tidal range is called a **spring** (*springen* = to rise up) **tide**,[6] because the tide is extremely large or "springs forth." When the Earth–moon–sun system is aligned, the moon is said to be in **syzygy** (*syzygia* = union).

When the moon is in either the first or third quarter[7] phase (Figure 9–9, *bottom*), the tide-generating force of the sun is working at right angles to the tide-generating

force of the moon. This results in *destructive interference*[8] between the lunar and solar tidal bulges. In this case, a small tidal range is experienced, characterized by *lower* high tides and *higher* low tides. This is called a **neap** (*nep* = scarcely or barely touching) **tide**,[9] and the moon is said to be in **quadrature** (*quadra* = four).

The time that elapses between successive spring tides (full moon and new moon) or neap tides (first quarter and third quarter) is one-half the monthly lunar cycle, which is about two weeks. The time that elapses between

---

[6] Spring tides have no connection with the spring season; they occur twice a month during the time when the Earth–moon–sun system is aligned.

[7] The third-quarter moon is often called the last-quarter moon, which is not to be confused with certain sports that have a fourth quarter.

---

[8] Destructive interference occurs when two waves (or, in this case, two tidal bulges) match up crest to trough and trough to crest.

[9] To help you remember a *neap* tide, think of it as one that has been "*nipped* in the bud," indicating a small tidal range.

a spring tide and a successive neap tide is one-quarter the monthly lunar cycle, which is about one week.

Other Factors

Besides the position of the moon and the sun around the spinning Earth, many other factors influence tides on Earth. Listed below are two of the most prominent factors that influence tides on Earth.

Declination of the Moon and Sun

Up to this point, we have assumed that the moon and sun always remain aligned over the Equator, but this is not the case. Most of the year, the sun and moon are either north or south of the Equator. This angular distance of the sun or moon above or below Earth's equatorial plane is called **declination**.

Earth revolves around the sun along an invisible ellipse in space. Imagine a plane in space that includes this ellipse. This plane is called the **ecliptic**. Earth is not perpendicular ("upright") on the ecliptic; it leans on its side by 23.5 degrees. This tilt is what causes Earth's seasons, as explained in Chapter 6.

To complicate matters further, the plane of the moon's orbit is at an angle of 5 degrees to the ecliptic. This 5 degrees supplements the 23.5 degrees of Earth's tilt, so that the declination of the moon's orbit relative to Earth's Equator can reach 28.5 degrees. The declination will change from 28.5 degrees south to 28.5 degrees north and back to 28.5 degrees south of the Equator in a period of one month.

The result of the declination is that tidal bulges rarely are aligned with the Equator. Instead, they occur mostly north and south of the Equator. Because the moon is the dominant force that creates tides in Earth's oceans, tidal bulges follow the moon as the moon shifts position throughout its monthly journey across the Equator, ranging from a maximum of 28.5 degrees north to a maximum of 28.5 degrees south of the Equator (Figure 9–10).

Effects of Elliptical Orbits

Additional considerations that affect the tide-generating force of the sun and moon on Earth are their shifting distances from Earth. Instead of a circular orbit around the sun, Earth follows an elliptical path (Figure 9–11). Thus, the distance between Earth and the sun varies by 2.5 percent [between 148.5 million kilometers (92.2 million miles) during the Northern Hemisphere winter and 152.2 million kilometers (94.5 million miles) during summer]. As Earth moves from its closest point (**perihelion**) to its most distant point (**aphelion**), the tidal range varies: Greater tidal ranges are experienced when Earth is near perihelion, which is in January each year.

Simultaneously, the Earth–moon distance varies by 8 percent [between 375,000 kilometers (233,000 miles) and 405,800 kilometers (252,000 miles)]. The moon moves from its closest point to Earth (**perigee**) to its most distant point (**apogee**). When the moon is near perigee, greater tidal ranges are experienced on Earth. The moon cycles between perigee, apogee, and back to perigee in a period of $27\frac{1}{2}$ days.

The elliptical orbits of Earth around the sun and the moon around Earth cause the orbital distances to change, thus affecting the tides. The net result is that spring tides have greater ranges during the Northern Hemisphere winter than in the summer, and spring tides have greater ranges when they coincide with perigee.

Idealized Tide Prediction

To predict tidal patterns for our idealized water-covered Earth, let us return to the effect of declination. The declination of the moon determines the position of the tidal bulges. The example illustrated in Figure 9–12 shows that the moon's declination is 28 degrees north of the Equator. Thus, the moon would be directly overhead at 28 degrees north latitude. If you stand at this latitude—for example, along a Florida beach near Cape

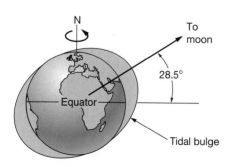

Figure 9–10 Maximum declination of tidal bulges from the Equator.

The center of the tidal bulges may lie at any latitude from the Equator to a maximum of 28.5 degrees on either side of the Equator, depending on the season of the year (solar angle) and the moon's position.

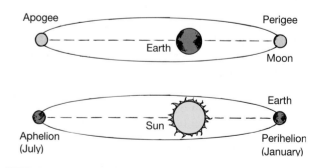

Figure 9–11 Effects of elliptical orbits.

Top: The moon moves from its closest point to Earth (perigee) to its most distant point (apogee). Greater tidal ranges are experienced when the moon is closest to Earth. *Bottom*: The Earth also moves from its closest point (perihelion) to its most distant point (aphelion). Greater tidal ranges are experienced when Earth is nearest the sun.

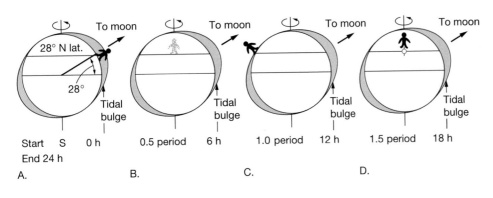

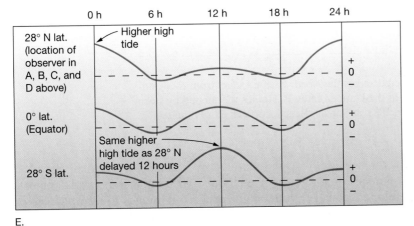

Figure 9–12 Predicted idealized tides.

A through **D.** Sequence showing the tide experienced at 28 degrees north latitude when the declination of the moon is 28 degrees north. **E.** Tide curves for 28 degrees north, 0 degrees, and 28 degrees south latitudes during the lunar day shown in the sequence above. The tide curves for 28 degrees north and 28 degrees south latitude show that the higher high tides occur 12 hours out of phase. All curves are for the same longitude.

Canaveral—your observation of the tides will be different from observations at the Equator.

Let's begin the observations when the moon is directly overhead. At this time, you observe that it is high tide (Figure 9–12A). Six lunar hours later (6 hours 12½ minutes solar time), you will experience low tide (Figure 9–12B). Six lunar hours later, it will be followed by another high tide, but it will be much lower than the initial high tide (Figure 9–12C). Six lunar hours later, another low tide is experienced (Figure 9–12D). Six hours later, at the end of a 24-lunar-hour period (24 hours and 50 minutes Earth time), you will have passed through a complete lunar-day cycle of high tide–low tide–high tide–low tide (two high tides and two low tides).

A representative curve of the type of tide you experience is shown in Figure 9–12E. Tide curves showing the heights of the same tides during one lunar day at the Equator and at 28 degrees south latitude are also provided. Note that the tide curves for 28 degrees north and

28 degrees south latitude have identically timed highs and lows, but the *higher* high tides and *lower* low tides are out of phase by 12 hours. This results from the fact that the bulges in the two hemispheres occur on opposite sides of Earth in relation to the moon. Table 9–1 shows a summary of characteristics of the tides on the idealized Earth.

Considering the great number of variables that are involved in predicting tides, how often are conditions right to produce the maximum tide-generating force? This occurs when the sun and moon are closest, and the moon is either between Earth and the sun (new moon) or opposite the sun (full moon), and when both the sun and moon have zero declination. This condition—which creates an absolute *maximum* spring tidal range—occurs only once every 1600 years; the next occurrence is predicted for the year 3300.

There are other times when conditions are right for producing large tide-generating forces. Are the spring tides

Table 9–1 Summary of characteristics of the tides on the idealized Earth.

- Any location will have two high tides and two low tides per lunar day.
- Neither the two high tides nor the two low tides are of the same height because of the declination of the moon and the sun (except for the rare occasions when the sun and moon are simultaneously above the Equator).
- Yearly and monthly cycles of tidal range are related to the changing distances of Earth from the moon and sun.
- Each week, there would be alternating spring and neap tides. Thus, in a lunar month, there are two spring tides and two neap tides.

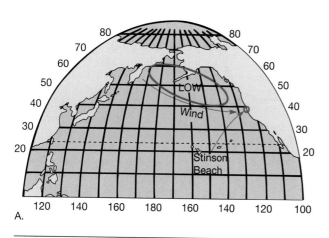

Figure 9–13 High tides of January 1983.

A. January 1983 storm winds blow uninterrupted across the North Pacific Ocean from the Kamchatka Peninsula of Russia to the west coast of the United States. **B.** Homes threatened by storm waves and unusually high tides on January 27, 1983, at Stinson Beach, north of San Francisco.

caused by these forces cause for concern? Yes, because even a less-optimized case of Earth–moon closeness that coincided with high spring tides caused substantial damage along the west coast of North America during early 1983. Slow-moving low-pressure cells developed over the Aleutian Islands in Alaska (Figure 9–13A). This caused strong northwest winds to blow across the ocean from Russia's Kamchatka Peninsula toward the U.S. coast, bringing with it many North Pacific storms. Averaging about 50 kilometers (30 miles) per hour, the winds produced a near fully developed 3-meter (10-foot) swell along the coast from Oregon to Baja California. The storm surge from this condition would have been trouble enough under average conditions, but the situation was made worse by two unusual events: high spring tides of 2.25 meters (7.4 feet), and a severe El Niño–Southern Oscillation event, which raised sea level by up to 20 centimeters (8 inches) due to thermal expansion of the warm water.

These unusually high spring tides occurred because Earth was still at its closest to the sun during its orbit (January 2), when the moon also was at its closest to Earth on January 28. Some of the largest waves came ashore January 26–28, causing over $100 million in damage (Figure 9–13B). At least a dozen lives were lost, 25 homes were destroyed, over 3500 homes were seriously damaged, and many commercial and municipal piers collapsed. With each such occurrence, more is learned about the foolishness of developing too close to the shore.

Tides in the Ocean

Tidal bulges are directed toward the moon and away from the moon on opposite sides of Earth. Thus, as Earth rotates, the bulges (more correctly called wave crests)

are separated by a distance of one-half Earth's circumference [about 20,000 kilometers (12,420 miles)]. Thus, one might expect the bulges to move across Earth at about 1600 kilometers (1000 miles) per hour.

However, the tides are an extreme example of shallow-water waves. As such, their speed is proportional to the water depth. For a tide wave to travel at 1600 kilometers (1000 miles) per hour, the ocean would have to be 22 kilometers (13.7 miles) deep! However, the average depth of the ocean is only 3.7 kilometers (2.3 miles). Thus, tidal bulges move as *forced waves*, with their velocity determined by ocean depth.

Based on the average ocean depth, the average speed at which tide waves can travel across the open ocean is only about 700 kilometers (435 miles) per hour. Thus, the idealized bulges that simply point toward and away from the tide-generating body cannot exist because they cannot keep up with the rotational speed of Earth. Instead, they break up into a number of cells.

In the open ocean, the crests and troughs of the tide wave actually rotate around a point near the center of each cell. This point is called an **amphidromic** (*amphi* = around, *dromus* = running) **point** and there is essentially no tidal range here. Radiating from this point, however, are **cotidal** (*co* = with, *tidal* = tide) **lines** that connect points along which high tide will occur simultaneously. Figure 9–14 shows cotidal lines, labeled to indicate the time of high tide in hours after the moon crosses the Greenwich Meridian.

The times shown on Figure 9–14 indicate that the rotation of the tide wave is counterclockwise in the Northern Hemisphere and clockwise in the Southern Hemisphere. The wave makes one complete rotation during the tidal period. The size of the cells is limited by

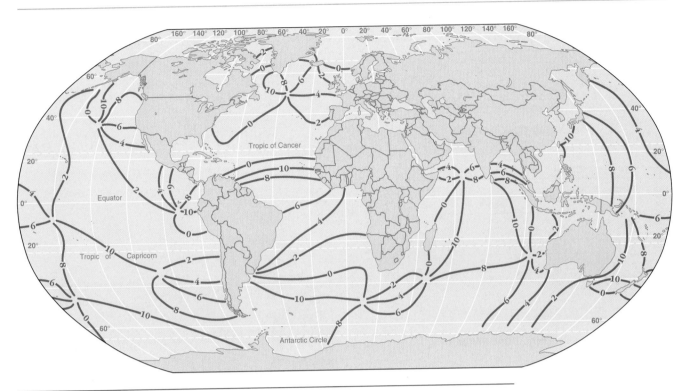

Figure 9–14 Cotidal map of the world.

Contour lines indicate times of the main lunar daily high tide in lunar hours after the moon has crossed the Greenwich Meridian (0 degrees longitude). Tidal ranges generally increase with increasing distance along cotidal lines away from the amphidromic points. Where cotidal lines terminate at both ends in amphidromic points, maximum tidal range will be near the midpoints of the lines.

the fact that the tide wave must make one complete rotation during the period of the tide (usually 12 lunar hours).

Within an amphidromic cell, low tide is six hours behind high tide. For example, if high tide is occurring along the cotidal line labeled "10," low tide occurs simultaneously along the cotidal line labeled "4."

Let's also consider the effect of the continents, which interrupt the free movement of the tidal bulges across the ideal unobstructed ocean surface considered earlier. The ocean basins between continents have freestanding waves set up within them. Their character modifies the forced astronomical tide waves that develop within the basin.

Studies of tidal phenomena reveal that over 150 different factors affect the tides at a particular coast. Even though the major factors have been discussed, there are simply too many other factors that influence the tides than can be adequately addressed here. For instance, high tide rarely occurs at the time the moon is at its highest point in the sky. In fact, the elapsed time between the passing of the moon across the meridian of an observer and the occurrence of high tides varies from place to place and is the result of the many factors that determine the tidal characteristics at a given location.

Tidal Patterns

Ideally, most areas on Earth should experience two high tides and two low tides of unequal heights during a lunar day. But due to modification from various depths, sizes, and shapes of ocean basins, tides in many parts of the world exhibit different patterns. The three tidal patterns are *diurnal* (*diurnal* = daily), *semidiurnal* (*semi* = twice, *diurnal* = daily), and *mixed*.[10] These are shown in Figure 9–15.

A **diurnal tidal pattern** has a single high and low water each lunar day. These tides are common in shallow inland seas such as the Gulf of Mexico and along the coast of Southeast Asia. Such tides have a tidal period of 24 hours 50 minutes.

A **semidiurnal tidal pattern** has two high and two low tides each lunar day. The heights of successive high tides and successive low tides are approximately the same.[11]

---

[10] Sometimes a *mixed* tidal pattern is referred to as *mixed semidiurnal.*

[11] Since tides are always growing higher or lower at any location due to the spring-neap tide sequence, successive high tides and successive low tides can never be *exactly* the same at any location.

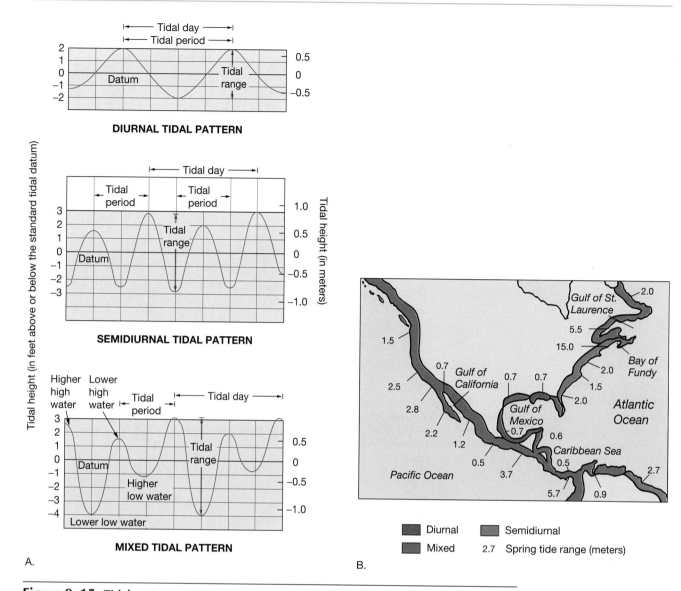

Figure 9–15 Tidal patterns.
A. Tide graphs showing two tidal days of the three tidal patterns. *Diurnal* (*top*) shows only one high and one low each lunar day. *Semidiurnal* (*middle*) shows two highs and lows of approximately equal heights during each lunar day. *Mixed* (*bottom*) shows two highs and lows of unequal heights during each lunar day. **B.** The types of tides observed along North and Central American coasts, showing the spring tidal range (in meters).

Among other places, semidiurnal tides are common along the Atlantic Coast of the United States. The tidal period is 12 hours 25 minutes.

A **mixed tidal pattern** may have characteristics of both diurnal and semidiurnal tides. Successive high tides and/or low tides will have significantly different heights, called *diurnal inequalities*. Mixed tides commonly have a tidal period of 12 hours 25 minutes, which is a semidiurnal characteristic, but they may also possess diurnal periods. This is the tide that is most common throughout the world and the type that is found along the U.S. Pacific Coast.

Figure 9–16 shows examples of monthly tidal curves for various coastal locations. Note that even though a tide at any particular location can be identified as having a single tidal pattern, it still may pass through stages of one or both of the other tidal patterns. Typically, however, the tidal pattern for a location will remain similar throughout the year.

Tides in Bays

When tide waves enter coastal waters, they are subject to reflection. In some cases, the standing waves set up by reflections may have periods near that of the forced tide wave. Under such conditions, constructive interference can produce significant increases in the tidal range.

Nova Scotia's **Bay of Fundy** is such a place, and it is here that the largest tidal range in the world is found.

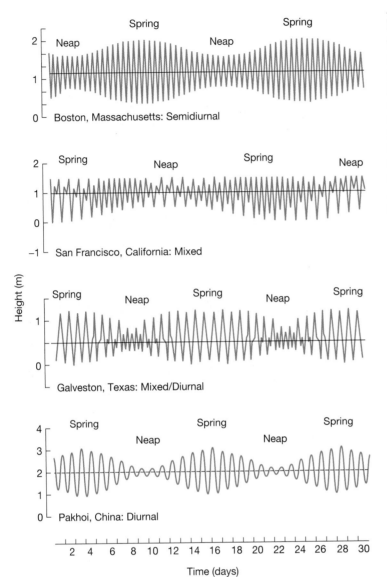

Figure 9–16 Examples of tidal curves.

Monthly tidal curves. *Top:* Boston, Massachusetts, showing semidiurnal tidal pattern. *Upper middle:* San Francisco, California, showing mixed tidal pattern. *Lower middle:* Galveston, Texas, showing mixed tidal pattern with strong diurnal tendencies. *Bottom:* Pakhoi, China, showing diurnal tidal pattern.

With a length of 258 kilometers (160 miles), the Bay of Fundy has a wide opening into the Atlantic Ocean. The Bay of Fundy splits into two narrow basins at its northern end, Chignecto Bay and Minas Basin (Figure 9–17A). In the Bay of Fundy, the period of free oscillation—the oscillation that occurs when a body is displaced and then released—is very nearly that of the tidal period. The resulting constructive interference—along with the narrowing and shoaling of the bay to the north—causes a buildup of tidal energy in the northern end of the bay. In addition, the bay is a right-curving bay in the Northern Hemisphere, so the Coriolis effect adds to the tidal range. These factors combine to produce maximum tidal ranges in the extreme northern end of Minas Basin.

During maximum spring tide conditions, the tidal range at the mouth of the bay (where it opens to the sea) is only about 2 meters (6.6 feet). However, the tidal range increases progressively from the mouth of the bay northward. In the northern end of Minas Basin, the maximum spring tidal range is 17 meters (56 feet). This extreme tidal range leaves boats high and dry during low tide (Figures 9–17B and 9–17C).

Coastal Tidal Currents

The current that accompanies the slowly turning tide crest in a Northern Hemisphere basin will turn in a counterclockwise direction, producing a **rotary current** in the open portion of the basin. Because of increased effects of friction in nearshore shoaling waters, the rotary current is changed to an alternating or **reversing current** that moves into and out of restricted passages along a coast.

Velocities of the rotary currents in the open ocean are usually well below 1 kilometer (0.6 mile) per hour. However, the reversing currents are of the greatest concern to coastal navigators, for in this setting they are known to reach velocities up to 44 kilometers (28 miles) per hour in restricted channels such as between islands of coastal British Columbia.

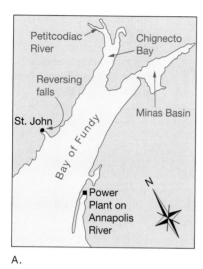

A.

B.

C.

Figure 9–17 Tidal range in the Bay of Fundy.
A. The largest tidal range in the world occurs in Nova Scotia's Bay of Fundy. Even thought the maximum spring tidal range at the mouth of the bay is only 2 meters (6.6 feet), amplification of tidal energy causes a maximum tidal range at the northern end of Minas Basin of 17 meters (56 feet), often stranding ships (**B** and **C**).

 Box 9–1
Tidal Bores: Boring Waves These Are Not!

A **tidal bore** (*bore* = crest or wave) is a wall of water that moves up certain low-lying rivers due to an incoming tide. Because it is a wave created by the tides, it may be considered to be a *true* tidal wave. Tidal bores form when an incoming tide rushes up a river, developing a steep forward slope due to resistance to the tide's advance by the river, which is flowing in the opposite direction (Figure 9C). This creates a tidal bore (Figure 9D), which may reach heights of 5 meters (16.4 feet) or more. Tidal bores move at speeds up to 22 kilometers (14 miles) per hour—faster than most people can run for any length of time—and sometimes trap unsuspecting people who venture onto tidal flats during low tide.

The conditions necessary for development of tidal bores include a large tidal range and a low-lying coastal river. When large tidal ranges occur, there is a greater potential for larg-

er tidal bores. Although tidal bores do not attain the size of some waves in the surf zone, tidal bores have been successfully surfed. The advantage over ocean waves is that a surfer can obtain a very long ride as the bore travels many kilometers upriver. Unfortunately, it is about half a day before the next tidal bore occurs, since it is associated with the incoming high tide, which occurs only twice a day.

The Amazon River probably possesses the longest estuary that is affected by oceanic tides. Tides can be measured as far as 800 kilometers (500 miles) from the river's mouth, although the effects are quite small at this distance. Tidal bores near the mouth of the Amazon River can be up to 5 meters (16.4 feet) in height and are locally called *pororocas* (waterfalls). Many other rivers also have notable tidal bores. These include the Chientang River in China, which has the largest

(continued)

tidal bores in the world, often reaching 8 meters (26 feet) in height; the Petitcodiac River in New Brunswick, Canada; the River Seine in France; the Trent River in England; and Cook Inlet near Anchorage, Alaska, where the largest tidal bore in the United States can be found.

It might seem logical that since the Bay of Fundy has the world's largest tidal range, it should have a large tidal bore. However, the tidal bore in the Bay of Fundy rarely exceeds 1 meter (3.3 feet), mostly because the bay is so wide. Still, the adventure of tidal bore rafting in the Bay of Fundy is promoted as a draw for tourists (Figure 9E).

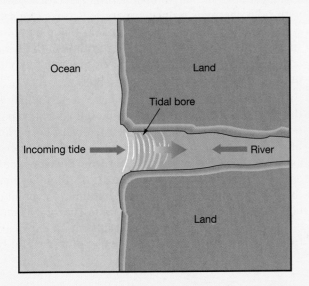

Figure 9C How a tidal bore develops.

Figure 9E Tidal bore rafting.

Figure 9D A tidal bore near Chignecto Bay, New Brunswick, Canada.

Reversing currents also exist in the mouths of bays (and some rivers) associated with the daily flow into and out of these bodies of water due to daily tidal changes. Figure 9–18 shows that a **flood current** is produced when water rushes into a bay (or river) with an incoming high tide; conversely, an **ebb current** is produced when water drains out of a bay (or river) due to an approaching low tide. No currents occur for several minutes during either **high slack water** (which occurs at the peak of each high tide) or during **low slack water** (at the peak of each low tide).

Reversing currents in bays can sometimes reach speeds of 40 kilometers (25 miles) per hour, producing a navigation hazard for ships. The benefits of these coastal tidal currents is that their daily flow into and out of bays often keeps sediment from closing off the bay. Also, the daily inflow and outflow of water keeps the bay supplied with new seawater and a fresh supply of ocean nutrients.

Even in deeper waters of the ocean, tidal currents can be significant. For example, strong tidal currents were encountered shortly after the discovery of the remains of the *Titanic* at a depth of 3795 meters (12,448 feet) on the continental slope south of Newfoundland's Grand Banks. These tidal currents were so strong that they forced re-

searchers to abandon the use of the camera-equipped tethered robot, *Jason Jr.* This experience demonstrated that significant tidal currents can occur at all depths in the ocean.

Some Considerations of Tidal Power

The most obvious benefit of generating electrical power with tidal energy is reduced operating costs compared with conventional thermal power plants that require fossil fuels or radioactive isotopes. Even though the initial cost of building a tidal power-generating plant may be higher, there would be no ongoing fuel bill.

A negative consideration involves the periodicity of the tides. Power could be generated only during a portion of a 24-hour day. People operate on a solar period, but tides operate on a lunar period; thus, the energy available through generating power from the tides would coincide with need only part of the time. To get around this, power would have to be distributed to the point of need somewhere on Earth at the moment, which could be a great distance away. This results in an expensive transmission problem. Alternatively, power could be

Figure 9–18 Reversing tidal currents.

As water drains out of a bay due to a low tide, outbound ebb currents are produced. Conversely, incoming high tides produce flood currents. No currents occur during either high slack water (*HSW*) or low slack water (*LSW*).

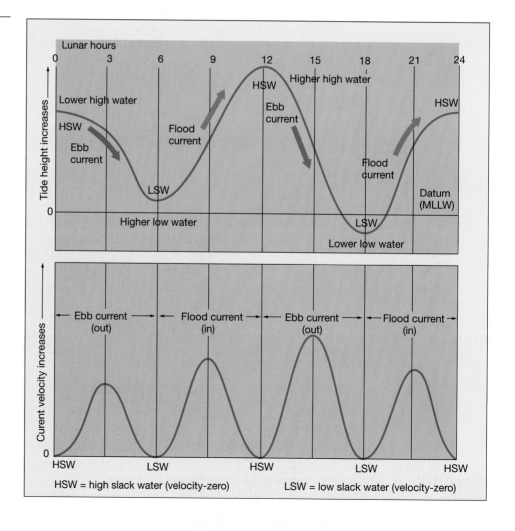

Box 9–2
Grunion: Doing What Comes Naturally on the Beach

Along the beaches of southern California and Baja California from March through September, a very unusual fish-spawning behavior can be observed. Shortly after the maximum spring tide has occurred, small silvery fish come ashore to bury their fertilized eggs in the sand. They are the **grunion** (*Leuresthes tenuis*), slender little fish 12 to 15 centimeters (4.7 to 6 inches) in length. Grunion are the only marine fish in the world that come completely out of water to spawn. The name "grunion" comes from the Spanish *gruñón*, which means "grunter." The early Spanish settlers gave the fish this name because of the faint noise they make during spawning.

The tidal pattern that occurs along southern California and Baja California beaches is a mixed tide. On most tidal days (24 hours and 50 minutes), there are two high and two low tides. There is usually a significant difference in the heights of the two high tides that occur each day. During the summer months, the higher high tide occurs at night. As the higher high tides become higher each night while the maximum spring-tide range is approached, sand is eroded from the beach (Figure 9F). After the maximum height of the spring tide has occurred, the higher high tide that occurs each night is a little lower than the one of the previous night. During this sequence of decreasing heights of the tides as neap-tide conditions approach, sand is deposited on the beach. The grunion depend greatly on this pattern of beach sand deposition and erosion for the success of their spawning.

Grunion spawn only after each night's higher high tide has peaked on the three or four nights following the night of the occurrence of the highest spring high tide. This behavior assures that the eggs will be covered deeply by sand deposited by the receding higher high tides each succeeding night. The fertilized eggs buried in the sand are ready to hatch nine days after spawning. By this time, another spring tide is approaching, and the higher high tide that occurs each night will be higher than that of the previous night. This condition causes the beach sand to erode, exposing the eggs to agita-

tion by waves that break ever higher on the beach. The eggs hatch about three minutes after being freed in the water. Tests done in laboratories have shown that the grunion eggs will not hatch until agitated in a manner that simulates the agitation of the eroding waves.

The spawning begins as the grunion come ashore immediately following an appropriate high tide, and it may last from one to three hours. Spawning activity usually peaks about an hour after its start and may last an additional 30 minutes to an hour. Thousands of fish may be on the beach at this time. During a run, the females, which are larger than the males, move high on the beach. If no males are near, a female may return to the water without depositing her eggs. In the presence of males, she will drill her tail into the semifluid sand until only her head is visible. The female continues to twist, depositing her eggs 5 to 7 centimeters (2 to 3 inches) below the surface.

The male curls around the female's body and deposits his milt against it (Figure 9G). The milt runs down the body of the female to fertilize the eggs. When the spawning is completed, both fish return to the water with the next wave.

Larger females are capable of producing up to 3000 eggs for each series of spawning runs, which are separated by the two-week period between spring-tide occurrences. As soon as the eggs are deposited, another group of eggs begins to form within the female. They will be deposited during the next spring tide run. Early in the spawning season, only older fish spawn, but by May, even the one-year-old females are in spawning condition.

Young grunion grow rapidly and are about 12 centimeters (5 inches) long when they are a year old and ready for their first spawning. They usually live two or three years, but four-year-olds have been recovered. The age of a grunion can be determined by the scales. After growing rapidly during the first year, they grow very slowly. There is no growth

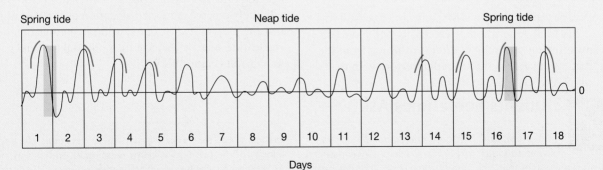

 Grunion deposit eggs in beach sand during early stages of the ebb of higher high tides on the three or four days following maximum spring tidal range.

Flood tides erode sand and free grunion eggs during higher high tide as maximum spring tidal range is approached.

Maximum spring tidal range

Figure 9F Grunion and the tidal cycle.

(continued)

at all during the 6-month spawning season, which causes marks to form on each scale that can be used to identify the grunion's age.

It is not known how the grunion are able to time their spawning behavior so precisely with the tides. Some investi-

gators believe the grunion are able to sense very small changes in the hydrostatic pressure caused by the changing level of the water associated with rising and falling sea level due to the tides. Certainly, a very dependable detection mechanism keeps the grunion accurately informed of the tidal conditions, because their survival depends on a spawning behavior precisely tuned to tidal motions.

Figure 9G Grunion spawning.

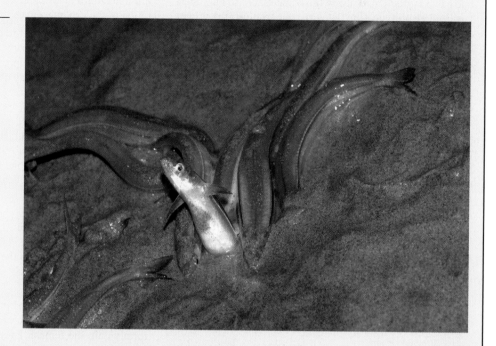

stored, but this presents a large and expensive technical problem.

To generate electricity effectively, electrical turbines (generators) need to run at a constant speed, which is difficult due to the variable flow of tidal currents in two directions (flood tide and ebb tide). This requires special design considerations that would allow both advancing and receding water to spin the turbine blades.

Another negative concern is that any use of coastal waters for the tidal generation of electricity will have certain unwanted environmental effects resulting from the modification of current flow. It will interfere with many

traditional uses of coastal waters, such as transportation and fishing.

Using tidal power as a source of renewable energy has waned in recent years because of the availability of inexpensive fossil fuels. However, there is a growing realization that these inexpensive fossil fuels will eventually run out. Whether or not tidal power stations are ever constructed on a large scale, this potential source of energy will likely receive increased attention as the cost of generating electricity by conventional means increases. Worldwide, many sites have the potential for tidal power generation (Figure 9–19).

? Students Sometimes Ask...

If the moon and Earth are gravitationally attracted to one another, why don't the two bodies collide?
Most of us are taught that "the moon orbits Earth," but it is not quite that simple. The two bodies actually rotate around a common center of mass [called a *barycenter* and located 1600 kilometers (1000 miles) beneath Earth's surface] as they travel through space. This can be visualized by imagining Earth and its moon as ends of a sledgehammer, flung into space, tumbling slowly end over end. It is because of this mutual orbit that the moon and Earth don't collide. As Earth accelerates toward the moon (and the moon accelerates toward Earth), the Earth accelerates

forward where the moon *was*, not where it *is going*. The angular velocity of the two bodies keeps them apart. In this way, orbits are established which keep objects at more-or-less fixed distances. This is also why the planets don't crash into the sun, even though all planets experience a strong gravitational attraction toward it.

I know that the sun is much more massive than the moon. Why is the sun's effect on the tide-generating force only about half that of the moon's?
Since the sun is 27 million times more massive than the moon, it should, solely on the basis of comparative mass, have a tide-

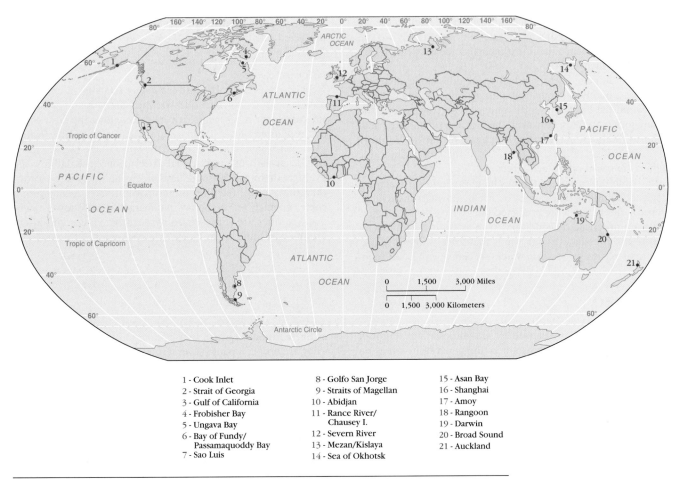

Figure 9–19 Sites with high potential for tidal power generation.

Twenty-one locations occur worldwide where tidal ranges are great enough to create potential for generating electricity. Where a large area exists for storing water behind a dam, a tidal range of 3 meters (10 feet) would suffice. Where smaller storage areas exist, greater tidal ranges would be required.

1 - Cook Inlet
2 - Strait of Georgia
3 - Gulf of California
4 - Frobisher Bay
5 - Ungava Bay
6 - Bay of Fundy/
  Passamaquoddy Bay
7 - Sao Luis
8 - Golfo San Jorge
9 - Straits of Magellan
10 - Abidjan
11 - Rance River/
  Chausey I.
12 - Severn River
13 - Mezan/Kislaya
14 - Sea of Okhotsk
15 - Asan Bay
16 - Shanghai
17 - Amoy
18 - Rangoon
19 - Darwin
20 - Broad Sound
21 - Auckland

generating force 27 million times greater than that of the moon. However, a more significant factor must be considered: the distances involved. The sun is 390 times farther from Earth than the moon (see the scale in Figure 9–8). As mentioned previously, tide-generating forces vary inversely as the *cube* of the distance between objects. Thus, the tide-generating force is reduced by the cube of 390, or about *59 million times* compared with that of the moon. These conditions result in the sun's tide-generating force being $^{27}/_{59}$ that of the moon, or 46 percent (about one-half).

Are there also tides in other objects, such as lakes and swimming pools?

Because the moon and the sun act on all objects that have the ability to flow, there most certainly are tides in many other objects. Tides in lakes are fairly easy to measure; tides affect water levels in wells; tides in swimming pools exist, but are very small. In fact, there are even extremely tiny tidal bulges in a glass of water! Of greater significance, however, are tides in the atmosphere and the "solid" Earth. Tides in the atmosphere—called *atmospheric tides*—can be miles high. The tides inside the Earth's interior—called *Earth tides*—cause a slight but measurable stretching of Earth's crust, typically only a few centimeters high.

I've heard of a blue moon. Is the moon really blue then?

No. It's just a phrase that has gained popularity and is synonymous with a rather unlikely occurrence. A blue moon is the second full moon of any calendar month. The $29^{1}/_{2}$-day lunar cycle must fall entirely within a 30- or 31-day month. Because the divisions between our calendar months were determined arbitrarily, a blue moon has no special significance, other than it occurs only once every 2.72 years. At that rate, it's certainly less common than a month of Sundays!

What are tropical tides?

Differences between successive high tides and succesive low tides occur each lunar day. Since these differences occur within a period of one day, they are called diurnal (daily) inequalities. These inequalities are at their greatest when the moon is at its maximum declination, and such tides are called *tropical tides* because the moon is over one of the tropic regions. When the moon is over the Equator, the inequality experienced within the tidal pattern is minimal; tides with this characteristic are called *equatorial tides*.

I know that tide tables are published years in advance. How can tide tables be so accurate?

Tide tables are handy references to allow one to determine high and low tidal occurrences along a coast. Often available free of charge at boating, fishing, and surf shops, tide tables indicate not only the exact time of all the high and low tides for a year, they also list the height of the tide within 0.25 centimeter (a tenth of an inch). What's more, tide tables are rarely wrong! Indeed, it does seem quite remarkable that the tides can be predicted with such accuracy.

Because so many factors influence the tides along a particular coast, a purely mathematical solution for predicting the tides is beyond the limits of marine science: There are simply too many local variables. However, there is a predictable periodicity to the tides, based mostly on an 18.6-year cycle that is, not surprisingly, related to the orbit of the moon. Thus, tide tables are based on historical records.

Still, tide tables can sometimes be in error. Local meteorological conditions that cannot be predicted years in advance cause a difference in the predicted tides. These *meteorological tides* are produced by conditions such as storm surge, low atmospheric pressure, or high winds that drive water on shore. Thermal expansion of warm water associated with ENSO conditions will also create higher sea level that is different than the predicted tide tables.

Some tide tables I've seen list negative tides. How can there ever be a negative tide?
Negative tides occur because the *datum* (starting point or reference point from which tides are measured) is an average of the tides over many years. For instance, along the west coast of the United States, the datum is Mean Lower Low Water (MLLW), which is the average of the *lower* of the two low tides that occur daily in a mixed tidal pattern. Since the datum is an average, there will be some days when the tide will be lower than the average (similar to the distribution of exam scores, some of which will be below the average). These lower-than-average tides are represented by negative values and often are the best times to visit local tide pool areas.

Summary

The tides of Earth are caused by gravitational attraction of the moon and sun, which produce tide-generating forces on Earth. According to a simplified model of tides that assumes an ocean of uniform depth and ignores the effects of friction, small horizontal forces (the tide-generating forces) tend to push water into two bulges on opposite sides of Earth: One directly facing the tide-generating body, and the other directly opposite. These tide-generating forces cause water to be pushed into bulges at regions on Earth facing toward and away from both the moon and the sun.

Despite its vastly smaller size, the moon has about twice the tide-generating effect of the sun because it is so much closer to Earth. Since the tidal bulges due to the moon's gravity (the lunar bulges) are dominant, the tides observed on Earth have periods dominated by lunar motions. However, the tides are modified by the changing position of the solar bulges relative to the lunar bulges. Earth's rotation is responsible for carrying us into and out of the various tidal bulges.

If Earth were a uniform sphere covered with an ocean of uniform depth, tides would be easily predicted. For most places on Earth, the time period between successive high tides would be 12 hours 25 minutes, or half a lunar day. The $29\frac{1}{2}$-day monthly tidal cycle would consist of tides with maximum tidal range (spring tides) that would occur each new moon and full moon, and tides with minimum range (neap tides) that would occur during the first- and third-quarter phases of the moon.

Since the moon may have a declination up to 28.5 degrees north or south of the Equator, and the sun is directly over the Equator only two times per year, tidal bulges usually are located to create two high tides of unequal height per lunar day. The same conditions apply for low tides. Tidal ranges are greater when Earth is nearest the sun and moon.

When friction and the true shape of ocean basins are included into the dynamics of tides, it is found that nowhere in the ocean are tides deep-water waves because the ocean basins are so shallow relative to the long wavelength of the wave crests created by tidal phenomena. The two bulges on opposite sides of Earth produced by the simplified model can not in fact exist. They are broken up into several rotating patterns where ocean tides rotate counterclockwise around an amphidromic point, a point of zero tidal range, in the Northern Hemisphere, and clockwise in the Southern Hemisphere. Tides on Earth are also influenced by a host of other factors, including the effects of continents on Earth and other coastal factors.

The three basic types of tidal patterns observed on Earth are diurnal (a single high and low tide each lunar day), semidiurnal (two high and two low tides each lunar day), and mixed (with characteristics of both). Mixed tidal patterns are usually dominated by semidiurnal periods and display significant diurnal inequality. Mixed tidal patterns are the most common type of tidal pattern in the world.

The effects of constructive interference and the shoaling and narrowing of coastal bays can be seen in the extreme tidal range at the northern end of Nova Scotia's Bay of Fundy. Here, the largest tidal range in the world [a maximum spring tidal range of 17 meters (56 feet)] is found.

Tidal currents follow a rotary pattern in open-ocean basins but are converted to reversing currents along continental margins. The maximum velocity of reversing currents occurs during flood and ebb currents when the water is halfway between high and low slack waters. Tidal bores are true tidal waves (a wave produced by the tides) that occur in certain rivers and bays due to an incoming high tide.

The tides are important to many marine organisms. For instance, grunion—small silvery fish that inhabit waters along the west coast of North America—time their spawning cycle in direct correspondence to the pattern of the tides.

Because tides can be used to generate power without need for fossil or nuclear fuel, the possibility of constructing such generating plants has always attracted interest. However, there are some significant drawbacks to creating a successful power plant using tidal power. Still, there are many sites worldwide that have the potential for tidal power generation.

Key Terms

Amphidromic point (p. 286)

Aphelion (p. 284)

Apogee (p. 284)

Bay of Fundy (p. 288)

Centripetal force (p. 279)

Cotidal line (p. 286)

Declination (p. 284)

Diurnal tidal pattern (p. 287)

Ebb current (p. 292)

Ebb tide (p. 281)

Ecliptic (p. 284)

Flood current (p. 292)

Flood tide (p. 281)

Full moon (p. 282)

Gravitational force (p. 278)

Grunion (*Leuresthes tenuis*) (p. 293)

High slack water (p. 292)

Low slack water (p. 292)

Lunar bulges (p. 281)

Lunar day (p. 281)

Mixed tidal pattern (p. 288)

Nadir (p. 278)

Neap tide (p. 283)

New moon (p. 282)

Newton, Sir Isaac (p. 277)

Perigee (p. 284)

Perihelion (p. 284)

Quadrature (p. 283)

Quarter moon (p. 282)

Resultant force (p. 279)

Reversing current (p. 289)

Rotary current (p. 289)

Semidiurnal tidal pattern (p. 287)

Solar bulges (p. 281)

Solar day (p. 281)

Spring tide (p. 283)

Syzygy (p. 283)

Tidal bore (p. 290)

Tidal period (p. 281)

Tidal range (p. 276)

Tide-generating force (p. 280)

Tides (p. 277)

Zenith (p. 278)

Questions And Exercises

1. Explain how a tidal power plant works, using as an example an estuary that has a mixed tidal pattern. Why does potential for usable tidal energy increase with an increase in the tidal range?

2. Explain why the sun's influence on Earth's tides is only 46 percent that of the moon's, even though the sun exerts a gravitational force on Earth 177 times greater than that of the moon.

3. Why is a lunar day 24 hours 50 minutes long, while a solar day is 24 hours long?

4. Which is more technically correct: The tide comes in and goes out; or Earth rotates into and out of the tidal bulges. Why?

5. From memory, draw the positions of the Earth–moon–sun system during a complete monthly tidal cycle. Indicate the tide conditions experienced on Earth, the phases of the moon, the time between those phases, and syzygy and quadrature.

6. Explain why the maximum tidal range (spring tide) occurs during new and full moon phases and the minimum tidal range (neap tide) at first-quarter and third-quarter moons.

7. What is declination? Discuss the degree of declination of the moon and sun relative to Earth's Equator. What are the effects of declination of the moon and sun on the tides?

8. Diagram the Earth–moon system's orbit about the sun. Label the positions on the orbit at which the moon and sun are closest to and farthest from Earth, stating the terms used to identify them. Discuss the effects of the moon's and Earth's positions on Earth's tides.

9. Are tides considered deep-water waves anywhere in the ocean? Why or why not?

10. Describe the number of high and low tides in a lunar day, the period, and any inequality of the following tidal patterns: diurnal, semidiurnal, and mixed.

11. Discuss factors that help produce the world's greatest tidal range in the Bay of Fundy.

12. Discuss the difference between rotary and reversing tidal currents.

13. Of flood current, ebb current, high slack water, and low slack water, when is the best time to enter a bay by boat? When is the best time to navigate in a shallow, rocky harbor?

14. Discuss at least two positive and two negative factors related to tidal power generation.

15. Describe the spawning cycle of grunion, indicating the relationship between tidal phenomena, where grunion lay their eggs, and the movement of sand on the beach.

16. Observe the moon from a reference location every night at about the same time for two weeks. Keep track of your observations about the phase of the moon and its position in the sky. Then, compare these to the reported tides in your area. How do the two compare?

References

Bowditch, N. 1995. *The American practical navigator: An epitome of navigation* (1995 edition). Bethesda, MD: Defense Mapping Agency Hydrographic/Topographic Center.

Clancy, E. P. 1969. *The tides: Pulse of the earth*. Garden City, NY: Doubleday.

Comins, N. F. 1993. *What if the moon didn't exist? Voyages to Earth that might have been*. New York: HarperCollins.

Defant, A. 1958. *Ebb and flow: The tides of earth, air, and water*. Ann Arbor, MI: University of Michigan Press.

Gill, A. E. 1982. *Atmosphere and ocean dynamics*. International Geophysics Series, Vol. 30. Orlando, FL: Academic Press.

Hauge, C. J. 1972. Tides, currents, and waves. *California Geology* 25:7, 147–160.

Le Provost, C., Bennett, A. F., and Cartwright, D. E. 1995. Ocean tides for and from TOPEX/POSEIDON. *Science* 267:5198, 639–642.

Pond, S., and Pickard, G. L. 1978. *Introductory dynamic oceanography*. Oxford: Pergamon Press.

Railsback, L. B. 1991. A model for teaching the dynamic theory of tides. *Journal of Geological Education* 39:1 15–18.

Sverdrup, H. U., Johnson, M. W., and Fleming, R. H. 1942. Renewal 1970. *The oceans: Their physics, chemistry, and biology*. Englewood Cliffs, NJ: Prentice-Hall.

von Arx, W. S. 1962. *An introduction to physical oceanography*. Reading, MA: Addison-Wesley.

Suggested Reading

Sea Frontiers

Canove, P. 1989. The reclamation of Holland. 35:3, 154–164. A comprehensive history of how the Dutch have reclaimed and protected coastal lands from the threat of rising water.

Holloway, T. 1989. Eling tide mill. 35:2, 114–119. The tide mill at Eling Toll Bridge near Southampton, England, has been in existence since at least 1086 A.D. It is now a working museum, and the article describes its history and how it works.

Sobey, J. C. 1982. What is sea level? 28:3, 136–142. The role of tides and other factors in changing the level of the ocean surface are discussed.

Zerbe, W. B. 1973. Alexander and the bore. 19:4, 203–208. An account of Alexander the Great's encounters with a tidal bore on the Indus River.

Scientific American

Goldreich, P. 1972. Tides and the earth-moon system. 226:4, 42–57. The tide-generating force of the sun and moon on Earth and the effect of transfer of angular momentum from Earth to the moon as a result of tidal friction are discussed. Also considered are theories of lunar origin.

Greenberg, D. A. 1987. Modeling tidal power. 257:3, 128–131. The effects of using the large tidal ranges experienced in the Bay of Fundy to generate electricity are modeled on a computer, and the idea of building a tidal power plant across the bay is explored.

Lynch, D. K. 1982. Tidal bores. 247:4, 146–157. Tidal bores can be spectacular walls of water rushing up rivers when the tide rises.

Oceanography on the Web

Visit the *Essentials of Oceanography* home page for on-line resources for this chapter. There you will find an on-line study guide with review exercises, and links to oceanography sites to further your exploration of the topics in this chapter. *Essentials of Oceanography* is at: **http://www.prenhall.com/thurman** (click on the Table of Contents menu and select this chapter).

CHAPTER 10
THE COAST: BEACHES AND SHORELINE PROCESSES

The Ultimate Protection:
The National Flood Insurance Program (NFIP)

In 1965, Hurricane Betsy devastated the east coast of the United States, killing 75 people and damaging property in excess of $1.4 million. The federal government launched a massive relief effort, but it was realized that the government would not be able to continue to lend such aid indefinitely. The federal government's response was to develop the National Flood Insurance Program (NFIP) in 1968. Its original goal was to encourage people to build safely inland of flood-threatened areas in return for federally subsidized flood insurance (in case some very unusual circumstance resulted in flood damage to property). The objective was to reduce deaths and property damage from flooding to the point that federal disaster relief would become less and less necessary. However, by the time the lobbyists for the banking, construction, and real estate industries had their concerns alleviated by members of Congress, the program turned into a vast subsidy that supported risky overdevelopment of the U.S. coastal region.

During the decade prior to passage of the program, hurricanes killed 186 people and did $2.2 billion in damage. Following the passage of the flood insurance program, coastal development boomed. Consequently, during the 10 years following passage of the program, there were 411 deaths and $4.7 billion in property damage from coastal storms—a 121 percent increase in deaths and a 114 percent increase in damage! One of the reasons for such drastic increases was that the NFIP had actually *encouraged* overdevelopment of the coastal areas that the original program was designed to protect, placing more people and buildings in harm's way (Figure 10A).

Problems with the NFIP continue to worsen. Studies conducted by the General Accounting Office indicate that the program operated from 1978 to 1987 at a $652 mil-

Figure 10A A house built too close to the beach.

lion deficit, which was made up by federal taxpayers. Studies into the soundness of the program showed that many insurance premiums should be increased by 800 percent. However, only new properties were subject to these

299

higher rates. Rates have increased in recent years, but the government still pays out billions with each flood disaster in the form of grants and loan subsidies—to cover damage that would not have happened if the U.S. government had a sensible policy on coastal development.

In addition to these benefits, which help owners replace structures built in locations that are clearly in danger of storm damage, the federal government subsidizes development of these areas by paying most of the cost of infrastructure development, including the construction of water systems, sewage treatment facilities, highways, and bridges. Not only does the NFIP cover flood damage, it also covers many other types of coastal damage unrelated to flooding, such as damage due to sea cliff erosion. Further, the federal government assumes from 50 to 70 percent of the cost of erosion-control programs. It is clear that U.S. taxpayers are subsidizing imprudent development decisions that benefit a small number of individuals who have chosen to live in areas prone to damage.

In 1982, Congress passed the Coastal Barrier Rezoning Act. It excludes construction in undeveloped coastal areas from receiving federal assistance. The act also encourages states to develop "setback" regulations that require construction to be set back at some distance from the shore, realistically incorporating the threat of storm damage into all coastal building programs.

After years of research, hearings, and investigations, the House of Representatives overwhelmingly passed a bill in 1991 to reform the program. A similar bill was introduced shortly thereafter in the Senate, but the bill died in 1992 because of special-interest pressure.

Rising sea level could exacerbate the problem. Many scientists—including members of the National Academy of Sciences—are warning of possible accelerated sea level rise in coming years. If this happens, the problem with existing homes and businesses could become immense in as few as 10 years. Still, insurance through the NFIP remains a real bargain. For instance, the owners of a $200,000 coastal home can insure their property by purchasing the equivalent of $18,000 in private insurance for about $1,000 from the NFIP.

What can be done to further reverse the direction of such unsound public policy? What general guidelines should be followed in developing a national program for conservative use of our coastal areas to preserve them for everyone for generations to come? Of course, politics is a major hurdle. However, an important first step is to understand processes operating in the coastal environment that make it one of the most dynamic places on the planet.

Throughout our history, humans have been attracted to the coastal regions of the world for their moderate climate, seafood, transportation, recreational opportunities, and commercial benefits. In the United States, our migration toward the Atlantic, Pacific, and Gulf coasts has continued; 80 percent of the U.S. population now lives within easy access of these regions. The coastal regions are enduring increasing stress that threatens to damage this national resource to the point that future generations may not have the opportunity to enjoy it as many people do today.

The coastal region is constantly changing. The mechanism involved in producing such a dynamic environment is ocean waves, which break along the shoreline and release their energy from distant storms. Waves crash along most shorelines more than 10,000 times a day. This wave activity causes many changes that occur hourly, daily, weekly, monthly, seasonally, and yearly. Every wave that breaks along the shore releases energy, causing erosion in some areas and deposition in others.

In this chapter, we shall examine the major features of the seacoast and shore and the processes that modify them. We shall also feature some examples of how people interfere with these processes, creating hazards to themselves and to the environment.

The Coastal Region

Traveling by boat in the ocean, as you approach land, you encounter the shore. The **shore** is a zone that lies between the lowest tide level (low tide) and the highest elevation on land that is affected by storm waves. From this point landward is the **coast**, which extends inland as far as ocean-related features can be found (Figure 10–1). The width of the shore varies between a few meters and hundreds of meters. The width of the coast may vary from less than a kilometer (0.6 mile) to many tens of kilometers. The **coastline**, which marks the boundary between the shore and the coast, is a line that connects points at which the highest effective wave action takes place.

Beach Terminology

The beach profile shown in Figure 10–1 shows some features characteristic of the shore. The shore is divided into the **backshore**, above the high-tide shoreline, which is covered with water only during storms, and the **foreshore**,[1] that portion exposed at low tide and submerged at high tide. The **shoreline** migrates back and forth with the tide

---

[1]The foreshore is often referred to as the *intertidal* or *littoral* (*litoralis* = the shore) *zone*.

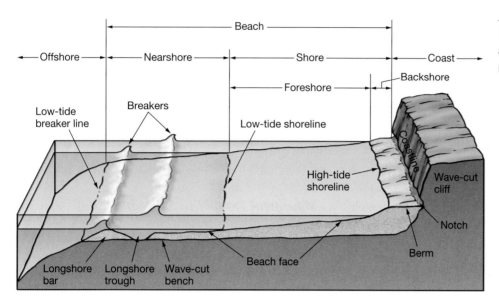

Figure 10–1 Landforms and terminology of coastal regions.

and is the water's edge. Never exposed to the atmosphere, but affected by waves that touch bottom, is the **nearshore**, which extends seaward to the low-tide breaker line. Beyond the low-tide breakers is the **offshore** zone, where depths are such that waves rarely affect the bottom.

A **beach** is a deposit of the shore area. It consists of wave-worked sediment that moves along the **wave-cut bench** (a flat, wave-eroded surface). A beach may continue from the coastline across the nearshore region to the line of breakers. Thus, the beach is defined as the entire active area of a coast that experiences changes due to breaking waves. Beaches might best be thought of as material in transit along the shoreline. The area of the beach above the shoreline is often referred to as the **recreational beach**.

The **berm** is the dry, gently sloping region at the foot of the coastal cliffs or dunes. The berm is often composed of sand, making it a favorite place of beachgoers. The **beach face** is the wet, sloping surface that extends from the berm to the shoreline. It is more fully exposed during low tide, and is also known as the **low tide terrace**. Offshore beyond the beach face is one or more sand bars that parallel the coast called **longshore bars**. A longshore bar may not always be present throughout the year, but when one is present, it can often be exposed during extremely low tides. The presence of a longshore bar can often cause waves to initially break. Separating the longshore bar from the beach face is a **longshore trough**.

Beach Composition

Beaches are composed of whatever material is locally available. In areas where this material—sediment—is provided by erosion of beach cliffs or nearby coastal mountains, beaches will be composed of mineral particles from these rocks and may be relatively coarse in texture.

If the sediment is provided primarily by rivers that drain lowland areas, sediment that reaches the coastal regions normally will be finer in texture. Often, mud flats develop along the shore because only tiny clay-sized and silt-sized particles are emptied into the ocean.

In low-relief, low-latitude areas such as southern Florida, where there are no mountains or other sources of rock-forming minerals nearby, most beach material is derived from the remains of the organisms that live in the coastal waters. Beaches in these areas are composed predominantly of shell fragments and the remains of microscopic organisms.

Many beaches on volcanic islands in the open ocean are composed of dark fragments of the basaltic lava that comprise the islands, or of coarse debris from coral reefs that develop around islands in low latitudes. Thus, around the world, there are beaches composed of crushed coral, beaches made of black (or even green) volcanic material, beaches composed of shell fragments, beaches made of quartz sand, beaches made of rock fragments, and even artificial beaches composed of scrap metal dumped at the beach.

Whatever the composition of the beach material, the material that comprises the beach does not stay in one place. Rather, it is constantly being moved by waves that crash along the shoreline.

Movement of Sand on the Beach

Material at the beach is in continual motion. The movement of sand on the beach occurs both perpendicular to the shoreline (*toward* and *away from* shore) and parallel to the shoreline (often referred to as *up-coast* and *down-coast*).

Movement Perpendicular to Shoreline
The mechanism involved in moving sand perpendicular to the shoreline is breaking waves. As each wave breaks, water rushes

up the beach face toward the berm. This movement of water from breaking waves is called **swash**. Some of the water soaks into the beach and eventually returns to the ocean. However, most of the water drains away from shore as **backwash**. As the water moves down the beach face as backwash, the next wave breaks and its swash comes in over the top of the previous wave's backwash.

Even standing in ankle-deep water at the shoreline, you can see that sediment is transported up and down the beach face perpendicular to the shoreline by swash and backwash. The process that dominates the transport system—either swash or backwash—determines whether sand is deposited or eroded from the berm.

In *light wave activity* (characterized by less energetic waves), the swash from breaking waves rushes up the beach face and much of it soaks into the beach. Consequently, there is a reduced amount of backwash. Thus, the swash dominates the transport system. Since the swash moves sediment up the beach face, it causes a net movement of the sand up the beach face toward the berm. This results in deposition of sand on the berm, causing the berm to become wide and well developed.

In *heavy wave activity* (characterized by high-energy waves), the beach is saturated with water from previous waves, causing very little of the swash to soak into the beach. There is much backwash, which is carrying a great deal of sediment with it as it drains down the beach face. Adding to this effect is the fact that when a wave breaks, the incoming swash comes *on top of* the previous wave's backwash, effectively protecting the beach from the incoming swash. For these reasons, the backwash dominates the transport system. This causes sand to be eroded from the berm, which results in a much narrower berm.

During heavy wave activity, where does the sand from the berm go? The shallow nature of orbital motion in waves prohibits the sand from moving very far offshore. Thus, the sand accumulates just beyond where the waves are breaking and forms one or more offshore sand bars (the longshore bars).

A balance exists at most beaches between light and heavy wave activity. As light and heavy wave activity alternate at a beach throughout the seasons, the characteristics of the beach will change (Table 10–1). For instance, light wave activity produces a wide berm at the expense of the longshore bar, which creates a **summer-time beach**: one that has a wide sandy berm and an overall steep beach face. Conversely, heavy wave activity destroys the berm and builds prominent longshore bars, creating a **wintertime beach**: one that has a narrow rocky berm and an overall flattened beach face. It is interesting to note that a wide berm that takes several months to build can be destroyed in just a few hours by high-energy wintertime storm waves. The differences between summertime and wintertime beaches can be dramatic (Figure 10–2).

Movement Parallel to Shoreline At the same time that movement occurs perpendicular to shore, movement parallel to shoreline also occurs. The mechanism involved in moving sand parallel to the shoreline is refracting (bending) waves. As discussed in Chapter 8, refraction causes waves to bend and line up *nearly* parallel to the shore. The slight angle that the waves approach shore is important in creating a current that moves sediment along the shoreline.

As waves strike the shore at a slight angle (Figure 10–3A), they set up a longshore movement of water, called a **longshore current**. This current moves parallel to the shore between the shoreline and the outer edge of the surf zone. With each breaking wave, the swash moves up onto the exposed beach at a slight angle. Gravity pulls the backwash straight down the beach face. As a result, water moves in a zigzag fashion along the shore. Longshore currents have speeds that have been measured at up to 4 kilometers (2.5 miles) per hour. The speed of longshore currents increase with increasing beach slope, with increasing angle at which breakers arrive at the beach, with increasing wave height, and with increasing frequency of waves.

Many beachgoers are inadvertently carried by the longshore current while they play within the waves in the surf zone. Often, they are surprised when they discover that they have been carried far from where they initially entered the water. This frequent occurrence demonstrates that longshore currents are strong enough to move people as well as a vast amount of sand in a zigzag fashion along the shore.

Longshore drift or **longshore transport** refers to the movement of *sediment* in a zigzag fashion by the longshore current (Figure 10–3B). Because a longshore current affects the entire surf zone, the resulting longshore drift also works within the entire surf zone as well.

Table 10–1   Characteristics of beaches affected by light and heavy wave activity

	Light wave activity	Heavy wave activity
Berm/longshore bars	Berm is built at the expense of the longshore bars	Longshore bars are built at the expense of the berm
Wave energy	Low wave energy (non-storm conditions)	High wave energy (storm conditions)
Time span	Long time span (weeks or months)	Short time span (hours or days)
Characteristics	Creates summertime beach: sandy, wide berm, steep beach face	Creates wintertime beach: rocky, narrow berm, flattened beach face

A.

B.

The amount of longshore drift in any coastal region is determined by an equilibrium between erosional and depositional forces. Any interference with the movement of sediment along the shore will destroy this equilibrium, causing a new erosional and depositional pattern to form. However, studies of the annual amount of longshore drift that occurs along coastal regions is on the order of millions of tons of sediment that pass by any particular point of a shore during a year.

A river is an interesting analog to the coastal zone: Both move water *and* sediment from one area (*upstream*) to another (*downstream*). The analogy is so close, in fact, that the beach has often been referred to as a "river of sand." The differences between a river and a beach are that a longshore current moves in a zigzag fashion (rivers flow mostly in a turbulent, swirling fashion), and the direction of flow of longshore currents along a shoreline can change (rivers always flow in the same direction, that is, downhill). The change of direction of the longshore current along a shoreline is due to the seasonal change in direction at which waves approach the beach. However, there is some consistency with which longshore currents move: The average direction of annual longshore current is *southward along both the Atlantic and Pacific shores of the United States.*

Figure 10–3 Longshore current and longshore drift.
A. Waves approaching the beach at a slight angle near Oceanside, California, producing a longshore current moving toward the right of the photo. **B.** A longshore current, caused by refracting waves, moves water in a zigzag fashion along the shoreline. This causes a net movement of sand grains, also in a zigzag fashion, from upstream to downstream ends.

A.

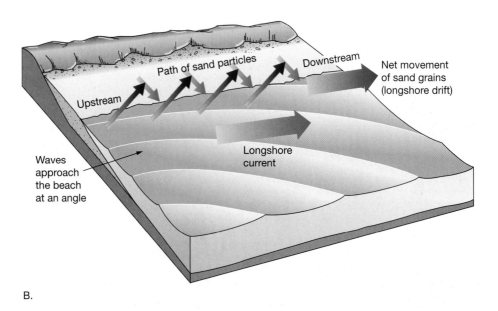

B.

Erosional- and Depositional-Type Shores

As waves beat against the shore, they cause erosion. This erosion produces sediment that is transported along the shore and deposited in areas where wave energy is low. Even though all shores experience some degree of both erosion and deposition, shores can often be identified primarily as one type or the other. Shores often described as primarily **erosional** have well-developed cliffs and generally are in areas where tectonic uplift of the coast is occurring. For example, much of the U.S. Pacific Coast is classified as erosional.

In contrast, along the U.S. southeastern Atlantic Coast and the Gulf Coast, sand deposits and offshore barrier islands are common. These coasts are experiencing a gradual subsidence of the shore. Although such shores are classified as primarily **depositional**, erosion can be a major problem when human development interferes with natural coastal processes. This can cause easily eroded deposits to be washed away.

Features of Erosional-Type Shores

Because of wave refraction, wave energy is concentrated on any **headlands** that jut out from the continent, while the amount of energy reaching the shore in bays is reduced. As the waves concentrate their energy on the headlands, erosion occurs and the shoreline retreats. Some of these erosional features are shown in Figure 10–4.

Box 10–1
Warning: Rip Currents... Do You Know What To Do?

In the development of longshore currents, water from breaking waves washes up onto the shore and then runs back into the ocean. This backwash of water generally finds its way into the open ocean as a flow of water across the ocean bottom and is commonly referred to as "sheet flow." However, some of this water flows back in surface **rip currents**. These currents typically occur perpendicular to the coast and move in the direction opposite to breaking waves. They happen where topographic lows in longshore bars or other conditions allow their formation.

Rip currents may be less than 25 meters (80 feet) wide and can attain velocities of 7 to 8 kilometers (4 to 5 miles) per hour—faster than most people can swim for any length of time [in fact, it is useless to swim for long against a current stronger than about 2 kilometers (1.2 miles) per hour]. Rip currents do not travel far from shore before they break up. If a light-to-moderate swell is breaking, numerous rip currents may develop, which are moderate in size and velocity.

A heavy swell will usually produce fewer, more concentrated rips. They can often be recognized by the way that they interfere with incoming waves, or by the sediment that is often held in suspension within the rip current (lighter areas in offshore water in Figure 10B).

The rip currents that occur during heavy swell are a significant hazard to coastal swimmers. However, swimmers who realize that they are caught in a rip current can escape by swimming parallel to the shore for a short distance (simply swimming out of the narrow rip current) and then riding the next wave in toward the beach. However, even excellent swimmers who panic or try to fight the current by swimming directly into it are eventually overcome by exhaustion and may drown. Even though most beaches have warning signs posted when rip currents occur in a coastal area and are patrolled by lifeguards, many people lose their lives each year because of rip currents.

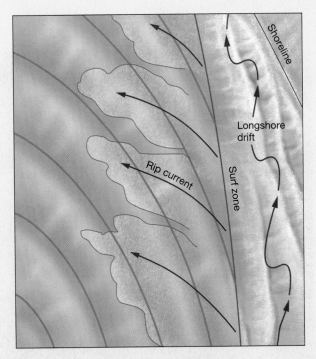

Figure 10B Rip currents.
Photograph and matching drawing of rip currents.

A **wave-cut cliff** is produced by wave action pounding relentlessly away at its base. The cliff develops as the upper portions collapse after being undermined by wave action. The undermining may be evident in the form of a notch at the base of the cliff, which may be characterized by **sea caves**—caves formed by wave attack along the base of a wave-cut cliff.

Further wave action eroding the softer portions of the rock outcrops may develop the caves into openings running through the headlands, called **sea arches**. Some sea arches are large enough to allow a boat to maneuver safely through them. Continued erosion and crumbling of the tops of these arches will produce **sea stacks** (Figure 10–5). Such remnants rise from the relatively smooth wave-cut

Figure 10–4 Features of erosional coasts.

Diagrammatic view of some features characteristic of erosional coasts.

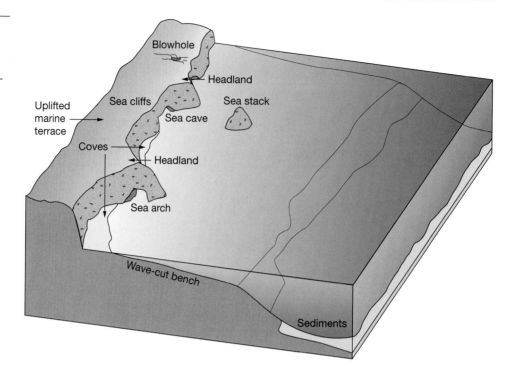

Figure 10–5 Sea arch and sea stack along the coast of Iceland.

When the roof of a sea arch (*left*) collapses, a sea stack (*right*) is formed.

bench cut into the bedrock by wave erosion. Uplift of the wave-cut bench creates a gently sloping **marine terrace** above sea level (Figure 10–6).

Rates of coastal erosion are influenced by the degree of exposure to waves, the amount of tidal range, and the composition of the coastal bedrock. Regardless of the erosion rate, all coastal regions follow the same developmental path. As long as there is no change in the elevation of the landmass relative to the ocean surface, the cliffs will continue to erode and retreat until the beaches widen sufficiently to prevent waves from reaching them. The eroded material is carried from the high-energy areas and deposited in the low-energy areas.

Figure 10–6 Wave-cut bench and marine terrace.
A wave-cut bench is exposed at low tide along the California coast at Bolinas Point near San Francisco. An elevated wave-cut bench, called a marine terrace, is shown at right.

Features of Depositional-Type Shores

Coastal erosion of sea cliffs produces large amounts of sediment. Adding to this sediment is additional material that is derived from the erosion of rocks and delivered to the shore by rivers. All of this sediment is distributed along the continental margin by wave activity.

Numerous depositional features exist that are partially or wholly separated from the shore. These features are primarily deposits of sand moved by longshore drift but are also modified by other coastal processes. Some of these features are shown in Figure 10–7 and are briefly described here.

A **spit** (*spit* = spine) is a linear ridge of sediment attached to land at one end. The other end points in the direction of longshore drift and ends in the open water. Spits are simply extensions of beaches into the deeper water near the mouth of a bay. The open-water end of the spit normally curves into the bay as a result of current action.

If tidal currents, or currents from river runoff, are too weak to keep the mouth of the bay open, the spit may eventually extend across the bay and connect to the mainland, thus completely cutting off the bay from the open ocean. The spit then has become a **bay barrier** or **bay-mouth bar** (Figure 10–8A). Although bay barriers are a buildup of sand usually less than 1 meter (3.3 feet) above sea level, bay barriers have become areas where permanent buildings have unwisely been constructed.

A **tombolo** (*tombolo* = mound) is a sand ridge that connects an island or sea stack to the mainland (Figure 10–8B). Tombolos can also be found between two adjacent islands. Formed in the wave-energy shadow of an island, tombolos are usually aligned perpendicular to the average direction of wave approach.

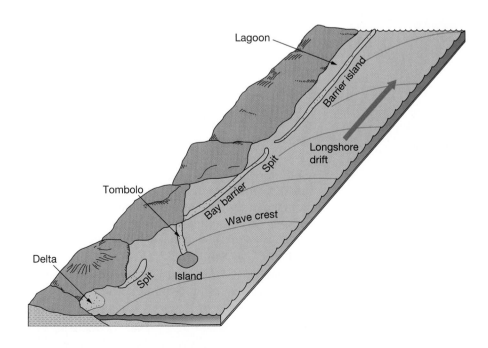

Figure 10–7 Features of depositional coasts.
Diagrammatic view of some features characteristic of depositional coasts.

A.

B.

Figure 10–8 Coastal depositional features.
A. Barrier coast, spit, and bay barrier along the coast of Martha's Vineyard, Massachusetts. **B.** Tombolo in the Gulf of California along Baja California.

Barrier Islands Extremely long offshore deposits of sand lying parallel to the coast are called **barrier islands** (Figure 10–9). They are well named, for they do constitute a barrier to storm waves that otherwise would severely assault the shore. Their origin is complex, and several explanations have been proposed for their existence. However, it appears that many barrier islands developed during the worldwide rise in sea level that began with the melting of the most recent major glaciers some 18,000 years ago.

Barrier islands are nearly continuous along the Atlantic Coast of the United States. They extend around Florida and along the Gulf of Mexico coast, where they exist well south of the Mexican border. Barrier islands may exceed 100 kilometers (60 miles) in length and have widths of several kilometers. They are separated from the mainland by a lagoon. Examples of barrier islands are Fire Island off the New York coast, North Carolina's Outer Banks, and Padre Island off the coast of Texas.

A typical barrier island has the following physiographic features, shown in Figure 10–10A. From the ocean landward, they are (1) ocean beach, (2) dunes, (3) barrier flat, (4) high salt marsh, (5) low salt marsh, and (6) lagoon between the barrier island and the mainland. Let's look briefly at each portion.

The **ocean beach** is typical of the beach environment discussed earlier. During the summer, as gentle waves carry sand to the beach, it widens and becomes steeper. Higher-energy winter waves carry sand offshore and produce a narrow, gently sloping beach.

Winds blow sand inland during dry periods to produce coastal **dunes**, which are stabilized by dune grasses. These plants can withstand salt spray and burial by sand. Dunes are the lagoon's primary protection against excessive flooding during storm-driven high tides. Numerous passes exist through the dunes, particularly along the southeastern Atlantic Coast, where dunes are less well developed than to the north.

Behind the dunes, the **barrier flat** forms as the result of deposition of sand driven through the passes during storms. These flats are quickly colonized by grasses. During storms, these low barriers are washed over by seawater. If for some reason the frequency of overwash by storms decreases, the plants will undergo natural biological succession, with the grasses successively being replaced by thickets, woodlands, and even forests.

Salt marshes typically lie inland of the barrier flat. They are divided into the *low marsh*, extending from about mean sea level to the high neap-tide line, and the *high marsh*, extending to the highest spring-tide line. Biologically, the low marsh is by far the most productive part of the salt marsh.

New marshland is formed as overwash carries sediment into the lagoon, filing portions so they become intermittently exposed by the tides. Marshes may be poorly

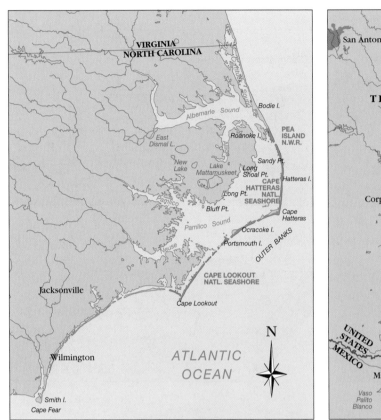

A.

B.

C.

Figure 10–9 Barrier islands.
A. Barrier islands of North Carolina's Outer Banks **B.** Barrier islands along the south Texas coast.
C. A portion of a heavily developed barrier island off the New Jersey coast.

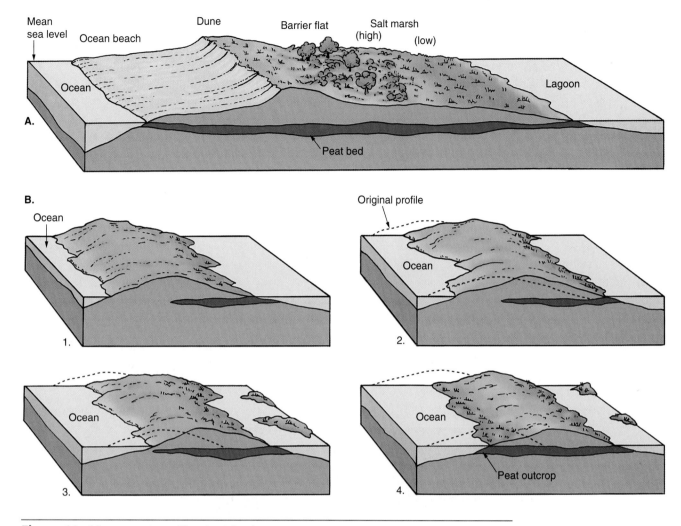

Figure 10–10 Formation of barrier islands.
A. Diagrammatic view through a barrier island, showing the major physiographic zones of a barrier island. The peat bed represents ancient marsh environments. MSL is mean sea level. **B.** Sequence (1–4) showing how a barrier island migrates and exposes peat deposits that have been covered by the island as it migrates toward the mainland in response to rising sea level.

developed on parts of the island that are far from flood-tide inlets. Their development is greatly restricted on barrier islands where people perform artificial dune enhancement and fill inlets, activities that prevent overwashing and flooding.

In response to a gradual rise of sea level that is occurring along the eastern North American coast, barrier islands are migrating landward (Figure 10–10B). The barrier island in essence slowly turns like a tractor's tread, moving toward the mainland as sea level rises. This is clearly visible to those who build structures on these islands. Other evidence for such migration can be seen in peat deposits, which are remnants of old marshes that lie beneath the barrier islands. Since the only barrier island environment in which peat forms is the salt marsh, the island must have moved inland over previous marsh development.

Deltas Some rivers carry more sediment than can be distributed by longshore currents. Such rivers dump sed-

iment at their mouths, developing a **delta** (*delta* = triangular) deposit. One of the largest such features on Earth is that produced by the Mississippi River where it empties into the Gulf of Mexico (Figure 10–11A). Deltas are fertile, flat, low-lying areas that are subject to periodic flooding.

Delta formation begins when a river has filled its mouth with sediment. Once the delta forms, it grows through the distribution of sediment by forming **distributaries**, which are branching channels that radiate out over the delta (Figures 10–11A). Distributaries lengthen as they deposit sediment and produce finger-like extensions to the delta. When the fingers get too long, they become choked with sediment. At this point, a flood may easily shift the distributary course and provide sediment to low-lying areas between the fingers. Where depositional processes dominate over coastal erosion and transportation processes, a branching "bird's foot" Mississippi-type delta results.

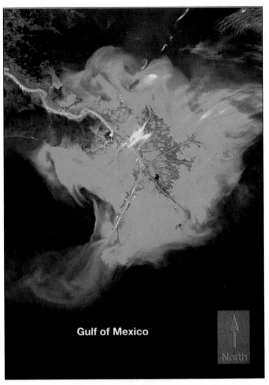

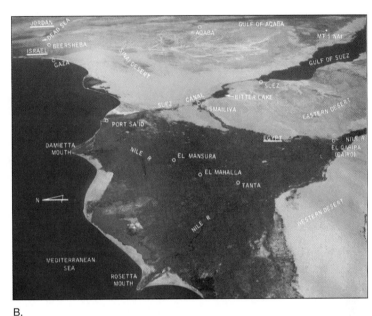

A.

B.

Figure 10–11 Deltas.
A. False-color infrared image of the branching "bird's foot" structure of the Mississippi River Delta. Red color is vegetation on land; light blue color is suspended sediment within the water.
B. The relatively smooth, curved shoreline of the Nile River Delta dominates this view. View is looking toward the southeast, with the Mediterranean Sea to the left.

Where erosion and transportation processes dominate (instead of deposition), a delta shoreline will be smoothed to a gentle curve, like that of the Nile River Delta in Egypt (Figure 10–11B). The Nile Delta is presently eroding due to the intervention of humans: Sediment is being entrapped behind the Aswan High Dam, completed in 1964. Prior to the building of the dam, generous volumes of sediment were carried by the Nile into the Mediterranean Sea.[2]

Erosion is only one negative effect of this dam. The problem of reduced sediment availability may become very significant along the eastern end of the delta, where subsidence (sinking) has occurred at a rate of 0.5 centimeter (0.2 inch) per year for 7500 years. At the present rate of subsidence, the sea may move inland 30 meters (100 feet) during the next century.

Beach Compartments **Beach compartments** consist of three components: a series of rivers that supply sand to a beach; the beach itself, where sand is moving due to longshore transport; and submarine canyons that lie offshore, where sand is drained away from the beach. Among the coasts that display beach compartments, one of the best examples is the coast of southern California, which contains four separate beach compartments (Figure 10–12).

Within an individual beach compartment, sand is supplied to the beach primarily by rivers, although coastal erosion can add a significant portion of the sand supply to the beach compartment. The sand is moved with the longshore current, in the case of southern California, to the south (Figure 10–12). This causes wider beaches to occur near the southern (*downstream*) end of each beach compartment. Although some sand is washed offshore along the way, most of the sand on the beach eventually moves near a head of a submarine canyon, where it is diverted away from the beach and onto the ocean floor. Once the sand is removed from the coastal environment, it is lost from the beach forever. To the south of this beach compartment, the beaches will not have very much sand, and will consequently be thin and rocky. However, the next beach compartment's rivers add their sediment, and, farther downstream, the beach widens and has an abundance of sand until this beach compartment's sand is also moved down a submarine canyon.

---

[2] On a positive note, however, this new, human-induced erosion may have made possible the discovery of the ruins of the ancient city of Alexandria beneath the Mediterranean Sea at the edge of the delta.

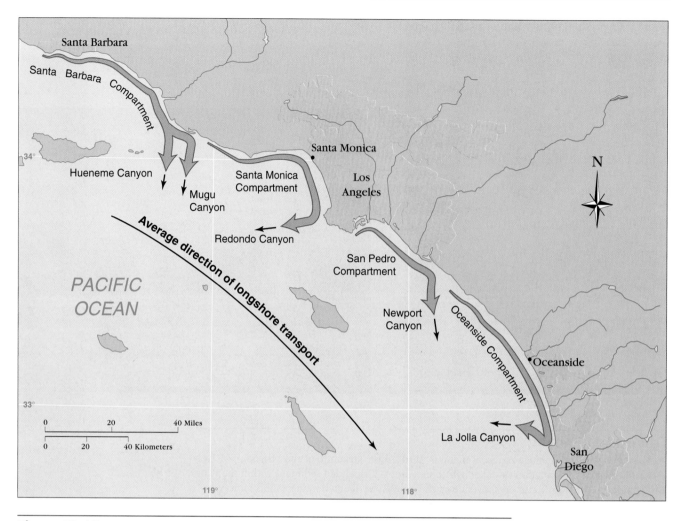

Figure 10–12 Beach compartments.
Southern California has several beach compartments, which include rivers that bring sediment to the beach, the beach that experiences longshore transport, and the submarine canyons that remove sand from the beaches. Average longshore transport is toward the south.

Human activities have altered the natural system of beach compartments. When a dam is built along one of the rivers that supplies sediment to the beach compartment, it deprives the beach of the sand that it would receive. Lining rivers with concrete for flood control further reduces the sediment load delivered to coastal regions. The process of longshore transport continues to operate (in a manner analogous to a broom sweeping the shoreline's sand into the submarine canyons), causing the beaches to become narrower. If all the rivers are blocked in this way, the beaches will nearly disappear. The interruption of sand supply and the resulting narrowing of beaches is called **beach starvation**.

What can be done to avoid the problem of beach starvation in beach compartments? One obvious solution is to eliminate the dams. Since most dams are useful for protecting areas from flooding, for water storage, and for the generation of hydropower, it is unlikely that the dams will be eliminated so that the beach compartments can return to a natural balance. Another option is **beach re-**plenishment (also called **beach nourishment**), which involves placing sand on the beach to replace the sand removed by dams. However, beach replenishment is an expensive prospect, since huge volumes of sand must be continually supplied to the beach. When dams are built, it is not often taken into account that they may have an effect on beaches great distances downstream. When the supply of sand to the beach is affected, it is soon realized that the rivers are parts of much larger systems that operate along the coast.

Emerging and Submerging Shorelines

Another way to classify shorelines is based on their position relative to sea level. However, it is important to note that *sea level has not remained at a fixed position throughout time*. Sea level change can be caused by a change in the level of the land, a change in the level of the sea, or a combination of the two. Shorelines that are

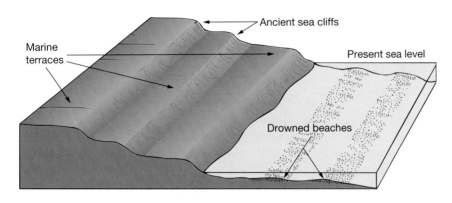

Figure 10–13 Evidence of ancient shorelines.

Marine terraces result from exposure of ancient sea cliffs and wave-cut benches above present sea level. Below sea level, drowned beaches indicate the sea level has risen relative to the land.

emerging above sea level are called **emerging shorelines**. Conversely, shorelines that are submerging below sea level are called **submerging shorelines**.

Many features characterize emerging shorelines. Along some coasts are flat platforms called marine terraces, which are backed by cliffs (Figure 10–13). Marine terraces are formed when a wave-cut bench is exposed above sea level. **Stranded beach deposits** and other evidence of marine processes may exist many meters above the present shoreline, indicating that the former shoreline has risen above sea level.

Submerging shorelines are characterized by features such as underwater **drowned beaches** (Figure 10–13) and **submerged dune topography**. These features, along with the **drowned river valleys** along the present shoreline, indicate that these shorelines are sinking below sea level.

What causes these changes in sea level that produce submerging and emerging shorelines? One main cause is tectonic and isostatic movements, which raise or lower the land surface relative to sea level. Another main cause is worldwide changes in sea level, which affect the sea itself. Let's examine each of these two main causes.

Tectonic and Isostatic Movements[3] of Earth's Crust

Changes in sea level relative to a continent may occur because of movement of the land, which is called *tectonic movement*. Such movement could mean large-scale uplift or large-scale subsidence of major portions of continents or ocean basins. Alternatively, this movement could mean a more localized deformation of the continental crust involving folding, faulting, and tilting. The most dramatic examples of sea level changes during the past 3000 years have been due to tectonic processes.

Earth's crust also responds isostatically, by *sinking* under the accumulation of heavy loads of ice, vast piles of sediment, or outpourings of lava; or by *rising* when such loads are removed. This response to the addition or

removal of weight on Earth's crust is called *isostatic adjustment*.

Most of the U.S. Pacific Coast is an emerging shoreline because of its active tectonic origin, typical of continental margins where plate collisions occur. Earthquakes, volcanic action, and mountain chains paralleling the coast are common characteristics of such active margins. In contrast, most of the U.S. Atlantic Coast is a submerging shoreline. When a continent is moving away from a spreading center (such as the Mid-Atlantic Ridge), its trailing edge subsides due to cooling. The additional weight of accumulating sediment also causes the continent to subside. Passive margins such as these are characterized by a low level of tectonic deformation, earthquakes, and volcanism, making the Atlantic Coast far more quiet and stable than the Pacific Coast.

Another way to accomplish a change in the elevation of the land involves glaciation. There is evidence that, during the last 2.5 to 3 million years, at least four major accumulations of glacial ice developed in high latitudes. Although Antarctica is still covered by a very large, thick glacial accumulation, much of the ice cover that once existed over northern Asia, Europe, and North America has disappeared. The most recent period of melting began about 18,000 years ago. Accumulations of ice that were up to 3 kilometers (2 miles) thick have disappeared from northern Canada and Scandinavia.

While these areas were beneath the thick ice sheet, the crust beneath them sank. Today, they are still recovering, slowly rebounding after the melting of the ice. By the time this isostatic rebound is finished, the floor of Hudson Bay, which is now about 150 meters (500 feet) deep, will be close to or above sea level. There is evidence in the Gulf of Bothnia, between Sweden and Finland, of 275 meters (900 feet) of isostatic rebound during the last 18,000 years. Many other areas worldwide that experienced glaciation during the previous Ice Age are also isostatically rebounding today.

Generally, tectonic and isostatic changes in the level of the shoreline are confined to a segment of the shoreline of a given continent. For a *worldwide* increase or decrease in sea level, mechanisms other than those mentioned here must occur.

---

[3]Tectonics and isostatic adjustment have been discussed in Chapter 2, "Plate Tectonics and the Ocean Floor."

Eustatic Changes in Sea Level

A worldwide change in sea level can be caused by:

- An increase or decrease of seawater volume; or
- An increase or decrease in ocean basin capacity.

Such a change in sea level is called **eustatic** (*eu* = good, *stasis* = standing).[4] For example, small eustatic changes in sea level can be created by formation or destruction of large inland lakes. When large inland lakes form, water that normally runs off the land into the ocean is trapped in these lakes, and sea level is subsequently lowered worldwide.

Changes in sea-floor spreading rates can cause eustatic sea level change by changing the capacity of the ocean basin. For instance, fast spreading produces larger rises, such as the East Pacific Rise, which displace more water than slow-spreading ridges such as the Mid-Atlantic Ridge. Thus, fast spreading produces a rise in sea level. Conversely, slower spreading produces lower sea level worldwide. Significant changes of sea level resulting from changes in spreading rate typically occur over hundreds of thousands to millions of years.

A more recent cause of eustatic sea level change is the effect of ice ages. As glaciers form during an ice age, they tie up a vast volume of water on land, causing a eustatic lowering of sea level. An analogy to this effect is a sink of water representing an ocean basin. To simulate an ice age, some of the water from the sink is removed and frozen, causing the water level of the sink to be lower. In a similar fashion, worldwide sea level is lower during an ice age. During interglacial stages (such as the one we are in at present), the glaciers melt and release great volumes of water that drain to the sea, causing a eustatic rise of sea level. This would be analogous to putting the frozen chunk of ice on the counter near the sink and letting the ice melt, causing the water to drain into the sink and raise "sink level."

During the Pleistocene Epoch,[5] there were many advances and retreats of glaciers on land near the poles. This caused the amount of water in the ocean basins to fluctuate considerably, as evidenced by the present elevations above and below sea level of Pleistocene Age marine features. Because the climate was colder during glacial times, some of the lowering of the shoreline might be accounted for by the thermal contraction of the ocean as its temperature decreased. The thermal contraction and expansion of seawater works in a similar fashion to how a mercury thermometer works: As the fluid inside the thermometer heats, it expands and rises into the thermometer; as the fluid cools, it contracts. Similarly, cooler seawater contracts and occupies less volume, thereby exposing more shoreline areas and producing a eustatic *drop* in sea level. Warmer seawater causes a eustatic *rise* in sea level.

Calculations reveal that for every 1 degree centigrade (1.8 degrees Fahrenheit) decrease in the average temperature of the ocean water, sea level will drop 2 meters (6.6 feet). Temperature indications derived from study of microfossils in Pleistocene ocean sediments suggest that ocean surface temperature may have been as much as 5 degrees centigrade (9 degrees Fahrenheit) lower than at present. Therefore, thermal contraction of the ocean water may have lowered sea level by about 15 meters (40 feet).

Although it is difficult to state definitely the range of shoreline fluctuation during the Pleistocene, evidence suggests that it was at least 120 meters (400 feet) below the present shoreline (Figure 10–14). It is also estimated that if *all* the remaining glacial ice on Earth were to melt, sea level would rise another 60 meters (200 feet). Thus, the *minimum* sea level change during the Pleistocene is on the order of 180 meters (600 feet), most of which must be explained through the capture and release of Earth's water by land-based glaciers.

During the last 18,000 years, ocean volume has increased due to expansion of seawater resulting from warmer temperatures plus the melting of polar sea ice and glacial ice. The combination of tectonic and eustatic changes in sea level is very complex and difficult to sort out, so it is difficult to classify coastal regions as purely emergent or submergent. Most coastal areas show evidence of having experienced *both* submergence and emergence in the recent past. It is believed, however, that sea level has not risen significantly as a result of melting glacial ice during the last 3000 years.

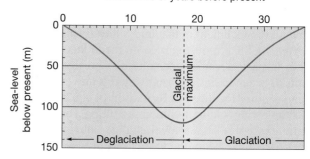

Figure 10–14 Sea level change during the most recent advance and retreat of Pleistocene glaciers.

Sea level dropped worldwide by about 120 meters (400 feet) as the last glacial advance removed water from the oceans and transferred it to continental glaciers. About 18,000 years ago, sea level began to rise as the glaciers melted and water was returned to the oceans.

---

[4] The term *eustatic* refers to a highly idealized situation in which all of the continents remain static (in *good standing*), while only the sea rises or falls.

[5] The Pleistocene Epoch of geologic time (also called the "Ice Age") is from 1.6 million to 10 thousand years ago (see the Geologic Time Scale, Figure 1–17).

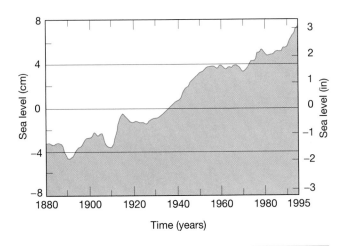

Figure 10–15 Sea level change from 1880 to 1995.
Tide-gauge data, averaged over five-year periods, show that
global mean sea level has increased about 10 centimeters
(4 inches) over the past 100 years.

There is at present much discussion about combined
gleanings from historical research. There is clear evidence
that carbon dioxide in the atmosphere has increased by
25 percent since 1958, when monitoring began. There has
also been an apparent global warming of about 0.5 degree centigrade (0.9 degree Fahrenheit) and a eustatic
rise in sea level of 10 centimeters (4 inches) since 1880
(Figure 10–15). Some researchers have documented a
current rate of sea level rise of 1.8 millimeters (0.07 inch)
per year based on analysis of worldwide tide gauge
records that extend for long periods of time.

The urgent question is this: Are all of these observations related, and are they the result of an increased
greenhouse effect? The answer is unknown. But the great
lesson for humankind from these observations is that we
cannot dominate nature; we must live within it. To do
otherwise is futile. Some of the consequences of a rising
sea level for U.S. coastal communities are discussed in
the following sections.

Characteristics of U.S. Coasts

Whether the dominant process along a coast is erosion
or deposition depends on the combined effect of many
variables. These variables include: degree of exposure
to ocean waves, tidal range, composition of coastal
bedrock, tectonic subsidence or emergence, isostatic subsidence or emergence, and eustatic sea level change.

In 1971 the U.S. Army Corps of Engineers reported on
erosion of the U.S. coastline, which is 135,870 kilometers (84,240 miles) long, including Alaska (but not
Hawaii). They found over 24 percent of it to be "seriously
eroding." Subsequent studies supported by the U.S.
Geological Survey along the shores of the contiguous
coastal states and Alaska produced the rates of shoreline change presented in Figure 10–16.

The Atlantic Coast

As shown in Figure 10–16, the U.S. Atlantic Coast includes a wide variety of complex coastal conditions:

- Most of the Atlantic coastline is exposed to storm
  waves from the open ocean. However, barrier island development from Massachusetts southward
  protects the mainland from large storm waves.

- Tidal ranges generally increase from less than
  1 meter (3.3 feet) along the Florida coast to more
  than 2 meters (6.5 feet) in Maine.

- Bedrock for most of Florida is a resistant type of
  sedimentary rock called *limestone*. From Florida
  northward through New Jersey, most bedrock is
  composed of nonresistant sedimentary rocks formed
  in the recent geologic past. As these rocks rapidly
  erode, they become sand sources for barrier islands
  and other depositional features common along the
  coast. The bedrock north of New York is composed
  of very resistant rock types.

- From New York northward, continental glaciers affected the coastal region directly. Many coastal features, including Long Island and Cape Cod, are
  glacial deposits (called *moraines*) left behind when
  the glaciers melted.

North of Cape Hatteras in North Carolina, the coast is
subject to very high-energy wave conditions during fall
and winter when storms called "nor'easters" (northeast-
ers) blow in from the North Atlantic. The great energy
of these storms is manifested in waves up to 6 meters
(20 feet) high, with a 1-meter (3.3-foot) rise in sea level
that follows the low pressure as it moves northward. Such
high-energy conditions significantly change coastlines
that are predominantly depositional and cause considerable erosion.

Most of the Atlantic Coast appears to be experiencing
a sea level rise of about 0.3 meter (1 foot) per century.
Evidence of submerging shorelines includes drowned
river valleys, which are common along the coast and form
large bays (Figure 10–17). The submergence of the coast
may be reversed in northern Maine because of isostatic
rebound due to the melting of the Pleistocene ice sheet.

The Atlantic Coast has an average annual rate of erosion of 0.8 meter (2.6 feet).[6] In everyday terms, this means
that the sea is migrating landward each year by a depth
equal to about the length of your legs! Thus, the majority of east coast states are slowly losing real estate.
Virginia leads the way, with a loss of 4.2 meters (13.7 feet)
per year, which is confined largely to barrier islands.

Erosion rates for Chesapeake Bay are about average
for the Atlantic Coast, but Delaware Bay has erosion

---

[6] Note that erosion values are shown as *negative* values in Figure 10–16, identifying them as separate from rates of deposition.

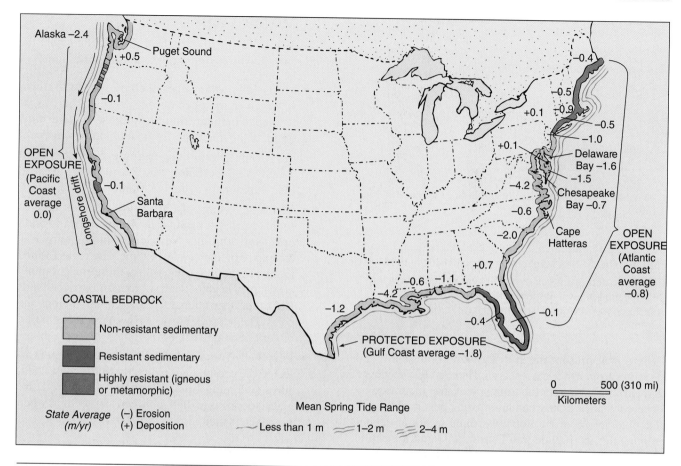

Figure 10–16 Coastal erosion and deposition, 1979–1983.

For the Pacific, Atlantic, and Gulf Coasts, this map shows coastal bedrock type (*red, yellow, and blue colors*), the mean spring-tide range (*light blue lines*), degree of exposure, and rates of erosion/deposition. Net erosion (–) or net deposition (+) in meters per year is shown for specific locations along with average condition for each coast.

rates [1.6 meters (5.2 feet) per year] that are about twice that of the Atlantic Coast average. Of the observations made along the Atlantic Coast, 79 percent showed some degree of erosion. Delaware, Georgia, and New York display depositional coasts despite serious erosion problems in these states as well.

The Gulf Coast

The erosion/deposition story of the Gulf Coast (often called America's "third coast") is simpler. The Louisiana-Texas coastline is dominated by the Mississippi River Delta, which is being deposited in an area that experiences a small tidal range—normally less than 1 meter (3.3 feet)—and generally low-energy conditions. Except during the hurricane season (June to November), wave energy is generally low. Tectonic subsidence is common throughout the Gulf Coast, and the average rate of sea level rise is similar to that of the southeast Atlantic coast, about 0.3 meter (1 foot) per century. Some areas of coastal Louisiana have experienced a 1-meter (3.3-foot) rise during the last century, due to consolidation of sediments deposited by the Mississippi River.

The average rate of erosion is 1.8 meters (6 feet) per year in the Gulf Coast. The Mississippi River Delta experiences the greatest rate, so the state of Louisiana is losing an average of 4.2 meters (13.7 feet) per year. Made worse by dredging of barge channels through marshlands, Louisiana has lost more than 1 million acres of delta since 1900. The Pelican State is now losing marshland at a rate exceeding 130 square kilometers (50 square miles) per year.

Although all Gulf states show a net loss of land, and the Gulf Coast has a greater erosion rate than the Atlantic Coast, only 63 percent of the shore is receding because of erosion. The high average rate of erosion reflects the heavy erosion loss in the Mississippi River Delta area.

The Pacific Coast

In general, the Pacific Coast is experiencing less erosion than the Atlantic and Gulf Coasts. Along the Pacific Coast, relatively young and easily eroded sedimentary rocks dominate the bedrock, with local outcrops of more resistant rock types. Tectonically, the coast is rising, as ev-

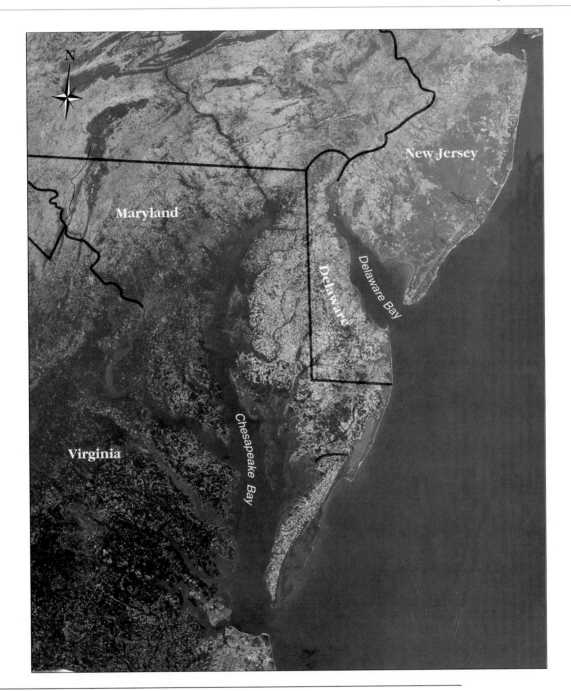

Figure 10–17 Drowned river valleys.
Drowned river valleys such as Chesapeake and Delaware Bays along the east coast of the United
States were formed by a relative rise in sea level that followed the end of the Ice Age.

idenced by marine (wave-cut) terraces (Figure 10–18).
Sea level still shows at least small rates of rise, except for
a segment along the coast of Oregon and the Alaskan
coast. The tidal range is mostly between 1 and 2 meters
(3.3 and 6.6 feet).

The Pacific Coast is fully exposed to large storm waves,
and is said to have *open exposure*. High-energy waves
may strike the coast in winter, with 1-meter (3.3-foot)
waves being typical. Frequently, the wave height will
increase to 2 meters (6.6 feet), and a few times per year
6-meter (20-foot) waves hammer the shore! These high-

energy waves erode sand from many beaches. The exposed beaches, which are composed primarily of pebbles and boulders during the winter months, regain their
sand as low-energy summer waves beat more gently
against the shore.

Many Pacific Coast rivers have been dammed for flood
control and hydroelectric power generation. As mentioned previously, this has reduced the amount of sediment supplied by rivers to the shoreline for longshore
transport. As a result, some areas now experience a severe erosion threat.

Figure 10–18 Marine (wave-cut) terraces.

Each marine terrace on San Clemente Island offshore of southern California was created by wave activity at sea level. Subsequently, each terrace has been exposed by tectonic uplift. The highest (oldest) terraces near the top of the photo are now about 400 meters (1320 feet) above sea level.

With an average erosion rate of only 0.005 meter (0.016 foot)[7] per year and only 30 percent of the coast showing erosion loss, the Pacific Coast appears to be under much less of an erosion threat than the Atlantic and Gulf coasts. Yet some areas along the coast experience high rates of erosion due to high wave energy and relatively soft rocks exposed along the coast. For example, some parts of Alaska are losing an average of 2.4 meters (7.9 feet) per year.

Of the Pacific states, only Washington shows a net sediment deposition. The long, protected Washington shoreline within Puget Sound helps skew the Pacific Coast values (Figure 10–16). Although the average erosion rate for California is only 0.1 meter (0.33 foot) per year, over 80 percent of the California coast is experiencing erosion, with local rates of up to 0.6 meter (2 feet) per year.

Effects of Hard Stabilization

People continually modify coastal sediment erosion/deposition in attempts to improve or preserve their property. These structures, built to protect a coast or to prevent the movement of sand along a beach, are known as **hard stabilization**. Hard stabilization can take many forms and often results in predictable yet unwanted outcomes.

One type of hard stabilization is a **groin** (*groin* = ground). Groins are barriers to sediment flow built perpendicular to a coastline (Figure 10–19). They are constructed of many types of material, but most commonly they are constructed of large blocky material called **rip-rap**. Sometimes, groins are even constructed of sturdy wood pilings (similar to a fence built out into the ocean).

A groin is constructed with the express purpose of trapping sand. When a groin is constructed along a shoreline, it does have the effect of trapping sand *upstream from the groin*. However, on the downstream side of the groin, erosion occurs because the sand that this segment of the beach would normally receive is trapped on the upstream side of the groin. To alleviate this situation, another groin is constructed downstream, which creates erosion downstream from it. More groins are needed to alleviate the beach erosion, and soon a **groin field** is created (Figure 10–20).

Does a groin (or a groin field) actually cause more sand to occur on the beach? Because groins eventually allow sand to migrate around their ends, there is no additional sand on the beach; it is only *distributed differently*. With proper engineering, an equilibrium may be reached that allows sufficient sand transport along the coast before excessive erosion occurs downstream from the last groin. However, some serious erosional problems have developed in many areas due to attempts to stabilize sand on the beach by the excessive use of groins.

Another type of hard stabilization is a **jetty** (*jettee* = to project). A jetty is similar to a groin in that it is a structure built perpendicular to the shore and is usually constructed of rip-rap. However, jetties are constructed to protect harbor entrances from wave action and only secondarily do they trap sand (Figure 10–21). Because jetties are often built in closely spaced pairs and are usually longer than groins, they can cause more pronounced upstream deposition and downstream erosion of shoreline sediment.

Many examples of beach destruction resulting from the use of jetties can be cited. Along the southern California coast, excessive beach erosion related to the development of many harbor facilities is especially common. California's longshore drift is predominantly southward,

---

[7] 0.005 meters is equal to 5 millimeters (0.2 inch).

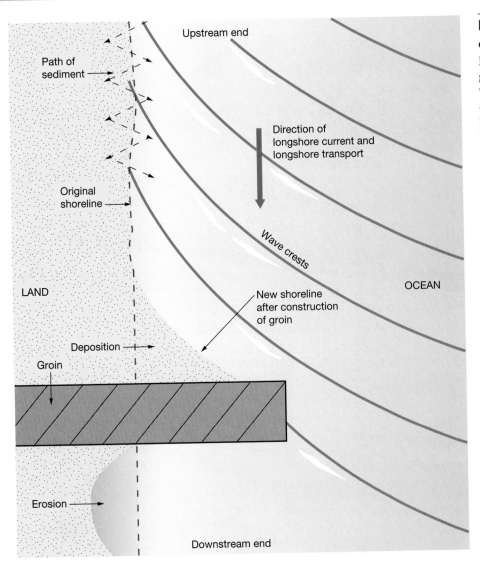

Figure 10–19 Interference of sand movement. Hard stabilization like the groin shown here interferes with the movement of sand along the beach, causing deposition of sand upstream of the groin and erosion immediately downstream.

Figure 10–20 Groin Field. A series of groins has been built along the shoreline north of Ship Bottom, New Jersey, in an attempt to trap sand, altering the distribution of sand on the beach. The view is toward the north, and the primary direction of longshore current is toward the bottom of the photo (toward the south).

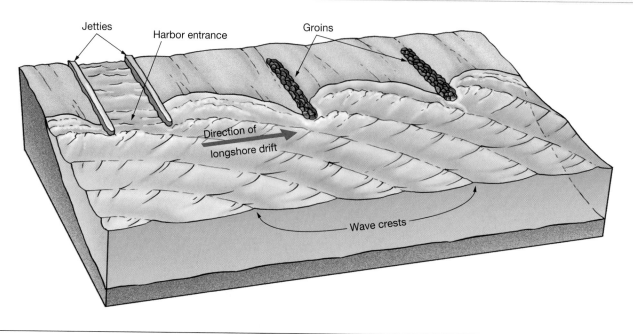

Figure 10–21 Jetties and groins.

Jetties and groins trap sand that would otherwise be moved along the shore by wave action.

Figure 10–22 Santa Barbara Harbor.

Construction of a breakwater at Santa Barbara Harbor interfered with the longshore drift, creating a broad beach. As the beach extended around the breakwater into the harbor, dredging operations were initiated to keep the harbor from being closed off by sand. The placement of dredged sand downstream of the harbor helped reduce coastal erosion there.

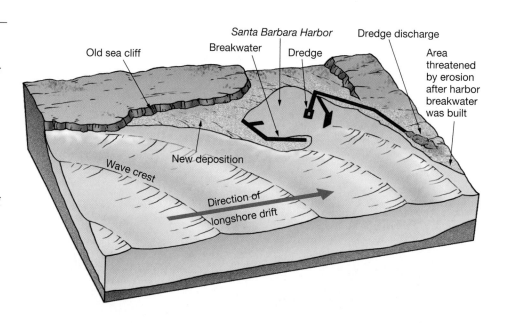

so construction of a jetty tends to trap sediment on its northern (upstream) side, which causes increased erosion on its southern (downstream) side.

The harbor at Santa Barbara, California, provides a unique example. Figure 10–22 shows the jetty constructed as a **breakwater**—hard stabilization built parallel to a shoreline—to protect the harbor. This jetty greatly disturbed the equilibrium established in this coastal region. The barrier created by the breakwater on the western side of the harbor caused an accumulation of sand that was migrating eastward along the coast. The beach to the west of the harbor continued to grow until finally the

sand moved around the breakwater and began to fill in the harbor (Figure 10–22).

While abnormal deposition occurred to the west, erosion proceeded at an alarming rate east of the harbor. The wave energy available east of the harbor was no greater than before, but the sand that had formerly moved down the coast was no longer available, because it was entrapped behind the breakwater.

A similar situation occurred in Santa Monica, California, where a breakwater was built to provide a boat anchorage. Shortly after the breakwater was built, a bulge in the beach formed behind the breakwater

A.

B.

Figure 10–23 Santa Monica Breakwater.
A. The shoreline and pier at Santa Monica as it appeared in 1931. **B.** The same area in 1949, showing that the construction of a breakwater to create a boat anchorage disrupted the longshore transport of sand and caused a bulge of sand in the beach.

(Figure 10–23). Downstream from the breakwater, severe erosion occurred. The breakwater interfered with the natural transport of sand by blocking the wave energy necessary to keep the sand moving. If something was not done to put energy back into the system, the breakwater would soon be attached by a tombolo of sand, and further erosion downstream might destroy coastal structures.

In Santa Barbara and Santa Monica, there was a need to compensate for erosion downstream from the breakwater and to keep the harbor or anchorage from being filled with sand. The solution in both cases was to constantly dredge sand from behind the breakwater to keep it from filling in. Dredged sand is pumped down the coast so it can reenter the longshore drift and replenish the eroded beach.

The dredging operation has stabilized the situation in Santa Barbara, but at a considerable (and ongoing) expense. In Santa Monica, sand dredging was conducted until the breakwater was largely destroyed during winter storms in 1982–1983. Shortly after the breakwater was destroyed, wave energy was able to move sand along the coast again, and the system restored itself to near-normal conditions. When people interfere with natural processes in the coastal region, they must provide the energy needed to replace what they have misdirected through modification of the shore environment.

One of the most destructive types of hard stabilization is the **seawall** (Figure 10–24). Instead of extending out into the longshore current, as do groins and jetties, seawalls are built parallel to the shore. The idea is to armor the coastline and protect landward developments from the action of ocean waves.

However, once waves begin breaking against a seawall, turbulence generated by the abrupt release of wave energy quickly erodes the sediment on its seaward side, causing it to collapse into the surf (Figure 10–24). Where seawalls have been used to protect property on barrier islands, the seaward slope of the island beach has steepened and the rate of erosion increased. In essence, seawalls destroy recreational beaches.

A well-designed seawall may last for many decades, but seawalls do not last forever. The constant pounding of wave energy eventually takes its toll (Figure 10–25). Eventually, the cost of repairing or replacing seawalls will be more than the property is worth, and the sea will claim more of the coast through the natural processes of coastal erosion. It's all just a matter of time, with homeowners who live too close to the coast gambling that it won't happen within their lifetime.

Although most state laws assert that beaches belong to the public, government action allows owners of shore property to destroy beaches in an effort to provide short-term protection to their property. To add to the public insult, taxpayers subsidize property owners in many shore developments by helping people rebuild their storm-damaged properties at great public expense through the National Flood Insurance Program described in the chapter-opening feature at the beginning of this chapter.

Figure 10–24 Seawalls. When a seawall is built along a beach to protect beachfront property (**A**), the first large storm will remove the beach from the seaward side of the wall and steepen its seaward slope (**B**). Eventually, the wall is undermined and falls into the sea (**C**). The property is lost (**D**) as the oversteepened beach slope advances landward in its effort to reestablish a natural slope angle.

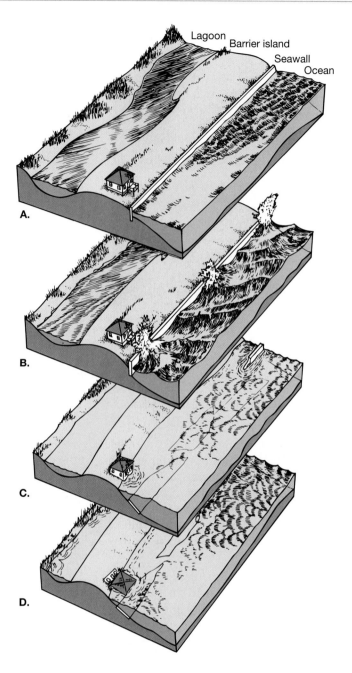

Lagoon
Barrier island
Seawall
Ocean

A.

B.

C.

D.

? Students Sometimes Ask...

What is the difference between a rip current and a rip tide?

Like tidal waves (tsunami), rip tides are a misnomer and have nothing to do with the tides. Rip tides are more correctly called rip currents. Perhaps it is because they occur suddenly (like an incoming tide) that rip currents have incorrectly been called rip tides.

What is an undertow? How is it different from a rip current?

An undertow is a flow of water along the ocean bottom away from shore. It resembles a rip current but is much wider, and is usually more concentrated along the ocean floor. An undertow is really a continuation of backwash that flows down the beach face. Undertows are stronger during heavy wave activity, when water stacks up at the beach and flows out away from the shore

along the bottom. Undertows can be strong enough to knock people off their feet, but fortunately, they are confined to the immediate floor of the ocean and only within the surf zone.

Can submarine canyons fill with sediment?

Yes. In many beach compartments, the submarine canyons that drain sand from the beach empty into deep basins offshore. However, given several million years of tons of sediments per year sliding down the submarine canyons, the offshore basins begin to fill up and can eventually be exposed above sea level. In fact, the Los Angeles basin in California was filled in by sediment derived from local mountains in this manner during the geologic past. Today, Los Angeles basin with its submarine canyons is completely filled in, and the entire basin is above present sea level.

Figure 10–25 Seawall damage.
This seawall in Leucadia, California, has been damaged and needs repair. Although seawalls appear to be sturdy, they can be destroyed by the continual pounding of high-energy storm waves.

I know that the positions of the continents have not always remained the same. Has the movement of the continents changed sea level?
Yes, but in an indirect way. If plate movement occurs such that large continental masses are located in polar regions, this provides a location for thick continental glaciation to occur (such as in Antarctica today). Glacial ice forms from water vapor in the atmosphere (in the form of snow) which originally is derived from the evaporation of seawater. Thus, water is removed from the oceans when continents assume more polar positions, resulting in a lowering of sea level.

Does beach starvation affect coastal cliff erosion?
Most certainly! When beaches become narrower, coastal cliffs are more prone to wave attack because they lack a wide, protective, energy-absorbing beach. Instead of storm waves expending their energy crashing against the beach, areas that have experienced beach starvation often receive storm waves that crash directly against a coastal cliff. The impressive power of these waves causes increased amounts of coastal cliff erosion.

How much does beach replenishment cost?
The cost of beach replenishment depends on many variables, including the type and quantity of material placed on the beach, how far the material must be transported, and how it is to be distributed on the beach. Most sand used for replenishment comes from offshore areas, but sand that is dredged from nearby rivers, drained dams, harbors, and lagoons is also used.
The average cost of sand used to replenish beaches is between $5 and $10 per 0.76 cubic meter (1 cubic yard). For comparative purposes, a typical top-loading trash dumpster holds about 2.3 cubic meters (3 cubic yards) of material, and a typical dump truck holds about 45 cubic meters (60 cubic yards) of material. The problem with replenishment projects is that a huge volume of sand is needed, and new sand must continually be supplied. A small beach replenishment project of several hundred cubic meters can cost around $10,000 per year. Larger projects that move several thousand cubic meters of sand cost several million dollars per year. As can be seen, this is an expensive (and recurring) prospect.
A new source for beach replenishment sand has been proposed. Recycled glass that is ground to sand size and spread on the beach may be a less expensive option, but the health and safety risks have not been fully explored.

I have the opportunity to live in a house at the edge of a coastal cliff where there is an incredible view along the entire coast. Is it safe from coastal erosion?
Based on what you've described, most certainly not! Geologists have long known that cliffs are naturally unstable. The property you describe most likely has significant danger of falling into the ocean. Even if the cliffs appear to be stable (or have been stable for a number of years), one significant storm can produce an amazing amount of cliff failure.
The number-one form of coastal erosion is direct wave attack undermining the support and causing failure of the coastal cliff. You might want to check the base of the cliff and examine the local bedrock to determine for yourself if you think it will be undamaged by the pounding of powerful storm waves that can move rocks weighing several tons. Other dangers include drainage runoff, weaknesses in the bedrock, slumps and landslides, seepage of water through the cliff, and even burrowing animals.
The state of California recommends a 12-meter (40-foot) setback from the edge of the cliff for all new buildings. Sometimes, even that distance isn't enough: Large sections of "stable" cliffs can fail all at once. For instance, several city blocks of real estate have been eroded from the cliff edges during the last 100 years in some areas of southern California. Even though the view sounds outstanding, you may suddenly find out that the house is built a little *too* close to the edge of the cliff!

Summary

The coastal region is characterized by change that occurs continuously. The region of contact between the oceans and the continents is marked by the shore, lying between the lowest low tides and the highest elevation on the continents affected by storm waves. The coast extends inland from the shore as far as marine-related features can be found. The coastline marks the boundary between the shore and the coast. The shore is divided into the foreshore, extending from low tide to high tide, and the backshore, extending beyond the high-tide line to the coastline. Seaward of the low-tide are the nearshore zone, extending to the breaker line, and the offshore zone beyond.

A beach is a deposit of the shore area that consists of wave-worked sediment that moves along a wave-cut bench. It includes the recreational beach, berm, beach face, low-tide terrace, one or more longshore bars, and longshore trough. Beaches are composed of whatever material is locally available.

Waves that break at the shore move sand perpendicular to shore (toward and away from shore). In light wave activity, swash dominates the transport system and sand is moved up the beach face toward the berm. In heavy wave activity, backwash dominates the transport system and sand is moved down the beach face away from the berm toward longshore bars. In a natural system, there is a balance between light and heavy wave activity, alternating between sand piled on the berm (summertime beach) and sand stripped from the berm (wintertime beach), respectively.

Movement of sand also occurs parallel to the shore. As waves break at an angle to the shore, a longshore current is created that results in a zigzag movement of sediment called longshore drift (longshore transport). Each year, millions of tons of sediment are moved from upstream to downstream ends of beaches. Most of the year, the direction of longshore drift is southward along both the Pacific and Atlantic shores of the United States.

Erosional-type shores are characterized by various erosional features such as headlands, wave-cut cliffs, sea caves, sea arches, sea stacks, and marine terraces (caused by uplift of a wave-cut bench). Wave erosion increases with greater exposure of the shore to the open ocean, decreasing tidal range, and decreasing strength of bedrock.

Depositional-type shores are characterized by depositional features such as beaches, spits, bay barriers, tombolos, barrier islands, deltas, and beach compartments. Viewed from ocean side to lagoon side, barrier islands commonly have an ocean beach, dunes, barrier flat, and salt marsh. Deltas form at the mouths of rivers that carry more sediment to the ocean than can be distributed by the longshore current. Many beach compartments suffer from beach starvation due to the interruption of sand supply. Beach replenishment (beach nourishment) is an expensive and temporary way to reduce beach starvation.

Shores can also be classified as emerging or submerging shorelines based on their position relative to sea level. A sea level drop relative to land may be indicated by ancient wave-cut cliffs and stranded beaches well above the present shoreline. A sea level rise relative to land may be indicated by old drowned beaches, submerged dunes, wave-cut cliffs, or drowned river valleys. Such sea level changes may result from tectonic processes causing local movement of the landmass or from eustatic processes changing the amount of water in the oceans or the capacity of ocean basins. Melting of continental ice caps during the past 18,000 years has caused a eustatic rise in sea level of about 120 meters (400 feet).

Along the Atlantic Coast, sea level is rising about 0.3 meter (1 foot) every century along most of the coast, and the average annual erosion rate is –0.8 meter (–2.6 feet). Along the Gulf Coast, sea level is rising at 0.3 meter (1 foot) per century, and the average rate of erosion is –1.8 meters (–6 feet) per year. The Mississippi River Delta is eroding at a rate of 4.2 meters (13.7 feet) per year, resulting in a large annual loss of wetlands. Along the Pacific Coast, the average erosion rate is only –0.005 meter (–0.016 foot) per year. Different shorelines erode at different rates depending on factors such as wave exposure, amount of uplift, and type of bedrock.

Hard stabilization is often constructed in an attempt to stabilize a shoreline. Types of hard stabilization include groins, jetties, breakwaters, and seawalls. Groins (built to trap sand) and jetties (built to protect harbor entrances) widen the beach by trapping sediment on their upstream side, but erosion usually becomes a problem downstream. Similarly, breakwaters (built parallel to a shore) trap sand behind the structure, but cause unwanted erosion downstream. Seawalls (built to armor a coast) often cause loss of the recreational beach. Eventually, the constant pounding of waves will destroy all types of hard stabilization.

Key Terms

Backshore (p. 300)	Beach starvation (p. 312)	Dune (p. 308)
Backwash (p. 302)	Berm (p. 301)	Emerging shoreline (p. 313)
Barrier flat (p. 308)	Breakwater (p. 320)	Erosional-type shore (p. 304)
Barrier island (p. 308)	Coast (p. 300)	Eustatic sea level change (p. 314)
Bay barrier (bay-mouth bar) (p. 307)	Coastline (p. 300)	Foreshore (p. 300)
Beach (p. 301)	Delta (p. 310)	Groin (p. 318)
Beach compartment (p. 311)	Depositional-type shore (p. 304)	Groin field (p. 318)
Beach face (p. 301)	Distributary (p. 310)	Hard stabilization (p. 318)
Beach replenishment (beach nourishment) (p. 312)	Drowned beach (p. 313)	Headland (p. 304)
	Drowned river valley (p. 313)	Jetty (p. 318)

Questions And Exercises

1. Why has the National Flood Insurance Program (NFIP) had a negative impact on the nation's shores?

2. To help reinforce your knowledge of beach terminology, construct and label your own diagram similar to Figure 10–1 from memory.

3. Describe differences between summertime and wintertime beaches. Explain why these differences occur.

4. What variables affect the speed of longshore currents?

5. What is longshore drift, and how is it related to a longshore current?

6. How is the flow of water in a stream similar to a longshore current? How are the two different?

7. Why does the direction of longshore current sometimes reverse in direction? Along both U.S. coasts, what is the primary direction of annual longshore current?

8. Describe the formation of rip currents. What is the best strategy to ensure that you won't drown if you are caught in a rip current?

9. Discuss the formation of such erosional features as wave-cut cliffs, sea caves, sea arches, sea stacks, and marine terraces.

10. Describe the origin of these depositional features: spit, bay barrier, tombolo, and barrier island.

11. Describe the response of a barrier island to a rise in sea level. Why do some barrier islands develop peat deposits running through them from the ocean beach to the salt marsh?

12. Discuss why some rivers have deltas and others do not. Also include the factors that determine whether a "bird's-foot" or a smoothly curved, Nile-type delta will form.

13. Describe all parts of a beach compartment. What will happen when dams are built across all of the rivers that supply sand to the beach?

14. Compare the causes and effects of tectonic versus eustatic changes in sea level.

15. List the two basic processes by which coasts advance seaward, and list their counterparts that lead to coastal retreat.

16. List and discuss four factors that influence the classification of a coast as either erosional or depositional.

17. Describe the tectonic and depositional processes causing subsidence along the Atlantic Coast.

18. Compare the Atlantic Coast, Gulf Coast, and Pacific Coast by describing the conditions and features of emergence-submergence and erosion-deposition that are characteristic of each.

19. List the types of hard stabilization and describe what each is intended to do.

20. Draw an aerial view of a shoreline to show the effect on erosion and deposition caused by constructing a groin, a jetty, a breakwater, and a seawall within the coastal environment.

References

Baltuck, M., Dickey, J., Dixon, T., and Harrison, C. G. A. 1996. New approaches raise questions about future sea level change. *EOS, Transactions of the AGU* 77:40, 385.

Bascom, W. 1980. *Waves and beaches: The dynamics of the ocean surface* (rev. ed.). New York: Anchor Books (Doubleday).

Bird, E. C. F. 1985. *Coastline changes: A global review.* Chichester, U.K.: Wiley.

Burk, K. 1979. The edges of the ocean: An introduction. *Oceanus* 22:3, 2–9.

Coates, R., ed. 1973. *Coastal geomorphology. Publications in Geomorphology.* Binghamton, NY: State University of New York.

Kuhn, G. G., and Shepard, F. P. 1984. *Sea cliffs, beaches, and coastal valleys of San Diego County: Some amazing histo-ries and some horrifying implications.* Berkeley: University of California Press.

Leatherman, S. P. 1983. Barrier dynamics and landward migration with Holocene sea-level rise. *Nature* 301:5899, 415–417.

May, S. K., Kimball, H., Grandy, N., and Dolan, R. 1982. The Coastal Erosion Information System. *Shore Beach* 50, 19–26.

Millemann, B. 1993. The National Flood Insurance Program. *Oceanus* 36:1, 6–8.

Peltier, W. R., and Tushingham, A. M. 1989. Global sea level rise and the greenhouse effect: Might they be connected? *Science* 244:4906, 806–810.

Roemmich, D. 1992. Ocean warming and sea level rise along the southwest U.S. coast. *Science* 257:5068, 373–375.

Sahagian D. L., Schwartz, F. W., and Jacobs, D. K. 1994. Direct anthropogenic contributions to sea levels rise in the twentieth century. *Nature* 367:6458, 54–56.

Shepard, F. P. 1973. *Submarine geology*, 3rd ed. New York: Harper & Row.

———. 1977. *Geological oceanography*. New York: Crane, Russak & Company.

Stanley, D. J., and Warne, A. G. 1993. Nile delta: Recent geological evolution and human impact. *Science* 260:5108, 628–634.

Suggested Reading

Earth

Eustis, M. 1992. The Nile. 1:6, 38–45. A photoessay about the Nile River and its delta, including photos of the Nile from space.

Flanagan, R. 1993. Beaches on the brink. 2:6, 24–33. A look at shoreline erosion problems encountered as barrier island beaches along the east and gulf coasts of the United States migrate with rising sea level.

Sea Frontiers

Carr, A. P. 1974. The ever-changing sea level. 20:2, 77–83. A discussion of the causes of sea level change is very well presented.

Emiliani, C. 1976. The great flood. 22:5, 256–270. An interesting discussion of the possible relationship of the rise in sea level resulting from the melting of glaciers 11,000 to 8000 years ago and biblical and other ancient accounts of a great flood.

Feazel, C. 1987. The rise and fall of Neptune's kingdom. 33:2, 4–11. A discussion of factors that change the level of the sea, including atmospheric conditions, currents, climate, ocean topography, and sea floor spreading.

Fulton, K. 1981. Coastal retreat. 27:2, 82–88. The problems of coastal erosion along the southern California coast are considered.

Grasso, A. 1974. Capitola Beach. 20:3, 146–151. The destruction of the beach of Capitola, California, shortly after the construction of a harbor by the Army Corps of Engineers at Santa Cruz to the north.

Mahoney, H. R. 1979. Imperiled sea frontier: Barrier beaches of the east coast. 25:6, 329–337. The natural alteration of barrier beaches is considered.

Pilkey, O. H. 1990. Barrier islands. 36:6, 30–39. The origin, stages of evolution, and types of barrier islands are explained by an advocate of barrier islands as living entities.

Schumberth, C. J. 1971. Long Island's ocean beaches. 17:6, 350–362. This is a very informative article on the nature of barrier islands. The specific problems observed on the Long Island barriers serve as examples.

Wanless, H. R. 1989. The inundation of our coastlines: Past, present and future with a focus on South Florida. 35:5, 264–267. This is an informative overview of the evidence that is useful in evaluating the past and future movements of the shorelines.

Wanless, H. R., and Tedesco, L. P. 1988. Sand biographies. 34:4, 224–232. Discusses how to tell where beach sand came from and how it traveled to its present location.

Westgate, J. W. 1983. Beachfront roulette. 29:2, 104–109. The problems related to the development of barrier islands are discussed.

Scientific American

Bascom, W. 1960. Beaches. 203:2, 80–97. A comprehensive consideration of the relationship of beach processes, both large and small scale, to release of energy by waves.

Broecker, W. S., and Denton, G. H. 1990. What drives glacial cycles? 262:1, 48–107. Recent research is included in this consideration of the causes of glacial cycles.

Dolan, R., and Lins, H. 1987. Beaches and barrier islands. 257:1, 68–77. A discussion dealing with the ultimate futility of trying to develop and protect structures on beaches and barrier islands.

Fairbridge, R. W. 1960. The changing level of the sea. 202:5, 70–79. A discussion of what is known of the causes of the changing level of the sea, which appears to be related mostly to the formation and melting of glaciers and changes in the ocean floor.

Schneider, D. 1997. The rising seas. 276:3, 112–117. An analysis of the true threat of rising sea level and its relationship to global warming.

Oceanography on the Web

Visit the *Essentials of Oceanography* home page for on-line resources for this chapter. There you will find an on-line study guide with review exercises, and links to oceanography sites to further your exploration of the topics in this chapter. *Essentials of Oceanography* is at: **http://www.prenhall.com/thurman** (click on the Table of Contents menu and select this chapter).

CHAPTER 11
THE COASTAL OCEAN

The Law of the Sea

Who owns the ocean? Who owns the sea floor? If a company wanted to drill for oil offshore between two different countries, must it obtain permission from either country? Extensive exploitation of the ocean floor for minerals and petroleum is imminent, necessitating laws that answer such questions. In the future, much exploration activity will occur beyond the jurisdiction of the country that has the nearest coastline. Furthermore, problems of overfishing and pollution are worsening. Are such problems covered by long-established laws? The answer to that question is yes... and no.

In 1609 Hugo Grotius, a Dutch jurist and statesman whose writings were important in the formulation of international law, established in his treatise *Mare liberum* (*mare* = sea, *liberum* = free) the doctrine that the oceans are free to all nations. Nevertheless, controversy continued over whether nations could control a *portion* of an ocean, such as the ocean adjacent to a nation's coastline.

Cornelius van Bynkershonk attempted to solve this problem in the work *De dominio maris* (*dominio* = domain, *maris* = sea), published in 1702. It provided for national domain over the sea out to the distance that could be protected by cannons from the shore, an area called the **territorial sea**. Just how far from shore did the territorial sea extend? The first official width of this territorial sea was probably established in 1672 when the British determined that cannon range encompassed a distance out to 1 league (3 nautical miles) from shore. Thus, every country with a coastline had a *three-mile territorial limit*.

Because of the rapid development of technology for drilling beneath the ocean, the first **United Nations Conference on the Law of the Sea**, held in Geneva, Switzerland, in 1958, produced a treaty regarding ownership as related to the continental shelf. The treaty stated that prospecting and mining of minerals on the continental shelf are under the control of the country that owns the nearest land. Since the continental shelf is that portion of the sea floor extending from the coastline out to where the slope markedly increases, the seaward limit of the shelf can be interpreted differently. Unfortunately, the continental shelf was not well defined in the treaty and led to disputes. In 1960, the second United Nations Conference on the Law of the Sea was also held in Geneva, but it made little progress toward an unambiguous and fair treaty concerning ownership of the coastal ocean.

Meetings of the third Law of the Sea Conference were held during 1973–1982 and came to a disappointing conclusion when a large majority of countries voted in favor of a new Law of the Sea treaty. The vote was 130 to 4, with 17 abstentions. Most developing nations that could benefit significantly from the treaty voted for its adoption; the opposition was led by the United States, and included Turkey, Israel, and Venezuela. The countries opposing the new treaty did so in support of private enterprises that are interested in sea floor mining operations because certain provisions of the treaty made such ventures unprofitable. The abstaining countries included the Soviet Union, Great Britain, Belgium, the Netherlands, Italy, and West Germany, all of which are interested in sea floor mining. The treaty was ratified by the required sixtieth nation in 1993, thus establishing it as international law. Negotiations removed the objections of nations interested in sea floor mining, and the United States signed the revised treaty in 1994, when the treaty was enacted as law.

The primary components of the treaty are as follows.

1. *Coastal nations jurisdiction.* The treaty established a uniform 12-mile (19-kilometer) territorial sea and a 200-nautical-mile (370-kilometer) **exclusive economic zone (EEZ)** from all land (including islands) within a nation. Each coastal nation has jurisdiction over mineral resources, fishing, and pollution regulation within its EEZ. If the continental shelf (defined geologically) extends beyond the 200-mile (322-kilometer) EEZ, the EEZ is extended out to 350 miles (563 kilometers) from shore.

2. *Ship passage.* The right of free passage for all vessels on the high seas is preserved. The right of free passage is also provided within territorial seas and through straits used for international navigation.

3. *Deep-ocean mineral resources.* Private exploitation of sea floor resources may proceed under the regulation of the International Seabed Authority (ISA), within which a mining company will be strictly controlled by the United Nations. This provision, which caused some industrialized nations to oppose ratification, required mining companies to fund two mining operations—their own and one operated by the regulatory United Nations. Recently, this portion of the law was modified to eliminate some of the reg-

ulatory components, thus favoring free market principles and development by private companies. Still, this portion of the Law of the Sea has been one of the most contentious issues in international law.

4. *Arbitration of disputes.* A United Nations Law of the Sea tribunal will arbitrate any disputes in the treaty or disputes concerning ownership rights.

Since the passage of the Law of the Sea, 42 percent of the world's oceans have come under the control of coastal nations. In the United States, the U.S. Geologic Survey has been involved in exploring the sea floor of the nation's EEZ (Figure 11A) to evaluate its economic potential.

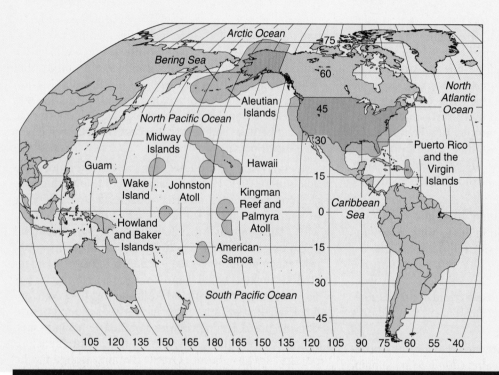

Figure 11A Exclusive Economic Zone (EEZ) of the United States.

The coastal ocean is a very busy place, filled with life, commerce, recreation, fisheries, and waste. Of the world fishery,[1] 95 percent is obtained within about 320 kilometers (200 miles) of shore. Indeed, it is these coastal waters that also support about 95 percent of the total mass of life in the oceans. In addition, these waters are the focal point of most shipping routes, oil and gas production, and recreational activities. Unfortunately, they are also the final destination of much of the waste products of those living on the adjacent land.

The human activity that is probably the most damaging to the ocean environment is commercial fishing. Since

1989, the mass of the world fishery has decreased from its peak production of 72.7 million metric tons (80.0 million short tons), caused primarily from overfishing by a vastly increased fishing effort. This decreased catch, along with increasing human population, has diminished the per-capita catch from 19.5 kilograms (43.0 pounds) in 1989 to 18.8 kilograms (41.4 pounds) in 1995, according to the United Nations' Food and Agriculture Organization. Beyond decreasing the fish yield, overfishing may strongly impact the overall ecology of the oceans.

Besides overfishing, there is another major worry: marine pollution. Marine pollution in coastal waters results from accidental spills of petroleum and the accumulation of sewage, certain chemicals (such as DDT and PCBs), and the element mercury. Such pollution is of great importance because these pollutants can have a se-

---

[1] The term *fishery* refers to fish caught from the ocean by commercial fishers.

vere deleterious effect on organisms living in the ocean. To better predict the effects of pollution on coastal waters, much more must be learned about the physical processes that give these waters their high biological productivity as well as an amazing resiliency to the onslaught of contamination.

Coastal Waters

The primary distinction between coastal waters and the open ocean is based on depth: **Coastal waters** are those relatively shallow-water areas that adjoin continents or islands. If nearshore regions are underlain by broad and shallow continental shelves, coastal waters can extend several hundred kilometers from land. However, if underwater bathymetry has significant relief or drops rapidly onto the deep-ocean basin, coastal waters will occupy a relatively thin band near the margin of the land.

Because of their proximity to land, coastal waters are directly influenced by processes that occur on or near land. For instance, river runoff and tidal currents have a far more significant effect on these shallow-water coastal waters than on the open ocean.

Salinity

Along the coast, river runoff reduces the salinity of the ocean's surface layer. The lower density of fresh water impedes the mixing of fresh water with seawater. Instead, the fresh water forms a wedge of water at the surface. Thus, a well-developed **halocline**[2] forms (Figure 11–1A). Eventually, the fresh water mixes with seawater, causing salinity throughout the water column to be reduced where mixing does occur (Figure 11–1C). There is no halocline here; instead, the water column is **isohaline** (*iso* = same, *halo* = salt).

Where precipitation on land is mostly rain, river runoff peaks in the same season that precipitation does. However, if runoff is fed largely by melting snow and ice, runoff always peaks in summer. In general, salinity is lower in coastal regions than in the open ocean because of the runoff of fresh water from the continents.

Counteracting the salinity-reducing effect of runoff in some coastal regions is the presence of prevailing offshore winds. As winds travel over a continent, they usually lose most of their moisture. Thus, dry offshore winds typically evaporate considerable water as they move across the surface of the coastal waters. The increase in evaporation rate in these areas tends to increase surface salinity, also resulting in the development of a halocline (Figure 11–1B), but with a reversed gradient from

the halocline developed from the input of fresh water (compare with Figure 11–1A).

Temperature

In coastal regions where the water is relatively shallow, large changes in temperature may occur over a year. Sea ice forms in many high-latitude coastal areas where temperatures are determined by the water's freezing point, generally warmer than –2 degrees centigrade (28.4 degrees Fahrenheit) (Figure 11–1D). In low-latitude coastal regions, where the water is restricted in its circulation with the open ocean and thus is protected from strong mixing, maximum surface temperature may approach 45 degrees centigrade (113 degrees Fahrenheit) (Figure 11–1E). In both high- and low-latitude coastal waters, **isothermal** (*iso* = same, *thermo* = heat) conditions prevail.

Seasonal changes in temperature can be most easily detected in coastal regions of the mid-latitudes, where surface temperatures are coolest in winter and warmest in late summer. Strong **thermoclines**[3] may develop from surface water being warmed during the summer (Figure 11–1F) and cooled during the winter (Figure 11–1G). In summer, very-high-temperature surface water may form a relatively thin layer. Mixing reduces the surface temperature by distributing the heat through a greater vertical column of water, thus pushing the thermocline deeper and making it less pronounced. In winter, cooling may cause surface water to sink by increasing its density, causing a well-mixed isothermal water column.

Prevailing winds can significantly affect surface temperatures, if they blow from the continent. These offshore winds, previously mentioned regarding salinity, usually have relatively high temperature during the summer. They increase the ocean surface temperature and seawater evaporation. During winter, they are much cooler than the ocean surface and absorb heat from the ocean surface, cooling surface water near shore. Mixing from strong winds may drive the thermoclines in Figures 11–1F and 11–1G deeper and even mix the entire water column, producing an isothermal condition. Another factor that influences the vertical mixing of shallow coastal waters is tidal currents, which can cause considerable vertical mixing.

Coastal Geostrophic Currents

As described in Chapter 7, geostrophic (*geo* = earth, *strophio* = turn) currents are currents that move in a circular path around the middle of a current gyre. Geostrophic currents also develop in coastal waters, resulting from two specific agents: wind and runoff. These currents are called **coastal geostrophic currents**.

---

[2]Remember that a halocline (*halo* = salt, *cline* = slope) is a zone of rapidly changing salinity, as discussed in Chapter 5.

---

[3] Remember that a thermocline (*themo* = heat, *cline* = slope) is a zone of rapidly changing temperature, as discussed in Chapter 5.

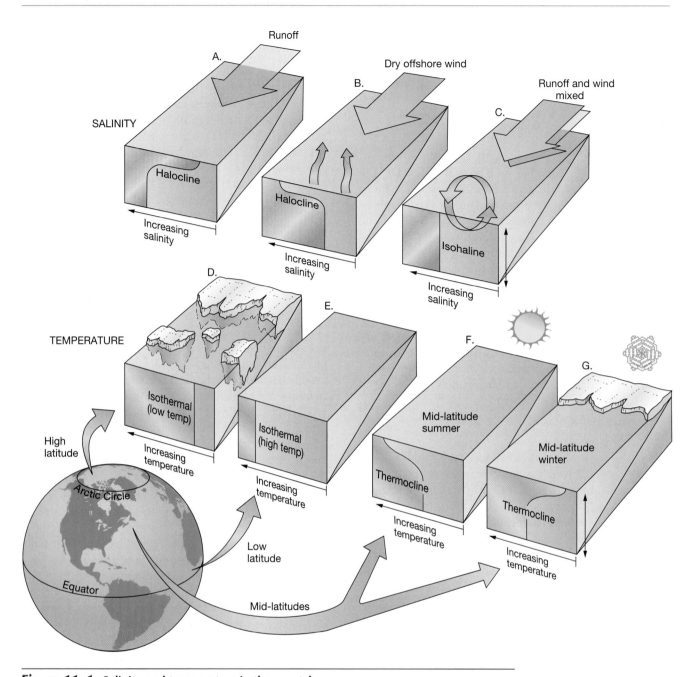

Figure 11–1 Salinity and temperature in the coastal ocean.

Changes in coastal salinity (*top row*) can be caused by the input of freshwater runoff (**A**), by dry offshore winds causing a high rate of evaporation (**B**), or by both (**C**). Changes in coastal temperature (*bottom row*) depend on latitude. In high latitudes (**D**), the temperature of coastal water remains uniformly near freezing. In low latitudes (**E**), coastal water may become uniformly warm. In the mid-latitudes, coastal surface water is significantly warmed during summer (**F**) and cooled during the winter (**G**).

Wind blowing parallel to the coast piles up water along the shore. Gravity eventually pulls this piled-up water back downslope toward the open ocean. As it runs downslope away from the shore, the Coriolis effect causes it to curve from its intended direction. In the Northern Hemisphere, the coastal geostrophic current curves to the right, or *northward* on the western coast and *southward* on the eastern coast of continents. In the Southern Hemisphere, the direction of the Coriolis effect is re-

versed, causing the movement of water flowing out from the coast to be reversed as well.

The other condition that produces coastal geostrophic flow along continental margins is the high-volume runoff of fresh water that gradually mixes with oceanic water. This produces a surface slope of water away from the shore (Figure 11–2). The seaward slope has salinity and density gradients because both properties increase sea-

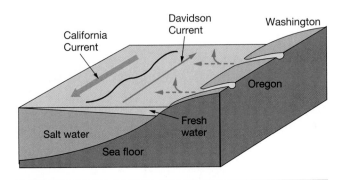

Figure 11–2 Coastal geostrophic current.
The north-flowing Davidson Current is a coastal geostrophic current that flows along the coast of Washington and Oregon. During the winter rainy season, runoff produces a freshwater wedge (*light blue*) that slopes away from shore. This causes a surface flow of low-salinity water toward the open ocean, which is acted upon by the Coriolis effect, curving to the right.

ward. These variable currents, which depend on the wind and the amount of runoff for their strength, are bounded on the ocean side by the steadier boundary currents resulting from the open-ocean gyres.

These local geostrophic currents frequently flow in a direction opposite to the nearby boundary current. Such is the case with the **Davidson Current**, which develops along the coast of Washington and Oregon during the winter (Figure 11–2). Heavy precipitation occurs in the Pacific Northwest during winter, and southwesterly winds are strongest then. Their combined effect produces a relatively strong northward-flowing coastal geostrophic current. It flows between the shore and the southward-flowing California Current, which is part of the North Pacific Gyre.

Estuaries

An **estuary** (*aestus* = tide) is defined as a partially enclosed coastal body of water in which salty ocean water is significantly diluted by fresh water from land runoff. The most common estuary is a river mouth, where the river empties into the sea. Many bays, inlets, gulfs, and sounds may be considered estuaries, too. All estuaries exhibit large temperature and/or salinity variations.

The mouths of large rivers form the most economically significant estuaries, because many are seaports and centers of ocean commerce—Baltimore, New York, San Francisco, Buenos Aires, London, Tokyo, and many others. Many estuaries support important commercial fisheries as well. Environmental changes caused by heavy commercial use of estuaries are a major concern as future development is planned.

Origin of Estuaries

Essentially all estuaries in existence today owe their origin to sea level rise over the past 18,000 years. Sea level has risen approximately 120 meters (400 feet) due to

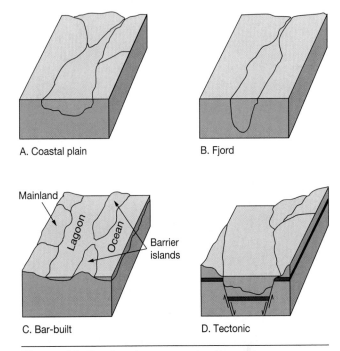

Figure 11–3 Classifying estuaries by origin.
Diagrammatic views of the four types of estuaries based on origin. **A.** Coastal plain estuary. **B.** Glacially carved fjord. **C.** Bar-built estuary. **D.** Tectonic estuary.

extensive melting of major continental glaciers. As described in Chapter 10, these glaciers covered portions of North America, Europe, and Asia during the Pleistocene Epoch, more commonly referred to as the Ice Age. Four major classes of estuaries can be identified on the basis of their origin (Figures 11–3):

1. A **coastal plain estuary** forms as rising sea level causes the oceans to invade existing river valleys. These estuaries are appropriately called **drowned river valleys**. Chesapeake Bay in Maryland and Virginia is an example (see Figure 10–17).

2. A **fjord**[4] is a glaciated valley that has been flooded by the sea, creating an estuary. Unlike a water-carved valley, which is V-shaped in profile, glacially carved fjords are U-shaped valleys with steep walls. Commonly, a shallowly submerged glacial deposit of debris (called a *moraine*) is located near the ocean entrance that marks the farthest extent of the glacier. Fjords are common along the coasts of Norway, Canada, Alaska, and New Zealand (Figure 11–4A).

3. A **bar-built estuary** is shallow and is separated from the open ocean by sand bars that are deposited parallel to the coast by wave action. Lagoons that separate **barrier islands** from the mainland are bar-built estuaries. Examples abound along the U.S.

---

[4] The Norwegian word *fjord* is pronounced "FEE-yord."

A.

B.

Figure 11–4 Estuaries.
A. This Norwegian fjord is a glacially formed estuary, where the deep U-shaped valley has been flooded by the sea. **B.** Aerial view of San Francisco Bay in California, which is a tectonic estuary that was created by faulting.

Gulf and East Coasts, including Laguna Madre along the Texas coast and Pamlico Sound in North Carolina (see Figure 10–9).

4. **A tectonic estuary** is produced by faulting or folding of rocks. This forms a restricted downdropped area into which rivers flow. San Francisco Bay is in part a tectonic estuary (Figure 11–4B), formed by movement along faults including the San Andreas Fault.

Water Mixing in Estuaries

Generally, fresh water runoff that flows into an estuary moves as an upper layer of low-density water across the estuary toward the open ocean. Beneath this upper layer, a layer of denser, salty seawater typically moves in the *opposite* direction to the surface water. Mixing takes place at the contact between these water masses.

Estuaries can be classified into four patterns based on mixing of fresh water and seawater, as shown in Figure 11–5:

1. **Vertically mixed estuary**—a shallow, low-volume estuary where the net flow always proceeds from the river head of the estuary toward the estuary's mouth. Salinity at any point in the estuary is uniform from surface to bottom because river water

mixes evenly with ocean water at all depths. Salinity simply increases from the head to the mouth of the estuary, as shown in Figure 11–5A. Note that the curved salinity lines at the edge of the estuary are caused by the inflow of seawater influenced by the Coriolis effect.

2. **Slightly stratified estuary**—a somewhat deeper estuary in which salinity increases from the head to the mouth at any depth, as in a vertically mixed estuary. However, two water layers can be identified: less saline, less dense upper water provided by the river, and denser, deeper seawater. These two layers are separated by a zone of mixing. An **estuarine circulation pattern** begins to develop in this type of estuary, which is characterized by a net surface flow of low-salinity water toward the ocean and an opposite net subsurface flow of seawater toward the head of the estuary (Figure 11–5B).

3. **Highly stratified estuary**—a deep estuary in which upper-layer salinity increases from the head to the mouth, reaching a value close to that of open-ocean water. The deep-water layer has a rather uniform open-ocean salinity at any depth throughout the length of the estuary. An estuarine circulation pattern is well developed in this type of estuary (Figure 11–5C). Net flow of the two layers is simi-

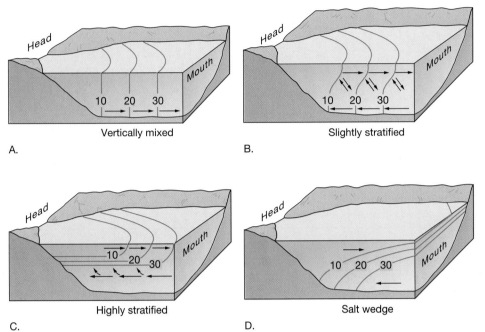

A. Vertically mixed

B. Slightly stratified

C. Highly stratified

D. Salt wedge

Figure 11–5 Classifying estuaries by mixing.

The basic flow pattern in an estuary is a surface flow of less-dense freshwater toward the ocean and an opposite flow in the subsurface of salty seawater into the estuary. Numbers represent salinity in ‰; arrows indicate flow directions. **A.** Vertically mixed estuary. **B.** Slightly stratified estuary. **C.** Highly stratified estuary. **D.** Salt wedge estuary.

lar to that in a slightly stratified estuary, but mixing at the interface of the upper water and the lower water is such that net movement is from the deepwater mass into the upper water. Less-saline surface water simply moves from the head toward the mouth of the estuary, growing more saline as water from the deep mass mixes with it. Relatively strong haloclines develop in such estuaries at the contact between the upper and lower water masses.

4. **Salt wedge estuary**—an estuary in which a saline wedge of water intrudes from the ocean beneath the river water; this is typical of the mouths of deep, high-volume rivers. No horizontal salinity gradient exists at the surface in these deep estuaries. Surface water is essentially fresh throughout the length of, and even beyond, the estuary (Figure 11–5D). There is, however, a *horizontal* salinity gradient at depth and a very pronounced vertical salinity gradient manifested as a strong halocline at any location throughout the length of the estuary. This halocline is shallower and more highly developed near the mouth of the estuary.

The mixing patterns described often cannot be applied to an estuary as a whole. Mixing within an estuary may change with location within an estuary, seasonal changes, or tidal conditions.

Estuaries and Human Activities

Estuaries are vital natural ecosystems that have evolved over millennia. Because estuaries are important breeding grounds and protective nurseries for many marine animals, the health of estuaries is essential to the world fishery and to coastal environments worldwide. However, they are exploited heavily by economic activities that do not depend on the health of the estuary. For instance, estuaries support shipping, logging, manufacturing, waste disposal, and other activities that can potentially damage the estuarine environment.

Obviously, estuaries are most threatened where human population is large and expanding. But even where population pressure is still modest, estuaries can be severely damaged. A typical example of human effects on estuaries is the damage done by development in the Columbia River estuary.

Columbia River Estuary The Columbia River, which forms most of the border between Washington and Oregon, has a long estuary at its entrance to the Pacific Ocean (Figure 11–6). This estuary is a salt-wedge estuary, caused by the strong flow of the river and tides that drive a salt wedge as far as 42 kilometers (26 miles) upstream and raise the river's water level by over 3.5 meters (12 feet). When the tide falls, the huge flow of fresh water [up to 28,000 cubic meters (1,000,000 cubic feet) per second] creates a vast fresh water wedge in the Pacific Ocean, sometimes extending for hundreds of kilometers away from shore.

Most rivers create floodplains along their lower courses, which have rich soil that can be used for growing crops. In the late nineteenth century, farmers and dairymen moved onto the rich floodplains along the Columbia River. Eventually, protective dikes were built along the river to prevent the yearly natural flooding. However, periodic flooding brings new nutrients to the floodplains and maintains soil richness. Ironically, protecting these lands from the natural ecosystem caused them to be deprived of the nutrients necessary to maintain agriculture. Consequently, the land was no longer suitable for most types of agriculture.

Figure 11–6 Columbia River estuary.

The long estuary at the mouth of the Columbia River has been severely affected from interference by floodplains that have been diked, by logging activities, and—most severely—by hydroelectric dams.

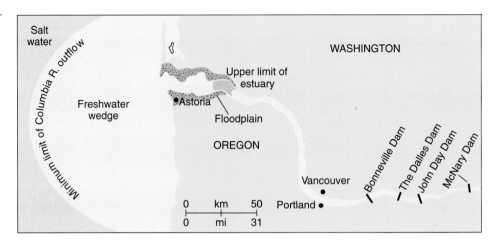

Today, the river has been the principal conduit for the logging industry, which has dominated the region's economy through most of its modern history. Fortunately, the river's ecosystem has largely survived the damage inflicted by the logging industry. However, the hydroelectric dams have dealt the ecosystem a more serious and permanent blow. Many of these dams did not include salmon ladders, which help fish "climb" in short vertical steps around the dams to reach the headwaters of their home streams. This closed off many areas previously used as spawning grounds. As is the case with most interactions between nature and humans, the dams were essential for the development of the region, but a high price was paid. On a more positive note, the Columbia River's large flow volume aids in reducing the effects of chemical pollution in the river.

At present, in an effort to diversify the local economy, shipping facilities are being developed by periodically dredging the river of sediment. Accompanying the dredging operations is an increased pollution potential. If such problems have developed at such sparsely populated areas as the Columbia River estuary, what must be the conditions at more highly populated estuaries?

Chesapeake Bay Estuary Chesapeake Bay is an example of a large coastal plain estuary that experiences classic estuarine circulation. The bay was produced by the drowning of the Susquehanna River and its tributaries (Figure 11–7). Most of the fresh water entering the bay enters along the western margin via rivers that drain the slopes of the Appalachian Mountains.

The dark blue lines in the bay in Figure 11–7 show the surface salinity pattern, which not only increases oceanward but also illustrates the Coriolis effect. The Coriolis effect causes seawater entering the bay to curve to the right in the Northern Hemisphere, so that seawater tends to hug the *eastern* side of the bay. Similarly, the Coriolis effect also acts on the fresh water flowing out of the bay, causing it to hug the *western* side of the bay. This pattern is represented by the near north–south orientation of the salinity lines in the middle part of Chesapeake Bay.

With maximum river flow in the spring, a strong halocline (and *pycnocline*[5]) develops, preventing mixing of the fresh surface water and saltier deep water. Beneath the pycnocline, which can be as shallow as 5 meters (16 feet), waters may become **anoxic** (*a* = without, *oxic* = oxygen) from May through August, as dead organic matter decays in the deep water. Major kills of commercially important blue crab, oysters, and other bottom-dwelling organisms occur during these events.

The degree of stratification and kills of bottom-dwelling animals have increased in this area since the early 1950s. Also during this time, the bay has experienced intense pressure from human activities. Studies suggest that increased nutrients from sewage and agricultural fertilizers have been added to the bay as a result of human activities. These increased nutrients have caused an increase in the production of microscopic algae (algal blooms), thereby increasing the deposition of organic matter on the bottom and increasing the possibility of the development of anoxic conditions.

Much is yet to be learned about estuarine circulation in Chesapeake Bay. This is illustrated by the following observations, made over an extended period in Chesapeake Bay.

- Researchers have recorded intervals of upstream surface flow, accompanied by downstream deep flow, which is exactly the opposite of normal estuarine circulation.

- Periods of downstream flow without any upstream flow at any level have been observed.

- Patterns that are even more complex are known to develop. Sometimes, a surface and bottom flow in one direction is separated by a mid-depth flow in the opposite direction. At other times, a landward flow occurs along both shores while a seaward flow occurs in the central portion of the estuary.

---

[5]Remember that a pycnocline (*pycno* = density, *cline* = slope) is a zone of rapidly changing density, as discussed in Chapter 5. A pycnocline is caused by a change in temperature and/or salinity.

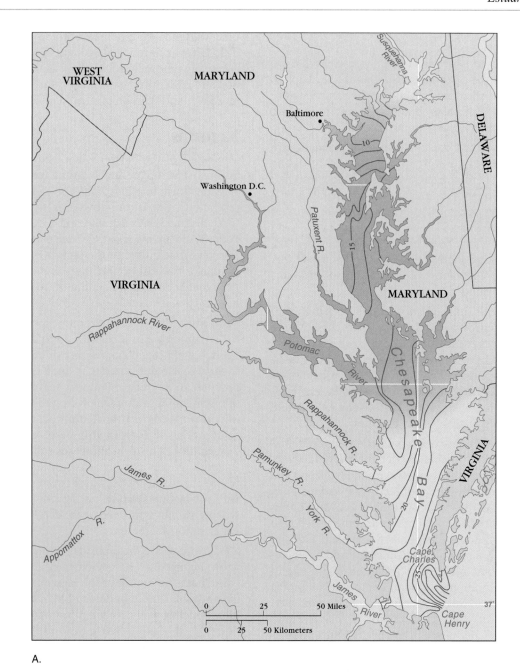

A.

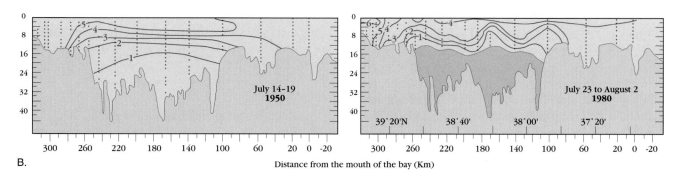

B.

Figure 11–7 Chesapeake Bay.

A. Map of the Chesapeake Bay region, showing average surface salinity (*blue lines*) in ‰. The area colored darker blue from Baltimore to the mouth of the Potomac River is the area of anoxic (oxygen-poor) waters. **B.** Profiles for comparison of dissolved oxygen levels for the summers of 1950 (*left*) and 1980 (*right*), showing the deep anoxic mid-bay waters in darker blue in 1980.

Wetlands

Wetlands border estuaries and other shore areas that are protected from the open ocean. Wetlands are strips of land that are delicately in tune with natural shore processes. The biological productivity of wetlands is one of the highest of any ecosystem on Earth.

Two types of wetlands exist: **salt marshes** and **mangrove swamps**. Both are intermittently submerged by ocean water and are characterized by oxygen-poor mud and peat deposits. Marshes, characteristically inhabited by a variety of grasses, are known to occur from the Equator to latitudes as high as 65 degrees (Figures 11–8A and 11–8B). Mangroves are restricted to latitudes below 30 degrees both north and south of the Equator (Figures 11–8A and 11–8C). Once mangroves colonize an area, they normally grow into and replace marsh grasses.

Research reveals that wetlands have a very high economic value when left alone. Salt marshes are believed to serve as nurseries for over half the species of commercially important fishes in the southeastern United States. Other fishes, such as flounder and bluefish, use marshes for feeding and protection during the winter. Fisheries of oysters, scallops, clams, and fishes such as eels and smelt are located directly in the marshes.

A very important characteristic of wetlands is their ability to cleanse polluted water. Wetlands remove inorganic nitrogen compounds (from sewage and fertilizers) and metals (from groundwater polluted by land sources), in essence "scubbing" the water of its pollution. Most of this removal is achieved when pollutants become attached to clay-sized particles in the wetlands. Some nitrogen compounds trapped in sediment are decomposed by bacteria that release the nitrogen to the atmosphere as gas. Many of the remaining nitrogen compounds support plant production in this environment, thus causing wetlands to have very high productivity. As plants die in marshes, their organic nitrogen compounds are either incorporated into the sediment to become peat or are broken up to become food for bacteria, fungi, or fish.

Serious Loss of Valuable Wetlands

With all the benefits that wetlands provide, it is indeed surprising that nearly 60 percent of the nation's wetlands have been lost. Of the original 215 million acres of wetlands that once existed in the United States (excluding Alaska and Hawaii), only about 90 million acres remain. Both marsh and mangrove wetlands have been filled in and developed for housing, industry, and agriculture. All of this is a consequence of people's strong desire to live near the oceans. Ironically, people often view wetlands as unproductive, useless land that harbors diseases.

To help prevent the loss of remaining wetlands, the U.S. Environmental Protection Agency established an Office of Wetlands Protection (OWP) in 1986. At that time, wetlands were being lost to development at a rate of 121,000 hectares (300,000 acres) per year! The OWP is actively enforcing regulations against wetlands pollution and is identifying the most valuable wetlands so that these wetlands can be protected or restored.

Lagoons

Landward of the barrier islands lie protected, shallow bodies of water called **lagoons** (see Figure 11–3C). Lagoons form in a bar-built type of estuary. Because of restricted circulation between lagoons and the ocean, three distinct zones usually can be identified within a lagoon. Typically, a *freshwater zone* exists near the mouths of rivers that flow into the lagoon. A *transitional zone* of brackish[6] water occurs near the middle of the lagoon. A *saltwater zone* is present close to the entrance (Figure 11–9A).

Salinity within a lagoon is influenced by the proximity to the lagoon's entrance, with the areas closer to the entrance exhibiting higher salinity (Figure 11–9B). Lagoon salinity is also determined by another factor other than nearness to the sea. In latitudes that have seasonal variations in temperature and precipitation, ocean water will flow through the entrance during a warm, dry summer to compensate for the volume of water lost through evaporation. This flow increases salinity in the lagoon. Lagoons actually may become *hypersaline*[7] in arid regions, where the inflow of seawater is too small to keep pace with the lagoon's surface evaporation. During the rainy season, the lagoon becomes much less saline as fresh water runoff increases.

Tidal effects are maximized near the entrance to the lagoon (Figure 11–9C). Inland from the saltwater zone that exists near the mouth of the lagoon, tidal effects diminish and are usually undetectable in the fresh water region, which is well protected from tidal effects.

Laguna Madre

Laguna Madre is one of the best-known and best-studied lagoons. It is located along the Texas coast between Corpus Christi and the mouth of the Rio Grande. This long, narrow body of water is protected from the open ocean by Padre Island, an offshore island 160 kilometers (100 miles) long. The lagoon probably formed about 6000 years ago as sea level was approaching its present height.

Much of the lagoon is less than 1 meter (3.3 feet) deep. The tidal range of the Gulf of Mexico in this area is about 0.5 meter (1.6 feet). The inlets at each end of the barrier

---

[6] Brackish water is water with salinity between that of fresh water and seawater.

[7] Hypersaline conditions are created when water becomes excessively salty.

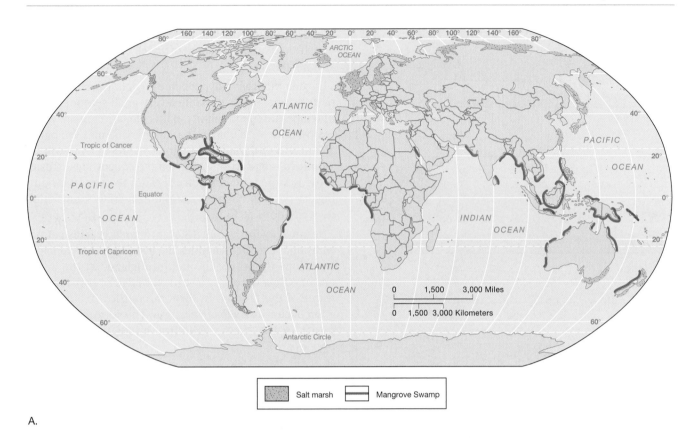

A.

B.

C.

Figure 11–8 Salt marshes and mangrove swamps.

A. Map showing the distribution of salt marshes (higher latitudes) and mangrove swamps (lower latitudes). **B.** Salt marsh along San Francisco Bay at Shoreline Park, California. **C.** Mangrove trees on Lizard Island, Great Barrier Reef, Australia.

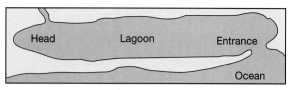

A. Geometry

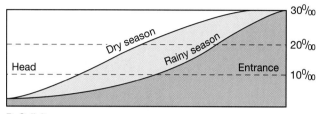

B. Salinity

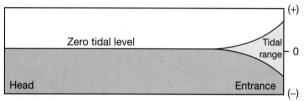

C. Tidal Effects

Figure 11–9 Lagoons.
Typical geometry (**A**), salinity (**B**), and tidal effects (**C**) of a lagoon.

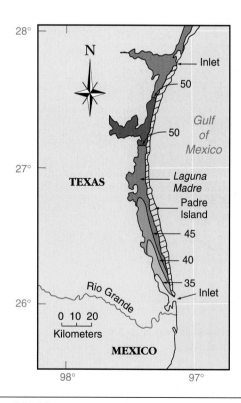

Figure 11–10 Laguna Madre, Texas.
Map showing geometry of Laguna Madre and typical summer surface salinity (in ‰).

island are quite narrow (Figure 11–10). There is, therefore, very little tidal interchange of water between the lagoon and the open sea.

Laguna Madre is a hypersaline lagoon. The shallowness of its waters makes possible a very great seasonal range of temperature and salinity in this semiarid region. Water temperatures are high in the summer and may fall below 5 degrees centigrade (41 degrees Fahrenheit) in winter. Salinities range from 2‰ when infrequent local storms provide large volumes of fresh water to over 100‰ during dry periods. High evaporation generally keeps salinity well above 50‰.[8]

Because even the salt-tolerant marsh grasses cannot withstand such high salinities, the marsh has been replaced by an open sand beach on Padre Island. At the inlets, ocean inflow occurs as a surface wedge *over* the denser water of the lagoon. In turn, the outflow from the lagoon occurs as a *subsurface* flow. This is exactly the opposite of the classic estuarine circulation described previously.

A Case Study: The Mediterranean Sea

The Mediterranean (*medi* = middle, *terra* = land) Sea is a most unusual body of water. It actually is a number of small seas connected by narrow necks of water into one larger sea. It is the remnant of an ancient sea (the Tethys Sea) that existed during the time when all the continents were combined several hundred million years ago. It is over 4300 meters (14,100 feet) deep, and is one of the few inland seas in the world underlain by oceanic crust.

The Mediterranean Sea has a very irregular coastline, which divides it into subseas such as the Aegean Sea and Adriatic Sea (Figure 11–11A). Each of these seas has a separate circulation pattern.

The Mediterranean is bounded by Europe and Asia Minor on the north and east and Africa to the south. As its name implies, it is surrounded by land except for two very shallow and narrow connections to other water bodies: to the Atlantic Ocean through the Strait of Gibraltar [about 14 kilometers (9 miles) wide], and to the Black Sea through the Bosporus [roughly 1.6 kilometers (1 mile) wide]. In addition, the Mediterranean Sea has a man-made passage to the Red Sea via the Suez Canal, a waterway 160 kilometers (100 miles) long that was completed in 1869.

The Mediterranean has two major basins. They are separated by an underwater ridge called a **sill** that extends from Sicily to the coast of Tunisia at a depth of 400 meters (1300 feet). This sill restricts the flow between the two basins, resulting in strong currents that run between Sicily and the Italian mainland through the Strait of Messina (Figure 11–11A).

---

[8] Remember that normal salinity in the open ocean averages 35‰.

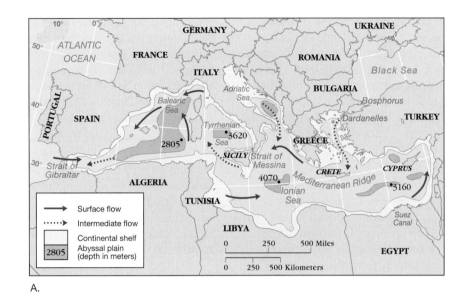

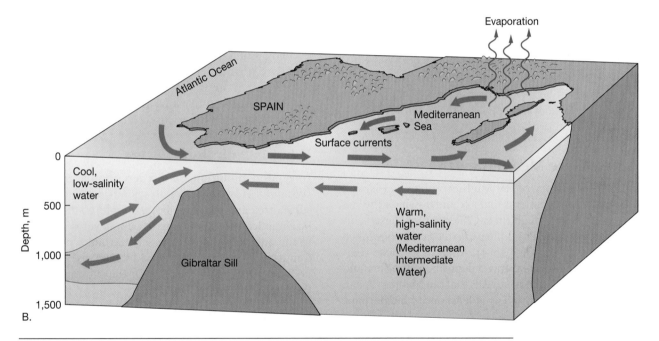

Figure 11–11 Mediterranean Sea.
A. Map of the Mediterranean Sea region showing its subseas, depths, sills (underwater ridges), surface flow, and intermediate flow. **B.** Diagrammatic view of Mediterranean circulation in the Gibraltar Sill area.

Mediterranean Circulation

Atlantic Ocean water enters the Mediterranean as a surface flow through the Strait of Gibraltar. It enters to replace water that rapidly evaporates in the very arid eastern end of the sea. The water level in the eastern Mediterranean is generally 15 centimeters (6 inches) lower than at the Strait of Gibraltar. The surface flow follows the northern coast of Africa throughout the length of the Mediterranean and spreads northward across the sea (Figure 11–11A).

The remaining Atlantic Ocean water continues eastward to Cyprus. Here, during winter, it sinks to form

what is called the *Mediterranean Intermediate Water*, which has a temperature of 15 degrees centigrade (59 degrees Fahrenheit) and a salinity of 39.1‰. This water flows westward at a depth of 200 to 600 meters (650 to 2000 feet) and returns to the North Atlantic, but this time as a *subsurface* flow through the Strait of Gibraltar (Figure 11–11B).

By the time Mediterranean Intermediate Water passes through Gibraltar, its temperature has dropped to 13 degrees centigrade (55 degrees Fahrenheit) and its salinity to 37.3‰. It is still denser than even Antarctic bottom water and much denser than water at this depth in the Atlantic Ocean, so it moves down the continental slope. While descending, it mixes with Atlantic Ocean water and becomes less dense. At a depth of about 1000 meters (3300 feet) it assumes a density equal to that of Atlantic Ocean water at that depth and spreads in all directions (Figure 11–11B). It remains detectable at depth over a broad area by its high salinity. In fact, this water mass has been detected in the far northern part of the Atlantic Ocean near Iceland.

Circulation between the Mediterranean Sea and the Atlantic Ocean is typical of closed, restricted basins where evaporation exceeds precipitation. Such restricted basins always lose water rapidly to surface evaporation, and this water must be replaced by surface inflow from the open ocean. Evaporation of inflowing water from the open ocean increases the sea's salinity to very high values. This denser water eventually sinks and returns to the open ocean as a subsurface flow.

This circulation pattern is opposite to the pattern characteristic of estuaries, where surface fresh water flow goes into the open ocean and saline subsurface flow enters the estuary. The Mediterranean Sea is the classic area for this type of circulation, which is called **Mediterranean circulation**. Conversely, typical estuarine circulation develops between a marginal water body and the ocean when fresh water input exceeds the water loss to evaporation.

History of the Mediterranean Sea

The geologic history of the Mediterranean Sea is well preserved in the sediments that are laid down on the sea floor beneath it through time. Collection of these sediments is accomplished by drilling into the layers to retrieve sediment cores (see Chapter 4). Analysis of these sediments (including unusual gravel deposits, evaporites, and shallow-water carbonates) reveals that the Mediterranean Sea must have nearly dried up at least once (and perhaps several times) in its history.

Based on evidence obtained from deep-sea cores, the Mediterranean Sea was cut off from the Atlantic Ocean at the narrow Strait of Gibraltar—in effect, creating a dam—about 6 million years ago. This may have been accomplished by tectonic activity or a drop in sea level. With the inflow of the Atlantic Ocean eliminated, high evaporation rates caused the Mediterranean Sea to nearly evaporate in just a few thousand years. As the seawater evaporated, the dissolved substances remained within the water and caused the water to have higher and higher salinity. In time, the dissolved substances began to precipitate out of water, resulting in the deposition of thick layers of evaporite minerals on the sea floor. Eventually, most of the water evaporated and a hot, salty, dry basin floor remained far below sea level. The thickness of evaporite minerals suggests that the basin may have partially filled and dried up several times during this period. At the same time, unusual gravel deposits washed in from the continents and shallow-water carbonate algal

mats called stromatolites also formed. Other supporting evidence for the drying of the Mediterranean Sea includes changes in climate, fossil evidence, and even deep gorges cut into the surrounding river valleys. The dry Mediterranean Sea basin would have been an impressive sight: as hot as Death Valley and three times deeper (and vastly wider) than the Grand Canyon.

A half million years later, erosion of the sill, further tectonic activity at Gibraltar, or a rise in sea level caused the dam to breach. Estimated to be 1000 times larger than the flow of all rivers in the world, the waterfall that spilled into the Mediterranean was probably the largest waterfall the world has ever seen. At that rate, the Mediterranean would have again been full of seawater in only 100 years.

Pollution in Coastal Waters

As the use of coastal areas has increased for residences, recreation, and commerce, pollution of coastal waters has increased as well. Coastal waters are in greater risk of being polluted than the open ocean for two reasons: (1) More pollution is dumped into coastal waters than into the open ocean, and (2) coastal waters are not as well circulated as the open ocean. Thus, many of the concerns of pollution are focused on coastal waters.

What Is Pollution?

In general, **pollution** can be described as *any harmful substance*. However, how do scientists determine which substances are harmful? For example, a substance may be esthetically unappealing to people yet is not harmful to the environment. Conversely, certain types of pollution cannot be easily detected by humans, yet they can do harm to the environment. Or the substance may not be immediately harmful, but it causes harm years, decades, or even centuries later. Also, to whom must this harm be done? For instance, some marine species thrive when exposed to a particular pollutant that is quite toxic to other species. It is interesting to note that some natural conditions in coastal waters may be considered "pollution" by some people. An example is dead seaweed on the beach. Nature may produce many conditions we do not like, but it does not pollute. Should conditions such as this be prevented from occurring? Moreover, the amount of a pollutant is also important: If a substance that causes pollution is present in extremely tiny amounts, can it still be characterized as a pollutant? All of these questions are difficult to answer.

In an effort to clarify the definition of pollution, the World Health Organization defines pollution of the marine environment as:

The introduction by man, directly or indirectly, of substances or energy into the marine environment, including estuaries, which results or is likely to result in such deleterious effects as harm to living resources and marine life, hazards to human health, hindrance to marine activities, including fishing and other legitimate uses of the sea, impairment of quality for use of sea water and reduction of amenities.

As far as the marine environment is concerned, however, there is great difficulty in establishing the degree to which pollution is occurring. In most cases, an area affected by pollution was not studied sufficiently before society's introduction of pollutants. Therefore, scientists cannot tell how the marine environment has been altered by these pollutants. In essence, there is no adequate baseline from which scientists can compare polluted versus nonpolluted regions. And since the marine environment is affected by decade- to century-long cycles, it is difficult to determine whether a change is due to a natural biologic cycle or whether it is caused by any number of introduced pollutants, many of which combine to produce new compounds.

To date, the most widely used technique for determining the level of pollutant concentration that negatively affects the living resources of the ocean is the **standard laboratory bioassay**. Regulatory agencies such as the Environmental Protection Agency (EPA) use a bioassay that determines the concentration of a pollutant that causes 50 percent mortality among the test organisms. If a pollutant exceeds a 50 percent mortality rate, then concentration limits are established for the discharge of the pollutant into coastal waters. One shortcoming of the bioassay is that it does not predict the long-term effect of sublethal doses on marine organisms. Another shortcoming is that is does not take into account the combining of a pollutant with other chemicals, which potentially creates new pollutants.

Another issue involves limiting the dumping of pollutants in coastal waters. If limits are placed on dumping wastes in coastal regions, where should the waste be dumped? For instance, proper waste disposal facilities on land (such as landfills) have capacities that are already being exceeded. One region that is being carefully scrutinized as a location for waste disposal is the open ocean. The open ocean is different than the coastal ocean in that mixing mechanisms (waves, tides, and currents) serve to distribute a pollutant over a wide area—even as far as an entire ocean basin. By diluting pollutants, they are often less harmful. It is an interesting proposition, but do we really want to distribute a pollutant across an entire ocean without knowing what the long-term effects might be?

Some experts believe that we should not dump *anything* in the ocean, while others believe that the ocean can be a repository for many of society's wastes, as long as proper monitoring is conducted. Certainly, there are no easy answers and the issues are complex. What is clear is that more research is needed to assess the impact of pollutants in the ocean. The following sections present some well-documented cases of chemical pollution of coastal waters by petroleum, sewage, halogenated hydrocarbons, and mercury.

Petroleum

Major oil (petroleum) spills into the ocean are a fact of our modern oil-powered economy. Some oil spills are the result of tanker accidents, which are caused by collisions, tankers running aground, or loading/unloading accidents. Other spills are caused by the blowout of undersea oil wells in the coastal ocean during drilling or while producing (Figure 11–12). Still others are intentional, such as during the Persian Gulf War of 1991, when 477 million liters (126 million gallons) of crude oil were discharged into the Persian Gulf. Although it is useful to know how oil enters the oceans and from what sources, let's focus on the effects of oil pollution on marine organisms and on the marine environment in general.

Oil is a **hydrocarbon**, meaning that it is a chemical composed of the elements *hydrogen* and *carbon*. Hydrocarbons are organic substances, and thus are broken down by microorganisms. A substance that can be biologically degraded in this way is said to be "biodegradable." Because hydrocarbons are biodegradable, it may be surprising that many experts consider oil to be among the *least* damaging pollutants introduced into the ocean! In fact, natural undersea oil seeps have occurred for millions of years, and the ocean ecosystem seems to be unaffected (or even, since oil is a source of energy, *enhanced*) by their presence.

Still, oil is a complex mixture of hydrocarbons and other substances, including the elements oxygen, nitrogen, and sulfur, along with various trace metals. When this

Figure 11–12 **Blowout from the Ixtoc #1 Well, Gulf of Mexico.**

Firefighting equipment is shown spraying water around the flaming oil directly above the wellhead that blew out in June 1979, creating the world's largest oil spill. Oil can be seen spreading across the ocean surface.

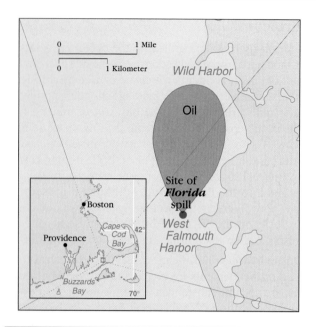

Figure 11–13 *Florida* oil spill at West Falmouth Harbor, Massachusetts.

When the barge *Florida* came ashore and ruptured, currents carried its load of number-2 fuel oil northward into Wild Harbor, where the most severe damage occurred.

complex chemical mixture is combined with seawater—another complex chemical mixture, which also contains organisms—the result can indeed be negative to those organisms. In addition, the sheer amount of oil spilled at one time is devastating to many forms of life that are killed outright by being coated with oil.

The Florida Spill in West Falmouth Harbor One of the best-studied oil spills in the United States occurred in September 1969 near West Falmouth Harbor in Buzzards Bay, Massachusetts. When the barge *Florida* came ashore and ruptured, it spilled about 680,000 liters (180,000 gallons) of number-2 fuel oil, which is similar to diesel oil and is used in home heating. The oil spread northward into Wild Harbor, where the most severe damage occurred (Figure 11–13).

How long does it take for a shore to recover from an oil spill? Because of the chemical complexity of petroleum, it is difficult to answer this question. In the case of the *Florida* spill, the initial kill was nearly absolute for marsh grasses and intertidal and subtidal[9] animals in the most severely oiled area. A sharp reduction in *species diversity* (the number of different species present) was accompanied by rapid increases in the population of a species that is resistant to oil: polychaete worms.[10] In

fact, a single small red polychaete worm species, *Capitella capiata*, accounted for up to 99.9 percent of the individuals collected in samples of the most severely oiled locations during the first year. Species diversity did not increase appreciably until well into the third year after the spill.

During a period from three to five years after the spill, marsh grasses and animals reentered the area. Amazingly, no visible damage could be seen after 10 years. After 20 years, there was virtually no oil in the subtidal sediments, and the intertidal marsh sediments were more than 99 percent free of oil from the *Florida* spill. However, at the most heavily oiled site in Wild Harbor, oil was still present in beaches at a depth of 15 centimeters (6 inches). In this region, there was enough oil 20 years after the spill to kill animals that burrow into the sediments.

Considering the extensive damage that was caused by the oil spill at Wild Harbor in 1969, it is clear that even in quiet and protected marsh environments, recovery from an oil spill can occur much more quickly than some researchers thought possible. Evidently, natural processes effectively biodegrade and remove oil from the marine environment, although it can take at least a couple of decades. Studies of other areas affected by oil spills indicate that the majority of these areas seem to experience no long-term damage to the environment.

The Argo Merchant *Spill off Nantucket Island* When oil spills do not come ashore, their effects are less obvious. The sinking of the *Argo Merchant* after it ran aground on Fishing Rip Shoals off Nantucket Island, Massachusetts, in December 1976 provides the best-studied effects of such a spill. The vessel held 29 million liters (7.7 million gallons) of number-6 fuel oil, which was spilled into the ocean 40 kilometers (25 miles) southeast of Nantucket Island (Figure 11–14). Fortunately, the winds were such that no oil came ashore. The surface slick moved eastward out to sea and was gone within a month of the spill.

Although the oil was not visible for long, it did significant damage to many organisms. The biological damage was investigated by scientists from Woods Hole Oceanographic Institution, the National Marine Fisheries Service, and local institutions.

The primary observable effect of the *Argo Merchant* spill on marine organisms was in planktonic (drifting) fish eggs—pollock and cod—that were collected shortly after the spill. Of the 49 pollock eggs recovered, 94 percent were coated with oil; and of the 60 cod eggs recovered, 60 percent were also oil fouled (Figure 11–15). In addition, 20 percent of the cod eggs and 46 percent of the pollock eggs were dead or dying. Little is known about the natural mortality rate of such fish eggs. For comparison, however, only 4 percent of cod eggs spawned in a laboratory under natural conditions were dead or dying at a similar developmental stage.

---

[9] The intertidal zone extends from high to low tide; the subtidal zone is below that.

[10] Polychaete worms are close relatives of segmented earthworms found on land.

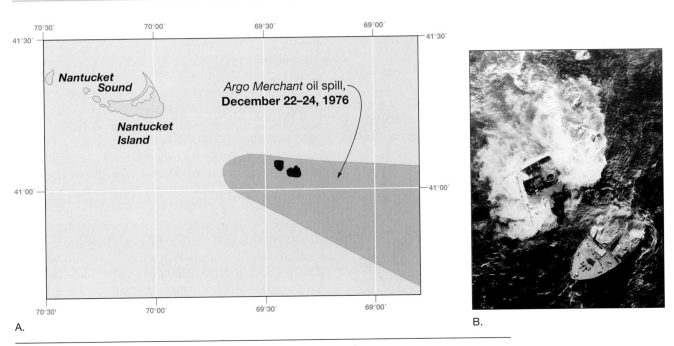

Figure 11–14 *Argo Merchant* oil spill off Nantucket, Massachusetts.
A. Map showing the ship's grounding site (*black*) and oiled area of the ocean (*gray*). **B.** The *Argo Merchant*, after breaking up and spilling much of its cargo into the Atlantic Ocean southeast of Nantucket Island, Massachusetts.

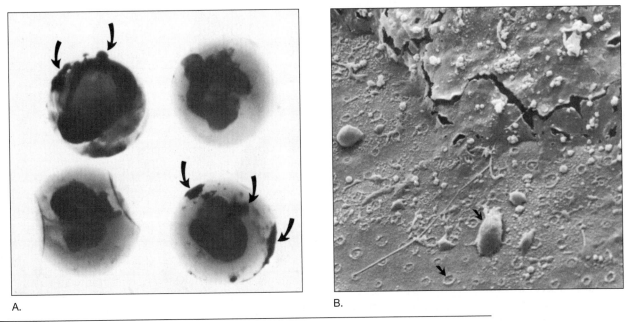

Figure 11–15 Pollock eggs affected by the *Argo Merchant* spill.
A. The eggs at the upper left and lower right show outer membranes that are contaminated with oil (*arrows*); the other two eggs have malformed or abnormal embryos. The eggs are about 1 millimeter (0.04 inch) in diameter. **B.** A photomicrograph of a portion of the surface of an oil-contaminated egg (magnified about 5000 times). The upper arrow points to one of many oil droplets; the lower arrow points to a membrane pore.

Because most of the oil floated as a surface slick, contamination in subsurface water samples did not exceed more than 250 parts per billion. Other than the described damage to fish eggs and significant damage to other plankton, little direct evidence of major biological damage was recorded. Numerous oiled birds, however, did wash ashore at Nantucket Island and Martha's Vineyard.

Studies indicate that each season, pollock females spawn about 225,000 eggs each, and that cod spawn about 1 million eggs. Because of these large numbers, and the fact that the fish spawn over a large area from New Jersey to Greenland, it is unlikely that a single oil spill would significantly affect these fisheries. Still, fishing communities of the northwest Atlantic are pleased about the poor results of oil exploration on Georges Bank. This is because it seems doubtful that the pollock and cod fisheries—both severely stressed by overexploitation—would survive the level of pollution brought about by even a small oil spill.

Cleaning Oil Spills Perhaps the best method of protecting areas from oil spills is to prevent oil spills from occurring in the first place. Because of our society's reliance on petroleum products, however, oil spills are a problem that will be with us for many years (Figure 11–16). It is likely that oil spills may become more common as petroleum reserves underlying the continental shelves of the world are increasingly exploited. Therefore, much attention has been focused on cleaning oil spills in the marine environment.

Once an oil spill occurs in the ocean, lower-density oil will initially float on top of higher-density water. This floating oil can be collected by skimming the surface with specially designed skimmers or it can be soaked into absorbent materials. However, the collected oil (or oiled materials) must still be disposed of elsewhere. In addition, the cleanup must begin immediately after the spill or the oil will begin to disperse and come ashore or form tar balls and sink. Because oil is a natural energy source, it is naturally biodegraded by microorganisms such as bacteria and fungi. The method of using these organisms to help clean oil spills is called **bioremediation**.

Certain types of bacteria and fungi are effective in breaking down only a particular variety of hydrocarbon, and none is effective against all forms. In 1980 Dr. A. M. Chakrabarty, now a microbiologist at the University of Illinois, worked at the General Electric Research and Development Center in Schenectady, New York. There he isolated a microorganism capable of breaking down

Figure 11–16 Who is at fault?

nearly two-thirds of the hydrocarbons in most crude oil spills.

Bioremediation can take the form of releasing bacteria directly into the marine environment. A strain of oil-degrading bacteria was released into the Gulf of Mexico to test its effectiveness in helping clean up about 15 million liters (4 million gallons) of crude oil spilled after a 1990 explosion disabled the tanker *Mega Borg*. Preliminary results indicate that the bacteria are effective at reducing the amount of oil, and no negative effects on the ecology of gulf waters due to the bacteria have been reported.

Virtually all marine ecosystems harbor naturally occurring bacteria that degrade hydrocarbons. Another way to clean oil spills using bioremediation is to provide conditions that will stimulate the growth of these bacteria. Encouraged by good results on small test plots, Exxon spent $10 million dollars to spread fertilizers rich in phosphorus and nitrogen on Alaskan shorelines to boost the development of indigenous oil-eating bacteria after the *Exxon Valdez* spill (Box 11–1). The application of these fertilizers resulted in a cleanup rate more than twice that under natural conditions. Bioremediation has proven to be one of the most successful methods for coping with the deleterious effects of oil spills.

Sewage Sludge

Sewage sludge is the semisolid material that remains as a result of typical sewage treatment. During the past several decades, at least 500,000 metric tons (550,000 short tons) of sewage sludge has been dumped through sewage outfalls into the coastal waters of southern California each year. This sludge contains a toxic brew of human waste, oil, zinc, copper, lead, silver, mercury, pesticides, and other chemicals. In the New York Bight between Long Island and the New Jersey shore, at least several million metric tons of sewage sludge has been dumped each year for the past several decades.

Although the Clean Water Act of 1972 prohibited dumping of sewage into the ocean after 1981, the high cost of treating and disposing of sewage sludge on land resulted in extension waivers being granted to many municipalities. In the summer of 1988, an unrelated event forced a change in the law. Nonbiodegradable debris including medical waste—probably carried into the ocean through storm drains by heavy rains—washed up on Atlantic coast beaches and caused a negative impact on the tourist business. Although this event was completely unrelated to sewage disposal at sea, it focused public awareness on ocean pollution and brought about the passage of new legislation to terminate sewage disposal in the ocean.

It now appears that political pressure to clean up coastal waters is sufficient to ensure that treatment and land disposal will be the primary disposal procedure of the future. For the present, Los Angeles has stopped pumping

Box 11–1
The *Exxon Valdez* Oil Spill: Not the Worst Spill Ever

Many oil spills of various sizes have occurred in the ocean. One of the most publicized oil spills in the last few decades was from the supertanker *Exxon Valdez*.

Crude oil produced from the North Slope of Alaska is carried by pipeline to the southern port of Valdez, Alaska. At this port, the oil is loaded onto supertankers like the *Exxon Valdez*, capable of holding almost 200 million liters (53 million gallons) when full. On March 29, 1989, the tanker left Valdez with a full load of crude oil and was headed toward refineries in California. She was only 40 kilometers (25 miles) out of Valdez when the ship's officers noted icebergs from nearby Columbia Glacier within the shipping channel. While maneuvering around the icebergs, the ship ran aground on a shallowly submerged rocky outcrop known as Bligh Reef (Figure 11B), rupturing eight of the ship's 11 cargo tanks. About 22 percent of her cargo—almost 44 million liters (about 11.6 million gallons) of oil—spilled into Prince William Sound.

Most of the volatile portions of the spilled oil (up to 25 percent of oil) evaporated within a few days, but the remainder formed a large oil slick at the surface that was moved by winds, waves, currents, and tides. This spill in the pristine waters of Prince William Sound spread to the Gulf of Alaska, where over 1775 kilometers (1000 miles) of shoreline were fouled when about 40 to 45 percent of the total amount of oil began to wash up and coat beaches (Figure 11C). For the floating oil, the heavier components of the oil eventually formed balls of varying sizes, some of which sank to the bottom and were assimilated by organisms or incorporated into

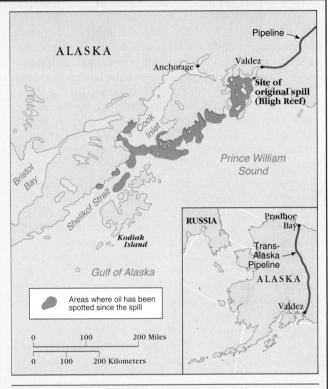

Figure 11B *Exxon Valdez* oil spill, Alaska.

A.

B.

Figure 11C Oil in the Ocean and on the Beach from the *Exxon Valdez* oil spill.

(continued)

sediments. Bacteria naturally decompose the tar balls and the oil that washes up on beaches, but the process may take several years to complete.

The U.S. Fish and Wildlife Service reported that at least 994 sea otters and 34,434 birds were killed outright by the spill. By some official estimates, the actual kill could have been 10 times that amount, because not all dead organisms are recovered. Due to the remote location and the size of the area affected by oil, the exact total death toll will never be known. As well, the long-term effect of the oil on organisms is largely unknown. It was predicted that affected waters would have a long, slow recovery, but the fisheries that closed in 1989 bounced back with *record* takes in 1990.

Exxon spent over $2.5 billion in cleaning the spill (and another $900 million in subsequent years for restoration). On the oil in the water, absorbent materials and skimming devices were used. One of the primary methods used to clean beaches was super hot water sprayed with high-pressure hoses. The hot water, at temperatures up to 160 degrees centigrade (40 degrees Fahrenheit), was blasted onto rocky beaches, which did help clean the beaches of oil but had the unwanted effect of killing most shoreline organisms. Recent studies show that the area's beaches appear to have largely recovered from the effects of the oil. Analysis of the clean-up effort, as compared to areas that were left to biodegrade naturally, indicates that perhaps the best—but clearly one of the most difficult—strategies to aid recovery is to do nothing at all.

The *Exxon Valdez* spill was the largest oil spill in the history of U.S. waters. Surprisingly, the spill was relatively small when compared to others around the world. In fact, it was only the *twenty-second* largest oil spill worldwide. The largest

oil spill on record occurred in June 1979, as a result of extracting oil from beneath the sea floor in the Gulf of Mexico. A Petroleus Mexicanos (PEMEX) oil-drilling station named Ixtoc #1 in the Bay of Campeche off the Yucatán peninsula, Mexico, blew out and caught fire. Before it was capped nearly 10 months later, it spewed 530 million liters (140 million gallons) of oil into the Gulf of Mexico (Figure 11D; see also Figure 11–12). The amount of oil from this spill was over *12 times greater* than the spill from the *Exxon Valdez*.

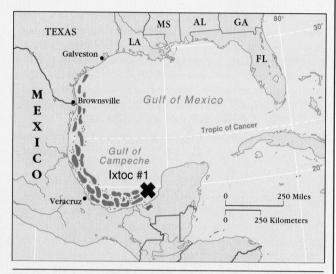

Figure 11D Location of the Gulf of Campeche blowout and oil slick that affected the Texas coast.

sewage into the sea. If the city is allowed to resume this practice, the sewage sludge pumped through the lines will have to be scrubbed of toxic chemicals and pathogens.

New York's Sewage Sludge Disposal at Sea

Off the East Coast, more than 8 million metric tons of sewage sludge has been transported offshore by barge and dumped in the ocean each year. This material was spread out over sites totaling 150 square kilometers (58 square miles) within the New York Bight Sludge Site and the Philadelphia Sludge Site (Figure 11–17).

The water depth at the New York Bight Sludge Site is about 29 meters (95 feet), and Philadelphia's site is 40 meters (130 feet) deep. In such shallow water, the water column displays relatively uniform characteristics from top to bottom. Even the smallest sludge particles reach the bottom without undergoing much horizontal transport, and the ecology of the dump site can be severely affected. At the very least, such concentration of organic and inorganic nutrients seriously disrupts the chemical cycling of nutrients. Greatly reduced species diversity results, and in some locations the environment becomes devoid of oxygen (anoxic).

In 1986, the shallow-water sites were abandoned and sewage was subsequently transported to a deep-water site 171 kilometers (106 miles) out to sea (Figure 11–17). At the deep-water site, beyond the continental shelf break, there is usually a well-developed density gradient that separates low-density, warmer surface water from high-density, colder deep water. Internal waves moving along this density gradient can retard the sinking rate of particles and may allow a horizontal transport rate 100 times greater than the sinking rate.

Local fishermen have been concerned about the deep-water dumping and reported adverse effects on their fisheries soon after it began. Also, there was concern that the sewage could be transported in eddies of the Gulf Stream (see Chapter 7) and transported great distances. This East Coast sewage dumping program was terminated in 1993, and municipalities must now find the money for land-based disposal.

Boston Harbor Sewage Project

Some 48 different communities in the greater Boston area have historically deposited their sewage into Boston Harbor, making it one of the most polluted water bodies in the country. The main

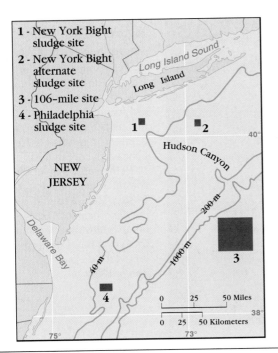

Figure 11–17 Atlantic sewage sludge disposal sites.
More than 8 million metric tons of sewage sludge was dumped by barge annually at the New York Bight Sludge Site (1) and the Philadelphia Sludge Site (4). After 1986, the new dump site is the larger and deeper-water 106-mile site (3).

problem with the disposal system was that it deposited the sewage at the entrance to Boston Harbor, where it could be swept back into the bay by tidal currents. A new system is being developed to carry the sewage through a tunnel 15.3 kilometers (9.5 miles) long into the deeper water of Massachusetts Bay (Figure 11–18A). By the year 2000, all of the effluent will be treated with bacteria-killing chlorine. To pay the $4 billion system cost, the average sewage bill for a Boston-area household will be about $1200.

Some fear that the project will degrade the environment in Cape Cod Bay and Stellwagen Bank, an important whale habitat (Figure 11–18B). The area's natural resources were acknowledged recently when the area was designated a National Marine Sanctuary, which may affect the feasibility of dumping there.

DDT and PCBs

The pesticide **DDT** (dichlorodiphenyltrichlorethane) and the industrial chemicals called **PCBs** (polychlorinated biphenyls) are now found throughout the marine environment. They are persistent, biologically active chemicals that have been put into the oceans entirely as a result of human activities.

DDT became a widely used pesticide during the 1950s and was responsible for the rapid improvement in crop production throughout developing countries for several decades. However, its extreme effectiveness and persistence as a toxin in the environment eventually wrought a host of environmental problems.

PCBs are industrial chemicals found in a wide variety of products from paint to plastics. They have been indicated as causes of spontaneous abortions in sea lions and the death of shrimp in Escambia Bay, Florida.

DDT and Eggshells A near-total ban on U.S. production of the pesticide DDT was emplaced in 1971. By that time, 2 billion kilograms (4.4 billion pounds) had been manufactured, most by the United States. Since 1972, the use of DDT in the Northern Hemisphere has virtually ceased.

The danger of excessive use of DDT and similar pesticides first became apparent in the marine environment when it affected marine bird populations. During the 1960s there was a serious decline in the brown pelican population of Anacapa Island off the coast of California (Figure 11–19). High concentrations of DDT in the fish eaten by the birds had caused them to produce excessively thin eggshells.

The osprey is a common bird of prey in coastal waters, similar to a large hawk. A decline in the osprey population of Long Island Sound began in the late 1950s and continued throughout the 1960s. This also was caused by thinning eggshells brought on by DDT contamination.

Studies showed a 1 percent increase in eggshell thickness for brown pelicans and ospreys from 1970 to 1976, as the concentration of DDT residue decreased. Because there was an increased egg hatching rate associated with these changes, it was concluded that DDT was the cause of the decline in the bird populations during the 1960s. Since the ban on DDT, those species that were affected by the chemical are making a remarkable comeback.

DDT and PCBs Linger in the Environment The main routes by which DDT and PCBs enter the ocean are through the atmosphere and river runoff. They are concentrated initially in the thin surface slick of organic chemicals at the ocean surface. Then they gradually sink to the bottom, attached to sinking particles. A study off the coast of Scotland indicated that open-ocean concentrations of DDT and PCBs are 10 and 12 times less, respectively, than in coastal waters. Long-term studies have shown that DDT residue in mollusks along the U.S. coasts peaked in 1968.

The pervasiveness of DDT and PCBs in the marine environment can best be demonstrated by the fact that Antarctic marine organisms contain measurable quantities of them. Obviously, there has been no agriculture or industry in Antarctica to explain their presence. Evidently, these substances have been transported from distant sources by winds and ocean currents.

Mercury and Minamata Disease

The metal **mercury** is a liquid at room temperatures, and has many industrial uses. Unfortunately, when it enters the ecosystem, mercury forms an organic compound that is toxic to most living things.

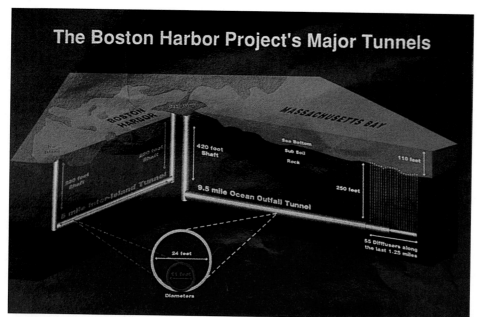

The Boston Harbor Project's Major Tunnels

A.

Figure 11–18 Boston Harbor Sewage Project. **A.** Diagrammatic view of Boston Harbor Project tunnels that will transport the sewage to an outfall 15 kilometers (9.5 miles) away at a depth of 76 meters (250 feet) beneath the ocean floor. **B.** Bathymetric map of the coastal ocean in the Boston–Cape Cod area, showing the proximity of the outfall to Stellwagen Bank and Cape Cod Bay. Depths in meters, vertical exaggeration = 100 ×.

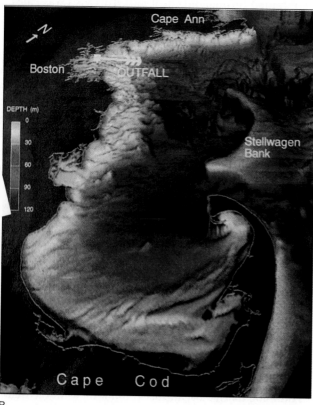

B.

Figure 11–19 Brown pelican (*Pelecanus occidentalis*) in breeding plumage.

The stage was set for the first tragic occurrence of mercury poisoning with the establishment of a chemical factory on Minamata Bay, Japan, in 1938. A product of this plant was acetaldehyde, which requires mercury in its manufacture. Mercury was discharged into Minamata Bay, where it was changed by bacterial action into a form that was later ingested and concentrated in the tissues of larger marine organisms. The first ecological changes in Minamata Bay were reported in 1950 and, because of in-gestion of contaminated seafood, human effects were noted as early as 1953. The mercury poisoning that is now known as Minamata disease became epidemic in 1956, when the plant was only 18 years old. **Minamata disease** is a degenerative neurologic disorder that affects the human nervous system. This was the first major human disaster resulting from ocean pollution.

It was not until 1968, however, that the Japanese government declared mercury the cause of the disease. The plant was immediately shut down, but by 1969, over

Box 11–2
From A to Z in Plastics: The Miracle Substance?

Even though plastic products have been in use for over a century, their commercial development occurred during World War II when shortages of rubber and other materials created great demand for alternative products. Plastic products have many properties that make them advantageous over other, similar materials: Plastic products are *lightweight*, *strong*, *durable*, and *inexpensive*. Thus, materials made of plastic were initially considered to be composed of a miracle substance. Since their introduction, the use of plastics has increased at a tremendous rate. Today, everything from airplane parts to zippers is made of plastic (Figure 11E). We wear plastics, drive in plastics, cook in plastics, and even carry plastic components inside us as artificial parts. The convenience of plastic items intended for one-time use has also contributed to the popularity of plastics.

However, what was once thought of as a miracle substance has several disadvantages. Disposing of this "throwaway" component of modern culture has already strained the capacity of land-based solid-waste disposal systems. Plastic waste is now an increasingly abundant component of oceanic flotsam (floating refuse). In fact, studies reveal that plastics constitute the vast majority of floating trash in all oceans worldwide. Unfortunately, the very same properties that make plastics so advantageous have also caused them to be unusually persistent and damaging when released into the marine environment:

- Their light weight causes them to float and be concentrated at the surface
- Their high strength results in entanglement of marine organisms
- Their durability means that they don't biodegrade easily, causing them to last almost indefinitely
- Their inexpensive cost allows them to be mass produced and used in almost everything

Some of the best-known negative effects of plastics on marine organisms have been documented in the strangulation of seals and birds caught in plastic netting and packing straps (Figure 11F). As well, marine turtles have been killed when they ingest plastic bags, evidently mistaking them for jellyfish or other transparent plankton on which they typically feed.

Small pellets ranging in size from a BB to a pea are used to produce essentially all plastic products. They are transported in bulk aboard commercial vessels and are found throughout the oceans. The pellets probably find their way into the oceans as a result of spillage at loading terminals. In coastal waters, plastic products used in fishing are commonly thrown overboard by recreational and commercial vessels. Plastic trash also finds its way into the open-ocean waters from careless people on land.

Even though plastic material does not sink or biodegrade, plastics eventually are removed from the ocean. This happens as ocean currents wash against the shorelines of islands and continents, depositing the plastic waste on beaches. Thus, even remote beaches are littered with plastic pellets and plastic trash. Studies conducted between 1984 and 1987 show that the plastic pellet content of beaches throughout the world is increasing. For instance, some Bermuda beaches have up to 10,000 pellets per square meter (10.7 square feet), and some beaches on Martha's Vineyard in Massachusetts yielded

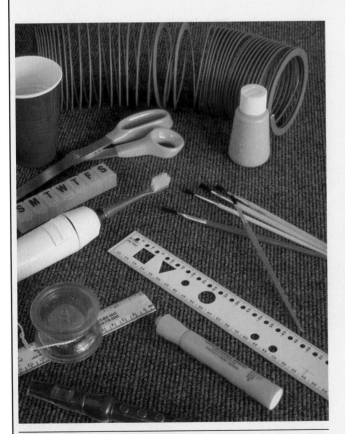

Figure 11E Plastic products.

100 people were known to suffer from the disease (Figure 11–20). Almost half of these victims died. A second occurrence of mercury poisoning resulted from pollution by another acetaldehyde factory, closed in 1965, in Niigata, Japan. Between 1965 and 1970, 47 fishing families contracted Minamata disease. Recent studies of the concentration of mercury in Minamata Bay have indicated that the bay no longer has unusually high mercury levels.

Evidently, there has been enough time for the mercury to be widely dispersed within the marine environment.

Bioaccumulation During the 1960s and 1970s mercury contamination in seafood received considerable attention. This is because certain marine organisms tend to concentrate within their tissues many substances found in minute concentrations in seawater. This process

(*continued*)

16,000 plastic spherules per square meter (10.7 square feet). Plastic trash is not limited to beaches: In the northern Sargasso Sea, researchers have counted more than 10,000 plastic pieces and 1500 pellets per square kilometer (0.4 square mile). The 1987 survey mentioned above found that since 1972, the concentration of plastic pellets had doubled! This suggests that there has also been a doubling of the amount of floating plastic trash in the ocean. Indeed, equipment deployed from research vessels often returns entangled in plastic trash such as 6-pack rings, styrofoam, and fishing lines, nets, and floats.

What can be done to limit the amount of plastic in the marine environment? People on land can help by limiting their use of disposable plastic, recycling plastic material, and disposing of their plastic trash properly. Internationally, plastics are the only substances that are prohibited from being dumped in the ocean. If people aboard ships follow this regulation, it would aid greatly in reducing the increasing concentration of plastics in the world's oceans.

Figure 11F Elephant seal with plastic packing strap around its neck.

Figure 11–20 A victim of Minamata disease.

is called **bioaccumulation**. Because mercury has been increasing in abundance in the ocean (mostly due to the mercury in disposable batteries being introduced into the environment), some seafood such as tuna and swordfish were thought to contain unusually high amounts of mercury.

Studies done on the amount of seafood consumed by various human populations led to the establishment of safe levels of mercury in fish to be marketed. To establish these levels, three variables were considered:

1. The fish consumption rate of each group of people under consideration

2. Mercury concentration in the fish being consumed by that population

3. The minimum ingestion rate of mercury that induces disease symptoms

Assuming that the data are dependable, and applying a safety factor of 10, a maximum allowable mercury concentration can be established. Thus, it can be asserted with a high degree of confidence that the health of the human population will be safeguarded from mercury poisoning if people do not exceed the intake of fish recommended.

Figure 11–21 shows how such data are used to derive a safe level of mercury in fish to be consumed by different groups of people. For the general populations of Japan, Sweden, and the United States, the average individual fish consumption is shown in Table 11–1. By com-

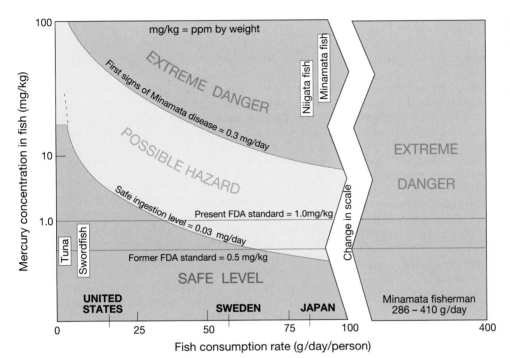

Figure 11–21 Mercury concentrations in fish versus consumption rates for various populations.

The graph shows that the first signs of Minamata disease can be expected at 0.3 mg/day. Safe ingestion level with a safety factor of 10 is 0.03 mg/day. The U.S. Food and Drug Administration has set a limit for mercury concentration in fish at 1.0 parts per million, making tuna and most swordfish safe to consume.

Table 11–1 **Fish consumed, mercury concentrations, and the amount of fish that can be safely consumed.**

Country	Average amount of fish consumed (grams/day)	Mercury poisoning symptoms occur when mercury concentrations in fish exceed (ppm)	Maximum concentration of mercury in fish that can be safely consumed[a] (ppm)
Japan	84	4	0.4
Sweden	56	6	0.6
United States	17	20	2.0

[a] Assumes a safety factor of 10 times.

parison, members of the Minamata fishing community averaged 286 and 410 grams of fish per day during winter and summer seasons, respectively.

The minimum level of mercury consumption that causes poisoning symptoms, as determined by Swedish scientists using Japanese data, is 0.3 milligrams per day over a 200-day period. The three populations shown in Figure 11–21 would begin to show symptoms if they ate fish with mercury concentrations shown in Table 11–1 (second column).

If a safety factor of 10 is applied, the maximum concentration of mercury in fish that could be safely consumed by these populations is also shown in Table 11–1 (third column). Although the U.S. Food and Drug Administration (FDA) initially established an extreme-

ly cautious limit of 0.5 part per million of mercury in fish, the present limit of 1 part per million adequately protects the health of U.S. citizens.

Essentially all tuna falls below this concentration, and most swordfish is acceptable.[11] The present limit corresponds to a highly conservative safety factor of 20 instead of 10. Thus, unless you eat an unusually large amount of tuna or swordfish, you should have little concern about contracting Minamata disease by consuming these fishes.

[11] Some swordfish has mercury concentration above the present 1.0 part per million limit imposed by the U.S. Food and Drug Administration.

? Students Sometimes Ask...

What is the strategy behind having such large ships for transporting oil? Wouldn't smaller ships be safer?
Theoretically, one larger ship has a lower chance of being damaged and spilling its load than several smaller ships. If a smaller ship spilled its oil, it would cause less damage, but there would be more spills over time because there would be more ship traffic. Imagine the problems that could potentially exist if there were 10 times the number of ships transporting oil! Perhaps a good way to think about this is to analyze the amount of *exposure* of a spill to the ocean. For example, if smaller ships are used, then more trips will be required to transport the same amount of oil. An increase in the number of trips increases ship traffic and thus increases the exposure of a spill to the ocean, just as frequent trips in your car increase your exposure to being involved in an accident. The strategy behind having larger ships is that it limits the exposure of a spill to the ocean by necessitating fewer journeys. If only a few trips are made (and during those trips, special caution is applied), then over time less oil will be spilled.

Legislation has been introduced that would require new tankers to be constructed with stronger, double-hulled designs. However, analysis of the *Exxon Valdez* spill indicates that even a double-hulled tanker would not have prevented the disaster. Tanker designs are also being modified to limit the amount of oil spilled should there be a hull rupture.

Can the influence of highly productive estuaries extend beyond the estuary itself?
Yes. Microscopic plankton (floating algal cells and animals), which are important in establishing a food web, abound in nutrient-rich estuary waters. Their influence extends well beyond the estuary itself, however. Because plankton float, they are carried by ocean currents from estuaries to the adjacent continental shelf areas. Studies conducted to determine the influence of estuaries on coastal plankton communities reveal that the abundance of populations of animal plankton may be explained by the fact that all were produced in estuaries and flushed into the shelf waters by estuarine outflow. Evidently, estuaries have a powerful influence on the plankton populations and the overall ecology of the adjacent continental shelf.

The story of the Mediterranean Sea drying up and refilling is intriguing. A waterfall that large must have created severe scouring at its base. Is there any evidence for this?
Imagine what an astonishing sight and sound a waterfall that large must have been! You're right in thinking that the waterfall must have created impressive scouring where it dumped into the dry Mediterranean basin. Several hundred meters of sediment is missing on the sea floor immediately to the east of the sill at the Strait of Gibraltar. The missing sediment extends to distances of up to 160 kilometers (100 miles) from Gibraltar, and in its place is a sequence of broken rock material and other disturbed sediment. The missing sediment is found in other parts of the basin, so the missing sediment was most likely caused from mechanical scouring by the world's largest waterfall.

I've seen signs on storm drains to the effect that any water dumped into the drain goes directly into the ocean. Don't storm drains receive treatment before emptying into the ocean?
Contrary to popular belief, water (and any other material) that goes down a storm drain does *not* receive any treatment before being emptied into a river or directly into the ocean. Sewage treatment plants receive enough waste to process without the additional runoff from storms. Thus, it is important to monitor carefully what is disposed into storm drains. For instance, some people discharge used motor oil into storm drains, thinking that the waste oil will be received and processed by a sewage plant. A good rule of thumb is this: *Don't put anything down a storm drain that you wouldn't put directly into the ocean itself.*

What is "non-point-source" pollution?
Non-point-source pollution, also called "poison runoff," is any type of pollution entering the surface water system from sources other than underwater pipelines. Mostly, non-point-source pollution arrives at the ocean via runoff from storm drains. Since non-point-source pollution comes from many different locations, it is difficult to pinpoint where the pollution originates, although the *cause* of the pollution may be readily apparent. Trash that is washed down a storm drain to the ocean is one such example. Other examples include pesticides and fertilizers from agriculture, and oil from automobiles that is washed to the ocean whenever it rains (the amount of road oil and improperly disposed oil regularly discharged each year into U.S. waters as non-point-source pollution is equivalent to as much as 26 times the amount of the *Exxon Valdez* oil spill!).

Non-point-source pollution is as difficult to prevent as it is to determine its exact point of origination. Since the general public is directly responsible for this type of pollution, perhaps the best method of prevention is through educating the public that their actions are directly responsible for this type of pollution. Then it is up to *all of us* to be involved in preventing non-point-source pollution.

Is dilution the solution to ocean pollution?
That's certainly a catchy phrase, but the implications are controversial. The idea suggests that the oceans can be used as a disposal area for society's wastes, as long as the wastes are properly spread out and diluted such that they no longer pose a threat to marine organisms (which in itself is difficult to determine). Because the oceans are vast in size and are composed mostly of a good solvent (water), the oceans appear to be ideally suited to this type of disposal strategy. In addition, the oceans have good mixing mechanisms (currents, waves, and tides), in effect being able to dilute many forms of pollution.

It is interesting to note that air pollution was once viewed in a similar manner. That is, disposal of pollutants into the atmosphere was thought to be acceptable as long as they were dispersed widely and high enough. Thus, tall smokestacks were constructed for this purpose. Over time, however, pollutants such as nitric acid and acid sulfates increased in the atmosphere to the point that acid rain is now a problem. The ocean, like the atmosphere, has a certain *holding capacity* for pollutants. The holding capacity for various pollutants dissolved in the ocean is not readily agreed upon even by experts.

As disposal sites on land begin to fill, the ocean is increasingly evaluated as an area for disposal of society's wastes. One thing that we can *all* do is to limit the amount of waste generated, thereby alleviating some of the problem of where to put waste. However, it is likely that the ocean will continue to be used as a dumping ground in the foreseeable future. In spite of many new ideas about disposal techniques, there doesn't yet appear to be any long-term "solution" to ocean pollution.

Summary

Coastal waters support about 95 percent of the total mass of life in the oceans, and they are important areas for commerce, recreation, fisheries, and the disposal of waste. Characteristics of the coastal ocean, such as temperature and salinity, vary over a greater range than the open ocean. This is because the coastal ocean is shallow and experiences river runoff, tidal currents, and seasonal changes in solar radiation. Coastal geostrophic currents are produced as a result of runoff of fresh water and coastal winds.

Estuaries are semienclosed bodies of water where freshwater runoff from the land mixes with ocean water. Estuaries are classified by their origin as coastal plain, fjord, bar built, or tectonic. Mixing patterns of fresh and saline water in estuaries include vertically mixed, slightly stratified, highly stratified, and salt wedge. Typical estuarine circulation consists of a surface flow of low-salinity water toward the mouth and a subsurface flow of marine water toward the head.

Estuaries provide important breeding and nursery areas for many marine organisms but often suffer because of human population pressures. The Columbia River Estuary is one such example, having become degraded because of agriculture, the logging industry, and from the construction of dams upstream. Another example is Chesapeake Bay, where an anoxic zone occurs during the summer which kills many commercially important species.

Wetlands are of two types: salt marsh and mangrove swamp. These biologically productive regions are alternately covered and exposed by the tides. Wetlands are important for scrubbing the water of land-derived pollutants before the water reaches the ocean. Despite their ecological and pollution-filtering importance, wetlands are rapidly disappearing from the world's coasts as they are destroyed by human activities.

Long offshore deposits called barrier islands protect marshes and lagoons. Some lagoons have restricted circulation with the ocean and may exhibit a great range of salinity and temperature conditions as a result of seasonal change.

The Mediterranean Sea has a classical circulation characteristic of restricted bodies of water in areas where evaporation greatly exceeds precipitation. This circulation is the reverse of estuarine circulation and is called Mediterranean circulation. Several lines of evidence suggest that the Mediterranean Sea was isolated from the waters of the Atlantic Ocean and almost completely evaporated between 5.5 and 6 million years ago.

Although marine pollution seems easily defined, an all-encompassing definition is quite detailed. There is also difficulty in establishing the degree to which pollution is occurring in most ocean areas. The most widely used technique for determining the effects of pollution is the standard laboratory bioassay, which determines the concentration of pollution that causes a 50 percent mortality rate among marine organisms. There is continuing debate about whether society's wastes should be dumped in the ocean.

Oil is a complex mixture of hydrocarbons and other substances, most of which are naturally biodegradable. Thus, many experts consider oil to be among the least damaging of all substances that are introduced into the marine environment. In areas that have experienced oil spills, recovery can be as rapid as a few years. Still, oil spills can cover large areas and kill many animals that are coated with oil.

Oil pollution reduces species diversity and persists longer on muddy bottoms than on sandy or rocky bottoms. Oil spills that do not come ashore, such as that of the *Argo Merchant*, do considerably less environmental damage than those that do. The greatest damage resulting from the *Argo Merchant* spill was probably to plankton, especially pollock and cod eggs. After the much-publicized *Exxon Valdez* oil spill in Alaska, many novel approaches were used to help clean the spilled oil, including the use of oil-eating bacteria as a form of bioremediation.

Millions of tons of sewage sludge has been dumped offshore in coastal waters. Although 1972 legislation required an end to dumping of sewage in the coastal ocean by 1981, exceptions continue to be made. Increased public concern resulted in new legislation to prohibit sewage dumping in the ocean.

Plastics were once thought to be composed of a miracle substance: lightweight, strong, durable, and inexpensive. Unfortunately, those same qualities create unwanted characteristics in the ocean in the form of floating plastic trash. The amount of plastic accumulating in the oceans has increased dramatically. Certain forms of plastic are known to be lethal to marine mammals and turtles, and national and international legislation has been enacted to ban the disposal of plastic in the oceans.

DDT and PCBs are persistent, biologically hazardous chemicals that have been introduced into the ocean by human activities. DDT pollution produced a decline in the Long Island osprey population in the 1950s and the brown pelican population of the California coast in the 1960s. Virtual cessation of DDT use in the Northern Hemisphere in 1972 allowed the recovery of both populations. The DDT thinned the eggshells and reduced the number of successful hatchings. PCBs have been implicated in causing health problems in sea lions and shrimp.

Mercury poisoning was the first major human disaster resulting from ocean pollution, which is now called Minamata disease after the bay in Japan where it first occurred in 1953. The toxic metal mercury bioaccumulates in the tissues of many large fish, most notably tuna and swordfish, and works its way up the food web. However, stringent mercury contamination levels have been established by the FDA that have been effective in preventing further poisonings.

Key Terms

Anoxic (p. 334)

Bar-built estuary (p. 331)

Barrier island (p. 331)

Bioaccumulation (p. 350)

Bioremediation (p. 344)

Coastal geostrophic current (p. 329)

Coastal plain estuary (p. 331)

Coastal water (p. 329)

Davidson Current (p. 331)

DDT (p. 347)

Drowned river valley (p. 331)

Estuarine circulation pattern (p. 332)

Estuary (p. 331)

Exclusive economic zone
(EEZ) (p. 327)

Fjord (p. 331)

Halocline (p. 329)

Highly stratified estuary (p. 332)

Hydrocarbon (p. 341)

Isohaline (p. 329)

Isothermal (p. 329)

Lagoon (p. 336)

Mangrove swamp (p. 336)

Mediterranean circulation (p. 340)

Mercury (p. 347)

Minamata disease (p. 348)

PCBs (p.347)

Pollution (p. 340)

Salt marsh (p. 336)

Salt-wedge estuary (p. 333)

Sewage sludge (p. 344)

Sill (p. 338)

Slightly stratified estuary (p. 332)

Standard laboratory bioassay (p. 341)

Tectonic estuary (p. 332)

Territorial sea (p. 327)

Thermocline (p. 329)

United Nations Conference on
the Law of the Sea (p. 327)

Vertically mixed estuary (p. 332)

Wetland (p. 336)

Questions And Exercises

1. Discuss possible reasons why less-developed nations believe that the open ocean is the common heritage of all, whereas the more developed nations believe that the open ocean's resources belong to those who recover them.

2. For coastal oceans where deep mixing does not occur, discuss the effect that offshore winds and fresh water runoff will have on salinity distribution. How will the winter and summer seasons affect the temperature distribution in the water column?

3. How does coastal runoff of low-salinity water produce a coastal geostrophic current?

4. Based on their origin, draw and describe the four major classes of estuaries.

5. Describe the difference between vertically mixed and salt wedge estuaries in terms of salinity distribution, depth, and volume of river flow. Which displays the more classical estuarine circulation pattern?

6. Discuss the factors that may explain why the surface salinity of Chesapeake Bay is greater on its east side, and periods of summer anoxia in the deep-water mass are increasing in severity with time.

7. Name the two types of wetland environments and the latitude ranges where each will likely develop. How do wetlands contribute to the biology of the oceans and the cleansing of polluted river water?

8. What factors lead to a wide seasonal range of salinity in Laguna Madre?

9. Describe the circulation between the Atlantic Ocean and the Mediterranean Sea, and explain how and why it differs from estuarine circulation.

10. Discuss the lines of evidence that indicate that the Mediterranean Sea dried up within the geologic past. When did these events occur?

11. Without consulting the textbook, define pollution. Then, consider these items and determine if each one is a pollutant based on your definition (and refine your definition as necessary):

- Dead seaweed on the beach
- Natural oil seeps
- A small amount of sewage
- Warm water dumped into the ocean

12. Do you agree or disagree with this statement: The solution to ocean pollution is dilution. Why or why not?

13. Why would many marine pollution experts consider oil among the *least* damaging pollutants in the ocean?

14. Describe the effect of oil spills on species diversity and recovery of bottom-dwelling organisms based on the experience at Wild Harbor.

15. Compare oil spills that wash ashore to those that do not, such as the *Argo Merchant*, in terms of destruction to marine life.

16. Discuss techniques used to clean oil spills. Why is it important to begin the cleanup immediately?

17. When and where was the world's largest oil spill? How many times larger was it than the *Exxon Valdez* oil spill?

18. How would dumping sewage in deeper water off the East Coast help reduce the negative effects to the ocean bottom?

19. What properties contributed to plastics being considered a miracle substance? How do those same properties cause them to be unusually persistent and damaging in the marine environment?

20. Discuss the animal populations that clearly suffered from the effects of DDT and the way in which this negative effect was manifested.

21. What causes Minamata disease? What are the symptoms of the disease in humans?

22. Refer to Figure 11–21 and discuss the desirability of setting a mercury contamination level for fish at the 0.5-milligram/kilogram level (instead of the present 1.0-milligram/kilogram level approved by the U.S. FDA) for the citizens of the United States, Sweden, and Japan.

References

Boincourt, W. C. 1993. Estuaries: Where the river meets the sea. *Oceanus* 36:2, 29–37.

Borgese, E. M., and Ginsburg, N., eds. *1988 Ocean Yearbook*, Vol. 7. Chicago: University of Chicago Press.

Bragg, J. R. Prince, R. C., Harner, E. J., and Atlas, R. M. 1994. Effectiveness of bioremediation for the *Exxon Valdez* oil spill. *Nature* 368:6470, 413–418.

Cherfas, C. 1990. The fringe of the ocean: Under siege from land. *Science* 248:4952, 163–165.

Clark, W. C. 1989. Managing planet earth. *Scientific American* 261:3, 47–54.

Farrington, J. W., Capuzzo, J. M., Leschine, T. M., and Champ, M. A. 1982. Ocean dumping. *Oceanus* 25:4, 39–50.

Fenichelli, S. 1996. *Plastic: The making of a synthetic century.* New York: Harper Business Press.

Fye, P. M. 1982. The law of the sea. *Oceanus* 25:4, 7–12.

Galt, J. A., Lehr, W. J., and Payton, D. L. 1996. Fate and transport of the *Exxon Valdez* oil spill, *in* Pirie, R. G., ed., *Oceanography: Contemporary Readings in Ocean Sciences*, 3rd ed., New York: Oxford University Press.

Hodgson, B. 1990. Alaska's big spill: Can the wilderness heal? *National Geographic* 177:1, 5–43.

Hsü, K. 1983. The Mediterranean was a desert: A voyage of the *Glomar Challenger*. Princeton, NJ: Princeton University Press.

Knauss, J. A. 1974. Marine science and the 1974 Law of the Sea Conference: Science faces a difficult future in changing Law of the Sea. *Science* 184:1335–1341.

Lees, D. C., Houghton, J. P., and Driskell, W. B. 1996. Short-term effects of several types of shoreline treatment on rocky intertidal biota in Prince William Sound. *American Fisheries Society Symposium* 18:329–348.

Manheim, T., and Butman, B. 1994. A crisis in waste management, economic vitality, and a coastal marine environment: Boston Harbor and Massachusetts Bay. *GSA Today* 4:7, 197–199.

Michel, J. 1990. The *Exxon Valdez* oil spill: Status of the shoreline. *Geotimes* 35:5, 20–22.

———. 1991. Prince William Sound, Alaska: The cleanup continues. *Geotimes* 36:3, 16–17.

Officer, C. B. 1976. Physical oceanography of estuaries. *Oceanus* 19:5, 3–9.

Officer, C. B., Briggs, R. B., Taft, J. L., Tyler, M. A., and Bovnton, W. R. 1984. Chesapeake Bay anoxia: Origin, development, and significance. *Science* 223:4631, 22–27.

O'Hara, K. J., Iudicello, S., and Bierce, R. 1988. *A citizen's guide to plastics in the ocean: More than a litter problem*, 2nd ed., Washington, DC: Center for Marine Conservation, Inc.

Pickard, G. L. 1964. *Descriptive physical oceanography: An introduction.* New York: Macmillan.

Stanley, D. J. 1990. Med desert theory is drying up. *Oceanus* 33:1, 14–23.

Stommel, H. M., ed. 1950. *Proceedings of the colloquium on "The flushing of estuaries."* Woods Hole, MA: Woods Hole Oceanographic Institution.

Teal, J. M. 1993. A local oil spill revisited. *Oceanus* 36:2, 65–70.

The Open University Course Team. 1991. *Case studies in oceanography and marine affairs.* Oxford: Pergamon Press.

Valiela, I., and Vince, S. 1976. Green borders of the sea. *Oceanus* 19:5, 10–17.

Wilber, R. J. 1987. Plastics in the North Atlantic. *Oceanus* 30:3, 61–68.

Zeder, J. B., and Powell, A. N. 1993. Managing coastal wetlands. *Oceanus* 36:2, 19–28.

Suggested Reading

Earth

Howell, D. G., Bird, K. J., and Gautier, D. L. 1993. Oil: Are we running out? 2:2, 26–33. A critical look at the status of the world's petroleum reserves and the potential of new discoveries.

McInnis, D. 1998. And the waters prevailed. 7:4, 46–54. Based on characteristic features of sediment cores, marine geologists have discovered that the Black Sea abruptly filled about 5600 B.C., perhaps inspiring the story of Noah's flood.

Sea Frontiers

Baird, T. M. 1983. Life in the high marsh. 29:6, 335–341. The ecology of the salt marsh is discussed.

Baker, R. D. 1972. Dangerous shore currents. 18:3, 138–143. A discussion of the hazards to swimmers of various types of currents found near the shore.

Cardoza, Y., and Hirsch, B. 1991. Tidal creeks and scuba gear. 37:4, 32–37. The life forms inhabiting Florida tidal creeks are discussed.

deCastro, G. 1974. The Baltic: To be or not to be? 20:5, 269–273. A review of the pollution threat to life in the Baltic Sea and what has been done to solve the problem.

Edwards, L. 1982. Oyster reefs: Valuable to more than oysters. 28:1, 23–25. The value of oyster reefs to various estuarine life forms is detailed.

Loeffelbein, B. 1983. Law of the Sea treaty: What does it mean to nonsigners (like the United States)? 29:6, 358–366. An overview of the Law of the Sea Convention and a discussion of the implications of the treaty not being signed by all nations.

Parker, P. A. 1990. Clearing the oceans of the plastic threat. 36:2, 18–27. An examination of the laws regarding the disposal of plastics at sea, and what can be done about the increasing amount of plastic in the ocean.

Sefton, N. 1981. Middle world of the mangrove. 27:5, 267–273. The life forms associated with Cayman Island mangrove swamps are described.

Smith, F. 1982. When the Mediterranean went dry. 28:2, 66–73. The role of global plate tectonics in causing the Mediterranean Sea to dry up some 5 million years ago.

Sousa Schaefer, F. 1993. To harvest the Chesapeake: Oysters, blue crabs, and softshell clams take a dive. 39:1, 36–49. The effects of excessive nutrients entering the bay from sewage and acid rain have depleted the biological resources of Chesapeake Bay.

White, I. C. 1985. An organization to help combat oil spills. 31:1, 15–21. The efforts of the International Tanker Owners Pollution Federation Limited to increase the efficiency of cleanup operations and reduce the negative effects of oil spills have resulted in more intelligent responses to oil spills.

Scientific American

Borgese, E. M. 1983. The law of the sea. 248:3, 42–49. One hundred nineteen nations sign a convention aimed at establishing a new international scheme of regulating the oceans.

Halloway, M. 1994. Nurturing nature. 270:4, 98–108. Focusing on the Florida Everglades, the potential to restore damaged wetlands is investigated.

———. 1996. Sounding out science. 275:4 106–112. An update on the recovery of Prince William Sound after the *Exxon Valdez* oil spill, with a focus on how science is used to analyze the recovery of affected areas.

Hsü, K. H. 1972. When the Mediterranean dried up. 227:6, 26–45. Evidence is presented that shows that the Mediterranean Sea was a dry basin 6 million years ago.

McDonald, I. R. 1998. Natural oil spills. 279:5, 56–61. A look at the effects of natural oil seeps in The Gulf of Mexico and the unique biologic communities that consume these hydrocarbons.

Oceanography on the Web

www.prenhall.com/thurman

Visit the *Essentials of Oceanography* home page for on-line resources for this chapter. There you will find an on-line study guide with review exercises, and links to oceanography sites to further your exploration of the topics in this chapter. *Essentials of Oceanography* is at: **http://www.prenhall.com/thurman** (click on the Table of Contents menu and select this chapter).

CHAPTER 12
THE MARINE HABITAT

Charles Darwin and the Voyage of HMS *Beagle*

In the early nineteenth century, an English naturalist named **Charles Darwin** (Figure 12A) entered the scientific scene. Darwin's interest, which was investigating the whole of nature, led him to make one of the most outstanding contributions to the field of biology: the theory of evolution by natural selection. Much of the background on which he based his conclusions was gained from observations made while aboard the vessel HMS *Beagle* during its five-year round-the-world journey (Figure 12B).

The *Beagle* sailed from Devonport, England, on December 27, 1831, under the command of Captain Robert Fitzroy. The major objective of the voyage was to complete a survey of the coast of Patagonia (Argentina) and Tierra del Fuego, and to make chronometric measurements. The voyage allowed the young naturalist the opportunity to study the plants and animals throughout the world. He studied closely the changes that occurred in the animal populations living in different environments and concluded that all animals change slowly over the immense period of time represented by the geologic past.

Darwin's observations led him to conclude that birds and mammals must have evolved from the reptiles and that the similar skeletal framework of the human, the bat, the horse, the giraffe, the elephant, the porpoise, and other vertebrates required that they be grouped together. Darwin felt that the superficial differences that could be observed between populations were the result of adaptation to different environments and modes of existence.

On his return to England, Darwin assimilated much of the data that he collected while on board the *Beagle*. In 1858 he and Alfred Russel Wallace simultaneously published summaries of their independently conceived notions of natural selection. A year later Darwin set forth the structure of his theory and evidence to support it in his controversial book, *The Origin of Species*, which dealt not so much with the origin of life but with the evolution of living things into their many forms. Although Darwin's ideas were highly controversial, many now have been widely accepted, supported by an impressive array of scientific observations.

Darwin's scientific interest was not confined to biology. In fact, prior to the publication of *The Origin of Species*, Darwin considered himself to be primarily a geologist. During the voyage of the *Beagle*, Darwin formulated a theory on the origin of coral reefs on volcanic islands, which progress through different stages of development (from fringing reef to barrier reef to atoll) as outlined in Chapter 2, "Plate Tectonics and the Ocean Floor." Prior to the publication of *The Origin of Species*, Darwin published his ideas about coral reef development in *The Structure and Distribution of Coral Reefs* in 1842.

Figure 12A Naturalist Charles R. Darwin.

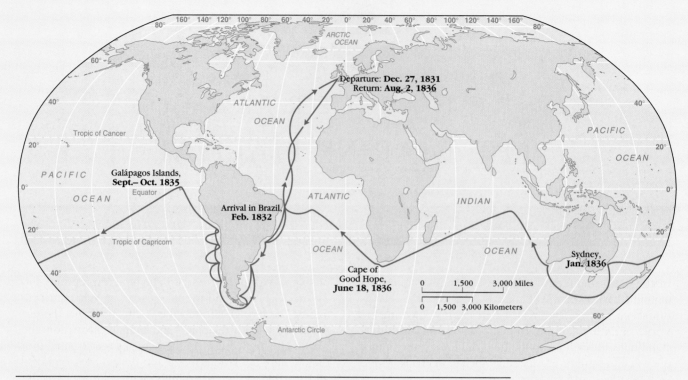

Figure 12B Voyage of HMS *Beagle.*

A wide variety of organisms inhabit the marine environment. These organisms range in size from microscopic bacteria and algae to the largest animal living in the world today: the blue whale, which is as long as three buses lined up end to end. Marine biologists have identified over 235,000 marine species, and this number is constantly increasing as new marine organisms are discovered.

Most marine life lives within the sunlit surface waters of the ocean. Strong sunlight supports photosynthesis by marine algae, which either directly or indirectly provide a source of food for the vast majority of marine organisms. All marine algae live near the surface because they need the sunlight; most marine animals live near the surface because this is where food can be obtained. In shallow water areas close to land, sunlight reaches all the way to the ocean floor, resulting in an abundance of marine life.

There are advantages and disadvantages to living in the marine environment. One advantage is that there is an abundance of water available, which is necessary for supporting all types of life. A disadvantage is that maneuvering around in this seemingly open and fluid environment can be difficult. The individual success of species depends on their ability to find food, to avoid predators, and to cope with the many physical variables that form barriers to their movement.

Classification of Living Things

All living things belong to one of three major domains (branches) of life: Archaea, Bacteria, and Eukarya (Figure 12–1). The domain **Archaea** (*archaeo* = ancient) is a group of simple microscopic bacteria-like creatures that includes methane producers and sulfur oxidizers that inhabit deep-sea vents and seeps,[1] as well as other forms—many of which prefer environments of extreme conditions of temperature and/or pressure. The domain **Bacteria** (*bakterion* = rod) includes simple life forms with cells that usually lack a nucleus, including purple bacteria, green nonsulfur bacteria, and cyanobacteria (blue-green algae). The domain **Eukarya** (*eu* = good, *karuon* = nut) includes complex organisms—protoctists, fungi, and multicellular plants and animals—with cells that usually contain a nucleus.

Within the three domains of life, a system of five kingdoms of life was first proposed by the ecologist and biologist Robert H. Whittaker in 1959. Since that time, there have been modifications to the kingdoms necessitated by the development of new biochemical and electron microscopy techniques. These five kingdoms of life are Monera, Protoctista, Fungi, Plantae, and Animalia.

Kingdom **Monera** (*monos* = single) includes some of the simplest of all organisms. These organisms are single-celled but lack a discrete nucleus. Instead, their nuclear

---

[1] A *seep* is an area where water trickles out of the sea floor.

358

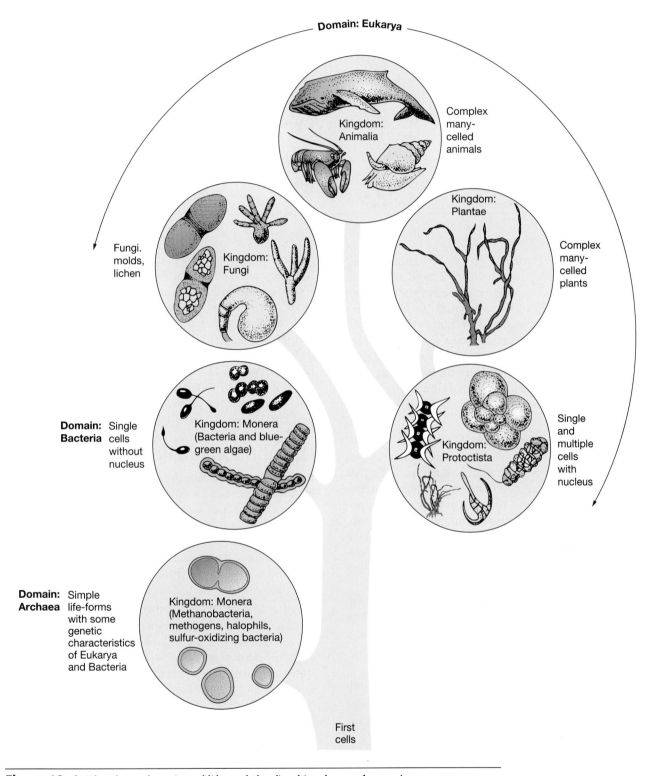

Figure 12–1 The three domains of life and the five kingdoms of organisms.

material is spread throughout the cell. Included in this kingdom are the cyanobacteria (blue-green algae), heterotrophic bacteria and archaea. Recent discoveries have shown bacteria to be a much more important part of marine ecology than previously believed. These organisms are found throughout the breadth and depth of the oceans.

Kingdom **Protoctista** (*protos* = first, *ktistos* = to establish) represents a higher stage of evolutionary development. It includes single- and multicelled organisms that have a nucleus. Protoctista include single-celled algae that produce food for most marine animals, single-celled animals called **protozoa** (*proto* = first, *zoa* = animal), and multicelled marine algae.

Table 12–1   Classification of selected organisms.

Category	Human	Common dolphin	Giant kelp
Kingdom	Animalia	Animalia	Protoctista
Phylum	Chordata	Chordata	Phaeophyta
Subphylum	Vertebrata	Vertebrata	
Class	Mammalia	Mammalia	Phaeophycae
Order	Primates	Cetacea	Laminariales
Family	Hominidae	Delphinidae	Lessoniaceae
Genus	*Homo*	*Delphinus*	*Macrocystis*
Species	*sapiens*	*delphis*	*pyrifera*

Kingdom **Fungi** (*fungus* = [probably from Greek *sp(h)ongos* = sponge]), which includes mold and lichen, appears to be poorly represented in the oceans. Less than one-half of 1 percent of the estimated 100,000 species of fungi are sea-dwellers. Although fungi exist throughout the marine environment, they are much more common in the intertidal zone. Here, they live in a relationship with cyanobacteria or green algae, creating what is called lichen. Other fungi remineralize organic matter and function primarily as decomposers in the marine ecosystem.

Kingdom **Plantae** (*planta* = plant) comprises the multicelled plants, all of which photosynthesize. In the ocean, only a few species of true plants—such as surf grass (*Phyllospadix*) and eelgrass (*Zostera*)—inhabit shallow coastal environments, and they are not a major component of the marine productive community. In the ocean, photosynthetic marine algae occupy the ecological niche of land plants. However, certain plants can be very important producers within coastal ecosystems, including mangrove swamps and salt marshes.

Kingdom **Animalia** (*anima* = breath) comprises the multicelled animals. Organisms from kingdom Animalia range in complexity from the simple sponges to complex vertebrates (animals with backbones), which also includes humans.

The five kingdoms are divided into increasingly specific groupings. The system of classification used today was introduced by Carl von Linné (Linnaeus) in 1758. It includes these major categories (Table 12–1):

Kingdom

Phylum (Division for plants)

Class

Order

Family

Genus

Species

The categories assigned to an individual species must be accepted by agreement of an international panel of experts. All organisms that share a common category (for instance, a family) have certain characteristics and evolutionary similarities (such as the familiar cat family). Every organism has a unique scientific name that includes its genus and species, which is usually italicized with the generic (genus) name capitalized—for example, *Delphinus delphis*. Most organisms also have a common name. For instance, *Delphinus delphis* is better known to many people as the common dolphin. If the scientific name is referred to repeatedly within a document, it is often shortened by abbreviating the genus name to its first letter. Thus, *Delphinus delphis* becomes *D. delphis*.

Distribution of Life in the Oceans

It is difficult to describe the degree to which the sea is inhabited because of its immense volume and the paucity of knowledge about marine organisms. As well, some populations fluctuate greatly each season. These conditions increase the difficulty of describing the extent to which the marine environment is populated. However, the marine and terrestrial environments may be compared by examining the number of marine and land species.

Figure 12–2 shows the distribution of Earth's known species. Well over 1,410,000 species are known from both terrestrial and marine environments.[2] However, only about 17 percent of all known species live in the ocean. It is interesting to note that there are many fewer species of marine organisms than terrestrial organisms.

Why Are There So Few Marine Species?

If the ocean is such a prime habitat for life, and if life originated in this environment, why is such a small percentage of the world's species living there? This lesser number may indeed result from the fact that the marine environment is more stable than the terrestrial environment: The relatively uniform conditions of the open ocean do not produce pressures for organisms to adapt, which is thought to be responsible for the creation of new

[2] Many biologists believe there may be from 3 to 10 million additional unnamed and undescribed species living on Earth, of which a large number probably inhabit the oceans.

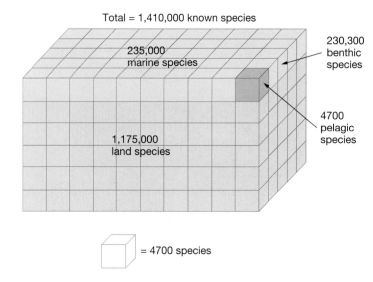

Total = 1,410,000 known species

235,000 marine species

1,175,000 land species

230,300 benthic species

4700 pelagic species

= 4700 species

Figure 12–2 Distribution of species on Earth. The entire large cube represents the 1,410,000 species known to be living on Earth. The blue-shaded portion indicates the proportion (16.7 percent) that live in the oceans. Of the marine species, 98% are benthic and live in or on the ocean floor (*light blue*). Only 2% are pelagic—either plankton or nekton—and live within the water column (*dark blue*).

species. In addition, temperatures are not only stable but also relatively low below the surface layers of the ocean. Thus, the rate of chemical reactions is slowed by this lower temperature, which in turn may reduce the tendency for variation to occur.

The great variety of organisms on the continents suggests that this development was the product of a less stable environment, one presenting many opportunities for natural selection to produce new species to inhabit varied new niches. At least 75 percent of all land organisms are insect species that have evolved the capability of inhabiting very restricted environmental niches. If the insects are ignored, the sea does possess over 45 percent of the remaining species living in marine and terrestrial environments.

Divisions of the Marine Environment

The oceans can be readily divided into two main environments. The ocean water itself is the **pelagic** (*pelagios* = of the sea) **environment**. Here, floaters and swimmers play out their lives in a complex food web. The ocean bottom constitutes the **benthic** (*benthos* = bottom) **environment**. Marine algae and animals that do not float or swim (or at least not very well) spend their lives here.

Based on the somewhat incomplete knowledge of marine species diversity, the distribution of the 235,000 known marine species shows that only about 4700 (about 2 percent) inhabit the pelagic environment (Figure 12–2). The other 98 percent are benthic, inhabiting the ocean floor. These numbers are certainly minimums, as recent discoveries indicate that many more benthic species may exist than previously thought.

Pelagic (Open Sea) Environment

There are a great many terms for the different divisions of the pelagic environment. Each division separates zones in the ocean that have distinctive biological characteristics, called **biozones**. These biozones and their depths are shown in Figure 12–3.

The pelagic environment is divided into two main provinces. The **neritic** (*neritos* = of the coast) **province** extends from the shore seaward, including all water overlying an ocean bottom that is less than 200 meters (660 feet) in depth.

Seaward, where depth increases beyond 200 meters (660 feet), is the **oceanic province** (Figure 12–3). It includes water with a very great range in depth, from the surface to the bottom of the deepest ocean trenches.

The oceanic province is further subdivided into four biozones:

1. The **epipelagic** (*epi* = top, *pelagios* = of the sea) **zone** from the surface to a depth of 200 meters (660 feet)

2. The **mesopelagic** (*meso* = middle, *pelagios* = of the sea) **zone** from 200 to 1000 meters (660 to 3280 feet)

3. The **bathypelagic** (*bathos* = depth, *pelagios* = of the sea) **zone** from 1000 to 4000 meters (3,300 to 13,000 feet)

4. The **abyssopelagic** (*a* = without, *byssus* = bottom, *pelagios* = of the sea) **zone** which includes all the deepest parts of the ocean below 4000 meters (13,000 feet) depth

The single most important factor in determining the distribution of life in the oceanic province is the availability of sunlight. Thus, in addition to the classification of the marine environment into four biozones, the distribution of life in the ocean is also divided into zones based on the availability of light as follows.

- The **euphotic** (*eu* = good, *photos* = light) **zone** extends from the surface to a depth where enough light still exists to support photosynthesis. This rarely is deeper than 100 meters (330 feet).

- Below this zone, there are small but measurable quantities of light within the **disphotic** (*dis* = apart

Figure 12–3 Oceanic biozones of the pelagic and benthic environments.

Pelagic environments are shown in blue; Benthic environments are shown in brown; Sea floor features and sunlight zones are shown in black lettering.

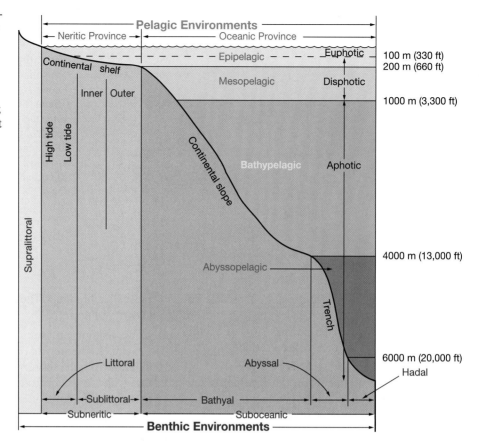

from, *photos* = light) **zone**, to a depth of about 1000 meters (3300 feet).

- Below about 1000 meters (3300 feet), no light exists in the **aphotic** (*a* = without, *photos* = light) **zone**.

Epipelagic Zone The upper epipelagic zone is the only oceanic biozone in which there is sufficient light to support photosynthesis. The boundary between it and the mesopelagic zone, at 200 meters (660 feet) depth, is also the approximate depth at which the level of dissolved oxygen begins to decrease significantly (Figure 12–4, *red curve*).

This decrease in oxygen results because no photosynthetic algae occurs below about 150 meters (500 feet), and dead organic tissue descending from the biologically productive upper waters is undergoing decomposition by bacterial oxidation. Nutrient content also increases abruptly below 200 meters (600 feet) (Figure 12–4, *green curve*). This depth also serves as the approximate bottom of the mixed layer, seasonal thermocline, and surface water mass.

Mesopelagic Zone Within the mesopelagic zone, a dissolved **oxygen minimum layer (OML)** occurs at a depth of about 700 to 1000 meters (2300 to 3280 feet) (Figure 12–4). The intermediate-water masses that move horizontally in this depth range often possess the highest levels of nutrients in the ocean.

Within the mesopelagic zone, although sunlight from the surface is very, very dim, there is evidence that animals sense this light. The mesopelagic zone is inhabited by fish that have unusually large and sensitive eyes, capable of detecting light levels 100 times lower than humans can sense.

Other important inhabitants of this zone are the **bioluminescent** organisms, which can produce light biologically, causing them literally to "glow in the dark." This group includes certain species of shrimp, squid, and fish. Approximately 80 percent of these organisms carry light-producing cells called **photophores**. These are glandular cells containing luminous bacteria surrounded by dark pigments. Some contain lenses to amplify the light radiation.

Bioluminescent light is produced by a chemical process involving the compound *luciferin*. Molecules of luciferin are excited and emit photons of light in the presence of oxygen. Only a 1 percent loss of energy is required to produce this illumination. This system is similar to that used by fireflies and glow worms.

Bathypelagic Zone and Abyssopelagic Zone The aphotic (lightless) bathypelagic and abyssopelagic zones represent over 75 percent of the living space in the oceanic province. In this region of total darkness, many completely blind fish exist. Bizarre-looking small predaceous species comprise the entire population of these fish.

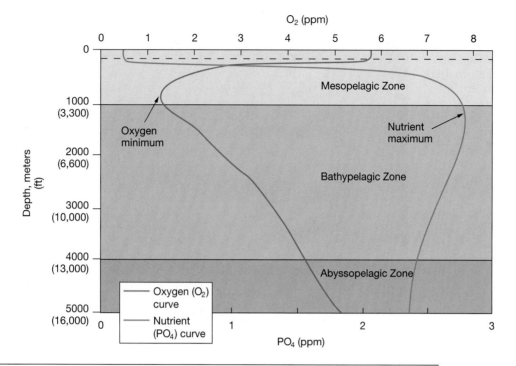

Figure 12–4 Distribution of oxygen and nutrients through the water column.
In surface water, oxygen (O_2) is abundant due to continual mixing with the atmosphere and continual plant photosynthesis, and nutrient content (phosphate, PO_4) is low due to continual uptake by marine algae. Below the surface euphotic zone, oxygen content abruptly decreases and nutrient content abruptly increases. At or near the base of the mesopelagic zone, there exists an oxygen minimum layer, which coincides with a nutrient maximum. Below that, nutrient levels remain high to the bottom and oxygen increases with depth (ppm = parts per million).

Many species of shrimp that normally feed on **detritus**[3] become predators at these depths, where the food supply is greatly reduced from that available in surface waters. Animals that live in these deep zones feed mostly upon one another. They have evolved impressive warning devices and unusual apparatuses to make them more efficient predators (Figure 12–5). These organisms have special adaptations to their predatory lifestyle, such as extremely large mouths relative to body size and very effective sets of teeth.

Below the oxygen minimum layer, oxygen content increases with depth because deep- and bottom-water masses carry oxygen from the cold surface waters where they formed to the deep ocean. The abyssopelagic zone is the realm of the bottom-water masses that commonly move in the opposite direction of the overlying deep-water masses of the bathypelagic zone.

Benthic (Sea Bottom) Environment

The benthic, or sea floor, environment is subdivided into two main units. These correspond to the previously discussed neritic and oceanic provinces of the pelagic environment.

- The **subneritic province** extends from the spring high tide shoreline to a depth of 200 meters (660 feet), approximately encompassing the continental shelf.
- The **suboceanic province** includes the entire benthic environment deeper than 200 meters (660 feet) depth.

Let's briefly examine the benthic environment, starting from the shore and moving out to sea. The transitional region from land to sea floor that is above the spring high-tide line is called the **supralittoral** (*supra* = above, *littoralis* = the shore) **zone**. Commonly called the *spray zone*, it is covered with water only during periods of extremely high tides and when tsunami or large storm waves break on the shore.

The *intertidal zone* (the zone between high and low tides) coincides with the **littoral** (*littoralis* = the shore) **zone**. From low-tide shoreline out to 200 meters (660 feet) water depth, the subneritic province is called the **sublittoral** (*sub* = below, *littoralis* = the shore) **zone**, or *shallow subtidal zone*, and is essentially the continental shelf.

The **inner sublittoral zone** includes the sublittoral to a depth of approximately 50 meters (160 feet). This seaward limit varies considerably because it is determined

---

[3] *Detritus* is a catchall term for dead and decaying organic matter, including waste products.

Figure 12–5 Deep-sea anglerfish.

The anglerfish, *Lasiognathus saccostoma,* attracts two deep-water shrimp to its bioluminescent "lure." The fish is about 10 centimeters (4 inches) long.

Figure 12–6 Benthic organisms.

As benthic organisms—such as this sea urchin and brittle star—move across or burrow through the ocean bottom, they often leave tracks on the sea floor such as those seen here.

by the depth at which no marine algae is found growing attached to the ocean bottom. This is controlled primarily by the amount of solar radiation that penetrates the surface water. When the seaward margin of the inner sublittoral region is reached, there are no more attached species of marine algae. All photosynthesis seaward of the inner sublittoral zone is carried on by microscopic algae floating in the waters above.

The **outer sublittoral zone** includes that portion of the sublittoral zone from the inner sublittoral zone out to a depth of 200 meters (660 feet) or the shelf break, which is the seaward edge of the continental shelf.

With increased depth in the suboceanic system, the **bathyal** (*bathus* = deep) **zone** is encountered. It extends from a depth of 200 to 4000 meters (660 to 13,000 feet) and corresponds generally to the continental slope.

From a depth of 4000 to 6000 meters (13,000 to 20,000 feet) stretches the **abyssal zone**, which includes over 80 percent of the benthic environment. The ocean floor representing the abyssal zone is covered by soft oceanic sediment, primarily abyssal clay. The tracks and burrows of animals that live in this sediment are frequently recorded in bottom photographs (Figure 12–6).

The **hadal** (*hades* = hell[4]) **zone** is a restricted environment, including all depths below 6000 meters (20,000 feet),

---

[4] The inhospitable high-pressure environment of the hadal zone is aptly named.

found only in deep trenches along the margins of continents. Isolation in these deep, linear depressions allows the development of animal communities unique to these trenches.

Classification of Marine Organisms

Marine organisms can be classified based on the portion of the ocean they inhabit (their environment) and the means by which they move (their mobility). Those organisms that inhabit the pelagic environment can be classified as either *plankton* (floaters) or *nekton* (swimmers). All other organisms are *benthos* (bottom dwellers).

Plankton (Floaters)

Plankton (*planktos* = wandering) include all organisms—algae, animals, and bacteria—that drift with ocean currents. An individual organism is called a **plankter**. The fact that plankters drift does not mean that all are unable to swim. Many plankters have this capacity but either move only weakly or are restricted to vertical movement and cannot determine their horizontal position within the ocean.

Among plankton, the algae (microscopic photosynthetic cells) are called **phytoplankton** (*phyto* = plant, *planktos* = wandering). The animals are **zooplankton** (*zoo* = animal, *planktos* = wandering). Representative members of each group are shown in Figure 12–7.

Plankton also include bacteria. It has recently been discovered that these free-living *bacterioplankton* are much more abundant in the plankton community than previously thought. Having an average dimension of only one-half of a micrometer[5] (0.00002 inch), they were missed in earlier studies because of their small size.

Most people have never heard of plankton, let alone seen them. However, plankton are unbelievably abundant and important within the marine environment. In fact, *most of Earth's* **biomass**—*the mass of living organisms— is found adrift in the oceans as plankton!* The volume of Earth's space inhabited by marine organisms that either drift or swim greatly exceeds that occupied by all animals that live on land or on the ocean's bottom. Even though 98 percent of marine *species* are bottom dwelling, the vast majority of the ocean's *biomass* is planktonic.

Plankton range greatly in size. They include large floating animals and algae, such as jellyfish and *Sargassum*.[6] These are called **macroplankton** (*macro* = large, *planktos* = wandering) and measure 2 to 20 centimeters (0.8 to 8 inches). At the other extreme are bacterioplankton that are too small to be filtered from the water even by a fine silk net and must be removed by other types of microfilters. These very tiny floaters are called **picoplankton** (*pico* = small, *planktos* = wandering) and measure 0.2 to 2 microns (0.000008 to 0.00008 inch).

Plankton are typically classified as either phytoplankton, zooplankton, or bacterioplankton. An additional scheme for classifying plankton is based on the portion of their life cycle spent within the plankton community. Organisms that spend their entire life as plankton are **holoplankton** (*holo* = whole, *planktos* = wandering).

Many organisms that are normally considered to be nekton or benthos (because they spend their adult life in one of these living modes) actually spend a portion of their life cycle as plankton. Many nekton and very nearly all benthos make the plankton community their home during their juvenile and/or larval stages (Figure 12–8). These organisms, which as adults either sink to the bottom to become benthos or begin to swim freely as nekton, are called **meroplankton** (*mero* = a part, *planktos* = wandering).

Nekton (Swimmers)

Nekton include all those animals capable of moving independently of the ocean currents, by swimming or other means of propulsion. They are not only capable of determining their own positions within relatively small areas of the ocean but are also capable, in many cases, of long migrations. Included in the nekton are most adult fish and squid, marine mammals, and marine reptiles (Figure 12–9). When you go ocean swimming, you become nekton, too.

Most freely moving nekton are unable to move at will throughout the breadth of the ocean. Certain nonvisible barriers effectively limit their lateral range. These barriers are gradual changes in temperature, salinity, viscosity, and availability of nutrients. As an example, the death of large numbers of fish frequently has been caused by temporary horizontal shifts of water masses in the ocean. The vertical range of nekton is normally controlled by water pressure.

Fish may appear to exist everywhere in the oceans, but normally they are more abundant near continents and islands and in colder waters. Some fish, such as salmon, ascend freshwater rivers to spawn. Many eels do just the reverse, growing to maturity in fresh water and then descending the streams to breed in the great depths of the ocean.

Benthos (Bottom Dwellers)

The term **benthos** describes organisms living on or in the ocean bottom. **Epifauna** (*epi* = upon, *fauna* = animal) live on the surface of the sea floor, either attached to rocks or moving along the bottom. **Infauna** (*in* = inside, *fauna* = animal) live buried in the sand, shells, or mud.

---

[5] One micrometer (also known as one *micron*) is 10^{-9} meter and is designated by the symbol μm.

[6] *Sargassum* is a floating type of brown macro marine algae commonly referred to as a "seaweed" that is particularly abundant in the Sargasso Sea.

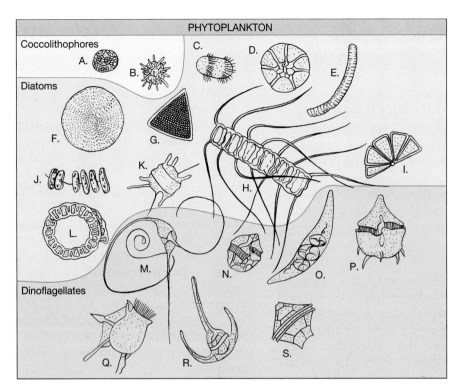

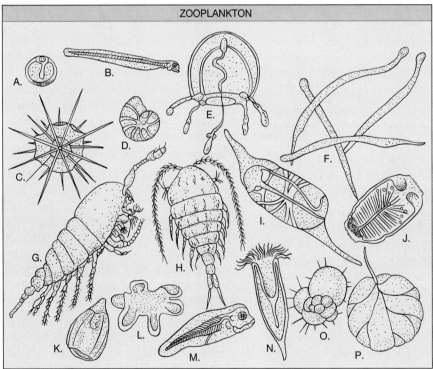

Figure 12–7 **Phytoplankton and zooplankton (floaters).**

Not drawn to scale; typical maximum dimension is in parentheses. **Phytoplankton: A and B.** Coccolithophoridae (15 μm, or 0.0006 in.). **C–L.** Diatoms (80 μm, or 0.0032 in.): **C.** *Corethron;* **D.** *Asteromphalus;* **E.** *Rhizosolenia;* **F.** *Coscinodiscus;* **G.** *Biddulphia favus;* **H.** *Chaetoceras;* **I.** *Licmophora;* **J.** *Thalassiorsira;* **K.** *Biddulphia mobiliensis;* **L.** *Eucampia.* **M–S.** Dinoflagellates (100 μm, or 0.004 in.): **M.** *Ceratium recticulatum;* **N.** *Goniaulax scrippsae;* **O.** *Gymnodinium;* **P.** *Goniaulax triacantha;* **Q.** *Dynophysis;* **R.** *Ceratium bucephalum;* **S.** *Peridinium.* **Zooplankton: A.** Fish egg (1 mm, or 0.04 in.). **B.** Fish larva (5 cm, or 2 in.). **C.** Radiolaria (0.5 mm, or 0.02 in.). **D.** Foraminifer (1 mm, or 0.04 in.). **E.** Jellyfish (30 cm, or 12 in.). **F.** Arrowworms (3 cm, or 1.2 in.). **G and H.** Copepods (5 mm, or 0.2 in.). **I.** Salp (10 cm, or 4 in.). **J.** Doliolum (10 cm, or 4 in.). **K.** Siphonophore (30 cm, or 12 in.). **L.** Worm larva (1 mm, or 0.04 in.). **M.** Fish larva (5 cm, or 2 in.). **N.** Tintinnid (1 mm, or 0.04 in.). **O.** Foraminifer (1 mm, or 0.04 in.). **P.** Dinoflagellate (*Noctiluca*) (1 mm, or 0.04 in.).

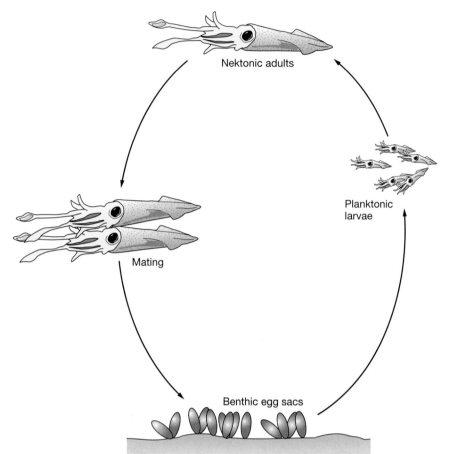

Nektonic adults

Mating

Planktonic larvae

Benthic egg sacs

Figure 12–8 Life cycle of a squid.

Adult squid are nekton, but their egg sacks are benthos, and their larval stage is planktonic. Thus, squid are considered to be meroplankton.

Some benthos not only live on the bottom but also move with relative ease by swimming through the water above the ocean floor or by crawling. These organisms are called **nektobenthos**. Examples of benthos organisms are shown in Figure 12–10.

The littoral (shore) zone and inner sublittoral zone have a great diversity of physical and nutritive conditions. Animal species have developed in great numbers within this nearshore benthic community as a result of the variations existing within the habitat. As one moves across the bottom from the shore into the deeper benthic environments, the *number of species* per square meter may remain relatively constant throughout the benthic environment. However, an inverse relationship can be observed between the *distance* from shore and the *biomass* of benthic organisms: the farther from shore, the lower the benthic biomass.

The littoral and inner sublittoral zones are the only areas where large marine algae (often called "seaweeds") attached to the bottom are found, because these are the only benthic zones to which sufficient light can penetrate.

Throughout most of the benthic environment, animals live in perpetual darkness, where photosynthetic production cannot occur. They must feed on each other, or on whatever outside nutrients fall from the productive zone near the surface. The deep-sea bottom is an environment of coldness, stillness, and darkness. Under these conditions, life progresses slowly. For animals that move around on the deep bottom, the streamlining that is so important to nekton is of little concern.

Organisms that live in the deep sea normally are widely distributed because physical conditions vary little on the deep-ocean floor, even over great distances. A few species appear to be extremely tolerant of pressure changes. In fact, members of a particular marine species are found in the littoral province and also at depths of several kilometers.

Hydrothermal Vent Biocommunities

In 1977, the first biocommunity at a hydrothermal vent was discovered in the Galápagos Rift off South America. This has shown us that high concentrations of deep-ocean benthos are possible. It appears that the primary limiting factor for life on the deep-ocean floor is the availability of food.

At these hydrothermal vents, food is abundant. It is produced not by photosynthesis, for no sunlight is available, but by chemosynthesis by archaeon (bacteria-like organisms). The size of individuals and the total biomass in the hydrothermal communities far exceed that previously known for the deep-ocean benthos. These biocommunities will be discussed in Chapter 15, "Animals of the Benthic Environment."

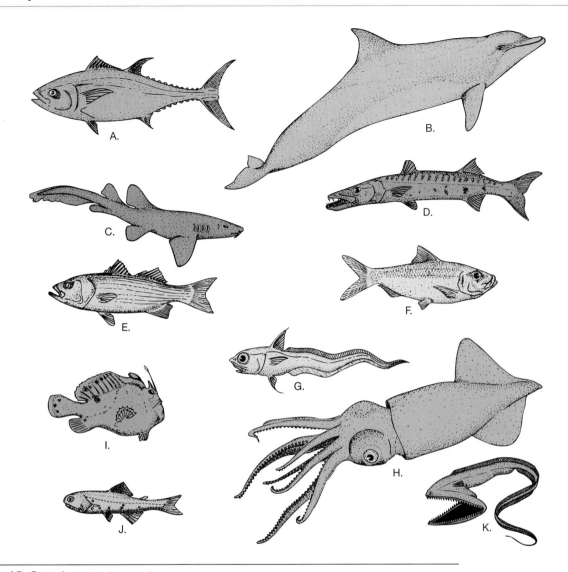

Figure 12–9 Nekton (swimmers).
Not drawn to scale; typical maximum dimension is in parentheses. **A.** Bluefin tuna (2 m, or 6.6 ft). **B.** Bottlenose dolphin (4 m, or 13 ft). **C.** Nurse shark (3 m, or 10 ft). **D.** Barracuda (1 m, or 3.3 ft). **E.** Striped bass (0.5 m, or 1.6 ft). **F.** Sardine (15 cm, or 6 in.). **G.** Deep-ocean fish (8 cm, or 3 in.). **H.** Squid (1 m, or 3.3 ft). **I.** Anglerfish (5 cm, or 2 in.). **J.** Lantern fish (8 cm, or 3 in.). **K.** Gulper (15 cm, or 6 in.).

Adaptations of Organisms to the Marine Environment

The ocean environment is far more stable than the terrestrial environment, particularly with regard to temperature. As a result of this stability, ocean-dwelling organisms generally have not developed highly specialized regulatory systems to combat sudden changes that might occur within their environment. They are therefore affected to various degrees by quite small changes in temperature, salinity, turbidity, pressure, and other environmental variables.

Water constitutes over 80 percent of the mass of **protoplasm**, the substance of living matter. Over 65 percent of your body's weight, and 95 percent of a jellyfish's weight, is simply water (Table 12–2). Water carries dissolved within it the gases and minerals needed by organisms. Water itself is a raw material used in photosynthesizing food, as is done by marine phytoplankton.

Land plants and animals have developed complex "plumbing systems" and devices to distribute water throughout their bodies and to prevent its loss. The threat of atmospheric desiccation (drying out) does not exist for the inhabitants of the open ocean, as they are abundantly supplied with water.

Need for Physical Support

One basic need of all plants and animals is for simple physical support. Land plants don't just sit on the ground; they have vast root systems that anchor the plant securely to the ground. In the animal kingdom, a number of support

Figure 12–10 Benthos (bottom dwellers)—representative intertidal and shallow subtidal forms.
Not drawn to scale; typical maximum dimension is in parentheses. **A.** Sand dollar (8 cm, or 3 in.).
B. Clam (30 cm, or 12 in.). **C.** Crab (30 cm, or 12 in.). **D.** Abalone (30 cm, or 12 in.). **E.** Sea urchins
(15 cm, or 6 in.). **F.** Sea anemone (30 cm, or 12 in.). **G.** Brittle star (20 cm, or 8 in.). **H.** Sponge
(30 cm, or 12 in.). **I.** Acorn barnacles (2.5 cm, or 1 in.). **J.** Snail (2 cm, or 0.8 in.). **K.** Mussels (25 cm,
or 10 in.). **L.** Gooseneck barnacles (8 cm, or 3 in.). **M.** Sea star (30 cm, or 12 in.). **N.** Brain coral
(50 cm, or 20 in.). **O.** Sea cucumber (30 cm, or 12 in.). **P.** Lamp shell (10 cm, or 4 in.). **Q.** Sea lily
(10 cm, or 4 in.). **R.** Sea squirt (10 cm, or 4 in.).

Table 12–2   Percent water content of selected organisms.

Organism	Percent water content
Human	65
Herring	67
Lobster	79
Jellyfish	95

systems are used. Each requires some combination of appendages—legs, arms, fingers, and toes—that must support the entire weight of the land animal.

In the ocean, this is not the case. The water of the oceans, as it so lavishly bathes the organisms with their needed gases and nutrients, serves as their physical support as well. Organisms that live in the open ocean depend primarily upon buoyancy and frictional resistance to sinking to maintain their desired position. Photosynthetic phytoplankton, which must live in the upper surface waters of the ocean, are an example.

This is not to say that these organisms maintain their positions without difficulty. Some organisms have developed special adaptations to increase their efficiency in positioning themselves, which will be discussed in this and succeeding chapters.

Water's Viscosity

Liquids (and some solids) flow with various degrees of ease. For example, syrup flows less readily than water; toothpaste flows even less readily than either one. This property is called **viscosity**, defined as *internal resistance to flow*. As described in Chapter 2, a substance that has high resistance to flow (high viscosity)—such as toothpaste—will not flow easily. Conversely, a substance that has low viscosity—such as water—will flow more easily. Viscosity is strongly affected by temperature (which helps explain why tar must be heated to decrease its viscosity before it can be spread onto roofs or roads).

The viscosity of ocean water is affected by two variables: temperature and salinity (which are discussed next). An increase in salinity increases seawater viscosity. Temperature has an even greater effect, with lower temperatures creating higher-viscosity seawater.

As a result of this temperature relationship, phytoplankton and animals that float in colder waters have less need for extensions to aid them in floating, because the water has a higher viscosity. In fact, members of the same species of floating crustaceans are very ornate, with featherlike appendages, where they occupy warmer waters, whereas these appendages are missing in colder,

more viscous environments (Figure 12–11). Thus, high-viscosity water benefits floating members of the marine biocommunity, helping them maintain their positions near the surface.

Viscosity and Streamlining Not all sea organisms are single-celled floaters. With increasing organism size and swimming ability, viscosity ceases to enhance survival and becomes an obstacle instead. This is particularly true of the large organisms that swim freely in the open ocean. They must pursue prey or flee predators, which means they must displace water to move forward. The faster they swim, the greater is the water stress on them. Not only must water be displaced ahead of the swimmer, but water also must move in behind it to occupy the space that the animal has vacated. These are important considerations in **streamlining**, which is producing a shape that offers the least resistance to fluid flow.

The familiar shape of free-swimming fish (and of mammals such as whales and dolphins) exemplifies the streamlining adaptations of organisms to move with minimum effort through water. A common shape to meet this need is a flattened body, which presents a small cross section at the front end and a gradually tapering back end to reduce the wake created by eddies (Figure 12–12). This form is characteristic of many fish because it reduces stress from movement through water and allows minimal energy to be expended in overcoming water's viscosity.

Temperature

Seawater temperature is one of the most important factors that affects marine life. The marine environment is much more stable than that on land, as can be seen by comparing the difference in temperature ranges (Figure 12–13). For example:

- *Extremes of ocean temperature have a far narrower range than temperatures on land.* The minimum surface temperature observed in the open ocean is seldom much below –2 degrees centigrade (28.4 degrees Fahrenheit). The maximum surface temperature seldom exceeds 32 degrees centigrade (89.6 degrees

Figure 12–11 Water temperature and appendages. **A.** Copepod (*Oithona*) displays the ornate plumage characteristic of warm-water varieties. **B.** Copepod (*Calanus*) displays the less ornate appendages found on temperate and cold-water forms.

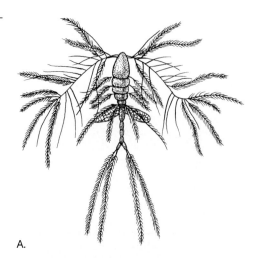

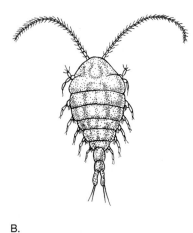

A.   B.

Fahrenheit), except in some shallow-water coastal regions. In contrast, record continental temperatures are –88 degrees centigrade (–127 degrees Fahrenheit) and 58 degrees centigrade (136 degrees Fahrenheit).

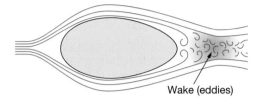
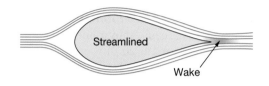

Figure 12–12 Streamlining.

To move through water efficiently, a body must produce as little stress as possible as it moves through and displaces the water. After the water has moved past the body, it must flow in behind the body with as little eddy action as possible.

- *Annual temperature variations of ocean water are small compared to those on land.* They range from 2 degrees centigrade (3.6 degrees Fahrenheit) at the Equator to 8 degrees centigrade (14.4 degrees Fahrenheit) between 35 and 45 degrees latitude, and decrease again in the higher latitudes. Annual variations of temperature in shallower coastal areas may be as high as 15 degrees centigrade (27 degrees Fahrenheit). In contrast, continental temperatures can vary dramatically over a year.

- *Daily temperature variations of the ocean surface are also small compared with those on land.* The daily range rarely exceeds 0.2 to 0.3 degree centigrade (0.4 to 0.5 degree Fahrenheit), although in shallower coastal waters they may vary daily by 2 to 3 degrees centigrade (3.6 to 5.4 degrees Fahrenheit). In contrast, continental temperatures vary much more widely on a daily basis (see Figure 5–6).

In addition, ocean temperatures change far more *slowly* than land temperatures, due largely to water's remarkable thermal properties (as discussed in Chapter 5, "Water and Seawater").

The temperature variations in surface waters are reduced in magnitude with depth. In the deep ocean, daily or seasonal temperature variations becomes insignificant.

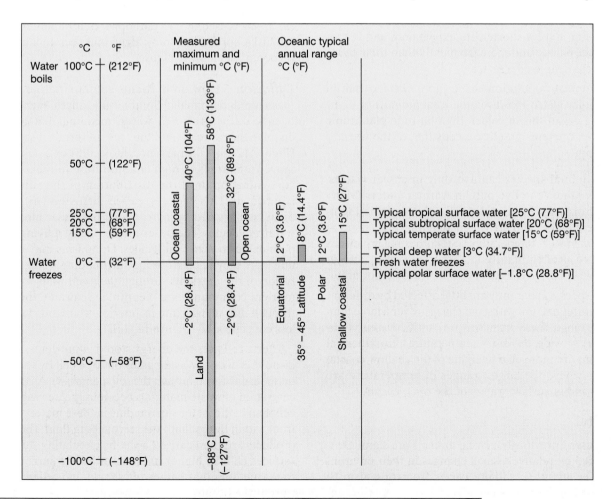

Figure 12–13 A comparison of land and ocean temperature ranges.

Throughout the deeper parts of the ocean, the temperature remains uniformly low. At the bottom of the ocean basin, where depth exceeds 1.5 kilometers (0.9 mile), temperatures hover around 3 degrees centigrade (37.4 degrees Fahrenheit) regardless of latitude.

Comparing Cold- and Warm-Water Species Physically, there are many differences between cold water and warm water. For instance, cold water is denser and has a higher viscosity than warm water. These factors, among others, profoundly influence marine life.

There are many differences between warm-water and cold-water species in the marine environment:

- In warmer waters, floating organisms are physically smaller than organisms in colder waters. This is probably related to the lower viscosity and density of seawater in lower-latitude waters. Thus, smaller tropical species can expose more surface area per unit of body mass.

- Warm-water species are characterized by ornate plumage to increase surface area, which is strikingly absent in the larger cold-water species (see Figure 12–11).

- Warmer temperatures increase the rate of biological activity, which more than doubles with an increase of 10 degrees centigrade (18 degrees Fahrenheit). Tropical organisms apparently grow faster, have a shorter life expectancy, and reproduce earlier and more frequently than their counterparts in colder water.

- Tropical populations have more *species* than in cooler water. However, the total *biomass* of floating organisms in colder, high-latitude planktonic environments greatly exceeds that of the warmer tropics.

Some animal species can live only in cooler waters, whereas others can live only in warmer waters. Many of these organisms can withstand only very small temperature changes and are called **stenothermal** (*steno* = narrow, *thermo* = temperature). Stenothermal organisms are found predominantly in the open ocean, and at depths where large ranges of temperature do not occur.

Other species apparently are little affected by different temperatures and can withstand changes over a large temperature range. These organisms are classified as **eurythermal** (*eury* = wide, *thermo* = temperature). Eurythermal organisms are more characteristic of the shallow coastal waters—where the largest ranges of temperature are found—and in surface waters of the open ocean.

Salinity

As discussed previously, marine animals are significantly affected by relatively small changes in their environment. Sensitivity to salinity varies from organism to organism. For example, some oysters live in estuarine environments at the mouths of rivers, and they are capable of withstanding a considerable range of salinity. In coastal areas, the daily rise and fall of the tides forces salty ocean water into river mouths and draws it out again, changing the salinity considerably. Further, during floods, salinity is extremely low. Consequently, oysters and most other organisms that inhabit coastal regions have evolved a tolerance for a wide range of salinity conditions, and are known as **euryhaline** (*eury* = broad, *halo* = salt).

By contrast, other marine organisms, particularly those that inhabit the open ocean, are seldom exposed to much salinity variation. Consequently, they are adapted to steady salinity and can withstand only very small changes. These organisms are called **stenohaline** (*steno* = narrow, *halo* = salt).

Extraction of Salinity Components Some phytoplankton and animals cause important alterations in the amount of dissolved material in ocean water. They do so by extracting these minerals to construct hard parts of their bodies, which serve as protective coverings. The compounds primarily used by phytoplankton and protozoans for this purpose are silica (SiO_2) and calcium carbonate ($CaCO_3$).

Silica is used by a very important phytoplankton population in the ocean—diatoms—and by microscopic protozoans such as radiolarians and silicoflagellates. *Calcium carbonate* is used by coccolithophores, foraminifers, most mollusks, corals, and some algae that secrete a calcium carbonate skeletal structure.

Diffusion How do nutrients get into marine phytoplankton and animals? Soluble substances, such as nutrients, diffuse through water, meaning that as they dissolve they spread until they are uniformly distributed (Figure 12–14A). In essence, molecules of a substance move from areas of *high* concentration to areas of *low* concentration until the distribution of the substance is uniform. This process, called **diffusion** (*diffuse* = dispersed), is caused by random molecular motion of water molecules. The outer membrane of a living cell is permeable to many molecules. Organisms may take in nutrients they need from the surrounding water by diffusion of the nutrients through their cell walls. Because nutrient compounds are plentiful in seawater, they pass through the cell wall into its interior, where nutrients are less concentrated (Figure 12–14B).

After a cell uses the energy stored in nutrients, it must dispose of waste. Waste passes out of a cell by the same method: diffusion through the cell membrane. As the concentration of waste materials becomes greater within the cell than in the water surrounding it, these materials pass from within the cell into the surrounding fluid. The waste products are then carried away by circulating fluid that services cells in higher animals, or by the surrounding water medium that bathes the simple one-celled organisms of the oceans.

DIFFUSION

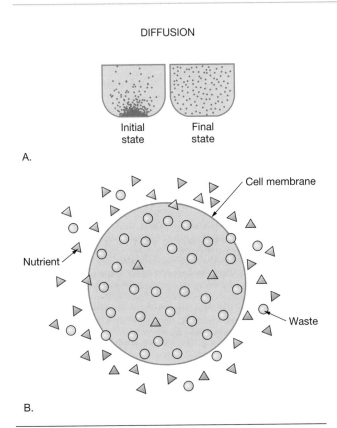

A.

B.

Figure 12–14 Diffusion.
A. A water-soluble substance that is added to the bottom of a container of water (*initial state*) will eventually become evenly distributed throughout the water by diffusion (*final state*).
B. Nutrients (*triangles*) are in high concentration outside the cell and diffuse into the cell through the cell membrane. Waste particles (*circles*) are in high concentration inside the cell and diffuse out of the cell through the membrane.

Osmosis When water solutions of unequal salinity are separated by a water-permeable membrane (such as the membrane around a living cell), water molecules diffuse through the membrane. They move from the *less* concentrated solution into the *more* concentrated solution. This process is called **osmosis** (*osmos* = push) (Figure 12–15). The membrane allows molecules of some substances to pass through while screening out others. **Osmotic pressure** is pressure that must be applied to the more concentrated solution to prevent passage of water molecules into it from a supply of pure water.

Osmosis is important to organisms that live in water, either marine or fresh. Their "skin," in whatever form, is a semipermeable membrane that separates internal body fluids from seawater. If the salinity of an organism's body fluid equals that of the ocean, it is said to be **isotonic** (*iso* = same, *tonos* = tension) and to have equal osmotic pressure. No net transfer of water will occur through the membrane in either direction.

If the seawater has a lower salinity than the body fluid within an organism's cells, seawater will pass through the cell walls into the cells (always toward the more

OSMOSIS

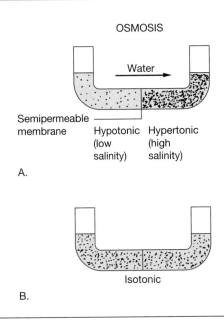

A.

B.

Figure 12–15 Osmosis.
A. The semipermeable membrane that separates two water solutions of different salinities allows water molecules to pass through it by the process of osmosis. However, the ions of dissolved substances do not pass through the membrane. Water molecules will diffuse through the membrane from the less concentrated (hypotonic) solution (*left*) into the more concentrated (hypertonic) solution (*right*). **B.** If the salinity of the two solutions is the same (isotonic), there is no net movement of water through the membrane.

concentrated solution). This organism is **hypertonic** (*hyper* = over, *tonos* = tension), which means that it is saltier relative to the seawater (Figure 12–15).

Should the opposite be true, with salinity in an organism's cells less than that of the surrounding seawater, water from the cells will pass through the cell membranes out into the seawater (toward the more concentrated solution). This organism is described as **hypotonic** (*hypo* = under, *tonos* = tension) relative to the external medium (Figure 12–15).

To simplify, consider osmosis simply as a diffusion process and examine the relative concentration of *water molecules* on either side of the semipermeable membrane. Thus, osmosis produces a net transfer of water molecules from the side where the *greatest concentration* of water molecules exists, through the membrane, to the side that contains a *lesser concentration* of water molecules.

It is interesting to note that three things can occur simultaneously across the cell membrane:

1. Water molecules move through the semipermeable membrane toward the lesser water concentration.

2. Nutrient molecules or ions move from where they are more concentrated into the cell, where they will be used to maintain the cell.

3. Waste molecules move from the cell into the surrounding seawater.

Keep in mind that during this process, molecules or ions of all the substances in the system are passing through the membrane in both directions. Nevertheless, there will be a net transport of molecules of a given substance from the side on which they are most highly concentrated to the side where the concentration is less, until equilibrium is attained.

In the case of marine invertebrates such as worms, mussels, and octopi, the body fluids and the seawater in which they are immersed are in a nearly isotonic state. Because no significant difference exists, these creatures have not had to evolve special mechanisms to maintain their body fluids at a proper concentration. Thus, they have an advantage over their fresh water relatives, whose body fluids are hypertonic in relation to the low-salinity fresh water.

An Example of Osmosis: Marine Fish Versus Freshwater Fish Marine fish have body fluids that are only slightly more than one-third as saline as ocean water. This low level may be the result of their having evolved in low-salinity coastal waters. They are, therefore, hypotonic (less salty) in relation to the surrounding seawater.

This salinity difference creates a situation in which saltwater fish, without some means of regulation, would lose water from their body fluids into the surrounding ocean by osmosis and eventually dehydrate. However, this loss is counteracted because marine fish drink ocean water and excrete the salts through special chloride-releasing cells located in their gills. These fish also help maintain their body water by discharging a very small amount of very highly concentrated urine (Figure 12–16A).

Alternatively, freshwater fish are hypertonic (internally more saline) in relation to the very dilute water in which they live. The osmotic pressure of the body fluids of such fish may be 20 to 30 times greater than that of the fresh water that surrounds them. A freshwater fish must avoid taking into its body cells excessive quantities of water through osmosis that would eventually rupture the cell walls.

To prevent this, the freshwater fish does not drink water. Its cells have the capacity to absorb salt. Water that may be gained as a result of body fluids being greatly hypertonic is disposed of with large volumes of very dilute urine (Figure 12–16B).

The Importance of Organism Size

Marine phytoplankton are quite simple compared with the specialized plant forms that exist on land. Why have phytoplankton remained as single-celled organisms instead of becoming multicellular? The major requirements of phytoplankton are that they stay in the upper portion of ocean water where solar radiation is available, that they have available necessary nutrients, that they efficiently take in these nutrients from surrounding waters, and that they expel waste materials. Let's examine how these requirements are satisfied by the ingenious design of phytoplankton, specifically their size and shape.

Phytoplankton live their entire lives in the upper layers of the ocean, yet they have no roots and no means of swimming. Interestingly, they are able to maintain their general position near the surface of the water. How they do so is closely related to the ratio between their surface area and body mass.

Greater surface area per unit of body mass means a greater resistance to sinking. This is because more surface area is in contact with the surrounding water, providing a greater amount of frictional resistance. An examination of Figure 12–17 will help to show how the surface area per unit of mass increases as an organism's size decreases. Although it is much smaller, cube A has 10 times the surface area *per unit of interior volume* as cube B. In practical terms, if cubes A and B were plankton, cube A would have 10 times the resistance to sinking *per unit of mass* as cube B. Therefore, it could stay afloat by exerting far less energy than cube B. Also, if cube A were a species of planktonic algae, it could take in nutrients and dispose of waste through its cell wall 10 times as efficiently as an alga with the dimensions of

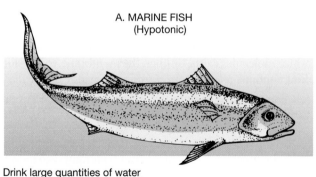

A. MARINE FISH
(Hypotonic)

Drink large quantities of water
Secrete salt through special cells
Small volume of highly concentrated urine

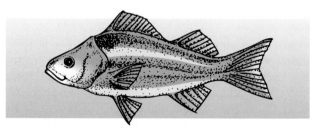

B. FRESHWATER FISH
(Hypertonic)

Do not drink
Cells absorb salt
Large volume of dilute urine

Figure 12–16 Marine fish versus freshwater fish.
Osmotic processes cause these two types of fish to have different adaptations to the environment.

CUBE A
Linear dimension = 1 cm
on each side

→|1cm|←

Area – 6 cm² (6 sides)
Volume – 1 cm³

Ration of surface area to volume:

$$\frac{6 \text{ cm}^2}{1 \text{ cm}^3} = 6 : 1$$

Cube A has 6 units of surface area
per unit of volume.

CUBE B
Linear dimension = 10 cm
on each side

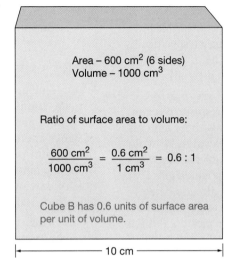

Area – 600 cm² (6 sides)
Volume – 1000 cm³

Ratio of surface area to volume:

$$\frac{600 \text{ cm}^2}{1000 \text{ cm}^3} = \frac{0.6 \text{ cm}^2}{1 \text{ cm}^3} = 0.6 : 1$$

Cube B has 0.6 units of surface area
per unit of volume.

|← 10 cm →|

Figure 12–17 Surface area per unit volume and size.

Although much smaller, cube A has 10 times the surface area *per unit of interior volume* as cube B.

cube B. Thus, one-celled organisms, which make up the bulk of photosynthetic marine life, are clearly benefited by being as small as possible. They are so small, in fact, that one needs a microscope to see them!

The efficiency with which photosynthetic cells take in nutrients from surrounding water and expel waste through their cell membranes also is related to the ratio of surface area to mass. Both the intake of nutrients and waste disposal depend to a large extent on diffusion, so there is a point at which increased size would reduce the ratio of surface area to body mass where the cell could not function properly and would die. This is why cells in all plants and animals are microscopic, regardless of the overall size of the organism, from bacteria to whales. Each cell must take in nutrients and dispose of wastes by diffusion.

Diatoms are one of the most important groups of phytoplankton. Diatoms often have unusual appendages, needle-like extensions, or even rings (Figure 12–18) to help increase their ratio of surface area to mass, which prevents them from sinking below the euphotic zone. Other types of microscopic marine organisms—especially warm-water species—display similar strategies to avoid sinking.

Some small organisms use another contrivance to stay in the upper layers of the ocean: They produce a tiny droplet of oil, which lowers their overall density and increases buoyancy. Accumulations of vast amounts of these organisms in sediment can produce offshore oil deposits if their droplets of oil are combined, sufficiently matured, and trapped in a reservoir.

Despite these adaptations to improve floating ability, the organisms still are denser than water, and tend to sink, if ever so slowly. However, this is not a serious handicap, for the tiny cells are carried readily with water

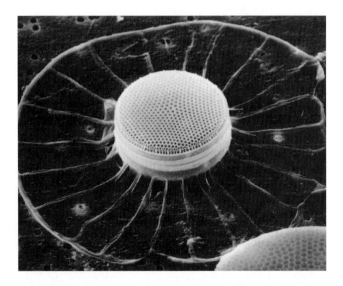

Figure 12–18 Warm-water diatom.

Scanning electron micrograph of the warm-water diatom *Planktoniella sol,* which has a prominent marginal ring to increase its surface area (diameter = 60 microns).

movements. Near the surface, wind causes considerable mixing and turbulence. Turbulence dominates the movement of these small organisms, keeping them positioned to bask in the solar radiation needed to photosynthesize, producing the energy used by essentially all other members of the marine community.

Dissolved Gases

You will recall that reducing the temperature of seawater increases its capacity to hold gases in solution; that is, colder water holds more dissolved gas in solution. This significantly affects organisms inhabiting the ocean. For instance, a direct consequence of the increased capacity of colder water to contain dissolved gases is seen in the vast phytoplankton communities that develop in high latitudes during summer, when solar energy becomes available for photosynthesis. A major factor in this phenomenon is the abundance of dissolved gases, specifically carbon dioxide (which phytoplankton need for photosynthesis) and oxygen (which is needed by all organisms to metabolize their food).

All animals that live in the ocean require oxygen. Except for air-breathing marine mammals, these animals must extract dissolved oxygen from seawater. How do they accomplish this? Most marine organisms have specially designed fibrous respiratory organs called **gills** that serve to exchange oxygen and carbon dioxide directly with seawater. For instance, fish extract oxygen from seawater by taking water in through their mouths (which gives them the appearance of "breathing" water), which is passed through their gills and exits through the gill slits on the sides of their bodies. Most fish need at least 4.0 parts per million (ppm) of oxygen per liter of seawater to survive for long periods, and even more for activity and rapid growth. That's why aquarium owners must use a pump to continually resupply oxygen to the water in the tank for most fish.

Gill structure and location vary among animals of different groups. In fishes, gills are located at the rear of the mouth and contain capillaries; in higher aquatic invertebrates, they protrude from the body surface and contain extensions of the vascular system; in mollusks, they are inside the mantle cavity; in aquatic insects, they occur as projections from the walls of the air tubes. In amphibians, gills are usually present only in the larval stage; in higher vertebrates (including humans), they occur merely as rudimentary, nonfunctional gill slits, which disappear during embryonic development.

Water's High Transparency

Water—including seawater—has relatively high transparency compared to many other substances, allowing sunlight to penetrate into the ocean. Although the depth of light penetration varies due to many factors such as the amount of turbidity (suspended sediment) in the water, the amount of plankton within the water, latitude, time of day, and the season, light typically penetrates to a depth of 1000 meters (3300 feet) in the open ocean.

Water's high transparency has resulted in well-developed eyesight in many marine organisms. A problem for organisms living in the ocean is that they can easily be seen and consumed by predators. To combat this, many marine organisms such as jellyfish are themselves nearly transparent, which helps them blend into their environment. Other organisms are colored so that they are well camouflaged (Figure 12–19A). Another strategy is for an organism to be colored differently on its top side than its bottom side, called **counter-shading**. Many fish—especially flat fish—are dark on top so that they cannot be easily seen against a dark background while looking into the ocean, and are light on bottom so that they blend into the sunlight when viewed from below (Figure 12–19B).

Box 12–1
The Mudskippers: Amphibious Fish

Mudskippers belonging to the genus *Periophthalmodon* are unique fish that inhabit tropical mud flats from West Africa to South Pacific islands. Specialized relatives of gobies, they spend most of their time skipping across mud flats, rocks, or mangrove roots (hence the common name, mudskippers). They have been observed climbing as much as 1 meter (3.3 feet) above water level on mangrove stalks. Possessing a frog-like head with bulging eyes set high on their head to search out even flying prey (Figure 12C), they eat mostly insects. The front of the body is supported by pectoral fins that have fleshy muscular bases. The muscular tail drags behind, but propels the fish on long leaps through the air.

Like all fish, mudskippers possess gills that allow them to obtain oxygen. They have chambers above their gills that they appear to refill with water periodically by dipping their heads into the water. Additionally, they may obtain oxygen by absorbing it through their skin.

Investigations in 1995 and 1996 have added to the knowledge of the nature of the mudskippers' life in their burrows, which were known to be used for refuge and spawning. The burrows of *P. schlosseri* observed in Malaysian mud flats are found in the high intertidal zone. A vertical shaft about 8 centimeters (3 inches) in diameter connects to one or two horizontal tunnels (Figure 12D). Always filled with water, the burrows studied reached a maximum depth of 125 centimeters (49 inches). The amount of oxygen in the burrow water varied from about 80 percent that present in surface water to less than 3 percent of that in surface water below a depth of 50 centimeters (20 inches).

A recent study analyzed gas concentrations in 28 mudskipper burrows from three locations, which are indicated by the triangles, green circles, and blue circles in Figure 12E. Basing the recency of use of mudskipper burrows by the color and texture of mud pellets ejected from the burrows by fish, it was determined that gas from more recently used burrows had higher partial pressure of oxygen (pO_2) and lower partial pressure of carbon dioxide (pCO_2). The study indicates that mudskippers can survive in low-oxygen/high-carbon dioxide conditions. For comparison, the red square in the lower right of Figure 12E represents typical values for normal air.

(continued)

The mudskipper *P. schlosseri* was observed to carry air into the burrow. Individuals were observed to surface from the burrow, open their mouths and gulp air to fill their cheek cavities (as the fish on the left in Figure 12C is doing). Laboratory studies have demonstrated that *P. schlosseri* regularly inflate their cheek cavities with air and, as long as air is available, they rarely use their gills to obtain oxygen.

The time that fish remained in their burrows varied, but the maximum length of time was more than 30 minutes. This behavior provides a sufficient oxygen reservoir for the fish and their developing embryos, which are within eggs that are commonly attached to the ceiling of burrow chambers. Studies are in progress that show that other species of mudskippers also have air in their spawning chambers.

Figure 12C Mudskippers.

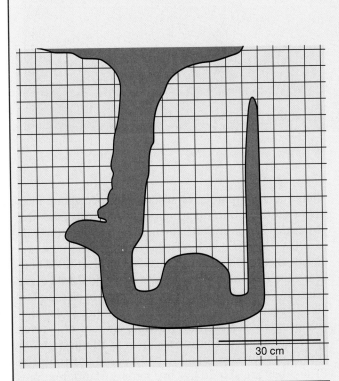

Figure 12D Geometry of Mudskipper burrows.

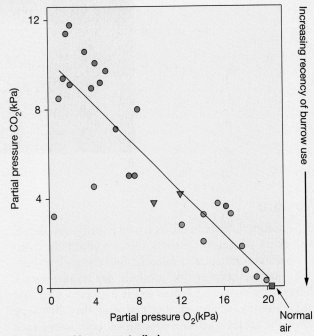

Increasing recency of burrow use →

Normal air

Locations of burrows studied:
Green circles: Burung River Estuary, Penang, Malaysia
Blue circles: Selangor River tributary, Kuala Selangor, Malaysia
Triangles: Pinang River Estuary, Penang, Malaysia

Figure 12E Mudskipper burrows and O_2.

Figure 12–19 Camouflage and counter-shading.
A. The head and eye of a well-camouflaged rock fish.
B. Halibut on a dock in Alaska show counter-shading.

A.

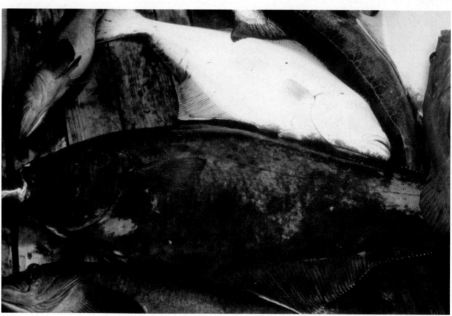

B.

Pressure

In the ocean, water pressure increases linearly with depth. With every 10 meters (33 feet) of water depth, the pressure increases by about 1 kilogram per square centimeter (1 atmosphere or 14.7 pounds per square inch). Humans are not well adapted to the high pressures that exist below the surface—as they find when they dive deep below the surface without specialized equipment (see chapter-opening feature in Chapter 1). Even when diving to the bottom of the deep end of a swimming pool, one can feel the dramatic increase in pressure in one's ears.

In the deep ocean, water pressure is on the order of several hundred kilograms per square centimeter (several hundred atmospheres, or several tons per square inch). How do deep-water marine organisms withstand pressures that can easily kill humans? The one thing that most marine organisms lack is large compressible air pockets inside their bodies. Thus, they do not have air-filled lungs, ear canals, or other passageways as we do. These organisms don't feel the high pressure pushing in on their bodies. Because water is nearly incompressible, their water-filled bodies have the same amount of pressure pushing outward. This causes them to be unaffected by the high pressures found in deep-ocean environments. Thus, they are able to live in a variety of different pressure conditions.

Box 12–2
The Deep Scattering Layer (DSL)

Within the mesopelagic layer, there exists a curious feature called the **deep scattering layer (DSL)**. It was discovered when the U.S. Navy was testing sonar equipment early in World War II to detect enemy submarines. On many of the sonar recordings, there was a prominent sound-reflecting surface that occurred at a relatively uniform depth in the ocean. This reflecting surface was much too shallow to be the ocean floor, but the cause of the echo was unknown. Often, it was referred to as a "false bottom" (see Figure 3–1). Interestingly, the *depth* of the layer changed during a day; the layer was deeper in midday and closer to the surface at night. This sound-reflecting layer occurred at a depth of about 100 to 200 meters (330 to 660 feet) during the night, but was as deep at 900 meters (3000 feet) during the day.

With the help of marine biologists, sonar specialists were able to determine what was producing this sound-reflecting layer. Marine biologists knew that copepods, which constitute a large proportion of planktonic animals, undertook daily vertical migrations by moving to a greater depth during the day and then closer to the surface at night (Figure 12F). Presumably, these organisms needed to come near the sur-

face to feed during the night but did not want to stay in the upper sunlit surface waters during the day, when they could be easily seen and captured as prey. Hence, they migrated to deeper depths during the day to remain hidden in the darker portions of the ocean.

Further investigation with plankton nets and submersibles revealed that the DSL contained various organisms, including large numbers of euphausids and lantern fish (family myctophidae) (Figure 12F). The deep scattering layer is presumably caused by sound reflecting from masses of migrating marine life, where even a small concentration of organisms (and especially fish) is sufficient to reflect sonar signals. Fish are predators, so organisms on which they prey—such as copepods—are responsible for the daily vertical migration of the DSL closer to and farther from the sunlit surface waters.

Today, sonar is still being used to track the movement of fish below the surface. On commercial and sport fishing trips, a sonar device (often called a "fish finder") is employed to help seek fish in the ocean, often with successful results.

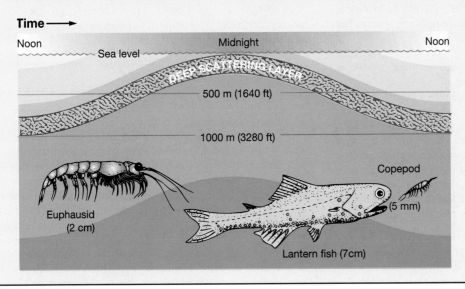

Figure 12F The deep scattering layer.

Students Sometimes Ask…

Why are there scientific names for all organisms? Wouldn't it be easier just to know the common name of an organism?

Since each individual species has a unique scientific name, using the scientific name is more specific in clearly identifying one particular species. Common names are often used for a variety of different species of organisms, which easily leads to confusion. For example, the common term "dolphin" is used to describe dolphins, porpoises, and even a type of fish! Most people would shudder at being served dolphin in a restaurant, but dolphin *fish* (also called mahi-mahi) is often a featured menu item.

Also, there can be more than one common name for the same species, which could lead to confusion. Another disadvantage of common names is that they vary with the language spoken. Because scientific names are Latin based, they are not translated to other languages. This allows, for example, a Chinese scientist to communicate effectively with a Greek scientist about a particular organism. It has been found that scientific names are quite useful, descriptive (if you know a bit about Latin terms and word roots), and unambiguous.

What is the difference between kelp, seaweed, and marine algae?

In common usage, they all refer to the same thing: large branching forms of photosynthetic marine organisms that contain pigments which give these organisms their color. However, there are differences between the terms.

"Seaweed" is an unfortunate term that was probably coined because these organisms were viewed by early mariners as having little use. For one thing, they could not be eaten. They also clogged harbors, entangled vessels, and washed up on the beach in great abundance after storms. Historically, all types of marine algae (except microscopic species) had similarities to weeds on land, which led to them becoming known collectively as "seaweeds."

Further study on the role of large marine algae has led to the awareness that they are a vital part of coastal ecosystems. Since the term "seaweed" implies that these marine organisms have no usefulness, marine biologists favor the term "marine algae." To differentiate from the microscopic planktonic species of algae, large marine algae are often called "marine macro algae." The branching types of brown algae (phylum Phaeophyta) are called "kelp." The various types of marine algae are discussed further in the next chapter.

I'm confused. You mentioned that 98 percent of marine species are benthic, yet the vast majority of the ocean's biomass is planktonic. How can this be true?
The key to understanding this apparent dilemma is to distinguish the difference between the *number of species* and the *amount of living organisms (biomass)*. For instance, a zoo has a high number of different species, but a low biomass of each. Conversely, a large cattle farm may have only one species, but a large biomass of that single species.

In benthic regions, there are many types of environments, such as sandy bottoms, rocky bottoms, muddy bottoms, upper tide zones, lower tide zones, high-energy shorelines, low-energy shorelines, and many others. This has led to a huge number of different species that have specific adaptations to the wide variety of benthic environments. Hence, there is high species diversity (many different types of species), but a smaller biomass (a smaller overall amount of organisms) than in sunlit surface waters. Conversely, the consistency of the pelagic environment has led to low species diversity (many individuals of the same species), but a large biomass (a large amount of living organisms, provided there are appropriate nutrient levels and sunlight conditions).

I know that some fish living in the ocean return to the same freshwater stream where they were born to spawn. How can fish adapted to saltwater survive in a freshwater environment?
It is true that some species of marine fish are born in fresh water. They spend most of their lives in the sea, only to return to the exact freshwater stream to spawn. (Although it is not known how these fish accurately return home, most investigators think the fishes' best guides are odors and currents.) Marine fish that return to a freshwater environment to spawn are called

anadromous (*ana* = up, *dromos* = a running) fish, which is the type of behavior that salmon of the Pacific and North Atlantic Oceans exhibit.

Since these marine fish are adapted to living in a saltwater environment, they cannot survive for long in a freshwater environment. While in fresh water, these fish take on much water through osmosis and are disinterested in food. The change in salinity causes the Pacific species to die after spawning. This results in many dead fish carcasses along the stream after a salmon run, which is a smelly mess but returns nutrients to the stream and, ultimately, to the ocean. Interestingly, Atlantic salmon are able to return to the ocean after as much as a week in freshwater streams.

Why do my fingers get all wrinkly when I stay in the water for a long time?
It's caused by water molecules passing through your skin (a semipermeable membrane) due to osmosis. Water-soluble atoms, molecules, and ions that can't move to the surface of your skin will draw water into your skin by way of osmosis. The water molecules outside your body—contained in either pure water or seawater—will flow into your skin cells in an attempt to dilute the dissolved particles inside those cells. After a short time in water, the cells of your skin will contain many more water molecules than before, which hydrates your skin and causes it to look wrinkly. Mostly, this affects appendages such as fingers and toes because those parts of your body have a high surface area per unit of volume. Fortunately, it's a temporary condition. Your body will absorb the excess water molecules once you are out of the water.

Why are tropical fish so brightly colored compared to fish in cooler waters?
That's an interesting question about which there has been much speculation. The bright coloration of tropical species may be an adaptation caused by the need for camouflage from predators in an environment where seawater has higher transparency than in high latitudes. However, many of these fish hardly blend into their environment: Their bright colors and distinctive markings actually make them *more* apparent, which may be advantageous for attracting mates. One possible explanation involves *disruptive coloration,* where large, bold patterns of contrasting colors tend to make an object blend in when viewed against an equally variable, contrasting background. This principle is used by certain animals (such as zebra and some snakes) and is also used for military camouflage. Studies of fish eyes reveal that some species of fish are able to distinguish more colors over a greater range of wavelengths than human eyes can. Evidently, there is some biological advantage to tropical fish having vivid colors—otherwise they wouldn't be so colorful. Scientists have yet to reach a widely accepted explanation of why tropical fish are so brightly colored. Perhaps these fish are brightly colored simply because they *can* be.

Summary

There are a wide variety of organisms living in the ocean, ranging in size from microscopic bacteria and algae to the largest animal on Earth today. All living things belong to one of three major domains (branches) of life: Archaea, simple microscopic bacteria-like creatures; Bacteria, simple life forms with cells that usually lack a nucleus; and Eukarya, which includes complex organisms including plants and animals.

Organisms are further divided into five kingdoms: Monera, single-celled organisms without a nucleus; Protoctista, single- and multi-celled organisms with a nucleus; Fungi, mold and lichen; Plantae, many-celled plants; and Animalia, many-celled animals. Classification of organisms involves dividing the kingdoms into increasingly specific groupings: phylum, class, order, family, genus, and species. Most organisms are known by their species name as well as by a common name.

In comparing marine and terrestrial environments, only about 17 percent of all known species inhabit the ocean. This is probably because the marine environment is much more stable than the terrestrial environment, which has led to lower species diversity in the ocean

The marine environment is divided into two basic units—the pelagic (open sea) and the benthic (sea bottom) environments. These regions, which are further divided primarily on the basis of depth, are inhabited by organisms that can be classified into three categories on the basis of lifestyle: plankton (free-floating forms with little power of locomotion), nekton (swimmers), and benthos (bottom dwellers). Most of the ocean's biomass is planktonic.

Marine organisms are well adapted to life in the ocean. The relatively stable marine environment is thought to have given rise to all living things. Those organisms that have established themselves on land have had to develop complex systems for support and for acquiring and retaining water. Pelagic marine organisms depend for support (and prevention of sinking) primarily on buoyancy and frictional resistance, often by exhibiting ornate appendages. However, the viscosity of seawater has necessitated nektonic organisms developing streamlined bodies.

Surface temperature variations of the world ocean do not exhibit as much change as those on land on a daily, seasonally, and yearly basis. Compared with life in colder-water regions, organisms living in warm water tend to be individually smaller, have ornate plumage, comprise a greater number of species, and constitute a much smaller total biomass. Warm-water organisms also tend to live shorter lives and reproduce earlier and more frequently than their cold-water counterparts.

Osmosis is the passing of water molecules from a region in which they are in higher concentration through a semipermeable membrane into a region where they are in lower concentration. If the body fluids of an organism and ocean water are separated by a membrane that allows water molecules to pass through, problems related to osmosis may result.

Many marine invertebrates are essentially isotonic, having body fluids with a salinity similar to that of ocean water. Most marine vertebrates are hypotonic, having body fluids with a salinity lower than that of ocean water, and tend to lose water through osmosis. Freshwater organisms are essentially all hypertonic, having body fluids much more saline than the water in which they live, so they must compensate for a tendency to take water into their cells through osmosis.

The algae that must stay in surface water to receive sunlight and the small animals that feed on them do not have effective means of locomotion. They depend, therefore, on their small size and other adaptations to give them a high ratio of surface area per unit of body mass, which results in a greater frictional resistance to sinking. Their small size also allows them to efficiently absorb nutrients and dispose of wastes.

Dissolved gases are important for organisms living in the ocean, and many marine animals extract oxygen through their gills. Water's high transparency has resulted in well-developed eyesight in many marine organisms. To avoid being seen and consumed by predators, many marine organisms are transparent, camouflaged, or counter-shaded. Although the increasing pressure with depth in the ocean affects humans, marine organisms do not have large internal air pockets that can be compressed. Thus, they are unaffected by the high pressure at depth in the ocean.

Key Terms

Abyssal zone (p. 364)

Abyssopelagic zone (p. 361)

Animalia (p. 360)

Aphotic zone (p. 362)

Archaea (p. 358)

Bacteria (p. 358)

Bathyal zone (p. 364)

Bathypelagic zone (p. 361)

Benthic environment (p. 361)

Benthos (p. 365)

Bioluminescence (p. 362)

Biomass (p. 365)

Biozone (p. 361)

Counter-shading (p. 376)

Darwin, Charles (p. 357)

Deep scattering layer (DSL) (p. 379)

Detritus (p. 363)

Diffusion (p. 372)

Disphotic zone (p. 361)

Epifauna (p. 365)

Epipelagic zone (p. 361)

Eukarya (p. 358)

Euphotic zone (p. 361)

Euryhaline (p. 372)

Eurythermal (p. 372)

Fungi (p. 360)

Gill (p. 376)

Hadal zone (p. 364)

Holoplankton (p. 365)

Hypertonic (p. 373)

Hypotonic (p. 373)

Infauna (p. 365)

Inner sublittoral zone (p. 363)

Isotonic (p. 373)

Littoral zone (p. 363)

Macroplankton (p. 365)

Meroplankton (p. 365)

Mesopelagic zone (p. 361)

Monera (p. 358)

Nektobenthos (p. 367)

Nekton (p. 365)

Neritic province (p. 361)

Oceanic province (p. 361)

Osmosis (p. 373)

Osmotic pressure (p. 373)

Outer sublittoral zone (p. 364)

Oxygen minimum layer (OML) (p. 362)

Pelagic environment (p. 361)

Photophore (p. 362)

Phytoplankton (p. 365)

Picoplankton (p. 365)

Plankter (p. 365)

Plankton (p. 365)

Plantae (p. 360)

Questions and Exercises

1. Discuss the accomplishments of the naturalist Charles Darwin.

2. List the three major domains of life and the five kingdoms of organisms. Describe the fundamental criteria used in assigning organisms to these divisions.

3. List the relative number of species of animals found in the terrestrial, pelagic, and benthic environments, and discuss the factors that may account for this distribution.

4. Describe how higher water temperatures in the tropics may account for the greater number of species in these regions compared with low-temperature, high-latitude areas.

5. Construct a table listing the subdivisions of the pelagic and benthic environments and the physical factors used in assigning their boundaries.

6. Describe the vertical distribution of oxygen and nutrients in the oceanic province, and discuss the factors that are responsible for this distribution.

7. What is bioluminescence? How are marine organisms able to produce light organically?

8. Describe the lifestyles of plankton, nekton, and benthos. Why is it true that plankton account for a much larger percentage of the ocean's biomass than the benthos and nekton?

9. List the subdivisions of plankton and benthos and the criteria used for assigning individual species to each.

10. Discuss the major differences between marine and terrestrial photosynthetic organisms, and explain the reasons why there is greater complexity in land plants.

11. Changes in water temperature significantly affect the density, viscosity of water, and ability of water to hold gases in solution. Discuss how decreased water temperature changes these variables and how these changes affect marine life.

12. List differences between cold- and warm-water species in the marine environment.

13. What do the prefixes *eury-* and *steno-* mean? Define the terms *eurythermal/stenothermal* and *euryhaline/stenohaline*. Where in the marine environment will organisms displaying a well-developed degree of each characteristic be found?

14. Describe the process of osmosis. How is it different from diffusion? What three things can occur simultaneously across the cell membrane during osmosis?

15. What is the problem requiring osmotic regulation that is faced by hypotonic fish in the ocean? How have these animals adapted to meet this problem?

16. List several reasons why it is advantageous to be of a small size in the ocean.

17. Compare the ability to resist sinking of an organism whose average linear dimension is 1 centimeter (0.4 inch) with that of an organism whose average linear dimension is 5 centimeters (2 inches). Discuss some adaptations other than size that are used by organisms to increase their resistance to sinking.

18. How does water temperature affect the water's ability to hold gases? How do marine organisms extract the dissolved oxygen from seawater?

19. What strategies are used by marine organisms to avoid being seen (and perhaps eaten) by predators?

20. Discuss the probable cause and composition of the deep scattering layer.

References

Borgese, E. M., Ginsburg, N., and Morgan, J. R., eds. 1991. *Ocean Yearbook 9*. Chicago: University of Chicago Press.

Bult, C., et al. 1996. Complete genome sequence of the methanogenic Archaeon, *Methanococcus jannaschii. Science* 273:5278, 1058–1072.

Coker, R. E. 1962. *This Great and Wide Sea: An Introduction to Oceanography and Marine Biology*. New York: Harper & Row.

Genin, A., Dayton, P. K., Lonsdale, P. F., and Spiess, F. N. 1986. Corals on seamount peaks provide evidence of current acceleration over deepsea topography. *Nature* 322:6074, 59–61.

Grassle, J. F., and Maciolek, N. J. 1992. Deep-sea species richness: Regional and local diversity estimates from quantitative bottom samples. *American Naturalist* 139:2, 313–341.

Hedpeth, J., and Hinton, S. 1961. *Common Seashore Life of Southern California*. Healdsburg, CA: Naturegraph.

Isaacs, J. D. 1969. The nature of oceanic life. *Scientific American* 221:65–79.

Ishimatsu, A., et al. 1998. Mudskippers store air in their burrows. *Nature* 391:6664, 237–238.

Lalli, C. M., and Parsons, T. R. 1993. *Biological Oceanography: An Introduction*. New York: Pergamon Press.

Margulis, L., and Schwartz, K. V. 1988, *Five Kingdoms: An Illustrated Guide to the Phyla of Life on Earth*, 2nd ed. New York: W. H Freeman.

Marshall, J. 1998. Why are reef fish so colorful? *Scientific American Presents: The Oceans* 9:3, 54–57.

May, R. M. 1988. How many species are there on Earth? *Science* 241:4872, 1441–1448.

Morell, V. 1996. Life's last domain. *Science* 273, 1043–1045.

Pimm, S. L., Russell, G. J., Gittlemen, J. L., and Brooks, T. M. 1995. The future of biodiversity. *Science* 269:5222, 347–350.

Sieburth, J. M. N. 1979. *Sea Microbes*. New York: Oxford University Press.

Sumich, J. L. 1976. *An Introduction to the Biology of Marine Life*. Dubuque, IA: Wm. C. Brown.

Sverdrup, H., Johnson, M., and Fleming R. 1942. Renewal 1970. *The Oceans*. Englewood Cliffs, NJ: Prentice-Hall.

Thorson, G. 1971. *Life in the Sea*. New York: McGraw-Hill.

Wilson, E. O. 1992. *The Diversity of Life*. Cambridge, MA: Belknap Press of Harvard University Press.

Wishner, K., Levin, L., Gowing, M., and Mullineaux, L. 1990. Involvement of the oxygen minimum in benthic zonation on a deep seamount. *Nature* 346:6279, 57–59.

Suggested Reading

Earth

Hecht, J. 1998. Tilt-a-world. 7:3, 34–37. A half billion years ago, the rapid movement of plates may have created marine habitats that led to the Cambrian explosion of life on Earth.

Sea Frontiers

Burton, R. 1977. Antarctica: Rich around the edges. 23:5, 287–295. The high level of biological productivity around the continent of Antarctica is the topic.

Grubner, M. 1970. Patterns of marine life. 16:4, 194–205. Many varieties of life in the ocean are discussed in terms of how their form and size fit them for life in a particular environmental niche.

Hammer, R. M. 1974. Pelagic adaptations. 16:1 2–12. A comprehensive discussion of the adaptations of pelagic organisms to reduce the energy required to maintain their position in the open ocean.

Patterson, S. 1975. To be seen or not to be seen. 21:1, 14–20. A discussion of the possible role of color in the protection and behavior of tropical fishes.

Perrine, D. 1987. The strange case of the freshwater marine fishes. 33:2, 114–119. Explains how marine crevalle jacks are able to inhabit the fresh waters of Crystal River, Florida.

Schellenger, K. 1974. Marine life of the Galápagos. 20:6, 322–332. A discussion of the unique life forms of the Galápagos Islands, 950 kilometers (590 miles) from South America.

Thresher, R. 1975. A place to live. 21:5, 258–267. An interesting discussion of how bottom-dwelling animals compete for space on the ocean floor.

Williams, L. B., and Williams, E. H., Jr. 1988. Coral reef bleaching: Current crisis, future warning. 34:2, 80–87. Corals and related reef animals underwent "bleaching" along the Central American coast in 1983 and in the Caribbean Sea in 1987.

Wu, N. 1990. Fangtooth, viperfish, and black swallower. 36:5, 32–39. Strange adaptations help fish survive in the food-scarce and dark waters below 1000-meter depths.

Scientific American

Denton, E. 1960. The buoyancy of marine animals. 203:1, 118–129. The means by which some marine animals reduce the energy expenditure required to live in the ocean water far above the ocean floor are discussed.

Eastman, J. T., and DeVries, A. L. 1986. Antarctic fishes. 255:5, 106–114. Explains how one group of fish survived when the Antarctic turned cold.

Herbert, S. 1986. Darwin as a geologist. 254:5, 116–123. Before publication of *The Origin of Species*, Charles Darwin considered himself to be primarily a geologist. This article outlines his contributions in this field, with special attention to his theory of coral reef formation.

Horn, M. H., and Gibson, R. N. 1988. Intertidal fishes. 258:1, 64–71. Intertidal fishes have undergone remarkable adaptation to survive this physically harsh environment.

Isaacs, J. D. 1969. The nature of oceanic life. 221:3, 146–165. A well-developed survey of the conditions for life in the ocean as they relate to the variety and distribution of marine life forms.

Isaacs, J. D., and Schwartzlose, R. A. 1975. Active animals of the deep-sea floor. 233:4, 84–91. A surprisingly large population of large fishes on the deep-sea floor is suggested by automatic cameras dropped to the ocean bottom.

Palmer, J. D. 1975. Biological clock and the tidal zone. 232:2, 70–79. This article investigates the mechanism of biological clocks set to rhythm of the tides, which are found in organisms from diatoms to crabs.

Partridge, B. L. 1982. The structure and function of fish schools. 246:6, 114–123. Schooling benefits and the means by which fish maintain contact with the school are considered.

Vogel, S. 1978. Organisms that capture currents. 239:2, 128–139. The manner in which sponges use ocean currents is an important part of this discussion.

Oceanography on the Web

Visit the *Essentials of Oceanography* home page for on-line resources for this chapter. There you will find an on-line study guide with review exercises, and links to oceanography sites to further your exploration of the topics in this chapter. *Essentials of Oceanography* is at: **http://www.prenhall.com/thurman** (click on the Table of Contents menu and select this chapter).

CHAPTER 13

BIOLOGICAL PRODUCTIVITY AND ENERGY TRANSFER

Baseline Studies in the California Current: The CalCOFI Program

When John Steinbeck's book *Cannery Row* was published in 1945, towns like Monterey, California, had a booming sardine fishery. By 1949, California faced the collapse of its sardine fishery, which was then the largest single-species fishery in the world. Scientists could not agree on whether the decrease in population had been caused by overfishing or by natural changes in the environment. To help solve this mystery, the state created the **California Cooperative Oceanic Fisheries Investigations (CalCOFI)**, a consortium of industry, universities, and state and federal agencies appointed to study the California current. The program continues today as a partnership among Scripps Institution of Oceanography, the California Department of Fish and Game, and the U.S. National Marine Fisheries Service. The program has been enlarged to include the waters along the West Coast of North America from the tip of Baja California, Mexico, to the Columbia River.

Early CalCOFI investigations proved the futility of studying a single species without examining its ocean environment and the cohabitants of that environment. Thus, the focus of the program was soon expanded to include all organisms of the eastern North Pacific Ocean as well as all source waters of the California Current. At the center of the investigations are the CalCOFI surveys, a shipboard monitoring program that makes seasonal measurements by using nets, temperature probes, and other instruments along precisely plotted pathways off the coast of California (Figure 13A). As a result, tens of thousands of measurements of zooplankton abundance, along with temperature, salinity, oxygen, nutrient, and chlorophyll levels, have been made in the California Current extending back almost 50 years.

This record allows oceanographers to define the "normal" patterns of the physical, chemical, and biological components of the California Current. Perhaps more importantly, it provides a baseline against which to identify any effects on organisms due to warming trends, such

as that predicted from a buildup of greenhouse gases in the environment.

Models and methods for studying fish such as sardines and anchovies developed by the CalCOFI program are

Figure 13A CalCOFI sampling bottles.

now used worldwide. Scientists in the program have developed new methods for estimating the size of adult populations, for determining the growth rates of larvae, and for estimating the number of larvae consumed by larger organisms.

Through the CalCOFI investigations, scientists came to understand that many environmental forces that have a minor effect on adult fish can have a tremendous effect on larvae. For example, young anchovies were found to be weak and vulnerable creatures that perish unless they find dense concentrations of phytoplankton for food. When the ocean is calm, phytoplankton often concentrate in narrow layers below the sea surface, providing a source of food on which larvae can feed and rapidly grow. Storms, however, can disrupt the layers of phytoplankton and prevent larvae from finding the high concentration of food needed for their survival. Thus, the strength and timing of storms can have a critical impact on the population of adult fish for several years.

In the end, an answer was found concerning the collapse of the sardine industry. CalCOFI investigators determined that the collapse had been caused by a combination of natural changes in the environment as well as by overfishing. The CalCOFI program highlights the success of long-term cooperative scientific studies between research institutions and state and federal government agencies. Today, oceanographers are becoming increasingly reliant on international cooperation and funding to support other such important studies.

Producers are plants and algae that photosynthesize their own food from carbon dioxide, water, and sunlight. Their ability to capture solar energy and bind it into their food sugars is the basis for all nutrition in the ocean (except near hydrothermal vents,[1] where chemosynthesis is the major source of "food" energy). The ocean's producers are the foundation of the ocean food web. The major primary producers of the oceans are marine algae. They capture nearly all of the solar energy used to support the entire marine biological community.

Although there are several large species of marine algae with which many people are familiar, these large species play only a minor part in the production of energy for the ocean population as a whole. Rather, marine organisms depend primarily on the small planktonic varieties of marine algae that inhabit the near-surface sunlit waters of the world's oceans. These algae are not so obvious: All are microscopic, and they are scattered throughout the breadth of the ocean's surface layer. They represent the largest community of **biomass**[2] in the marine environment—the *phytoplankton*.

In addition to the food production of algae living near the ocean surface, another food-production method operates in the deep ocean. In the total darkness of the deep sea, where no measurable sunlight penetrates, certain archaeon (bacteria-like organisms) use energy released by oxidizing hydrogen sulfide or methane to synthesize food. This food-production method supports a vast array of unusually large deep-sea benthos.

The chemical energy stored by both of these producers of organic matter—surface phytoplankton and deep-sea archaeon—is passed to the various populations of animals that inhabit the oceans through a series of feeding relationships called food chains and food webs.

Primary Productivity

Primary productivity is the amount of carbon fixed by organisms through the synthesis of organic matter using energy derived from solar radiation or chemical reactions. The major process through which primary productivity occurs is **photosynthesis**.

Recently, new knowledge has been acquired on the role of **chemosynthesis** (chemically converting the energy in gases such as hydrogen sulfide gas and methane) in supporting hydrothermal vent communities along oceanic spreading centers. However, chemosynthesis is much less significant in worldwide marine primary production than is photosynthesis. Because of this, let's concentrate our discussion on photosynthesis.

Photosynthetic Productivity

Photosynthesis is a chemical reaction in which energy from the sun is stored in organic molecules (Figure 13–1). The total amount of organic matter produced by photosynthesis per unit of time is the **gross primary production** of the oceans. Algae use some of this organic matter for their own maintenance, through respiration. What remains is **net primary production**,[3] which is manifested as growth and reproduction products. It is the net primary production that supports the rest of the marine population.

---

[1]Hydrothermal vent biocommunities are discussed in more detail in Chapter 15, "Animals of the Benthic Environment."

[2]Biomass is the mass of living organisms.

[3]The difference between gross and net primary production is similar to the difference between your gross pay (your earnings before taxes) and your net pay (the amount you take home after taxes).

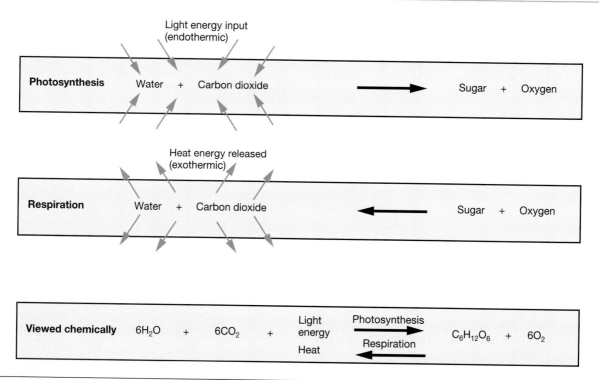

Figure 13–1 Photosynthesis and respiration.

Gross primary production has two components, new production and regenerated production. **New production** is that part supported by nutrients brought in from outside the local ecosystem by processes such as upwelling. The higher the ratio of new production to gross primary production in an ecosystem, the greater is its ability to support animal populations that are depended on for fisheries, such as pelagic fishes and benthic scallops. **Regenerated production** results from nutrients being recycled within the ecosystem.

Photosynthetic productivity can be measured by a variety of methods. One of the most direct ways is to physically measure the amount of phytoplankton in the ocean by capturing plankton in cone-shaped nylon nets called **plankton nets** (Figure 13–2). These nets—which resemble windsocks at airports—are used to filter plankton from the ocean by being towed at a specific depth by research vessels. Analysis of the types of organisms captured by plankton nets reveals much about the productivity of the area.

Other measurement methods include lowering specially designed bottles into the ocean, laboratory analysis using radioactive carbon, or determining chlorophyll levels based on seawater surface color from orbiting satellites. One such satellite, **SeaWiFS** (Sea-viewing Wide Field-of-View Sensor), began operating in 1997 and replaced the Coastal Zone Color Scanner, which operated between 1978 and 1986. SeaWiFS measures the color of the ocean with a radiometer and provides global coverage of ocean chlorophyll levels as well as land productivities every two days.

Availability of Nutrients

The distribution of life throughout the ocean's breadth and depth depends mainly on the availability of nutrients for phytoplankton, such as nitrate, phosphorous, iron, and silica. Where the physical conditions are right for supplying large quantities of nutrients, marine populations reach their greatest concentration. Where are these areas in which the greatest biomass of marine organisms is found? To answer this question, the *sources* of nutrients must be considered.

Water in the form of runoff has an important role in eroding the continents, carrying the eroded material to the oceans, and depositing it as sediment at the margins of the continents. During this process, water dissolves and transports many substances, including nitrate and phosphate compounds.[4] Once in the ocean, these nutrients serve as the basic nutrient supply for phytoplankton.

Through photosynthesis, phytoplankton combine these nutrients with carbon dioxide and water to produce the carbohydrates, proteins, and fats that store the produced energy (see Figure 13–1). The rest of the ocean's biological community depends upon this storehouse of energy.

Because the continents are the major sources of these nutrients, one might expect the greatest concentrations of marine life to be at the continental margins, and this

---

[4]Nitrates and phosphates are the basic ingredients in garden and farm fertilizer.

Figure 13–2 Plankton nets.
These large, cone-shaped, fine mesh plankton nets being washed during a CalCOFI cruise are lowered into the water and towed behind a research vessel to collect plankton.

is exactly the case. As one travels from the continental margins to the open sea, a progressively *lower* concentration of marine life is experienced. This is due to the vast depth of the world's oceans and the great distance between the open ocean and the coastal regions where nutrients are concentrated.

Recent studies of ocean productivity have indicated that some oceanic regions may be limited by a lack of iron.[5] Studies completed in the waters near Antarctica and in equatorial surface waters near the Galápagos Islands reveal that photosynthetic production is low even though the concentration of all nutrients—except iron—is high. Production is high only in regions of shallow water downcurrent from islands or landmasses where iron from rocks and sediments was dissolved into the water to significant levels.

---

[5]The idea of fertilizing the ocean with iron to stimulate productivity and increase the amount of CO_2 gas absorbed by the ocean is discussed in Chapter 6, "Air–Sea Interaction."

Availability of Solar Radiation

To use these nutrients, phytoplankton must meet another requirement. Photosynthesis cannot proceed unless it is powered by light energy, provided by solar radiation. Sunlight penetrates the atmosphere quite readily, so plants on the continents would seem to have an advantage over algae in the ocean. There is almost always an abundance of solar radiation to drive the photosynthetic reactions for these land-based plants.

By contrast, ocean water is a significant barrier to the penetration of solar radiation. In the clearest water, solar energy may be detected to depths of about 1 kilometer (0.6 mile). However, the amount of solar energy reaching these depths is extremely small and is not adequate to allow organisms to photosynthesize.

Photosynthesis in the ocean is restricted to the uppermost surface waters and those shallow areas of the sea floor that receive adequate light penetration. The depth at which net photosynthesis becomes zero is called the **compensation depth for photosynthesis**. The layer of the ocean that includes water from the surface to the compensation depth for photosynthesis is called the **euphotic zone**. In the open ocean, the euphotic zone extends to a depth of approximately 100 meters (330 feet). Near the coast, the euphotic zone may occur at a depth of less than 20 meters (66 feet) because the water contains more suspended inorganic material (turbidity) or microscopic organisms that limit light penetration.

How do the two factors necessary for photosynthesis—the supply of nutrients and the presence of solar radiation—differ between coastal and open ocean areas? In the open ocean, away from the continental margins, the column of water that has solar energy available extends deeper, but this column contains only a small concentration of nutrients. Conversely, in coastal regions, where the water is more turbid, light penetrates to much shallower depths, but the nutrient supply is quite rich. Because the coastal zone is by far more productive, it is clear that nutrient availability is the most important factor affecting the distribution of life in the oceans.

Margins of the Oceans

As mentioned previously, the stability of the ocean environment makes it ideal for the continuation of life processes. However, it is ironic that the richest concentration of marine organisms is in the very margins of the oceans, where conditions are least stable. The coastal region has many physical conditions that appear at first to be deterrents to the establishment of life:

- Water depths are shallow, allowing seasonal variations in temperature and salinity that are much greater than would be found in the open ocean.
- The thickness of the water column varies in the nearshore region as a result of tide movements that

periodically cover and uncover a thin strip of land along the margins of the continents.

- The surf condition represents a sudden release of energy that has been carried for great distances across the open ocean.

Each of these conditions appears to provide a stressful environment to organisms. Such a great concentration of biomass in an environment that contains so many stress factors highlights the importance of the *evolutionary process* in developing new species by *natural selection*. Over billions of years of geologic time, new life forms have developed to fit every imaginable biological niche. Many of these life forms have adapted to live under adverse conditions within which life can exist so long as nutrients are available.

Along continental margins, some areas have more abundant life than in other areas. What characteristics create such an uneven distribution of life? Again, only those basic requirements for the production of food need be considered. If one were to measure the properties of the water along continental coasts and compare these measurements from place to place, one could see that *areas that have the greatest biomass are those with lower water temperatures.* This is because cold water holds in solution larger amounts of the essential gases—oxygen and carbon dioxide—than warmer waters do. Of particular importance is the increased availability of carbon dioxide, which stimulates phytoplankton growth. In essence, the increased availability of one of the basic requirements of phytoplan—nutrient supply—profoundly affects the distribution of life in the oceans.

Upwelling Brings Nutrients to the Surface

In certain areas of the coastal margins, an additional factor enhances conditions for life—upwelling. **Upwelling**, as the name implies, is a flow of subsurface water toward the surface. Most notably, this phenomenon occurs along the western margins of continents where surface currents are moving toward the Equator (Figure 13–3A).

Along the western margins of continents in both the Northern and Southern Hemispheres, Ekman transport (as described in Chapter 7) moves surface water away from the coast. As this happens, water rises to replace it, from depths of 200 to 1000 meters (660 to 3300 feet) (Figure 13–3B).

This water comes up from depths below where photosynthesis occurs. It is rich in nutrients for phytoplankton because there are no phytoplankton in deeper waters to use these nutrients. This constant replenishing of nutrients at the surface enhances the conditions for life in areas of upwelling. Upwelling water is usually of low temperature, which produces the additional benefit of having a high capacity for dissolved gases (Figure 13–3C).

The combined availability of solar radiation and nutrients results in maximum biomass concentrations in shallow coastal waters. As has been discussed, biomass decreases with greater distance from shore and with greater water depth.

Water Color and Life in the Oceans

Areas of relatively high and low biological production can be determined in the ocean based on the water's color. Satellites can easily detect differences in water color, which are influenced by changing concentrations of photosynthetic pigment (Figure 13–4). The amount of photosynthetic pigment varies directly with the amount of primary productivity. In Figure 13–4, highly productive areas (such as shallow-water coastal regions and areas of upwelling) stand out in marked contrast to low-productivity, tropical open-ocean regions. Areas that have low productivity are termed **oligotrophic** (*oligo* = few, *tropho* = nourishment), whereas regions of high productivity are termed **eutrophic** (*eu* = good, *tropho* = nourishment).

Coastal waters are almost always greenish in color because they contain more suspended particulate matter (turbidity) due to runoff from land. This disperses solar radiation in such a way that the wavelengths most scattered are those for greenish or yellowish light. This condition is also partly the result of yellow-green microscopic marine algae in these coastal waters.

In the open ocean, where particulate matter is relatively scarce and marine life exists in low concentration, the water appears blue. This is due to the scattering of blue wavelengths of light caused by the extremely small size and hydrogen bonding of water molecules. Similarly, the scattering of blue light wavelengths by the atmosphere is responsible for producing the blue color of a clear sky.

Green color in water usually indicates the presence of a lush biological population, such as might correspond to the tropical forests on the continents. The deep indigo blue of the open oceans, particularly between the tropics, usually indicates an area that lacks abundant marine life. In fact, these areas are often considered to be biological deserts!

Photosynthetic Marine Organisms

Many types of marine organisms photosynthesize. Mostly, they are represented by marine algae, which occupy the equivalent ecological niche of plants on land. Let's first examine the true plants, then the macroscopic and microscopic types of algae.

Seed-Bearing Plants (Spermatophyta)

Marine plants are generally confined to coastal areas. The only members of kingdom Plantae observed in the marine environment belong to the highest group of plants, the seed-bearing Spermatophyta. Two common seed-bearing plants found in the marine environment are eelgrass (*Zostera*) and surf grass (*Phyllospadix*). Eelgrass, a grasslike plant with true roots, exists primarily in the quiet waters of bays and estuaries from the low-tide zone to a depth of some 6 meters (20 feet). Surf grass (Figure 13–5), which is also a seed-bearing plant with true roots, is

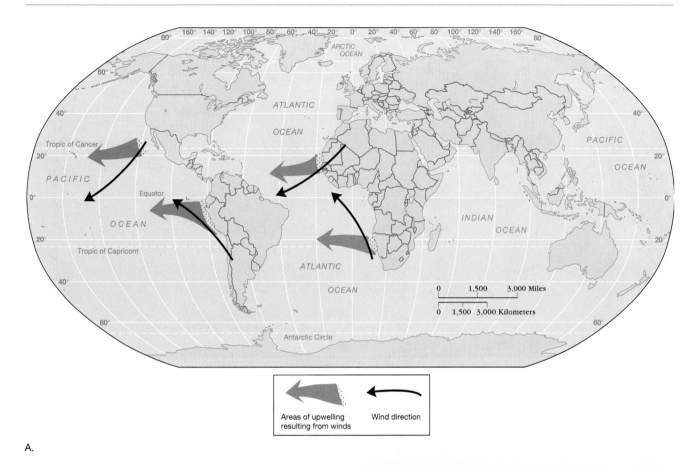

A.

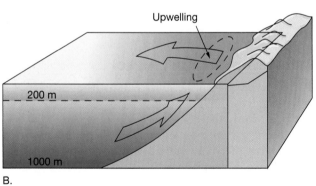

B.

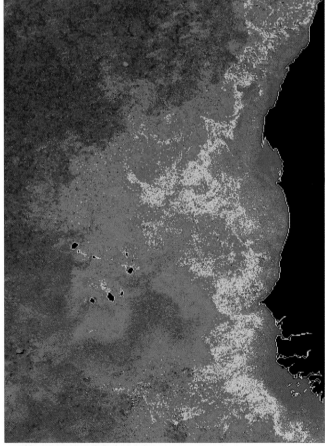

Figure 13–3 Upwelling.

A. Coastal winds (*black arrows*), aided by Ekman transport, drive surface water away from the western margins of continents (*blue arrows*). **B.** As the surface water moves away from the continent, water upwells and replaces it. Because this water comes from below the zone of photosynthesis, it is cold (unheated by sunlight) and rich in nutrients. **C.** False-color Coastal Zone Color Scanner (CZCS) image from the *Nimbus*-7 satellite of the Canary Islands and northwestern Africa, where wind-driven upwelling results in high phytoplankton biomass and productivity. The highest productivity is indicated by the red and orange colors and decreases through yellow, green, and blue.

C.

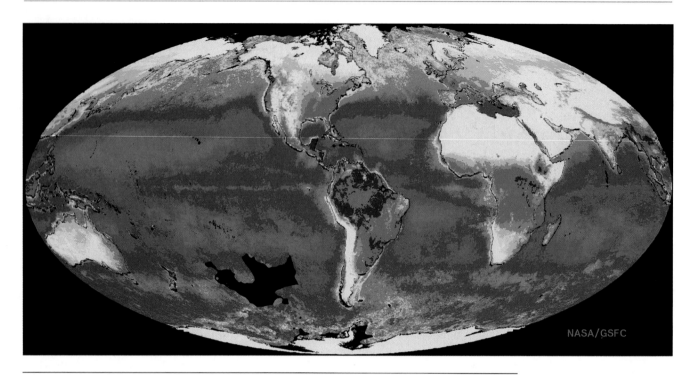

NASA/GSFC

Figure 13–4 Photosynthetic production in the world ocean.

Data from the Coastal Zone Color Scanner (CZCS) aboard the satellite *Nimbus-7* between November 1978 and June 1981 were used to produce this false-color image. The CZCS senses changes in seawater color caused by changing concentrations of photosynthetic pigment, which varies with the photosynthetic productivity. Low productivity values [below 0.1 milligram per cubic meter (mg/m^3)] are represented by the magenta color. Progressively higher values are shown with different colors, with orange indicating values exceeding 10.0 mg/m^3. Black areas indicate insufficient data.

Figure 13–5 Surf grass (*Phyllospadix*).

typically found in the high-energy environment of exposed rocky coasts from the intertidal zones down to a depth of 15 meters (50 feet).

Other seed-bearing plants are found in salt marshes and include grasses (mostly of the genus *Spartina*), whereas mangrove swamps contain primarily mangroves (genera *Rhizophora* and *Avicennia*). All of these plants are important sources of food and protection for the marine animals that inhabit these nearshore environments.

Macroscopic (Large) Algae

Various types of marine macro algae (the "seaweeds") are typically found in shallow waters along the ocean margins. Mostly, these large algae are attached to the bottom, but a few are floating species. Algae are classified based partly on color criteria, which is a result of the pigment they contain (Figure 13–6).

Brown Algae The brown algae of the phylum Phaeophyta (*phaeo* = dusky, *phytum* = a plant) include the largest members of the attached (not free-floating) species of the marine algal community. Their color ranges from very light brown to black. Brown algae occur primarily in temperate and cold-water areas. Their sizes range widely. The smallest is a black encrusting patch of *Ralfsia* of the upper and middle intertidal zones. The largest is bull kelp (*Pelagophycus*), which may grow in water deeper than 30 meters (100 feet) and extend to the surface. Other types of brown algae include *Sargassum* (Figure 13–6A) and *Macrocystis* (Figure 13–6B).

Green Algae Although green algae of phylum Chlorophyta (*chloro* = green, *phytum* = a plant) are common in freshwater environments, they are not well represented in the ocean. Most marine species are intertidal or grow in shallow bay waters. They contain the pigment chlorophyll, which makes most of them "grass

Figure 13–6 Macroscopic algae.

A. Brown algae, *Sargassum*. This attached form is similar to the floating form that is the namesake of the Sargasso Sea. **B.** Brown algae, *Macrocystis*, a major component of kelp beds. **C.** Green algae, *Codium fragile*, also known as sponge weed or dead man's fingers. **D.** Red algae, *Lithothamnion*, an encrusting form that produces a pink coating on these rounded cobbles.

green" in color. They grow only to moderate size, seldom exceeding 30 centimeters (12 inches) in the largest dimension. Forms range from finely branched filaments to flat thin sheets.

Various species of sea lettuce (*Ulva*), a thin membranous sheet only two cell layers thick, are widely scattered throughout colder-water areas. Sponge weed (*Codium*), a two-branched form more common in warm waters, can exceed 6 meters (20 feet) in length (Figure 13–6C).

Red Algae Red algae of phylum Rhodophyta (*rhodo* = red, *phytum* = a plant) are the most abundant and widespread of marine macroscopic algae. Over 4000 species occur from the very highest intertidal levels to the outer edge of the inner sublittoral zone. Many are attached to the bottom, either as branching forms or as forms that encrust surfaces (Figure 13–6D). They are very rare in fresh water. Red algae range from just barely visible to the unaided eye to lengths of 3 meters (10 feet). While found in both warm- and cold-water areas, the warm-water varieties are relatively small.

The color of red algae varies considerably depending on its depth in the intertidal or inner sublittoral zones. In upper, well-lighted areas, it may be green to black or purplish. In deeper-water zones, where less light is available, it may be brown to pinkish red.

The bulk of marine photosynthetic productivity is believed to occur within the surface layer of the ocean to a depth of 100 meters (330 feet), which corresponds to the depth of the euphotic zone. At this depth, the amount of light is reduced to 1 percent of that available at the surface. However, a red alga has been documented growing at a depth of 268 meters (880 feet) on a seamount near San Salvadore, Bahamas. Available light at this sighting was thought to be only 0.05 percent of the light available at the ocean's surface.

Microscopic (Small) Algae

Microscopic forms of algae produce over 99 percent of the food supply for marine animals. Most of these organisms constitute phytoplankton—photosynthetic algae that live in the upper surface waters and drift with currents—although some live on the bottom in the nearshore environment, where sunlight reaches the shallow ocean floor.

Golden Algae   Containing predominantly the yellow pigment **carotin**, the golden algae of phylum Chrysophyta (*chrysus* = golden, *phytum* = a plant) store food in the form of carbohydrates and oils. There are two types of golden algae: diatoms and coccolithophores, both of which have been described in Chapter 4, "Marine Sediments."

Diatoms   The **diatoms** (*diatoma* = cut in half) are a class of algae that are contained in a **test** (microscopic shell). The test is composed of opaline silica, which means that it is like the semiprecious mineral opal, con-

taining considerable water locked into its silicate structure ($SiO_2 \cdot nH_2O$). These silica housings are important geologically because they accumulate on the ocean bottom and produce a siliceous sediment called **diatomaceous earth**. Some deposits of diatomaceous earth, now elevated onto land by tectonic forces, are mined and used in filtering devices and numerous other applications (see Box 4–1). In terms of productivity, diatoms are the most important group of marine algae.

The tests of diatoms resemble a microscopic pillbox, with a top and bottom half that fits together, and can have a variety of shapes (Figure 13–7A). The protoplasm of the single cell is contained within this test, and it exchanges nutrients and waste with the surrounding water through slits or holes in its shell.

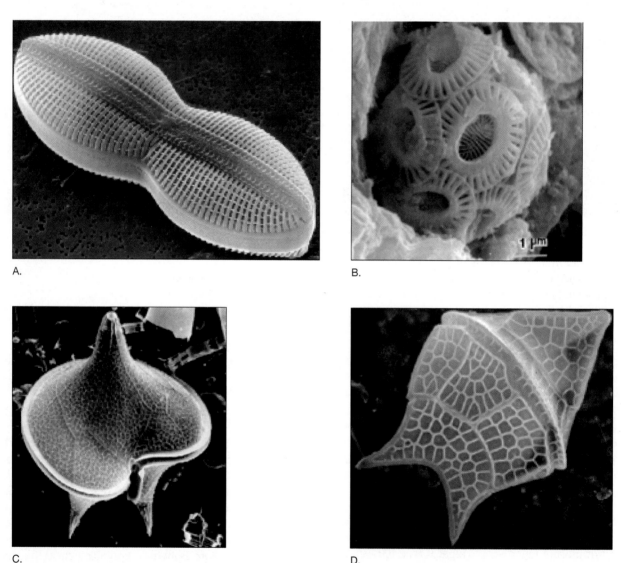

A.

B.

C.

D.

Figure 13–7 Microscopic algae.
A. Peanut-shaped diatom *Diploneis* (length = 50 microns, or 0.002 inch). **B.** Coccolithophore *Emiliania huxleyi*, showing disk-shaped calcium carbonate ($CaCO_3$) plates—called coccoliths—that cover the organism (bar scale = 1 micron, or 0.00004 inch). **C.** Dinoflagellate *Protoperidinium divergnes* (length = 70 microns, or 0.003 inch). **D.** Leaf-like tropical dinoflagellate *Heterodinium whittingae* (length = 100 microns, or 0.004 inch).

Coccolithophores. **Coccolithophores** (*coccus* = berry; *lithos* = stone; *phorid* = carrying) are covered with small calcareous plates called **coccoliths**, made of calcium carbonate ($CaCO_3$) (Figure 13–7B). The individual plates are about the size of a bacterium, and the entire organism is too small to be captured in plankton nets. Coccolithophores contribute significantly to calcareous deposits in all the temperate and warmer oceans.

Dinoflagellates The **dinoflagellates** (*dino* = whirling, *flagellum* = a whip) belong to the phylum Pyrrophyta (*pyrro* = red, *phytum* = a plant) (Figures 13–7C and D). In terms of marine productivity, they are second in importance only to the diatoms. They possess **flagella** (small, whip-like structures) for locomotion, giving them a slight capacity to move into areas that are more favorable for photosynthetic productivity. Dinoflagellates are rarely important geologically because their tests are made of the biodegradable material cellulose, which is not preserved in the fossil record. Many dinoflagellates have bioluminescent capabilities and they sometimes exist in great abundance, producing "red tides" (Box 13–1). In addition, many of the 1100 species undergo structural changes in response to changes in their environment.

Regional Productivity

Primary photosynthetic production in the oceans varies dramatically from place to place (see Figure 13–4). Typical units of photosynthetic production are in weight of carbon (*grams of carbon*) per unit of area (*square meter*) per unit of time (*year*), which is abbreviated as gC/m^2/yr. Values range from as low as 1 gC/m^2/yr in some areas of the open ocean to as much as 4000 gC/m^2/yr in some highly productive coastal estuaries (Table 13–1). This variability is the result of the uneven distribution of nutrients throughout the photosynthetic zone and seasonal changes in the availability of solar energy.[6]

About 90 percent of the biomass generated in the euphotic (sunlit) zone of the open ocean is decomposed into inorganic nutrients before descending below this zone. Approximately 10 percent of this organic matter sinks into deeper water, where most of it is decomposed. About 1 percent of this material reaches the deep-ocean floor, where it accumulates. This process of removing material from the euphotic zone and concentrating the material on the sea floor is called a **biological pump**, because it removes ("pumps") carbon dioxide and nutrients from the upper ocean and concentrates them in deep-sea waters and sea floor sediments.

Throughout much of the subtropical gyres, a permanent **thermocline** (and resulting **pycnocline**[7]) develops. It forms a barrier to vertical mixing and prevents the re-supply of nutrients to the sunlit surface layer. In the mid-latitudes, a thermocline develops only during the summer season, whereas it does not occur in polar regions. The degree to which waters develop a thermocline profoundly affects the patterns of biological production observed at different latitudes in the world ocean. Let's examine some oceanic regions of different latitudes and look at how productivity is influenced by nutrient supply and the development of a thermocline.

Productivity in Polar Oceans

The Barents Sea is a good example of a polar region that experiences dramatic seasonal changes in productivity. The Barents Sea is a part of the Arctic Ocean north of the Arctic Circle, off the northern coast of Europe. In this region above 70 degrees north latitude, there is continuous darkness for about three months of winter and continuous illumination for about three months during summer. Diatom productivity peaks here during May (Figure 13–8A), when the sun rises high enough in the sky to cause deep penetration of sunlight. As soon as this diatom food supply develops, the zooplankton population—mostly small crustaceans (Figure 13–8D)—begins feeding on it. The zooplankton biomass peaks in June and continues at a relatively high level until winter darkness begins in October.

In the Antarctic region, particularly at the southern end of the Atlantic Ocean, productivity is somewhat greater. The most likely explanation is the continual upwelling of water that has sunk in the North Atlantic. This North Atlantic Deep Water sinks and moves southward below the surface. Hundreds of years later, it finally rises to the surface near Antarctica, carrying with it high concentrations of nutrients (Figure 13–8B). When the summer sun provides sufficient radiation, there is an explosion of biological productivity.

As an illustration of the very great productivity that occurs during the short summer season in polar oceans, consider the growth rate of blue whales. The largest of all whales, the blue whale (see Figure 14–21) migrates through temperate and polar oceans at times of maximum zooplankton productivity. This perfect timing enables the whales to develop and support large calves [following a gestation of 11 months, the calves can exceed 7 meters (23 feet) length at birth].

The mother blue whale suckles the calf for 6 months with a teat that actually pumps the youngster full of rich

---

[6]For a review of Earth's seasons, see Chapter 6, "Air-Sea Interaction."

[7]Recall that a thermocline is a rapidly changing zone of temperature, and a pycnocline is a rapidly changing zone of density. Development of ocean thermoclines and pycnoclines is discussed in Chapter 5, "Water and Seawater."

Box 13–1
Red Tides: Was Alfred Hitchcock's *The Birds* Based on Fact?

Sometimes conditions in the oceans stimulate the productivity of certain dinoflagellates. During these times, up to 2 million dinoflagellates may be found in 1 liter (about 1 quart) of water, giving the water a reddish color and causing what is known as a **red tide** (Figure 13B) that can affect nearly any body of water. Red tides are by no means a new phenomenon. In fact, the Old Testament makes reference to waters turning blood red, which is most likely how the Red Sea got its name.

Although many red tides are harmless to marine animals and humans, red tides are sometimes associated with mass die-offs of marine organisms. This occurs when a great abundance of dinoflagellates die and decay, which removes oxygen from the seawater and literally suffocates marine life. In addition, two genera of dinoflagellates that are responsible for some red tides are *Ptychodiscus* and *Gonyaulax*, both of which produce water-soluble toxins. Certain filter-feeding shellfish—various clams, mussels, and oysters—strain the dinoflagellates from the water as a food source. *Ptychodiscus* toxin kills fish and shellfish. *Gonyaulax* toxin is not poisonous to shellfish, but it concentrates in their tissues and is poisonous to humans who eat the shellfish, even after the shellfish are cooked. This malady is called **paralytic shellfish poisoning (PSP)**.

Symptoms of PSP in humans can occur only 30 minutes after ingesting contaminated shellfish. The symptoms are similar to those of drunkenness, including incoherent speech, uncoordinated movement, dizziness, and nausea. Documented cases throughout the world include 300 deaths and 1750 nonfatal cases. Every few years a serious outbreak occurs somewhere in the world. There is no known antidote for the toxin, which attacks the human nervous system, but the critical period usually passes within 24 hours.

April through September are particularly dangerous months for red tides in the Northern Hemisphere. In most areas, quarantine prohibits harvesting those shellfish that feed on toxic microscopic organisms. Poisons may be concentrated in these shellfish and thus reach levels that are dangerous to humans.

Red tides are occurring more frequently, more species that are toxic are being recognized, and larger areas are being affected. As well, more marine species, including fish, dolphins, and humpback whales, are succumbing to the toxins. In fact, toxic species of dinoflagellates have been implicated in several group strandings of marine mammals. Unfortunately, human activities may be contributing to the increasing frequency of red tides. For example, sewage and fertilizer that make their way into coastal waters can produce dangerous phytoplankton blooms.

A new type of poisoning caused by domoic acid—a toxin produced by a diatom—occurred on Prince Edward Island off the east coast of Canada in 1987. One hundred people were poisoned by eating mussels; of these, four people died and 10 suffered permanent memory loss. Poisoning by domoic acid has been called **amnesic shellfish poisoning** because of the memory loss suffered by a number of its victims. Domoic acid appears to be capable of spreading throughout the food web, to zooplankton, fish, mussels, clams, anchovies, birds, and even humans.

This type of poisoning may be responsible for birds attacking humans in California's Monterey Bay area in 1961. The incident—reported to have inspired Alfred Hitchcock's thriller movie, *The Birds* (Figure 13C)—apparently was triggered after the birds consumed domoic acid produced by a

Figure 13B Red Tide near La Jolla, California.

(continued)

bloom of diatoms. During this time, birds smashed into structures during flight and pecked eight people. In the same area in September 1991, brown pelicans and Brandt's cormorants began exhibiting unusual behavior: The birds acted drunk, swam in circles, and made loud squawking sounds. Research revealed that toxic diatoms (*Pseudonitzschia australis*) were ingested by unaffected anchovies that were subsequently eaten by birds, causing more than 100 birds to wash up dead at the shore.

These and other recent outbreaks of red tides have led some scientists to question whether the phenomenon is occurrring more frequently around the globe and if the red tides are becoming potentially more threatening. It may, however, be too early to tell whether red tides are truly on the increase or if the system for reporting them has simply improved.

Figure 13C Actress Tippi Hedren fights off a seagull in Hitchcock's classic, *The Birds*.

Table 13–1   Values of net primary productivity for various ecosystems.

	Ecosystem	Range (gC/m²/yr)	Average (gC/m²/yr)
Oceanic	Algae beds and coral reefs	1000–3000	2000
	Estuaries	500–4000	1800
	Upwelling zone	400–1000	500
	Continental shelf	300–600	360
	Open ocean	1–400	125
Land	Freshwater swamp and marsh	800–4000	2500
	Tropical rainforest	1000–5000	2000
	Mid-latitude forest	600–2500	1300
	Cultivated land	100–4000	650

milk that is high in fat content. By the time the calf is weaned, it is over 16 meters (50 feet) long. In two years, it will attain a length of 23 meters (75 feet). After about three years, a 60-ton blue whale has developed. This phenomenal growth rate gives some indication of the enormous biomass of copepods and krill upon which these large mammals feed.[8]

Polar waters have little density or temperature changes with depth (Figure 13–8C). Consequently, the waters are considered **isothermal** (*iso* = same, *thermo* = temperature). In most polar areas, therefore, there is no barrier to mixing between surface waters and deeper, nutrient-rich waters. However, some density segregation of water

masses occurs when summer ice melts. This lays down a thin, low-salinity layer that does not readily mix with the deeper waters. Such stratification is crucial to summer production, because it helps prevent phytoplankton from being carried into deeper, darker waters. The result is that plankton are concentrated instead in the sunlit surface waters. Without this density barrier, the summer bloom could not develop.

It is becoming increasingly clear that high levels of biological productivity occur only under these specific conditions: when periods of deep mixing (which create high nutrient levels in the sunlit surface waters) are followed by periods of density stratification.

Nutrient concentrations (phosphates and nitrates) usually are adequate in high-latitude surface waters. Thus, photosynthetic productivity in these areas is more commonly limited by solar energy availability than by nutrients. The productive season in these waters is relatively short but outstandingly abundant.

Productivity in Tropical Oceans

In direct contrast with high productivity during the summer season in the polar seas, low productivity is the rule in tropical regions of the open ocean. This may sound contradictory, for warm, sunny tropical waters sound very productive—but there is a key limiting factor. It is true that light penetrates much deeper into the open tropical ocean than into temperate and polar waters, and this produces a very deep compensation depth for photosynthesis. However, in the tropical ocean a permanent thermocline produces a stratification of water masses. This prevents mixing between surface waters and nutrient-rich deeper waters, effectively eliminating

---

[8]As an analogy, consider how many ants you would have to eat as a child to grow to adult size.

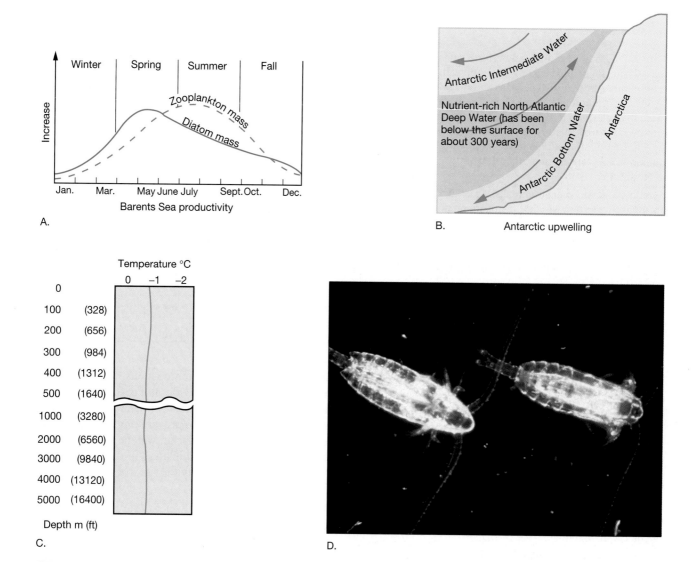

Figure 13–8 Productivity in polar oceans.
A. A springtime increase of diatom mass is followed closely by an increase in zooplankton abundance. **B.** The continuous upwelling of North Atlantic Deep Water keeps Antarctic waters rich in nutrients **C.** Polar water shows nearly uniform temperature with depth (an isothermal water column). **D.** Copepods of the genus *Calanus*, each about 8 millimeters (0.3 inch) in length.

any supply of nutrients from deeper waters below (Figure 13–9).

At about 20 degrees north and south latitude, phosphate and nitrate concentrations are commonly less than $^1/_{100}$ of the concentrations of these nutrients in temperate oceans during winter. In fact, nutrient-rich waters within the tropics lie for the most part below 150 meters (500 feet), and the highest nutrient concentration occurs between 500 and 1000 meters (1640 and 3300 feet) depth.

Generally, primary productivity in tropical oceans is at a steady but rather low rate. When the total annual productivity of tropical oceans is compared with that of the more productive temperate oceans, it reveals that tropical productivity is only about half of that found in the temperate region, averaged over an annual basis.

Within low-productivity tropical regions, three environments have unusually high productivity. These include regions of equatorial upwelling, regions of coastal upwelling, and coral reef ecosystems.

- *Equatorial upwelling.* Where trade winds drive westerly equatorial currents on either side of the Equator, surface water diverges as a result of the Ekman spiral. The surface water that moves off toward higher latitudes is replaced by nutrient-rich water that surfaces from depths of up to 200 meters (660 feet). This condition of equatorial upwelling is best developed in the eastern Pacific Ocean.

- *Coastal upwelling.* Where the prevailing winds blow toward the Equator and along western continental

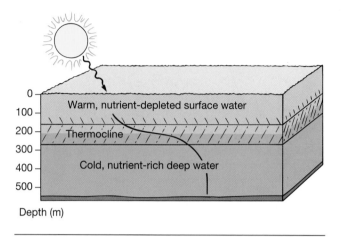

Figure 13–9 Productivity in tropical oceans.
In most tropical regions, deep penetration of sunlight produces a deep compensation depth for photosynthesis. A permanent thermocline serves as an effective barrier to mixing of surface and deep water. As plants consume nutrients in the surface layer, productivity is impeded because the thermocline prevents replenishment of nutrients from deeper water.

margins, surface waters are driven away from the coast. They are replaced by nutrient-rich waters from depths of 200 to 900 meters (660 to 2950 feet). This nutrient-rich upwelling promotes high primary productivity in these areas, which supports large fisheries. In the Pacific, such conditions exist along the southern coast of California and the southwestern coast of Peru; in the Atlantic, they exist along the northwestern coast of Morocco and the southwestern coast of Africa.

- *Coral reefs.* Organisms that comprise and live among coral reefs are superbly adapted to low-nutrient conditions, similar to how certain organisms on land are adapted to desert life. Symbiotic algae living within the tissues of coral and other species allow coral reefs to be highly productive ecosystems. As well, coral reefs tend to hold and concentrate what little nutrients exist. Coral reef ecosystems are discussed further in Chapter 15.

Productivity in Temperate Oceans

So far, we have described general productivity in the polar regions (where productivity is limited by available sunlight) and in the tropical low-latitude areas (where the limiting factor is nutrient supply). In the temperate (mid-latitude) regions, a combination of these two limiting factors controls productivity in a more complex pattern (Figure 13–10A). Let's examine how productivity varies in temperate regions with the change of seasons.

Winter Productivity in temperate oceans is very low during winter, despite high nutrient concentrations in the surface layers. Ironically, nutrient concentration is *highest* during winter (Figure 13–10A). However, the limit-

ing factor to productivity during the winter is the amount of solar energy. In Figure 13–10B (*winter*), the sun is at its lowest elevation above the horizon during this season. A high percentage of solar energy is reflected, leaving a smaller percentage to be absorbed into surface waters. During this time, the water column is isothermal, which is similar to what is seen in polar waters. Consequently, nutrients are distributed throughout the water column. During winter months, the compensation depth for photosynthesis is so shallow that it does not allow for much growth of phytoplankton. Contributing to this low rate of production is the absence of a thermocline, which allows algal cells to be carried down beneath the euphotic zone for extended periods by turbulence associated with winter waves.

Spring In Figure 13–10B (*spring*), the sun rises higher in the sky, causing the compensation depth for photosynthesis to deepen as solar radiation is transmitted deeper into surface waters. A "spring bloom" of algae occurs as a result of the availability of solar energy, nutrients, and the development of a seasonal thermocline (due to increased solar heating) that traps the algal cells in the euphotic zone (Figure 13–10A). This places a tremendous demand on the nutrient supply in the euphotic zone, and the supply of nutrients becomes limited, causing the productivity to sharply decrease. In most Northern Hemisphere areas, decreases in phytoplankton population occur in April, due to insufficient nutrients.

Summer In Figure 13–10B (*summer*), as the sun rises higher, surface waters in temperate parts of the ocean continue to warm. This causes a strong seasonal thermocline to develop at a depth of about 15 meters (50 feet). As a result, very little exchange of water occurs across this boundary and nutrients that are depleted from surface waters cannot be replaced by those from deep waters. Throughout summer, the phytoplankton population remains at a relatively low level (Figure 13–10A). Even though the compensation depth for photosynthesis is at its maximum, phytoplankton can actually become scarce in late summer.

Fall In Figure 13–10B (*fall*), solar radiation diminishes as the sun drops lower in the sky. This lowers surface temperature, and the summer thermocline breaks down. A return of nutrients to the surface layer occurs as increased wind strength mixes it with the deeper water mass in which the nutrients have been trapped throughout the summer months. These conditions create a "fall bloom" of phytoplankton, which is much less spectacular than that of the spring (Figure 13–10A). This bloom is very short-lived, and the phytoplankton population declines rapidly. The limiting factor in this case is the opposite of that which reduced the population of the spring phytoplankton bloom. In the case of the spring bloom, solar radiation was readily available, and the decrease in *nutrient supply* was the limiting factor. During the fall bloom,

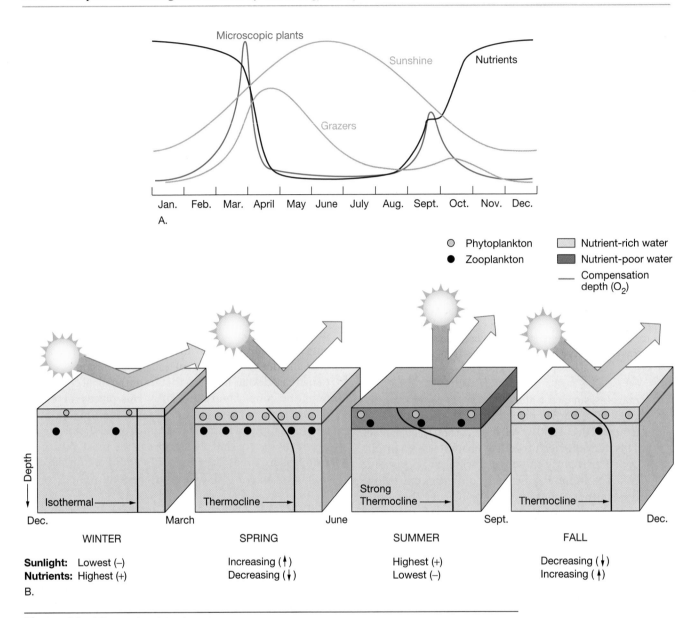

Figure 13–10 Productivity in temperate oceans.
A. Relationship among phytoplankton (*microscopic plants*), zooplankton (*grazers*), amount of sunshine, and nutrient levels for surface waters in temperate latitudes. **B.** The seasonal cycle of sunlight affects the depth of a seasonal thermocline, which affects the availability of nutrients. This, in turn, affects the abundance of phytoplankton and other organisms that feed on phytoplankton—such as zooplankton.

nutrients are increasing as the thermocline lessens, and the decrease in *sunlight* is the limiting factor.

Energy Flow

Before considering the cycles that transfer organic and inorganic matter, let's consider the flow of energy in general. It is important to understand that energy flow is not a *cycle* but a *unidirectional flow*, which originates from the constant input of solar energy and continually dissipates until no energy remains. Most energy in ecosystems begins as solar energy and enters a biotic community through algae. From the algae, the energy is diminished continually, culminating in **entropy**, which

is a state where energy is so dissipated that it no longer can do work.

Energy Flow in Marine Ecosystems

The term **biotic community** refers to the assemblage of organisms that live together within some definable area. An **ecosystem** is a larger unit that includes the biotic community plus the environment with which it interacts in the exchange of energy and chemical substances. For instance, a kelp forest biotic community includes all organisms living within or near the kelp and receiving some benefit from it. In comparison, a kelp forest ecosystem includes all those organisms plus the surrounding sea-

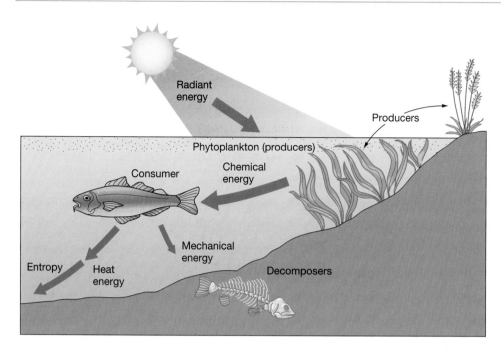

Figure 13–11 Energy flow through a photosynthetic ecosystem.

Energy enters the ecosystem as radiant solar energy and is converted to chemical energy through photosynthesis by producers. Metabolism in the fish (a consumer) then releases the chemical energy for conversion to mechanical energy. Energy also is lost from the biotic community (the algae and fish) as heat, which increases the entropy of the ecosystem. Decomposers work to break down the remaining energy after an organism dies.

water, hard substrate onto which the kelp is attached, and atmosphere where gases are exchanged.

In an algae-supported biotic community (Figure 13–11), energy enters the system as solar energy, which is absorbed by the algae. Photosynthesis converts this solar energy into a form of chemical energy, which is used for the algae's respiration. This energy is also passed on to animal consumers for their growth and other life functions. Energy is expended by the animals as mechanical and heat energy, which are progressively less recoverable forms of energy. Finally, the residual energy becomes biologically useless as entropy increases.

Within an ecosystem, there are generally three basic categories of organisms: **producers**, **consumers**, and **decomposers** (Figure 13–11). Algae, archaeons, and certain bacteria are the **autotrophic** (*auto* = self, *tropho* = nourishment) producers, with the capacity to nourish themselves through chemosynthesis or photosynthesis. The consumers and the decomposers are **heterotrophic** (*hetero* = different, *tropho* = nourishment) organisms that depend on the organic compounds produced by the autotrophs for their food supply.

The decomposers, such as bacteria, break down the organic compounds that comprise **detritus** (dead and decaying remains and waste products of organisms) for their own energy requirements. They characteristically release simple inorganic salts that are used by algae as nutrients.

Animals may be divided into three categories: **herbivores** (*herba* = grass, *vora* = eat), which feed directly on plants or algae; **carnivores** (*carni* = meat, *vora* = eat), which feed only on other animals; and **omnivores** (*omni* = all, *vora* = eat), which feed on both. As the role of ocean-dwelling bacteria becomes better understood, a fourth category of organisms, the **bacteriovores**, which

feed only on bacteria, have been recognized as an important component of the marine ecosystem.

Biogeochemical Cycling

Having discussed the noncyclic, unidirectional flow of energy through the biotic community, let us now consider the flow of nutrients, which is cyclic. These are the **biogeochemical cycles**.[9] In these cycles, matter does not dissipate but is *cycled* from one chemical form to another by the various members of the community.

In biogeochemical cycles, the chemical components of organic matter enter the biological system through photosynthesis (or through bacteria chemosynthesis at hydrothermal vents). These chemical components are passed on to animal populations through feeding. When algae and animals die, their organic remains are converted to inorganic form by bacterial or other decomposition processes. In this form, they are again available for uptake by plants and algae.

Of the many biogeochemical cycles, three cycles are especially important: carbon, nitrogen, and phosphorus. A fourth cycle, the silicon cycle, also has important implications in the marine environment.

Carbon, Nitrogen, and Phosphorus Cycles

The element carbon is the basic component of all organic compounds (including carbohydrates, protein, and fats). There is no scarcity of carbon for photosynthetic productivity: only about 1 percent of the total carbon in the

---

[9]Biogeochemical cycles are so named because they involve biological, geological (Earth processes), and chemical components.

Box 13–2
Symbiosis: The Story of Cohabitation

Symbiosis (*sym* = together, *bios* = life) is a relationship in which two or more organisms are closely associated in a way that benefits at least one of the participants, and sometimes both. Such relationships are classified as commensalism, mutualism, or parasitism.

In **commensalism**, a smaller or less dominant participant benefits without harming its host, which affords subsistence or protection to the other. An example is the remora, a fish that attaches itself to a shark or other fish to obtain food and transportation without harming its host (Figure 13D).

In **mutualism**, both participants benefit. Such a relationship exists between the clown fish and the sea anemone (Figure 13E). The clown fish receives protection by swimming among the stinging tentacles of the sea anemone. This relationship is believed to be mutual, because the anemones benefit by the clown fish serving as bait to draw other fish within reach of anemone tentacles.

In **parasitism**, one participant, the parasite, benefits at the expense of the host. Many fish are hosts to isopods, which attach to the fish and derive their nutrition from the body fluids of the fish, thereby robbing the host of some of its energy supply (Figure 13F). Usually, the parasite does not rob enough energy to kill the host because if the host dies, so does the parasite.

Another interesting type of symbiosis was discovered by observing how one small marine organism protects itself from predators. In 1990, a type of pelagic amphipod was observed to capture pteropods (sea butterflies) and carry them around on its back beneath the Antarctic ice (Figure 13G). Further investigation revealed that the 1.25-centimeter (0.5-inch) amphipod *Hyperiella dilatata* uses the 0.65-centimeter (0.25-inch) pteropod *Clione limacina* as a chemical defense against being eaten by predatory fish. The pteropod contains a chemical that tastes awful, and while the amphipod holds it captive, fish will not eat it.

Figure 13D Commensalim: A remora attaches itself to a Caribbean grouper.

Figure 13E Mutualism: The clown fish and the sea anemone.

sea is involved in photosynthetic productivity.[10] Thus, carbon does not limit productivity.

However, nitrates and phosphates need to be considered in more detail because they do, under many conditions, limit marine productivity. Comparatively, nitrogen compounds involved in phytoplankton productivity may be 10 times the total nitrogen compound concentration that can be measured as a yearly average. This level implies that the soluble nitrogen compounds must be cycled completely up to 10 times per

year. Available phosphates may be turned over up to four times per year.

The ratio of carbon to nitrogen to phosphorus in dry weights of diatoms is in the proportion of 41:7:1 (C:N:P). This ratio is also observed in the zooplankton that feed on the diatoms, and in ocean water samples taken worldwide. Thus, phytoplankton take up nutrients in the ratio in which they are available in the ocean water and pass them on to zooplankton in the same ratio. When these plankton and animals die, carbon, nitrogen, and phosphorus are restored to the water in this ratio.

The Carbon Cycle Figure 13–12 shows the carbon cycle. This cycle involves the uptake of carbon dioxide by algae and plants for their use in the photosynthetic

[10]The remainder of carbon atoms are in seawater as carbonate ions or are bound into calcium carbonate shells.

(continued)

The fish that prey on the amphipods depend on sight to find their food. Therefore, it is not surprising that up to 75 percent of the amphipods observed at depths of less than 9 meters (28 feet) were carrying pteropods on their backs. In the darker waters at 50 meters (160 feet), only 6 percent of the amphipods observed were protected by their foul-tasting captives.

This pattern of actively *capturing* another organism for the purpose of using its chemical defense against predators does not fit clearly into any of the traditional categories of symbiotic relationships. It is believed that the pteropods are eventually released unharmed, although they may experience only a short reprieve until they once again become hostages to a different amphipod.

Figure 13F Parasitism: A parasitic isopod on the head of a blackbar soldierfish.

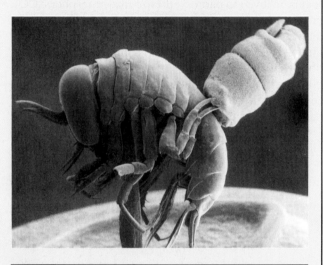

Figure 13G Scanning electron micrograph of a captive pteropod (*upper right*) that protects pelagic amphipod (*center*) from predators.

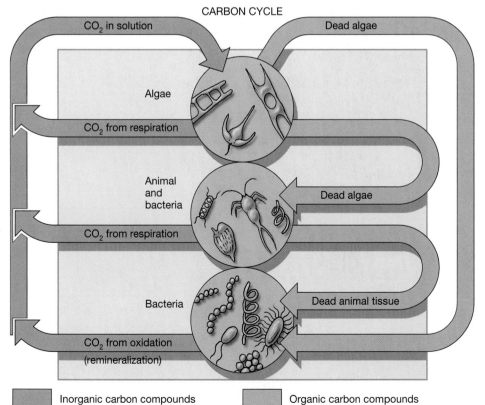

CARBON CYCLE

CO₂ in solution

Dead algae

Algae

CO₂ from respiration

Animal and bacteria

Dead algae

CO₂ from respiration

Bacteria

Dead animal tissue

CO₂ from oxidation (remineralization)

Inorganic carbon compounds   Organic carbon compounds

Figure 13–12 The Carbon Cycle.

There is a large supply of inorganic carbon (*purple*) in the oceans. Only about 1 percent of the carbon dissolved in the ocean is involved in photosynthetic productivity. Algae store carbon dioxide through photosynthesis. They pass this carbon on to animals that eat them. Bacteria decompose dead algae and animals to release carbon dioxide. All release carbon dioxide through respiration.

process. Carbon dioxide is returned to the ocean water primarily through respiration of algae, animals, and bacteria and secondarily by breakdown of dissolved organic materials.

The oceans are believed to remove about 25 to 50 billion metric tons of carbon from the atmosphere each year. This process is referred to as the "biological pump" that is believed to photosynthesize this atmospheric CO_2 into organic matter and then transport it to ocean sediments, where it may be held for millions of years. However, the magnitude of this "biological pump" is poorly understood. With the threat of an increased greenhouse effect resulting in part from increasing CO_2 concentration in the atmosphere, the marine carbon cycle has become an important topic of current research.

The Nitrogen Cycle Figure 13–13 shows the nitrogen cycle. Nitrogen is essential in producing amino acids, the building blocks of proteins that are synthesized by algae. These photosynthetic products are consumed by animals and free-living bacteria. They are then passed on to the decomposing bacteria, along with dead algae tissue, dead animal tissue, and excrement. The decomposing bacteria gain energy from breaking down these materials. This breakdown liberates inorganic compounds, such as nitrates, that are the basic nutrients used by algae.

Different bacteria involved in the nitrogen cycle make it somewhat complicated. Although some bacteria consume dissolved organic matter or convert organic compounds into inorganic substances, others can bind molecular nitrogen (N_2) into useful nutrients. These are

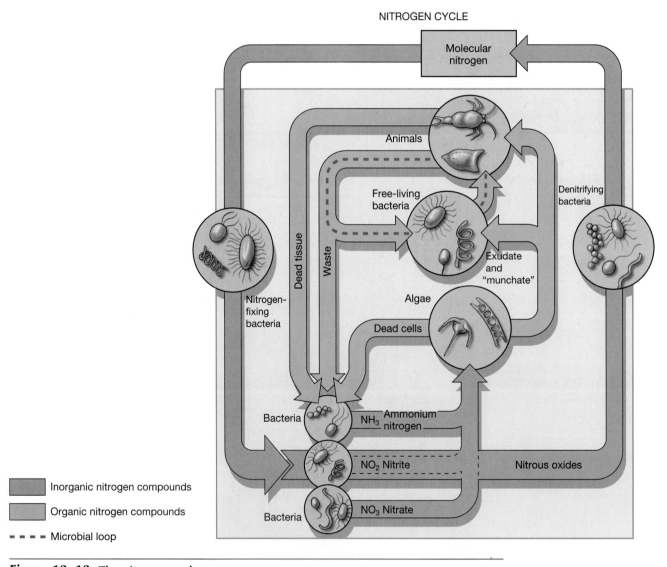

Figure 13–13 The nitrogen cycle.
The total nitrogen fixed into organic molecules (*tan*) at any given time may be 10 times as great as the yearly average of soluble nitrogen compounds dissolved in ocean water. Therefore, each nitrogen atom must be recycled biogeochemically about 10 times per year. Also, the decomposition of organic nitrogen compounds back into the preferred inorganic form of nitrogen, nitrate (NO_3), requires three steps of bacterial decomposition. Nitrogen is considered the nutrient most likely to limit biological productivity as a result of its depletion.

called **nitrogen-fixing bacteria**. Another special group is **denitrifying bacteria**, whose metabolism depends upon the breakdown of nitrates and the liberation of molecular nitrogen.

Studies of nutrient cycles clearly indicate that nitrogen availability is a limiting factor in productivity, especially during summer. One reason is that the process of converting organic substances to useful nutrients through bacterial action may require up to three months. The conversion begins in the lower portion of the euphotic (sunlit) zone as particulate matter sinks toward the ocean bottom.

By the time this conversion is completed, the nitrogen compounds usually are below the euphotic zone and thus are unavailable for photosynthesis. They cannot readily be returned to the euphotic zone during summer due to a strong thermocline that has developed throughout much of the ocean surface. Thus, nitrogen compounds become depleted and restrict productivity by algae. For nitrates to become available again to phytoplankton, the thermocline must disappear, allowing upwelling and mixing during winter.

The Phosphorus Cycle   Figure 13–14 shows the phosphorus cycle. The phosphorus cycle is simpler than the nitrogen cycle, primarily because the bacterial action involved in breaking down organic phosphorus compounds

is simpler. This difference can be studied by comparing Figures 13–13 and 13–14.

The rate at which the organic phosphate compounds can be decomposed into useful nutrients is much faster than for nitrogen. Consequently, much of it can be completed above the compensation depth for photosynthesis. Thus, phosphorus is made available for photosynthesis within the photosynthetic zone. Lack of phosphorus is rarely a limiting factor of algae productivity.

The Silicon Cycle   A fourth cycle, the silicon cycle, is also important. Although silicon is technically not considered a nutrient, the shells of diatoms are composed of silica (silicon dioxide, SiO_2). Availability of silica can be a limiting factor in the productivity of diatoms, but it is rarely a limiting factor of total primary productivity, because not all phytoplankton require silica as a protective covering. In some regions, however, primary production is known to be limited by the availability of silicon even when nitrates are in great abundance. One such region that experiences this phenomenon is the upwelling zone that occurs in the equatorial eastern Pacific Ocean.

Silicon concentrations range from barely measurable up to 400 milligrams per cubic meter (306 milligrams per cubic yard). Fluctuations in concentration coincide roughly with those observed in nitrogen and phosphorus.

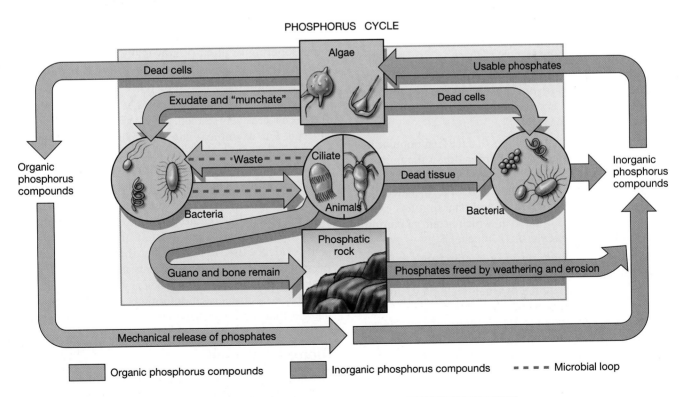

PHOSPHORUS CYCLE

Figure 13–14 The phosphorus cycle.
Each phosphorus atom may need to be recycled up to four times per year to maintain biological productivity in the oceans, yet phosphorus is seldom depleted to where it limits biological productivity. Organic phosphorus is quickly returned as a usable nutrient to algae through breakdown and the single step of bacterial decomposition.

However, the fluctuation displays a much greater amplitude than those for nitrogen and phosphorus. This condition is probably because silica does not undergo bacterial decay. Instead, it is dissolved directly into seawater.

Trophic Levels and Biomass Pyramids

As producers make food (organic matter) available to the consuming animals of the ocean, it passes from one feeding population to the next. Only a small percentage of the energy taken in at any level is passed on to the next because of energy consumption and energy loss at each level. As a result, the biomass of producers in the ocean is many times greater than the mass of the top consumers, such as sharks or killer whales. It is certainly surprising that the total biomass of large animals such as killer whales is much less than the total biomass of microscopic diatoms, but the following discussion of trophic levels should make it clear why this is true.

Trophic Levels

Chemical energy that is stored in the mass of the ocean's algae (the "grass of the sea") is transferred to the animal community mostly through feeding. Zooplankton eat diatoms and other microscopic marine algae, and in doing so, are *herbivores*, like cows.[11] Larger animals feed on the larger macroscopic algae and marine plants that grow attached to the ocean bottom near shore.

In turn, the herbivores are eaten by larger animals, the *carnivores*. They in turn are eaten by another population of larger carnivores, and so on. Each of these feeding levels is called a **trophic level**.

Generally, individual members of a feeding population are larger—but not too much larger—than the individuals in the population they eat. However, there are conspicuous exceptions to this condition. For instance, the blue whale—possibly the largest animal that has existed on Earth—feeds upon krill, which attain maximum lengths of *only 6 centimeters* (2.4 inches)!

Remember that the transfer of energy from one population to another represents a *continuous flow* of energy. This flow is interrupted by small-scale recycling and storage of this energy, which slows the process of converting potential (chemical) energy to kinetic energy, then to heat energy, and finally to entropy. However, despite this cycling of energy, all energy that enters the organic community ultimately dissipates and increases the overall entropy of the system.

[11]In fact, zooplankton might be considered the miniature drifting "cows of the sea."

Transfer Efficiency

In the transfer of energy between trophic levels, some energy is always lost due to inefficiency of the transfer process. Researchers have observed considerable variability in the efficiency of different algal species. As an average, the percentage of light energy absorbed by algae and ultimately synthesized into food made available to herbivores is only about 2 *percent*, which is highly inefficient!

The **gross ecological efficiency** at any trophic level is *the ratio of energy passed on to the next higher trophic level divided by the energy received from the trophic level below.* The ecological efficiency of herbivorous anchovies would be, for example, the energy consumed by carnivorous tuna that feed on the anchovies divided by the energy represented by the phytoplankton that the anchovies consumed.

Figure 13–15 shows that some of the chemical energy taken in as food by herbivores is passed by the animal as feces and the rest is assimilated by the animal. Of the assimilated chemical energy, much is quickly converted through respiration to kinetic energy for maintaining life, and what remains is available for growth and reproduction. Only a portion of this mass is passed on to the next trophic level through feeding.

Figure 13–16 shows the passage of energy between trophic levels through an entire ecosystem, from the solar energy assimilated by phytoplankton through all trophic levels to the mass of the ultimate carnivore—humans. Because of energy losses at each trophic level, it is indeed sobering to think of how many marine organisms are involved in producing a *single* fish that is so easily consumed during dinner!

The efficiency of energy transfer between trophic levels has been intensively studied. Many variables must be considered. For example, young animals display a higher growth efficiency than older animals. In addition, when food is plentiful, animals expend more energy in digestion and assimilation than when food is not readily available.

Most efficiencies range between 6 and 15 percent. It is well accepted that ecological efficiencies in natural ecosystems average approximately 10 percent. There is some evidence that, in populations that are important to present fisheries, this efficiency may run as high as 20 percent. The true value of this efficiency is of practical importance to us because it determines the size of the fish harvest that can be anticipated from the oceans.

Biomass Pyramid

Considering the great amount of energy loss that occurs between each feeding population, it is obvious that there must be a limit to the number of feeding populations in an ecosystem. It there were too many levels, there would not be enough energy to support the organisms that depend on the other levels below them because of

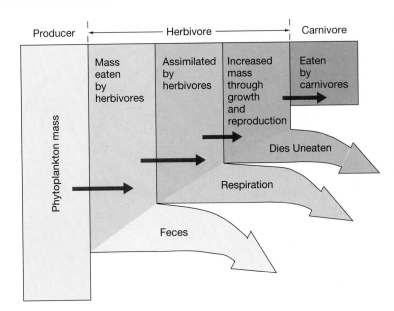

Figure 13–15 Passage of energy through a trophic level.

Because of utilization and losses of energy by the processes shown, only a small percentage (about 10 percent) of the food mass consumed by herbivores is available for consumption by carnivores that feed upon it.

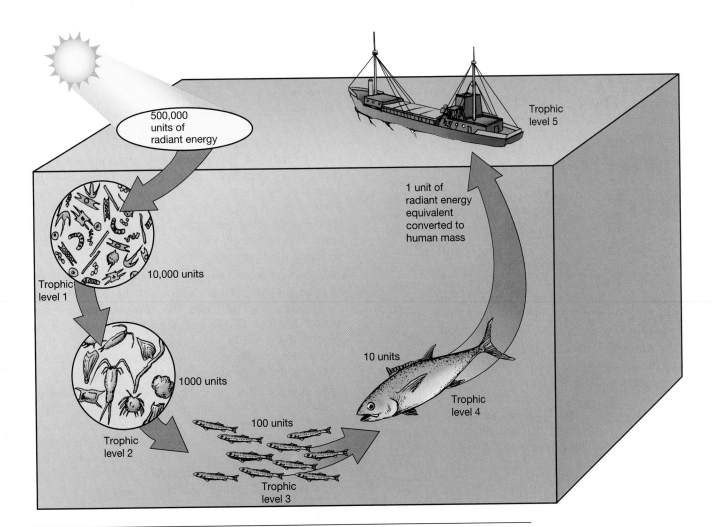

Figure 13–16 Ecosystem energy flow and efficiency.

This diagram shows the effects of energy transfer within an ecosystem. One unit of mass is added to the fifth trophic level (humans) for every 500,000 units of radiant energy input available to the producers (phytoplankton). This value is based on 2 percent efficiency of transfer by phytoplankton and 10 percent efficiency at all other levels.

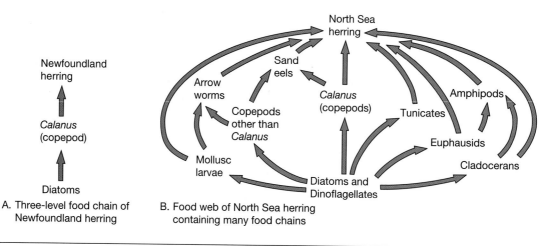

A. Three-level food chain of Newfoundland herring

B. Food web of North Sea herring containing many food chains

Figure 13–17 A Food chain and a food web.
A. A food chain is the passage of energy along a single path, such as from diatoms to copepods to Newfoundland herring. **B.** The food sources of the North Sea herring involve multiple paths—a food web—where this fish may be at the third or fourth trophic level.

the great amount of energy loss at each level. It also is evident that each feeding population necessarily must have less mass than the population it eats. Logically, individual members of a feeding population generally are *larger in size* and *less numerous* than their prey. Let's examine how these factors can be conceptualized by food chains and food webs.

Food Chains

A **food chain** is a sequence of organisms through which energy is transferred, starting with the primary producer, through the herbivore, and through successive carnivores, up to the "top carnivore," which is not usually preyed upon by any other organism.

Because of the inefficiency of energy transfer between trophic levels, it is beneficial for fishermen to choose a population that feeds as close to the primary producing population as possible. This increases the biomass available for food and the number of individuals available to be taken by the fishery.

An example of an animal population that is an important fishery, and which usually represents the third trophic level in a food chain, is herring that live off the coast of Newfoundland. Although some herring populations are involved in longer food chains, the Newfoundland herring feed primarily on a population of small crustaceans (copepods) that in turn feed upon diatoms (Figure 13–17A).

Food Webs

It is rather uncommon to see feeding relationships as simple as that of the Newfoundland herring.

More commonly, top carnivores in a food chain feed on a number of different animals, each of which has its own simple or complex feeding relationships. This constitutes a **food web** (Figure 13–17B).

One benefit to animals that feed through a food web rather than a linear chain is their greater likelihood of survival, because they have alternative foods to eat should others fail. Those animals involved in food webs, such as the North Sea herring shown in Figure 13–17B, are less likely to starve should one of their food sources diminish in quantity or even become extinct. Conversely, the Newfoundland herring eat only copepods, so extinction of copepods would have a catastrophic effect on this herring population.

The Newfoundland herring population does, however, have an advantage over its relatives that feed through the broader-based food web. The Newfoundland herring is more likely to have a larger biomass to eat, because the herring are only two steps removed from the producers, whereas the North Sea herring is at the fourth level in some of the food chains within its web.

The ultimate effect of energy transfer between trophic levels can be seen in the **biomass pyramid** in Figure 13–18. It depicts the *progressive decrease in numbers of individuals and total biomass* at successive trophic levels resulting from decreased amounts of available energy. The figure also depicts the progressive *increase in organism size* at successive tropic levels up the food web.

? Students Sometimes Ask...

I know that the tropics are an area of abundance on land—the number and variety of tropical species is astounding. I don't understand how the tropical oceans can have such low productivity.

Life on land does not necessarily correspond to life in the ocean! You correctly point out that tropical rain forests are amazingly verdant in terms of species diversity as well as biomass. However, oceanic conditions are quite different from

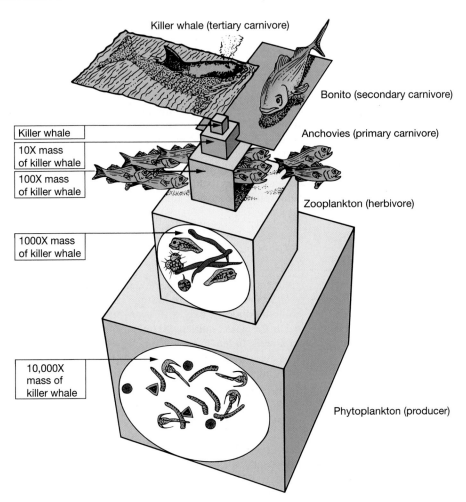

Killer whale (tertiary carnivore)

Bonito (secondary carnivore)

Anchovies (primary carnivore)

Zooplankton (herbivore)

Phytoplankton (producer)

| Killer whale |
| 10X mass of killer whale |
| 100X mass of killer whale |
| 1000X mass of killer whale |
| 10,000X mass of killer whale |

Figure 13–18 **Biomass pyramid.**
The higher on the biomass pyramid an organism lives, the physically larger it is as an individual. However, the total biomass represented by a population that is high on the pyramid is less than that for a population at a lower level.

those on land. In the tropical ocean, the development of a strong, permanent thermocline limits the availability of nutrients that are necessary for the growth of phytoplankton. Without abundant phytoplankton, not much else can live in the ocean. It is ironic that the clear blue water of the tropics so prominently displayed in tourist brochures indicates seawater that is biologically quite sterile!

I've heard of the "dead zone" that occurs periodically along the coast of the Gulf of Mexico. What is it?
A large hypoxic (*hypo* = under, *oxid* = oxygen) "dead zone" along the Gulf Coast has occurred each summer near the mouth of the Mississippi River since the record-breaking Midwest floods of 1993. Within that zone, which has reached the size of the state of New Jersey, oxygen levels drop from above 5.0 ppm to below 2.0 ppm. This oxygen level is lower than most marine animals can tolerate. As a consequence, fish, crustaceans, and other animals die and wash up on beaches bordering the Gulf.

Although small hypoxic zones have occurred during summer in the region for several decades, they have grown larger and more intense in recent years. These "dead zones" appear to be related to runoff of nutrients from agricultural activities on land that cause ocean **eutrophication**—the enrichment of waters by a previously scarce nutrient. Nutrients in the form of nitrate and phosphate fertilizers are washed down the Mississippi River, eventually reaching the Gulf, where they stimulate algal blooms. Once these algae die and sink to the bottom, bacteria feed on them and on the fecal matter of other organisms, resulting in depletion of oxygen along the ocean bottom. Some of the more

mobile organisms can escape the hypoxic conditions, but for those that cannot, the situation can be deadly.

Eutrophication plagues such nitrogen-laden bodies as New York State's Long Island Sound, California's San Francisco Bay, and the Baltic Sea in Europe. Fertilizer runoff from agricultural fields in Queensland also threatens parts of Australia's Great Barrier Reef. In the United States, proposals to combat the spread of the "dead zone" include filtering some river water through wetlands before it enters the Gulf, planting buffer strips along certain waterways, altering the timing of fertilizer applications, improving crop rotation, and enforcing existing antipollution regulations.

Why does a red tide glow green at night?
Many of the species of dinoflagellates that produce red tides (most notably those of genus *Gonyaulax*) also have bioluminescent capabilities—that is, they are able to produce light organically. When the organisms are disturbed, they emit a faint green glow. Thus, when waves break during a red tide at night, the waves are often spectacularly illuminated by millions of bioluminescent dinoflagellates. During these times, one can easily observe marine animals moving through the water because their bodies are silhouetted by bioluminescent dinoflagellates that light up as they pass over the animal's body.

How likely am I to see a whale during a whale-watching trip?
It depends on the area and the time of the year, but generally your chances are quite low. That's because large marine organisms—such as whales, sharks, and other large marine animals—comprise

such a small proportion of marine life. In terms of total oceanic biomass, these nektonic (swimming) organisms have been estimated to comprise only *one-tenth of 1 percent* of all organisms in the sea! With your knowledge of food pyramids, it should be no surprise that the majority of the ocean's biomass is comprised of one group—the microscopic drifting photosynthetic phytoplankton. Perhaps enough interest could be generated for commercial boat operators to conduct a *plankton*-watching trip! Everyone would be guaranteed to see dozens of different species of plankton—all that would be required is a plankton net and a microscope.

My friends are planning an ocean-fishing trip. To get to good fishing grounds that haven't been fished out, one of my friends has suggested that we go far offshore. Should we expect to find more fish there?

Generally, no. Even though few people have fished those waters, it is unlikely that you'll discover a good fishing area. That's because coastal waters have high nutrient levels and are highly productive, while most of the open ocean has a low nutrient level and correspondingly low productivity. Far offshore, there just isn't enough food for many fish to live (see, for example, the distribution of phytoplankton shown in Figure 13–4). Thus, the farther you go from land, the less likely there would be good fishing, especially past the continental shelf. However, there are some exceptions to this: Shallowly submerged banks and areas of upwelling do exist at great distances from shore.

Summary

Microscopic planktonic algae that photosynthesize represent the largest biomass in the ocean and, as such, they are the foundation of the ocean's food web. In addition to this primary photosynthetic productivity, organic biomass is produced through bacterial chemosynthesis. Chemosynthesis, observed on oceanic spreading centers in association with hydrothermal springs, is based on the release of chemical energy by the oxidation of hydrogen sulfide.

Photosynthetic productivity of the oceans is limited by the availability of nutrients and the amount of solar radiation. Nutrients—such as nitrate, phosphorous, iron, and silica—are most abundant in coastal areas, due to runoff and upwelling. The depth to which sufficient light penetrates to allow photosynthesis to produce only that amount of oxygen required for their respiration is the oxygen compensation depth. Algae cannot live successfully below this depth, which may occur at less than 20 meters (65 feet) in turbid coastal waters or to 100 meters (330 feet) in the open ocean.

There is an abundance of marine life along continental margins—where optimum conditions of nutrient supply and sunlight occur—that decreases away from the continents and with increased depth. In addition, cool water typically supports abundant life because of the water's ability to hold more dissolved gases necessary for life. Areas of upwelling bring cold, nutrient-rich water to the surface and have some of the highest productivities. The color of the oceans is an indicator of the abundance of life: It ranges from green in highly productive regions to blue in areas of low productivity.

There are many different types of photosynthetic marine organisms. The seed-bearing Spermatophyta are represented by a few genera of nearshore plants such as eelgrass (*Zostera*), surf grass (*Phyllospadix*), marsh grass (*Spartina*), and mangrove trees (genera *Rhizophora* and *Avicennia*). Macroscopic algae include brown algae (Phaetophyta, which includes kelp), green algae (Chlorophyta), and red algae (Rhodophyta). Microscopic algae include diatoms and coccolithophores (Chrysophyta), and dinoflagellates (Pyrrophyta, which are responsible for red tides).

In high-latitude (polar) areas thermoclines are generally absent, so upwelling can readily occur, and productivity is commonly limited more by the availability of solar radiation than by lack of nutrients. In low-latitude (tropical) regions, where a strong thermocline may exist year-round, productivity is limited by the lack of nutrients except in areas of upwelling or near coral reefs. In temperate regions, where distinct seasonal patterns are developed, productivity peaks in the spring and fall and is limited by lack of solar radiation in the winter and lack of nutrients in the summer.

Radiant energy captured by algae is converted to chemical energy and passed through the biotic community. It is expended as mechanical and heat energy and ultimately reaches a state of entropy, where it is biologically useless. There is, however, no loss of mass. The mass used as nutrients by algae is converted to biomass. Upon the death of organisms, the mass is decomposed to an inorganic form ready again for use as nutrients for algae.

Marine ecosystems are composed of populations of organisms called producers (which photosynthesize or chemosynthesize), consumers (which eat producers), and decomposers (which break down detritus). Animals can be divided into the herbivores (eat plants), the carnivores (eat animals), the omnivores (eat both), or bacteriavores (eat bacteria). Some of these organisms live closely together in various types of symbiotic relationships.

Biogeochemical cycles involve the cycling of nutrients and other chemicals by various organisms. Important biogeochemical cycles include the carbon, nitrogen, phosphorous, and silicon cycles. Of the nutrients required by algae, compounds of nitrogen are most likely to be depleted and restrict productivity, particularly during summer.

As energy is transferred from algae to herbivore and the various carnivore feeding levels, an average of only about 10 percent of the mass taken in at one feeding level is passed on to the next. The ultimate effect of this decreased amount of energy that is passed between trophic levels higher in the food chain is a decrease in the number of individuals, which are larger in size. It also shows a decrease in the total biomass of populations higher in the food chain toward the top of the biomass pyramid.

Key Terms

Amnesic shellfish poisoning (p. 394)

Autotrophic (p. 399)

Bacteriovore (p. 399)

Biogeochemical cycle (p. 399)

Biological pump (p. 393)

Biomass (p. 385)

Biomass pyramid (p. 406)

Biotic community (p. 398)

California Cooperative Oceanic Fisheries Investigations (CalCOFI) (p. 384)

Carnivore (p. 399)

Carotin (p. 392)

Chemosynthesis (p. 385)

Coccolith (p. 393)

Coccolithophore (p. 393)

Commensalism (p. 400)

Compensation depth for photosynthesis (p. 387)

Consumer (p. 399)

Decomposer (p. 399)

Denitrifying bacteria (p. 403)

Detritus (p. 399)

Diatom (p. 392)

Diatomaceous earth (p. 392)

Dinoflagellate (p. 393)

Ecosystem (p. 398)

Entropy (p. 398)

Euphotic zone (p. 387)

Eutrophic (p. 388)

Eutrophication (p. 407)

Flagella (p. 393)

Food chain (p. 406)

Food web (p. 406)

Gross ecological efficiency (p. 404)

Gross primary production (p. 385)

Herbivore (p. 399)

Heterotrophic (p. 399)

Isothermal (p. 395)

Mutualism (p. 400)

Net primary production (p. 385)

New production (p. 386)

Nitrogen-fixing bacteria (p. 403)

Oligotrophic (p. 388)

Omnivore (p. 399)

Paralytic shellfish poisoning (PSP) (p. 394)

Parasitism (p. 400)

Photosynthesis (p. 385)

Plankton net (p. 386)

Primary productivity (p. 385)

Producer (p. 399)

Pycnocline (p. 393)

Red tide (p. 394)

Regenerated production (p. 386)

SeaWiFS (p. 386)

Symbiosis (p. 400)

Test (p. 392)

Thermocline (p. 393)

Trophic level (p. 404)

Upwelling (p. 388)

Questions And Exercises

1. Describe the mission of the CalCOFI program. How did the CalCOFI program help solve the mystery it intended to answer?

2. Discuss chemosynthesis as a method of primary productivity. How does it differ from photosynthesis?

3. How does gross primary production differ from net primary production? What are the two components of gross primary production, and how do they differ?

4. An important variable in determining the distribution of life in the oceans is the availability of nutrients. How are the following variables related: proximity to the continents, availability of nutrients, and the concentration of life in the oceans?

5. Another important determinant of productivity is the availability of solar radiation. Why is biological productivity relatively low in the tropical open ocean, where the penetration of sunlight is greatest?

6. Discuss the characteristics of the coastal ocean where unusually high concentrations of marine life are found.

7. What factors create the color difference between coastal waters and the less-productive open-ocean water?

8. Compare the macroscopic algae in terms of color, maximum depth in which they grow, common species, and size.

9. The golden algae include two classes of important phytoplankton. Compare their composition and the structure of their tests and explain their significance in the fossil record.

10. Discuss and compare the contributions of the Pyrrophyta genera *Ptychodiscus* and *Gonyaulax* to red tide development.

11. How does paralytic shellfish poisoning (PSP) differ from amnesic shellfish poisoning? What types of microorganisms create each?

12. Describe how a biological pump works. What percentage of organic material from the euphotic zone accumulates on the sea floor?

13. Compare the biological productivity of polar, temperate, and tropical regions of the oceans. Consider seasonal changes, the development of a thermocline, the availability of nutrients, and solar radiation.

14. Generally, the productivity in tropical oceans is rather low. What are three environments that are exceptions to this, and what factors contribute to their higher productivity?

15. Describe the flow of energy through the biotic community and include the forms into which solar radiation is converted. How does this flow differ from the manner in which mass is moved through the ecosystem?

16. What are the three types of symbiosis, and how do they differ? What new type of symbiosis has been discovered?

17. What are the proportions by weight of carbon, nitrogen, and phosphorus in ocean water, phytoplankton, and zooplankton? Suggest how these amounts may support or refute the idea that life originated in the oceans.

18. Explain why nitrogen is much more likely than phosphorus to be a limiting factor in marine productivity.

19. What is the average efficiency of energy transfer between trophic levels? Use this efficiency to determine how much phytoplankton mass is required to add 1 *gram* of new mass to a killer whale, which is a third-level carnivore. Include a diagram that shows the different trophic levels and the relative size and abundance of organisms at different levels. How would your answer change if the efficiency were half the average rate, or twice the average rate?

20. Describe the probable advantage to the top carnivore of the food web over a single food chain as a feeding strategy.

References

Carpenter, E. J., and Romans, K. 1991. Major role of the cyanobacterium, Trichodesmium, in nutrient cycling in the North Atlantic Ocean. *Science* 254:5036, 1356–1358.

Culotta, E. 1996. Red menace in the world's oceans, *in* Pirie, R. G., ed., *Oceanography: Contemporary Readings in Ocean Sciences,* 3rd ed., New York: Oxford University Press.

Ducklow, H. W. 1983. Production and fate of bacteria in the oceans. *Bioscience* 33:8, 494–501.

Dugdale, R. C., and Wilkerson, F. P. 1998. Silicate regulation of new production in the equatorial Pacific upwelling. *Nature* 391:6664, 270–273.

Falkowski, P. G., Barber, R. T., and Smetacek, V. 1998. Biogeochemical controls and feedbacks on ocean primary production. *Science* 281:5374, 200–206.

George D., and George, J. 1979. *Marine life: An Illustrated Encyclopedia of Invertebrates in the Sea.* New York: Wiley-Interscience.

Grassle, J. F., et al. 1979. Galápagos '79: Initial findings of a deep-sea biological quest. *Oceanus* 22:2, 2–10.

Howard, J. 1995. Vanishing act: Critical link in marine food chain may be at risk. *Scripps Institution of Oceanography Explorations* 2:2, 11–17, Scripps Institution of Oceanography, University of California, San Diego.

———. 1996. Red tides rising: Local plankton bloom possible sign of growing global threat. *Scripps Institution of Oceanography Explorations* 2:3, 2–9, Scripps Institution of Oceanography, University of California, San Diego.

Jenkins, W. J. 1982. Oxygen utilization rates in North Atlantic subtropical gyre and primary production in oligotrophic systems. *Nature* 300, 246–248.

Lalli, C. M., and Parsons, T. R. 1993. *Biological Oceanography: An Introduction.* New York: Pergamon Press.

Little, M. M., Little, D. S., Blair, S. M., and Norris, J. N. 1985. Deepest known plant life discovered on an uncharted seamount. *Science* 227:4683, 57–59.

Malakoff, D. 1998. Death by suffocation in the Gulf of Mexico. *Science* 281:5374, 190–192.

Martinez, L., Silver, M., King, J., and Alldredge, A. 1983. Nitrogen fixation by floating diatom mats: A source of new nitrogen to oligotrophic ocean waters. *Science* 221:4066, 152–154.

McClintock, J. B., and Janssen, J. 1990. Pteropod abduction as a chemical defense in a pelagic Antarctic amphipod. *Nature* 346:6283, 462–464.

More, F. M. M., Reinfelder, J. R., Roberts, S. B., Chamberlain, C. P. Lee, J. G. and Yee, D. 1994. Zinc and carbon co-limitation of marine phytoplankton. *Nature* 369:6483, 740–742.

Paerl, H. W., and Bebout, B. M. 1988. Direct measurement of O_2-depleted microzones in marine Oscillatoria: Relation to N_2 fixation. *Science* 241:4864, 442–445.

Parsons, T. R., Takahashi, M., and Hargrave, B. 1974. *Biological Ooceanographic Processes,* 3rd ed. New York: Pergamon Press.

Pimm, S. L., Lawton, J. H., and Cohen, J. E. 1991. Food web patterns and their consequences. *Nature* 350:6320, 669–674.

Platt, T., and Sathyendranath, S. 1988. Oceanic primary production: Estimation by remote sensing at local and regional scales. *Science* 241:4873, 1613–1619.

Platt, T., Subba Rao, D. V., and Irwin, B. 1983. Photosynthesis of picoplankton in the oligotrophic ocean. *Nature* 310:5902, 702–704.

Russell-Hunter, W. D. 1970. *Aquatic Productivity.* New York: Macmillan.

Sathyendranath, S., Platt, T., Horne, E. P. W., Harrison, W. G., Ulloa, O., Outerbridge, R., and Hoepffner, N. 1991. Estimation of new production in the ocean by compound remote sensing. *Nature* 353:6340, 129–133.

Seliger, H. H. 1996. Bioluminescence: Excited states under cover of darkness, *in* Pirie, R. G., ed., *Oceanography: Contemporary Readings in Ocean Sciences,* 3rd ed. New York: Oxford University Press.

Sherr, B. F., Sherr, E. B., and Hopkinson, C. S. 1988. Trophic interactions within pelagic microbial communities: Indications of feedback regulation of carbon flow. *Hydrobiologia* 159:1, 19–26.

Shulenberger, E., and Reid, J. L. 1981. The Pacific shallow oxygen maximum, deep chlorophyll maximum, and primary productivity reconsidered. *Deep Sea Research* 28A:9, 901–919.

Strahler, A. H., and Strahler, A. N. 1992. *Modern Physical Geography,* 4th ed. New York: Wiley.

Sullivan, C. W., Arrigo, K. R., McClain, C. R., Comiso, J. C., and Firestone, J. 1993. Distributions of phytoplankton blooms in the Southern Ocean. *Science* 262:5141, 1832–1836.

Trefil, J. 1997. Nitrogen. *Smithsonian* 28:7, 70–78.

Suggested Reading

Earth

Zaburunov, S. A. 1992. As the world breathes: The carbon dioxide cycle. 1:1, 26–33. An analysis of the carbon dioxide cycle, including a look at why there is so little carbon dioxide and so much oxygen in Earth's atmosphere.

Sea Frontiers

Arehart, J. L. 1972. Diatoms and silicon. 18:2, 89–94. A very readable description of the important role of silicon and other elements in the ecology of diatoms, including microphotographs showing the varied forms of diatoms.

Coleman, B. A., Doetsch, R. N., and Sjblad, R. D. 1986. Red tide: A recurrent marine phenomenon. 32:3, 184–191. The problem of periodic red tides along the Florida, New England, and California coasts is discussed.

Cox, V. 1994. It's no snow job. 40:2. The search for marine snow conducted by biologist Alice Alldredge is the focus of this story. This fall of biological debris to the ocean floor plays an important role in helping remove carbon dioxide from the atmosphere and depositing in deep-sea sediments.

Idyll, C. P. 1971. The harvest of plankton. 17:5, 258–267. An interesting discussion of the potential of zooplankton as a major fishery.

Jensen, A. C. 1973. Warning—red tide. 19:3, 164–175. An informative discussion of what is known of the cause, nature, and effect of red tides.

Johnson, S. 1981. Crustacean symbiosis. 27:6, 351–360. A description of various symbiotic relationships entered into by tropical shrimps and crabs.

McFadden, G. 1987. Not-so-naked ancestors. 33:1, 46–51. The nature of the coverings of marine phytoplankton cells is revealed by the electron microscope.

Oremland, R. S. 1976. Microorganisms and marine ecology. 22:5, 305–310. The role of such microorganisms as phytoplankton and bacteria in cycling matter in the oceans is discussed.

Philips, E. 1982. Biological sources of energy from the sea. 28:1, 36–46. The potential for converting marine biomass to energy sources useful to society is discussed.

Scientific American

Anderson, D. M. 1994. Red tides. Dense blooms of algae are becoming more frequent in coastal waters. They are also involving species of algae and animals not previously known to be associated with red tide occurrences.

Benson, A. A. 1975. Role of wax in oceanic food chains. 232:3, 76–89. A report on the findings from observations made of the content of wax in the bodies of many marine animals from copepods to small deep-water fishes and their implications.

Childress, J. J., Feldback, H., and Somero, G. N. 1987. Symbiosis in the deep sea. 256:5, 114–121. Deep-sea hydrothermal vent animals have a symbiotic relationship with sulfur-oxidizing bacteria that allows them to live in the darkness of the deep ocean.

Coleman, G., and Coleman, W. J. 1990. How plants make oxygen. 262:2, 50–67. The process of oxygen production by plants is explained.

Levine, R. P. 1969. The mechanism of photosynthesis. 221:6, 58–71. Reveals what is known of the process by which energy is captured by plants and converted to useful forms of chemical energy while freeing oxygen to the atmosphere.

Pettit, J., Drucker, S., and Knox, B. 1981. Submarine pollination. 244:3, 134–144. Discusses the pollination of sea grasses by wave action.

Oceanography on the Web

Visit the *Essentials of Oceanography* home page for on-line resources for this chapter. There you will find an on-line study guide with review exercises, and links to oceanography sites to further your exploration of the topics in this chapter. *Essentials of Oceanography* is at: **http://www.prenhall.com/thurman** (click on the Table of Contents menu and select this chapter).

C H A P T E R 1 4
ANIMALS OF THE PELAGIC ENVIRONMENT

Alexander Agassiz: Advancements in Ocean Sampling

Biological oceanography was given strong support in America by the activities of Alexander Agassiz (Figure 14A). The son of the great Swiss scientist Louis Agassiz, he immigrated to the United States in 1846 with his parents. He graduated from Harvard University in 1855, became superintendent of a large copper mining operation in 1866, and was a multimillionaire by age 40. Agassiz was a friend of C. Wyville Thompson of the *Challenger* Expedition, and his successor Sir John Murray, who helped prepare the reports from the voy-

age of the *Challenger*. In fact, Agassiz had a prominent role in processing the organisms collected during the *Challenger* Expedition.

Although Agassiz contributed substantial money and effort to the development of U.S. ocean study, he got on the wrong end of a couple of controversies—both of which he initiated. He took a position adverse to Darwin's theory of coral reef development, and he vigorously opposed the view that there was an extensive midwater plankton community. Later, both of these ideas

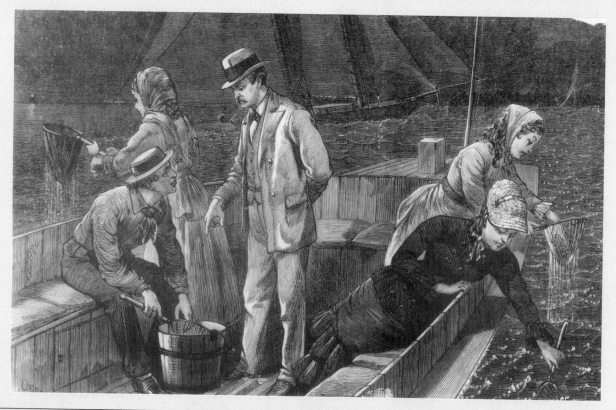

Figure 14A Alexander Agassiz (*standing*) and associates collecting specimens.

became well documented. Even though he was not always correct, Alexander Agassiz was credited by his contemporaries as the most important individual in marine research as it entered the twentieth century. He used his money freely to finance scientific voyages and to support the Museum of Comparative Zoology at Harvard, where valuable analysis of marine organisms has been conducted ever since.

He was the first to use stronger steel cables for deep-sea dredging during voyages between 1877 and 1880 aboard the steamer *Blake* and the first ship especially for oceanographic research, the steamer *Albatross*. He conducted many oceanographic research cruises in the Caribbean Sea and the Gulf of Mexico and off the Atlantic Coast of Florida on board these ships. Truly, he was widely acknowledged as the driving force that brought oceanography recognition as a science. Not only did Agassiz initiate the scheme that first brought oceanographic research strong financial and institutional support, but he was also noted for his ability to organize efficient and effective research voyages.

Because of his training as a mining engineer, he also developed ingenious devices—the prototypes for many modern oceanographic sampling devices in use today—that improved the quantitative value of biological samples recovered. One such device, a dredge used to collect organisms living on the bottom, was designed to work equally well no matter which way it landed on the ocean floor. It worked so well, in fact, that one haul from 3219 meters (10,560 feet) brought up more specimens of deep-sea fishes than the *Challenger* had collected in its entire three-and-a-half-year expedition!

Pelagic animals include all animals living suspended in seawater (not on the ocean floor), which comprises the vast majority of the ocean's **biomass**.[1] Because phytoplankton—directly or indirectly the source of food for most marine organisms—live within the surface waters of the ocean, most marine animals live there as well. The immense depth of the oceans provides a challenge for these organisms to remain afloat.

Phytoplankton depend primarily on their small size to provide a high degree of frictional resistance to sinking below the sunlit surface waters. Most animals, however, possess bodies that are usually more dense than ocean water, and have less surface area per unit of body mass. Therefore, they tend to sink more rapidly into the ocean.

To remain in surface waters where the food supply is greatest, pelagic marine animals must have adaptations that increase their buoyancy or they must continually swim. Various animals apply one or both of these strategies that allow them to remain in surface waters. This results in a wonderful variety of adaptations and lifestyles that make it possible for these animals to live successfully in the oceans.

Adaptations for Staying Above the Ocean Floor

Some animals depend on increased buoyancy to maintain themselves in near-surface waters. Their buoyancy results either from gas containers they possess, which significantly reduce their average density, or from soft bodies void of hard parts that are high in density. Larger animals with bodies denser than seawater must exert more energy to propel themselves through the water as swimmers. Let's look at some examples of how various marine animals employ these strategies.

Gas Containers

At sea level, air is approximately 1000 times less dense than water. As a result, even a small amount of air inside an organism can dramatically increase its buoyancy. To take advantage of this, some animals, such as the cephalopods (*cephalo* = the head, *podium* = a foot), have rigid gas containers in their bodies. For instance, the genus *Nautilus* have an external shell, whereas the cuttlefish *Sepia*[2] and deep-water squid *Spirula* have an internal chambered structure (Figure 14–1). These animals are neutrally buoyant, meaning that the amount of air in their bodies regulates their density so they can remain within the water column at a particular depth of their choosing.

Because the air pressure in their air chambers is always 1 kilogram per square centimeter (1 atmosphere, or 14.7 pounds per square inch), these creatures are limited in the depth to which they may venture. The *Nautilus* must stay above a depth of approximately 500 meters (1640 feet) to prevent collapse of its chambered shell, because the external pressure approaches 50 kilograms per square centimeter (50 atmospheres or 735 pounds per square inch). The *Nautilus* rarely ventures below a depth of about 250 meters (800 feet).

Neutral buoyancy is achieved by some slow-moving fish through by an internal organ called a **swim bladder** with gases (Figure 14–2). The swim bladder is normally not present in very active swimmers, such as the tuna, or in fish that live on the bottom, because neither need it. Some fish have a pneumatic duct that connects the swim bladder to the esophagus (Figure 14–2). These fish can

---

[1] Remember that *biomass* is the mass of living organisms.

[2] Many species of cephalopods have an inking response. The ink of the cuttlefish *Sepia* was used as writing ink (with the brand name Sepia) before alternatives were developed.

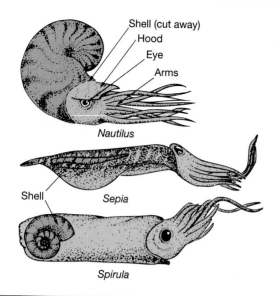

Figure 14–1 Gas containers in cephalopods.
The *Nautilus* has an external chambered shell, while *Sepia* and *Spirula* have rigid internal chambered structures that can be filled with gas to provide buoyancy.

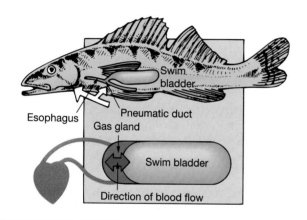

Figure 14–2 Swim bladder.
Some bony fishes have a swim bladder, which is connected to the esophagus by the pneumatic duct, allowing air to be added or removed rapidly. In fish with no pneumatic duct, all gas must be added or removed through the blood, which requires more time.

add or remove air through this duct. In other fish, the gases of the swim bladder must be added or removed more slowly by an interchange with the blood.

Because change in depth will cause the gas in the swim bladder to expand or contract, the fish removes or adds gas to the bladder to maintain a constant volume. Those fish that do not have the pneumatic duct are limited in the rate at which they can make these adjustments and, therefore, cannot withstand rapid changes in depth.

The composition of gases in the swim bladders of shallow-water fishes is similar to that of the atmosphere. With increasing depth, the concentration of oxygen in the swim bladder gas increases from the 20 percent common near

the surface to more than 90 percent. Fish with swim bladders have been captured from as deep as 7000 meters (23,000 feet), where the pressure is 700 kilograms per square centimeter (700 atmospheres, or 10,300 pounds per square inch). The high pressure at this depth causes gas to be compressed to a density of 0.7 gram per cubic centimeter.[3] This is approximately the same density as fat, so it is not surprising that many deep-water fish have special organs for buoyancy that are filled with fat instead of compressed gas.

Floating Organisms (Zooplankton)

Organisms floating at the surface range in size from microscopic plankton to some relatively large plankton, such as the familiar jellyfish. These floating organisms—collectively called zooplankton—comprise the second largest biomass in the ocean after the phytoplankton. Microscopic forms have a hard shell or test. Many larger forms have soft, gelatinous bodies with little if any hard tissue, which reduces their density and allows them to stay afloat.

Microscopic Zooplankton Microscopic zooplankton are incredibly abundant in the ocean. They are miniature planktonic organisms that are considered *primary consumers* because they are the first level of organisms that eat the abundant microscopic phytoplankton that comprises the majority of biomass in the ocean.

There are many different types of microscopic zooplankton. As with most microscopic planktonic organisms, they need to increase the surface area of their bodies (or shells) so that they won't sink below the sunlit surface waters too quickly, since they need to remain near their food source. Many zooplankton are herbivores and graze on diatoms, dinoflagellates, and other phytoplankton. Other zooplankton species are omnivores and eat other zooplankton in addition to phytoplankton.

Three of the most important groups of zooplankton are the radiolarians, the foraminifers (both of which have been discussed in Chapter 4, "Marine Sediments"), and the copepods.

The **radiolarians** (*radio* = a spoke or ray) are single-celled, microscopic organisms that build their hard shells out of silica (Figure 14–3). Their microscopic shells—called *tests*—have intricate ornamentation including long projections. Although this appears to be a defense mechanism from predators, the spikes and spines are used to increase the test's surface area so that the organism won't sink through the water column.

Foraminifers (*foramen* = an opening) are microscopic to (barely) macroscopic single-celled animals, many of which are planktonic. Foraminifers produce a hard test made of calcium carbonate in which the organism lives

---

[3] For comparison, note that the density of water is 1.0 gram per cubic centimeter.

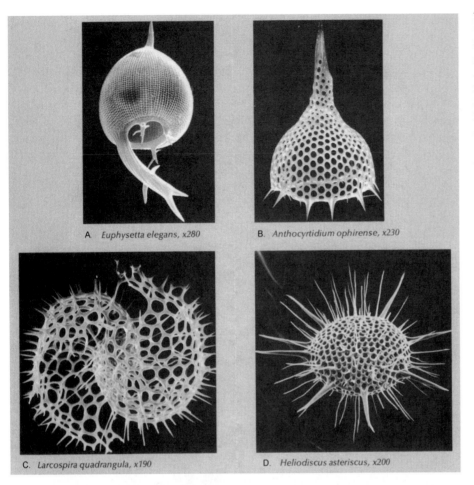

A.   *Euphysetta elegans, x280*

B.   *Anthocyrtidium ophirense, x230*

C.   *Larcospira quadrangula, x190*

D.   *Heliodiscus asteriscus, x200*

Figure 14–3 Radiolarians.
Scanning electron micrographs of various radiolarians. **A.** *Euphysetta elegans* (magnified 280 times). **B.** *Anthocyrtidium ophirense* (magnified 230 times). **C.** *Larcospira quadrangula* (magnified 190 times). **D.** *Heliodiscus asteriscus* (magnified 200 times).

Figure 14–4
Foraminiferans.
Scanning electron micrographs of various foraminifers, clockwise from upper left: *Eggerella bradyi* (magnified 200 times); *Pyrgo murrhyna* (magnified 193 times); *Uvigerina proboscidea* (magnified 228 times); and *Cibicides bradyi* (magnified 388 times).

(Figure 14–4). Many of their tests appear segmented or chambered, and all have a prominent opening in one end. The tests of both radiolarians and foraminifers are common components of deep-sea sediment.

Copepods (*kope* = oar, *pod* = a foot) are microscopic shrimp-like animals of the subphylum Crustacea, which also includes larger crustaceans such as shrimps, crabs, and lobsters. Like other crustaceans, copepods have a hard exoskeleton (*exo* = outside) and a segmented body with jointed legs (Figure 14–5). In terms of biomass, copepods probably represent the majority of the ocean's zooplankton and are an important link in many marine food webs. They have special adaptations for filtering their tiny floating food from seawater.

Macroscopic Zooplankton There are many types of zooplankton that are large enough to be seen without the aid of a microscope. One important group is **krill** (genus *Euphausia*), which are similar to copepods but larger in size. Krill and copepods are both members of the subphylum Crustacea, so they share similar characteristics with each other and with other crustaceans. In spite of their name—which means "young fry of fish" in Norwegian—krill resemble mini-shrimp (Figure 14–6).

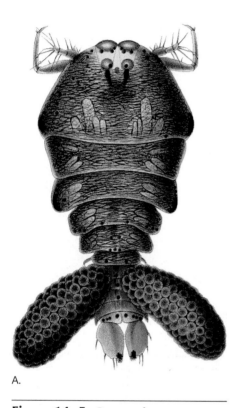

A.

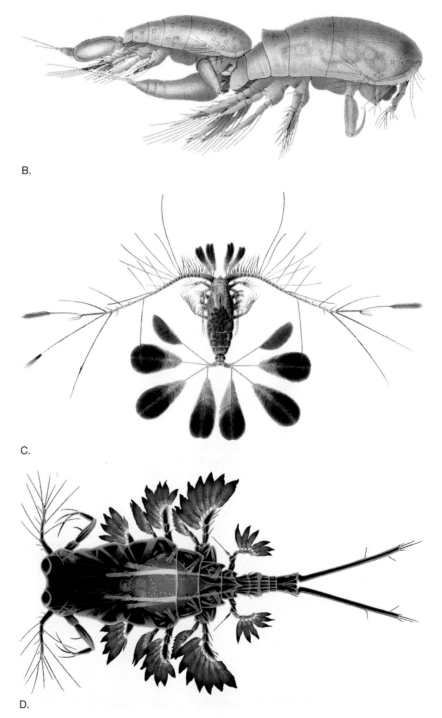

B.

C.

D.

Figure 14–5 Copepods.
Line drawings of various copepods (shown many times their actual size) from Wilhelm Giesbrecht's 1892 book on the flora and fauna of the Gulf of Naples. **A.** The adult female *Sapphirina auronitens* carries a pair of lobe-like egg sacks. **B.** Copulating pair of *Oncaea conifera*. **C.** *Calocalanus pavo* showing elaborate feathery appendages that are characteristic of warm-water species. **D.** *Copilia vitrea* uses its appendages to cling to large particles in the water column or to larger zooplankton.

A.

B.

Figure 14–6 Krill.
A. Close-up view of the krill *Meganyctiphanes norvegica*, which is about 3.8 centimeters (1.5 inches) long. **B.** A large aggregation of krill and smaller organisms.

There are over 1500 species of krill, most of which have lengths of 5 centimeters (2 inches) or less. They are abundant near Antarctica and form a critical link in the food web there, supplying a source of food for many organisms from sea birds to whales. In fact, krill are the favorite food of many of the largest whales in the world.

Coelenterates (*coel* = hollow, *entreon* = the intestines) are characterized by soft bodies that are more than 95 percent water. There are two basic types, the siphonophores and the scyphozoans (jellyfish).

Siphonophore (*siphon* = tube, *phoros* = bearing) coelenterates are represented in all oceans by the Portuguese man-of-war (genus *Physalia*) and "by-the-wind sailor" (genus *Velella*). Their gas floats, called pneumatophores, serve as floats and sails that allow the wind to push these creatures across the ocean surface (Figure 14–7A, *left*). Sometimes, large numbers of these organisms are pushed by the wind onto a beach, where they wash up and die. In the living organism, a colony of tiny individuals is sus-

pended beneath the float. Portuguese man-of-war tentacles may be many meters long and possess nematocysts long enough to penetrate the skin of humans; they have been known to inflict a painful and occasionally dangerous neurotoxin poisoning.

Jellyfish, or **scyphozoan** (*skuphos* = cup, *zoa* = animal) coelenterates, are individuals that have a bell-shaped body with a fringe of tentacles and a mouth at the end of a clapperlike extension hanging beneath the bell-shaped float (Figure 14–7). Ranging in size from nearly microscopic to 2 meters (6.6 feet) in diameter, with long tentacles (60 meters, or 200 feet), most jellyfish have bells with a diameter of less than 0.5 meter (1.6 feet).

Jellyfish move by muscular contraction. Water enters the cavity under the bell and is forced out by contractions of muscles that circle the bell, jetting the animals ahead in short spurts. To allow the animal to swim generally in an upward direction, sensory organs are spaced around the outer edge of the bell. These may be light sensitive or gravity sensitive. This orientation ability is important because the jellyfish feed by swimming to the surface and sinking slowly through the life-rich surface waters.

Animals called **tunicates** (*tunicatus* = to clothe with a tunic) are generally transparent and barrel shaped. They have openings on each end for current to flow in or out (Figure 14–8A). Individual tunicates may be 7 centimeters (3 inches) long, and often form chains of individuals that may reach great lengths (see Figure 14–8B). They move by a feeble form of jet propulsion, created by contraction of bands of muscles that force water into the incurrent opening and out of the excurrent opening.

The genus *Pyrosoma* is luminescent and colonial. Individual members have their incurrent openings facing the outside surface of a tube-shaped colony that may be a few meters long. One end of the tube is closed, and the excurrent openings of the thousands of individuals all empty into the tube. Muscular contraction forces water out the open end to provide propulsion.

Ctenophores (*kten* = comb, *phoros* = bearing) are animals closely related to the coelenterates. The body form of most ctenophores is spherical, with eight rows of cilia spaced evenly around the sphere. It is from these structures that the name "comb bearing" is derived; when magnified, they look like miniature combs. Often called "comb jellies," this group of animals is entirely pelagic and confined to the marine environment.

If a ctenophore possesses tentacles, there will be two that contain adhesive organs instead of stinging cells, to capture prey. Sea gooseberries (genera *Pleurobranchia*) (Figure 14–8C) and the pink, elongated *Beroe* (Figure 14–8D) range from gooseberry size to over 15 centimeters (6 inches) in the latter.

The **chaetognaths** (also called the *arrowworms*) are transparent and difficult to see, although they may grow to more than 5 centimeters (2 inches) in length

Figure 14–7 Planktonic cnidarians.

A. Portuguese man-of-war (*Physalia*) and jellyfish. **B.** A medusa jellyfish.

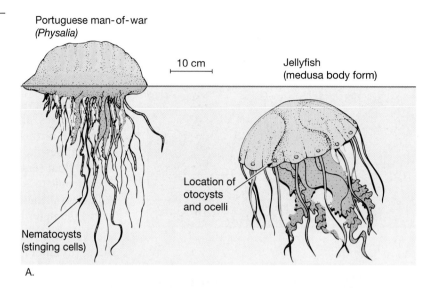

Portuguese man-of-war (*Physalia*)

10 cm

Jellyfish (medusa body form)

Location of otocysts and ocelli

Nematocysts (stinging cells)

A.

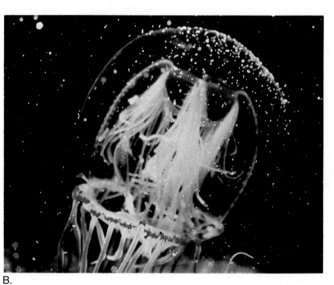

B.

(Figure 14–9). The name *chaetognath* (*chaeta* = bristle, *gnathos* = jaw) refers to the hairlike attachments around their mouths used to grasp prey while they devour it. They are voracious feeders, eating primarily small zooplankton. In turn, they are eaten by fishes and larger planktonic animals such as jellyfish. These exclusively marine, hermaphroditic animals are usually more abundant in the surface waters some distance from shore.

Swimming Organisms (Nekton)

Many larger pelagic animals have substantial powers of locomotion and can swim easily against currents. These organisms include rapid-swimming invertebrate squids, fish, and marine mammals. Because of their good swimming ability, some of these organisms undertake long migrations.

Swimming squid include the common squid (genus *Loligo*), flying squid (*Ommastrephes*), and giant squid (*Architeuthis*). Active predators of small fish, the smaller squid varieties have long, slender bodies with paired fins

(Figure 14–10). Unlike the less active *Sepia* and *Spirula* shown in Figure 14–1, they have no hollow chambers in their bodies and therefore require more energy to remain in the upper water of the oceans without sinking.

These invertebrates can swim about as fast as any fish their size, and do so by trapping water in a cavity between their soft body and pen-like shell and forcing it out through a siphon. To capture prey, they use two long arms with pads containing suction cups at the ends (Figure 14–10). Eight shorter arms with suckers convey the prey to the mouth, where it is crushed by a mouthpiece that resembles a parrot's beak.

Locomotion in fish is more complex than in the squid because of body motion and the role of fins. The basic movement of a swimming fish is the passage of a wave of lateral body curvature from the front to the back of the fish. This is achieved by the alternate contraction and relaxation of muscle segments along the sides of the body. These muscle segments are called **myomeres** (*myo* = a muscle, *merous* = parted). The backward pressure of the fish's

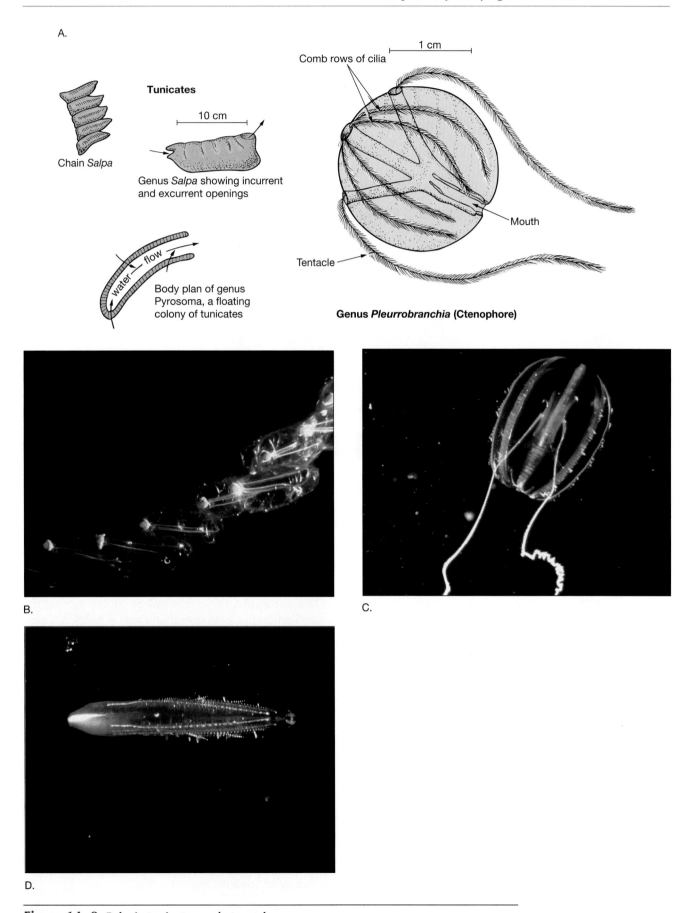

A. General body structure of tunicates (*left*) and a ctenophore (*right*). B. Chain of salps (tunicates). C. Ctenophore, *Pleurobranchia*. D. Ctenophore, *Beroe*.

Figure 14–8 Pelagic tunicates and ctenophores.

A.

Tunicates

Chain *Salpa*

10 cm

Genus *Salpa* showing incurrent and excurrent openings

water flow

Body plan of genus Pyrosoma, a floating colony of tunicates

Comb rows of cilia

1 cm

Mouth

Tentacle

Genus *Pleurrobranchia* (Ctenophore)

B.

C.

D.

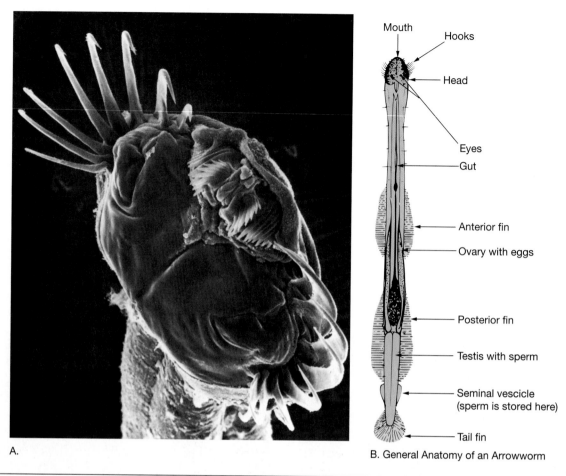

A.

B. General Anatomy of an Arrowworm

Figure 14-9 Arrowworm.

A. Scanning electron micrograph of the head of the chaetognath (arrowworm) *Sagitta tenuis,* magnified 161 times. **B.** General anatomy of an arrowworm.

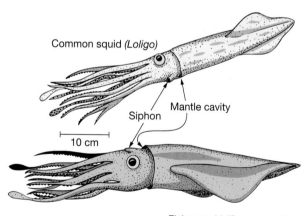

Figure 14-10 Squid.

Squid move by trapping water in their mantle cavity between their soft body and pen-like shell. They then jettison the water through their siphon for rapid propulsion.

body and fins produced by the movement of this wave provides the forward thrust (Figure 14–11).

Fin Designs in Fish Most active swimming fish have two sets of paired fins used in maneuvers such as turn-

ing, braking, and balancing. These are the *pelvic fins* and *pectoral (pectoralis* = breast) *fins* shown in Figure 14–11. When not in use, these fins can be folded against the body. Vertical fins, both *dorsal (dorsum* = back) and *anal,* serve primarily as stabilizers.

The fin that is most important in propelling the high-speed fish is the tail fin, or *caudal (cauda* = tail) *fin.* Caudal fins flare vertically to increase the surface area available to develop thrust against the water.[4] The increased surface area also increases frictional drag. The efficiency of the design of a caudal fin depends on its shape.

There are five basic shapes of caudal fins, illustrated in Figure 14–12 and keyed by letter to the following description.

A. The rounded fin is flexible and useful in accelerating and maneuvering at slow speeds.

B. & C. The somewhat flexible truncate tail (B) and forked tail (C) are found on faster fish and still may be used for maneuvering.

---

[4] This is equivalent to humans donning swimming flippers on their feet to enable them to swim more efficiently.

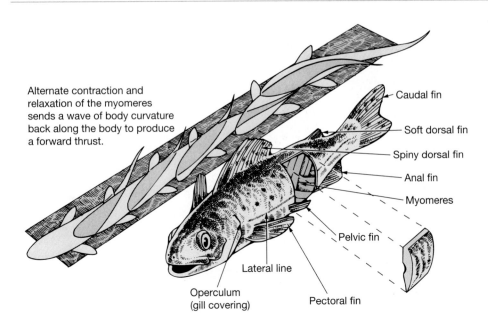

Figure 14–11 Fish: swimming motions and fins.

Alternate contraction and relaxation of the myomeres sends a wave of body curvature back along the body to produce a forward thrust.

Caudal fin

Soft dorsal fin

Spiny dorsal fin

Anal fin

Myomeres

Pelvic fin

Lateral line

Operculum (gill covering)

Pectoral fin

D. The lunate caudal fin is found on the fast-cruising fishes such as tuna, marlin, and swordfish; it is very rigid and useless for maneuverability, but very efficient for propelling.

E. The heterocercal (*hetero* = uneven, *cercal* = tail) fin is asymmetrical, with most of its mass and surface area in the upper lobe.

The heterocercal fin produces a significant lift to sharks as it is moved from side to side. This lift is important because sharks have no swim bladder and tend to sink when they stop moving. To aid this lifting, the pectoral (chest) fins are large and flat. Positioned on the shark's body like airplane wings, they in fact function like aircraft wings—or a hydrofoil boat—to lift the front of the shark's body to balance the rear lift supplied by the caudal fin. The shark gains tremendous lift but loses a lot in maneuverability as a result of this adaptation of the pectoral fins. This is why sharks tend to swim in broad circles, like a circling airplane, and are not marine acrobats.

Adaptations for Seeking Prey

Several factors affect each species' adaptation for capturing food. These include mobility (lunging versus cruising), speed, body length, body temperature, and circulatory system.

Lungers Versus Cruisers

Some fish spend most of their time waiting patiently for prey and exert themselves only in short bursts as they lunge at the prey. Others cruise relentlessly through the water, seeking prey. There is a marked difference in the musculature of fishes that use these different styles of obtaining food.

Lungers sit and wait. Figure 14–13A shows a grouper, representative of the lungers. It has a truncate caudal

fin for speed and maneuverability, and almost all its muscle tissue is white.

Cruisers, such as the tuna in Figure 14–13B, actively seek prey. Less than half of a cruiser's muscle tissue is white; most is red.

What is the significance of red versus white muscle tissue? *Red muscle fibers* are much smaller in diameter (25 to 50 microns, or 0.01 to 0.02 inch). *White muscle fibers* are much larger (135 microns, or 0.05 inch) and contain lower concentrations of **myoglobin** (*myo* = muscle, *globus* = sphere), a red pigment with an affinity for oxygen. These qualities allow the red fibers to obtain a much greater oxygen supply than is possible for white fibers. This supports a metabolic rate six times that of white fibers, which is needed for endurance. The red muscle tissue is abundant in cruisers that swim constantly.

Lungers can get along quite well with little red tissue because they do not need to move continually. White tissue, which fatigues much more rapidly than red tissue, is used by the tuna for short periods of acceleration while on the attack. It is also quite adequate for propelling the grouper and other lungers during their quick bursts of speed to obtain prey.

Speed and Body Size

The swimming speed of a fish depends on whether it is cruising (slow), hunting for prey (fast), or escaping from predators (fastest). Generally, the swimming speed of a fish is closely related to its body size: the larger the fish, the faster it can swim. For tuna, which are well adapted for sustained cruising and short bursts of high-speed swimming, cruising speed averages about three body lengths per second. They can maintain a maximum speed of about 10 body lengths per second, but only for one second. A yellowfin tuna, *Thunnus albacares*, has been clocked at an amazing speed of 74.6 kilometers (46 miles)

Figure 14–12 Caudal fin shapes.

A. Rounded fin on a queen angel (other examples: sculpin, flounder). **B.** Truncate fin on a gray angelfish (also on salmon, bass). **C.** Forked fin on a goatfish (also on herring, yellowtail). **D.** Lunate fin on a blue marlin (also on bluefish, tuna). **E.** Heterocercal fin on a silvertip shark (also on many other types of sharks).

per hour! This speed is more than 20 body lengths per second, although the tuna can maintain this burst of speed only for a *fraction* of a second.

Theoretically, a 4-meter (13-foot) bluefin tuna, *Thunnus thynnus*, can reach speeds up to about 144 kilometers

(90 miles) per hour.[5] Besides fish, many of the toothed whales are known to be capable of high rates of speed.

---

[5] Imagine how difficult it would be to clock a bluefin tuna accurately in the ocean at this speed!

Box 14–1
Some Myths and Facts About Sharks

Sharks (Figure 14B) are the fish most feared by humans. The strength, large size, sharp teeth, and unpredictable nature of sharks are enough to keep some people from *ever* entering the ocean. The publicity generated by the occasional shark attack on humans has led to many myths about sharks.

- *Myth #1: All sharks are dangerous.* Of the more than 350 shark species, about 80 percent are unable to hurt people or rarely encounter people. For instance, the largest shark—also the largest fish in the world—is the whale shark, which reaches lengths of up to 15 meters (50 feet) but eats only plankton.

- *Myth #2: Sharks are voracious eaters that must eat continuously.* Like other large animals, sharks eat periodically depending upon their metabolism and the availability of food. Humans are not a primary food source of any shark, and many large sharks prefer the higher fat content of seals and sea lions.

- *Myth #3: Most people attacked by a shark are killed.* There is a high survival rate from unprovoked shark attacks on humans: Of every 100 people attacked by sharks, over 85 survive. Many large sharks commonly attack by biting their prey to immobilize it before trying to eat it. Consequently, many potential prey escape and survive.

- *Myth #4: Many people are killed by sharks each year.* The chances of a person being killed by a shark are quite low (Table 14–1). Each year, an average of only 5-15 people are killed by sharks worldwide. However, the number of sharks killed by humans each year is as many as 100 million (mostly as "by-catch" from fishing activities). Compared to many other fishes, sharks have low re-

production rates and grow slowly, which may result in many sharks being designated as endangered species in the future.

- *Myth #5: The great white shark is a common, abundant species found off most beaches.* Great white sharks are relatively uncommon, large predators that prefer cooler waters. At most beaches, great whites are rarely encountered.

- *Myth #6: Sharks are not found in fresh water.* A specialized osmoregulatory system enables some species (such as the bull shark) to cope with dramatic changes in salinity, from the high salinity of seawater to the low salinity of freshwater rivers and lakes.

- *Myth #7: All sharks need to swim constantly.* Some sharks can remain at rest for long periods on the bottom and obtain enough oxygen through pumping water over their gills by opening and closing their mouths. Typically, sharks swim very slowly—typical cruising speeds are less than 9 kilometers (6 miles) per hour—but can swim at burst of over 37 kilometers (23 miles) per hour.

- *Myth #8: Sharks have poor vision.* The lens of a shark's eye is up to seven times more powerful than that of a human's. In fact, sharks can even distinguish color.

- *Myth #9: Eating shark meat makes one aggressive.* There is no indication that eating shark meat will alter a person's temperament. The firm texture, white flesh, low fat content, and mild taste of shark meat have made it a favorite type of seafood in many countries.

- *Myth #10: No one would ever want to enter water filled with sharks.* Long regarded with fear and suspicion, sharks are more recently receiving attention as subjects of fascination and wonder. Diving tours specializing in close encounters with sharks are becoming increasingly popular.

Figure 14B Great white shark (*Carcharodon carcharias*).

Table 14–1 Types and number of occurrences in the United States.

Occurrence to people in U.S.	Avg. number per year
Highway fatalities	2611
Struck by lightning	352
Killed by lightning	94
Bit by a squirrel in NY City	88
Bit by a shark	10
Killed by a shark	0.4

For instance, the porpoise *Stenella* has been clocked at 40 kilometers (25 miles) per hour, and it is believed that the top speed of killer whales may exceed 55 kilometers (34 miles) per hour.

Cold-Blooded Versus Warm-Blooded

As is true with many chemical reactions, metabolic processes occur more rapidly at higher temperatures. Fish are mostly **cold-blooded** or **poikilothermic**

A.

B.

Figure 14–13 Feeding styles—lungers and cruisers.
A. Lungers, such as this tiger grouper, sit patiently on the bottom and capture prey with quick, short lunges. **B.** Cruisers, such as these yellowfin tuna, swim constantly in search of prey and capture it with short periods of high-speed swimming.

(*poikilos* = spotted, *thermos* = heat), having body temperatures that are nearly the same as their environment. Usually, these fish are not fast swimmers. Nevertheless, there are some fast swimmers with body temperatures slightly above that of the surrounding water. For example, the mackerel (*Scomber*), yellowtail (*Seriola*), and bonito (*Sarda*) have body temperature elevations of only 1.3, 1.4, and 1.8 degrees centigrade (2.3, 2.5, and 3.2 degrees Fahrenheit), respectively, above adjacent seawater.

Other fast swimmers that have much higher temperature elevations than their environment are members of the mackerel shark genera, *Lamna* and *Isurus*, as well as the tuna, *Thunnus*. Bluefin tuna have been observed to maintain a body temperature of 30 to 32 degrees centigrade (86 to 90 degrees Fahrenheit) regardless of water temperature, which is characteristic of **warm-blooded** or **homeothermic** (*homeo* = alike, *thermos* = heat) organisms. Although these tuna are more commonly found in warmer water, where the temperature difference between fish and water is no more than 5 degrees centigrade (9 degrees Fahrenheit), body temperatures of 30 degrees centigrade (86 degrees Fahrenheit) have been measured in bluefin tuna swimming in 7 degrees centigrade (45 degrees Fahrenheit) water.

Why do these fish exert so much energy to maintain their body temperatures at high levels, when other fish do quite well with ambient body temperatures? Their mode of behavior is that of a cruiser, and any adaptation (high temperature and high metabolic rate) that increases the power output of their muscle tissue helps them seek and capture prey.

Circulatory System Modifications

Mackerel sharks and tuna are aided in maintaining their high body heats by a modified circulatory system (Figure 14–14). Most fish have a **dorsal aorta** located just beneath the vertebral column that provides blood to the swimming muscles. Some fish, such as mackerel sharks and tuna, have additional **cutaneous** (*cutane* = skin) **arteries** just beneath the skin on either side of the body. As cool blood flows into red muscle tissue, its temperature is increased by heat generated by muscle metabolism (muscle contractions). A fine network of tiny blood vessels within the muscle tissue is designed to minimize heat loss.

The vessels that return the blood to the **cutaneous vein**, parallel to the cutaneous artery along the side of the fish, are all paired with small vessels carrying blood into the muscle tissue. In this way, the warm blood leaving the tissue helps to heat the cooler blood entering from the cutaneous artery.

Adaptations to Avoid Being Prey

Obtaining food occupies most of the time of many inhabitants of the open ocean. Some animals are fast and agile, and obtain food through active predation. Other animals move more leisurely as they filter small food particles from the water. Examples of predators and **filter feeders** (organisms that feed by filtering their small prey from seawater) can be found in populations of pelagic animals, from the tiny zooplankton to the massive whales.

Many animals have unique adaptations to avoid being captured as prey and eaten. We have already discussed in this chapter the speed of various fish, which can be

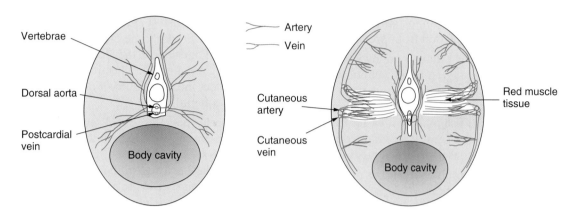

Figure 14–14 Circulatory modifications in fishes.
A. Most cold-blooded (poikilothermic) fishes have major blood vessels arranged so that blood flows to muscle tissue from the dorsal aorta and returns to the postcardial vein beneath the vertebral column. **B.** Warm-blooded (homeothermic) fishes such as the bluefin tuna have cutaneous arteries and veins that help maintain high blood temperature by using heat energy generated by contracting muscle tissue.

used to avoid predators. In Chapter 12 ("The Marine Habitat"), the adaptations of camouflage and countershading were presented. Here, we will consider an additional adaptation: schooling.

Schooling

Although vast populations of phytoplankton and zooplankton may be highly concentrated in certain areas of the ocean, these patches are not usually referred to as schools. The term **school** is usually reserved for large numbers of fish, squid, or shrimp that form well-defined social groupings.

The number of individuals in a school can vary from a few larger predaceous fish (such as bluefin tuna) to hundreds of thousands of small filter feeders (such as anchovies). Within the school, individuals of the same size move in the same direction with equal spacing between each. This spacing probably is maintained through visual contact and, in the case of fish, by use of the lateral line system (see Figure 14–11) that detects vibrations of swimming neighbors. The school can turn abruptly or reverse direction as individuals at the head or rear of the school assume leadership positions (Figure 14–15).

What is the advantage of schooling? The advantage might seem obvious from the reproductive point of view: During spawning, it ensures that there will be males to release sperm to fertilize the eggs shed into the water or deposited on the bottom by females. However, most

Figure 14–15 Schooling.
A school of soldier fish near a reef in the Maldives.

investigators believe the most important function of schooling in small fish is protection from predators.

At first, it may seem illogical that schooling would be protective. Any predator lunging into a school would surely catch something, just as land predators run a herd of grazing animals until one weakens and becomes a meal. So, aren't the smaller fish making it easier for the predators by forming a large target? Based more on conjecture than research, the consensus of scientists is no.

How schooling protects a group of organisms may grow out of the following considerations.

1. If members of a species form schools, they reduce the percentage of ocean volume in which a cruising predator might find one of their kind.

2. Should a predator encounter a large school, it is less likely to consume the entire unit than if it encounters a small school or an individual.

3. The school may appear as a single large and dangerous opponent to the potential predator and prevent some attacks.

4. Predators may find the continually changing position and direction of movement of fish within the school confusing, making attack particularly difficult for predators, who can attack only one fish at a time.

In support of schooling's advantage is the fact that over 2000 fish species are known to form schools and that half of all fishes join schools during a portion of their lives. This suggests that schooling behavior offers an advantage to species that form schools, especially for those organisms with no other means of defense.

There may be other, more subtle reasons for schooling. For instance, schooling may also aid in swimming as each schooling fish gets a boost from the vortex created by the fish swimming in front of it, thus helping schooling fish travel greater distances than individuals. Certainly, the behavior of schooling must enhance species survival because it is so widely practiced among pelagic animals.

Deep-Water Nekton

Living below the surface water but still above the ocean floor are deep-water nektonic (swimming) species—mostly various species of fish—that are specially adapted to the deep-water environment, where it is very still and completely dark. Their food source is one of two things: **detritus**—dead and decaying organic matter and waste products that slowly settle through the water column from the life-rich surface waters above—or each other. Thus, there are a number of predatory species. The lack of abundant food limits the *number* of organisms (total biomass) and also the *size* of these organisms. Thus, there are small populations of these organisms and most of these organisms are less than 30 centimeters (1 foot) long. As well, many of these organisms have low metabolic rates to conserve energy.

These **deep-sea fish** (Figure 14–16) have special adaptations to allow them to be efficient at finding and collecting food. For example, they have good sensory devices such as long antennae or sensitive lateral lines that are used to detect the movement of other organisms within the water column.

Many species have large and sensitive eyes—perhaps 100 times more sensitive to light than our own—that enable them to see potential prey. To avoid being prey, most species are dark colored so that they blend into the environment. Other species are blind and rely on their other senses to obtain food. For instance, the sense of smell is well developed in some species, which allows certain fish to track down prey items by following a scent trail.

Well over half of deep-sea fish have the capability to **bioluminesce**, that is, to produce light organically with specially designed structures or cells called **photophores**. In a world of darkness, the ability to produce light has many uses, including:

- Attracting prey [such as how the deep-sea anglerfish (Figure 14–16F) uses its specially modified dorsal fin as a bioluminescent lure]
- Staking out territory by constantly patrolling an area
- Communicating or seeking a mate by sending signals
- Escaping from predators by using a flash of light to temporarily blind a predator

Other adaptations to the deep-sea environment include large sharp teeth, expandable bodies, hinged jaws that can disarticulate, and huge mouths in proportion to their small bodies (Figure 14–17). These adaptations allow deep-sea fish to process even large food items efficiently whenever food is located. With these adaptations, many of these deep-sea fish are able to ingest other species that are larger than they are.

Marine Mammals

Perhaps it is because they are some of our closest marine relatives that marine mammals capture the imagination of so many people. Although marine mammals live in the marine environment, they have quite different characteristics than fish. In fact, all organisms in class Mammalia (including marine mammals) share these common characteristics:

- They are warm-blooded
- They breathe air
- They have hair (or fur) in at least some stage of their development
- They bear live young (except for a few egg-laying mammals of Australia from the subclass Prototheria, which include the duck-billed platypus and the spiny anteater)

Figure 14–16 Deep-sea fishes.
A. A hatchet fish. **B.** A lantern fish. **C.** A stomiatoid. **D.** A hatchet fish. **E.** A gulper eel.
F. A female deep-sea anglerfish, with attached parasitic male (**G**).

Figure 14–17 Adaptations of deep-sea fishes.
A. Large teeth, hinged jaw, and swallowing mechanism of the deep-sea viper fish *Chauliodus sloani*. **B.** Ingestion capability of *Chiasmodon niger*, with a curled-up fish in its stomach that is longer than itself.

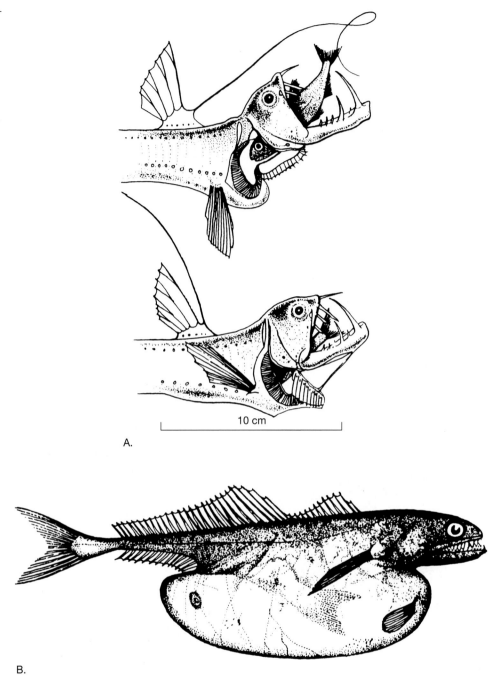

10 cm

A.

B.

• The females of each species have mammary glands—the namesake of the mammals—that are used to produce milk for their young

Recent discoveries of fossils of ancient whales found in Egypt and China provide strong evidence that marine mammals evolved from mammals on land about 50 million years ago. In fact, many of the so-called missing links in the evolution of marine mammals provide some of the strongest evidence for the transition back to the sea. For instance, fossils of ancient whales recovered in Egypt in 1989 have small, unusable hind legs, suggesting that the land mammal predecessor had no need for its hind legs with the development of a large paddle-shaped struc-

ture for a tail that was used to swim through water. As well, there are many anatomical similarities between land mammals and marine mammals. It is thought that mammals—which originally evolved from organisms that inhabited the sea millions of years ago—returned to the sea because of more abundant food sources. Truly, air-breathing marine mammals are well adapted to life in the sea.

Representatives of marine mammals include 116 species, which fall mostly into these three orders: order Pinnipedia, order Sirenia, and order Cetacea. Let's examine some of the characteristics of marine mammals within these three orders.

Order Pinnipedia

Marine mammals within order **Pinnipedia** (*pinni* = feather, *ped* = a foot) include walruses, seals, sea lions, and fur seals (Figure 14–18). The name pinniped describes these organisms' prominent skin-covered flippers, which are well adapted for propelling the pinnipeds through water.

Walruses are characterized by large body size and prominent canine teeth in adults, which are represented as elongated ivory tusks up to 1 meter (3 feet) long (Figure 14–18A). Both males and females have tusks, which are used for territorial fighting and for hauling themselves onto icebergs.

Seals—also called the *earless seals* or *true seals*—are common marine mammals. They differ from the **sea lions** and **fur seals**—also called the *eared seals*—by several characteristics:

- Seals lack prominent ear flaps that are specific to sea lions and fur seals (compare Figures 14–18B and 14–18C).
- Seals have smaller and less-prominent front flippers (called *fore flippers*) than sea lions and fur seals.
- Seals have prominent claws that extend from their flippers that sea lions and fur seals lack (Figure 14–19).

A.

B.

C.

Figure 14–18 Marine mammals of order Pinnipedia.
A. Walruses. **B.** Harbor seals. **C.** California sea lions.

Figure 14–19 Skeletal and morphological differences between sea lions and seals.

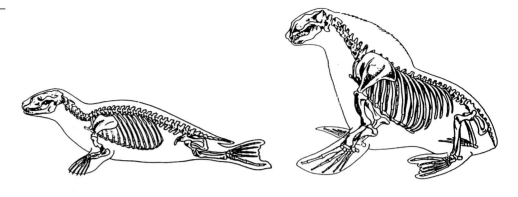

Skeleton of a typical seal, genus *Phoca*

Skeleton of the Steller sea lion

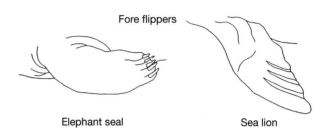

Fore flippers

Elephant seal

Sea lion

- Seals have a different hip structure than sea lions and fur seals. Thus, seals cannot move their rear flippers underneath their bodies as sea lions and fur seals can (Figure 14–19).

- Seals do not walk on land very well, and can only slither along like a caterpillar; sea lions and fur seals use their rear flippers under their bodies and their large front flippers to walk easily on land, and can even ascend steep slopes, climb stairs, and do other acrobatic tricks.

- Seals propel themselves through the water by a back-and-forth motion of the rear flippers (similar to a wagging tail), while sea lions and fur seals swim by flapping their large front flippers.

Order Sirenia

Animals of order **Sirenia** (*siren* = a mythical mermaid-like creature with an enticing voice) include the manatees and dugongs, collectively (and aptly) known as "sea cows." The manatees are concentrated in coastal areas of the tropical Atlantic Ocean, while the dugongs populate the tropical regions of the Indian and Western Pacific Oceans. Both of these animals have a paddle-like tail and rounded front flippers (Figure 14–20). The land-dwelling ancestors of sirenians were elephant-like, and sirenians today retain some of these unique characteristics, such as their large body size and, for manatees, the presence of toenails. The sirenians have very few hairs, which are bristly and concentrated around the mouth.

Sirenians eat only shallow-water grasses and are thus the only vegetarian marine mammals. They spend most of their lives in heavily traveled shallow coastal waters, which has led to their demise. These areas are heavily populated and heavily traveled, resulting in much competition for space. In addition, there have been many accidents, with boats running over these large animals that move slowly and cannot be easily seen. The populations of manatees and dugongs have been decreasing, and both are considered endangered.

Order Cetacea

The order **Cetacea** (*cetus* = a whale) includes the whales, dolphins, and porpoises (Figure 14–21). The cetacean body is more or less cigar shaped and insulated with a thick layer of blubber. Cetacean forelimbs are modified into flippers that move only at the "shoulder" joint. The hind limbs are vestigial (rudimentary), not attached to the rest of the skeleton, and are usually not visible externally. The characteristics that all cetaceans share in common are:

- An elongated (telescoped) skull
- Blowholes on top of the skull
- Very few hairs
- A horizontal tail fin called a *fluke* that is used for propulsion by vertical movements

These characteristics generally improve the streamlining of cetacean's bodies, allowing them to be excellent swimmers.

sue. The soft layer tends to decrease the pressure differences at the skin–water interface by compressing under regions of higher pressure and expanding in regions of low pressure, thus reducing turbulence and drag.

Modifications to Allow Deep Diving
Humans can free-dive to a maximum recorded depth of 130 meters (428 feet) and hold their breaths in rare instances for up to six minutes. In contrast, the sperm whale *Physeter macrocephalus* is known to dive deeper than 2200 meters (7200 feet), and the northern bottlenose whale *Hyperoodon ampullatus* has been known to stay submerged for up to two hours.

What adaptations do cetaceans have to permit such long, deep dives? One reason is that cetaceans can alternate between periods of normal breathing and cessation of breathing. The cessation periods occur while the animal is submerged. To understand how some cetaceans are able to go for long periods without breathing requires a general knowledge of their lungs and associated structures.

Cetacean Breathing Figure 14–22 shows internal modifications that allow cetaceans to remain submerged for extended periods. Inhaled air finds its way to tiny terminal chambers, the alveoli. Alveoli are lined by a thin alveolar membrane that is in contact with a dense bed of capillaries; the exchange of gases between the inhaled air and the blood (oxygen in, carbon dioxide out) occurs across the alveolar membrane. Some cetaceans have an exceptionally large concentration of capillaries surrounding the alveoli (Figure 14–22B), which have muscles that move air against the membrane by repeatedly contracting and expanding.

Cetaceans take from one to three breaths per minute while resting, compared with about 15 in humans. Because they hold the inhaled breath much longer, and because of the large capillary mass in contact with the alveolar membrane and the circulation of the air by muscular action, cetaceans can extract much more oxygen from each breath. In fact, marine mammals can extract almost 90 percent of the oxygen in each breath, compared with only 4 to 20 percent extracted by terrestrial mammals.

To use this large amount of oxygen efficiently during long periods under water, cetaceans may apply two strategies: (1) storing the oxygen and (2) reducing oxygen use. The storage of so much oxygen is possible because prolonged divers have a much greater blood volume per unit of body mass than those that dive for only short periods.

Compared with terrestrial animals, some cetaceans have twice as many red blood cells per unit of blood volume and up to nine times as much myoglobin in the muscle tissue. Thus, large supplies of oxygen can be chemically stored in the **hemoglobin** (*hemo* = blood, *globus* = sphere) of the red blood cells and the myoglobin of the muscles.

Figure 14–20 Marine mammals of order sirenia. **A.** Manatees. **B.** Dugong.

Modifications to Increase Swimming Speed
Cetaceans' muscles are not vastly more powerful than those of other mammals, so it is believed that their ability to swim at high speed must result from modifications that reduce frictional drag. To illustrate the importance of streamlining in reducing the energy requirements of swimmers, a small dolphin would require muscles five times more powerful than it has to swim at 40 kilometers (25 miles) per hour in turbulent flow.

In addition to a streamlined body, cetaceans are believed to modify the flow of water around their bodies and produce uniform flow conditions with the aid of a specialized skin structure. Their skin is composed of two layers: a soft outer layer that is 80 percent water and has narrow canals filled with spongy material, and a stiffer inner layer composed mostly of tough connective tis-

NORTHERN RIGHT WHALE (*Eubalaena glacialis*)
Baleen in northern right whales can reach 2.8 m (9 ft) in length and is used to strain copepods and krill from surface waters. A similar species, the southern right whale, is found in high southern latitudes. Length is to 18 m (60 ft).

SPERM WHALE (*Physeter macrocephalus*)
Found mostly in tropical waters, these deep-diving toothed whales have a huge snout that contains a large amount of oil. Length of male is to 18 m (59 ft); length of female is to 10.5 m (35 ft).

Figure 14–21 **Marine mammals of order cetacea.**
A composite drawing of representatives of the two whale suborders, drawn to relative scale. The toothed whales (Odontoceti) form complex social communities and include the bottlenose dolphin, killer whale, narwhal, and sperm whale. The baleen whales (Mysticeti) are the largest of all whales and include the right whale, gray whale, humpback whale, and blue whale.

BOTTLENOSE DOLPHIN (*Tursiops truncatus*)
Found in all oceans, this is one of the many species of
oceanic dolphins. Length is to 3 m (10 ft).

KILLER WHALE (*Orcinus orca*)
Cosmopolitan in distribution, this whale is unique in that it not only
feeds on fish but also on seals, birds, and other whales. The male
has a larger dorsal fin than the female. Length is to 9 m (31 ft).

NARWHAL (*Monodon monoceros*)
The scientific name, which means "one tooth, one horn" is accurate for
the male of this Arctic whale. Length is to 6 m (20 ft).

GRAY WHALE (*Eschrichtius robustus*)
This bottom-feeding baleen whale has a small head and very
reduced dorsal fin. It stays close to shore, and even enters
the surf zone. Length is to 14 m (45 ft).

HUMPBACK WHALE (*Megaptera novaeangliae*)
These baleen whales produce some of the most complex
songs of any marine mammal and are known for their energetic
and athletic behaviors, such as breaching. The genus name is
derived from their extraordinary large wing-like flippers.
Length is to 16 m (52 ft).

BLUE WHALE (*Balaenoptera musculus*)
Probably the largest animal to ever inhabit the Earth, these baleen whales weigh up to
150 tons and can consume over 5 tons of krill per day. Length is to 30 m (100 ft).

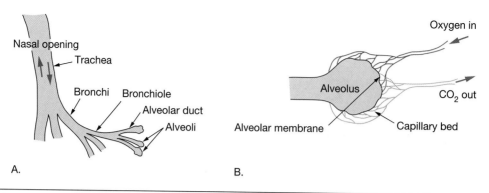

Figure 14–22 Cetacean modifications to allow prolonged submergence.
A. Basic lung design. Air enters the lung through the trachea, and oxygen is absorbed into the blood through the walls of the alveoli. **B.** Oxygen exchange in the alveolus. A dense mat of capillaries receives oxygen through the alveolar membrane, allowing whales to extract as much as 90 percent of oxygen from each breath.

Additionally, the muscles that start a dive with a significant oxygen supply can continue to function through anaerobic respiration when the oxygen is used up. The muscle tissue is relatively insensitive to high levels of carbon dioxide.

Research has shown that cetaceans' swimming muscles can function without oxygen during a dive. This suggests that these muscles and other organs, such as the digestive tract and kidneys, may be sealed off from the circulatory system by constriction of key arteries. The circulatory system might then be required to service only essential components, such as the heart and brain. Because of the decreased circulatory requirements, the heart rate can be reduced by 20 to 50 percent of its normal rate. However, other research has shown that no such reduction in heart rate occurs during dives by the common dolphin (*Delphinus delphis*), the white whale (*Delphinapterus leucas*), or the bottlenose dolphin (*Tursiops truncatus*).

Nitrogen Narcosis Another difficulty with deep and prolonged dives is the absorption of compressed gases into the blood of animals. When humans make dives using compressed air, the prolonged breathing of compressed air—which includes nitrogen and oxygen—can result in nitrogen narcosis or decompression sickness (the bends).[6] The effect of nitrogen narcosis is similar to drunkenness, and it can occur when a diver either goes too deep or stays too long at depths greater than 30 meters (98 feet).

If a diver surfaces too rapidly, the lungs cannot remove the excess gases fast enough, and the reduced pressure may cause small bubbles to form in the blood and tissue.

The bubbles interfere with blood circulation, and the resulting decompression sickness can cause excruciating pain, severe physical debilitation, or even death.

Cetaceans and other marine mammals do not suffer from these difficulties. Their main defense against absorbing too much nitrogen seems to be a more flexible rib cage. By the time a cetacean has reached a depth of 70 meters (230 feet), the rib cage has collapsed under the 8 kilograms per square centimeter (8 atmospheres, or 118 pounds per square inch) of pressure. The lungs within the rib cage also collapse, removing all air from the alveoli. Because most absorption of gases by the blood occurs across the alveolar membrane, the blood cannot absorb additional gases, and the problem of nitrogen narcosis is avoided.

It is possible, however, that the collapsible rib cage is not the main defense against the bends. An experiment was conducted that put enough nitrogen into the tissue of a dolphin to give a human a severe case of the bends. Remarkably, the dolphin suffered no ill effects. This dolphin, as well as other marine mammal species, may have simply evolved an insensitivity to nitrogen gas.

Suborders Mysticeti and Odontoceti Members of order Cetacea can be divided into two suborders. Suborder **Odontoceti** (*odonto* = a tooth, *cetus* = a whale)—also known as the toothed whales—includes the killer whale, sperm whale (which DNA studies suggest is more closely related to baleen whales than to other toothed whales), porpoises, and dolphins. Suborder **Mysticeti** (*mystic* = a moustache, *cetus* = a whale)—also known as the baleen whales—includes the world's largest whales such as the blue whale, finback whale, and humpback whale; and also the gray whale. The baleen whales have no teeth; instead, they have **baleen**, which are plates of fibrous material that hang from the whale's upper jaw and resemble a moustache (except that the baleen is on the *inside* of their mouths). Toothed whales

---

[6] The effects on human physiology in response to diving into the marine environment are discussed in the chapter-opening feature in Chapter 1, "Introduction to Planet 'Earth.'

have one external nasal opening (blowhole), while baleen whales have two external openings.

Suborder Odontoceti All members of the toothed whales posses prominent teeth. The toothed whales are predators that feed mostly on smaller fish and squid, although the killer whale is known to feed on a variety of larger animals, including other whales. The toothed whales form complex and long-lived social groups. Although both toothed and baleen whales can emit and receive sounds, the ability to use sound is best developed in the toothed whales.

It has long been known that cetaceans make a variety of sounds, despite their lack of vocal cords. Sounds are emitted from the blowhole or, in sperm whales (which have the most highly evolved sound ability), near a special structure called the *museau du singe* ("monkey's muzzle") (Figure 14–23). Contractions of muscles in

these structures produce sound that reflects off the front of the skull, which is bowl shaped and resembles a radar dish. The sound passes through an organ called the **melon** (or the **spermaceti organ** in sperm whales), in which sound is concentrated by forming various shapes and sizes of lenses. In essence, the organ acts as an acoustical lens, focusing the sound. Speculations about the sounds' purpose range from **echolocation**—using sound to determine the direction and distance of objects—(clearly true) to a highly developed language (doubtful). In fact, what marine biologists know about cetacean use of sound is limited.

All marine mammals have good vision, but conditions often limit its effectiveness. In coastal waters (where suspended sediment and dense plankton blooms make the water turbid) and in deeper waters (where light is limited or absent), echolocation surely would assist pursuit of prey or location of objects.

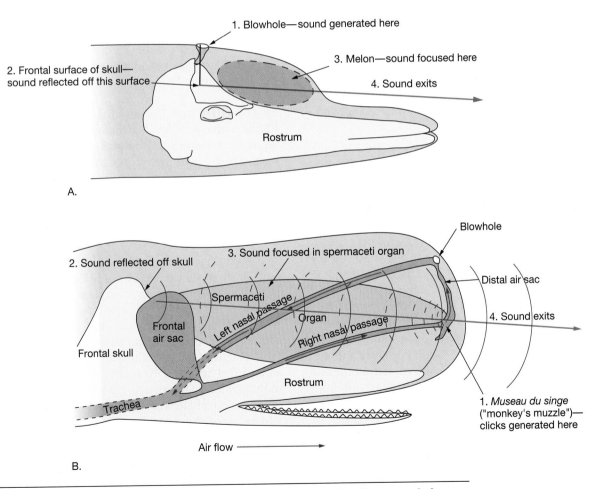

Figure 14–23 Generation of echolocation clicks in small toothed and sperm whales.
A. In small toothed whales, clicks may be generated within the blowhole mechanism, reflected off the frontal surface of the skull, and then focused by the fatty melon into a forward-directed beam. **B.** In the sperm whale, air passes from the trachea through the right nasal passage across the *museau du singe* ("monkey's muzzle"), where clicks are generated. The air passes along the distal air sac, past the closed blowhole, and returns to the lungs along the left nasal passage. The sounds are reflected off the frontal surface of the skull and are focused by the spermaceti organ before leaving the whale.

Figure 14–24 Echolo-cation.

Clicking sound signals are generated by cetaceans and bounced off objects in the ocean to determine their size, shape, distance, movement, internal structure, and density.

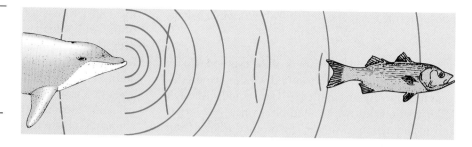

Because water transmits sound so well, toothed whales use echolocation to determine the distances of objects underwater. Some toothed whales use sound in frequencies as high as 150,000 hertz.[7] The clicks of the bottlenose dolphin (*Tursiops truncatus*) are of a frequency range that is partly audible to humans. However, some clicks are at ultrasonic frequencies (above human hearing) and are repeated up to 800 times per second.

Using lower-frequency clicks at great distance and higher frequency at closer range, the bottlenose dolphin can detect a school of fish at distances exceeding 100 meters (330 feet). It can pick out an individual fish 13.5 centimeters (5.3 inches) long at a distance of 9 meters (30 feet). It has been estimated that sperm whales can detect their main prey, squid, from a distance of up to 400 meters (1300 feet) by use of their low-frequency scanning clicks.

To locate an object, toothed whales send sound signals through the water, some of which are reflected from various objects and are returned to the animal (Figure 14–24). To interpret these signals, these animals perform the equivalent of a remarkable mathematical feat: They determine the distance to an object by quickly multiplying the velocity at which sound travels to and returns from the object by the time required for travel, and divide this product by 2. Because sound is able to penetrate objects, the echolocation ability of toothed whales can be used to determine much more about an object than can be determined by eyesight alone. For example, an interpretation of sound received back from echolocation can be used to produce a three-dimensional image of the object's internal structure and density. Recent research indicates that the whales may also use a sharp burst of sound from a distance to stun their prey.

Once the sound hits an object, some of the sound is reflected depending on the density of the object. How the returning sound is received is not yet fully understood. In most mammals, the bony housing of the inner ear structure is fused to the skull. When submerged, sounds transmitted through the water are picked up by the skull and travel to the hearing structure from many directions. This makes it impossible for such mammals to locate accurately the source of the sounds. Thus, such a hearing structure would not work for an animal that depends on echolocation to find objects accurately in water.

All cetaceans have evolved structures that insulate the inner ear housing from the rest of the skull. In toothed whales, the inner ear is separated from the rest of the skull and surrounded by an extensive system of air sinuses (cavities). The sinuses are filled with an insulating emulsion of oil, mucus, and air and are surrounded by fibrous connective tissue and venous networks. In many toothed whales, it is believed that sound is picked up by the thin, flaring jawbone and passed to the inner ear via the connecting oil-filled body.[8]

Suborder Mysticeti Baleen whales are much larger than toothed whales, which is primarily a result of their different food sources. Baleen whales eat lower on the food web, mostly devouring zooplankton such as krill and small nektonic organisms, which are relatively abundant in the marine environment. The main obstacle that baleen whales face in obtaining food is how to concentrate their small prey items and how to separate them from seawater.

To accomplish this, baleen whales do not have any teeth. Instead, they have rows of baleen (Figure 14–25A) in their mouth that hang from their upper jaw. Baleen is made of keratin—the same substance of which human nails and hair is composed—and, depending on the species of baleen whale, can be up to 4.3 meters (14 feet) long[9] (Figure 14–25B). The fibrous plates of baleen act as an ideal strainer. To feed, baleen whales fill their mouths with water that includes their prey items, allowing their pleated lower jaw to balloon in size. The whales force the water out between the baleen slats, trapping small fish, krill, and other plankton on the inside of their mouths. Mostly, baleen whales feed at or near the surface, sometimes working together coopera-

---

[7] One hertz equals one cycle per second. Youthful human ears are sensitive to frequencies from about 16 to 20,000 hertz.

---

[8] To simulate this, try pushing the end of a vibrating tuning fork into your chin. The sound is transmitted through your jaw directly to your ear.

[9] Baleen (also called whalebone) was used for such items as buggy whips and corset stays before synthetic materials were substituted.

Box 14–2
Killer Whales: A Reputation Deserved?

Killer whales (*Orcinus orca*) are sleek and powerful predatory animals that inhabit all oceans of the world. Once considered ferocious maneaters, it has been discovered that killer whales are social animals that don't harm people. So, how did they receive the name "killer"?

Killer whales prey upon many kinds of sea life. The killer whales' diet is comprised mainly of fish. When fish are not plentiful, some feed on other dolphins and even large whales, as well as on penguins, birds, squid, turtles, seals, or sea lions (Figure 14C). It is likely that the name "killer" comes from how other marine mammals have been hunted by killer whales. Killer whales often hunt in groups, employing various strategies, including deception, to trap their prey. In fact, their distinctive black-and-white coloration—a type of **disruptive coloration**—and white eye patch often confuse their prey as killer whales close in for the kill. Killer whales have often been called "wolves of the sea" because of their group hunting tactics.

Another interesting behavior of killer whales is that they often play with their food before killing it. Similar to the way a cat plays with a mouse, killer whales have been known to hone their hunting skills by releasing their prey only to catch it over and over again. Sometimes, they will throw their prey completely out of the water. Certainly, this behavior appears cruel, and it undoubtedly contributed to the whale's common name.

Killer whales are not maneaters, despite the fact that they have rammed and sunk boats many times and have even been documented to tug on divers' fins. In fact, there is only one documented case of a killer whale ever killing a human. This occurred in 1991 when trainer Keltie Byrne of the marine park Sealand of the Pacific in Victoria, Canada, accidentally slipped into the killer whale pool. As the 20-year-old champion swimmer attempted to climb out, she was pulled back into the water by a killer whale. She was then tossed about in the pool for 10 minutes. Finally, one whale carried her underwater in its mouth for longer than she could hold her breath, causing her to drown. Evidently, the killer whales saw her as a new play object and inadvertently killed her. It is interesting to note that in the wild, there has never been a documented instance where killer whales have maliciously attacked and killed a human being.

Figure 14C Killer whale (*Orcinus orca*) hunting sea lions near Valdes, Argentina.

tively in large groups (Figure 14–26). However, the gray whale has short baleen slats and feeds by filtering sediment from the shallow bottom of its North Pacific and Arctic feeding grounds, straining out a diet of benthic organisms.

Baleen whales include three families:

1. The **gray whale**, which has short, coarse baleen, no dorsal fin, and only two to five ventral grooves on its lower jaw

2. The **rorqual**[10] **whales**, which have short baleen, many ventral grooves, and are divided into two subfamilies:

   a. The *balaenopterids*, which have long, slender bodies, small sickle-shaped dorsal fins, and

---

[10] The term "rorqual" refers to the longitudinal groves on the lower jaw, called rorqual folds. The term rorqual is from the Old Norse term *raudhr*, meaning "red."

Figure 14–25 Baleen.

A. Parallel rows of these fibrous baleen plates from a gray whale hang from its upper jaw, where it is used to strain mud and water from bottom-dwelling organisms. **B.** Individual slat of baleen from a northern right whale.

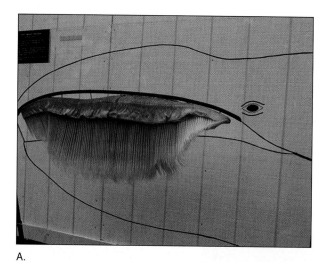

A.

B.

Figure 14–26 Cooperative feeding by humpback whales.

A large group of humpback whales surfaces with open mouths during cooperative feeding behavior. Evidently, the whales communicate with each other and work as a team to capture prey.

flukes with smooth edges (minke, Baird's, Bryde's, sei, fin, and blue whales)

b. The *megapterids*, or humpback whales, which have a more robust body, long flippers, flukes with uneven trailing edges, tiny dorsal fins, and tubercles (a row of large bumps) on the head

3. The **right whales**[11], which have long, fine baleen, broad triangular flukes, no dorsal fin, and no ventral grooves. The northern right whale is the baleen

[11] The right whales are so named because, from a whaler's point of view, they were the "right" whales to take.

whale that is most threatened with extinction. The southern right whale (Southern Hemisphere) and the bowhead whale that remains near the Arctic pack-ice edge are the other members of this family.

Baleen whales also produce sound, but at much different frequencies than toothed whales. Baleen whales produce sounds at frequencies generally below 5000 hertz. Gray whales produce pulses, possibly for echolocation, and moans that may maintain contact with other gray whales. Rorqual whales produce moans that last from one to many seconds. These sounds are extremely low in frequency, in the 10- to 20-hertz range, and are probably used to communicate over distances of up to 50 kilometers (31 miles). Blue whales produce sounds that may travel along the SOFAR channel across entire ocean basins. Moans of fin whales may be related to reproductive behavior, but much more observation is required to ascertain the message content of all these sounds.

An Example of Migration: Gray Whales

Many oceanic animals undertake migrations of various magnitudes. For instance, fish, sea turtles, and marine mammals all are known to migrate seasonally. Some of the longest migrations known in the open ocean are the seasonal migrations of baleen whales. Let's examine the migration of one of the best-studied whales—the gray whale (*Eschrichtius robustus*, often called the "California" gray whale).

Gray whales are moderate-sized whales, reaching lengths of 15 meters (50 feet) and weighing over 30 metric tons (33 short tons). They are unique among baleen whales in that they do not feed by straining pelagic crustaceans and small fish from the water; instead, they stir up bottom sediment with their snouts and strain bottom-dwelling organisms (mostly amphipods and shellfish) from the sea floor. They often take a large gulp of bottom sediment that contains their food items and wash the mud out through their short baleen plates, trapping their food inside their mouths.

Historically, the migratory routes of commercially important baleen whales such as gray whales have been well known since the mid–1800s. The paths of these and other air-breathing mammals are easy to observe, because they must surface periodically for air. In particular, gray whales spend their entire lives in nearshore areas, allowing them to be easily tracked. More recently, radio tracking of individual whales has helped to further track individuals during their migratory journey. This has allowed the documentation of the route and timing of the migration of gray whales.

Why Gray Whales Migrate   Gray whales undertake the longest known migration of any mammal, representing as much as a 22,000-kilometer (13,700-mile) round-trip journey every year. Gray whales feed in

colder high-latitude waters in the coastal Arctic Ocean and the far northern Pacific Ocean near Alaska; they breed and give birth in warm tropical lagoons along the west coast of Baja California and mainland Mexico (Figure 14–27). Feeding occurs during summer when the long hours of sunlight radiate the nutrient-rich waters to produce a vast feast of crustaceans and other bottom-dwelling organisms in the shallow Bering, Chukchi, and Beaufort Seas. Only this bountiful food enables the whales to sustain themselves during the long migrations and the mating and calving season, when feeding is minimal.

Initially, gray whales were thought to migrate so far because the physical environment of their cold-water feeding grounds does not meet the needs of young grey whales. However, recent research on the physiology of newborn gray whales indicates that gray whale calves have the ability to survive in much colder water. So why do they migrate? One new idea suggests that the migration of gray whales is a relict from the Ice Age, when sea level was lower. During this time, the present productive feeding grounds of the gray whales were above sea level. Hence, gray whales could not feast on the abundant

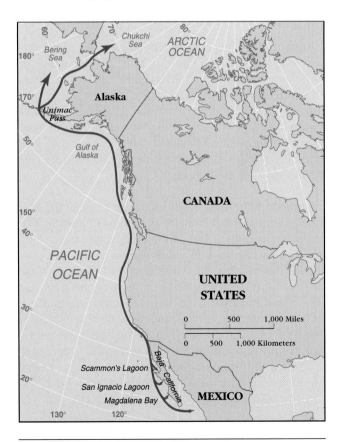

Figure 14–27 Migration route of the gray whale.
Gray whales undertake the longest annual migration of any mammal, which many be as much as 22,000 kilometers (13,700 miles) round-trip. They migrate from summer feeding grounds in the Bering and Chukchi Seas to winter breeding and calving grounds in coastal lagoons along Baja California and mainland Mexico.

Figure 14–28 A "friendly" gray whale approaches a boat full of people.

food and probably gave birth to smaller calves that could not survive in the colder water. This necessitated the migration to warmer-water regions, which continues to this day in spite of more abundant food supply.

An alternative explanation for leaving the colder waters to calve is to avoid killer whales. Killer whales are more numerous in colder waters and are a major threat to young whales. This may also help explain why only those lagoons with shallow entrances in Mexico are used as calving areas. However, killer whales have been seen near the lagoons and have also been observed to feed in water so shallow that they come completely out of the water (see Figure 14C).

Timing of Migration The migration route of the gray whale demonstrates how closely whale migration patterns are linked to physical oceanographic conditions. The whales' migration usually begins in September, after the fall bloom in productivity has peaked. Because of the abundant food supply in their feeding grounds, the whales are able to store enough fat reserves to last them until they return to high-latitude waters. However, gray whales have been observed to feed during their migration when the opportunity presents itself. When pack ice begins to form over the continental shelf areas that are their feeding grounds, the whales move south.

First to leave are the pregnant females. They are followed by a procession of nonpregnant mature females,

immature females, mature males, and then immature males. After navigating through passages between the Aleutian Islands, they follow the coast throughout their southern journey. Traveling at an average rate of about 200 kilometers (125 miles) per day, most reach the lagoons of Baja California by the end of January.

In these warm-water lagoons, the pregnant females give birth to 2-ton calves. The calves nurse from their mothers, whose milk is almost half butterfat and has the consistency of cheese, allowing the calves to put on weight quickly during the next two months. While the calves are nursing, the mature males breed with the mature females that did not bear calves. Because of the energy demands faced by female baleen whales in producing large offspring (the gestation period is up to one year) and providing them with fat-rich milk for several months, it is not uncommon for them to mate only once every two or three years. Also, it is in these lagoons that the well-known "friendly" behavior of gray whales is often observed, with whales and their calves closely approaching boats that venture into their breeding grounds (Figure 14–28).

Late in March, they return to the feeding grounds, with the procession order reversed. Most of the whales are back in the feeding grounds by the end of June, which coincides with the beginning of the spring bloom in productivity. The whales feed on prodigious quantities of bottom-dwelling organisms to replenish their depleted store of fat and blubber before the next trip south.

? Students Sometimes Ask…

Those deep-sea fish look frightening. Do they ever come to the surface? Are they related to piranhas?
Deep-sea fish live in a world of darkness and never come to the surface. That is fortunate for us, because if they ever did come

to the surface, people would be foolish to enter the ocean. Although quite small, deep-sea fish are vicious predators. However, they are only distantly related to piranhas (their only commonality is that they are within the same group of bony fishes). Their similarly adapted large, sharp teeth may be a good example of *convergent evolution*: the evolution of similar characteristics on different organisms independent of one another yet adapted to the same problem (in this case, a low food supply).

In a battle between a killer whale and a great white shark, which one would win?

Many people are fascinated with large and powerful wild animals, so this question is commonly posed. Up until recently, there was much speculation but little evidence to support either animal's claim as the "apex" or top carnivore in the ocean. In waters off northern California in 1997, the battle between a 6-meter (20-foot) juvenile killer whale and a 3.6-meter (12-foot) adult great white shark was captured on film—the first documented evidence of such a fight. The remarkable video shows that the killer whale completely severed the shark's head! If this is representative of the way these two animals interact in the wild, then the killer whale is king. It is believed that the killer whale's superior maneuverability and use of echolocation helped it conquer the shark.

Which other marine organisms are included within class Mammalia?

The only other marine mammals not within the orders Pinnipedia, Sirenia, and Cetacea are a few members of the order Carnivora (*carni* = meat, *vora* = eat). Animals that belong to order Carnivora are characterized by prominent canine teeth and include the cat and dog families on land. In the marine environment, there are two representatives: the sea otter and the polar bear.

The sea otter lacks an insulating layer of blubber but, instead, has extremely dense fur. This fur was highly sought for pelts, causing sea otters to be nearly hunted to extinction in the late 1800s. Fortunately, they have made a remarkable comeback and now inhabit most areas where they were formerly hunted. Sea otters are some of the smallest marine mammals, and seem particularly playful because they continually scratch themselves, which serves to clean their fur and also adds an insulating layer of air. They eat various shellfish and crustaceans, and often use a rock to break open the shells of their food while floating at the surface on their backs. Sea otters are a common inhabitant of kelp beds.

The polar bear has massive webbed paws that act as large paddles that allow the polar bear to be excellent swimmers. The polar bear's thick fur is hollow to allow for better insulation properties. Polar bears also have large teeth and sharp claws, which are used for prying and killing. Their main diet consists of seals, which are often captured by polar bears that patiently wait for seals to come up for a breath of air through holes in the Arctic ice.

What is the difference between a dolphin and a porpoise?

Both dolphins and porpoises are small-sized toothed whales, but there are several morphological differences. One difference is that porpoises are somewhat smaller and have a more "stout" (bulky and robust) body shape as compared to the more elongated and streamlined dolphins. Generally, porpoises have a blunt snout (*rostrum*) while dolphins have a longer rostrum. From a distance, a dolphin can be differentiated from a porpoise based on the shape of their dorsal fins: Porpoises have a smaller and more triangular (or, on one species, no) dorsal fin, whereas dolphins have a dorsal fin that is more sickle-shaped in profile and appears hooked and curved backwards (called "*falcate*").

The surest way to tell the two apart is by the shape of their teeth. The teeth of dolphins end in points, while the teeth of porpoises are blunt or flat (shovel shaped) on their ends and resemble our incisors (front teeth).

I've seen dolphins swimming and playing around just in front of the bow of a moving ship. Are they in danger of being run over by the ship?

Not really. When dolphins or porpoises are swimming in front of a moving ship, they are actually surfing the "bow wave" that forms as the ship plies through the water. Studies on trained dolphins have been conducted to monitor their respiratory rate during this activity. These studies reveal that these animals are just barely above a resting rate of activity. In essence, they are being pushed along and catching a free ride from the ship! The dolphins or porpoises will stay with the ship for as long as the ship is traveling in the direction they're going—or until a food source is located—which may be for an hour or more. Not only do dolphins and porpoises ride the bow waves of ships, they have also been observed to ride the bow waves of large baleen whales, which is a spectacular sight!

Sometimes, large groups of dolphins or porpoises have been observed to ride a ship's bow wave, moving back and forth in front of the ship and using echolocation to keep track of each other. Since these marine mammals have the ability to swim much faster than ships, they are in no real danger of being run over by the ship. However, there have been a few recorded instances where they have been inadvertently hit by a ship. This is probably due to the animals becoming either too brave or too careless.

How intelligent are toothed whales?

There may not be a definitive answer to that question. Certainly, the following facts about toothed whales (suborder Odontoceti) imply a certain level of intelligence:

- They communicate with each other by using sound.
- They have large brains relative to their body size.
- The amount of convolution of their brain is quite high—a characteristic shared by many organisms that are considered to have highly developed intelligence (such as humans and other primates).
- Some dolphins have been reported to assist drowning humans in the wild.
- Some dolphins have been trained to respond to hand signals and do certain tricks on command (such as retrieve objects).

It is clear that odontocetes have remarkable abilities, but this does not necessarily imply intelligence. For instance, pigeons—animals not known for having high intelligence—have also been trained to retrieve objects by using hand signals. Perhaps many of us would like to think that whales and dolphins are more intelligent than they really are because humans feel an attachment to these playful, seemingly ever-smiling, air-breathing creatures. It is interesting to note that even experts in the field of animal intelligence disagree on how to assess *human* intelligence accurately, let alone that of a marine mammal.

If the large brain that odontocetes have is not an indication of intelligence, then why is their brain so large? Leading whale researches don't exactly know, but it might be because odontocetes need a large brain to process the wealth of information they receive from the sound echoes they transmit. Because intelligence is difficult to measure, perhaps it is best to say that animals of suborder Odontoceti are tremendously well adapted to the marine environment.

Table 14–2   World whale populations.

Species	Pre-whaling population	Present population	Percent of initial population
Right	100,000	2,000	2
Blue	>196,000	>9,000	<5
Bowhead	>55,000	8,000	<15
Humpback	150,000	30,000	20
Sei	>105,000	25,000	<24
Fin	>464,000	119,000	<26
Minke	175,000	86,000	<50
Sperm	2,770,000	1,810,000	65
Gray	<20,000	22,000	110

Are gray whales still an endangered species?
Although many other whales are considered endangered (Table 14–2), the North Pacific gray whale was removed from the endangered species list in 1993 when their numbers exceeded 20,000, which surpassed the estimated size of their population prior to whaling. Other populations of gray whales were not so fortunate. For instance, the gray whales that used to inhabit the North Atlantic Ocean were hunted to extinction several centuries ago, and the population of gray whales that live in the waters near Japan may also have recently gone extinct.

These slow-moving whales of suborder Mysticeti spend most of their lives in coastal waters, which made them easy targets for whalers. In the mid–1800s, they were traced to their birthing and breeding lagoons and were hunted to the brink of extinction. A common strategy was to harpoon a calf and then its mother when she came to its rescue. At this time, gray whales were known as "devilfish" because they would often capsize small whaling boats when the adults came to the protection of their young. By the late 1800s, the number of gray whales had diminished to the point where they were difficult to find during their annual migration. Fortunately, these low numbers also made it difficult for whalers to hunt them successfully.

In 1938, the International Whaling Treaty banned the taking of gray whales, which were thought to be nearly extinct. This protection has allowed them steadily to increase in number to this day and to become the first marine creature to be removed from endangered status. Their repopulation is truly one of the most impressive success stories of how protecting animals can ensure their continued survival. What is perhaps most surprising is the friendly behavior that they now show toward people in boats (see Figure 14–28) in the same lagoons where they were hunted over 150 years ago. Devilfish, indeed!

Summary

Pelagic animals that comprise the majority of the ocean's biomass remain mostly within the upper surface waters of the ocean, where their primary food source exists. Those animals that are not planktonic (floating forms such as microscopic zooplankton) depend on buoyancy or their ability to swim to help them remain in food-rich surface waters.

Rigid gas containers found in some cephalopods and the expandable swim bladders of some fishes are adaptations that help increase buoyancy. Other organisms have adaptations to maintain their positions near the surface that include gas-filled floats (such as those of the Portuguese man-of-war) and soft bodies that lack high-density hard parts (such as the jellyfish).

Nekton—squid, fish, and marine mammals—are strong swimmers that depend on their swimming ability to avoid predators and to obtain food. Squid swim by trapping water in their body cavities and forcing it out through a siphon. Most fish swim by creating a wave of body curvature that passes from the front to the back of the fish and provides a forward thrust.

The caudal (rear) fin is the most important in providing thrust, whereas the paired pelvic and pectoral (chest) fins are used for maneuvering. The dorsal (back) and anal fins serve primarily as stabilizers. A rounded caudal fin is flexible and can be used for maneuvering at slow speeds. The lunate fin is rigid, is of little use in maneuvering, and is very efficient in producing thrust for fast swimmers such as tuna.

Lungers (such as groupers) are fish that sit motionless and make short, quick passes at passing prey. They have mostly white muscle tissue that has low concentrations of myoglobin. Alternatively, cruisers (such as tuna) are fish that swim constantly in search of prey and possess mostly red, myoglobin-rich muscle tissue. Red muscle fibers have a greater affinity for oxygen and tire less rapidly than white fibers, enabling tuna to maintain rapid cruising speeds.

Fish swim at different rates depending on whether they are cruising, hunting for prey, or escaping from predators. Although most fish are cold blooded, the fast-swimming tuna, *Thunnus*, is homeothermic, meaning that it maintains a body temperature well above water temperature.

A large number of marine organisms such as fish, squid, and crustaceans exhibit a behavior called schooling. It is likely that schooling is an adaptation that provides a better chance of avoiding predation than swimming alone.

The small biomass of deep-water nekton feed on detritus and each other. They have special adaptations—such as good sensory devices and bioluminescence—that allow them to survive in this still and completely dark environment.

The characteristics that differentiate marine mammals from fish are that marine mammals are warm blooded; they breathe air; they have hair or fur; they bear live young; and the females have mammary glands. There is good fossil evidence that indicates marine mammals evolved from land-dwelling animals about 50 million years ago. Marine mammals fall mostly into three orders: order Pinnipedia, order Sirenia, and order Cetacea.

Marine mammals within order Pinnipedia have prominent skin-covered flippers and include walruses, seals, sea lions, and fur seals. Representatives of order Sirenia have toenails, very few hairs, a vegetarian diet, and include manatees and dugongs (the "sea cows").

Perhaps the best-adapted mammals for life in the open ocean are those of the order Cetacea (whales, dolphins, and porpoises). Cetaceans have special adaptations that allow them to have highly streamlined bodies so that they can be fast swimmers. Other adaptations—such as being able to absorb 90 percent of the oxygen in air they inhale, store large quantities of oxygen, reduce the use of oxygen by noncritical organs, and collapse their lungs below depths of 100 meters (330 feet)—allow them to achieve deep dives without the effects of nitrogen narcosis.

Cetaceans are divided into two suborders: suborder Odontoceti (the toothed whales) and suborder Mysticeti (the baleen whales). Odontocetes are highly skilled in echolocation, which is used for finding their way through the ocean and locating prey. The clicking sounds emitted by whales are bounced off objects, and the animal can determine the size, shape, internal structure, and distance of the objects by the nature of the returning signals and the time elapsed.

Mysticetes include the largest whales in the world, which eat by separating their small prey from seawater using their baleen plates as a strainer. Baleen whales include the gray whale, the rorqual whales, and the right whales.

Gray whales migrate from their cold-water summer feeding grounds in the Arctic to warm, low-latitude lagoons in Mexico during winter for breeding and birthing purposes. This behavior may have evolved to allow their young to be born into warm water during the Ice Age, when lower sea level eliminated highly productive Arctic feeding areas.

Key Terms

Baleen (p. 434)

Bioluminesce (p. 426)

Biomass (p. 413)

Cetacea (p. 430)

Chaetognath (p. 417)

Coelenterate (p. 417)

Cold-blooded (p. 423)

Copepod (p. 416)

Cruiser (p. 421)

Ctenophore (p. 417)

Cutaneous artery (p. 424)

Cutaneous vein (p. 424)

Deep-sea fish (p. 426)

Detritus (p. 426)

Disruptive coloration (p. 437)

Dorsal aorta (p. 424)

Echolocation (p. 435)

Filter feeder (p. 424)

Foraminifer (p. 414)

Fur seal (p. 429)

Gray whale (p. 437)

Hemoglobin (p. 431)

Homeothermic (p. 424)

Krill (p. 416)

Lunger (p. 421)

Melon (p. 435)

Myoglobin (p. 421)

Myomere (p. 418)

Mysticeti (p. 434)

Odontoceti (p. 434)

Photophore (p. 426)

Pinnipedia (p. 429)

Poikilothermic (p. 423)

Radiolarian (p. 414)

Right whale (p. 438)

Rorqual whale (p. 437)

School (p. 425)

Scyphozoan (p. 417)

Sea lion (p. 429)

Seal (p. 429)

Siphonophore (p. 417)

Sirenia (p. 430)

Spermaceti organ (p. 435)

Swim bladder (p. 413)

Tunicate (p. 417)

Walrus (p. 429)

Warm-blooded (p. 424)

Questions And Exercises

1. What contributions did Alexander Agassiz make to advancing the study of the marine environment?

2. Discuss how the rigid gas chamber in cephalopods may be more effective in limiting the depth to which they can descend than the flexible swim bladders of some fish.

3. Draw and describe different types of microscopic zooplankton.

4. Describe the body form and lifestyle of the following macroscopic plankton: krill, Portuguese man-of-war (*Physalia*), jellyfish, tunicates, ctenophores, and arrowworms.

5. Name and describe the different types of fins that fish exhibit. What are the five basic shapes of caudal fins, and what are their uses?

6. Do humans have more to fear about being attacked by a shark, or do sharks have more to fear about being killed by a human?

7. What are the major structural and physiological differences between the fast-swimming cruisers and lungers that patiently lie in wait for their prey?

8. Are most fast swimming fish cold-blooded or warm-blooded? What circulatory system modifications do these fish have to minimize heat loss?

9. What are several benefits of schooling?

10. What are the two food sources of deep-water nekton? List several adaptations of deep-water nekton that allow them to survive in their environment.

11. What common characteristics do all organisms in class Mammalia share?

12. Describe marine mammals within the order Pinnipedia, including their adaptations for living in the marine environment.

13. How can true seals be differentiated from the eared seals (sea lions and fur seals)?

14. Describe the marine mammals within the order Sirenia, including their distinguishing characteristics.

15. List the modifications that are thought to allow some cetaceans the ability to (a) increase their swimming speed; (b) dive to great depths without suffering the bends; and (c) stay submerged for long periods.

16. Describe differences between cetaceans of the suborder Odontoceti (toothed whales) with those of the suborder Mysticeti (baleen whales). Be sure to include examples from each suborder.

17. Describe the process by which the sperm whale may produce echolocation clicks.

18. Discuss how sound may reach the inner ear of toothed whales.

19. Describe the mechanism by which baleen whales feed.

20. Discuss possible reasons why gray whales leave their cold-water feeding grounds during the winter season.

References

Bonner, N. 1989. *Whales of the World.* New York: Facts-on-File.

Carey, F. G. 1973. Fishes with warm bodies. *Scientific American* 228:2, 36–44.

Carwardine, M. 1995. *Whales, Dolphins and Porpoises: A Visual Guide to All the World's Cetaceans.* New York: Dorling Kindersley.

Denton, E. J., and Shaw, T. I. 1962. The buoyancy of gelatinous marine animals. *Journal of Physiology* 161:14P–15P.

Doak, W. 1989. *Encounters with Whales and Dolphins.* Dobbs Ferry, NY: Sheridan House.

Ellis, R. 1996. *Deep Atlantic.* New York: Alfred A. Knopf.

George, D., and George, J. 1979. *Marine Life: An Illustrated Encyclopedia of Invertebrates in the Sea.* New York: Wiley-Interscience.

Giesbrecht, W. 1892. *Fauna und Flora des Golfes von Neapel und der Angrenzenden Meeres—Abschnitte.* Berlin: Verlag Von R. Friedländer and Sohn.

Gingerich, P. D. 1994. The whales of Tethys. *Natural History* 103:4, 86–88.

Idyll, C. P., Ed. 1969. *The Science of the Sea: A History of Oceanography.* London: Thomas Y. Crowell.

Kanwisher, J. W., and Ridgway, S. H. 1983. The physiological ecology of whales and porpoises. *Scientific American* 248:6, 110–121.

Klimley, P A. 1996. The predatory behavior of the white shark, *in* Pirie, R. G., ed., *Oceanography: Contemporary Readings in Ocean Sciences,* 3rd ed. New York: Oxford University Press.

Kunzig, R. 1990. Invisible garden. *Discover* 11:4, 66–74.

Lalli, C. M., and Parsons, T. R. 1993. *Biological Oceanography: An Introduction.* New York: Pergamon Press.

Leatherwood, S., and Reeves, R. R. 1983 *The Sierra Club Handbook of Whales and Dolphins.* San Francisco: Sierra Club Books.

MacGinitie, G. E., and MacGinitie, N. 1968. *Natural History of Marine Animals.* 2nd ed. New York: McGraw-Hill.

Mrosovsky, N., Hopkins-Murphy, S. R., and Richardson, J. I. 1984. Sex ratio of sea turtles: Seasonal changes. *Science* 225:4663, 739–740.

Norris, K. S., and Harvey, G. W. 1972. A theory for the function of the spermaceti organ of the sperm whale (*Physeter catodon* L), *in Animal Orientation and Navigation,* 397–417. Washington, DC: National Aeronautics and Space Administration.

Obee, B. 1992. The great killer whale debate: Should captive orcas be set free? *Canadian Geographic* 112:1, 20–31.

Pike, G. C. 1962. Migration and feeding of the gray whale (*Eschrichtius gibbosus*). *Journal of the Fisheries Research Board of Canada* 19:815–838.

Reeves, R. R., Stewart, B. S., and Leatherwood, S. 1992. *The Sierra Club Handbook of Seals and Sirenians.* San Francisco: Sierra Club Books.

Thewissen, J. G. M., Hussain, S. T., and Arif, M. 1994. Fossil evidence for the origin of aquatic locomotion in archaeocete whales. *Science* 263:5144, 210–212.

Thorson, G. 1971. *Life in the Sea.* New York: McGraw-Hill.

Waller, G., ed. 1996. *SeaLife: A Complete Guide to the Marine Environment.* Washington, DC: Smithsonian Institution Press.

Würsig, B. 1989. Cetaceans. *Science* 244:4912, 1550–1557.

Suggested Reading

Earth

Livermore, B. 1997. Life behind the cold wall. 6:6, 50–56. Examines evidence that suggests that fish near Antarctica developed a form of antifreeze when Antarctica approached a position near the South Pole millions of years ago.

Sea Frontiers

Bachand, R. G. 1985. Vision in marine animals. 31:2, 68–74. An overview of the types of eyes possessed by marine animals.

Bleecker, S. E. 1975. Fishes with electric know-how. 21:3, 142–148. A survey of fishes that use electrical fields to navigate, to capture prey, and to defend themselves.

Bushnell, P. G., and Holland, K. N. 1989. Tunas: Athletes in a can. 35:1, 42–48. A discussion of the physiology of various fast-swimming tuna species.

Hersh, S. L. 1988. Death of the dolphins: Investigating the East Coast die-off. 34:4, 200–207. The 1987–1988 East Coast die-off of dolphins is discussed by a marine mammalogist.

Kleen, S. 1989. The diving seal: A medical marvel. 35:6, 370–374. The physiology of seals that allows them to dive deep and stay submerged for long periods of time is covered.

Klimley, A. P. 1976. The white shark: A matter of size. 22:1 2–8. Describes procedure used by Dr. John E. Randall for determining the size of sharks from the perimeter of the upper jaw and height of teeth.

Lineaweaver, T. H., III. 1971. The hotbloods. 17:2, 66–71. Discusses the physiology of the bluefin tuna and sharks that have high body temperatures.

Maranto, G. 1988. The Pacific walrus. 34:3, 152–159. A summary of the natural history of walruses of the Bering Sea and the role of humans as their predators.

McAuliffe, K. 1994. When whales had feet. 40:1, 20–33. Details the work of paleontologists in the Egyptian desert, where they found the remains of a 45-million-year-old whale that still possessed feet.

O'Feldman, R. 1980. The dolphin project. 26:2, 114–118. A description of the response of Atlantic spotted dolphin to musical sounds.

Reeve, M. 1971. The deadly arrowworm. 17:3, 175–183. This important group of plankton is described in terms of body structure and lifestyle.

Volger, G., and Volger, S. 1988. Northern elephant seals. 34:6, 342–347. This article deals primarily with the history of the elephant seal populations off the coast of California and Baja California.

Scientific American

Denton, E. 1960. The buoyancy of marine animals. 303:11, 118–128. How various marine animals use buoyancy to maintain their position in the water column.

Gosline, J. M., and DeMont, M. E. 1985. Jet propelled swimming in squids. 252:1, 96–103. By using jet propulsion resulting from expelling water through their siphons, squids can move as fast as the speediest fishes.

Gray, J. 1957. How fishes swim. 197:2, 48–54. The roles of musculature and fins in the swimming of fishes.

Horn, M. H., and Gibson, R. N. 1988. Intertidal fishes. 258:1, 64–71. A discussion of a wide variety of fishes that inhabit tide pools.

Kooyman, G. L. 1969. The Weddell seal. 221:2, 100–107. The lifestyle and problems related to this mammal's living in water permanently covered with ice.

Martin, K., Shariff, K., Psarakos, S., and White, D. J. 1996. Ring bubbles of dolphins. 275:2, 82–87. Shows how dolphins emit bubble-shaped rings from their blowholes and use them as play toys.

O'Shea, T. J. 1994. Manatees. 271:1, 66–73. Presents information about how manatees are thought to have evolved from the same ancestors as elephants and aardvarks and discusses threats to their survival.

Rudd, J. T. 1956. The blue whale. 195:6, 46–65. A description of the ecology of the largest animal that ever lived.

Sanderson, S. L., and Wassersug, R. 1990. Suspension-feeding vertebrates. 263:3, 96–102. A discussion of the ecology of filter-feeding fishes and cetaceans.

Shaw, E. 1962. The schooling of fishes. 206:6, 128–136. A consideration of theories seeking to explain the schooling behavior of fishes.

Whitehead, H. 1985. Why whales leap. 252:3, 84–93. An attempt to answer the enigmatic question, "Why do whales breech?"

Würsig, B. 1988. The behavior of baleen whales. 258:4, 102–107. A summary of the facts concerning the behavior of baleen whales.

Zapol, W. M. 1987. Diving adaptations of the Weddell seal. 256:6, 100–107. The physiological adaptations that enable the Weddell seal to make deep, long dives are discussed.

Oceanography on the Web

www.prenhall.com/thurman

Visit the *Essentials of Oceanography* home page for on-line resources for this chapter. There you will find an on-line study guide with review exercises, and links to oceanography sites to further your exploration of the topics in this chapter. *Essentials of Oceanography* is at: **http://www.prenhall.com/thurman** (click on the Table of Contents menu and select this chapter).

CHAPTER 15
ANIMALS OF THE BENTHIC ENVIRONMENT

The Debate on Life in the Deep Ocean: The Rosses and Edward Forbes

One of the first major controversies in marine science developed around the work of Sir John Ross (1777–1856), his nephew Sir James Clark Ross (1800–1862), and Edward Forbes (1815–1854)—all British naturalists and scientists. The Rosses were two of the earliest successful sounders of the deep oceans. Baffin Bay in Canada was explored by Sir John Ross in 1817 and 1818, and he was able to measure depths. Ross also collected samples of bottom-dwelling organisms and sediments with a device of his own design called a *deep-sea clamm,* which resembles a pair of metal jaws used to retrieve "bites" of the bottom mud. He recovered worms in mud samples from a depth of 1.8 kilometers (1.1 miles) and even retrieved a starfish from this depth, which had become entangled in the line that was used to lower the clamm to the bottom.

Sir James Clark Ross extended the soundings to greater depths on voyages to the Antarctic during 1839–1843 (Figure 15A). A 7-kilometer (4.3-mile) sounding line was used on these voyages, and even this great length was at times insufficient to reach the bottom of the ocean. Sir James observed that the animals he recovered from the cold waters of the Antarctic were the same species his uncle had recovered from the Arctic area, and they were found to be very sensitive to temperature increase. This discovery led him to the conclusion that the waters making up the deep ocean must be of uniformly low temperature.

During this same time, Edward Forbes was conducting oceanographic shipboard experiments to study the distribution of marine life. Forbes was interested in determining the vertical distribution of life in the ocean, and after repeated observations, he divided the sea into specific life-depth zones. He came to the conclusion that plant life is limited to the zone near the surface. Forbes also found that the animal concentration was greater near the surface and decreased with increasing depth until

Figure 15A The Ross Expedition in Antarctica.

only a small trace of life, if any, remained in the deepest waters.

Based on his sampling, Forbes announced that the abundance of animals in the ocean decreased steadily with depth. Forbes's followers either misinterpreted this statement or decided to apply their own logic to it and concluded that no life whatsoever existed in the deep ocean. They believed that life would be impossible in the deep ocean because of the high pressure that exists there and the absence of light and oxygen. Ignoring the find-ings of the Rosses in the high southern and northern latitudes, they persisted in their belief that no life exist-ed in the great depths of the ocean.

Today, it is well established that various life forms do indeed exist at all levels in the ocean, and those that are found in the deep ocean are well adapted to the condi-tions found there. As Sir James Ross expressed, "from however great a depth we may be able to bring the mud and stones of the bed of the ocean, we shall find them teeming with animal life."

Of the 235,000-plus known species that inhabit the ocean, more than 98 percent (about 230,000) live in or on the ocean floor. Ranging from the rocky, sandy, and muddy environments of the intertidal zone to the muddy deposits of the deepest ocean trench-es, the ocean floor provides a tremendously varied en-vironment. This environment is home to a diverse group of specially adapted organisms that comprise the ben-thic community.

Living at or near the interface of the ocean floor and seawater, an organism's success is closely tied to its abil-ity to cope with the physical conditions of the water, the ocean floor, and other members of the biological com-munity. The vast majority of known benthic species live on the continental shelf, where the water is often shal-low enough to allow sunlight to penetrate to the ocean bottom.

One of the most prominent variables affecting species diversification may be temperature. It can be observed that a significant difference in the number of benthic species is found at similar latitudes on opposite sides of an ocean basin because of the effect of ocean currents on coastal water temperature. For example, along the coast of Europe, where the Gulf Stream warms the water from the Spanish coast to the northern tip of Norway, over three times the number of benthic species exist than thrive along the similar latitudinal range of the Atlantic coast of North America, where the Labrador Current cools water as far south as Cape Cod, Massachusetts.

The distribution of benthic **biomass**[1] (Figure 15–1) closely mirrors the distribution of photosynthetic pro-ductivity in the surface waters (compare with Fig-ure 13–4). This suggests that life on the ocean floor is very much dependent upon the primary photosyn-thetic productivity within the ocean's surface waters. However, there are some exceptions to this: Coral reefs possess the greatest known diversity in the marine en-vironment, yet they exist in waters with the most lim-ited supply of nutrients. Let's examine some of the benthic environments that provide such an abundance of diversity.

Rocky Shores

Rocky shorelines are characterized by organisms that live on the surface of the ocean floor, called **epifauna** (*epi* = upon, *fauna* = animal), which are either perma-nently attached to the bottom (such as marine algae) or moving over it (such as crabs). These organisms have spe-cial adaptation to withstand the rigors of life on rocky shores (Table 15–1). In spite of many adverse conditions, rocky shores are typically teeming with life, as is often seen in tide pools.

A typical rocky shore (Figure 15–2A) can be divided into different environmental zones. The **spray zone**, above the spring high-tide line, is covered by water only during storms. The **intertidal zone** lies between the high and low tidal extremes. Along most shores, the intertidal zone can be clearly separated into three subzones (Figure 15–2):

- The **high tide zone**, which is relatively dry and is covered only by the highest high tides
- The **middle tide zone**, which is alternately covered by all high tides and exposed during all low tides
- The **low tide zone**, which is usually wet but is ex-posed during the lowest low tides

Along rocky shores, these divisions of the intertidal zone can be clearly delineated based on the sharp bound-aries between populations of organisms that attach them-selves to the bottom. Because each centimeter of the rocky shore has a significantly different character from the centimeter above and below it, evolution has pro-duced organisms with abilities to withstand very specif-ic degrees of exposure to the atmosphere. Consequently, the most finely delineated biozones in the marine envi-ronment can be found along rocky shores.

Diversity of species that inhabit rocky shores varies widely. Overall, rocky intertidal ecosystems have a mod-erate diversity of species compared to other benthic en-vironments. Interestingly, the greatest animal diversity is

---

[1] Remember that *biomass* is the mass of living organisms.

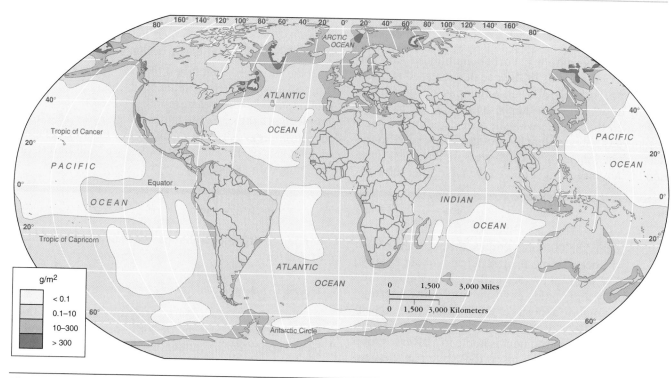

Figure 15–1 Distribution of benthic biomass.

The distribution of benthic biomass (in grams of biomass per square meter) is similar to that of photosynthetic pigment distribution (see Figure 13–4), suggesting that most of the benthic community depends directly on photosynthetic production in surface waters for nutrition. The lowest benthic biomass is beneath the centers of subtropical gyres, and the highest values are in high-latitude continental shelf areas.

Table 15–1   Adverse conditions of rocky intertidal zones and adaptations for adverse conditions

Adverse conditions of rocky intertidal zones	Adaptations for adverse conditions	Examples
Drying out during low tide	Ability to seek shelter or withdraw into shells Thick exterior or exoskeleton to prevent water loss	Sea slugs, snails, crabs
Strong wave activity	Strong holdfasts (in algae) to prevent being washed away Strong attachment threads, a muscular foot, multiple legs, or hundreds of tube feet (in animals) to allow them to attach firmly to bottom Hard structures adapted to withstand wave energy	Kelp, snails, sea stars, sea urchins
Predators occupy area during low tide	Firm attachment Stinging cells Camouflage Inking response Ability to break off body parts and regrow them later (regenerative capability)	Mussels, sea anemones, sea slugs, octopi, sea stars
Difficulty finding mates for attached species	Release of large numbers of egg/sperm into the water column during reproduction	Abalones, sea urchins
Rapid changes in temperature salinity, pH, and oxygen content	Ability to withdraw into shells to minimize exposure to rapid changes in environment Ability to exist in varied temperature, salinity, pH, and low-oxygen environments for extended periods of time	Snails, barnacles
Lack of abundant attachment sites	Organisms attach to others	Bryozoans, coral

at lower (tropical) latitudes, while the diversity of algae is greater in the mid-latitudes, which is probably due to a better nutrient supply.[2]

Spray (Supralittoral) Zone

Within the spray zone [also known as the **supralittoral** (*supra* = above, *littora* = the seashore) **zone**], one of the primary environmental conditions to which organisms are adapted is exposure out of water. Organisms that live here must avoid drying out during the long periods when this zone is above water level. Consequently, many animals that inhabit this zone have shells or other protective means to prevent drying, and there are few species of marine algae.

Throughout the world, one of the most obvious inhabitants of the rocky spray zone is the periwinkle snail (Figure 15–2B). The spray zone can easily be identified by locating the middle of the periwinkle belt, which is the boundary between the intertidal zone and the spray zone. Periwinkles of genus *Littorina* feed on marine algae by using a rasp-like structure in their mouth.

Hiding among the cobbles and boulders that typically cover the floors of sea caves well above the high tide line are rock lice or sea roaches (isopods of the genus *Ligia*). Neither name is particularly flattering for these little scavengers, which reach lengths of 3 centimeters (1.2 inches) and scurry about at night feeding on organic debris (Figure 15–2C). During the day, these creatures spend most of their time hiding in crevices.

Another indicator of the spray zone is a distant relative of the periwinkle snail, a limpet (genus *Acmaea*) with a flattened conical shell and a muscular foot with which it clings tightly to rocks (Figure 15–2D). Limpets feed on marine algae in a manner similar to periwinkles.

High Tide Zone

In the high tide zone, drying out remains a condition to which organisms must adapt. Similar to animals within the spray zone, most animals that inhabit the high tide zone have a protective covering. For example, periwinkles have a protective shell that serves them well in both the spray zone and the high tide zone. However, many other organisms cannot venture onto drier land. Unlike the periwinkles, the landward limit for buckshot barnacles (Figure 15–2E) is the high-tide shoreline. Although barnacles also have a protective shell, they are confined for two reasons that require them to be in water on a regular basis: They filter-feed from seawater, and their larval forms are planktonic.

The most conspicuous algae in the high tide zone are rock weeds, members of the genus *Fucus* in colder lati-

tudes (Figure 15–2F) and *Pelvetia* in warmer latitudes. Both have thick cell walls to reduce water loss during periods of low tide.

On a clean, rocky shore, the rock weeds are among the first organisms to colonize an area. Later, **sessile** (*sessil* = sitting on) animal forms—those that are attached to the bottom such as barnacles and mussels—begin to establish themselves, creating competition for attachment sites with the rock weeds.

Middle Tide Zone

In the middle tide zone, the constant bathing by seawater that most organisms receive results in more types of marine algae and more soft-bodied animals. Rock weeds continue to flourish in the middle tide zone, where the variety of life forms is much greater than in the high tide zone. Not only does the variety increase, but also the total biomass is much greater. There is, therefore, greater competition for rock space among sessile forms, which attach themselves to the bottom.

Among the shelled organisms that inhabit the middle tide zone are acorn barnacles (Figure 15–2G). However, the barnacle most characteristic of this zone is the goose-necked barnacle (*Pollicipes*) (Figure 15–2H), which attaches itself to the rock surface by a long, muscular neck. Even more successful than barnacles in the middle tide-zone space competition are various mussels (genera *Mytilus* and *Modiolus*). They attach to bare rock, algae, or barnacles during their planktonic stage and remain in place by means of tough threads called *byssus* threads.

Mussels are fed upon by predators such as carnivorous snails and sea stars. Two common genera of sea stars are the *Pisaster* and *Asterias*. To pry open the mussel shell and expose the edible tissue inside, sea stars exert a continuous pull on the shells by using alternating suction with their tube-like feet so that some are always pulling while others are resting. The mussel eventually becomes fatigued and can no longer hold its shell halves closed. When the shell opens ever so slightly, the sea star turns its stomach inside out, slips it through the crack in the mussel shell, and digests the mussel without having to take it into its mouth (Figure 15–3).

Perhaps the most dominant feature of the middle tidal zone along most rocky coasts is a mussel bed that thickens toward the bottom until it reaches an abrupt bottom limit. This may be so pronounced that it appears as though an invisible horizontal plane has prevented the mussels from growing below this depth. Protruding from the mussel bed will be numerous goose-necked barnacles, and concentrated in the lower levels of the bed will be sea stars browsing on the mussels. Less-conspicuous forms common to mussel beds are varieties of algae, worms, clams, and crustaceans.

Where the rock surface flattens out within the middle tidal zone, tide pools trap water as the tide ebbs. These pools support interesting microecosystems containing a

---

[2] As discussed in Chapter 13, "Biological Productivity and Energy Transfer," this increased nutrient supply is a result of the lack of a permanent thermocline in mid-latitude regions.

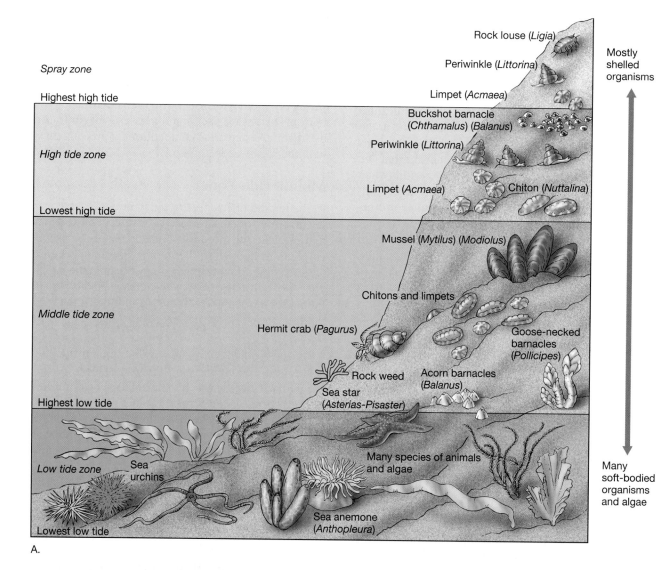

Spray zone

Highest high tide

High tide zone

Lowest high tide

Middle tide zone

Highest low tide

Low tide zone

Lowest low tide

Rock louse (*Ligia*)

Periwinkle (*Littorina*)

Limpet (*Acmaea*)

Buckshot barnacle (*Chthamalus*) (*Balanus*)

Periwinkle (*Littorina*)

Limpet (*Acmaea*)   Chiton (*Nuttalina*)

Mussel (*Mytilus*) (*Modiolus*)

Chitons and limpets

Hermit crab (*Pagurus*)

Goose-necked barnacles (*Pollicipes*)

Rock weed   Acorn barnacles (*Balanus*)

Sea star (*Asterias-Pisaster*)

Many species of animals and algae

Sea urchins

Sea anemone (*Anthopleura*)

Mostly shelled organisms

Many soft-bodied organisms and algae

A.

B.

C.

D.

E.

F.

G.

H.

I.

Figure 15–2 **The rocky shore.**

A. Zonations and typical organisms of a rocky shore (not to scale). **B.** Periwinkles (*Littorina*), spray zone to upper high-tide zone. **C.** Rock louse (*Ligia*), spray zone. **D.** Rough keyhole limpet (*Diodora aspera*) with encrusting red algae (*Lithothamnion*), high tide zone. **E.** Buckshot barnacles (*Chthamalus*), high tide zone. **F.** Rock weed (*Fucus filiformes*), middle tide zone. **G.** Acorn barnacles (*Balanus*), middle tide zone. **H.** Goose-necked barnacles (*Pollicipes*) and blue mussels (*Mytilus*), middle tide zone. **I.** Sea anemone (*Anthopleura*), low tide zone.

Figure 15–3 Sea star feeding on a mussel.

An ochre sea star (*Pisaster*) pulls apart the two halves of a mussel's (*Mytilus*) shell with its tube feet. The star then turns its own stomach inside out and forces it through the opening between the mussel's shells, where it digests the mussel's soft tissue in place inside the mussel's shell.

wide variety of organisms. The most conspicuous member of this community is often the sea anemone (see Figure 15–2I).

Sea anemones are relatives of jellyfish and live attached to the bottom. Shaped like a sack, anemones have a flat foot disk that provides a suction attachment to the rock surface. Directed upward, the open end of the sack is the only opening to the gut cavity, the mouth, surrounded by rows of tentacles (Figure 15–4). The tentacles are covered with cells that contain a stinging needle-like cell called a *nematocyst* (Figure 15–4, *inset*), which injects its victim with a potent neurotoxin. It is automatically released to penetrate any organism that brushes against its tentacles.

One of the most amusing inhabitants of tide pools is the hermit crab, *Pagurus*. With a well-armored pair of claws and upper body, hermit crabs have a soft, unprotected abdomen. They "adopt" protection for it, usually by moving into an abandoned snail shell (Figure 15–5A), which they drag around with them. They can often be seen scurrying around the tide pool area or fighting for new shells with other hermit crabs. Their abdomen has even evolved a curl to the right to make it fit properly into snail shells. Once in the snail shell, the crab can protect itself by closing off the shell's opening with its large claws.

In tide pools near the lower limit of the middle tide zone, sea urchins may be found feeding on algae (Figure 15–5B). Sea urchins have a five-toothed mouth structure centered on the bottom side of their hard spherical covering that supports many spines. The hard protective covering is composed of fused calcium carbonate plates that are perforated to allow tube feet and water to pass through. Resembling a pincushion, sea urchins have numerous spines that help protect the organism and

are also used to scrape out protective holes for themselves in rocks.

Low Tide Zone

Unlike the upper and middle tide zones, the low tide zone is dominated by algae rather than animals. This is because this zone is almost always submerged, favoring the growth of algae. A diverse community of animals does exist here, but they are less obvious because they are hidden by the great variety of marine algae and surf grass (*Phylospadix*) (Figure 15–6). The encrusting red algae (*Lithothamnion*), which are also seen in middle-zone tide pools, becomes very abundant in the lower tide pools (see Figure 15–2D). In temperate latitudes, moderate-sized red and brown algae provide a drooping canopy beneath which much of the animal life can hide during low tide.

Scampering from crevice to crevice and in and out of tide pools across the full range of the intertidal zone are various species of shore crabs (Figure 15–7). These scavengers help keep the shore clean. Shore crabs spend most of the daylight hours hiding in cracks or beneath overhangs. At night they engage in most of their eating activity, shoveling in algae as rapidly as they can tear them from the rock surface with their large front claws (chelae). Their hard exoskeleton allows these creatures to spend long periods of time out of water.

Sediment-Covered Shores

The sediment-covered shore ranges from steep boulder beaches, where wave energy is high, to the mud flats of quiet, protected embayments. However, the sediment-covered shore with which most people are familiar is the sandy beach typical of areas where wave energy usually is moderate.

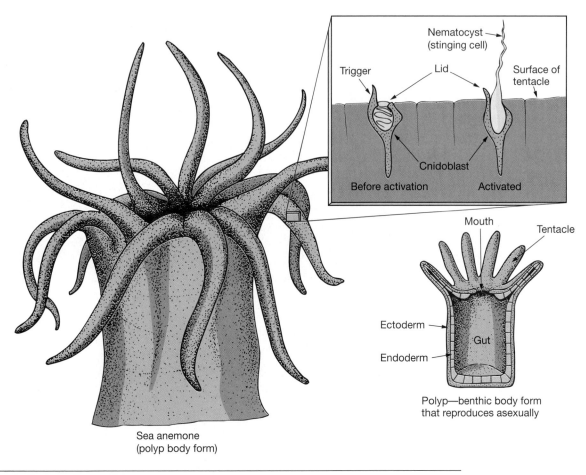

Figure 15–4 Sea anemone.

Basic body plan of a sea anemone and detail of its stinging nematocyst.

Mostly, the organisms that inhabit sediment-covered shores have the ability to burrow into the shore, and are called **infauna** (*in* = inside, *fauna* = animal). Many sediment-covered shores have a similar intertidal zonation as a rocky shore. There is much less species diversity (fewer species), but what species there are exist in great numbers (large biomass).

The Sediment

The sediment-covered shore includes *beaches, salt marshes,* and *mud flats,* which represent progressively lower-energy environments and are composed of progressively finer sediment. As the energy level diminishes,[3] particle size becomes smaller. This causes the sediment slope to be reduced, which, in turn, increases sediment stability. Thus, the sediment on a sandy beach is less stable than the fine-grained sediment found in mud flats.

You may have noticed that a large quantity of water from breaking waves is absorbed into sandy beaches. This water rapidly sinks into the sand and brings a continual supply of nutrients and oxygen-rich water for animals that live in the sediment. This readily available oxygen supply also enhances bacterial decomposition of dead tissue. Salt marshes and mud flats, however, have sediment that is not nearly so rich in oxygen. Consequently, decomposition occurs more slowly in these environments.

Intertidal Zonation

A faunal distribution exists across the intertidal range of sediment-covered shore, similar to that observed on rocky shores. The species of animals that are found in each of the corresponding intertidal zones are appropriately different. However, life forms across both the intertidal rocky and sediment-covered shores have two common characteristics: The maximum *number* of species and the greatest *biomass* are found near the low-tide shoreline, and both decrease toward the high-tide shoreline.

This zonation, shown in Figure 15–8, is best developed on steeply sloping, coarse-sand beaches and is less readily identified on the more gentle sloping, fine-sand beaches. On a mud flat, the tiny clay-sized particles form a deposit with essentially no slope, which eliminates the possibility of zonation in this protected, low-energy environment.

---

[3] The energy level is related to the strength of waves and the strength of longshore currents.

Figure 15–5 Hermit crab and sea urchins.

A. Hermit crab (*Pagurus*) has taken up residence in a *Maxwellia gemma* shell.
B. Sea urchins (*Echinus*) burrowed into the bottom of a tide pool in the lower middle-tide zone.

A.

B.

Life in the Sediment

Life on and in the sediment requires very different adaptations than on the rocky coast. The sandy beach supports fewer species than the rocky shore, and mud flats fewer still; however, the total *number* of individuals may be as high. In the low tide zone of some beaches and on mud flats, as many as 5000 to 8000 burrowing clams have been counted in only 1 square meter (10.8 square feet).

Because burrowing is the most successful adaptation for life in sediment-covered shores, life here is hidden beneath the sand and is much less obvious than in other environments. By burrowing only a few centimeters beneath the surface, organisms encounter a much more stable environment where they are not bothered by fluctuations of temperature and salinity or the threat of drying out.

Burrowing in the sediment does not prevent animals from suspension feeding or straining plankton from the clear water above the sediment. Using various techniques, the water above the sediment surface can be strained of its plankton by buried animals. Two common methods used for feeding by animals of sediment-covered shores are:

1. **Suspension feeding,**[4] where organisms that are buried in sediment use specially designed structures to filter their planktonic food items from seawater (Figure 15–9A). For instance, clams bury themselves in sediment and extend siphons through the surface. They pump in overlying water and filter suspended plankton and other organic matter from it.

2. **Deposit feeding**, where organisms feed on food items that occur as deposits. These deposits in-

---

[4] Suspension feeding is also known as *filter feeding*.

Figure 15–6 Algal colony of the low tide zone exposed during extremely low tide.

The dark-colored sea palms (a brown alga) and green surf grass of this California low-tide zone provide protection for many organisms.

A.

B.

Figure 15–7 Shore crabs.

A. Coral crab, or queen crab (Bonaire Island, Netherlands Antilles). **B.** Shore crab, *Pachygrapsis crassipes*. This female is carrying eggs (dark oval structure curled under her abdomen).

clude detritus—dead and decaying organic matter and waste products—and the sediment itself, which is coated with organic matter. Some deposit feeders, such as the segmented worm *Arenicola* (Figure 15–9B), feed by ingesting sediment and extracting organic matter from it. Others, such as the amphipod *Orchestoidea* (Figure 15–9C), feed on more concentrated deposits of organic matter (detritus) on the sediment surface.

Another method of feeding is **carnivorous feeding**. An organism that displays this feeding strategy is the sand star *Astropecten* (Figure 15–9D). It cannot climb rocks the way its sea star relatives can, but it can burrow rapid-ly into the sand, where it feeds voraciously on crustaceans, mollusks, worms, and other echinoderms.

The Sandy Beach

When you go to the beach, you do not see as many animals exposed at the surface as you do along a rocky shore. This is because there is no stable, fixed surface onto which creatures can attach, so most of them burrow into the sand and are safely hidden from view. Let's study some examples.

Bivalve Mollusks A **bivalve** (*bi* = two, *valva* = a valve) is an animal having two hinged shells, such as a clam or a mussel. A **mollusk** is a member of the phylum Mollusca

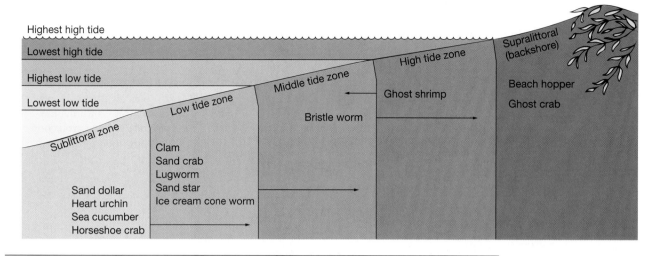

Figure 15–8 Intertidal zonation on a sediment-covered shore.
Zonation is best displayed on coarse sand beaches with steep slopes. As the sediment becomes
finer and the beach slope decreases, zonation becomes less distinct and disappears entirely on
mud flats.

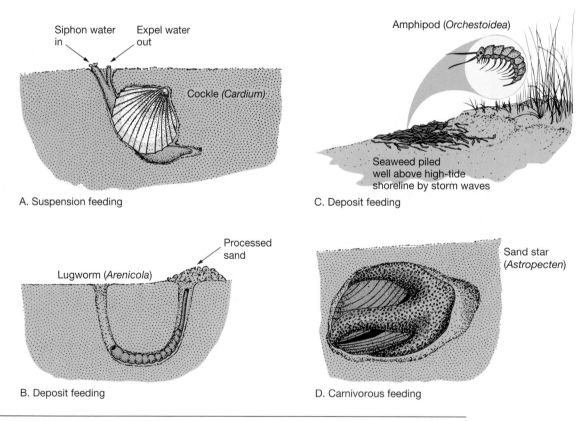

Figure 15–9 Modes of feeding along sediment-covered shores.
A. Suspension feeding by the clam (cockle) *Cardium*, which uses its siphon to filter plankton and
other organic matter that is suspended in the water. **B.** Deposit feeding by the segmented worm
Arenicola, which feeds by ingesting sediment and extracting organic matter from it. **C.** Deposit feeding by the amphipod *Orchestoidea*, which feeds on more concentrated deposits of organic matter (detritus) on the sediment surface. **D.** Carnivorous feeding of a clam by the sand star *Astropecten*.

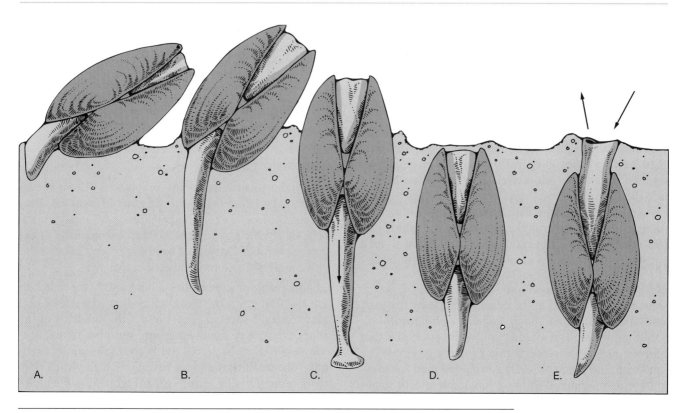

Figure 15–10 How a clam burrows.

A clam burrows into the sediment by (**A**) extending its pointed foot into the sediment and (**B**) forcing the foot deeper into the sediment and using this increasing leverage to bring the exposed, shell-clad body toward vertical. When the foot has penetrated deeply enough, a bulbous anchor forms at the tip (**C**), and a quick muscular contraction pulls the entire animal into the sediment (**D**). The siphons are then pushed up above the sediment to pump in water, from which the clam extracts food and oxygen (**E**).

(*molluscus* = soft), characterized by having a soft body and either an internal or external hard shell. Bivalve mollusks possess a soft body, a portion of which secretes calcium carbonate valves (shells) that hinge together.

Bivalve mollusks—such as clams—are well adapted for life in the sediment. A single foot digs into the sediment to pull the creature down into the sand. Their siphon extends vertically through the sediment for feeding (Figure 15–9A). The procedure used by a clam to bury itself by using its muscular foot is shown in Figure 15–10.

How deeply a bivalve can bury itself depends on the length of its siphons; they must reach above the sediment surface to pull in water from which plankton will be filtered. Oxygen is also extracted in the gill chamber before the water is expelled. Undigestible matter is forced back out the siphon periodically by quick muscular contractions. The greatest biomass of clams is burrowed into the low-tide region of sandy beaches and it decreases where the sediment becomes muddier.

Annelid Worms   A variety of **annelids**, or segmented worms, are also well adapted for life in the sediment. Most common of the sand worms is the lugworm (*Arenicola* species). It lives in a U-shaped burrow (see

Figure 15–9B), the walls of which are strengthened with mucus. The worm moves forward to feed and extends its proboscis (snout) up into the head shaft of the burrow to loosen sand with quick pulsing movements. A cone-shaped depression forms at the surface over the head end of the burrow as sand continually slides into the burrow and is ingested by the worm. As the sand passes through its digestive tract, the organic content is digested, and the processed sand is deposited at the surface (see Figure 15–9B).

Crustaceans   Staying high on the beach and feeding on kelp cast up by storm or high tide waves are numerous **crustaceans**[5] (*crusta* = shell) called *beach hoppers*. Beach hoppers are known to jump distances more than 2 meters (6.6 feet). A common genus is *Orchestoidea*, which usually ranges from 2 to 3 centimeters (0.8 to 1.2 inches) in length (see Figure 15–9C). Laterally flattened, they

---

[5] Crustaceans—such as crabs, lobsters, shrimps, and barnacles—include predominately aquatic animals that are characterized by a segmented body, a hard exoskeleton, and paired, jointed limbs.

usually spend the day buried in the sand or hidden in the kelp on which they feed. They become particularly active at night, when large groups many hop at the same time and form large clouds above the piles of seaweed on which they are feeding.

Sand crabs (*Emerita*) (Figure 15–11) are another crustacean that is a common inhabitant on many sandy beaches. Ranging in length from 2.5 to 8 centimeters (1 to 3 inches), they move within the sand up and down the beach near the shoreline. They bury their bodies into the sand and leave their long, curved, V-shaped antennae pointing up the beach slope. These little crabs filter food particles from the water, and can be located by looking in the lower intertidal zone for a V-shaped pattern in the swash as it runs down the beach face.

Echinoderms Echinoderms (*echino* = spiny, *derma* = skin) are represented in the beach deposits by the sand star (*Astropecten*) and heart urchins (*Echinocardium*). Sand stars are well adapted to prey on invertebrates that burrow into the low-tide region of sandy beaches. The sand star (see Figure 15–9D) is well designed for moving through sediment, with five tapered legs with spines and a smooth back.

More flattened and elongated than the sea urchins of the rocky shore, heart urchins live buried in the sand near the low-tide line. They gather sand grains into their mouths, where the coating of organic matter is scraped off and ingested (Figure 15–12).

Meiofauna Meiofauna (*meio* = lesser, *fauna* = animal) are small organisms that live in the spaces between sediment particles. These organisms, only 0.1 to 2 millimeters (0.004 to 0.08 inch) in length, feed primarily on bacteria removed from the surface of sediment particles. The meiofauna population, composed primarily of polychaetes, mollusks, arthropods, and nematodes (Figure 15–13), are found in sediment from the intertidal zone to the deep-ocean trenches.

The Mud Flat

Two widely distributed plants associated with mud flats are eelgrass (*Zostera*) and turtle grass (*Thalassia*). They occupy the low tide zone and adjacent shallow sublittoral regions bordering the flats. Numerous openings at the surface of mud flats attest to a large population of bivalve mollusks and other invertebrates.

Quite visible and interesting inhabitants of the mud flats are the fiddler crabs (*Uca*), which live in burrows that may exceed 1 meter (3 feet) in depth. Relatives of the shore crabs, they usually measure no more than 2 centimeters (0.8 inch) across the body. Male fiddler crabs have one small claw and one oversized claw, which is up to 4 centimeters (1.6 inches) long (Figure 15–14). Fiddler crabs get their name because this large claw is waved around in a manner such that the crab seems to be playing an imaginary fiddle. The females have two normal-sized claws. The large claw of the male is used to court females and to fight competing males.

Shallow Offshore Ocean Floor

Extending from the spring low-tide shoreline to the seaward edge of the continental shelf is an environment that is mainly sediment covered, although rocky exposures may occur locally near shore. On rocky exposures, many types of marine algae exist, which have adaptations for

Figure 15–11 Sand crab.

The head of a sand crab (*Emerita*) emerges from the sand. Sand crabs can often be located just beneath the surface.

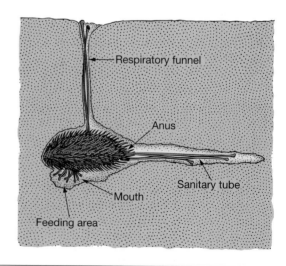

Figure 15–12 Heart urchin.

Feeding and respiratory structures of a heart urchin (*Echinocardium*), which feeds on the film of organic matter that covers sand grains.

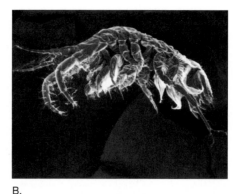

A.

B.

C.

Figure 15–13 Scanning electron micrographs of meiofauna.
A. Nematode head, magnified 804 times. The projections and pit on the right side may be sensory structures. **B.** Amphipod, magnified 20 times. This 3-millimeter-long organism builds a burrow of cemented sand grains. **C.** A 1-millimeter long polychaete (magnified 55 times) with its proboscis extended.

Figure 15–14 Life on the mud flat.

Eel-grass (*Zostera*) and fiddler crab (*Uca*) at Cape Hatteras, North Carolina.

reaching from the shallow sea floor to near the sunlit surface waters, such as gas-filled floats.

The sediment-covered shelf has moderate to low species diversity. Surprisingly, the diversity of organisms inhabiting the benthic environment is *lowest* beneath upwelling regions. This is because upwelling carries nutrients to the surface, making pelagic production great, which produces an excess of dead organic matter. When this material rains down on the bottom and decomposes, consuming oxygen, the oxygen supply can be locally depleted. A specialized shallow-water community with higher diversity is the giant kelp bed associated with rocky bottoms.

The Rocky Bottom (Sublittoral)

A rocky bottom within the shallow inner sublittoral region usually will be covered with algae. Along the North American Pacific coast, the giant brown bladder kelp (*Macrocystis*) attaches to rocks as deep as 30 meters

(100 feet) if the water is clear enough to allow sunlight to support plant growth at this depth. The kelp can extend for another 30 meters (100 feet) along the surface, which allows for good exposure to sunlight for these photosynthetic organisms.

The giant brown bladder kelp and another fast-growing kelp (bull kelp, *Nereocystis*) often form beds called **kelp forests** along the Pacific coast (Figure 15–15A). Smaller tufts of red and brown algae are found on the bottom, living on the kelp fronds. All these types of marine algae have a root-like structure called a *holdfast* that securely anchors the algae to the bottom. The holdfast is so strong that only during large storm waves do algae break free and wash up onto shore.

Kelp forests are highly productive ecosystems that provide shelter for a wide variety of smaller life forms that make their home within or directly on the larger kelp species. These organisms serve as an important food source for many of the animals living in and near the kelp

A.

B.

Figure 15–15 Life in the sublittoral zone.
A. Giant brown bladder kelp (*Macrocystis*). **B.** Sea hares (*Aplysia californica*) and sea urchins (*Echinus*) in a kelp forest.

A.

B.

Figure 15–16 Spiny and american lobsters.

A. Spiny lobster (*Panulirus interruptus*). **B.** American lobster (*Homarus americanus*).

forest. Such organisms include mollusks, sea stars, fishes, octopus, lobsters, and even marine mammals. Surprisingly, very few animals feed directly on the living kelp plant. Among those that do are the large sea hare (*Aplysia*) and sea urchins (Figure 15–15B).

Lobsters Large crustaceans—including lobsters and crabs—are common along rocky bottoms. The spiny lobsters are named for their spiny covering and have two very large, spiny antennae (Figure 15–16A). These antennae serve as feelers and are equipped with noise making devices near their base that are used in protection. The genus *Panuliris*, considered a delicacy, lives deeper than 20 meters (65 feet) along the European coast and reaches lengths to 50 centimeters (20 inches). For unknown reasons, the Caribbean species *Panulirus argus* sometimes migrates single-file across the sea floor in lines that are several kilometers long.

Panulirus interruptus is the spiny lobster of the American West Coast. All spiny lobsters are taken for food, but none are as highly regarded as the so-called true lobsters (genus *Homarus*), which include the American lobster, *Homarus americanus* (Figure 15–16B). Although they are scavengers like their spiny relatives, the true lobsters also feed on live animals, including mollusks, crustaceans, and members of their own species.

Oysters Oysters are thick-shelled sessile (anchored) bivalve mollusks found in estuarine environments. They prefer a steady flow of clean water to provide plankton and oxygen.

Oysters are food for a variety of sea stars, fishes, crabs, and boring snails that drill through the shell and rasp away the soft tissue of the oyster (Figure 15–17). In fact, this may be one of the main reasons that oysters have such a thick shell.[6] Oysters also have great commercial importance to humans throughout the world as a food source.

Oyster beds are composed of empty shells of many previous generations that are cemented to a hard substrate or to one another, with the living generation on top. Each female produces many millions of eggs each year, which become planktonic larvae when fertilized. After a few weeks as plankton, the larvae settle to attach themselves to the bottom. As a material upon which to anchor, the oyster larvae prefer (in order) live oyster shells, dead oyster shells, and rock.

Coral Reefs

Individual corals—called **polyps**—are small benthic marine animals that feed with stinging tentacles and are related to jellyfish. Most species of corals are about the size of an ant, live in large colonies, and construct hard calcium carbonate structures as protection. Coral species are found throughout the ocean, but coral accumulations that are classified as **coral reefs** are restricted to shallow warmer-water regions.

---

[6] It is interesting to note that for every type of armor or defense possessed by one species, a weapon is evolved by another species to defeat it.

Figure 15–17 An oyster drill snail feeding on an oyster.

An oyster drill snail drills through the shell of an oyster with a rasp-like mouthpiece to feed on the soft tissue beneath the shell.

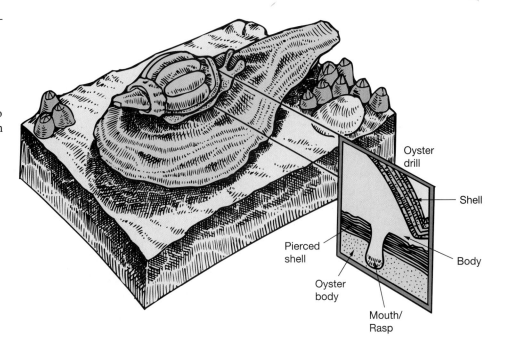

To survive, corals need water where the average monthly temperature exceeds 18 degrees centigrade (64 degrees Fahrenheit) throughout the year (Figure 15–18). Corals are very temperature sensitive: Water that is warmer than 30 degrees centigrade (86 degrees Fahrenheit) will cause many corals to die. Warmer-than-normal sea-surface water temperatures occur during El Niño events, which often decimate corals in the eastern tropical Pacific and in other locations.

Warm water temperatures that support coral growth are found primarily within the tropics. However, reefs also grow to latitudes approaching 35 degrees north and south on the western margins of ocean basins, where warm-water currents move into high-latitude areas and raise average sea surface temperatures (Figure 15–18).

Also shown in Figure 15–18 is the greater diversity of reef-building corals on the western side of ocean basins. More than 50 genera of corals thrive in a broad area of the western Pacific Ocean and a narrow belt of the western Indian Ocean. Fewer than 30 genera occur in the Atlantic Ocean, with the greatest diversity occurring in the Caribbean Sea. The explanation for this pattern of diversity may be related to past or present ocean circulation patterns.

Besides warm water, other environmental conditions that allow for coral growth include:

- Strong sunlight (not for the corals themselves, which are animals and can exist in deeper water, but for a symbiotic photosynthetic type of algae called **zooxanthellae** that live within the coral's tissues[7])

- Strong wave or current action (to bring nutrients and oxygen)

- Lack of turbidity (Suspended particles in the water tend to interfere with the coral's filter-feeding capability and absorb radiant energy, so corals are not usually found close to areas where major rivers drain to the sea.)

- Salt water (Corals will die if the water is too fresh, which is an additional reason why coral reefs are absent near the mouths of freshwater rivers.)

- A hard substrate for attachment (Corals cannot attach onto a muddy bottom, so coral often build upon the hard skeletons of their dead ancestors, creating coral reef structures that are several kilometers in thickness.)

Because of changes in wave energy, salinity, water depth, temperature, and other less obvious factors, there is a well-developed vertical and horizontal zonation of the reef slope (Figure 15–19). These zones can be readily identified based on the types of coral present and the assemblages of other organisms found in and near the reef.

Symbiosis of Coral and Algae "Coral reefs" are more than just coral. Algae, mollusks, and foraminifers make important contributions to the reef structure. Individual reef-building corals are **hermatypic** (*herma* = secret, *typi* = type). This means that they have a mutualistic relationship[8] with a type of microscopic algae called zooxanthellae, which live within the tissue of the coral polyp. The algae provide their coral host with a continual supply

---

[7] It is zooxanthellae (*zoo* = animal, *xantho* = yellow, *ella* = small) algae that give corals their distinctive bright coloration (which can be many colors besides yellow).

[8] See Box 13–2 for a discussion of the types of symbiosis, including mutualism.

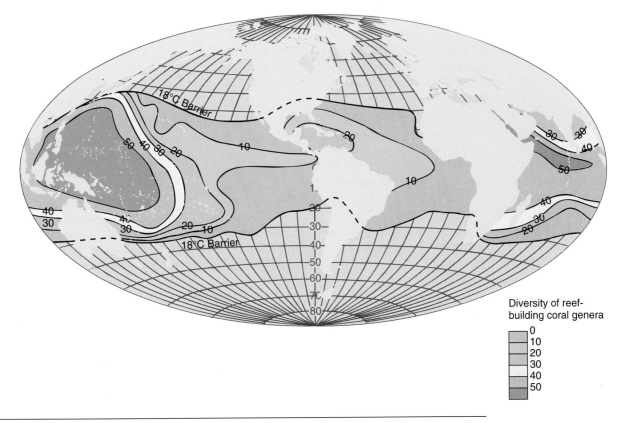

Diversity of reef-
building coral genera

0
10
20
30
40
50

Figure 15–18 Coral reef distribution and diversity.

Coral reef development is restricted to the low-latitude area between the two 18°C (64°F) temperature lines shown. In each ocean basin, the coral reef belt is wider and the diversity of coral genera is greater on the western side, which may be related to past or present ocean circulation patterns.

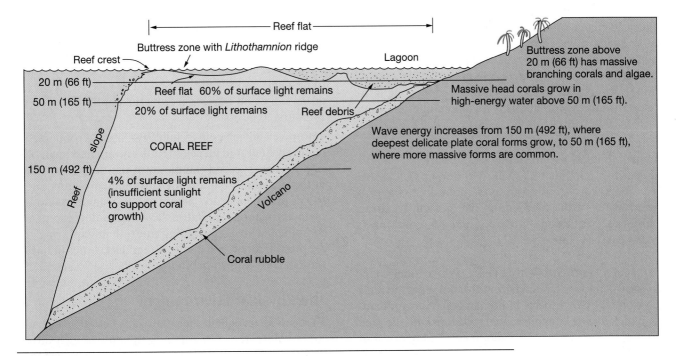

Figure 15–19 Coral reef zonation.

As depth increases, wave energy decreases and light intensity decreases. Massive branching corals occur above 20 meters (66 feet), where wave energy is great. Corals become more delicate with increasing depth, until a depth of around 150 meters (500 feet), where there is too little solar radiation to allow the survival of their symbiotic zooxanthellae algae.

of food, and the corals contribute to the mutual relationship by providing nutrients to the zooxanthellae. Although coral polyps capture tiny planktonic food with their stinging tentacles, most reef-building corals receive the majority of their nourishment from their symbiotic zooxanthellae algae. In this way, corals are able to survive in the nutrient-poor waters characteristic of the tropical oceans.

Other reef animals also have a symbiotic relationship with various types of marine algae similar to that exhibited by reef-building corals. Those that derive part of their nutrition from their algae partners are called **mixotrophs** (*mixo* = mix, *tropho* = nourishment), and include coral, foraminifers, sponges, and mollusks (Figure 15–20). The algae not only nourish the coral but also may contribute to their calcification capability by extracting carbon dioxide from the coral's body fluids.

Coral reefs actually contain up to three times as much algal biomass as animal biomass. The zooxanthellae account for less than 5 percent of the reef's overall algal mass; most of the rest is filamentous green algae. However, zooxanthellae account for up to 75 percent of the biomass of reef-building corals and provide the coral with up to 90 percent of its nutrition.

Because of the sunlight requirement for the algae, the greatest depth to which active coral growth extends is 150 meters (500 feet), below which there is not enough sunlight. Water motion is less at these depths, so relatively delicate plate corals can live on the outer slope of the reef from 150 meters (500 feet) up to about 50 meters (165 feet), where light intensity exceeds 4 percent of the surface intensity (see Figure 15–19).

From 50 meters (164 feet) to about 20 meters (66 feet), the strength of water motion from breaking waves increases on the side of the reef facing into the prevailing current flow. Correspondingly, the mass of coral growth and the strength of the coral structure supporting it increase toward the top of this zone, where light intensity exceeds 20 percent of surface value.

The reef flat may have a water depth of a few centimeters to a few meters at low tide. A variety of colorful reef fish inhabits this shallow water, as well as sea cucumbers, worms, and a variety of mollusks. In the protected water of the reef lagoon live gorgonian coral, anemones, crustaceans, mollusks, and echinoderms of great variety (Figure 15–21).

Coral Reefs and Nutrient Levels

When human populations increase in lands adjacent to coral reefs, the reefs deteriorate. Many aspects of human behavior can damage a reef: fishing, trampling, boat collisions with the reef, sediment increase due to development, and removal of reef inhabitants by visitors. One of the more subtle effects is the inevitable increase in the nutrient levels of the reef waters due to sewage discharge and farm fertilizers.

As nutrient levels increase in reef waters, the dominant benthic community changes:

- At low nutrient levels, hermatypic corals and other reef animals that contain algal symbiotic partners thrive.
- As nutrient levels in the water increase, moderate nutrient levels favor development of fleshy benthic plants, and high nutrient levels favor suspension feeders such as clams.
- At high nutrient levels, the phytoplankton mass exceeds the benthic algal mass, so benthic populations that are tied to the phytoplankton food web dominate.

The clarity of water is reduced by increased phytoplankton biomass. The fast-growing members of the phytoplankton-based ecosystem destroy the reef structure in two ways: by overgrowing the slow-growing coral and through **bioerosion**, which is erosion of the reef by organisms. Bioerosion by sea urchins and sponges is particularly damaging to many coral reefs.

The Deep-Ocean Floor

The vast majority of the ocean floor lies permanently submerged below several kilometers of water. Less is known about life in the deep ocean than about life in any of the shallower nearshore environments because of the great difficulty and expense associated with investigating the deep sea. Just to obtain samples from the deep-ocean floor requires a specially designed submersible or a properly equipped research vessel that has a spool of high-strength cable at least 12 kilometers (7.5 miles) long.

Collecting samples by submersible or with a biological dredge is a time-consuming process. Because of the need for oxygen, manned submersibles can stay submerged for only 12 hours, and it may take eight of those hours to descend and ascend. To send a dredge to the deep-ocean floor and retrieve it from the depths takes about 24 hours.

However, advances in technology—such as robotics and remotely operated vehicles (ROVs)—are making it possible to observe and sample even the deepest reaches of the ocean more easily. Because they are unmanned, ROVs are cheaper to operate and can stay beneath the surface for months if necessary. These developments, combined with a continuing interest in exploring the deep-ocean floor, will allow further discoveries in one of Earth's least-known habitats.

The Physical Environment

The deep-ocean floor includes the bathyal, abyssal, and hadal zones.[9] Here, the physical environment is much different than at the surface, but quite stable and homogeneous.

---

[9] The bathyal, abyssal, and hadal zones are described in Chapter 12, "The Marine Habitat."

A.

Figure 15–20 Coral reef inhabitants that depend on symbiotic algae.

A. Coral polyps, which are nourished by internal zooxanthellae algae and also by extending their tentacles to capture tiny planktonic organisms from the surrounding water. **B.** The blue-gray sponge *Niphates digitalis* (*left*), and the brown sponge *Angelas* (*right*), which contain symbiotic algae or bacteria. **C.** A giant clam, *Tridacna gigas*, which depends on symbiotic algae living within its mantle tissue.

B.

C.

A.

B.

Figure 15–21 Non-reef building inhabitants.
A. This puffer (*Arothron*) is one of the greatly varied, color-ful, and interesting fishes of coral reef and other shallow-water tropical environments. Feeble swimmers, puffers can't quickly escape a predator, but expand their bodies to pro-duce a large, spherical shape. **B.** Many members of the coral family do not secrete calcium carbonate (such as reef builders do). An example is this soft gorgonian coral, shown with feeding (*purple*) polyps extending from its branches.

Box 15–1
How White I Am: Coral Bleaching and Other Diseases

Loss of color in coral reef organisms causes them to turn white, a phenomenon known as **coral bleaching**. The cause of the bleaching is due to the removal or expulsion of the coral's symbiotic partner, the zooxanthellae algae. Essentially all reef-building corals and some other reef mixotrophs are nourished by these algae, which live within their tissue and give these organisms vivid colors. The loss of this source of nourishment will turn the coral white, giving the coral a bleached appearance. Figure 15B shows the bleaching of a round starlet coral (*Siderastrea siderea*) on Enrique Reef, Puerto Rico. Normally a rusty brown color, this coral has been bleached across the left side of this picture.

Coral bleaching has occurred locally numerous times in the past. For instance, Florida's coral reefs have recorded eight widespread bleaching episodes since 1911. However, a mass mortality of at least 70 percent of the corals along the Pacific Central American coast occurred as a result of a bleaching episode associated with the severe El Niño–Southern Oscillation (ENSO) event of 1982–1983. This eastern Pacific bleaching was probably caused by water temperatures that were much higher than normal. For instance, coral reefs around the Galápagos Islands thrive in ocean water at or below 27 degrees centigrade (81 degrees Fahrenheit). But if the water is even 1 or 2 degrees centigrade (2 or 4 degrees Fahrenheit) warmer for an extended period of time, the coral tissue may expel the algae, in effect "bleaching" itself. The warming during 1982–1983 was so severe and long-lived that two species of Panamanian coral became extinct during this event. The bleaching episode of 1987 was especially severe in Florida and throughout the Caribbean, and also affected other coral reefs worldwide. Since then, widespread episodes of bleaching have been occurring with increasing frequency and intensity. For instance, the warming associated with the ENSO event of 1997–1998 created water temperatures several degrees higher than normal has been blamed for causing coral bleaching in the equatorial eastern Pacific Ocean, as well as the Yucatàn coast, the Florida Keys, and the Netherlands Antilles.

Scientists are still trying to understand why the symbiotic alga and coral become separated. There seems to be a high correlation between abnormally high surface water temperatures (such as El Niño events) and coral bleaching, but other reasons have also been advanced to explain the phenomenon. For instance, it has been suggested that the algae may leave their host due to some undetermined condition, such as pollution, elevated ultraviolet radiation levels, changes in salinity, invasion of disease, or a combination of factors.

Still, the single factor that is most consistently linked with coral bleaching is high water temperatures. Some researchers believe that during times of warm temperatures, excess oxygen builds up in the coral's tissues and becomes toxic. This causes the algae to be expelled or, perhaps, the algae leave with dead tissue. The strong correlation between coral bleaching and elevated water temperature concerns scientists, some of whom believe that coral bleaching may be one of the first oceanic indications of global warming. Whatever the reason, experts agree that coral bleaching indicates that the coral is suffering severe environmental stress.

Bleaching can occur as quickly as overnight. The coral does not die immediately, but is weakened and does not grow. In this state, the coral may be more susceptible to other diseases. Recovery from a minor bout of bleaching can take as little as four weeks, but severe bouts can take corals as long as four years to recover. If the coral does not regain its symbiotic zooxanthellae algae, it will eventually die.

John Porter—a coral reef ecologist at the University of Georgia—and his colleagues are studying diseases that affect corals and have discovered many new diseases. They set up 160 stations along the Florida Keys in 1995 to study the health of the reef as part of the Environmental Protection Agency's Coral Reef Monitoring Program.

A disease called *white plague* had first been observed in the 1970s. It was thought to have disappeared but it has returned in a more deadly form (Figure 15C). Twelve other diseases with ominous-sounding names have also been indentified. From 1996 to 1997 the number of stations with infected corals increased from 25 to 94, and of 44 species monitored, those affected increased from nine to 28. One disease called *black band disease* (Figure 15D) appears to be associated with a cyanobacterium. In addition, a fungus (*Aspergillus*) attacks soft corals such as *Gorgonia flabellum* (Figure15E). Other diseases that have appeared in recent years are *white band disease* (Figure 15F) and *white pox* (Figure 15G), both of which were discovered by Porter in 1996. Other diseases include *yellow-band disease* (*yellow-blotch disease*), *patchy necrosis*, and *rapid wasting disease*. In most cases, once the coral polyps are killed by a disease, only the white calcium carbonate skeleton of the coral remains.

The cause of most of these diseases is still being investigated, and it is not know if the diseases are the result of living organisms—bacteria, viruses, or fungi—or environmental stresses. As human population has increased along the Florida Keys, the coral reefs of the Keys have begun to show signs

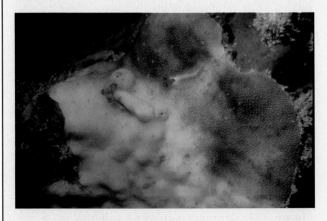

Figure 15B Bleached coral.

(continued)

of stress, thus making them more susceptible to a host of diseases. It is suspected that the increased nutrient levels and water turbidity resulting from soil runoff and improper sewage disposal from as far north as Fort Myers, Florida, are contributing to the problem. The Keys, in fact, have over 20,000 illegal septic tanks in operation.

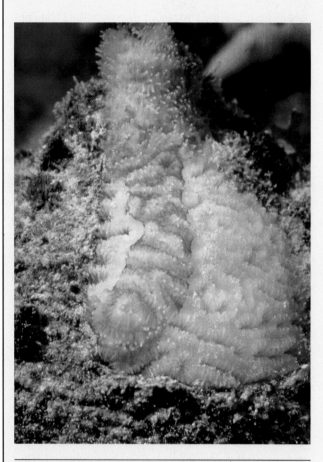

Figure 15C *Dendrogyra cylindrus* suffering from white plague disease.

Figure 15E Fungus *Aspergillus* attacks *Gorgonia flabellum*.

Figure 15D *Montastrea annularis* suffering from black band disease.

Figure 15F *Acropora cervicornis* suffering from white band disease.

(continued)

Figure 15G *Acropora palmata* suffering from white pox disease.

Light is present in only the lowest concentrations down to a maximum of 1000 meters (3300 feet), and absent below this depth. Everywhere the temperature is low, rarely exceeding 3 degrees centigrade (37 degrees Fahrenheit) and falling as low as –1.8 degrees centigrade (28.8 degrees Fahrenheit) in the high latitudes. Salinity remains at slightly less than 35‰.[10] Oxygen content is constant and relatively high. Pressure exceeds 200 kilograms per square centimeter (200 atmospheres or 2940 pounds per square inch) on the oceanic ridges, ranges between 300 and 500 kilograms per square centimeter (300 and 500 atmospheres or 4410 and 7350 pounds per square inch) on the deep-ocean abyssal plains, and attains a value of over 1000 kilograms per square centimeter (1000 atmospheres or 14,700 pounds per squre inch) in the deepest trenches.[11] Bottom currents are generally slow but more variable than once believed. For instance, **abyssal storms** created by warm- and cold-core eddies of surface currents affect certain areas, lasting several weeks and causing bottom currents to reverse and/or increase in speed.

Much of the deep-ocean floor is covered by at least a thin layer of sediment. On abyssal plains and in deep trenches, sediment is composed of mud-like abyssal clay deposits. The accumulation of oozes—composed of dead planktonic organisms that have sunk through the water column—occurs on the flanks of oceanic ridges and rises. On the continental rise, there may be some coarse sediment from nearby land sources. Sediment may be absent near the crests of the oceanic ridges and rises and down the slopes of seamounts and oceanic islands, where basaltic ocean crust forms the bottom.

Food Sources and Species Diversity

Because of the lack of light, primary productivity by photosynthesis cannot occur. With the exception of chemosynthetic productivity that occurs around hydrothermal vents, all benthic organisms receive their food from the surface waters above. Thus, the low availability of food—not low temperature, low oxygen concentration, or high pressure—is what limits deep-sea benthic biomass. It has been estimated that only 1 to 3 percent of the food produced in the euphotic zone is transferred to the deep-ocean floor. However, there may be some seasonal variability in the supply of food associated with spring or fall blooms at the surface. The potential food sources for deep-sea organisms are outlined in Figure 15–22.

Because of environmental conditions much different than near the surface, it was believed for many years that the species diversity of the deep-ocean floor was quite low compared with that of shallow-water communities. However, a 1989 study of sediment-dwelling animals in the North Atlantic revealed an unexpectedly large diversity of species. An area of 21 square meters (225 square

---

[10] Remember that average surface seawater salinity is 35 parts per thousand (‰).

[11] The pressure is 1 atmosphere (1 kilogram per square centimeter) at the ocean surface and increases by 1 atmosphere for each 10 meters (33 feet) of depth. Thus, a pressure of 1000 atmospheres is 1000 times that at the ocean's surface

Box 15–2
How Long Would Your Remains Remain on the Sea Floor?

What happens to people who are buried at sea? How long do their remains remain on the sea floor? How long do the remains of a large organism such as a whale remain on the sea bottom? Oceanographers who study deep-sea biocommunities have conducted experiments in the deep sea to help answer questions such as these.

One such experiment was conducted along the sea floor in the Philippine Trench at a depth of 9600 meters (31,500 feet) in 1975. Several whole fish were used as bait and lowered onto the sea floor at the end of a long cable. An underwater camera positioned above the bait took a picture every few minutes to observe how long the fish remains remained on the sea floor (Figure 15H). After only a few hours on the sea floor, the bait had been discovered by *Hirondellea gigas*, a scavenging benthic amphipod (a shrimp-like animal). After nine hours, the bait was swarming with amphipods. The time-lapse photos show that the bait was stripped of flesh in as little as 16 hours! Other studies reveal similar results and imply similar-sized organisms—including humans—would be expected to have their soft tissue devoured within a day on the deep ocean floor.

For deep-sea organisms, large food falls represent an unpredictable but intense nutrient supply that arrives periodically on the sea floor. Evidently, deep-sea scavengers such as amphipods, hagfish, and sleeper sharks are well equipped with special chemoreceptive sensory devices that allow them to identify and quickly locate food items on the sea floor. It is believed that a common type of food fall is a dead whale. The frequency of whale falls remains a mystery, but the carcasses of whales discovered on the sea floor have been recently found to support a thriving ecosystem of benthic organisms, including some species that inhabit hydrothermal vents.

To test how long a whale's remains remain on the sea floor, researchers used two dead gray whales that washed up on the beach in southern California in 1996 and 1997. With permission from the National Marine Fisheries Service, the whales were intentionally weighted and sunk in the San Diego Trough. At regular intervals, the whale carcasses were visited by researchers in a deep-diving submersible. This study revealed that the 5000-kilogram (5.5-ton) gray whales were completely stripped of flesh by flesh-eating scavengers in four months. Other studies indicate that even the largest baleen whales—blue whales, which can weigh more than 25 times that of a gray whale—are stripped of flesh in as little as six months after arriving on the deep ocean floor, leaving just bones.

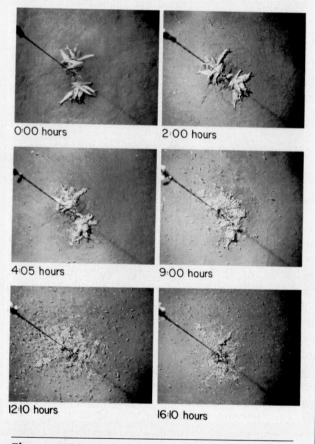

Figure 15H Time-sequence photography of fish remains on the deep-ocean floor.

feet) contained 898 species, of which 460 of these were new to science. After analyzing 200 samples, new species were being revealed at a rate that suggested millions of deep-sea species!

It has recently been established that there is high species diversity in the deep sea, especially with regard to small infaunal deposit feeders. As more samples have been obtained from the deep sea and more new species have been described, it becomes more apparent that these deep-sea areas seem to rival the species diversity present in highly diverse terrestrial environments such as tropical rain forests. However, it also appears that the distribution of deep-sea life is patchy and varies to a large degree on the presence of certain microenvironments.

Deep-Sea Hydrothermal Vent Biocommunities

An active hydrothermal (*hydro* = water, *thermo* = heat) vent field on the ocean floor was visited for the first time in 1977 during a dive of the submersible *Alvin*. The field exists in complete darkness in water depths below 2500 meters (8200 feet) in the Galápagos Rift, near the Equator in the eastern Pacific Ocean (Figures 15–23 and 15–24). Water temperature in the vicinity of the vents was 8 to 12 degrees centigrade (46 to 54 degrees Fahrenheit), whereas normal bottom-water temperature at these depths is about 2 degrees centigrade (36 degrees Fahrenheit).

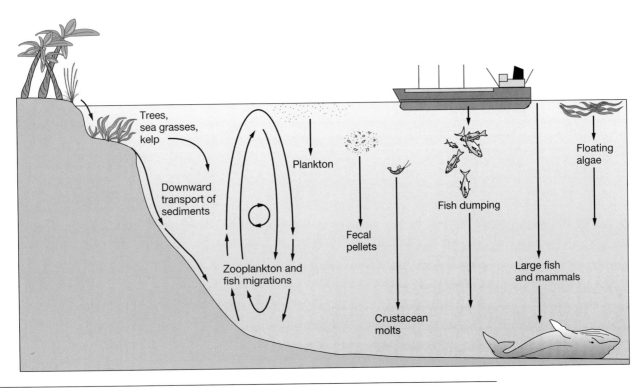

Figure 15–22 Potential food sources for deep-sea organisms.

Figure 15–23 *Alvin* approaches a hydrothermal vent community.

Schematic view of a hydrothermal vent area, showing lava pillows and a black smoker that spews hot (350°C, or 662°F), sulfide-rich water from its metallic sulfide chimney. Organisms (counterclockwise from *Alvin*) include the grenadier fish (or rattail fish), octacoras, a sea anemone, white brachyuran crabs, large clams (*Calypotogena*), and tube worms (*Riftia*).

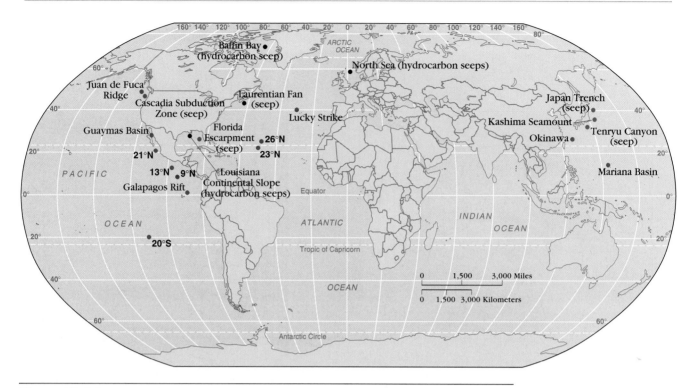

Figure 15–24 Vents and seeps known to support biocommunities.
Locations of selected hydrothermal vents (*red*), cold seeps (*blue*), and hydrocarbon seeps (*black*).

What was perhaps the most surprising discovery was that these vents were found to support the first known **hydrothermal vent biocommunities**, consisting of unusually large organisms for these depths. Most of these organisms were unknown to science. The most prominent organisms are tube worms over 1 meter (3.3 feet) long, giant clams up to 25 centimeters (10 inches) in length, large mussels, two varieties of white crabs, and extensive microbial mats. These hydrothermal vents biocommunities had up to 1000 times more biomass than the normal deep-ocean floor. In a region of scarce nutrients and small populations of organisms, these hydrothermal vents are truly the oases of the deep ocean.

At 21 degrees north latitude on the East Pacific Rise, south of the tip of Baja California, tall underwater chimneys were found to belch hot vent water (350 degrees centigrade or 662 degrees Fahrenheit) so rich in metal sulfides that it colored the water black. These chimney vents, first observed in 1979, were composed primarily of sulfides of copper, zinc, and silver and came to be called **black smokers**. A new species of vent fish was first observed here.

The most important members of these hydrothermal vent biocommunities are the microscopic archaea, which are simple bacteria-like life forms. Through chemosynthesis of sulfur compounds, they manufacture organic molecules that feed the community and are the base of the food web around the vents. The archaea use the chemical energy released by the oxidation of sulfur to produce carbohydrates from water and carbon dioxide, much as marine algae use the sun's energy to carry on photosynthesis. In essence, these archaea produce their own food chemosynthetically by combining inorganic nutrients dissolved in the deep-ocean water. Although some animals feed directly on archaea and larger prey, many of them depend primarily on a symbiotic relationship with archaea. For instance, tube worms and giant clams found here depend entirely on sulfur-oxidizing archaea that live symbiotically within their tissues (Figure 15–25).

In 1981, humans in a submersible first visited the Juan de Fuca Ridge biocommunity offshore of Oregon. Although vent fauna at this site are less abundant than at the Galápagos Rift and on the East Pacific Rise, the metallic sulfide deposits from the vents aroused much interest because they are the only active hydrothermal vent deposits in U.S. waters.

In 1982, the first hydrothermal vents beneath a thick layer of sediment were discovered during a submersible dive in the Guaymas Basin of the Gulf of California. In this region, a spreading center is actively working to rift apart the sea floor as it is being covered with sediment. Sediment samples recovered in this region were high in sulfide and saturated with hydrocarbons, which may have entered the food chain through bacterial uptake. The abundance and diversity of life discovered here may exceed that of the rocky bottom vents where these communities were first described, including those along the East Pacific Rise.[12]

---

[12] The reason the East Pacific Rise is so rocky and lacks sediment is because the sea floor is so young. In essence, there hasn't been enough time for sediment to accumulate there.

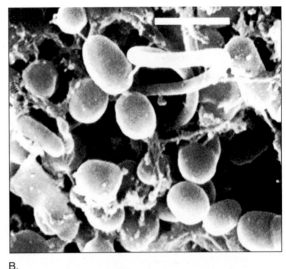

B.

C.

Figure 15–25 Chemosynthetic life.
A. Tube worms up to 1 meter long (3.3 feet) are found at the Galápagos Rift and other deep-sea hydrothermal vents. **B.** Sulfur-oxidizing bacteria (enlarged 20,000 times; white bar at top is 1 micron) that live symbiotically within the tissue of tube worms, clams, and mussels found at hydrothermal vents. **C.** A hydrothermal vent community from the Mariana Back-Arc Basin includes a new genus and species of sea anemone (*Marianactis bythios*), a new family, genus, and species of gastropod (*Alviniconcha hessleri*, the first known snail to contain chemosynthetic bacteria), and the galatheid crab (*Munidopsis marianica*).

Geologically, a situation similar to that of the Guaymas Basin occurs in the Mariana Basin of the western Pacific: a small spreading-center system beneath a sediment-filled basin. A research dive in a submersible in 1987 revealed many new species of hydrothermal vent organisms (Figure 15–25C). Subsequent exploration has revealed numerous hydrothermal vent biocommunities in other parts of the Pacific Ocean, including one in the Southern Hemisphere on the East Pacific Rise at 20 degrees south latitude (see Figure 15–24).

In 1985, the first active hydrothermal vents with associated biocommunities in the Atlantic Ocean were discovered at depths below 3600 meters (11,800 feet) near the axis of the Mid-Atlantic Ridge between 23 and 26 degrees north latitude. The predominant fauna of these vents consists of shrimp that have no eye lens but can detect levels of light emitted by the black smoker chimneys that are not visible to the human eye (Figure 15–26).

In 1993, a hydrothermal vent community was discovered on a flat-topped volcano rising to 1525 meters (5000 feet)—well above the walls of the Mid-Atlantic Ridge rift valley. Called the "Lucky Strike" vent field, it is about 1000 meters (3300 feet) shallower than most other sites. It is the only Mid-Atlantic Ridge site to possess the mussels common at many other vent sites and is the only location where a new species of pink sea urchin has been found.

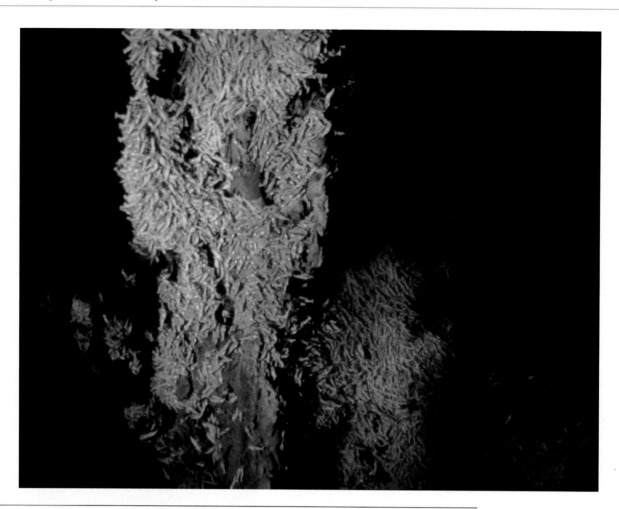

Figure 15–26 Atlantic Ocean hydrothermal vent organisms.
Swarm of particulate-feeding shrimp, the predominant animals observed at hydrothermal vents near 26 degrees north latitude on the Mid-Atlantic Ridge.

Life Span of Hydrothermal Vents Recent evidence indicates that changes occur quickly at hydrothermal vents and that vent fields are sometimes short-lived. For instance, a hydrothermal vent field called the Coaxial Site along the Juan de Fuca Ridge offshore of Washington that had been active was revisited a few years later and found to be inactive. Inactive sites such as this one are identified by an accumulation of large numbers of dead hydrothermal vent organisms. It appears that when the vent becomes inactive and the hydrogen sulfide that serves as the source of energy for the community is no longer available, organisms of the community die if they cannot move elsewhere.

Other sites indicate an increase in the amount of volcanic activity. For example, at a site along the East Pacific Rise known as Nine-Degrees North, a large number of tube worms were essentially cooked by lava flowing into their midst in what has been described as a "tube-worm barbecue." The discovery of newly formed and ancient vent areas along spreading centers indicates that hydrothermal vents can suddenly appear or cease to operate.

Although some hydrothermal vent fields may remain active for several decades, the life span of some hydrothermal vents may be only a few years or even as short as a few months. This relatively short time span is not surprising, considering the intense tectonic activity associated with mid-ocean ridge spreading centers. Evidently, hydrothermal vent areas die out and new ones appear quite regularly.

The lifestyles of many of these hydrothermal vent organisms are well adapted to the temporary nature of hydrothermal vents. For example, most organisms have high metabolic rates, which cause them to mature rapidly so that they can reproduce quickly while the vent is still active.

Studies of several hydrothermal vent sites suggest that there is low species diversity here. In fact, only a little more than 300 animal species have been found to date within this very short-lived environment. Interestingly, many species are common to widely separated hydrothermal vent fields. It is well established that hydrothermal vent animals typically disperse as water-borne larvae that drift with currents. However, it remains a mystery how these organisms are able to move between areas that are separated by as much as several hundred kilometers of environment that is highly inhospitable to them.

One idea, called the "dead whale hypothesis," suggests that when large animals such as whales die they may sink onto the deep-ocean floor and provide a temporary stepping-stone between hydrothermal vent fields. Decomposition of a whale carcass provides an energy source for these organisms, which breed and release their larvae, some of which make it to the next hydrothermal vent field. Other researchers believe that deep-ocean currents are strong enough to transport drifting larvae to new sites. Still others have suggested that the rift valleys of mid-ocean ridges may have acted as passageways along which drifting larvae of hydrothermal vent organisms have traversed to inhabit new vent fields. By whatever means these organisms travel, they are able to colonize new hydrothermal vents soon after the vents are created. An example of this was documented with the discovery of a newly formed hydrothermal vent in 1989 along the Juan de Fuca Ridge. Although no life forms were apparent at the time, a visit to the same site in 1993 revealed that tube worms and other life forms had already established themselves.

Hydrothermal Vents and the Origin of Life
Interest in exploring hydrothermal vents has intensified recently because of the possibility that these areas may have been some of the first regions where life became established on Earth. Life is thought to have begun in the oceans, and environments similar to that of the vents must have been present in the early history of the planet. The uniformity of conditions and abundant energy source of hydrothermal vents has lead some scientists to speculate that hydrothermal vents would have provided an ideal habitat for the origin of life. In fact, hydrothermal vents may represent one of the oldest life-sustaining environments, since hydrothermal activity occurs wherever there are both volcanoes and water. The recent discovery of a new domain of life at hydrothermal vents—the archaea, which have ancient genetic makeup—helps support this idea. Indeed, the process that gives birth to new ocean floor may also be responsible for the development of one of Earth's most unique characteristics: that of life itself.

Low-Temperature Seep Biocommunities

Three additional submarine seep environments—locations where water trickles out of the sea floor—have been found that chemosynthetically support biocommunities similar to hydrothermal vent communities.

Hypersaline Seeps
In 1984, a hypersaline seep was studied in water depths below 3000 meters (9800 feet) at the base of the Florida Escarpment in the Gulf of Mexico (Figure 15–27A). The water from this seep had a salinity of 46.2‰ but its temperature was not warmer than normal. Researchers discovered a **hypersaline seep biocommunity** similar in many respects to the hydrothermal vent communities. The seeping water appears to flow from fractures at the base of a limestone escarpment (Figure 15–27B) and move out across the clay deposits of the abyssal plain at a depth of about 3200 meters (10,500 feet).

The hydrogen sulfide-rich waters support a number of white microbial growths called mats, which support chemosynthesis in a fashion similar to archaea at hydrothermal vents. These and other chemosynthetic archaea may provide most of the sustenance for a diverse community of animals. The community includes sea stars, shrimp, snails, limpets, brittle stars, anemones, tube worms, crabs, clams, mussels, and a few species of fish (Figure 15–27C).

Hydrocarbon Seeps
Also observed in 1984 were dense biological communities associated with oil and gas seeps on the Gulf of Mexico continental slope (Figure 15–28). Trawls at depths of between 600 and 700 meters (2000 and 2300 feet) recovered fauna similar to those observed at hydrothermal vents and at the hypersaline seep locality in the Gulf of Mexico mentioned above. Subsequent investigations identified seeps with associated communities to depths of 2200 meters (7300 feet) on the continental slope.

Carbon-isotope analysis indicates that these **hydrocarbon seep biocommunities** are based on a chemosynthetic productivity that derives its energy from hydrogen sulfide and/or methane. Microbial oxidation of methane results in the production of calcium carbonate slabs found here and at other hydrocarbon seeps (see Figure 15–24).

Subduction Zone Seeps
In 1984, a third type of submarine seep biocommunity was discovered—a **subduction zone seep biocommunity**—during one of the *Alvin's* dives to study folding of the sea floor in a subduction zone. The seep is located near the Cascadia subduction zone of the Juan de Fuca Plate at the base of the continental slope off the coast of Oregon (Figure 15–29). Here the trench is filled with sediments, which are folded into a ridge at the seaward edge of the slope. At the crest of this ridge, water slowly flows from the 2-million-year-old folded sedimentary rocks into a thin overlying layer of soft sediment on the sea floor. Eventually, the water is released from the sediment through seeps on the ocean bottom.

At a depth of 2036 meters (6678 feet), the seeps produce water that is only slightly warmer (about 0.3 degree centigrade, or 0.5 degree Fahrenheit) than seawater at that depth. The vent water contains methane that is

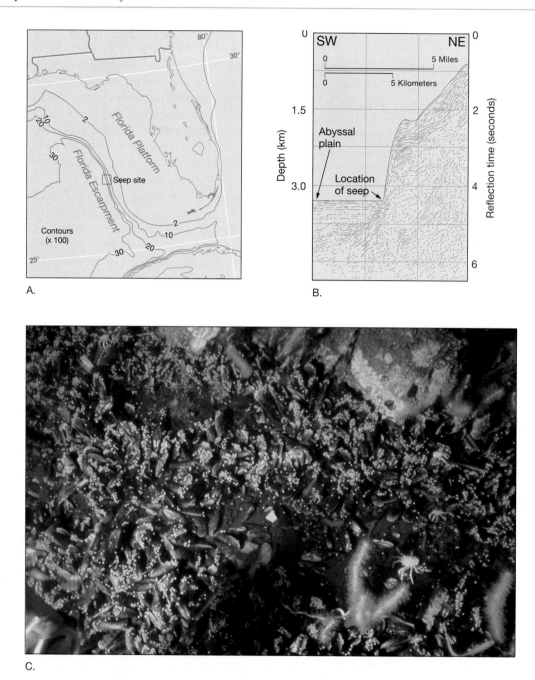

Figure 15–27 **Hypersaline seep biocommunity at base of Florida Escarpment.**
A. Location of seep and biocommunity. **B.** Seismic reflection profile of Florida Escarpment and
abyssal sediments at its base. Arrow marks location of seep. **C.** Florida Escarpment seep biocommunity of dense mussel beds. White dots are small gastropods on mussel shells. Tube worms
(*lower right*) are covered with hydrozoans and galatheid crabs.

probably produced by decomposition of organic material in the sedimentary rocks. The methane is the source
of energy for microbes that oxidize it and chemosynthetically produce food for themselves and the rest of the
community, which contains many of the same genera
found at other vent and seep sites.

Since the detection of subduction zone seeps, similar
communities have been discovered in subduction zones
in other locations, including the Japan Trench and the
Peru-Chile Trench. All these subduction zone seeps are
located on the landward side of the trenches at depths
from 1300 to 5640 meters (4265 to 18,500 feet).

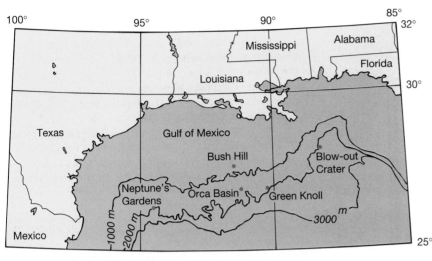

Figure 15–28 Hydrocarbon seeps on the continental slope of the Gulf of Mexico.
A. Locations of known hydrocarbon seeps with associated biocommunities.
B. Chemosynthetic mussels and tube worms from the Bush Hill seep.
C. Alaminos Canyon site (Neptune's Gardens), discovered in 1990, contains a new species of mussel and two new species of tube worms.

B.

C.

Students Sometimes Ask…

What is an urchin barren?

An urchin barren is created when the population of sea urchins goes unchecked and the urchins devour entire areas of the giant brown bladder kelp (*Macrocystis*), one of the main types of algae that create kelp forests. The sea urchins crawl across the ocean floor to new areas in search of food, and set the kelp adrift by chewing through the holdfast structure that the kelp needs to remain attached to the ocean floor. In California, the elimination of species that prey upon sea urchins (such as the wolf eel and the sea otter) has resulted in an upset in the natural balance in oceanic food webs. Consequently, sea urchins have proliferated and urchin barrens now exist where there were once lush kelp beds.

I've been at a tide pool and seen sea anemones. When I put my finger on one, it tends to gently grab my finger. Why does it do that?

The sea anemone is trying to kill you and wants to eat you (seriously!). Disguised as a harmless flower, the sea anemone is actually a vicious predator that will attack any unsuspecting animal (even a human) that is entrapped by its stinging tentacles. Fortunately, the skin on our hands is thick enough to resist the stinging nematocyst and its neurotoxin. However, a couple of people have found out the hard way how potent the neurotoxin released by sea anemones really is. Thinking that it would be fun to see if the sea anemone grabbed other things with its tentacles, they put their *tongue* into a sea anemone. After a short time, their throats swelled almost completely closed, and they had to be rushed to a hospital. They lived, but the moral of this story is: *NEVER* put your tongue into a sea anemone!

I know that coral reefs are pretty, but do they have any practical uses?

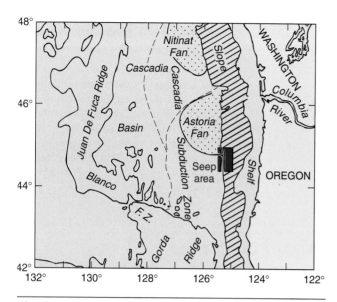

Figure 15–29 Locations of vent communities off the coast of Oregon.

These communities are associated with the Cascadia subduction zone of the Juan de Fuca Plate. Sediment filling the trench is folded into a ridge with vents at its crest.

Many. Coral reefs are amazing creations, some of which are the largest hard structures on Earth [such as the Great Barrier Reef, which stretches over 2000 kilometers (1250 miles) in length]. Coral reefs foster a diversity of species that surpasses even that of the tropical rain forests. Reefs provide shelter, food, and breeding grounds for an estimated 35,000 to 60,000 species worldwide, including almost a third of the world's estimated 12,000 kinds of marine fishes. Other species that inhabit reefs include creatures as diverse as anemones, sea stars, crabs, eels, sea slugs, clams, sharks, and sponges.

From a human perspective, they are also important to the economies of countries that have coral reefs with their waters. Many of these tropical countries receive over 50 percent of their gross national product as tourism related to reefs, which provides a much-needed incentive for these countries to protect coral reefs. In addition, pharmacologists and marine chemists have discovered a storehouse of new medical compounds that fight maladies such as cancer and infections in humans. One unique use of coral is that some human bone grafts have been fashioned from the hard calcium carbonate skeletons of coral.

I've heard that the Great Barrier Reef is plagued by an organism called the crown-of-thorns. What is it?
The crown-of-thorns (*Acanthaster planci*) is a sea star (Figure 15I). Since 1962, its proliferation has caused destruction of living coral on many reefs throughout the western Pacific Ocean. The sea star moves across reefs and eats the coral polyps. Normally, the coral can grow back provided it has enough time to do so. However, vast numbers of crown-of-thorns sea stars upset the natural balance, causing decimation of coral communities. Initially, divers were employed to smash

the crown-of-thorns, but sea stars (which have tremendous regenerating capabilities) can easily produce new individuals from various body parts, so it only made the problem worse.

Some investigators believe the proliferation of the crown-of-thorns sea star is a modern phenomenon brought about by the activities of humans. However, there is little evidence to point to such a cause. A 1989 study of the Great Barrier Reef indicated that, during the past 80,000 years, the crown-of-thorns sea star has been even more abundant on the reefs studied than it is today. If this is true, the sea star may be an integral part of the reef ecology in this region. Thus, their increase may be part of a long natural cycle rather than a destructive upstart taking advantage of human actions that have modified the reef in some way favorable to its proliferation.

What happens to the zooxanthellae algae that are released by corals that become bleached?
This has not been thoroughly researched, but it is believed that zooxanthellae algae that leave or are expelled from corals enter the stream of phytoplankton, which is largely consumed by primary consumers such as copepods and other zooplankton. It is not believed that they can exist outside of their host, so those that are not eaten probably die.

The mussel beds near hydrothermal vents are certainly extensive. Are those organisms edible?
Unfortunately, not for humans. This is because hydrogen sulfide gas (which has a characteristic "rotten egg" odor) is used as a source of energy by microbes which are at the base of the food web. Thus, sulfide—a substance that is poisonous to most organisms even at low levels—tends to be concentrated within the tissues of these organisms. Although organisms within the hydrothermal biocommunitites can ingest sulfide and have unique adaptations for getting rid of it, hydrothermal vent organisms such as mussels and clams are toxic to humans. Even if they were edible, the mussels would be difficult to harvest economically because they live in such deep water.

Figure 15I Crown-of-thorns sea star.

Summary

Over 98 percent of the more than 235,000 known marine species live in diverse environments within or on the ocean floor. Species diversity of these benthic organisms depends on their ability to adapt to the conditions of their environment, particularly temperature. With few exceptions, the biomass of benthic organisms closely matches that of photosynthetic productivity in surface waters above.

Many adverse conditions exist in the intertidal zone of rocky shores, but organisms have adaptations to allow them to densely populate these environments. Influenced by the tides, rocky shores can be divided into the high tide zone (mostly dry), middle tide zone (equally wet and dry), and the low tide zone (mostly wet). The intertidal zone is bounded by the supralittoral (covered only by storm waves) and the sublittoral, which extends below the low-tide shoreline.

Each of these zones contains characteristic types of life. The supralittoral zone is characterized by periwinkle snails, rock louses, and limpets. Sessile organisms are characteristic of the high-tide zone, especially buckshot barnacles. Algae become more abundant in the middle tide zone, and the diversity and abundance of the flora and fauna in general increase toward the lower intertidal zone. A common assemblage of rocky middle-tide zones includes acorn barnacles, goose-necked barnacles, mussels, and sea stars. The middle tide zone also houses sea anemones, fishes, hermit crabs, and sea urchins. The low tide zone in temperate latitudes is characterized by a variety of moderately sized red and brown algae providing a drooping canopy for the animal life.

Along sediment-covered shores, many varieties of burrowing infauna are common. However, compared with rocky shores, the diversity of species in sediment-covered shores is reduced. Sediment-covered shores include beaches, salt marshes, and mud flats. As is true for the rocky shore, the diversity of species and abundance of life on the sediment-covered shore increases toward the low-tide shoreline.

In more protected segments of the shore, lower levels of wave energy allow deposition of sand and mud. Sand deposits are usually well oxygenated as compared to the low oxygen content of mud deposits. An intertidal zonation exists in sediment-covered shores that is similar to that of rocky shores.

Two common methods for feeding on sediment-covered shores include suspension feeding (filtering planktonic organisms from the water) and deposit feeding (ingesting sediment and detritus). Another method is carnivorous feeding (preying directly upon other organisms). Organisms characteristic of sandy beaches include bivalve (two-shelled) mollusks, lugworms, beach hoppers, sand crabs, sand stars, and heart urchins. Organisms characteristic of mud flats include eelgrass, turtle grass, bivalve mollusks, and fiddler crabs.

Attached to the rocky sublittoral bottom just beyond the shoreline is a band of algae that often creates kelp forests. Kelp forests are the home of many organisms, including other varieties of algae, mollusks, sea stars, fishes, octopus, lobsters, marine mammals, sea hares, and sea urchins.

Spiny lobsters are common to rocky bottoms in the Caribbean and along the West Coast, and the American lobster is found from Labrador to Cape Hatteras. Oyster beds found in estuarine environments consist of individuals that attach themselves to the bottom or to the empty shells of previous generations.

Coral reefs consist of large colonies of coral polyps and many other species that need specific environmental conditions to live, such as warm water and strong sunlight. Mostly, coral reefs are found within nutrient-poor tropical waters. Reef-building corals and other mixotrophs are hermatypic, containing algae symbiotic partners (zooxanthellae) in their tissues. Delicate varieties are found at 150 meters, and they become more massive near the surface, where wave energy is higher. The potentially lethal "bleaching" of coral reefs is caused by the removal or expulsion of symbiotic algae, probably under stress of elevated temperature. Many other lethal diseases of coral are known.

The physical conditions of the deep ocean floor are much different from those of shallow water. The primary food source is from the surface waters above, which limits deep-sea benthic biomass. Even though it was believed that species diversity in the deep ocean was low, recent studies indicate that it is much higher than was previously thought.

The important discovery of hydrothermal vent communities made in 1977 on the Galápagos Rift has shown that near black smokers, chemosynthesis is an important means of primary productivity. Some evidence suggests that hydrothermal vents may have been some of the first regions where life became established on Earth, in spite of the short life span of individual vents. Similar chemosynthesis has also been discovered in low-temperature seep biocommunities near hypersaline, hydrocarbon, and subduction zone seeps.

Key Terms

Abyssal storm (p. 469)

Annelid (p. 457)

Bioerosion (p. 464)

Biomass (p. 447)

Bivalve (p. 455)

Black smoker (p. 472)

Carnivorous feeding (p. 455)

Coral bleaching (p. 467)

Coral reef (p. 461)

Crustacean (p. 457)

Deposit feeding (p. 454)

Echinoderm (p. 458)

Epifauna (p. 447)

Hermatypic (p. 462)

High tide zone (p. 447)

Hydrocarbon seep biocommunity (p. 475)

Hydrothermal vent biocommunity (p. 472)

Hypersaline seep biocommunity (p. 475)

Infauna (p. 453)

Intertidal zone (p. 447)

Kelp forest (p. 459)

Low tide zone (p.447)

Meiofauna (p. 458)

Middle tide zone (p. 447)

Mixotroph (p. 464)

Mollusk (p. 455)

Polyp (p. 461)

Sessile (p. 449)

Spray zone (p. 447)

Subduction-zone seep biocommunity (p. 475)

Supralittoral zone (p. 449)
Suspension feeding (p. 454)

Zooxanthellae (p. 462)

Questions And Exercises

1. What controversy developed over the observations of the Rosses and Edward Forbes? How has the controversy been resolved?

2. What are some adverse conditions of rocky intertidal zones? What are some organism's adaptations for those adverse conditions? Which conditions seem to be most important in controlling the distribution of life?

3. Draw a diagram of the zones within the rocky-shore intertidal region and list characteristic organisms of each zone.

4. One of the most dominant features of the middle tide zone along rocky coasts is a mussel bed. Describe general characteristics of mussels, and include a discussion of other organisms that are associated with mussels.

5. Describe how sandy and muddy shores differ in terms of energy level, particle size, sediment stability, and oxygen content.

6. Describe the two types of feeding styles (other than predation) that are characteristic of the rocky, sandy, and muddy shores. One of these feeding styles is rather well represented in all these environments; name it and give an example of an organism that uses it in each environment.

7. How does the diversity of species on sediment-covered shores compare with that of the rocky shore? Suggest at least one reason why this occurs.

8. In which intertidal zone of a steeply sloping, coarse sand beach would you find each of the following organisms: clams; beach hoppers; ghost shrimp; sand crabs; and heart urchins?

9. Discuss the dominant species of kelp, their epifauna, and animals that feed on kelp in Pacific coast kelp forests.

10. Describe the environmental conditions required for development of coral reefs.

11. Describe the zones of the reef slope, the characteristic coral types, and the physical factors related to zonation.

12. What is coral bleaching? How does it occur? What other diseases affect corals?

13. What are some problems that limit the study of the deep-ocean floor? How are advances in technology overcoming these problems?

14. As one moves from the shoreline to the deep-ocean floor, what changes in the physical environment are experienced?

15. How long would human remains remain on the sea floor? What happens to the flesh? How long would a whale's remains remain on the sea floor?

16. What is the primary food source for organisms living on the deep-ocean floor? How does this affect benthic biomass?

17. Is species diversity on the deep-ocean floor relatively high or low? Explain.

18. Describe the characteristics of hydrothermal vents. What evidence suggests that hydrothermal vents have short life spans?

19. What is the "dead whale hypothesis"? What other ideas have been suggested to help explain how organisms from hydrothermal vent biocommunities populate new vent sites?

20. What are the major differences between the conditions and biocommunities of the hydrothermal vents and the cold seeps? How are they similar?

References

Brandon, J. L, and Rokip, F. J. 1985. *Life Between the Tides: The Natural History of the Common Seashore Life of Southern California.* San Diego, CA: American Southwest Publishing Company of San Diego.

Broad, W. J. 1997. *The Universe Below: Discovering the Secrets of the Deep Sea.* New York: Simon & Schuster.

Bruckner, A. W., and Bruckner, R. J., 1997. Emerging infections on the reefs. *Natural History* 106:11, 48.

Chadwick, W. W., Embley, R. W., and Fox, C. G. 1991. Evidence for volcanic eruption on the southern Juan de Fuca ridge between 1981 and 1987. *Nature* 350, 416–418.

Childress, J. J., Fisher, C. R., Brooks, J. M., Kennicutt, M. C., II, Bidigare, R., and Anderson, A. E. 1986. A methanotrophic marine molluscan (Bivalvia, Mytilidae) symbiosis: Mussels fueled by gas. *Science* 233:4770, 1306–1308.

Dawson, E. Y., and Foster, M. S. 1982. *Seashore Plants of Southern California.* Berkeley: University of California Press.

Fisher, C. R. 1990. Chemoautrophic and methanotrophic symbioses in marine invertebrates. *Reviews in Aquatic Sciences* 2:3 and 4, 399–436.

George, D., and George, J. 1979. *Marine Life: An Illustrated Encyclopedia of Invertebrates in the Sea.* New York: Wiley-Interscience.

Grassle, F. J., and Maciolek, N. J. 1992. Deep-sea species richness: Regional and local diversity estimates from quantitative bottom samples. *American Naturalist* 139:2, 313–341.

Hallock, P., and Schlager, W. 1986. Nutrient excess and the demise of coral reefs and carbonate platforms. *Palaios* 1:389–398.

Hessler, R. R., Ingram, C. L., Yayanos, A. A., and Burnett, B. R. 1978. Scavenging amphipods from the floor of the Philippine Trench. *Deep-Sea Research* 25:1029–1047.

Hessler, R. R., and Lonsdale, P. F. 1991. Biogeography of Mariana Trough hydrothermal vent communities. *Deep-Sea Research* 38:2, 185–199.

Huyghe, P. 1990. The storm down below. *Discover* 11:11, 70–76.

Kennicutt, M. C., II, Brooks, J. M., Bidigare, R. R., Fay, R. R., Wade, T. L., and McDonald, T. J. 1985. Vent-type taxa in a hydrocarbon seep region on the Louisiana slope. *Nature* 317:6035, 351–353.

Klum, L. D., et al. 1986. Oregon subduction zone: Venting fauna and carbonates. *Science* 231:4738, 561–566.

Lalli, C. M., and Parsons, T. R. 1993. *Biological Oceanography: An Introduction.* New York: Pergamon Press.

Lutz, R. A., and Kennish, M. J. 1996. Ecology of deep-sea hydrothermal vent communities: A review, *in* Pirie, R. G., ed.,

Oceanography: Contemporary Readings in Ocean Sciences, 3rd ed., New York: Oxford University Press.

MacGinitie, G. E., and MacGinitie, N. 1968. *Natural History of Marine Animals,* 2nd ed. New York: McGraw-Hill.

Milliman, J. D., ed. 1998. Deep sea biodiversity: A compilation of recent advances in honor of Robert R. Hessler. *Deep Sea Research* 45:1–2.

Ricketts, E. F., Calvin, J., and Hedgpeth, J. 1968. *Between Pacific Tides.* Stanford, CA: Stanford University Press.

Rona, P. A., Klinkhammer, G., Nelson, T. A., Trefry, J. H., and Elderfield, H. 1986. Black smokers, massive sulphides and vent biota at the Mid-Atlantic Ridge. *Nature* 321:6065, 33–37.

Smith, S. 1996. Life after death on the seafloor: Aerobic oases of whales' bones. *Pacific Discovery* 49:1, 32–35.

Thorne-Miller, B., and Catena, J. 1991. *The Living Ocean: Understanding and Protecting Marine Biodiversity.* Washington, DC: Island Press.

Walbran, P. D., Henderson, R. A., Jull, A. J. T., and Head, M. J. 1989. Evidence from sediments of long-term *Acanthaster planci* predation on corals of the Great Barrier Reef. *Science* 245:4920, 847–850.

Waller, G., ed. 1996. *SeaLife: A Complete Guide to the Marine Environment.* Washington, DC: Smithsonian Institution Press.

Williams, E. H., Jr., and Bunkley-Williams, L. 1990. The worldwide coral reef bleaching cycle and related sources of coral mortality. *Smithsonian Atoll Research Bulletin* 335.

Williams, E. H., Jr., Goenaga, C., and Vicente, V. 1987. Mass bleaching on Atlantic coral reefs. *Science* 238:4830, 877–878.

Yonge, C. M. 1963. *The Sea Shore.* New York: Atheneum.

Zenevitch, L. A., Filatove, A., Belyaev, G. M., Lukanove, T. S., and Suetove, I. A. 1971. Quantitative distribution of zoobenthos in the world ocean. *Bulletin der Moskauer Gen der Naturforscher, Abt. Biol.* 76, 27–33.

ZoBell, C. E. 1968. Bacterial life in the deep sea, in Proceedings of the U.S.–Japan Seminar on Marine Microbiology, August 1966, Tokyo. *Bulletin Misaki Marine Biology, Kyoto Inst. Univ.* 12:77–96.

Suggested Reading

Earth

Flannagan, R. 1993. Corals under siege. 2:3, 26–35. An examination of the causes of coral bleaching.

Hart, S. 1997. Tubeworm travels. 6:3, 48–53. Examines how organisms of hydrothermal vent biocommunities may be able to travel from one hydrothermal vent field to another.

Sea Frontiers

Alper, J. 1990. Methane eaters. 36:6, 22–29. An account of the exploration of hydrocarbon seep biocommunities in the Gulf of Mexico.

Coleman, N. 1974. Shell-less molluscs. 20:6, 338–342. A description of nudibranchs, gastropods without shells.

George, J. D. 1970. The curious bristle-worms. 16:5, 291–300. The variety of worms belonging to the class *Polychaeta* of phylum Annelida is described. The discussion includes locomotion, feeding, and reproductive habits of the various members of the class.

Gibson, M. E. 1981. The plight of *Allopora* 27:4, 211–218. The reason the author believes the unusual California hydrocoral is headed for the endangered species list is the topic of this article.

Humann, P. 1991. Loving the reef to death. 37:2, 14–21. The many ways divers in particular and humans in general degrade the coral reef environment are discussed.

McClintock, J. 1994. Out of the oyster. 40:3, 18–23. The history of our interest in pearls as jewelry and the development of the industry of culturing pearls are discussed.

Ruggiero, G. 1985. The giant clam: Friend or foe? 31:1, 4–9. The ecology and behavior of the giant clam (*Tridacna gigas*) is discussed with reference to whether it is a danger to divers.

Shinn, E. A. 1981. Time capsules in the sea. 27:6, 364–374. The method by which geologists determine past climatic and environmental conditions by studying coral reefs is discussed.

Viola, F. J. 1989. Looking for exotic marine life? Don't leave the dock. 35:6, 336–341. Marine life living on and near pilings for a boat dock is described, along with good color photos.

Winston, J. E. 1990. Intertidal space wars. 36:1, 47–51. Florida intertidal life is discussed.

Scientific American

Caldwell, R. L., and Dingle, H. 1976. Stomatopods. 234:1, 80–89. Presents the ecology of these interesting crustaceans that have appendages specialized for spearing and smashing prey.

Feder, H. A. 1972. Escape responses in marine invertebrates. 227:1, 92–100. Discusses the surprisingly rapid movements and other responses made by invertebrates to the presence of predators. Some interesting photographs accompany the text, which describes the escape responses of limpets, snails, clams, scallops, sea urchins, and sea anemones.

Martini, F. H. 1998. Secrets of the slime hag. 279:4, 70-75. An examination of the biology ad life history of hag fishes, scavengers of the deep sea.

Wicksten, M. K. 1980. Decorator crabs. 242:2, 146–157. Describes how species of spider crabs use materials from their environment to camouflage themselves.

Younge, C. M. 1975. Giant clams. 232:4, 96–105. The distribution and general ecology of the tridacnid clams, some of which grow to lengths well over 1 meter (3.3 feet), are investigated.

Oceanography on the Web

Visit the *Essentials of Oceanography* home page for on-line resources for this chapter. There you will find an on-line study guide with review exercises, and links to oceanography sites to further your exploration of the topics in this chapter. *Essentials of Oceanography* is at: **http://www.prenhall.com/thurman** (click on the Table of Contents menu and select this chapter).

AFTERWORD

At the end of our journey through this book together, it seems appropriate to examine people's perceptions of the ocean. If asked to describe the ocean, people's responses commonly include the words "powerful," "awe inspiring," "moving," "serene," "abundant," and "majestic." Also included are words that describe its huge extent, such as "vast," "infinite," or "boundless." These are equally appropriate descriptions, because even though humans have been exploring and studying the ocean for centuries, the ocean still holds many secrets. Recent findings of newly discovered life forms in the ocean bear testimony to how limited our knowledge of the ocean really is.

In spite of the ocean's impressive size, it is beginning to feel the effects of human influence. For instance, every ocean contains floating plastic trash and even remote beaches are littered with trash. Organisms living in the ocean are also feeling the effect of humankind's use of the ocean. One example is that many great whale populations were pushed to near extinction by whaling in the nineteenth and twentieth centuries. Because of restrictions placed on whaling and the development of substitutes for whale products, it appears that whales will survive this particular threat. The major threat to whales now may be related to habitat destruction in the form of disruption of their feeding and breeding grounds. Another example is overfishing, which has been the single most devastating activity with regard to altering the marine ecosystem. Although the amount of fish caught by the world fishery peaked in 1989, the fishing effort has not decreased significantly. These examples illustrate that perhaps the ocean is not quite as "vast," "infinite," and "boundless" as most people believe.

Whether it is coral bleaching, increased concentrations of greenhouse gases in the lower atmosphere, depletion of upper-atmosphere ozone, the increased frequency of red tides and "dead zones," or polluting the ocean with petroleum, plastics, sewage, chemicals, or toxin-laden sediment, there is little doubt that human-induced changes in the environment are broad in scope and beginning to reach worldwide proportions. All of these may be symptoms of a worldwide pathology—sickness—that will demand major changes to human behavior before it can be cured. To draw awareness to these and other problems, the United Nations designated 1997 as the International Year of the Coral Reef and 1998 as the International Year of the Ocean.

One step in the right direction was initiated in the 1970s when the U.S. Congress began establishing National Marine Sanctuaries. Today, there are 12 National Marine Sanctuaries in various places such as the Florida Keys, the Stellwagen Bank off Massachusetts, the Channel Islands off southern California, the Flower Garden Banks in the Gulf of Mexico, and the Hawaiian Islands. The purpose of marine sanctuaries is to help protect vital pockets of the ocean from further degradation caused by human influence. Other countries are also realizing the benefit of protecting ocean resources: For those countries with coral reefs offshore, a healthy coral reef can be worth more as a tourist draw than it might be as a producer of seafood.

It is important to keep in mind that although the ocean is feeling the effects of humans, the ocean has a tremendous ability to withstand change. Remarkably, the vast majority of ocean water is still relatively unpolluted. The exceptions to this are shallow-water coastal areas near large population centers or near the mouths of major rivers. As well, natural processes that occur in the ocean tend to disperse and eventually remove many types of pollutants.

Because the ocean has traditionally been considered so incredibly large—and capable of absorbing many substances—it has been used as a dumping ground for many of society's wastes. Even today, humans are adding pollutants to the ocean at staggering rates. What can be done? Each of us as an individual can do several things that will not only help the environment in general, but will also help reduce human impact on the ocean:

- Minimize your impact on the environment. Reduce the amount of waste you generate by making wise consumer choices. Avoid products with excessive packaging and support companies that have good environmental records of accomplishment. Reuse and recycle items, then help close the recycling loop by buying goods made of recycled materials.

- Be responsible. Don't assume that someone else will take care of your trash. Be aware that storm drains lead directly to the ocean or to a river that leads to the ocean. If you throw something on the street, it may very well find its way to a river—and ultimately to the ocean—after the next rainstorm.

- Become politically aware. Many ocean-related issues come before the public and require a majority of voters to approve a proposal before they are enacted into law. This is true for local as well as national and international issues. For instance, within our lifetimes, we may very well decide on such issues as whether to spend large amounts of money to add finely ground iron to the ocean to reduce the amount of carbon dioxide in the atmosphere. Many political issues in the future will involve the ocean.

- Educate yourself about how the ocean works. A recent poll indicated that over 90 percent of the American public consider themselves to be scientifically illiterate. With our society becoming more scientifically advanced, people need to understand how science operates. Science is not meant to be comprehended by an elite few. Rather, science is for everyone. As a student of the ocean, you have completed one of the first steps toward that end: By studying oceanography, you have gained an understanding into how the ocean works. It is our hope that this book has piqued your curiosity about the ocean and its workings to the point that you will continue to be a life-long student of the ocean.

Figure Aft–1 Sunset at the ocean.

APPENDIX I
METRIC AND ENGLISH UNITS COMPARED

How many inches are there in a mile? How many cups in a gallon? How many pounds in a ton? In our daily lives, we often need to convert between units. Worldwide, the metric system of measurement is the most widely used system. Besides the United States, only *two other countries in the world* still use the English system as their primary system of measurement: Liberia and Myanmar (formerly Burma). The metric system of measurement has many advantages over the English system. It is simple, logical, and is easy to convert between units. For those of us in the United States, it is only a matter of time before the change to the metric system occurs.

On December 23, 1975, President Gerald R. Ford signed the Metric Conversion Act of 1975. It defined the metric system as being the International System of Units (officially called the Système International d'Unités, or SI) as interpreted in the United States by the secretary of commerce. The Trade Act of 1988 and other legislation declared the metric system the preferred system of weights and measures for U.S. trade and commerce, called for the federal government to adopt metric specifications, and mandated the Commerce Department to oversee the program.

Still, the metric system has not become the system of choice for most Americans' daily use, and there is great resistance to using the metric system. Indeed, it appears to be a daunting task to switch over to an unfamiliar system of units, no matter how logical that system may be. It is interesting to note that Benjamin Franklin proposed that the new colonies of the United States change over to the metric system in the late 1700s. To be competitive in world markets, the United States will need to switch to the metric system. Leading the way in this changeover are many of the sciences. Oceanographers all over the world have been using the metric system for years.

The English System

The English system actually consists of two related systems—the U.S. Customary System,[1] used in the United States and dependencies, and the British Imperial

System. Great Britain, the originator of the latter system, has now largely converted to the metric system. The basic unit of length in the English System is the *yard* (yd); the basic unit of weight (not mass[2]) is the *pound* (lb).

In the English system, the units of length were initially based rather arbitrarily on dimensions of the body. For example, the yard as a measure of length can be traced to the early Saxon kings. They wore a sash or belt around the waist that could be removed and used as a convenient measuring device. Thus, the word *yard* comes from the Saxon word *gird* (like a girdle), in reference to the circumference of a person's waist. Since the circumference of a person's waist varies from person to person—and even varies from time to time on the *same* person—its usefulness was limited. There had to be some standardization to make it a useful measurement. Tradition holds that King Henry I of England decreed that the yard should be the distance from the tip of his nose to the end of his thumb! Somehow, this stuck, and the yard as defined today is that distance.

Other English units have a history that is also based on antique measurements. For instance, why are there 5280 feet in a mile (*mille passuum* = 1000 paces), rather than a round number (such as 5000)? Ironically, the mile was initially defined as an even 5000 feet by the Romans. However, a *furlong* (or "furrow-long") was established as 220 yards by early Tudor rulers in England based on the length of agricultural fields. To facilitate the conversion between miles and furlongs, Queen Elizabeth I declared in the sixteenth century that the traditional Roman mile of 5000 feet would be forever replaced by one of 5280 feet, making the mile exactly 8 furlongs. Even though it made the conversion between miles and other units more difficult, this provided a convenient relationship between miles and furlongs, which were previously ill-related measures. Today, a furlong is an archaic unit of measurement, but we still must remember that a mile is exactly 5280 feet.

It was due largely to this standardization of units and that the British had a superior navy—thus leading them to colonize and trade with many nations—that led to the English system of units being used around the world by

---

[1] The names of the units and the relationships between them are generally the same in the U.S. Customary System and the British Imperial System, but the sizes of the units differ, sometimes considerably.

---

[2] Although the basic unit of mass in the English system is rarely used, it is the *slug*.

the eighteenth century. However, it remained difficult to convert between English units and other English units. For example, the use of the number 12 as a base instead of 10 makes for awkward conversions. The well-known 12 inches to a foot illustrates this problem.

The Metric System

The metric system is a system of weights and measures planned in France and adopted there in 1799. Now used by most of the countries of the world, it is based on a unit of length called the *meter* (m) and a unit of mass called the *kilogram* (kg). Originally defined as one ten-millionth the distance from the North Pole to the Equator, the meter is now defined as the distance light travels through a vacuum in 1/299,792,458 of a second. The kilogram was originally related to the volume of one cubic meter of water. It is now defined as the mass of the International Prototype Kilogram, a platinum-iridium cylinder kept at Sèvres (near Paris), France. Other metric units can be defined in terms of the meter and the kilogram.

Fractions and multiples of the metric units are related to each other by powers of 10, allowing conversion from one unit to a multiple of it simply by shifting the decimal point. This avoids the lengthy arithmetical operations required with English units of measurement. Even the names of the metric units indicate how many are in a larger unit. For instance, how many cents are there in a dollar? It is the same number as the number of *centi*grams in a gram. The prefixes listed in the following table, "Common Prefixes for Basic Metric Units" have been accepted for designating multiples and fractions of the meter, the gram (which is equal to $1/1000$ of a kilogram), and other units.

In spite of many fears that changing to the metric system will be a large economic burden on businesses of the United States, the conversion to the metric system has already begun. For example, you use 35-millimeter film, purchase 2-liter soft drink bottles or 750-milliliter wine bottles, buy prescription drugs in milligrams, work on computers with gigabytes of memory, and many people participate in 10-kilometer (10-K) runs. Cars are one example of a manufactured item that is now produced metrically in order to compete with metric-based countries. Most of us in the United States will experience the increasing use of the metric system in our daily lives. Some ways to ease yourself into this unfamiliar (yet sensible) system of units include noticing distances on road signs in kilometers, using a meterstick instead of a yardstick, and charting your weight from your bathroom scale in kilograms.

Temperature

The Celsius (centigrade) temperature scale was constructed based on water's characteristics at one atmosphere of pressure (standard sea level pressure). The scale was invented in 1742 by a Swedish astronomer named Anders Celsius. The scale was designed so that the freezing point of pure water—the temperature at which it changes state from a liquid to a solid—is set as 0 degrees. Similarly the boiling point of pure water—the temperature at which it changes state from a liquid to a gas—is set at 100 degrees. Doing so made the liquid range of water—0 degrees to 100 degrees—an even 100 units, giving the scale its name: centigrade (*cent* = 100, *grade* = graduations).

In contrast, the commonly used Fahrenheit temperature scale sets water's freezing and boiling points at 32 degrees and 212 degrees, for a spread of 180 units. This spread, along with the freezing and boiling points of water set at irregular numbers, makes the Fahrenheit scale difficult to use. This scale was devised by a German-born physicist named Gabriel Daniel Fahrenheit, who invented the mercury thermometer in 1714. It is interesting to note that he initially designed the scale so that normal body temperature was set at an even 100 degrees Fahrenheit. However, because of inaccuracies in early thermometers or how the scale was initially devised, normal human body temperature is well known as an uneven 98.6 degrees Fahrenheit (37 degrees centigrade).

Scientific Notation

Common Prefixes for Basic Metric Units

Factor	Name	Prefix Name
$10^{12} = 1,000,000,000,000$	trillion	tera
$10^{9} = 1,000,000,000$	billion	giga
$10^{6} = 1,000,000$	million	mega
$10^{3} = 1,000$	thousand	kilo
$10^{2} = 100$	hundred	hecto
$10^{1} = 10$	ten	deka
$10^{-1} = 0.1$	tenth	deci
$10^{-2} = 0.01$	hundredth	centi
$10^{-3} = 0.001$	thousandth	milli
$10^{-6} = 0.000001$	millionth	micro
$10^{-9} = 0.000000001$	billionth	nano
$10^{-12} = 0.000000000001$	trillionth	pico

Conversion Tables

Length

1 micrometer (μm)	0.001 millimeter
	0.0000394 inch
1 millimeter (mm)	1,000 micrometers
	0.1 centimeter
	0.001 meter
	0.0394 inch
1 centimeter (cm)	10 millimeters
	0.01 meter
	0.394 inch

1 meter (m)	100 centimeters
	39.4 inches
	3.28 feet
	1.09 yards
	0.547 fathom
1 kilometer (km)	1,000 meters
	1,093 yards
	3,280 feet
	0.62 statute mile
	0.54 nautical mile
1 inch (in)	25.4 millimeters
	2.54 centimeters
1 foot (ft)	12 inches
	30.5 centimeters
	0.305 meter
1 yard (yd)	3 feet
	0.91 meter
1 fathom (fm)	6 feet
	2 yards
	1.83 meters
1 statute mile (mi)	5,280 feet
	1,760 yards
	1,609 meters
	1.609 kilometers
	0.87 nautical mile
1 nautical mile (nm)	1 minute of latitude
	6,076 feet
	2,025 yards
	1,852 meters
	1.15 statute miles
1 league (lea)	5,280 yards
	15,840 feet
	4,805 meters
	3 statute miles
	2.61 nautical miles

Area

1 square centimeter (cm^2)	0.155 square inch
	100 square millimeters
1 square meter (m^2)	10,000 square centimeters
	10.8 square feet
1 square kilometer (km^2)	100 hectares
	247.1 acres
	0.386 square mile
	0.292 square nautical mile

1 square inch (in^2)	6.45 square centimeters
1 square foot (ft^2)	144 square inches
	929 square centimeters

Volume

1 cubic centimeter (cc; cm^3)	1 milliliter
	0.061 cubic inch
1 liter (l)	1,000 cubic centimeters
	61 cubic inches
	1.06 quarts
	0.264 gallon
1 cubic meter (m^3)	1,000,000 cubic centimeters
	1,000 liters
	264.2 gallons
	35.3 cubic feet
1 cubic kilometer (km^3)	0.24 cubic mile
	0.157 cubic nautical mile
1 cubic inch (in^3)	16.4 cubic centimeters
1 cubic foot (ft^3)	1,728 cubic inches
	28.32 liters
	7.48 gallons

Mass

1 gram (g)	0.035 ounce
1 kilogram (kg)	2.2 pounds[a]
	1,000 grams
1 metric ton (mt)	2,205 pounds
	1,000 kilograms
	1.1 U.S. short tons
1 pound[a] (lb)	16 ounces
	454 grams
	0.454 kilogram
1 U.S. short ton (ton; t)	2,000 pounds
	907.2 kilograms
	0.91 metric ton

[a] The pound is a weight unit, not a mass unit, but is often used as such.

Pressure

1 atmosphere (atm)	760 millimeters of mercury
(at sea level)	14.7 pounds per square inch
	29.9 inches of mercury
	33.9 feet of fresh water
	33 feet of seawater

Speed

1 centimeter per second (cm/s)	0.0328 foot per second
1 meter per second (m/s)	2.24 statute miles per hour
	1.94 knots
	3.28 feet per second
	3.60 kilometers per hour
1 kilometer per hour (kph)	27.8 centimeters per second
	0.62 mile per hour
	0.909 foot per second
	0.55 knot
1 statute mile per hour (mph)	1.61 kilometers per hour
	0.87 knot
1knot (kt)	1 nautical mile per hour
	51.5 centimeters per second
	1.15 miles per hour
	1.85 kilometers per hour

Oceanographic data

Velocity of sound in 34.85‰ seawater	4,945 feet per second
	1,507 meters per second
	824 fathoms per second
Seawater with 35 grams of dissolved substances per kilogram of seawater	3.5 percent
	35 parts per thousand
	35,000 parts per million
	35,000,000 parts per billion

Temperature

Exact Formula	Approximation (easy way)
$°C = \dfrac{(°F - 32)}{1.8}$	$°C = \dfrac{(°F - 30)}{2}$
$°F = (1.8 × °C) + 32$	$°F = (2 × °C) + 30$

Some useful equivalent temperatures:

100°C	= 212°F	(boiling point of pure water)
40°C	= 104°F	(heat wave conditions)
37°C	= 98.6°F	(normal body temperature)
30°C	= 86°F	(very warm—almost hot)
20°C	= 68°F	(room temperature)
10°C	= 50°F	(a warm winter day)
3°C	= 37°F	(average temperature of deep water)
0°C	= 32°F	(freezing point of pure water)

In degrees centigrade:

Thirty is hot, twenty is pleasing;

Ten is not, and zero is freezing.

APPENDIX II
LATITUDE AND LONGITUDE ON EARTH

Suppose that you find a great fishing spot or a shipwreck on the sea floor. How would you remember the spot—far from any sight of land—so that you might return? Furthermore, how do sailors navigate a ship when they are at sea? Even using sophisticated equipment such as the global positioning system (GPS), which uses satellites to determine location, how is this information reported?

To solve problems like these, a navigational grid—a series of intersecting lines across a globe—is used. Once starting places are fixed for a grid, a location can be identified based on its position within the grid. An example of this is how certain cities use the regular numbering of streets as a grid. Even if you have never been to the intersection of 5th Avenue and 42nd Street in New York City, you would be able to locate it on a map, or know which way to travel if you were at 10th Avenue and 42nd Street. Although many grid (or coordinate) systems are in use today, the one most universally accepted uses *latitude* and *longitude*.

A glance at any globe will reveal a series of north–south and east–west lines that together comprise a grid system that can be used to locate points on Earth's surface whether at sea or on land. The north–south lines of the grid are called *meridians* and extend from pole to pole (Figure A2–1). Notice that the meridian lines converge at the poles and are spaced farther apart at the Equator. The east–west lines of the grid are called *parallels*. True to their name, they are parallel to one another. Notice that the longest parallel is the *Equator* (so called because it equates the globe into two equal hemispheres), and that the parallels at the poles are a single point (Figure A2–1).

Latitude and Longitude

Latitude is defined as the angular distance (in degrees of arc) measured north or south of the starting line of latitude (the Equator) from the center of Earth. Parallels are used to show latitude. Since all points that lie along the same parallel are an identical distance from the

Figure A2–1 Earth's grid system.

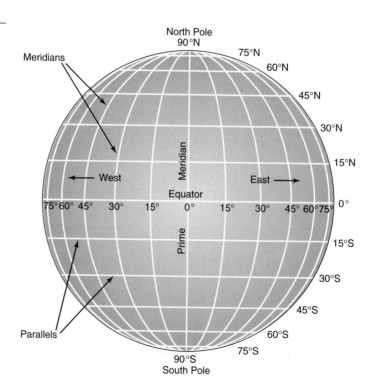

Equator, they all have the same latitude designation. The latitude of the Equator is 0 degrees, and the North and South Poles lie at 90 degrees north and 90 degrees south, respectively. Figure A2–2 shows that the latitude of New Orleans is an angular distance of 30 degrees north from the Equator.

Longitude is defined as the angular distance (in degrees of arc) measured east or west of the starting line of longitude from the center of Earth. Since all meridians are identical (none is longer or shorter than the others, and they all pass through both poles), the choice of a starting line of longitude has been chosen arbitrarily. Before it was agreed upon by all nations, different countries used different starting points for 0 degrees longitude. Some of the starting points that have been used for longitude include the Canary Islands, the Azores, Rome, Copenhagen, Jerusalem, St. Petersburg, Pisa, Paris, Philadelphia, and others. At the 1884 International Meridian Conference, it was agreed that the meridian that passes through the Royal Observatory at Greenwich, England, would be used as the zero or starting point for the measurement of longitude. This zero degree line of longitude is also called the *prime meridian*. Thus, the longitude for any place on Earth is measured east or west from this line. Longitude can vary from 0 degrees along the prime (Greenwich) meridian to 180 degrees (either east or west), which is halfway around the globe and is known as the *International Date Line*. For the example shown in Figure A2–2, New Orleans is 90 degrees west of the prime meridian.

A location given in latitude and longitude defines a unique location on Earth. However, the direction from the starting point (for latitude, either north or south; for longitude, either east or west) must also be given for it to be unique. For instance, a location of 42 degrees north and 120 degrees west defines a unique spot on Earth, different from 42 degrees south and 120 degrees west and also different from 42 degrees north and 120 degrees east.

A degree of latitude or longitude can be divided into smaller units. One degree (°) of arc (angular distance) is equal to 60 minutes (′) of arc. One minute of arc is equal to 60 seconds (″) of arc. When using a small-scale map or a globe, it may be difficult to estimate latitude and longitude to the nearest degree or two. Conversely, when using a large-scale map that shows much detail of an area, it is often possible to determine the latitude and longitude to the nearest fraction of a minute.

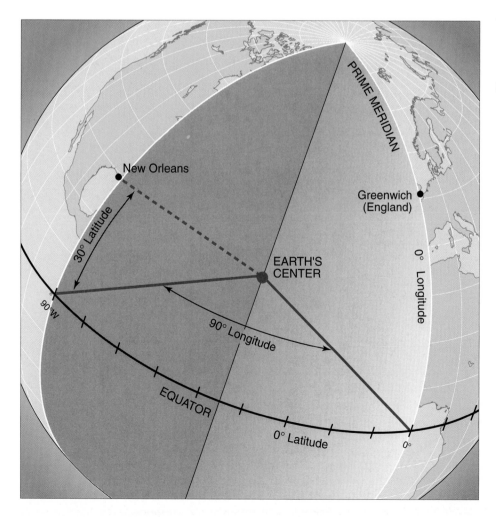

Figure A2–2 New Orleans is located at 30 degrees north latitude and 90 degrees west longitude.

Determination of Latitude and Longitude

Today, latitude and longitude can be determined very precisely by using satellites that remain in orbit around Earth in fixed positions. How did navigators determine their position before this new technology was available?

The method of determining latitude was based on the positions of particular stars. Initially, navigators in the Northern Hemisphere would measure the angle between the horizon and the North Star (Polaris), which is directly above the North Pole. The latitude north of the Equator is the angle between the two sightings (Figure A2–3). Similar determinations may be made in the Southern Hemisphere by using the Southern Cross, which is directly overhead at the South Pole. Later, the angle of the sun above the horizon, corrected for the date, was also used.

For millennia, there was no method of determining longitude. The method of determining longitude based on time was developed only at the end of the eighteenth century. As Earth turns on its rotational axis, it moves through 360 degrees of arc every 24 hours (one complete rotation on its axis) (Figure A2–4A). Since Earth rotates through 15 degrees of arc or longitude per hour

($360° \div 24$ hours = 15 degrees/hour), a navigator needed only to know what time it was at the prime meridian at the exact time that the sun was at its highest point (local noon) at the ship's location. In this way, a navigator aboard a ship could calculate the ship's longitude each day at noon. That's why the development of John Harrison's chronometer (see Box 1–1 in Chapter 1) was such an important step in determining longitude and helping sailors know where they were at sea.

Let's look at an example of how ships used the difference in local noon versus Greenwich noon to determine their longitude. A ship sets sail west across the Atlantic Ocean from Europe, checking its longitude each day at noon local time. One day when the sun is at the noon position (noon local time), the navigator checks the chronometer, which is faithfully recording time based on Greenwich time (Figure A2–4B). The clock reads 16:18 hours (0:00 is midnight and 12:00 is noon). What is the ship's longitude (Figure A2–4C)?

If the clock is keeping good time (which Harrison's chronometer did), then we know that the ship is 4 hours and 18 minutes behind (west of) Greenwich time. To determine the longitude, we must convert time into longitude. Each hour represents 15 degrees of longitude based on the rotation of Earth. Thus, the four hours rep-

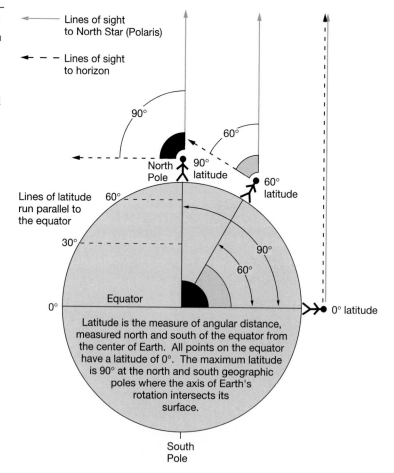

Figure A2–3 Determining latitude based on the North Star (Polaris).

Latitude can be determined by noting the angular difference between the horizon and the North Star, which is directly over the North Pole of Earth.

resents 60 degrees of longitude (4 hours × 15 degrees of longitude = 60 degrees). Since one degree is divided into 60 minutes of arc, Earth rotates through $\frac{1}{4}$ degree (15′) of arc per minute of time. Thus, 18 minutes of time multiplied by 15 minutes of arc per minute of time equals 270 minutes of arc. To convert 270 minute of arc into degrees, it must be divided by 60 minutes per degree, which gives 4.5 degrees of longitude. Therefore, the answer is 60 degrees plus 4.5 degrees, or 64.5 degrees west longitude.

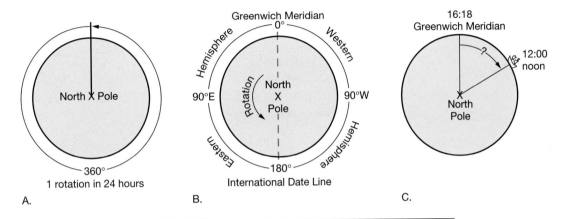

Figure A2–4 Determining longitude based on time.
View of Earth as seen from above the North Pole. **A.** Earth rotates 360 degrees of arc every 24 hours. **B.** The Greenwich Meridian is set as 0 degree longitude, which divides the globe into Eastern and Western Hemispheres. The International Date Line is 180 degrees from the Greenwich Meridian. **C.** An example of how a ship at sea can determine its longitude using time.

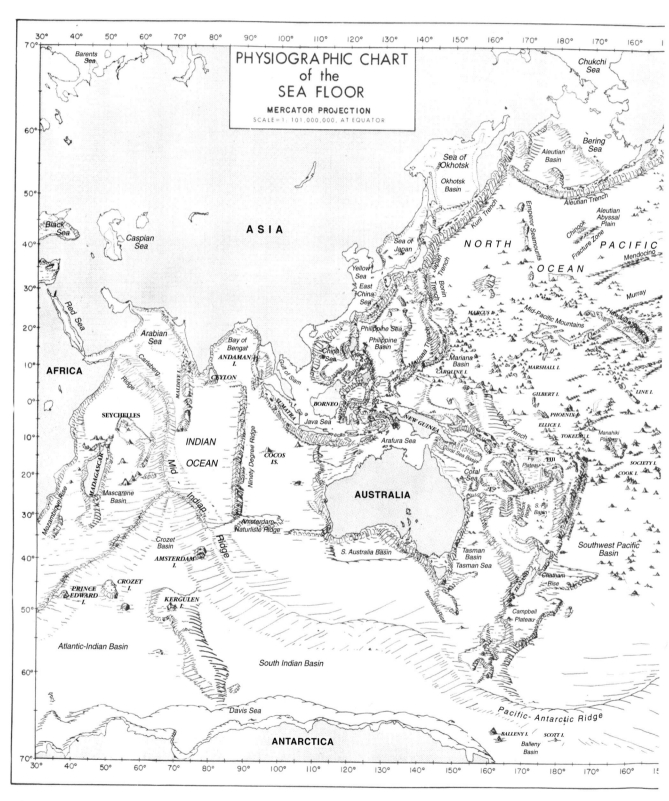

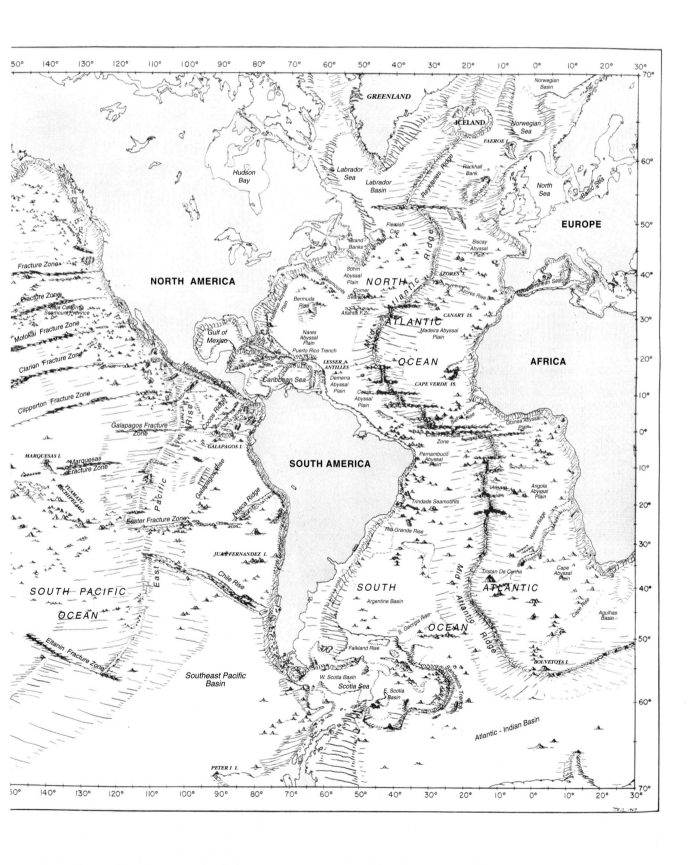

A P P E N D I X   I V
Careers in Oceanography

Many people think of a career in oceanography as consisting of swimming with marine animals at a marine life park or snorkeling in crystal-clear tropical waters studying coral reefs. In reality, these kinds of jobs are extremely rare and there is intense competition for the few jobs that do exist. Most oceanographers work in fields that use science to solve a particular question about the ocean. Examples of these questions include:

- What is the role of the ocean in limiting the greenhouse effect?
- What kinds of pharmaceuticals can be found naturally in marine organisms?
- How does sea floor spreading relate to the movement of tectonic plates?
- What economic deposits are there on the sea floor?
- Can rogue waves be predicted?
- What is the role of longshore transport in the distribution of sand on the beach?
- How does a particular pollutant affect organisms in the marine environment?

Preparation for a Career in Oceanography

Preparing yourself for a career in oceanography is probably one of the most interesting and rewarding (yet difficult) paths to travel. The study of oceanography is typically divided into four different academic disciplines (or subfields) of study:

- *Geological oceanography* is the study of the structure of the sea floor and how the sea floor has changed through time; the creation of sea floor features; and the history of sediments deposited on it.
- *Chemical oceanography* is the study of the chemical composition and properties of seawater; how to extract certain chemicals from seawater; and the effects of pollutants.
- *Physical oceanography* is the study of waves, tides, and currents; the ocean–atmosphere relationship that influences weather and climate; and the transmission of light and sound in the oceans.

- *Biological oceanography* is the study of the various oceanic life forms and their relationships to one another; adaptations to the marine environment, and developing ecologically sound methods of harvesting seafood.

Other disciplines include ocean engineering, marine archaeology, and marine policy. Since the study of oceanography often examines in detail all the different disciplines of oceanography, it is often described as being an *interdisciplinary* science, or one covering all the disciplines of science as they apply to the oceans. Thus, some of the most exciting work and best employment opportunities combine two or more of these disciplines.

Individuals in oceanography and marine-related fields need a good background in at least one area of basic science (for example, geology, physics, chemistry, or biology) or engineering. In almost all cases, mathematics is required as well. Marine archaeology requires a background in archaeology or anthropology; marine policy studies require a background in at least one of the social sciences (such as law, economics, or political science).

The ability to speak and write clearly—as well as critical thinking skills—are prerequisites for any career. Fluency in computers—preferably PC systems, not Macintosh—is rapidly becoming a necessity. Because many job opportunities in oceanography require trips on research vessels, any shipboard experience is also desirable. Mechanical ability (the ability to fix equipment while on board a vessel without having to return to port) is a plus. Depending on the type of work that is required, other traits that may be desirable include: the ability to speak one or more foreign languages; certification as a scuba diver; the ability to work for long periods of time in cramped conditions; physical stamina; physical strength; and, of course, a high tolerance to motion sickness.

Since oceanography is a relatively new discipline of science (with much room left for discovery), most people enter the field with an advanced degree (master's or doctorate). One exception to this is to work as a marine technician, which usually requires a bachelor's degree or applicable experience. It does take a large commitment to achieve an advanced degree, but, in the end, the journey itself is what makes all the hard work worthwhile.

Job Duties of Oceanographers

There are many job opportunities for oceanographers with scientific research institutions (universities) and various government agencies. Private companies that are engaged in searching for economic sea floor deposits, investigating areas for sea farming, and evaluating natural energy production from waves, currents, and tides also hire oceanographers. The job duties of oceanographers vary from place to place, but can be generally described as follows:

- *Geological oceanographers* and *geophysicists* explore the ocean floor and map submarine geological structures. Studies of the physical and chemical properties of rocks and sediments give us valuable information about Earth's history. The results of their work help us understand the processes that created the ocean basins and the interactions between the ocean and the sea floor.

- *Chemical oceanographers* and *marine geochemists* investigate the chemical composition of seawater and its interaction with the atmosphere and the sea floor. Their work may include analysis of seawater components, desalination of seawater, and studying the effects of pollutants. They also examine chemical processes operating within the marine environment and work with biological oceanographers on studies of living systems. Their study of trace chemicals in seawater helps us understand how ocean currents move seawater around the globe, and how the ocean affects climate.

- *Physical oceanographers* investigate such ocean properties as temperature, density, wave motions, tides, and currents. They study ocean–atmosphere relationships that influence weather and climate, the transmission of light and sound through water, and the ocean's interactions with its boundaries at the sea floor and the coast.

- *Biological oceanographers, marine biologists,* and *fisheries scientists* study marine plants and animals. They are interested in how marine organisms develop, relate to one another, adapt to their environment, and interact with it. Their work includes developing ecologically sound methods of harvesting seafood and studying biological responses to pollution. New fields associated with biological oceanography include marine biotechnology (the use of natural marine resources in the development of new industrial and biomedical products) and molecular biology (the study of the structure and function of bioinformational molecules—such as DNA, RNA, and proteins—and the regulation of cellular processes at the molecular level). Because marine biology is the most well-known oceanographic field (and because the larger marine animals have such wide appeal), it is currently the most competitive sector of oceanography.

- *Marine* and *ocean engineers* apply scientific and technical knowledge to practical uses. Their work ranges from designing sensitive instruments for measuring ocean processes to building marine structures that can withstand ocean currents, waves, tides, and severe storms. Subfields include acoustics, robotics, electrical, mechanical, civil, and chemical engineering as well as naval architecture. They often use highly specialized computer techniques.

- *Marine archaeologists* are involved in the systematic recovery and study of material evidence, such as shipwrecks, graves, buildings, tools, and pottery remaining from past human life and culture that is now covered by the sea. Marine archaeologists use state-of-the-art technology to locate various underwater archaeological sites.

- *Marine policy experts* combine their knowledge of oceanography and social sciences, law, or business to develop guidelines and policies for the wise use of the ocean and coastal resources. Marine policy requires a knowledge of at lest one of the other disciplines as well as a sound understanding of oceanographic issues.

Other job opportunities for oceanographers include work as science journalists specializing in marine science, teachers at various grade levels, and aquarium and museum curators.

Report from a Student/Oceanographer

One of the true pleasures of being a teacher is that some of your students become so interested in the subject matter you teach that they themselves pursue a career in your field of interest. One of Al Trujillo's former students, Joe Cooney, works as a facilities manager for the Hubbs Sea World Research Institute's Sea Bass Fish Hatchery in Carlsbad, California (Figure A4–1). His job responsibility is to provide maintenance and upkeep of equipment involved in raising sea bass for release to the ocean to restore the former abundance of this sportfish in southern California waters. Joe writes:

> After completing coursework in oceanography at a community college, I wanted more information about what types of jobs were available and what degree would be most valuable. I was not certain if I wanted to be involved in work like marine biologists I had seen, or look more towards physical oceanography as a career. Fortunately, I had the opportunity to get a summer job at a marine fish hatchery, which led to an offer of a permanent position. I am now involved daily in a marine sciences working environment where my primary responsibility is to monitor and maintain the seawater systems for the hatchery.
>
> Unlike a traditional fish hatchery that uses water diverted from a river to provide large amounts of clean water at

virtually no cost, a saltwater hatchery must pump in its entire water supply from the sea. Also, conditions must be maintained so that open-ocean fish will live and reproduce, and that their larvae will survive and grow.

Water is pumped to recirculating and experimental systems, as well as to traditional flow-through raceways where juvenile fish are briefly held before being released into the ocean. Incoming water is diverted throughout a hatchery building via computer-controlled valves, pumps, filters and other automated systems. Some water is temperature regulated and its gas content manipulated for experimental use. Controlled light and temperature conditions are maintained in larger tanks to achieve regulated spawning for year-round fish production. Much like the automated system that runs a car engine, sensors throughout the hatchery's system allow a computer to maintain desired conditions, and signal an alarm if conditions vary from prescribed limits. The system contains many components, and maintenance of it requires knowledge of everything from computer hardware to large industrial pumps. In addition, the saltwater is very corrosive and constant maintenance is needed to keep everything running smoothly. Much of what is done is experimental, which presents daily challenges to design, build, and run systems that will achieve the desired results. Both the animal caretakers and the machines must work correctly to successfully produce fish.

Lacking an advanced college degree, I was able to draw on past work experience to get the position that I have. In order to be considered for most jobs in marine science, a degree is a virtual necessity. However, experience in any marine-related field is good not only to make you more valuable to an employer, but also to allow you to better choose your goals. My position has exposed me to current developments, which in turn reinforces my desire to continue working in this field and obtain further schooling.

As humans place increasing demands on the ocean, I believe there will be an even larger need for more people to work in many different marine-related areas. Even in such specialized operations such as aquaculture, public aquariums, and research facilities, there are many jobs in animal care, research, lab work, and system operation. The ocean is only recently being studied extensively, and I am looking forward to the investigation into what is found, which should continue for a long time to come.

Sources of Information

- Consult the catalog of any college or university that offers a curriculum in oceanography or marine science.

- The Oceanography Society publishes an excellent brochure entitled *Careers in Oceanography and Marine-Related Fields*. The Oceanography Society can be contacted at 4052 Timber Ridge Drive, Virginia Beach, VA 23455, and their telephone number is (804) 464-0131.

- The National Sea Grant College Program of NOAA publishes a comprehensive brochure entitled *Marine Science Careers: A Sea Grant Guide to Ocean Opportunities,* which includes interviews with working oceanographers. The Sea Grant College can be reached c/o NOAA, SSMC3 Room 11606, 1315 East-West Highway, Silver Springs, MD 20910, and their telephone number is (301) 713-2431.

- The Scripps Institution of Oceanography at the University of California, San Diego, publishes an

Figure A4–1
Joe Cooney at work in the Hubbs Sea World Sea Bass Fish Hatchery in Carlsbad, California.

informative brochure aimed at perspective students entitled *Preparing for a Career in Oceanography*. General information about Scripps can be obtained by contacting the Scripps Communication Office at the Scripps Institution of Oceanography, University of California San Diego, 9500 Gilman Drive, Department 0233, La Jolla, CA 92093-0233, and their telephone number is (619) 534-3624.

- Sea World manages marine and adventure parks and sometimes hires educators and animal trainers—especially those interested in training marine mammals. Their Education Department publishes a booklet on working in a marine life park entitled *The Sea World/Bush Gardens Guide to Zoological Park Careers*, available from Sea World of California Education Department, 1720 South Shores Road, San Diego, CA. 92109-7995, and their telephone number is (619) 226-3834.

- For more information about graduate school, The National Academy of Sciences has published a book entitled *Careers in Science and Engineering: A Student Planning Guide to Graduate School and Beyond* (National Academy Press, 1996). The National Academy of Sciences can be contacted at 2101 Constitution Ave., NW Washington, DC. 20418, and their telephone number is (800) 624-6242.

Some WWW sites that contain oceanography career information on-line include:

- The International Oceanographic Foundation at: http://www.rsmas.miami.edu/iof/

- The U.S. Navy's web site on careers in oceanography at:
  http://www.cnmoc.navy.mil/educate/career-o.html
  or
  http://www.oc.nps.navy.mil/careers.html

- The Office of Naval Research's web site, which includes The Oceanography Society's brochure entitled *Careers in Oceanography and Marine-Related Fields* at:
  http://www.onr.navy.mil/onr/careers/default.htm

- A comprehensive list of information about careers in oceanography, marine science, and marine biology is available through the Scripps Institution of Oceanography Science Library's web site at: http://scilib.ucsd.edu/sio/guide/career.html

- The popular "So You Want to Become A Marine Biologist" web site at:
  http://www-siograddept.ucsd.edu/Web/To_Be_A_Marine_Biologist.html

- A listing of marine laboratories and institutions is available at:
  http://life.bio.sunysb.edu/marinebio/mblabs.html

GLOSSARY

Abiotic environment The nonliving components of an ecosystem.

Abyssal clay Deep-ocean (oceanic) deposits containing less than 30 percent biogenous sediment. Often oxidized and red in color, thus commonly termed red clay.

Abyssal hill Volcanic peaks rising less than 1 kilometer (0.6 mile) above the ocean floor.

Abyssal hill province Deep-ocean regions, particularly in the Pacific Ocean, where oceanic sedimentation rates are so low that abyssal plains do not form and the ocean floor is covered with abyssal hills.

Abyssal plain A flat depositional surface extending seaward from the continental rise or oceanic trenches.

Abyssal storm Storm-like occurrences of rapid current movement affecting the deep-ocean floor. They are believed to be caused by warm- and cold-core eddies of surface currents.

Abyssal zone The benthic environment between 4000 and 6000 meters (13,000 and 20,000 feet).

Abyssopelagic Open-ocean (oceanic) environment below 4000 meters (13,000 feet) in depth.

Acid A substance that releases hydrogen ions (H^+) in solution.

Acoustic Thermometry of Ocean Climate (ATOC) The measurement of ocean-wide changes in water properties such as temperature by transmitting and receiving low-frequency sound signals.

Active margin A continental margin marked by a high degree of tectonic activity, such as those typical of the Pacific rim. Types of active margins include convergent active margins (marked by plate convergence) and transform active margins (marked by transform faulting).

Adiabatic Pertaining to a change in the temperature of a mass resulting from compression or expansion. It requires no addition of heat to or loss of heat from the substance.

Agulhas Current A warm current that carries Indian Ocean water around the southern tip of Africa and into the Atlantic Ocean.

Air mass A large area of air that has a definite area of origin and distinctive characteristics.

Algae One-celled or many-celled plants that have no root, stem, or leaf systems. Simple plants.

Alkaline A substance that releases hydroxide ions (OH^-) in solution. Also called basic.

Amino acid One of more than 20 naturally occurring compounds that contain NH_2 and COOH groups. They combine to form proteins.

Amnesic shellfish poisoning Partial or total loss of memory resulting from poisoning caused by eating shellfish with high levels of domoic acid, a toxin produced by a diatom.

Amphidromic point A nodal or "no-tide" point in the ocean or sea around which the crest of the tide wave rotates during one tidal period.

Amphipoda Crustacean order containing laterally compressed members such as the "beach hoppers."

Anadromous Pertaining to a species of fish that spawns in fresh water and then migrates into the ocean to grow to maturity.

Anaerobic respiration Respiration carried on in the absence of free oxygen (O_2). Some bacteria and protozoans carry on respiration this way.

Andesite A gray, fine-grained volcanic rock composed chiefly of plagioclase and feldspar.

Animalia Kingdom of many-celled animals.

Annelida Phylum of elongated segmented worms.

Anomalistic month The time required for the moon to go from perigee to perigee, 27.5 days.

Anoxic Without oxygen.

Antarctic Bottom Water A water mass that forms in the Weddell Sea, sinks to the ocean floor, and spreads across the bottom of all oceans.

Antarctic Circle The latitude of 66.5 degrees south.

Antarctic Convergence The zone of convergence along the northern boundary of the Antarctic Circumpolar Gyre where the southward-flowing boundary currents of the subtropical gyres converge on the cold Antarctic waters.

Antarctic Divergence The zone of divergence separating the westward-flowing East Wind Drift and the eastward-flowing West Wind Drift.

Antarctic Intermediate Water Antarctic zone surface water that sinks at the Antarctic convergence and flows north at a depth of about 900 meters (2950 feet) beneath the warmer upper-water mass of the South Atlantic subtropical gyre.

Anticyclonic flow The flow of air around a region of high pressure, clockwise in the Northern Hemisphere.

Antilles Current A warm current that flows north seaward of the Lesser Antilles from the north equatorial current of the Atlantic Ocean to join the Florida Current.

Antinode Zone of maximum vertical particle movement in standing waves, where crest and trough formation alternate.

Aphelion The point in the orbit of a planet or comet where it is farthest from the sun.

Aphotic zone Without light. The ocean is generally in this state below 1000 meters (3280 feet).

Apogee The point in the orbit of the moon or an artificial satellite that is farthest from Earth.

Aragonite A form of $CaCO_3$ that is less common and less stable than calcite. Pteropod shells are usually composed of aragonite.

Archaea One of the three major domains of life. The domain consists of simple microscopic bacteria-like creatures (including methane producers and sulfur oxidizers that inhabit deep-sea vents and seeps) and other microscopic life forms, many of which prefer environments of extreme conditions of temperature and/or pressure.

Archipelago A large group of islands.

Arctic Circle The latitude of 66.5 degrees north.

Arctic Convergence A zone of converging currents similar to the Antarctic Convergence but located in the Arctic.

Arrowworm A member of the phylum Chaetognatha. This organism averages about 1 centimeter (0.4 inch) in length and is an important type of zooplankton.

Aschelminthes Phylum of worm-like pseudocoelomates.

Aspect ratio The index of propulsive efficiency obtained by dividing the square of the height by fin area.

Asthenosphere A plastic layer in the upper mantle 80 to 200 kilometers (50 to 124 miles) deep that may allow lateral movement of lithospheric plates and isostatic adjustments.

Atoll A ring-shaped coral reef growing upward from a submerged volcanic peak. It may have low-lying islands composed of coral debris.

Atom A unit of matter, the smallest unit of an element, having all the characteristics of that element and consisting of a dense, central, positively charged nucleus surrounded by a system of electrons.

Autolytic decomposition Decomposition of organic matter that is achieved by enzymes present in the tissue. The enzymes are triggered to begin their work shortly after death of the tissue.

Autotroph Plants and bacteria that can synthesize organic compounds from inorganic nutrients.

Autumnal equinox The passage of the sun across the Equator as it moves from the Northern Hemisphere into the Southern Hemisphere, approximately September 23. During this time, all places in the world experience equal lengths of night and day. Also called the fall equinox.

Backshore The inner portion of the shore, lying landward of the mean spring-tide high-water line. Acted upon by the ocean only during exceptionally high tides and storms.

Backwash The flow of water down the beach face toward the ocean from a previously broken wave.

Bacteria One of the three major domains of life. The domain includes unicellular, prokaryotic microorganisms that vary in terms of morphology, oxygen and nutritional requirements, and motility.

Bacterioplankton Bacteria that live as plankton.

Bacteriovore Organisms that feed on bacteria.

Baleen The fibrous material forming the fringed plates that hang from the upper jaw of baleen whales (suborder Mysticeti) and serve to strain plankton from water.

Bar-built estuary A shallow estuary (lagoon) separated from the open ocean by a bar deposit such as a barrier island. The water in these estuaries usually exhibits vertical mixing.

Barrier flat Lying between the salt marsh and dunes of a barrier island, it is usually covered with grasses and even forests if protected from overwash for sufficient time.

Barrier island A long, narrow, wave-built island separated from the mainland by a lagoon.

Barrier reef A coral reef separated from the nearby landmass by open water.

Basalt A dark-colored volcanic rock characteristic of the ocean crust. Contains minerals with relatively high iron and magnesium content.

Base See *Alkaline*.

Bathyal zone The benthic environment between the depths of 200 and 4000 meters (660 and 13,000 feet). It includes mainly the continental slope and the oceanic ridges and rises.

Bathymetry The measurement of ocean depth.

Bathypelagic zone The pelagic environment between the depths of 1000 and 4000 meters (3300 and 13,000 feet).

Bathyscaphe A specially designed deep-diving submersible.

Bathysphere A specially designed deep-diving submersible that resembles a sphere.

Bay barrier A marine deposit attached to the mainland at both ends and extending entirely across the mouth of a bay, separating the bay from the open water. Also known as a bay-mouth bar.

Beach Sediment seaward of the coastline through the surf zone that is in transport along the shore and within the surf zone.

Beach compartment A series of rivers, beaches, and submarine canyons involved in the movement of sediment to the coast, along the coast, and down one or more submarine canyons.

Beach face The wet, sloping surface that extends from the berm to the shoreline. Also known as the low-tide terrace.

Beach replenishment The addition of beach sediment to replace lost or missing material. Also called beach nourishment.

Beach starvation The interruption of sediment supply and resulting narrowing of beaches.

Benguela Current The cold eastern boundary current of the South Atlantic subtropical gyre.

Benthic Pertaining to the ocean bottom.

Benthos The forms of marine life that live on the ocean bottom.

Berm The dry, gently sloping region on the backshore of a beach at the foot of the coastal cliffs or dunes.

Berm crest The area of a beach that separates the berm from the beach face. The berm crest is often the highest portion of a berm.

Bicarbonate ion (HCO₃⁻) An ion that contains the radical group HCO_3^-.

Bioaccumulation The accumulation of a substance, such as a toxic chemical, in various tissues of a living organism.

Bioerosion Erosion of reef or other solid bottom material by the activities of organisms.

Biogenous sediment Sediment containing material produced by plants or animals, such as coral reefs, shell fragments, and tests of diatoms, radiolarians, foraminifers, and coccolithophores.

Biogeochemical cycle The natural cycling of compounds among the living and nonliving components of an ecosystem.

Biological pump The movement of CO_2 that enters the ocean from the atmosphere through the water column to the sediment on the ocean floor by biological processes—photosynthesis, secretion of shells, feeding, and dying.

Bioluminescence Light produced organically by a chemical reaction. Found in bacteria, phytoplankton, and various fishes (especially deep-sea fish).

Biomass The total mass of a defined organism or group of organisms in a particular community or in the ocean as a whole.

Biomass pyramid A representation of trophic levels that illustrates the progressive decrease in total biomass at successive higher levels of the pyramid.

Bioremediation The technique of using microbes to assist in cleaning toxic spills.

Biotic community The living organisms that inhabit an ecosystem.

Biozone A region of the environment that has distinctive biological characteristics.

Bivalve A mollusk, such as an oyster or a clam, that has a shell consisting of two hinged valves.

Black smoker A hydrothermal vent on the ocean floor that emits a black cloud of hot water filled with dissolved metal particles.

Body wave A longitudinal or transverse wave that transmits energy through a body of matter.

Boiling point The temperature at which a substance changes state from a liquid to a gas at a given pressure.

Bolide An exploding or exploded meteor or meteorite.

Bore A steep-fronted tide crest that moves up a river in association with an incoming high tide.

Boundary current The northward- or southward-flowing currents that form the western and eastern boundaries, respectively, of the subtropical circulation gyres.

Brackish Low-salinity water caused by the mixing of fresh water and saltwater.

Brazil Current The warm western boundary current of the South Atlantic subtropical gyre.

Breaker zone Region where waves break at the seaward margin of the surf zone.

Breakwater Any artificial structure constructed roughly parallel to shore and designed to protect a coastal region from the force of ocean waves.

Brittle Descriptive term for a substance that is likely to fracture when force is applied to it.

Bryozoa Phylum of colonial animals that often share one coelomic cavity. Encrusting and branching forms secrete a protective housing (zooecium) of calcium carbonate or chitinous material. Possess lophophore feeding structure.

Buffering The process by which a substance minimizes a change in the acidity of a solution when an acid or base is added to the solution.

Buoyancy The ability or tendency to float or rise in a liquid.

Calcareous Containing calcium carbonate.

Calcite The mineral of the chemical formula $CaCO_3$.

Calcite compensation depth (CCD) The depth at which the amount of calcite ($CaCO_3$) produced by the organisms in the overlying water column is equal to the amount of calcite the water column can dissolve. There will be no calcite deposition below this depth, which, in most parts of the ocean, is at a depth of 4.5 kilometers (2.8 miles).

Calcium carbonate (CaCO₃) A chalk-like substance secreted by many organisms in the form of coverings or skeletal structures.

California Current The cold eastern boundary current of the North Pacific subtropical gyre.

Calorie Unit of heat, defined as the amount of heat required to raise the temperature of 1 gram of water 1 degree centigrade.

Calving The process by which a glacier breaks at an edge, so that a portion of the ice separates and falls from the glacier.

Canary Current The cold eastern boundary current of the North Atlantic subtropical gyre.

Capillarity The action by which a fluid, such as water, is drawn up in small tubes as a result of surface tension.

Capillary wave An ocean wave whose wavelength is less than 1.74 centimeters (0.69 inch). The dominant restoring force for such waves is surface tension.

Carapace Chitinous or calcareous shield that covers the cephalothorax of some crustaceans. Dorsal portion of a turtle shell.

Carbohydrate An organic compound containing the elements carbon, hydrogen, and oxygen with the general formula $(CH_2O)_n$.

Carbonate ion (CO₃)⁻² An ion that contains the radical group CO_3^-.

Caribbean Current The warm current that carries equatorial water across the Caribbean Sea into the Gulf of Mexico.

Carnivore An animal that depends on other animals solely or chiefly for its food supply.

Carotin A red to yellow pigment found in plants.

Celerity Speed of travel of a wave form.

Centigrade temperature scale (°C) A temperature scale based on the freezing point (0 degrees centigrade = 32 degrees Fahrenheit) and boiling point (100 degrees centigrade = 212 degrees Fahrenheit) of pure water. Also known as the Celsius scale after its founder.

Centripetal force A center-seeking force that tends to make rotating bodies move toward the center of rotation.

Cephalopoda A class of the phylum Mollusca with a well-developed pair of eyes and a ring of tentacles surrounding the mouth. The shell is absent or internal on most members. The class includes the squid, octopus, and nautilus.

Cetacea An order of marine mammals that includes the whales, dolphins, and porpoises.

Chaetognatha A phylum of elongate, transparent, worm-like pelagic animals commonly called arrowworms.

Chalk A soft, compact form of calcite, generally gray-white or yellow-white in color and derived chiefly from microscopic fossils.

Chemical energy A form of potential energy stored in the chemical bonds of compounds.

Chemosynthesis A process by which bacteria synthesize organic molecules from inorganic nutrients using chemical energy released from the bonds of some chemical compound by oxidation.

Chloride ion, Cl⁻ A chlorine atom that has become negatively charged by gaining one electron.

Chlorinity The amount of chloride ion and ions of other halogens in ocean water expressed in parts per thousand (‰) by weight.

Chlorophyll A group of green pigments that make it possible for plants to carry on photosynthesis.

Chlorophyta Green algae. Characterized by the presence of chlorophyll and other pigments.

Chondrite A stony meteorite composed primarily of silicate rock material and containing chondrules (spheroidal granules). They are the most commonly found meteorites.

Chronometer An exceptionally precise timepiece.

Chrysophyta An important phylum of planktonic algae that includes the diatoms. The presence of chlorophyll is masked by the pigment carotin, which gives the plants a golden color.

Cilium A short, hair-like structure common on lower animals. Beating in unison, cilia may create water currents that carry food toward the mouth of an animal or may be used for locomotion.

Circadian rhythm Behavioral and physiological rhythms of organisms related to the 24-hour day. Sleeping and waking patterns are an example.

Circular orbital motion The motion of water particles caused by a wave as the wave is transmitted through water.

Circumpolar Current Eastward-flowing current that extends from the surface to the ocean floor and encircles Antarctica.

Clay 1. A particle size between silt and colloid. **2.** Any of various hydrous aluminum silicate minerals that are plastic, expansive, and have ion-exchange capabilities.

Climate The meteorological conditions, including temperature, precipitation, and wind, that characteristically prevail in a particular region; the long-term average of weather.

Cnidoblast Stinging cell of phylum Coelenterata that contains the stinging mechanism (nematocyst) used in defense and capturing prey.

Coast A strip of land that extends inland from the coastline as far as marine influence is evidenced in the landforms.

Coastal geostrophic current See *Geostrophic current*.

Coastal plain estuary An estuary formed by rising sea level flooding a coastal river valley.

Coastal upwelling The movement of deeper nutrient-rich water into the surface water mass as a result of wind-blown surface water moving offshore.

Coastal waters Those relatively shallow water areas that adjoin continents or islands.

Coastline Landward limit of the effect of the highest storm waves on the shore.

Coccolith Tiny calcareous disks averaging about 3 micrometers (0.00012 inch) in diameter that form the cell wall of coccolithophores.

Coccolithophore A microscopic planktonic form of algae encased by a covering composed of calcareous disks (coccoliths).

Coelenterata Phylum of radially symmetrical animals that includes two basic body forms, the medusa and the polyp. Includes jellyfish (medusoid) and sea anemones (polypoid). Preferred name is now *Cnidaria*.

Cohesion The intermolecular attraction by which the elements of a body are held together.

Cold-core ring A circular eddy of a surface current that contains cold water in its center and rotates counterclockwise.

Cold front A weather front in which a cold air mass moves into and under a warm air mass. It creates a narrow band of intense precipitation.

Cold-blooded See *ectothermic*.

Colonial animals Animals that live in groups of attached or separate individuals. Groups of individuals may serve special functions.

Columbia River Estuary An estuary at the border between the states of Washington and Oregon that has been most adversely affected by the construction of hydroelectric dams.

Comb jelly Common name for members of the phylum Ctenophora. (See *Ctenophora*.)

Commensalism A symbiotic relationship in which one party benefits and the other is unaffected.

Compensation depth for photosynthesis The depth at which net photosynthesis becomes zero and photosynthetic organisms can no longer survive. This depth is greater in the open ocean (up to 100 meters or 330 feet) than near the shore, due to factors that limit light penetration in coastal regions.

Condensation The conversion of water from the vapor to the liquid state. When it occurs, the energy required to vaporize the water is released into the atmosphere. This is about 585 calories per gram of water at 20 degrees centigrade.

Conduction The transmission of heat by the passage of energy from particle to particle.

Conjunction The apparent closeness of two heavenly bodies. During the new-moon phase, Earth and the moon are in conjunction on the same side of the sun.

Constant proportions, principle of A principle that states that the major constituents of ocean water salinity are found in the same relative proportions throughout the ocean water volume, independent of salinity.

Constructive interference A form of wave interference in which two waves come together in phase, for example, crest to crest, to produce a greater displacement from the still water line than that produced by either of the waves alone.

Consumer An animal within an ecosystem that consumes the organic mass produced by the producers.

Continent About one-third of Earth's surface that rises above the deep-ocean floor to be exposed above sea level. Continents are composed primarily of granite, an igneous rock of lower density than the basaltic oceanic crust.

Continental accretion Growth or increase in size of a continent by gradual external addition of crustal material.

Continental arc An arc-shaped row of active volcanoes produced by subduction that occurs along convergent active continental margins.

Continental borderland A highly irregular portion of the continental margin that is submerged beneath the ocean and is characterized by depths greater than those characteristic of the continental shelf.

Continental drift A term applied to early theories supporting the possibility the continents are in motion over Earth's surface.

Continental margin The submerged area next to a continent comprising the continental shelf, continental slope, and continental rise.

Continental rise A gently sloping depositional surface at the base of the continental slope.

Continental shelf A gently sloping depositional surface extending from the low water line to the depth of a marked increase in slope around the margin of a continent or island.

Continental slope A relatively steeply sloping surface lying seaward of the continental shelf.

Convection Heat transfer in a gas or liquid by the circulation of currents from one region to another.

Convection cell A circular-moving loop of matter involved in convective movement.

Convergence The act of coming together from different directions. There are polar, tropical, and subtropical regions of the oceans where water masses with different characteristics come together. Along these lines of convergence, the denser mass will sink beneath the others.

Convergent plate boundary A lithospheric plate boundary where adjacent plates converge, producing ocean trench-island arc systems, ocean trench-continental volcanic arcs, or folded mountain ranges.

Copepoda An order of microscopic to nearly microscopic crustaceans that are important members of the zooplankton in temperate and subpolar waters.

Coral A group of benthic anthozoans that exist as individuals or in colonies and secrete $CaCO_3$ external skeletons. Under the proper conditions, corals may produce reefs composed of their external skeletons and the $CaCO_3$ material secreted by varieties of algae associated with the reefs.

Coral bleaching The loss of color in coral reef organisms that causes them to turn white. The bleaching is due to the removal or expulsion of the coral's symbiotic partner, the zooxanthellae algae, in response to high water temperatures or other adverse conditions.

Coral reef A calcareous organic reef composed significantly of solid coral and coral sand. Algae may be responsible for more than half of the $CaCO_3$ reef material. Found in waters where the minimum average monthly temperature is 18 degrees centigrade (64 degrees Fahrenheit) or higher.

Core 1. The core of Earth is composed primarily of iron and nickel. It has a liquid outer portion 2270 kilometers (1410 miles) thick and a solid inner core with a radius of 1216 kilometers (756 miles). **2.** A cylinder of sediment and/or rock material.

Coriolis effect An apparent force resulting from Earth's rotation causes particles in motion to be deflected to the right in the Northern Hemisphere and to the left in the Southern Hemisphere from a frame of reference on Earth's surface.

Cosmogenous sediment All sediment derived from outer space.

Cotidal lines Lines connecting points where high tide occurs simultaneously.

Counter-shading Protective coloration in an animal or insect, characterized by darker coloring of areas exposed to light and lighter coloring of areas that are normally shaded.

Covalent bond A chemical bond formed by the sharing of one or more electrons, especially pairs of electrons, between atoms.

Crest (wave) The portion of an ocean wave that is displaced above the still water level.

Cruiser Fish, such as the bluefin tuna, that constantly cruise the pelagic waters in search of food.

Crust 1. Unit of Earth's structure that is composed of basaltic ocean crust and granitic continental crust. The total thickness of the crustal units may range from 5 kilometers (3 miles) beneath the ocean to 50 kilometers (30 miles) beneath the continents. **2.** A hard covering or surface layer of hydrogenous sediment.

Crustacea A class of phylum Arthropoda that includes barnacles, copepods, lobsters, crabs, and shrimp.

Crystalline rock Igneous or metamorphic rocks. These rocks are made up of crystalline particles with orderly molecular structures.

Ctenophora A phylum of gelatinous organisms that are more or less spheroidal with biradial symmetry. These exclusively marine animals have eight rows of ciliated combs for locomotion, and most have two tentacles for capturing prey.

Current A horizontal movement of water.

Cutaneous artery The artery that runs down both sides of some cruiser-type fish to help maintain a constant elevated temperature in the myomere musculature used for swimming.

Cutaneous vein The vein that runs down both sides of some cruiser-type fish to help maintain a constant elevated temperature in the myomere musculature used for swimming.

Cyclone An atmospheric system characterized by the rapid, inward circulation of air masses about a low-pressure center, usually accompanied by stormy, often destructive, weather. Cyclones circulate counterclockwise in the Northern Hemisphere and clockwise in the Southern Hemisphere.

Cyclonic flow The flow of air around a region of low pressure, counterclockwise in the Northern Hemisphere.

Davidson Current A northward-flowing current along the Washington-Oregon coast that is driven by geostrophic effects on a large freshwater runoff.

DDT An insecticide that caused damage to marine bird populations in the 1950s and 1960s. Its use is now banned throughout most of the world.

Declination The angular distance of the sun or moon above or below the plane of Earth's Equator.

Decomposers Primarily bacteria that break down nonliving organic material, extract some of the products of decomposition for their own needs, and make available the compounds needed for plant production.

Deep boundary current Relatively strong, deep currents flowing across the continental rise along the western margin of ocean basins.

Deep current Density-driven circulation that is initiated at the ocean surface by temperature and salinity conditions that produce a high-density water mass, which sinks and spreads slowly beneath surface waters.

Deep-ocean basin Areas of the ocean floor that have deep water, are far from land, and are underlain by basaltic crust.

Deep scattering layer (DSL) A layer of marine organisms in the open ocean that scatter signals from an echo sounder. It migrates daily from depths of slightly over 100 meters (330 feet) at night to more than 800 meters (2600 feet) during the day.

Deep-sea fan A large fan-shaped deposit commonly found on the continental rise seaward of such sediment-laden rivers as the Amazon, Indus, or Ganges-Brahmaputra. Also known as a submarine fan.

Deep-sea fish Any of a large group of fishes that lives within the aphotic zone and has special adaptations for finding food and avoiding predators in darkness.

Deep-sea system Includes all benthic environments beneath the littoral (sublittoral, bathyal, abyssal, and hadal).

Deep-water The water beneath the permanent thermocline (and resulting pycnocline) that has a uniformly low temperature.

Deep-water wave Ocean wave traveling in water that has a depth greater than one-half the average wavelength. Its speed is independent of water depth.

Delta A low-lying deposit at the mouth of a river, usually having a triangular shape as viewed from above.

Denitrifying bacteria Bacteria that reduce oxides of nitrogen to produce free nitrogen (N_2).

Density Mass per unit volume of a substance. Usually expressed as grams per cubic centimeter (g/cm^3). For ocean water with a salinity of 35‰ at 0 degrees centigrade, the density is 1.028 g/cm^3.

Density stratification Layering based on density, where the highest-density material occupies the lowest space.

Deposit feeder An organism that feeds on food items that occur as deposits, including detritus and various detritus-coated sediments.

Depositional-type shores Shorelines dominated by processes that form deposits (such as sand bars and barrier islands) along the shore.

Desalination The removal of salt ions from ocean water to produce pure water.

Destructive interference A form of wave interference in which two waves come together out of phase, for example, crest to trough, and produce a wave with less displacement than the larger of the two waves would have produced alone.

Detritus 1. Any loose material produced directly from rock disintegration. **2.** Material resulting from the disintegration of dead organic remains.

Diatom Member of the class Bacillariophyceae of algae that possesses a wall of overlapping silica valves.

Diatomaceous earth A deposit composed primarily of the tests of diatoms mixed with clay. Also called diatomite.

Dipolar Having two poles. The water molecule possesses a polarity of electrical charge, with one pole being more positive and the other more negative in electrical charge.

Diffraction Any change in the direction or intensity of a wave after passing an obstacle that cannot be interpreted as refraction or reflection.

Diffusion A process by which fluids move through other fluids from areas of high concentration to areas in which they are in lower concentrations by random molecular movement.

Dinoflagellates Single-celled microscopic organisms that may possess chlorophyll and belong to the plant phylum Pyrrophyta (autotrophic) or may ingest food and belong to the class Mastigophora of the animal phylum Protozoa (heterotrophic).

Discontinuity An abrupt change in a property such as temperature or salinity at a line or surface.

Disphotic zone The dimly lit zone, corresponding approximately to the mesopelagic, in which there is not enough light to carry on photosynthesis; sometimes called the twilight zone.

Disruptive coloration A marking or color pattern that confuses prey.

Dissolved oxygen Oxygen that is dissolved in ocean water.

Distillation A method of purifying liquids by heating them to their boiling point and condensing the vapor.

Distributary A small stream flowing away from a main stream. Such streams are characteristic of deltas.

Diurnal inequality The difference in the heights of two successive high tides or two successive low tides during a lunar (tidal) day.

Diurnal tidal pattern A tidal pattern exhibiting one high tide and one low tide during a tidal day; a daily tide.

Divergence A horizontal flow of water from a central region, as occurs in upwelling.

Divergent plate boundary A lithospheric plate boundary where adjacent plates diverge, producing an oceanic ridge or rise (spreading center).

Doldrums A belt of light variable winds within 10 to 15 degrees of the Equator, resulting from the vertical flow of low-density air within this belt.

Dolphin 1. A brilliantly colored fish of the genus *Coryphaena*. **2.** The name applied to the small, beaked members of the cetacean family Delphinidae.

Dorsal Pertaining to the back or upper surface of most animals.

Dorsal aorta For most fish, this is the only major artery that runs the length of the fish through openings in the vertebrae and supplies blood. Some pelagic cruisers also have cutaneous arteries.

Downwelling In the open or coastal ocean where Ekman transport causes surface waters to converge or impinge on the coast, surface water will be carried down beneath the surface.

Drift bottle Any equipment used to study current movement by drifting with currents.

Drifts Thick sediment deposits on the continental rise produced where the deep boundary current slows and loses sediment when it changes direction to follow the base of the continental slope.

Drowned beach An ancient beach now beneath the coastal ocean because of rising sea level.

Drowned river valley The lower part of a river valley that has been submerged by rising sea level.

Dunes Coastal deposits of sand lying landward of the beach and deriving their sand from onshore winds that transport beach sand inland.

Dynamic topography A surface configuration resulting from the geopotential difference between a given surface and a reference surface of no motion. A contour map of this surface is useful in estimating the nature of geostrophic currents.

Earthquake A sudden motion or trembling in Earth, caused by the sudden release of slowly accumulated strain by faulting (movement along a fracture in Earth's crust) or volcanic activity.

East Australian Current The warm western boundary current of the South Pacific subtropical gyre.

East Pacific Rise A fast-spreading divergent plate boundary extending southward from the Gulf of California through the eastern South Pacific Ocean.

East Wind Drift The coastal current driven in a westerly direction by the polar easterly winds blowing off Antarctica.

Eastern boundary current Equatorward-flowing cold drifts of water on the eastern side of all subtropical gyres.

Ebb current The flow of water seaward during a decrease in the height of the tide.

Ebb tide A receding or outgoing tide.

Echinodermata Phylum of animals that have bilateral symmetry in larval forms and usually a five-sided radial symmetry as adults. Benthic and possessing rigid or articulating exoskeletons of calcium carbonate with spines, this phylum includes sea stars, brittle stars, sea urchins, sand dollars, sea cucumbers, and sea lilies.

Echolocation A sensory system in odontocete cetaceans in which usually high-pitched sounds are emitted and their echoes interpreted to determine the direction and distance of objects.

Echosounder A device that transmits sound from a ship's hull to the ocean floor where it is reflected back to receivers. The speed of sound in the water is known, so the depth can be determined from the travel time of the sound signal.

Ecliptic The plane of the center of the Earth–moon system as it orbits around the sun.

Ecosystem All the organisms in a biotic community and the abiotic environmental factors with which they interact.

Ectothermic Of or relating to an organism that regulates its body temperature largely by exchanging heat with its surroundings; cold-blooded.

Eddy A current of any fluid forming on the side of a main current. It usually moves in a circular path and develops where currents encounter obstacles or flow past one another.

Ekman spiral A theoretical consideration of the effect of a steady wind blowing over an ocean of unlimited depth and breadth and of uniform viscosity. The result is a surface flow at 45 degrees to the right of the wind in the Northern Hemisphere. Water at increasing depth will drift in directions increasingly to the right until at about 100 meters (330 feet) depth it is moving in a direction opposite to that of the wind. The net water transport is 90 degrees to the wind, and speed decreases with depth.

Ekman transport The net transport of surface water set in motion by wind. Due to the Ekman spiral phenomenon, it is theoretically in a direction 90 degrees to the right and 90 degrees to the left of the wind direction in the Northern Hemisphere and Southern Hemisphere, respectively.

El Niño A southerly flowing warm current that generally develops off the coast of Ecuador around Christmastime. Occasionally it will move farther south into Peruvian coastal waters and cause the widespread death of plankton and fish.

El Niño–Southern Oscillation (ENSO) The correlation of El Niño events with an oscillatory pattern of pressure change in a persistent high-pressure cell in the southeastern Pacific Ocean and a persistent low-pressure cell over the East Indies.

Electrical conductivity The ability or power to conduct or transmit electricity.

Electrolysis A separation process by which salt ions are removed from saltwater through water-impermeable membranes toward oppositely charged electrodes.

Electromagnetic energy Energy that travels as waves or particles with the speed of light. Different kinds possess different properties based on wavelength. The longest wavelengths belong to radio waves, up to 100 kilometers (60 miles) in length. At the other end of the spectrum are cosmic rays with greater penetrating power and wavelengths of less than 0.000001 micrometer.

Electromagnetic spectrum The spectrum of radiant energy emitted from stars and ranging between cosmic rays with wavelengths of less than 10 to 11 centimeters (4 to 4.3 inches) and very long waves with wavelengths in excess of 100 kilometers (60 miles).

Electron A subatomic particle that orbits the nucleus of an atom and has a negative electric charge.

Electrostatic force A force caused by electric charges at rest.

Element One of a number of substances, each of which is composed entirely of like particles—atoms—that cannot be broken into smaller particles by chemical means.

Emerging shoreline A shoreline resulting from the emergence of the ocean floor relative to the ocean surface. It is usually rather straight and characterized by marine features usually found at some depth.

Endothermic reaction A chemical reaction that absorbs energy. For example, energy is stored in the organic products of the chemical reaction photosynthesis.

Entropy A thermodynamic quantity that reflects the degree of randomness in a system. It increases in all natural systems.

Environment The sum of all physical, chemical, and biological factors to which an organism or community is subjected.

Epicenter The point on Earth's surface that is directly above the focus of an earthquake.

Epifauna Animals that live on the ocean bottom, either attached or moving freely over it.

Epipelagic zone A subdivision of the oceanic province that extends from the surface to a depth of 200 meters (660 feet).

Equator The imaginary great circle around the Earth's surface, equidistant from the poles and perpendicular to the Earth's axis of rotation. It divides the Earth into the Northern Hemisphere and the Southern Hemisphere.

Equatorial Pertaining to the equatorial region.

Equatorial countercurrent Eastward-flowing currents found between the north and south equatorial currents in all oceans, but particularly well developed in the Pacific Ocean.

Equatorial current Westward-flowing currents that travel along the Equator in all ocean basins and caused by the trade winds. They are called North or South Equatorial Currents depending on their position north or south of the Equator.

Equatorial low A band of low atmospheric pressure that encircles the globe along the Equator.

Erosion The group of natural processes, including weathering, dissolution, abrasion, corrosion, and transportation, by which material is worn away from the Earth's surface.

Erosional-type shores Shorelines dominated by processes that form erosional features (such as cliffs and sea stacks) along the shore.

Estuarine circulation pattern A flow pattern in an estuary characterized by a net surface flow of low-salinity water toward the ocean and an opposite net subsurface flow of seawater toward the head of the estuary.

Estuary A partially enclosed coastal body of water in which salty ocean water is significantly diluted by fresh water from land runoff. Examples of estuaries include river mouths, bays, inlets, gulfs, and sounds.

Eukarya One of the three major domains of life. The domain includes single-celled or multicellular organisms whose cells usually contain a distinct membrane-bound nucleus.

Euphotic zone A layer that extends from the surface of the ocean to a depth where enough light exists to support photosynthesis, rarely deeper than 100 meters (330 feet).

Euryhaline Descriptive term for organisms with a high tolerance for a wide range of salinity conditions.

Eurythermal Descriptive term for organisms with a high tolerance for a wide range of temperature conditions.

Eustatic sea level change A worldwide raising or lowering of sea level.

Eutrophic A region of high productivity.

Eutrophication The enrichment of waters by a previously scarce nutrient.

Evaporation To change from a liquid to the vapor state.

Evaporite A sedimentary deposit that is left behind when water evaporates. Evaporite minerals include gypsum, calcite, and halite.

Exclusive Economic Zone (EEZ) A coastal zone 200 nautical miles wide over which the coastal nation has jurisdiction over mineral resources, fishing, and pollution. If the continental shelf extends beyond 200 miles, the EEZ may be up to 350 miles (about 560 kilometers) in width.

Exothermic reaction A chemical reaction that liberates energy. For example, the energy stored in the products of photosynthesis is released by the chemical reaction respiration.

Extrusive rocks Igneous rocks that flow out onto Earth's surface before cooling and solidifying (lavas).

Eye **1.** An organ of vision or of light sensitivity. **2.** The circular area of relative calm at the center of a hurricane.

Fahrenheit temperature scale (°F) A temperature scale whereby the freezing point of water is 32 degrees and the boiling point of water is 212 degrees.

Falkland Current A northward-flowing cold current found off the southeastern coast of South America.

Fall equinox See *Autumnal equinox.*

Fan A gently sloping, fan-shaped feature normally located near the lower end of a canyon. Also known as a submarine fan.

Fat An organic compound formed from alcohol, glycerol, and one or more fatty acids; a lipid, it is a solid at atmospheric temperatures.

Fathom (fm) A unit of depth in the ocean, commonly used in countries using the English system of units. It is equal to 1.83 meters (6 feet).

Fault A fracture or fracture zone in Earth's crust along which displacement has occurred.

Fault block A crustal block bounded on at least two sides by faults. Usually elongate; if it is downdropped, it produces a graben; if it is uplifted, it is a horst.

Fauna The animal life of any particular area or of any particular time.

Fecal pellet Excrement of planktonic crustaceans that assist in speeding up the descent rate of sedimentary particles by combining them into larger packages.

Ferrel cell The large atmospheric circulation cell that occurs between 30 and 60 degrees latitude in each hemisphere.

Ferromagnesium Minerals rich in iron and magnesium.

Fetch 1. Pertaining to the area of the open ocean over which the wind blows with constant speed and direction, thereby creating a wave system. **2.** The distance across the fetch (wave-generating area) measured in a direction parallel to the direction of the wind.

Filter feeder An organism that obtains its food by filtering seawater for other organisms. Also known as suspension feeder.

Fjord A long, narrow, deep, U-shaped inlet that usually represents the seaward end of a glacial valley that has become partially submerged after the melting of the glacier.

Flagellum A whip-like living structure used by some cells for locomotion.

Flood current A tidal current associated with increasing height of the tide, generally moving toward the shore.

Flood tide An incoming or rising tide.

Flora The plant life of any particular area or of any particular time.

Florida Current A warm current flowing north along the coast of Florida. It becomes the Gulf Stream.

Folded mountain range Mountain ranges formed as a result of the convergence of lithospheric plates. They are characterized by masses of folded sedimentary rocks that formed from sediments deposited in the ocean basin that was destroyed by the convergence.

Food chain The passage of energy materials from producers through a sequence of a herbivore and a number of carnivores.

Food web A group of interrelated food chains.

Foraminifer An order of planktonic and benthic protozoans that possess protective coverings, usually composed of calcium carbonate.

Forced wave A wave that is generated and maintained by a continuous force such as the gravitational attraction of the moon.

Foreshore The portion of the shore lying between the normal high and low water marks; the intertidal zone.

Fossil Any remains, print, or trace of an organism that has been preserved in Earth's crust.

Fracture zone An extensive linear zone of unusually irregular topography of the ocean floor, characterized by large seamounts, steep-sided or asymmetrical ridges, troughs, or long, steep slopes. Usually represents ancient, inactive transform fault zones.

Free wave A wave created by a sudden rather than a continuous impulse that continues to exist after the generating force is gone.

Freeze separation The desalination of seawater by multiple episodes of freezing, rinsing, and thawing.

Freezing The process by which a liquid is converted to a solid at its freezing point.

Freezing point The temperature at which a liquid becomes a solid under any given set of conditions. The freezing point of water is 0 degrees centigrade at one atmosphere pressure.

Fringing reef A reef that is directly attached to the shore of an island or continent. It may extend more than 1 kilometer (0.6 mile) from shore. The outer margin is submerged and often consists of algal limestone, coral rock, and living coral.

Full moon The phase of the moon that occurs when the sun and moon are in opposition, that is, they are on opposite sides of Earth. During this time, the lit side of the moon faces Earth.

Fully developed sea The maximum average size of waves that can be developed for a given wind speed when it has blown in the same direction for a minimum duration over a minimum fetch.

Fungi Any of numerous eukaryotic organisms of the kingdom Fungi, which lack chlorophyll and vascular tissue and range in form from a single cell to a body mass of branched filamentous structures that often produce specialized fruiting bodies. The kingdom includes the yeasts, molds, lichens, and mushrooms.

Fur seal Any of several eared seals of the genera *Callorhinus* or *Arctocephalus*, having thick, soft underfur.

Fusion reaction A type of nuclear reaction where hydrogen atoms are converted to helium atoms, thereby releasing large amounts of energy.

Galápagos Rift A divergent plate boundary extending eastward from the Galápagos Islands toward South America. The first deep-sea hydrothermal vent biocommunity was discovered here in 1977.

Galaxy One of the billions of large systems of stars that make up the universe.

Gas hydrate A lattice-like compound composed of water and natural gas (usually methane) formed in high-pressure and low-temperature environments such as those found in deep-ocean sediments. Also known as clathrates because of their cage-like chemical structure.

Gaseous state A state of matter in which molecules move by translation and interact only through chance collisions.

Gastropoda A class of mollusks, most of which possess an asymmetrical, spiral one-piece shell and a well-developed flattened foot. A well-developed head will usually have two eyes and one or two pairs of tentacles. Includes snails, limpets, abalone, cowries, sea hares, and sea slugs.

Geostrophic current A current that grows out of Earth's rotation and is the result of a near balance between gravitational force and the Coriolis effect.

Gill A thin-walled projection from some part of the external body or the digestive tract used for respiration in a water environment.

Glacial deposit A sedimentary deposit formed by a glacier and characterized by poor sorting.

Glacier A large mass of ice formed on land by the recrystallization of old, compacted snow. It flows from an area of accumulation to an area of wasting where ice is removed from the glacier by melting.

Glauconite A group of green hydrogenous minerals consisting of hydrous silicates of potassium and iron.

Global Positioning System (GPS) A system of satellites that transmit microwave signals to Earth, allowing people at the surface to locate themselves accurately.

Globigerina ooze An ooze that contains a large percentage of the calcareous tests of the foraminifer Globigerina.

Goiter An enlargement of the thyroid gland, visible as a swelling at the front of the neck, that is often associated with iodine deficiency.

Gondwanaland A hypothetical protocontinent of the Southern Hemisphere named for the Gondwana region of India. It included the present continental masses Africa, Antarctica, Australia, India, and South America.

Graded bedding Stratification in which each layer displays a decrease in grain size from bottom to top.

Gradient The rate of increase or decrease of one quantity or characteristic relative to a unit change in another. For example, the slope of the ocean floor is a change in elevation (a vertical linear measurement) per unit of horizontal distance covered. Commonly measured in meters per kilometer.

Grain size The average size of the grains of material in a sample. Also known as fragment or particle size.

Granite A light-colored igneous rock characteristic of the continental crust. Rich in nonferromagnesian minerals such as feldspar and quartz.

Gravitational force The force of attraction that exists between any two bodies in the universe that is proportional to the product of their masses and inversely proportional to the distance between the centers of their masses.

Gravity wave A wave for which the dominant restoring force is gravity. Such waves have a wavelength of more than 1.74 centimeters (0.69 inch), and their speed of propagation is controlled mainly by gravity.

Gray whale A baleen whale (*Eschrichtius robustus*) of northern Pacific waters, having grayish-black coloring with white blotches.

Greenhouse effect The heating of Earth's atmosphere that results from the absorption by components of the atmosphere such as water vapor and carbon dioxide of infrared radiation from Earth's surface.

Groin A low artificial structure built perpendicular to the shore and designed to interfere with longshore transportation of sediment so that it traps sand and widens the beach on its upstream side.

Groin field A series of closely spaced groins.

Gross ecological efficiency The amount of energy passed on from a trophic level to the one above it divided by the amount it received from the one below it.

Gross primary production The total carbon fixed into organic molecules through photosynthesis or chemosynthesis by a discrete autotrophic community.

Grunion A small fish (*Leuresthes tenuis*) of coastal waters of California and Mexico that spawns at night along beaches during the high tides of spring and summer. Grunion time their reproductive activities to coincide with tidal phenomena and are the only fish that come completely out of water to spawn.

Gulf Stream The high-intensity western boundary current of the North Atlantic Ocean subtropical gyre that flows north off the east coast of the United States.

Guyot See *Tablemount.*

Gypsum A colorless, white, or yellowish evaporite mineral, $CaSO_4 \cdot 2H_2O$.

Gyre A large, horizontal, circular-moving loop of water. Used mainly in reference to the circular motion of water in each of the major ocean basins centered in subtropical high-pressure regions.

Habitat A place where a particular plant or animal lives. Generally refers to a smaller area than environment.

Hadal Pertaining to the deepest ocean environment, specifically that of ocean trenches deeper than 6 kilometers (3.7 miles).

Hadal zone Pertaining to the deepest ocean benthic environment, specifically that of ocean trenches deeper than 6 kilometers (3.7 miles).

Hadley cell The large atmospheric circulation cell that occurs between the Equator and 30 degrees latitude in each hemisphere.

Half-life The time required for half the atoms of a sample of a radioactive isotope to decay to an atom of another element.

Halite A colorless or white evaporite mineral, NaCl, which occurs as cubic crystals and is used as table salt.

Halocline A layer of water in which a high rate of change in salinity in the vertical dimension is present.

Hard stabilization Any form of artificial structure built to protect a coast or to prevent the movement of sand along a beach. Examples include groins, breakwaters, and seawalls.

Headland A steep-faced irregularity of the coast that extends out into the ocean.

Heat Energy moving from a high-temperature system to a lower-temperature system. The heat gained by the one system may be used to raise its temperature or to do work.

Heat budget The equilibrium that exists on the average between the amount of heat absorbed by Earth and its atmosphere in one year and the amount of heat radiated back into space in one year.

Heat capacity Usually defined as the amount of heat required to raise the temperature of 1 gram of a substance by 1 degree centigrade.

Heat energy Energy of molecular motion. The conversion of higher forms of energy such as radiant or mechanical energy to heat energy within a system increases the heat energy within the system and the temperature of the system.

Heat flow (flux) The quantity of heat flow to Earth's surface per unit of time.

Hemoglobin A red pigment found in red blood corpuscles that carries oxygen from the lungs to tissue and carbon dioxide from tissue to lungs.

Herbivore An animal that relies chiefly or solely on plants for its food.

Hermatypic coral Reef-building corals that have symbiotic algae in their ectodermal tissue. They cannot produce a reef structure below the euphotic zone.

Heterotroph Animals and bacteria that depend on the organic compounds produced by other animals and plants as food. Organisms not capable of producing their own food by photosynthesis.

High slack water The period of time associated with the peak of high tide when there is no visible flow of water into or out of bays and rivers.

High tide zone That portion of the littoral zone that lies between the lowest high tides and highest high tides that occur in an area. It is, on the average, exposed to desiccation for longer periods each day than it is covered by water.

High water (HW) The highest level reached by the rising tide before it begins to recede.

Higher high water (HHW) The higher of two high waters occurring during a tidal day where tides exhibit a mixed tidal pattern.

Higher low water (HLW) The higher of two low waters occurring during a tidal day where tides exhibit a mixed tidal pattern.

Highly stratified estuary A relatively deep estuary in which a significant volume of marine water enters as a subsurface flow. A large volume of freshwater stream input produces a widespread low-surface-salinity condition that produces a well-developed halocline throughout most of the estuary.

Holoplankton Organisms that spend their entire life as members of the plankton.

Homeothermic Of or relating to an animal that maintains a precisely controlled internal body temperature using its own internal heating and cooling mechanisms; warm blooded.

Horse latitudes The latitude belts between 30 and 35 degrees north and south latitude where winds are light and variable, since the principal movement of air masses at these latitudes is one of vertical descent.

The climate is hot and dry, resulting in the creation of the major continental and maritime deserts of the world.

Hotspot The relatively stationary surface expression of a persistent column of molten mantle material rising to the surface.

Hurricane A tropical cyclone in which winds reach speeds in excess of 120 kilometers (74 miles) per hour. Generally applied to such storms in the North Atlantic Ocean, eastern North Pacific Ocean, Caribbean Sea, and Gulf of Mexico. Such storms in the western Pacific Ocean are called typhoons, and those in the Indian Ocean are known as cyclones.

Hydration The condition of being surrounded by water molecules, as when sodium chloride dissolves in water.

Hydrocarbon Any organic compound consisting only of hydrocarbon and carbon. Crude oil is a mixture of hydrocarbons.

Hydrocarbon seep biocommunity Deep bottom-dwelling community of organisms associated with a hydrocarbon seep from the ocean floor. The community depends on methane and sulfur-oxidizing bacteria as producers. The bacteria may live free in the water, on the bottom, or symbiotically in the tissues of some of the animals.

Hydrogen bond An intermolecular bond that forms within water because of the dipolar nature of water molecules.

Hydrogenous sediment Sediment that forms from precipitation from ocean water or ion exchange between existing sediment and ocean water. Examples are manganese nodules, phosphorite, glauconite, phillipsite, and montmorillonite.

Hydrologic cycle The cycle of water exchange among the atmosphere, land, and ocean through the processes of evaporation, precipitation, runoff, and subsurface percolation. Also called the water cycle.

Hydrothermal spring Vents of hot water found primarily along the spreading axes of oceanic ridges and rises.

Hydrothermal vent Ocean water that percolates down through fractures in recently formed ocean floor is heated by underlying magma and surfaces again through these vents. They are usually located near the axis of spreading along mid-ocean ridges.

Hydrothermal vent biocommunity Deep bottom-dwelling community of organisms associated with a hydrothermal vent. The hot-water vent is usually associated with the axis of a spreading center, and the community is dependent on sulfur-oxidizing bacteria that may live free in the water, on the bottom, or symbiotically in the tissue of some of the animals of the community.

Hypersaline Waters that are highly or excessively saline.

Hypersaline lagoon Shallow lagoons such as Laguna Madre, which may become hypersaline due to little tidal flushing and seasonal variability in freshwater input. High evaporation rates and low freshwater input can result in very high salinities.

Hypersaline seep At the base of the Florida Escarpment there is a biocommunity that depends on methane and sulfur-oxidizing bacteria as producers. The bacteria may live free in the water, on the bottom, or symbiotically in the tissue of some of the animals within the biocommunity.

Hypersaline seep biocommunity Bottom-dwelling community of organisms associated with a hypersaline seep.

Hypertonic Pertaining to the property of an aqueous solution having a higher osmotic pressure (salinity) than another aqueous solution from which it is separated by a semipermeable membrane that will allow osmosis to occur. The hypertonic fluid will gain water molecules through the membrane from the other fluid.

Hypothesis A tentative explanation that accounts for a set of facts and can be tested by further investigation.

Hypotonic Pertaining to the property of an aqueous solution having a lower osmotic pressure (salinity) than another aqueous solution from which it is separated by a semipermeable membrane that will allow osmosis to occur. The hypotonic fluid will lose water molecules through the membrane from the other fluid.

Hypsographic curve A curve that displays the relative elevations of the land surface and depths of the ocean.

Ice Age The most recent glacial period, which occurred during the Pleistocene epoch.

Ice rafting The movement of trapped sediment within or on top of ice by flotation.

Ice sheet An extensive, relatively flat accumulation of ice.

Iceberg A massive piece of glacier ice that has broken from the front of the glacier (calved) into a body of water. It floats with its tip at least 5 meters (16 feet) above the water's surface, and at least four-fifths of its mass submerged.

Igneous rock One of the three main classes into which all rocks are divided (igneous, metamorphic, and sedimentary). Rock that forms from the solidification of molten or partly molten material (magma).

In situ In place; in situ density of a sample of water is its density at its original depth.

Inertia Newton's first law of motion. It states that a body at rest will stay at rest and a body in motion will remain in uniform motion in a straight line unless acted on by some external force.

Infauna Animals that live buried in the soft substrate (sand or mud).

Infrared radiation Electromagnetic radiation lying between the wavelengths of 0.8 micrometer (0.000003 inch) and about 1000 micrometers (0.04 inch). It is bounded on the shorter-wavelength side by the visible spectrum and on the long side by microwave radiation.

Inner sublittoral Pertaining to the inner continental shelf, above the intersection with the euphotic zone, where attached plants grow.

Insolation The rate at which solar radiation is received per unit of surface area at any point at or above Earth's surface.

Interface A surface separating two substances of different properties, such as density, salinity, or temperature. In oceanography, it usually refers to a separation of two layers of water with different densities caused by significant differences in temperature and/or salinity.

Interface wave An orbital wave that moves along an interface between fluids of different density. An example is ocean surface waves moving along the interface between the atmosphere and the ocean, which is 1000 times more dense.

Interference (wave) The overlapping of different wave groups, either in phase (constructive interference, which results in larger waves), out of phase (destructive interference, which results in smaller waves), or some combination of the two (mixed interference).

Intermolecular bond A relatively weak bond that forms between molecules of a given substance. The hydrogen bond and the van der Waals bonds are intermolecular bonds.

Internal wave A wave that develops below the surface of a fluid, the density of which changes with increased depth. This change may be gradual or may occur abruptly at an interface.

Intertidal zone The ocean floor within the foreshore region that is covered by the highest normal tides and exposed by the lowest normal tides, including the water environment of tide pools within this region.

Intertropical Convergence Zone (ITCZ) Zone where northeast trade winds and southeast trade winds converge. Averages about 5 degrees north latitude in the Pacific and Atlantic oceans and 7 degrees south latitude in the Indian Ocean.

Intraplate feature Any feature that occurs within a tectonic plate and not along the plate's boundaries.

Intrusive rocks Igneous rocks such as granite that cool slowly beneath Earth's surface.

Invertebrate Animal without a backbone.

Ion An atom that becomes electrically charged by gaining or losing one or more electrons. The loss of electrons produces a positively charged cation, and the gain of electrons produces a negatively charged anion.

Ionic bond A chemical bond formed as a result of the electrical attraction.

Irminger Current A warm current that branches off from the Gulf Stream and moves up along the west coast of Iceland.

Iron hypothesis, the A hypothesis that states that an effective way of increasing productivity in the ocean is to fertilize the ocean by adding the only nutrient that appears to be lacking—iron. Adding iron to the ocean also increases the amount of carbon dioxide removed from the atmosphere.

Irons A meteorite consisting essentially of iron but that may also contain up to 30 percent nickel.

Island arc A linear arrangement of islands, many of which are volcanic, usually curved so that the concave side faces a sea separating the islands from a continent. The convex side faces the open ocean and is bounded by a deep-ocean trench.

Island mass effect As surface current flows past an island, surface water is carried away from the island on the downcurrent side. This water is replaced in part by upwelling of water on the downcurrent side of the island.

Isohaline Of the same salinity.

Isopoda An order of dorsoventrally flattened crustaceans that are mostly scavengers or parasites on other crustaceans or fish.

Isostasy A condition of equilibrium, comparable to buoyancy, by which Earth's brittle crust floats on the plastic mantle.

Isostatic adjustment The adjustment of crustal material due to isostasy.

Isostatic rebound The upward movement of crustal material due to isostasy.

Isotherm A line connecting points of equal temperature.

Isothermal Of the same temperature.

Isotonic Pertaining to the property of having equal osmotic pressure. If two such fluids were separated by a semipermeable membrane that will allow osmosis to occur, there would be no net transfer of water molecules across the membrane.

Isotope One of several atoms of an element that has a different number of neutrons, and therefore a different atomic mass, than the other atoms, or isotopes, of the element.

Jellyfish Free-swimming, umbrella-shaped medusoid members of the coelenterate class Scyphozoa. Also frequently applied to the medusoid forms of other coelenterates.

Jet stream An easterly moving air mass at an elevation of about 10 kilometers (6 miles). Moving at speeds that can exceed 300 kilometers (185 miles) per hour, the jet stream follows a wavy path in the midlatitudes and influences how far polar air masses may extend into the lower latitudes.

Jetty A structure built from the shore into a body of water to protect a harbor or a navigable passage from being closed off by the deposition of longshore drift material.

Juan de Fuca Ridge A divergent plate boundary off the Oregon-Washington coast.

K–T event An extinction event marked by the disappearance of the dinosaurs that occurred 65 million years ago at the boundary between the Cretaceous (K) and Tertiary (T) Periods of geological time.

Kelp Large varieties of Phaeophyta (brown algae).

Kelp forest An extensive bed of various species of macroscopic brown algae that provides a habitat for many other types of marine organisms.

Key A low, flat island composed of sand or coral debris that accumulates on a reef flat.

Kinetic energy Energy of motion. It increases as the mass or speed of the object in motion increases.

Knot (kt) Unit of speed equal to 1 nautical mile per hour, approximately 51 centimeters (1.67 feet) per second.

Krill A common name frequently applied to members of crustacean order Euphausiacea (euphausids).

Kuroshio Current The warm western boundary current of the North Pacific subtropical gyre.

La Niña An event where the surface temperature in the waters of the eastern South Pacific fall below average values. It usually occurs at the end of an El Niño–Southern Oscillation event.

Labrador Current A cold current flowing south along the coast of Labrador in the northeastern Atlantic Ocean.

Lagoon A shallow stretch of seawater partly or completely separated from the open ocean by an elongate narrow strip of land such as a reef or barrier island.

Laguna Madre A hypersaline lagoon located landward of Padre Island along the south Texas coast.

Laminar flow Flow in which water, or any fluid, flows in parallel layers or sheets. The direction of flow at any point does not change with time. Nonturbulent flow.

Land breeze The seaward flow of air from the land caused by differential cooling of Earth's surface.

Langmuir circulation A cellular circulation set up by winds that blow consistently in one direction with speeds in excess of 12 kilometers (7.5 miles) per hour. Helical spirals running parallel to the wind direction are alternately clockwise and counterclockwise.

Larva An embryo that has a different form before it assumes the characteristics of the adult of the species.

Latent heat The quantity of heat gained or lost per unit of mass as a substance undergoes a change of state (such as liquid to solid) at a given temperature and pressure.

Latent heat of evaporation The heat energy that must be added to one gram of a liquid substance to convert it to a vapor at a given temperature below its boiling point. For water, it is 585 calories at 20 degrees centigrade.

Latent heat of melting The heat energy that must be added to one gram of a substance at its melting point to convert it to a liquid. For water, it is 80 calories.

Latent heat of vaporization The heat energy that must be added to one gram of a substance at its boiling point to convert it to a vapor. For water, it is 540 calories.

Lateral line system A sensory system running down both sides of fishes to sense subsonic pressure waves transmitted through ocean water.

Latitude Location on Earth's surface based on angular distance north or south of the Equator. Equator = 0 degrees; North Pole = 90 degrees north latitude; South Pole = 90 degrees south latitude.

Laurasia An ancient landmass of the Northern Hemisphere. The name is derived from Laurentia, pertaining to the Canadian Shield of North America, and Eurasia, of which it was composed.

Lava Fluid magma coming from an opening in Earth's surface, or the same material after it solidifies.

Law of gravitation See *Gravitational force.*

Law of the Sea, The International law that defines jurisdiction, rights, passage through, and arbitration over oceanic waters.

Leeuwin Current A warm current that flows south out of the East Indies along the western coast of Australia.

Leeward Direction toward which the wind is blowing or waves are moving.

Levee 1. Natural levees are low ridges on either side of river channels that result from deposition during flooding. **2.** Artificial levees are built by human beings.

Light Electromagnetic radiation that has a wavelength in the range from about 4000 (violet) to about 7700 (red) angstroms and may be perceived by the normal unaided human eye.

Light-year The distance traveled by light during one year at a speed of 300,000 kilometers (186,000 miles) per second. It equals 9.8 trillion kilometers (6.2 trillion miles).

Limestone A class of sedimentary rocks composed of at least 80 percent carbonates of calcium or magnesium. Limestones may be either biogenous or hydrogenous.

Limpet A mollusk of the class Gastropoda that possesses a low conical shell that exhibits no spiraling in the adult form.

Liquid state A state of matter in which a substance has a fixed volume but no fixed shape.

Lithify Process by which sediment becomes hardened into sedimentary rock.

Lithogenous sediment Sediment composed of mineral grains derived from the weathering of rock material and transported to the ocean by various mechanisms of transport, including running water, gravity, the movement of ice, and wind.

Lithosphere The outer layer of Earth's structure, including the crust and the upper mantle to a depth of about 200 kilometers (124 miles). It is this layer that breaks into the plates that are the major elements of the theory of plate tectonics.

Lithothamnion ridge A feature common to the windward edge of a reef structure, characterized by the presence of the red algae, *Lithothamnion.*

Littoral zone The benthic zone between the highest and lowest spring-tide shorelines; the intertidal zone.

Lobster Large marine crustacean considered a delicacy. *Homarus americanus* (American lobster) possesses two large chelae (pincers) and is found off the New England coast. *Panulirus sp.* (spiny lobsters or rock lobsters) have no chelae but possess long, spiny antennae that are effective in warding off predators. *P. argue* is found off the coast of Florida and in the West Indies, whereas *P. interruptus* is common along the coast of southern California.

Longitude Location on Earth's surface based on angular distance east or west of the Prime (Greenwich) Meridian (0 degrees longitude). 180 degrees longitude is the International Date Line.

Longitudinal wave A wave phenomenon where particle vibration is parallel to the direction of energy propagation.

Longshore bar A deposit of sediment that forms parallel to the coast within or just beyond the surf zone.

Longshore current A current located in the surf zone and running parallel to the shore as a result of waves breaking at an angle to the shore.

Longshore drift The load of sediment transported along the beach from the breaker zone to the top of the swash line in association with the longshore current. Also called longshore transport or littoral drift.

Longshore trough A low area of the beach that separates the beach face from the longshore bar.

Lophophore Horseshoe-shaped feeding structure bearing ciliated tentacles characteristic of the phyla Bryozoa, Brachiopoda, and Phoronidea.

Low slack water The period of time associated with the peak of low tide when there is no visible flow of water into or out of bays and rivers.

Low tide terrace See *Beach face.*

Low tide zone That portion of the intertidal zone that lies between the lowest low-tide shoreline and the highest low-tide shoreline.

Low water (LW) The lowest level reached by the water surface at low tide before the rise toward high tide begins.

Lower high water (LHW) The lower of two high waters occurring during a tidal day where tides exhibit a mixed tidal pattern.

Lower low water (LLW) The lower of two low waters occurring during a tidal day where tides exhibit a mixed tidal pattern.

Lunar day The time interval between two successive transits of the moon over a meridian, approximately 24 hours and 50 minutes of solar time.

Lunar hour One-twenty-fourth of a lunar day; about 62.1 minutes.

Lunar tide The part of the tide caused solely by the tide-producing force of the moon.

Lunger Fish such as groupers that sit motionless on the ocean floor waiting for prey to appear. A quick burst of speed over a short distance is used to capture prey.

Macroplankton Plankton larger than 2 centimeters (0.8 inch) in their smallest dimension.

Magma Fluid rock material from which igneous rock is derived through solidification.

Magnetic anomaly Distortion of the regular pattern of Earth's magnetic field resulting from the various magnetic properties of local concentrations of ferromagnetic minerals in Earth's crust.

Magnetic dip The dip of magnetite particles in rock units of Earth's crust relative to sea level. It is approximately equivalent to latitude. Also called magnetic inclination.

Magnetic field A condition found in the region around a magnet or an electric current, characterized by the existence of a detectable magnetic force at every point in the region and by the existence of magnetic poles.

Magnetic inclination See *Magnetic dip.*

Magnetite The mineral form of black iron oxide, Fe_3O_4, that often occurs with magnesium, zinc, and manganese and is an important ore of iron.

Manganese nodules Concretionary lumps containing oxides of iron, manganese, copper, and nickel found scattered over the ocean floor.

Mangrove swamp A marsh-like environment that is dominated by mangrove trees. They are restricted to latitudes below 30 degrees.

Mantle The relatively plastic zone rich in ferromagnesian minerals between the core and crust of Earth.

Mantle plume A rising column of molten magma from Earth's mantle.

Marginal sea A semienclosed body of water adjacent to a continent and floored by submerged continental crust.

Mariculture The application of the principles of agriculture to the production of marine organisms.

Marine terrace Wave-cut benches that have been exposed above sea level by a drop in sea level.

Marsh An area of soft, wet, flat land that is periodically flooded by saltwater and is common in portions of lagoons.

Maturity A texture of lithogenous sediment, where increasing maturity (caused by increased time of transport) is indicated by decreased clay content, increased sorting, and increased rounding of the grains within the deposit.

Mean high water (MHW) The average height of all the high waters occurring over a 19-year period.

Mean low water (MLW) The average height of the low waters occurring over a 19-year period.

Mean sea level (MSL) The mean surface water level determined by averaging all stages of the tide over a 19-year period, usually determined from hourly height observations along an open coast.

Mean tidal range The difference between mean high water and mean low water.

Meander A sinuous curve, bend, or turn in the course of a current.

Mechanical energy Energy manifested as work being done; the movement of a mass through some distance.

Mediterranean circulation Circulation characteristic of bodies of water with restricted circulation with the ocean that results from an excess of evaporation as compared to precipitation and runoff similar to the Mediterranean Sea. Surface flow is into the restricted body of water with a subsurface counterflow as exists between the Mediterranean Sea and the Atlantic Ocean.

Medusa Free-swimming, bell-shaped coelenterate body form with a mouth at the end of a central projection and tentacles around the periphery.

Meiofauna Small species of animals that live in the spaces among particles in a marine sediment.

Melon A fatty organ located forward of the blowhole on certain odontocete cetaceans that is used to focus echolocation sounds.

Melting point The temperature at which a solid substance changes to the liquid state.

Mercury A silvery-white poisonous metallic element, liquid at room temperature and used in thermometers, barometers, vapor lamps, and batteries and in the preparation of chemical pesticides.

Meridian of longitude Great circles running through the North and South Poles.

Meroplankton Planktonic larval forms of organisms that are members of the benthos or nekton as adults.

Mesopelagic zone That portion of the oceanic province 200 to 1000 meters (660 to 3300 feet) deep. Corresponds approximately with the disphotic (twilight) zone.

Mesosaurus An extinct, presumably aquatic reptile that lived about 250 million years ago. The distribution of its fossil remains helps support the theory of plate tectonics.

Mesosphere The middle region of Earth below the asthenosphere and above the core.

Metal sulfide A compound containing one or more metals and sulfur.

Metamorphic rock Rock that has undergone recrystallization while in the solid state in response to changes of temperature, pressure, and chemical environment.

Meteor A bright trail or streak that appears in the sky when a meteoroid is heated to incandescence by friction with the Earth's atmosphere. Also called a falling or shooting star.

Meteorite A stony or metallic mass of matter that has fallen to the Earth's surface from outer space.

Microplankton Net plankton. Plankton not easily seen by the unaided eye, but easily recovered from the ocean with the aid of a silk-mesh plankton net.

Mid-Atlantic Ridge A slow-spreading divergent plate boundary running north–south and bisecting the Atlantic Ocean.

Mid-ocean ridge A linear, volcanic mountain range that extends through all the major oceans, rising 1 to 3 kilometers (0.6 to 2 miles) above the deep-ocean basins. Averaging 1500 kilometers (930 miles) in width, rift valleys are common along the central axis. Source of new oceanic crustal material.

Middle tide zone That portion of the intertidal zone that lies between the highest low-tide shoreline and the lowest high-tide shoreline.

Migration Long journeys undertaken by many marine species for the purpose of successful feeding and reproduction.

Minamata Bay, Japan The site of the occurrence of human poisoning in the 1950s by mercury contained in marine organisms that were consumed by victims.

Minamata disease A degenerative neurological disorder caused by poisoning with a mercury compound found in seafood obtained from waters contaminated with mercury-containing industrial waste.

Mineral An inorganic substance occurring naturally on Earth and having distinctive physical properties and a chemical composition that can be expressed by a chemical formula. The term is also sometimes applied to organic substances such as coal and petroleum.

Mixed interference A pattern of wave interference in which there is a combination of constructive and destructive interference.

Mixed surface layer The surface layer of the ocean water mixed by wave and tide motions to produce relatively isothermal and isohaline conditions.

Mixed tidal pattern A tidal pattern exhibiting two high tides and two low tides per tidal day with a marked diurnal inequality. Coastal locations that experience such a tidal pattern may also show alternating periods of diurnal and semidiurnal patterns. Also called mixed semidiurnal.

Mixotroph An organism that depends on a combination of autotrophic and heterotrophic behavior to meet its energy requirements. Many coral reef species exhibit such behavior.

Mohorovicic discontinuity (Moho) A sharp compositional discontinuity between the crust and mantle of Earth. It may be as shallow as 5 kilometers (3 miles) below the ocean floor or as deep as 60 kilometers (37 miles) beneath some continental mountain ranges.

Molecular motion Molecules move in three ways: vibration, rotation, and translation.

Molecule A group of two or more atoms that stick together by sharing or trading electrons.

Mollusca Phylum of soft, unsegmented animals usually protected by a calcareous shell and having a muscular foot for locomotion. Includes snails, clams, chitons, and octopi.

Monera Kingdom of organisms that do not have nuclear material confined within a sheath but spread throughout the cell. Includes bacteria, archaebacteria, and blue-green algae.

Mononodal Pertaining to a standing wave with only one nodal point or nodal line.

Monsoon A name for seasonal winds derived from the Arabic word for season, *mausim*. The term was originally applied to winds over the Arabian Sea that blow from the southwest during summer and from the northeast during winter.

Moraine A deposit of unsorted material deposited at the margins of glaciers. Many such deposits have become important economically as fishing banks after being submerged by the rising level of the ocean.

Mud Sediment consisting primarily of silt and clay-sized particles smaller than 0.06 millimeter (0.002 inch).

Mutualism A symbiotic relationship in which both participants benefit.

Myoglobin A red, oxygen-storing pigment found in muscle tissue.

Myomere A muscle fiber.

Mysticeti The baleen whales.

Nadir The point on the celestial sphere directly opposite the zenith and directly beneath the observer.

Nanoplankton Plankton less than 50 micrometers (0.002 inch) in length that cannot be captured in a plankton net and must be removed from the water by centrifuge or special microfilters.

Nansen bottle A device used by oceanographers to obtain samples of ocean water from beneath the surface.

Natural selection The process in nature by which, according to Darwin's theory of evolution, only the organisms best adapted to their environment tend to survive and transmit their genetic characters in increasing numbers to succeeding generations, while those less well adapted tend to be eliminated.

Nauplius A microscopic free-swimming larval stage of crustaceans such as copepods, ostracodes, and decapods. Typically has three pairs of appendages.

Neap tide Tides of minimal range occurring about every two weeks when the moon is in either first- or third-quarter phase.

Nearshore The zone of a beach that extends from the low tide shoreline seaward to where breakers begin forming.

Nebula A diffuse mass of interstellar dust and/or gas.

Nebular hypothesis A model that describes the formation of the solar system by contraction of a nebula.

Nektobenthos Those members of the benthos that can actively swim and spend much time off the bottom.

Nekton Pelagic animals such as adult squids, fish, and mammals that are active swimmers to the extent that they can determine their position in the ocean by swimming.

Nematath A linear chain of islands and/or seamounts that are progressively older in one direction. It is created by the passage of a lithospheric plate over a hotspot.

Nematocyst The stinging mechanism found within the cnidoblast of members of the phylum Cnidaria (Coelenterata).

Neritic province That portion of the pelagic environment from the shoreline to where the depth reaches 200 meters (660 feet).

Neritic sediment That sediment composed primarily of lithogenous particles and deposited relatively rapidly on the continental shelf, continental slope, and continental rise.

Net primary production The primary production of plants after they have removed what is needed for their metabolism.

Neutral A state in which there is no excess of either the hydrogen or the hydroxide ion.

Neutron An electrically neutral subatomic particle found in the nucleus of atoms that has a mass approximately equivalent to that of a proton.

New moon The phase of the moon that occurs when the sun and the moon are in conjunction, that is, they are both on the same side of Earth. During this time, the dark side of the moon faces Earth.

New production Photosynthetic production supported by nutrients supplied from outside the immediate ecosystem by upwelling or other physical transport.

Niche The ecological role of an organism and its position in the ecosystem.

Niigata, Japan The site of mercury poisoning of humans in the 1960s by ingestion of contaminated seafood.

Nitrogen fixation Conversion by bacteria of atmospheric nitrogen (N_2) to oxides of nitrogen (NO_2, NO_3) usable by plants in primary production.

Nitrogen-fixing bacteria Bacteria that perform nitrogen fixation.

Node The point on a standing wave where vertical motion is lacking or minimal. If this condition extends across the surface of an oscillating body of water, the line of no vertical motion is a nodal line.

North Atlantic Current The northernmost current of the North Atlantic current gyre.

North Atlantic Deep Water A deep-water mass that forms primarily at the surface of the Norwegian Sea and moves south along the floor of the North Atlantic Ocean.

North Pacific Current The northernmost current of the North Pacific current gyre.

Northeast Monsoon A northeast wind that blows off the Asian mainland onto the Indian Ocean during the winter season.

Norwegian Current A warm current that branches off from the Gulf Stream and flows into the Norwegian Sea between Iceland and the British Isles.

Nucleus The positively charged central region of an atom, composed of protons and neutrons and containing almost all of the mass of the atom.

Nudibranch Sea slug. A member of the mollusk class Gastropoda that has no protective covering as an adult. Respiration is carried on by gills or other projections on the dorsal surface.

Nutrients Any number of organic or inorganic compounds used by plants in primary production. Nitrogen and phosphorus compounds are important examples.

Occam's razor A test that states that an accepted hypothesis should be the simplest hypothesis that explains any given phenomenon.

Ocean The entire body of saltwater that covers more than 71 percent of Earth's surface.

Ocean acoustical tomography A method by which changes in water temperature may be determined by changes in the speed of transmission of sound. It has the potential to help map ocean circulation patterns over large ocean areas.

Ocean beach The beach on the open-ocean side of a barrier island.

Ocean current A mass of oceanic water that flows from one place to another.

Ocean Drilling Program In 1983, this program replaced the Deep Sea Drilling Project. It focuses more on drilling the continental margins using the drill ship *JOIDES Resolution*.

Ocean thermal energy conversion (OTEC) A technique of generating energy by using the difference in temperature between surface waters and deep waters in low latitude regions.

Oceanic common water Deep water that is common to all ocean basins of the world except the Arctic Ocean.

Oceanic crust A mass of rock with a basaltic composition that is about 5 kilometers (3 miles) thick and may or may not extend beneath the continents.

Oceanic province That division of the pelagic environment where the water depth is greater than 200 meters (660 feet).

Oceanic ridge A portion of the global mid-ocean ridge system that is characterized by slow spreading and steep slopes.

Oceanic rise A portion of the global mid-ocean ridge system that is characterized by fast spreading and gentle slopes.

Oceanic sediment The inorganic abyssal clays and the organic oozes that accumulate slowly on the deep-ocean floor.

Oceanic spreading center The axes of oceanic ridges and rises that are the locations at which new lithosphere is added to lithospheric plates. The plates move away from these axes in the process of sea floor spreading.

Ocelli Light-sensitive organ around the base of many medusoid bells.

Odontoceti Toothed whales.

Offset Separation that occurs due to movement along a fault.

Offshore The comparatively flat submerged zone of variable width extending from the breaker line to the edge of the continental shelf.

Oligotrophic Areas such as the midsubtropical gyres where there are low levels of biological production.

Omnivore An animal that feeds on both plants and animals.

Oolite A deposit formed of small spheres from 0.25 to 2 millimeters (0.01 to 0.08 inch) in diameter. They are usually composed of concentric layers of calcite.

Ooze A pelagic sediment containing at least 30 percent skeletal remains of pelagic organisms, the balance being clay minerals. Oozes are further defined by the chemical composition of the organic remains (siliceous or calcareous) and by their characteristic organisms (examples: diatomaceous ooze, foraminifer ooze).

Opal An amorphous form of silica ($SiO_2 \cdot nN_2O$) that usually contains from 3 to 9 percent water. It forms the shells of radiolarians and diatoms.

Opposition The separation of two heavenly bodies by 180 degrees relative to Earth. The sun and moon are in opposition during the full-moon phase.

Orbital wave A wave phenomenon in which energy is moved along the interface between fluids of different densities. The wave form is propagated by the movement of fluid particles in orbital paths.

Orthogonal lines Lines constructed perpendicular to wave fronts and evenly spaced so that the energy between lines is equal at all times. Orthogonals are used to help determine how energy is distributed along the shoreline by breaking waves.

Orthophosphate Phosphoric oxide (P_2O_5) can combine with water to produce orthophosphates ($3H_2O \cdot P_2O_5$ or H_3PO_4) that may be used by plants as nutrients.

Osmosis Passage of water molecules through a semipermeable membrane separating two aqueous solutions of different solute concentration. The water molecules pass from the solution of lower solute concentration into the other.

Osmotic pressure A measure of the tendency for osmosis to occur. It is the pressure that must be applied to the more concentrated solution to prevent the passage of water molecules into it from the less concentrated solution.

Osmotic regulation Physical and biological processes used by organisms to counteract the osmotic effects of differences in osmotic pressures of their body fluids and the water in which they live.

Ostracoda An order of crustaceans that are minute and compressed within a bivalve shell.

Otocyst Gravity-sensitive organs around the bell of a medusa.

Outer sublittoral zone The continental shelf below the intersection with the euphotic zone where no plants grow attached to the bottom.

Outgassing The process by which gases are removed from within Earth's interior.

Oxygen compensation depth The depth in the ocean at which marine plants receive just enough solar radiation to meet their basic metabolic needs. It marks the base of the euphotic zone.

Oxygen minimum layer (OML) A zone of low dissolved oxygen concentration that occurs at a depth of about 700 to 1000 meters (2300 to 3280 feet).

Pacific Ring of Fire An extensive zone of volcanic and seismic activity that coincides roughly with the borders of the Pacific Ocean.

Paleoceanography The study of the physical and biological changes of the oceans brought about by the changing shapes and positions of the continents.

Paleogeography The study of the historical changes of shapes and positions of the continents and oceans.

Paleomagnetism The study of Earth's ancient magnetic field.

Pangaea An ancient supercontinent of the geological past that contained all Earth's continents.

Panthalassa A large, ancient ocean that surrounded Pangaea.

Paralytic shellfish poisoning (PSP) Paralysis resulting from poisoning caused by eating shellfish contaminated with the dinoflagellate *Gonyaulax*, a toxin.

Parasitism A symbiotic relationship between two organisms in which one benefits at the expense of the other.

Parts per thousand (‰) A unit of measurement used in reporting salinity of water, equal to the number of grams of dissolved substances in 1000 grams of water (typically seawater). One ‰ is equivalent to 0.1 percent or 1000 parts per million.

Passive margin A continental margin that lacks a plate boundary and is marked by a low degree of tectonic activity, such as those typical of the Atlantic Ocean.

PCBs A group of industrial chemicals used in a variety of products; responsible for several episodes of ecological damage in coastal waters.

Pelagic environment The open-ocean environment, which is divided into the neritic province (water depth 0 to 200 meters or 656 feet) and the oceanic province (water depth greater than 200 meters or 656 feet).

Pelecypoda A class of mollusks characterized by two more or less symmetrical lateral valves with a dorsal hinge. These filter feeders pump water through the filter system and over gills through posterior siphons. Many possess a hatchet-shaped foot used for locomotion and burrowing. Includes clams, oysters, mussels, and scallops.

Perigee The point on the orbit of an Earth satellite (moon) that is nearest Earth.

Perihelion That point on the orbit around the sun of a planet or comet that is closest to the sun.

Permeability Capacity of a porous rock or sediment for transmitting fluid.

Peru Current The cold eastern boundary current of the South Pacific subtropical gyre.

Petroleum A naturally occurring liquid hydrocarbon.

pH scale A measure of the acidity or alkalinity of a solution, numerically equal to 7 for neutral solutions, increasing with increasing alkalinity and decreasing with increasing acidity. The pH scale commonly in use ranges from 0 to 14.

Phaeophyta Brown algae characterized by the carotenoid pigment fucoxanthin. Contains the largest members of the marine plant community.

Phosphate Any of a number of phosphorous-bearing compounds.

Phosphorite A sedimentary rock composed primarily of phosphate minerals.

Photic zone The upper ocean in which the presence of solar radiation is detectable. It includes the euphotic and disphotic zones.

Photophore One of several types of light-producing organs found primarily on fishes and squids inhabiting the mesopelagic and upper bathypelagic zones.

Photosynthesis The process by which plants produce carbohydrate food from carbon dioxide and water in the presence of chlorophyll, using light energy and releasing oxygen.

Phycoerythrin A red pigment characteristic of the Rhodophyta (red algae).

Phytoplankton Plant plankton. The most important community of primary producers in the ocean.

Picoplankton Small plankton within the size range of 0.2 to 2.0 micrometers (0.000008 to 0.00008 inch) in size. Composed primarily of bacteria.

Pillow basalt A basalt exhibiting pillow structure. See *Pillow lava*.

Pillow lava A general term for those lavas that display discontinuous pillow-shaped masses (pillow structure), caused by the rapid cooling of lava as a result of underwater eruption of lava or lava flowing into water.

Ping A sharp, high-pitched sound made by the transmitting device of many sonar systems.

Pinnipedia A suborder of marine mammals that includes the sea lions, seals, and walruses.

Plankter Informal term for *plankton*.

Plankton Passively drifting or weakly swimming organisms that are not independent of currents. Includes mostly microscopic algae, protozoa, and larval forms of higher animals.

Plankton bloom A very high concentration of phytoplankton, resulting from a rapid rate of reproduction as conditions become optimal during the spring in high-latitude areas. Less-obvious causes produce blooms that may be destructive in other areas.

Plankton net Plankton-extracting device that is cone shaped and typically of a silk material. It is towed through the water or lifted vertically to extract plankton down to a size of 50 micrometers (0.0002 inch).

Plantae Kingdom of many-celled plants.

Plastic 1. Capable of being shaped or formed. **2.** Composed of plastic or plastics.

Plate tectonics Theory of global dynamics having to do with the movement of a small number of semirigid sections of the Earth's crust, with seismic activity and volcanism occurring primarily at the margins of these sections. This movement has resulted in continental drift and changes in the shape and size of ocean basins and continents.

Plume A rising column of molten mantle material that is associated with a hotspot when it penetrates Earth's crust.

Plunging breaker Impressive curling breakers that form on moderately sloping beaches.

Pneumatic duct An opening into the swim bladder of some fishes that allows rapid release of air into the esophagus.

Poikilotherm An organism whose body temperature varies with and is largely controlled by its environment.

Polar Pertaining to the polar regions.

Polar cell The large atmospheric circulation cell that occurs between 60 and 90 degrees latitude in each hemisphere.

Polar easterly winds Cold air masses that move away from the polar regions toward lower latitudes.

Polar emergence The emergence of low- and mid-latitude temperature-sensitive deep-ocean benthos onto the shallow shelves of the polar regions, where temperatures similar to that of their deep-ocean habitat exist.

Polar front The boundary between the prevailing westerly and polar easterly wind belts that is centered at about 60 degrees latitude in each hemisphere.

Polar high The region of high atmospheric pressure that surrounds the poles.

Polar wandering curve A curve that shows the change in the position of a pole through time.

Polarity Intrinsic polar separation, alignment, or orientation, especially of a physical property (such as magnetic or electrical polarity).

Pollution (marine) The introduction of substances that result in harm to the living resources of the ocean or humans who use these resources.

Polychaeta Class of annelid worms that includes most of the marine segmented worms.

Polyp A single individual of a colony or a solitary attached coelenterate.

Population A group of individuals of one species living in an area.

Porifera Phylum of sponges. Supporting structure composed of $CaCO_3$ or SiO_2 spicules or fibrous spongin. Water currents created by flagella-waving choanocytes enter tiny pores, pass through canals, and exit through a larger osculum.

Porosity The ratio of the volume of all the empty spaces in a material to the volume of the whole.

Precession Describes the change in the attitude of the moon's orbit around Earth as it slowly changes its direction. The cycle is completed every 18.6 years and is accompanied by a clockwise rotation of the plane of the moon's orbit that is completed in the same time interval.

Precipitation In a meteorological sense, the discharge of water in the form of rain, snow, hail, or sleet from the atmosphere onto Earth's surface.

Prevailing westerlies The air masses moving from subtropical high-pressure belts toward the polar front. They are southwesterly in the Northern Hemisphere and northwesterly in the Southern Hemisphere.

Primary productivity The amount of organic matter synthesized by organisms from inorganic substances within a given volume of water or habitat in a unit of time.

Prime Meridian The meridian of longitude 0 degrees used as a reference for measuring longitude that passes through the Royal Observatory at Greenwich, England. Also known as the Greenwich Meridian.

Producer The autotrophic component of an ecosystem that produces the food that supports the biocommunity.

Progressive wave A wave in which the waveform moves progressively.

Propagation The transmission of energy through a medium.

Protein A very complex organic compound made up of large numbers of amino acids. Proteins make up a large percentage of the dry weight of all living organisms.

Protoctista A kingdom of organisms that includes any of the unicellular eukaryotic organisms and their descendant multicellular organisms. Includes single- and multicelled marine algae as well as single-celled animals called protozoa.

Protoearth The young, early developing Earth.

Proton A positively charged subatomic particle found in the nucleus of atoms that has a mass approximately equivalent to that of a neutron.

Protoplanet Any planet that is in its early stages of development.

Protoplasm The self-perpetuating living material making up all organisms, consisting mostly of the elements carbon, hydrogen, and oxygen combined into various chemical forms.

Protozoa Phylum of one-celled animals with nuclear material confined within a nuclear sheath.

Pseudopodia An extension of protoplasm in a broad, flat, or long needle-like projection used for locomotion or feeding. Typical of amoeboid forms such as foraminifers and radiolarians.

Pteropoda An order of pelagic gastropods in which the foot is modified for swimming and the shell may be present or absent.

Pycnocline A layer of water in which a high rate of change in density in the vertical dimension is present.

Pyrrophyta A phyla of dinoflagellates that possess flagella for locomotion.

Quadrature The state of the moon during the first- and third-quarter moon phases (at right angles to one another relative to Earth).

Quarter moon First- and third-quarter moon phases, which occur when the moon is in quadrature about one week after the new moon and full moon phases, respectively. The third-quarter moon phase is also known as the last-quarter moon phase.

Quartz A very hard mineral composed of silica, SiO_2.

Radiata A grouping of phyla with primary radial symmetry—phyla Coelenterata and Ctenophora.

Radioactivity The spontaneous breakdown of the nucleus of an atom resulting in the emission of radiant energy in the form of particles or waves.

Radiolaria An order of planktonic and benthic protozoans that possess protective coverings usually made of silica.

Radiometric age dating The use of radioisotope half-lives to determine the age of rock units in years before present within 2 or 3 percent.

Ray A cartilaginous fish in which the body is dorsoventrally flattened, eyes and spiracles are on the upper surface, and gill slits are on the bottom. The tail is reduced to a whip-like appendage. Includes electric rays, manta rays, and stingrays.

Recreational beach The area of a beach above shoreline, including the berm, berm crest, and the exposed part of the beach face.

Red clay See *Abyssal clay*.

Red muscle fiber Fine muscle fibers rich in myoglobin that are abundant in cruiser-type fishes.

Red tide A reddish-brown discoloration of surface water, usually in coastal areas, caused by high concentrations of microscopic organisms, usually dinoflagellates. It probably results from increased availability of certain nutrients. Toxins produced by the dinoflagellates may kill fish directly; decaying plant and animal remains or large populations of animals that migrate to the area of abundant plants may also deplete the surface waters of oxygen and cause asphyxiation of many animals.

Reef A strip or ridge of rocks, sand, coral, or humanmade objects that rises to or near the surface of the ocean and creates a navigational hazard.

Reef flat A platform of coral fragments and sand on the lagoon side of a reef that is relatively exposed at low tide.

Reef front The upper seaward face of a reef from the reef edge (seaward margin of reef flat) to the depth at which living coral and coralline algae become rare, 16 to 30 meters (50 to 100 feet).

Reflection The process in which a wave has part of its energy returned seaward by a reflecting surface.

Refraction The process by which the part of a wave in shallow water is slowed down, causing the wave to bend and align itself nearly parallel to underwater contours.

Regenerated production The portion of gross primary production that is supported by nutrients recycled within an ecosystem.

Relict beach A beach deposit laid down and submerged by a rise in sea level. It is still identifiable on the continental shelf, indicating that no deposition is presently taking place at that location on the shelf.

Relict sediment A sediment deposited under a set of environmental conditions that remains unchanged although the environment has changed, and it remains unburied by later sediment. An example is a beach deposited near the edge of the continental shelf when sea level was lower.

Residence time The average length of time a particle of any substance spends in the ocean. It is calculated by dividing the total amount of the substance in the ocean by the rate of its introduction into the ocean or the rate at which it leaves the ocean.

Respiration The process by which organisms utilize organic materials (food) as a source of energy. As the energy is released, oxygen is used and carbon dioxide and water are produced.

Restoring force A force such as surface tension or gravity that tends to restore the ocean surface displaced by a wave to that of a still water level.

Resultant force The difference between the gravitational force of various bodies and the centripetal force on Earth. The horizontal component of the resultant force is the tide-producing force.

Reverse osmosis A method of desalinating ocean water that involves forcing water molecules through a water-permeable membrane under pressure.

Reversing current The tide current as it occurs at the margins of landmasses. The water flows in and out for approximately equal periods of time separated by slack water where the water is still at high and low tidal extremes.

Rhodophyta Phylum of algae composed primarily of small encrusting, branching, or filamentous plants that receive their characteristic red color from the presence of the pigment phycoerythrin. With a worldwide distribution, they are found at greater depths than other algae.

Rift valley A deep fracture or break, about 25 to 50 kilometers (15 to 30 miles) wide, extending along the crest of a mid-ocean ridge.

Rifting The movement of two plates in opposite directions such as along a divergent boundary.

Right whales Surface-feeding baleen whales of the family Balaenidae, which were the favorite targets of early whalers.

Rip current A strong, narrow surface or near-surface current of short duration and high speed flowing seaward through the breaker zone at nearly right angles to the shore. It represents the return to the ocean of water that has been piled up on the shore by incoming waves.

Rip-rap Any large, blocky material used to armor coastal structures.

Rogue wave An unusually large wave that usually occurs unexpectedly amid other waves of smaller size. Also known as a superwave.

Rorqual whales Large baleen whales with prominent ventral groves (rorqual folds) of the family Balaenopteridae: the minke, Baird's, Bryde's, sei, fin, blue, and humpback whales.

Rotary current Tidal current as observed in the open ocean. The tidal crest makes one complete rotation during a tidal period.

Rounding A sediment texture where well-rounded sediment is characterized as having grains that lack sharp corners.

Salinity A measure of the quantity of dissolved solids in ocean water. Formally, it is the total amount of dissolved solids in ocean water in parts per thousand (‰) by weight after all carbonate has been converted to oxide, the bromide and iodide to chloride, and all the organic matter oxidized. It is normally computed from conductivity, refractive index, or chlorinity.

Salinometer An instrument that is used to determine the salinity of seawater by measuring its electrical conductivity.

Salpa Genus of pelagic tunicates that are cylindrical, transparent, and found in all oceans.

Salt Any substance that yields ions other than hydrogen or hydroxyl. Salts are produced from acids by replacing the hydrogen with a metal.

Salt marsh A relatively flat area of the shore where fine sediment is deposited and salt-tolerant grasses grow. One of the most biologically productive regions of Earth.

Salt wedge estuary A very deep river mouth with a very large volume of freshwater flow beneath which a wedge of saltwater from the ocean invades. The Mississippi River is an example.

San Andreas Fault A transform fault that cuts across the state of California from the northern end of the Gulf of California to Point Arena north of San Francisco.

Sand Particle size of 0.0625 to 2 millimeters (0.002 to 0.08 inch). It pertains to particles that lie between silt and granules on the Wentworth scale of grain size.

Sargasso Sea A region of convergence in the North Atlantic lying south and east of Bermuda where the water is very clear, deep blue in color, and contains large quantities of floating *Sargassum*.

Sargassum A brown alga characterized by a bushy form, substantial holdfast when attached, and a yellow-brown, green-yellow, or orange color. Two species, *S. fluitans* and *S. natans,* make up most of the macroscopic vegetation in the Sargasso Sea.

Saturation diving The technique of saturating the bloodstream with different mixtures of gases so that divers are able to undertake prolonged stays underwater.

Scarp A linear steep slope on the ocean floor separating gently sloping or flat surfaces.

Scavenger An animal that feeds on dead organisms.

Schooling Large, well-defined groups of fish, squid, and crustaceans that apparently aid the members to survive.

Scientific method The principles and empirical processes of discovery and demonstration considered characteristic of or necessary for scientific investigation, generally involving the observation of phenomena, the formulation of a hypothesis concerning the phenomena, experimentation to demonstrate the truth or falseness of the hypothesis, and a conclusion that validates or modifies the hypothesis.

Scyphozoa A class of coelenterates that includes the true jellyfish, in which the medusoid body form predominates and the polyp is reduced or absent.

Scuba An acronym for self-contained underwater breathing apparatus, it is a portable device containing compressed air that is used for breathing under water.

Sea 1. A subdivision of an ocean, generally enclosed by land and usually composed of saltwater. Two types of seas are identifiable and defined. They are the Mediterranean seas, where a number of seas are grouped together collectively as one sea, and adjacent seas, which are connected individually to the ocean. **2.** A portion of the ocean where waves are being generated by wind.

Sea anemone A member of the class Anthozoa whose bright color, tentacles, and general appearance make it resemble flowers.

Sea arch An opening through a headland caused by wave erosion. Usually develops as sea caves are extended from one or both sides of the headland.

Sea breeze The landward flow of air from the sea caused by differential heating of Earth's surface.

Sea cave A cavity at the base of a sea cliff formed by wave erosion.

Sea cow See *Sirenia.*

Sea cucumber A common name given to members of the echinoderm class Holotheuroidea.

Sea floor spreading A process producing the lithosphere when convective upwelling of magma along the oceanic ridges moves the ocean floor away from the ridge axes at rates between 2 to 12 centimeters (0.8 to 5 inches) per year.

Sea ice Any form of ice originating from the freezing of ocean water.

Sea lion Any of several eared seals with a relatively long neck and large front flippers, especially the California sea lion, *Zalophus californianus,* of the northern Pacific. Along with the fur seals, these marine mammals are known as eared seals.

Sea otter A seagoing otter that has recovered from near extinction along the North Pacific coasts. It feeds primarily on abalone, sea urchins, and crustaceans.

Sea snake A reptile belonging to the family Hydrophiidae with venom similar to that of cobras. Sea snakes are found primarily in the coastal waters of the Indian Ocean and the western Pacific Ocean.

Sea stack An isolated, pillar-like rocky island that is detached from a headland by wave erosion.

Sea turtle Any of the reptilian order Testudinata found widely in warm water.

Sea urchin An echinoderm belonging to the class Echinoidea possessing a fused test (external covering) and well-developed spines.

Seaknoll See *Abyssal hill.*

Seal 1. Any of several earless seals with a relatively short neck and small front flippers. Also known as true seals.

2. A general term that describes any of the various aquatic, carnivorous

marine mammals of the families *Phocidae* and *Otariidae* (true seals and eared seals), found chiefly in the Northern Hemisphere and having a sleek, torpedo-shaped body and limbs that are modified into paddle-like flippers.

Seamount An individual volcanic peak extending over 1000 meters (3300 feet) above the surrounding ocean floor.

Seasonal thermocline A thermocline that develops due to surface heating of the oceans in mid- to high-latitudes. The base of the seasonal thermocline is usually above 200 meters (656 feet).

Seawall A wall built parallel to the shore to protect coastal property from the waves.

SeaWiFS A satellite launched in 1997 that measures the color of the ocean with a radiometer and provides global coverage of ocean chlorophyll levels as well as land productivities every two days.

Secchi disk A light-colored, disk-shaped device that is lowered into water in order to measure the water's clarity.

Sediment Particles of organic or inorganic origin that accumulate in loose form.

Sediment maturity A condition in which the roundness and degree of sorting increase and clay content decreases within a sedimentary deposit.

Sedimentary rock A rock resulting from the consolidation of loose sediment, or a rock resulting from chemical precipitation, such as sandstone and limestone.

Seep An area where water of various temperature trickles out of the sea floor.

Seiche A standing wave of an enclosed or semienclosed body of water that may have a period ranging from a few minutes to a few hours, depending on the dimensions of the basin. The wave motion continues after the initiating force has ceased.

Seismic Pertaining to an earthquake or Earth vibration, including those that are artificially induced.

Seismic moment magnitude A scale used for measuring earthquake intensity based on energy released in creating very long-period seismic waves.

Seismic sea wave See *Tsunami.*

Seismic surveying The use of sound-generating techniques to identify features on or beneath the ocean floor.

Semidiurnal tidal pattern A tidal pattern exhibiting two high tides and two low tides per tidal day with small inequalities between successive highs and successive lows; a semidaily tide.

Sessile Permanently attached to the substrate and not free to move about.

Sewage sludge Semisolid material precipitated by sewage treatment.

Shallow-water wave A wave on the surface having a wavelength of at least 20 times water depth. The bottom affects the orbit of water particles and speed is determined by water depth: Speed (meters per second) = 3.1 $\sqrt{\text{Water depth (meters)}}$.

Shelf break The depth at which the gentle slope of the continental shelf steepens appreciably. It marks the boundary between the continental shelf and continental rise.

Shelf ice Thick shelves of glacial ice that push out into Antarctic seas from Antarctica. Large tabular icebergs calve at the edge of these vast shelves.

Shoaling To become shallow.

Shore Seaward of the coast, extends from highest level of wave action during storms to the low-water line.

Shoreline The line marking the intersection of water surface with the shore. Migrates up and down as the tide rises and falls.

Shoreline of emergence Shorelines that indicate a lowering of sea level by the presence of stranded beach deposits and marine terraces above it.

Shoreline of submergence Shorelines that indicate a rise in sea level by the presence of drowned beaches or submerged dune topography.

Side-scan sonar A method of mapping the topography of the ocean floor along a strip up to 60 kilometers (37 miles) wide using computers and sonar signals that are directed away from both sides of the survey ship.

Silica Silicon dioxide (SiO_2).

Silicate A mineral whose crystal structure contains SiO_4 tetrahedra.

Siliceous A condition of containing abundant silica (SiO_2).

Sill A submarine ridge partially separating bodies of water such as fjords and seas from one another or from the open ocean.

Silt A particle size of 0.008 to 0.0625 millimeter (0.0003 to 0.002 inch). It is intermediate in size between sand and clay.

Siphonophora An order of hydrozoan coelenterates that forms pelagic colonies containing both polyps and medusae. Examples are *Physalia* and *Velella.*

Sirenia An order of large, vegetarian marine mammals that includes dugongs and manatees, which are also know as sea cows.

Slack water Occurs when a reversing tidal current changes direction at high or low water. Current speed is zero.

Slick A smooth patch on an otherwise rippled surface caused by a monomolecular film of organic material that reduces surface tension.

Slightly stratified estuary An estuary of moderate depth in which marine water invades beneath the freshwater runoff. The two water masses mix so that the bottom water is slightly saltier than the surface water at most places in the estuary.

SOFAR channel Sound fixing and ranging channel, which is a low-velocity sound travel zone that coincides with the permanent thermocline in low and mid-latitudes.

Solar day The 24-hour period during which Earth completes one rotation on its axis.

Solar distillation A process by which ocean water can be desalinated by evaporation and the condensation of the vapor on the cover of a container. The condensate then runs into a separate container and is collected as fresh water. Also called solar humidification.

Solar humidification See *Solar distillation.*

Solar system The sun and the celestial bodies, asteroids, planets, and comets that orbit around it.

Solar tide The partial tide caused by the tide-producing forces of the sun.

Solid state A state of matter in which the substance has a fixed volume and shape. A crystalline state of matter.

Solstice The time during which the sun is directly over one of the tropics. In the Northern Hemisphere the summer solstice occurs on June 21 or 22 as the sun is over the Tropic of Cancer, and the winter solstice occurs on December 21 or 22 when the sun is over the Tropic of Capricorn.

Solute A substance dissolved in a solution. Salts are the solute in saltwater.

Solution A state in which a solute is homogeneously mixed with a liquid solvent. Water is the solvent for the solution that is ocean water.

Solvent A liquid that has one or more solutes dissolved in it.

Somali Current This current flows north along the Somali coast of Africa during the southwest monsoon season.

Sonar An acronym for sound navigation and ranging, it is a method by which objects may be located in the ocean.

Sorting A texture of sediments, where a well-sorted sediment is characterized by having great uniformity of grain sizes.

Sounding A measured depth of water beneath a ship.

Southwest Monsoon A southwest wind that develops during the summer season. It blows off the Indian Ocean onto the Asian mainland.

Southwest Monsoon Current During the southwest monsoon season, this eastward-flowing current replaces the west-flowing North Equatorial Current in the Indian Ocean.

Space dust Micrometeoroid space debris.

Species diversity The number or variety of species found in a subdivision of the marine environment.

Specific gravity The ratio of density of a given substance to that of pure water at 4 degrees centigrade and at one atmosphere pressure.

Specific heat The quantity of heat required to raise the temperature of 1 gram of a given substance by 1 degree centigrade. For water it is 1 calorie.

Spermaceti organ A large fatty organ located within the head region of sperm whales (*Physeter macrocephalus*) that is used to focus echolocation sounds.

Spermatophyta Seed-bearing plants.

Spherule A cosmogenous microscopic globular mass composed of silicate rock material (tektites) or of iron and nickel.

Spicule A minute, needle-like calcareous or siliceous projection found in sponges, radiolarians, chitons, and echinoderms that acts to support the tissue or provide a protective covering.

Spilling breaker A type of breaking wave that forms on a gently sloping beach, which gradually extracts the energy from the wave to produce a turbulent mass of air and water that runs down the front slope of the wave.

Spit A small point, low tongue, or narrow embankment of land commonly consisting of sand deposited by longshore currents and having one end attached to the mainland and the other terminating in open water.

Splash wave A long-wavelength wave created by a massive object or series of objects falling into water; a type of tsunami.

Sponge See *Porifera.*

Spray zone The shore zone lying between the high-tide shoreline and the coastline. It is covered by water only during storms.

Spreading center A divergent plate boundary.

Spreading rate The rate of divergence of plates at a spreading center.

Spring equinox See *Vernal equinox.*

Spring tide Tide of maximum range occurring about every two weeks when the moon is in either new or full moon phase.

Stack An isolated mass of rock projecting from the ocean off the end of a headland from which it has been detached by wave erosion.

Standard laboratory bioassay An standard assessment technique that determines the concentration of a pollutant that causes 50 percent mortality among selected test organisms.

Standard seawater Ampules of ocean water for which the chlorinity has been determined by the Institute of Oceanographic Services in Wormly, England. The ampules are sent to laboratories all over the world so that equipment and reagents used to determine the salinity of ocean water samples can be calibrated by adjustment until they give the same chlorinity as shown on the ampule label.

Standing wave A wave, the form of which oscillates vertically without progressive movement. The region of maximum vertical motion is an antinode. On either side are nodes, where there is no vertical motion but maximum horizontal motion.

Stenohaline Pertaining to organisms that can withstand only a small range of salinity change.

Stenothermal Pertaining to organisms that can withstand only a small range of temperature change.

Stick chart A device made of sticks or pieces of bamboo that was used by early navigators at sea.

Still water level Halfway between crest and trough, it is the level where the water would reside if there were no waves. Also known as zero energy level.

Storm An atmospheric disturbance characterized by strong winds accompanied by precipitation and often by thunder and lightning.

Storm surge A rise above normal water level resulting from wind stress and reduced atmospheric pressure during storms. Consequences can be more severe if a storm surge occurs in association with high tide.

Strait of Gibraltar The narrow opening between Europe and Africa through which the waters of the Atlantic Ocean and Mediterranean Sea mix.

Stranded beach An ancient beach deposit found above present sea level because of a lowering of sea level.

Streamlining The shaping of an object so it produces the minimum of turbulence while moving through a fluid medium. The teardrop shape displays a high degree of streamlining.

Structural style The type and degree of deformation exhibited by a group of rocks.

Subduction The process by which one lithospheric plate descends beneath another as they converge.

Subduction zone A long, narrow region beneath Earth's surface in which subduction takes place.

Subduction-zone seep biocommunity Animals that live in association with seeps of pore water squeezed out of deeper sediments. They depend on sulfur-oxidizing bacteria that act as producers for the ecosystem.

Sublittoral zone That portion of the benthic environment extending from low tide to a depth of 200 meters (660 feet); considered by some to be the surface of the continental shelf.

Submarine canyon A steep, V-shaped canyon cut into the continental shelf or slope.

Submarine fan See *Deep-sea fan.*

Submerged dune topography Ancient coastal dune deposits found submerged beneath the present shoreline because of a rise in sea level.

Submerging shoreline A shoreline formed by the relative submergence of a landmass in which the shoreline is on landforms developed under subaerial processes. It is characterized by bays and promontories and is more irregular than a shoreline of emergence.

Subneritic province The benthic environment extending from the shoreline across the continental shelf to the shelf break. It underlies the neritic province of the pelagic environment.

Suboceanic province Benthic environments seaward of the continental shelf.

Subpolar Pertaining to the oceanic region that is covered by sea ice in winter. The ice melts away in summer.

Subpolar low A region of low atmospheric pressure located at about 60 degrees latitude.

Substrate The base on which an organism lives and grows.

Subsurface current A current that usually flows below the pycnocline, generally at slower speed and in a different direction from the surface current.

Subtropical Pertaining to the oceanic region poleward of the tropics (about 30 degrees latitude).

Subtropical convergence The zone of convergence that occurs within all subtropical gyres as a result of Ekman transport driving water toward the interior of the gyres.

Subtropical gyre The trade winds and westerly winds initiated in the subtropical regions of all ocean basins, with the influence of the Coriolis effect, set large regions of ocean water in motion. They rotate clockwise in the Northern Hemisphere and counterclockwise in the Southern Hemisphere, and they are centered in the subtropics.

Subtropical high A region of high atmospheric pressure located at about 30 degrees latitude.

Sulfur A yellow mineral composed of the element sulfur. It is commonly found in association with hydrocarbons and salt deposits.

Sulfur-oxidizing bacteria Bacteria that support many deep-sea hydrothermal vent and cold-water seep biocommunities by using energy released by oxidation to synthesize organic matter chemosynthetically.

Summer solstice In the Northern Hemisphere, it is the instant when the sun moves north to the Tropic of Cancer before changing direction and moving southward toward the Equator, approximately June 21.

Summertime beach A beach that is characteristic during summer months. It typically has a wide sandy berm and a steep beach face.

Superwave See *Rogue wave.*

Supralittoral zone The splash or spray zone above the spring high-tide shoreline.

Surf beat An irregular wave pattern caused by mixed interference that results in a varied sequence of larger and smaller waves.

Surf zone The region between the shoreline and the line of breakers where most wave energy is released.

Surface tension The tendency for the surface of a liquid to contract owing to intermolecular bond attraction.

Surging breaker A compressed breaking wave that builds up over a short distance and surges forward as it breaks. It is characteristic of abrupt beach slopes.

Suspension feeder See *Filter feeder.*

Suspension settling The process by which fine-grained material that is being suspended in the water column slowly accumulates on the sea floor.

Sverdrup (sv) A flow rate of 1 million cubic meters per second, named after Swedish oceanographer Otto Sverdrup.

Swash A thin layer of water that washes up over exposed beach as waves break at the shore.

Swell A free ocean wave by which energy put into ocean waves by wind in the sea is transported with little energy loss across great stretches of ocean to the margins of continents where the energy is released in the surf zone.

Swim bladder A gas-containing, flexible, cigar-shaped organ that aids many fishes in attaining neutral buoyancy.

Symbiosis A relationship between two species in which one or both benefit or neither one is harmed. Examples are commensalism, mutualism, and parasitism.

Syzygy Either of two points in the orbit of the moon (full or new moon phase) when the moon lies in a straight line with the sun and Earth.

Tablemount A conical volcanic feature on the ocean floor resembling a seamount except that its top is truncated to a relatively flat surface.

Tectonic estuary An estuary, the origin of which is related to tectonic deformation of the coastal region.

Tectonics Deformation of Earth's surface by forces generated by heat flow from Earth's interior.

Tektite See *Spherule.*

Temperate Pertaining to the oceanic region where pronounced seasonal change occurs (about 40 to 60 degrees latitude). Also known as the mid-latitudes.

Temperature A direct measure of the average kinetic energy of the molecules of a substance.

Temperature of maximum density The temperature at which a substance reaches its highest density. For water, it is 4 degrees centigrade.

Terrigenous sediment. Sediment similar to lithogenous sediment but derived exclusively from a landmass (not including islands).

Territorial sea A strip of ocean, 12 nautical miles wide, adjacent to land over which the coastal nation has control over the passage of ships.

Test The supporting skeleton or shell (usually microscopic) of many invertebrates.

Tethys Sea An ancient body of water that separated Laurasia to the north and Gondwanaland to the south. Its location was approximately that of the present Alpine-Himalayan mountain system.

Texture The general physical appearance of an object.

Theory Systematically organized knowledge applicable in a relatively wide variety of circumstances, especially a system of assumptions, accepted principles, and rules of procedure devised to analyze, predict, or otherwise explain the nature or behavior of a specified set of phenomena.

Thermal contraction To reduce in size during times when temperature is lowered.

Thermocline A layer of water beneath the mixed layer in which a rapid change in temperature can be measured in the vertical dimension.

Thermohaline circulation The vertical movement of ocean water driven by density differences resulting from the combined effects of variations in temperature and salinity; produces deep currents.

Tidal bore A steep-fronted wave that moves up some rivers when the tide rises in the coastal ocean.

Tidal bulges The mounds of water on both sides of Earth caused by the relative positions of the moon (lunar tidal bulges) and the sun (solar tidal bulges).

Tidal period The time that elapses between successive high tides. In most parts of the world, it is 12 hours and 25 minutes.

Tidal range The difference between high-tide and low-tide water levels over any designated time interval, usually one lunar day.

Tide Periodic rise and fall of the surface of the ocean and connected bodies of water resulting from the gravitational attraction of the moon and sun acting unequally on different parts of Earth.

Tide-generating force The magnitude of the centripetal force required to keep all particles of Earth having identical mass moving in identical circular paths required by the movements of the Earth–moon system is identical. This required force is provided by the gravitational attraction between the particles and the moon. This gravitational force is identical to the required centripetal force only at the center of Earth. For ocean tides, the horizontal component of the small force that results from the difference between the required and provided forces is the tide-generating force on that individual particle. These forces are such that they tend to push the ocean water into bulges toward the tide-generating body on one side of Earth and away from the tide-generating body on the opposite side of Earth.

Tide wave The long-period gravity wave generated by tide-generating forces and manifested in the rise and fall of the tide.

Tissue An aggregate of cells and their products developed by organisms for the performance of a particular function.

Tombolo A sand or gravel bar that connects an island with another island or the mainland.

Topography The configuration of a surface. In oceanography it refers to the ocean bottom or the surface of a mass of water with given characteristics.

Trade winds The air masses moving from subtropical high-pressure belts toward the Equator. They are northeasterly in the Northern Hemisphere and southeasterly in the Southern Hemisphere.

Transform fault A fault characteristic of mid-ocean ridges along which they are offset. Oceanic transform faults occur wholly on the ocean floor, while continental transform faults occur on land.

Transform plate boundary The boundary between two lithospheric plates formed by a transform fault.

Transitional wave A wave moving from deep water to shallow water that has a wavelength more than twice the water depth but less than 20 times the water depth. Particle orbits are beginning to be influenced by the bottom.

Transporting media An agent that causes the movement of sediment, such as running water, gravity, the flow of ice, or wind.

Transverse wave A wave in which particle motion is at right angles to energy propagation.

Trench A long, narrow, and deep depression on the ocean floor, with relatively steep sides.

Trophic level A nourishment level in a food chain. Plant producers constitute the lowest level, followed by herbivores and a series of carnivores at the higher levels.

Tropic of Cancer The latitude of 23.5 degrees north, which is the farthest location north that receives vertical rays of the sun.

Tropic of Capricorn The latitude of 23.5 degrees south, which is the farthest location south that receives vertical rays of the sun.

Tropical Pertaining to the regions of the tropics (at or near 23.5 degrees latitude).

Tropical cyclone See *Hurricane*.

Tropical tide A tide occurring twice monthly when the moon is at its maximum declination north and south of the Equator. It is in the tropical regions where tides display their greatest diurnal inequalities.

Tropics The region of Earth's surface lying between the Tropic of Cancer and the Tropic of Capricorn. Also known as the Torrid Zone.

Troposphere The lowermost portion of the atmosphere, which extends from Earth's surface to 12 kilometer (7 miles). It is where all weather is produced.

Trough (wave) The part of an ocean wave that is displaced below the still water level.

Tsunami Seismic sea wave. A long-period gravity wave generated by a submarine earthquake or volcanic event. Not noticeable on the open ocean but builds up to great heights in shallow water.

Tunicates Members of the chordate subphylum Urochordata, which includes sack-like animals. Some are sessile (sea squirts), whereas others are pelagic (salps).

Turbidite deposit A sediment or rock formed from sediment deposited by turbidity currents characterized by both horizontally and vertically graded bedding.

Turbidity A state of reduced clarity in a fluid caused by the presence of suspended matter.

Turbidity current A gravity current resulting from a density increase brought about by increased water turbidity. Possibly initiated by some sudden force such as an earthquake, the turbid mass continues under the force of gravity down a submarine slope.

Turbulent flow Flow in which the flow lines are confused heterogeneously due to random velocity fluctuations.

Typhoon See *Hurricane*.

Ultraplankton Plankton for which the greatest dimension is less than 5 micrometers (0.0002 inch). They are very difficult to separate from the water.

Ultrasonic Sound frequencies above those that can be heard by humans (above 20,000 hertz).

Ultraviolet radiation Electromagnetic radiation shorter than visible radiation and longer than X rays. The approximate range is 1 to 400 nanometers.

Upper water Includes the mixed layer and the permanent thermocline. It is approximately the top 1000 meters (3300 feet) of the ocean.

Upwelling The process by which deep, cold, nutrient-laden water is brought to the surface, usually by diverging equatorial currents or coastal currents that pull water away from a coast.

Valence The combining capacity of an element measured by the number of hydrogen atoms with which it will combine.

van der Waals force Weak attractive force between molecules resulting from the interaction of one molecule and the electrons of another.

Vent An opening on the ocean floor that emits hot water and dissolved minerals.

Ventral Pertaining to the lower or under surface.

Vernal equinox The passage of the sun across the Equator as it moves from the Southern Hemisphere into the Northern Hemisphere, approximately March 21. During this time, all places in the world experience equal lengths of night and day. Also known as the spring equinox.

Vertebrata Subphylum of chordates that includes those animals with a well-developed brain and a skeleton of bone or cartilage; includes fish, amphibians, reptiles, birds, and land animals.

Vertically mixed estuary Very shallow estuaries such as lagoons in which fresh water and marine water are totally mixed from top to bottom so that the salinity at the surface and the bottom is the same at most places within the estuary.

Viscosity A property of a substance to offer resistance to flow caused by internal friction.

Volcanic arc An arc-shaped row of active volcanoes directly above a subduction zone. Can occur as a row of islands (island arc) or as mountains on land (continental arc).

Walker circulation cell The pattern of atmospheric circulation that involves the rising of warm air over the East Indies low-pressure cell and its descent over the high-pressure cell in the southeastern Pacific Ocean off the coast of Chile. It is the weakening of this circulation that accompanies an El Niño event, which has led to the development of the term *El Niño–Southern Oscillation event.*

Walrus A large marine mammal (*Odobenus rosmarus*) of Arctic regions belonging to the order Pinnipedia and having two long tusks, tough wrinkled skin, and four flippers.

Warm blooded See *Homeothermic*.

Warm front A weather front in which a warm air mass moves into and over a cold air mass, producing a broad band of gentle precipitation.

Water cycle See *Hydrologic cycle*.

Water mass A body of water identifiable from its temperature, salinity, or chemical content.

Water spout A tornado-like structure occuring over water and resulting in a funnel-shaped whirling column of air and spray.

Wave A disturbance that moves over the surface or through a medium with a speed determined by the properties of the medium.

Wave base The depth at which circular orbital motion becomes negligible; exists at a depth of one-half wavelength, measured from still water level.

Wave-cut beach A gently sloping surface produced by wave erosion and extending from the base of the wave-cut cliff out under the offshore region.

Wave-cut cliff A cliff produced by landward cutting by wave erosion.

Wave dispersion The separation of waves as they leave the sea area by wave size. Larger waves travel faster than smaller waves and thus leave the sea area first, followed by progressively smaller waves.

Wave frequency (f) The number of waves that pass a fixed point in a unit of time (usually one second). A wave's frequency is the inverse of its period.

Wave height (H) The vertical distance between a crest and the adjoining trough.

Wave period (T) The elapsed time between the passage of two successive wave crests (or troughs) past a fixed point. A wave's period is the inverse of its frequency.

Wave speed (S) The rate at which a wave travels. It can be calculated by dividing a wave's wavelength (L) by its period (T).

Wave steepness Ratio of wave height (H) to wavelength (L). If a 1:7 ratio is ever exceeded by the wave, then the wave breaks.

Wave train A series of waves from the same direction. Informally known as a wave set.

Wavelength (L) The horizontal distance between two corresponding points on successive waves, such as from crest to crest.

Weather The state of the atmosphere at a given time and place, with respect to variables such as temperature, moisture, wind velocity, and barometric pressure.

Weathering A process by which rocks are broken down by chemical and mechanical means.

Wentworth scale A logarithmic scale for size classification of sediment particles.

West Australian Current This cold current forms the eastern boundary current of the Indian Ocean subtropical gyre. It is separated from the coast by the warm Leeuwin Current except during El Niño–Southern Oscillation events, when the Leeuwin Current weakens.

West Wind Drift The surface portion of the Antarctic Circumpolar Gyre driven in an easterly direction around Antarctica by the strong prevailing westerly wind belt.

Westerly winds The air masses moving away from the subtropical high-pressure belts toward higher latitudes. They are southwesterly in the Northern Hemisphere and northwesterly in the Southern Hemisphere.

Western boundary current Poleward-flowing warm currents on the western side of all subtropical gyres.

Western boundary undercurrent (WBUC) A bottom current that flows along the base of the continental slope eroding sediment from it and redepositing the sediment on the continental rise. It is confined to the western boundary of deep-ocean basins.

Westward intensification Pertaining to the intensification of the warm western portion of the subtropical gyre currents that is manifested in higher velocity, faster flow, and deeper flow compared with the cold eastern boundary currents that drift leisurely toward the Equator.

Wetlands Biologically productive regions bordering estuaries and other protected coastal areas. They are usually salt marshes at latitudes greater than 30 degrees and mangrove swamps at lower latitudes.

White muscle fiber Thick muscle fibers with relatively low concentrations of myoglobin that make up a large percentage of the muscle fiber in lunger-type fishes.

White smoker Similar to a black smoker, but emits water of a lower temperature that is white in color.

Wind-driven circulation Any movement of ocean water that is driven by winds. This includes most horizontal movements in the surface waters of the world's oceans.

Windward The direction from which the wind is blowing.

Winter solstice The instant the southward-moving sun reaches the Tropic of Cancer before changing direction and moving north back toward the Equator, approximately December 21.

Wintertime beach A beach that is characteristic during winter months. It typically has a narrow rocky berm and a flat beach face.

Zenith That point on the celestial sphere directly over the observer.

Zooplankton Animal plankton.

Zooxanthellae A form of algae that lives as a symbiont in the tissue of corals and other coral reef animals and provides varying amounts of their required food supply.

CREDITS AND ACKNOWLEDGMENTS

Hole Oceanographic Institution. **Figure 3–10** Courtesy of Daniel J. Fornari, Lamont-Doherty Geological Observatory, Columbia University. Reprinted with permission of the American Geophysical Union. **Figure 3–11** After Tarbuck, E. J., and Lutgens, F. K., *The Earth: An Introduction to Physical Geology,* 5th Ed. (Fig. 19.7), Prentice-Hall, 1996. **Figure 3–13** Courtesy of Aluminum Company of America (Alcoa). **Figure 3–14A** Courtesy of Woods Hole Oceanographic Institution. **Figure 3–14B** Courtesy of Br. Robert McDermott, S.J. **Figure 3–15** Courtesy of A. E. J. Engel, Scripps Institution of Oceanography, University of California, San Diego. **Figure 3–16** After Tarbuck, E. J., and Lutgens, F. K., *The Earth: An Introduction to Physical Geology,* 5th Ed. (Fig. 21.23), Prentice-Hall, 1996. Photo by Dr. Fred N. Spiess, Scripps Institution of Oceanography, University of California, San Diego.

Chapter 4

Figure 4A Courtesy of the Ocean Drilling Program. **Figure 4B** Courtesy of the Ocean Drilling Program. **Figure 4C** Reprinted by permission from Hallegraeff, G. M., *Plankton: A Microscopic World,* 1988 (p. 20). Courtesy of E. J. Brill, Inc. **Figure 4D** Photo courtesy of World Minerals, Inc., Lompoc, California. **Figure 4F** Courtesy of the Ocean Drilling Program, Texas A&M University **Figure 4–1** Courtesy of the RISE Project Group, F. N. Spiess et al., Scripps Institution of Oceanography, University of California, San Diego. **Figure 4–2** APT photo. **Figure 4–3** After Tarbuck, E. J., and Lutgens, F. K., *The Earth: An Introduction to Physical Geology,* 5th Ed. (Fig. 5.11), Prentice-Hall, 1996. **Figure 4–4A–D** APT photos. **Figure 4–5** Courtesy of Walter N. Mack, Michigan State University. **Figure 4–6** After Leinen, M., et al. 1986. Distribution of biogenic silica and quartz in recent deep-sea sediments. *Geology* 14:3, 199–203. **Figure 4–8A** Reprinted by permission from Hallegraeff, G. M., *Plankton: A Microscopic World,* 1988 (p. 46). Courtesy of E. J. Brill, Inc. **Figure 4–8B** Courtesy of Warren Smith, Scripps Institution of Oceanography, University of California, San Diego. **Figure 4–8C** Photo courtesy of World Minerals, Inc., Lompoc, California (sample from Celite Corporations Diatomite Mine in Lompoc, California). **Figure 4–9A** Reprinted by permission from Hallegraeff, G. M., *Plankton: A Microscopic World,* 1988 (p. 8). Courtesy of E. J. Brill, Inc. **Figure 4–9B** Reprinted by permission from Hallegraeff, G. M., *Plankton: A Microscopic World,* 1988 (p, 16). Courtesy of E. J. Brill, Inc. **Figure 4–9C** Courtesy of Memorie Yasuda, Scripps Institution of Oceanography, University of California, San Diego. **Figure 4–9D** Courtesy of the Deep Sea Drilling Project, Scripps Institution of Oceanography, University of California, San Diego. **Figure 4–10** APT photo. **Figure 4–11** Courtesy of Jimmy Greenslate, Scripps Institution of Oceanography, University of California, San Diego. **Figure 4–12** APT photo. **Figure 4–13** Reprinted by permission of The Open University Course Team, *Ocean Chemistry and Deep-Sea Sediments,* Butterworth-Heinemann, 1989. **Figure 4–16** After Biscaye, P. E. et al., 1976; Berger, W. H. et al., 1976; and Kolla V. and Biscaye, P.E., 1976. **Figure 4–18** Courtesy of Joseph Holliday, El Camino College. **Figure 4–19** After Sverdrup, H. U. et al., 1942. **Figure 4–20** Courtesy of Susumu Honjo, Woods Hole Oceanographic Institution. **Figure 4–21** Photo by Earl Roberge, courtesy Photo Researchers, Inc.. **Figure 4–22** Courtesy of deep Sea Ventures, Inc. **Figure 4–23** After Cronan, D. S. 1977. Deep sea nodules: Distribution and geochemistry, *in* Glasby, G. P., ed., *Marine Manganese Deposits,* Elsevier Scientific Publishing Co. **Table 4–1** After Deen and Trujillo, 1996. **Table 4–2** Source: Wentworth, 1992; After Udden, 1898.

Chapter 5

Figure 5A From C. W. Thompson and Sir John Murray, Report on the Scientific Results of the Voyage of H.M.S. Challenger, Vol. 1. Great Britain: Challenger Office, 1895, Plate 1. **Figure 5C** Peter Arnold, Inc. **Figure 5D** Courtesy of Dr. Walter Munk, Scripps Institution of Oceanography, University of California, San Diego. **Figure 5–1** After Tarbuck, E. J., and Lutgens, F. K., *The Earth: An Introduction to Physical Geology,* 5th Ed. (Fig. 2.4), Prentice-Hall, 1996. **Figure 5–3** Used with permission from R. W. Christopherson,

Geosystems, 2nd Ed., Macmillan Publishing Company, 1994, Figure 7–7, p. 186. **Figure 5–6** After data from Moustafa T. Chahine, Jet Propulsion Laboratory, California Institute of Technology. **Figure 5–11** Photo by Electron Microscopy Laboratory, ARS, USDA. **Figure 5–15** Reprinted by permission from Tarbuck, E. J., and Lutgens, F. K., *The Earth: An Introduction to Physical Geology,* 4th Ed. (Fig. 10.2), Macmillan Publishing Company, 1993. **Figure 5–16** After G. L. Pickard, *Descriptive Physical Oceanography,* © 1963. By permission of Pergamon Press Ltd., Oxford, England. **Figure 5–17** After Sverdrup, H. U. et al., 1942. **Figure 5–19** After G. L. Pickard, *Descriptive Physical Oceanography,* © 1963. By permission of Pergamon Press Ltd., Oxford, England. **Figure 5–20A and B** After G. L. Pickard, *Descriptive physical oceanography,* © 1963. By permission of Pergamon Press Ltd., Oxford, England. **Figure 5–22** Courtesy of Patty Deen, Palomar College.

Chapter 6

Figure 6A Hutton Getty/Liaison Agency, Inc. **Figure 6B** Photos courtesy of Woods Hole Oceanographic Institution, Emory Kristof/National Geographic Society. **Figure 6C** Courtesy of Morgan P. Sanger, The Columbus Foundation. **Figure 6G** Courtesy of J. H. Golden, NOAA/OAR. **Figure 6–4** After Gross, M. G., *Oceanography,* 6th Ed. (Fig. 5–2), Prentice-Hall, 1993. **Figure 6–14** After Gross, M. G., *Oceanography,* 6th Ed. (Fig. 5–17), Prentice-Hall, 1993. **Figure 6–15** Reprinted by permission from Lutgens, F. K., and Tarbuck, E. J., *The Atmosphere,* 6th Ed. (Fig. 8.2), Prentice-Hall, 1995. **Figure 6–19B** Photo by Bob Stovall, Bruce Coleman, Inc. **Figure 6–20** Courtesy of National Hurricane Center, NOAA. **Figure 6–24** Data courtesy Charles Keeling and Timothy Whorf, Scripps Institution of Oceanography, University of California, San Diego. **Tables 6–4 and 6–5** Source: After Rodhe, H., 1990.

Chapter 7

Figure 7A Courtesy of U.S. Navy. **Figure 7B** Courtesy of *EOS, Transactions* 75:37, 425 (1994). **Figure 7C** Courtesy of *EOS, Transactions* 73:34, 361 (1992). **Figure 7D** Photo courtesy of The Fram Museum, Oslo, Norway. **Figure 7–1A** Courtesy of Douglas Alden, Scripps Institution of Oceanography, University of California, San Diego. **Figure 7–1B** Courtesy of Aanderaa Instruments. **Figure 7–2** Courtesy of NASA. **Figure 7–8A** NASA and NOAA. **Figure 7–16A** Image courtesy of Dr. Charles McLain at the Rosenstiel School of Marine and Atmospheric Science, University of Miami. **Figure 7–19** Map courtesy of the International Research Institute for Climate Prediction, Lamont-Doherty Earth Observatory, Columbia University. **Figure 7–20** Courtesy Paul C. Fiedler at the National Marine Fisheries Service. **Figure 7–26** Courtesy of Woods Hole Oceanographic Institution. **Figure 7–27** Courtesy of U.S. Department of Energy. **Table 7–3** Data modified from Quinn, et al. 1987.

Chapter 8

Figure 8A Courtesy of *California Geology.* **Figure 8B** Courtesy of *California Geology.* **Figure 8C** Courtesy of the Ocean Drilling Program. **Figure 8–1B** Photo courtesy of NASA. **Figure 8–2** After Kinsman, B., 1965. **Figure 8–3B** From *The Tasa Collection: Shorelines.* Published by Macmillan Publishing Co., New York. Copyright © 1986 by Tasa Graphic Arts, Inc. All rights reserved. **Figure 8–11** Courtesy of Jet Propulsion Laboratory, NASA. **Figure 8–13** Official photograph U.S. Navy. **Figure 8–16** After Gross, M.G. *Oceanography.* 6th Ed. (Fig. 8–4), Prentice–Hall, 1993. **Figure 8–18** From *The Tasa Collection: Shorelines.* Published by Macmillan Publishing, New York. Copyright © 1986 by Tasa Graphic Arts, Inc. All rights reserved. **Figure 8–19A** © Peter Arnold, Inc. **Figure 8–19B** © Tony Arruza/Bruce Coleman, Inc. **Figure 8–19C** Photo © Woody Woodworth, Creation Captured. All rights reserved. **Figure 8–21** Adapted from Tarbuck, E. J., and Lutgens, F. K., *Earth Science,* 6th Ed. New York: Macmillan Publishing Company, 1991. **Figure 8–24B, C, D** Photos courtesy Kydo News Agency. **Figure 8–25** Map constructed from data in U.S. Navy Summary of Synoptic Meteorological Observations, SSMO. Adapted from *Sea Frontiers* (July–August 1987), "Sea Secrets," International Oceanographic Foundation (Vol. 33,

No. 4), pp. 260–261; produced by The National Climatic Data Center with support from the U.S. Department of Energy.

Chapter 9
Figure 9B Courtesy of Phototeque/Electricite de France. **Figure 9D** Courtesy New Brunswick Department of Tourism. **Figure 9E** Courtesy Shubenacadie Tidal Bore Park and Upriver Rafting, Nova Scotia, Canada. **Figure 9G** Photo by Eda Rogers. **Figure 9–8** From *The Tasa Collection: Shorelines.* Published by Macmillan Publishing Co., New York. Copyright © 1986 by Tasa Graphic Arts, Inc. All rights reserved. **Figure 9–12** Tom Strickland, Courtesy of Wide World Photos. **Figure 9–13** After von Arx, W. S., 1962; original by H. Poincaré 1910, Leçons de Mécanique Céleste, a Gauther-Crofts, Vol. 3. **Figure 9–14** After C. Hauge, Tides, currents, and waves. *California Geology*, July 1972. **Figure 9–16B and C** Photos courtesy of Nova Scotia Department of Tourism.

Chapter 10
Figure 10A Photo courtesy of Patty Deen, Palomar College. **Figure 10B** Photo courtesy of Scripps Institution of Oceanography, University of California, San Diego. **Figure 10–2** APT photos. **Figure 10–3A** Photo by John S. Shelton. **Figure 10–3B** After Tarbuck, E. J., and Lutgens, F. K., *The Earth: An Introduction to Physical Geology,* 4th Ed. (Fig. 14.8), Macmillan Publishing Company, 1993. **Figure 10–5** Photo by Bruce F. Molnia, Terra-photographics/BPS. **Figure 10–6** Photo by John S. Shelton. **Figure 10–8A** Photo by USDA-ASCS. **Figure 10–8B** APT photo. **Figure 10–9A and B** After Tarbuck, E. J., and Lutgens, F. K., *The Earth: An Introduction to Physical Geology,* 4th Ed. (Fig. 14.12), Macmillan Publishing Company, 1993. **Figure 10–9C** Photo by USDA-ASCS. **Figure 10–11A** Photo by GEOPIC®, Earth Satellite Corporation. **Figure 10–11B** Photo courtesy of NASA. **Figure 10–15** After V. Gornits, S. Lebedeff, and J. Hausen, 1982. *Science* 215:1611–1614. **Figure 10–18** Photo by John S. Shelton. **Figure 10–20** Photo by John S. Shelton. **Figure 10–21** Reprinted by permission from Tarbuck, E. J., and Lutgens, F. K., *Earth science,* 5th Ed., Merrill Publishing Company, 1988. **Figure 10–22** From U.S. Coast and Geodetic Survey Chart 5161. **Figure 10–23A** Courtesy of The Fairchild Aerial Photography Collection at Whittier College, California. Flight C–1670, Frame 4 (9/18/31). **Figure 10–23B** Courtesy of The Fairchild Aerial Photography Collection at Whittier College, California. Flight C–14180, Frame 3:61 (10/21/49). **Figure 10–25** APT photos.

Chapter 11
Figure 11C Photos courtesy of NOAA. **Figure 11E** © Tony Freeman/Photo Edit. **Figure 11F** Photo by Wayne Perryman, NMFS. **Figure 11–4A** Photo courtesy of The Stock Market. **Figure 11–4B** Courtesy NASA. **Figure 11–8B** Photo © Eda Rogers. **Figure 11–7** After Officer et al., 1984. **Figure 11–8C** Photo © R. N. Mariscal/Bruce Coleman, Inc. **Figure 11–11A** Adapted from *Encyclopedia of Oceanography*, edited by Rhodes Fairbridge, © 1966. Reprinted by permission of Dowden, Hutchinson, & Ross, Inc., Stroudsburg, PA. **Figure 11–11B** After Judson, S. et al. *Physical Geology*, 7th Ed. Prentice-Hall, 1987. **Figure 11–12** Photo courtesy of NOAA. **Figure 11–14B** Photo courtesy of U.S. EPA/EMSL. **Figure 11–15** Courtesy of A. Crosby Longwell. **Figure 11–16** Courtesy of John Trever, *Albuquerque Journal*. **Figure 11–18A** Courtesy of Massachusetts Water Resources Authority. **Figure 11–18B** Courtesy of U.S. Geological Survey, Woods Hole, MA. **Figure 11–19** Photo by Thomas D. Mangelsen, courtesy Peter Arnold, Inc. **Figure 11–20** Minamata, Tokomo is bathed by her mother, ca 1972. Photograph by W. Eugene Smith. Collection, Center for Creative Photography, The University of Arizona. © Aileen M. Smith, Courtesy Black Star, Inc., New York.

Chapter 12
Figure 12A Photo courtesy Corbis–The Bettmann Archive. **Figure 12C–12E** Photo and graphs courtesy of Atsushi Ishimatsu. Permission granted by Atsushi Ishimatsu and *Nature*. **Figure 12–5** Photo by Peter Arnold, Inc. **Figure 12–11** After Sverdrup, H. U.

et al., 1942. **Figure 12–18** Reprinted by permission from Hallegraeff, G. M., *Plankton: A Microscopic World*, 1988 (p. 21). Courtesy of E. J. Brill, Inc. **Figure 12–19A** Courtesy of Deeanne Edwards, Scripps Institution of Oceanography, University of California, San Diego. **Figure 12–19B** APT photo.

Chapter 13
Chapter Opening Feature Reprinted with permission and light editing from Howard, J. 1995. Vanishing act: Critical link in marine food chain may be at risk. *Scripps Institution of Oceanography Explorations* 2:2, p. 13, Scripps Institution of Oceanography, University of California, San Diego. **Figure 13A** Courtesy of Dr. Elizabeth Venrick, Scripps Institution of Oceanography, University of California, San Diego. **Figure 13B** Courtesy of Susan R. Green, Scripps Institution of Oceanography, University of California, San Diego. **Figure 13C** Copyright © 1999 by Universal City Studios, Inc. Courtesy of Universal Studios Publishing Rights, a division of Universal Studios Licensing, Inc. All Rights Reserved. Photo Courtesy of the Academy of Motion Picture Arts and Sciences. **Figure 13D** Photo © Marty Snyderman. **Figure 13E** Photo by Christopher Newbert. **Figure 13F** Photo © Marty Snyderman. **Figure 13G** Photo courtesy of James B. McClintock with permission from *Nature*, Vol. 346, 462–464, 1990. Photo by Phil Oshal. **Figure 13–2** Courtesy of Dr. Elizabeth Venrick, Scripps Institution of Oceanography, University of California, San Diego. **Figure 13–3C** Courtesy of NASA/Goddard Space Flight Center. **Figure 13–4** Courtesy of Jane A. Elrod and Gene Feldman, NASA/Goddard Space Flight Center. **Figure 13–5** APT photos. **Figure 13–6A, B, and D** APT photos. **Figure 13–7A** Reprinted by permission from Hallegraeff, G. M., *Plankton: A Microscopic World*, 1988 (p. 43). Courtesy of E. J. Brill, Inc. **Figure 13–7B** Courtesy of Dr. Wuchang Wei, Scripps Institution of Oceanography, University of California, San Diego. **Figure 13–7C** Reprinted by permission from Hallegraeff, G. M., *Plankton: A Microscopic World*, 1988 (p. 91). Courtesy of E. J. Brill, Inc. **Figure 13–7D** Reprinted by permission from Hallegraeff, G. M., *Plankton: A Microscopic World*, 1988 (p. 81). Courtesy of E. J. Brill, Inc. **Figure 13–8D** Photo courtesy of Scripps Institution of Oceanography, University of California, San Diego. **Table 13–1** Data after Strahler, A. H. and Strahler, A. N., *Modern Physical Geograph*, 4th Ed., Wiley, 1992 (Table 25.1).

Chapter 14
Figure 14A Photo courtesy of Corbis–The Bettmann Archive. **Figure 14B** Photo by Kelvin Aitken, courtesy Peter Arnold, Inc. **Figure 14C** Photo by Francois Gohier, courtesy of Photo Researchers, Inc. **Figure 14–3** Courtesy of Kozo Takahashi, Kyushu University, Fukuoka, Japan. **Figure 14–4** Courtesy of Susan Burke, Scripps Institution of Oceanography, University of California, San Diego. **Figure 14–5** From Wilhelm Giesbrecht's 1892 book *Fauna und Flora des Golfes von Neapel und der Angrenzenden Meeres–Abschnitte*, Berlin: Verlag Von R. Friedländer and Sohn. Courtesy of *Scripps Institution of Oceanography Explorations*, Scripps Institution of Oceanography, University of California, San Diego. **Figure 14–6A** Photo by Pam Blades-Eckelbarger, Harbor Branch Oceanographic Institution, Inc. **Figure 14–6B** Photo by Marsh Youngbluth, Harbor Branch Oceanographic Institution, Inc. **Figure 14–7B** Photo by Larry Ford. **Figures 14–8B, C, and D** by James M. King, Graphic Impressions. **Figure 14–9A** Photo courtesy of Howard J. Spero. **Figure 14–12A** Photo © Marty Snyderman. **Figure 14–12B** Photo © Wayne and Karen Brown. **Figure 14–12C** Photo © Greg Ochocki/The Stock Market. **Figure 14–12D** Photo © Bob Gomel/The Stock Market. **Figure 14–12E** Photo © Valerie Taylor/Peter Arnold, Inc. **Figure 14–13A** Photo © Fred Bavendam/Peter Arnold, Inc. **Figure 14–13B** Photo courtesy of National Marine Fisheries Service. **Figure 14–15** Photo © Thomas Ives/The Stock Market. **Figure 14–16** After Lalli and Parsons, 1993 (Fig. 6.5). **Figure 14–17** After Lalli and Parsons, 1993 (Fig. 6.6b and 6.7a). **Figure 14–18A** Photo © Ralph Lee Hopkins/Wilderness Images. **Figure 14–18B** Photo © C. Allan Morgan/Peter Arnold, Inc. **Figure 14–18C** APT photo. **Figure 14–20A** Photo © Fred Bavendam/Peter Arnold, Inc. **Figure 14–20B**

D. Fleetham/Animals Animals/Earth Scenes. **Figure 14–25A** APT photos, courtesy Sea World of California. **Figure 14–25B** APT photos, courtesy Sheldon Jackson Museum, Sitka, Alaska. **Figure 14–26** Photo © Ralph Lee Hopkins, Wilderness Images. **Figure 14–28** Photo © Ralph Lee Hopkins, Wilderness Images. **Table 14–1** Data from the International Shark Attack File. **Table 14–2** Data from the U.S. National Marine Fisheries Service.

Chapter 15

Figure 15A Reprinted by permission of The Bettman Archive. **Figure 15B** Photo by Dr. Lucy Bunkley-Williams. **Figures 15C–15G** Courtesy of John Porter, University of Georgia, Athens. **Figure 15H** Photos courtesy of Dr. Robert Hessler, Scripps Institution of Oceanography, University of California, San Diego. Reprinted from *Deep-Sea Research,* Vol. 25, Hessler, R. R., Ingram, C. L., Yayanos, A. A., and Burnett, B. R., Scavenging amphipods from the floor of the Phillipine Trench, Fig. 1, Copyright 1978, with permission from Elsevier Science. **Figure 15I** Courtesy of Deeanne Edwards, Scripps Institution of Oceanography, University of California, San Diego. **Figure 15–1** After Zenkevitch, L. A. et al., 1971. **Figure 15–2D** Photo © Norbet Wu/Peter Arnold, Inc. **Figure 15–2F** Photo © Breck P. Kent. **Figures 15–2I** APT photos. **Figure 15–3** Photo © Joy Sparr/Bruce Coleman, Inc. **Figure 15–5A** Photo by James McCullagh. **Figure 15–7A** Photo © Fred Bavendam/Peter Arnold, Inc. **Figure 15–7B** Photo © Eda Rogers. **Figure 15–13** Courtesy of Howard J. Spero, University of South Carolina. **Figure 15–14** Photo by Stephen J. Kraseman © Peter Arnold, Inc. **Figure 15–15A** Photo by Mia Tegner. **Figure 15–15B** Photo by James McCullagh. **Figure 15–16A** Photo by B. Kiwala. **Figure 15–16B** Photo by Harold W. Pratt/Biological Photo Service. **Figure 15–18** After Stehli, F. G. and Wells, J. H. 1971. Diversity and age patterns in hermatypic corals. *Systematic Zoology* 20, 115 126. **Figure 15–20A** Photo by Christopher Newbert. **Figure 15–20B** Photo by C. R. Wilkerson, Australian Institute of Marine Science. **Figure 15–20C** Photo by Ken Lucas/Biological Photo Service. **Figure 15–21A and B** Photos by Christopher Newbert. **Figure 15–22** After Lalli and Parsons, 1993 (Fig. 8.16). **Figure 15–25A** Photo courtesy of Woods Hole Oceanographic Institution. **Figure 15–25B** SEM photomicrograph courtesy of Woods Hole Oceanographic Institution. **Figure 15–25C** Photo courtesy of Robert Hessler, Scripps Institution of Oceanography, University of California, San Diego. **Figure 15–26** Courtesy of Peter A. Rona, NOAA. **Figure 15–27** Courtesy of C. K. Paull, Scripps Institution of Oceanography, University of California, San Diego. **Figures 15–28B and C** Courtesy of Charles R. Fisher, Penn State University. **Figure 15–29** Courtesy of L. D. Klum, Oregon State University.

Afterword

Figure Aft–1 APT photo.

Appendices

AIII Reprinted with permission of Hubbard Scientific, Inc. Physiographic Chart of the Sea Floor © Hubbard Scientific. **Figure A4–1** APT photo.

INDEX